Developmental Biology Protocols

METHODS IN MOLECULAR BIOLOGY™

John M. Walker, SERIES EDITOR

METHODS IN MOLECULAR BIOLOGY™

Developmental Biology Protocols
Volume I

Edited by

Rocky S. Tuan

Thomas Jefferson University, Philadelphia, PA

and

Cecilia W. Lo

University of Pennsylvania, Philadelphia, PA

Humana Press ✳ Totowa, New Jersey

To Chuck and our parents

© 2000 Humana Press Inc.
999 Riverview Drive, Suite 208
Totowa, New Jersey 07512

This publication is printed on acid-free paper. ∞
ANSI Z39.48-1984 (American Standards Institute)
Permanence of Paper for Printed Library Materials.

Cover design by Patricia F. Cleary

Cover illustration taken from Fig. 1 in Chapter 20. It is a composite three-dimensional view of epifluorescence micrographs of representative stages of the developing rat embryo coated with acridine orange and imaged in toto by confocal laser-scanning microscopy.

For additional copies, pricing for bulk purchases, and/or information about other Humana titles, contact Humana at the above address or at any of the following numbers: Tel.: 973-256-1699; Fax: 973-256-8341; E-mail: humana@humanapr.com; or visit our Website: http://humanapress.com

Printed in the United States of America. 10 9 8 7 6 5 4 3 2 1

Library of Congress Cataloging in Publication Data

Main entry under title: Developmental biology protocols/edited by Rocky S. Tuan and Cecilia W. Lo.

Methods in molecular biology™.

Developmental biology protocols, vol. I/edited by Rocky S. Tuan and Cecilia W. Lo.
    p.   cm.—(Methods in molecular biology; v. 135, 136, 137)
    Includes bibliographical references and indexes.
    ISBN 0-89603-852-1 (v. 1: alk. paper) (hardcover), ISBN 0-89603-574-3 (paper); ISBN 0-89603-853-X (v. 2: alk. paper) (hardcover), ISBN 0-89603-575-1 (paper); ISBN 0-89603-854-8 (v. 3: alk. paper) (hardcover), ISBN 0-89603-576-X (paper); ISBN 0-89603-855-6 (hc: alk paper); ISBN 0-89603-578-6 (comb: alk paper)
    1.   Developmental biology—Laboratory manuals.   2. Molecular biology—Laboratory manuals.   I.   Tuan, Rocky S.   II. Lo, Cecilia W.   III.   Series.
QH491.D4625 2000
571.8—dc21
                                                   99-23532
                                                     CIP

Preface

Developmental biology is one of the most exciting and fast-growing fields today. In part, this is so because the subject matter deals with the innately fascinating biological events—changes in form, structure, and function of the organism. The other reason for much of the excitement in developmental biology is that the field has truly become the unifying melting pot of biology, and provides a framework that integrates anatomy, physiology, genetics, biochemistry, and cellular and molecular biology, as well as evolutionary biology. No longer is the study of embryonic development merely "embryology." In fact, development biology has produced important paradigms for both basic and clinical biomedical sciences alike.

Though modern developmental biology has its roots in "experimental embryology" and the even more classical "chemical embryology," the recent explosive and remarkable advances in developmental biology are critically linked to the advent of the "cellular and molecular biology revolution." The impressive arsenal of experimental and analytical tools derived from cell and molecular biology, which promise to continue to expand, together with the exponentially developing sophistication in functional imaging and information technologies, guarantee that the study of the developing embryo will contribute one of the most captivating areas of biological research in the next millennium.

There is a demonstrated need for students of developmental biology to be knowledgeable of the breadth and depth of the available experimental methodologies, by necessity derived from multiple disciplines, which are applicable to the study of the developing embryo. In particular, because developmental biology deals with multiple model systems, from organismal to tissue and cell levels, as well as a wide range of "change"-related biological activities, the investigator is often frustrated as to how his/her findings relate to those obtained in another model system and/or by using different reagents or functional markers. Compared to other more strictly defined fields of biological research, the number of "reference" publications that deal specifically with the practical aspects of experimental developmental biology are, however, relatively scarce.

Developmental Biology Protocols grows out of the need for a comprehensive laboratory manual that provides the readers the principles, background, rationale, as well as the practical protocols, for studying and analyzing the events of embryonic development. This three-volume set, consisting of 142 chapters, is intentionally broad in scope, because of the nature of modern developmental biology. Information is grouped into the following topics: (1) systems—production, culture, and storage; (2) developmental pattern and morphogenesis; (3) embryo structure and function; (4) cell lineage analysis; (5) chimeras; (6) experimental manipulation of embryos; (7) application of viral vectors; (8) organogenesis; (9) abnormal development and teratology;

(10) screening and mapping of novel genes and mutations; (11) transgenesis production and gene knockout; (12) manipulation of developmental gene expression and function; (13) analysis of gene expression; (14) models of morphogenesis and development; and (15) in vitro models and analysis of differentiation and development.

Throughout *Developmental Biology Protocols*, the authors have consistently striven for a balanced presentation of both background information and actual laboratory details. It is believed that this highly practical format will permit readers to bring the concepts and principles we present into their personal research practices in a most efficient manner. Specifically, the wide range of model systems and multidisciplinary experimental techniques presented here should lower the "activation energy" for the student of developmental biology to become a contributing member of this exciting scientific discipline. In addition, teachers of developmental biology at all levels should also readily find relevant and useful information to enrich the experience of their students.

The practice of developmental biology is currently in a state of constant change, reflecting the close relationship of the field to other rapidly developing fields of biological research, particularly cell and molecular biology, and imaging and information technology. The materials presented in this three-volume set are therefore the beginning of a project that will involve continuous update and upgrade to reach and enhance the scientific endeavors of developmental biologists at large.

The production of *Developmental Biology Protocols* would not have been possible without the outstanding work of the contributing authors who share here with the readers the hands-on wisdom they have earned in the laboratory. We are grateful for their intellectual contributions as well as their remarkable tolerance to our constant reminders. Tom Lanigan and his staff at Humana Press worked diligently on the project to ensure a final product of the highest quality. Chuck, our young son, persevered throughout the gestation period of the project, and constantly demonstrated to us the meaning of "developmental biology."

Our final, heartfelt thanks go to Lynn Stierle, who expertly and single-handedly maintained the massive organization of the manuscripts and the correspondence (snail-mail and e-mail), as well as the sanity of the editors! Michelle Levinski also provided valuable assistance in proofreading the final production.

Finally, we hope that these volumes will find their place on the laboratory shelves, with their pages well soiled and their contents tried and tested, and prove their utility as an everyday resource for the students of developmental biology, the most exciting discipline of biology for many decades to come!

Rocky S. Tuan, PhD
Cecilia W. Lo, PhD

List of Color Plates

Color plates 1–13 appear as an insert following p. 258.

Contents

TABLES OF CONTENTS FROM VOLUMES II AND III

VOLUME II

VOLUME III

Contributors

TAKASHI ARIIZUMI • *CREST, Japan Science and Technology Corp., The University of Tokyo, Tokyo, Japan*

MAKOTO ASASHIMA • *Department of Life Sciences (Biology), Graduate School of Arts and Sciences, The University of Tokyo, Tokyo, Japan*

H. SCOTT BALDWIN • *Department of Pediatric Cardiology, Children's Hospital of Philadelphia, Philadelphia, PA*

JEAN BENNETT • *Department of Department of Ophthalmology, Scheie Eye Institute, University of Pennsylvania School of Medicine, Philadelphia, PA*

STEPHEN A. BOPPART • *Department of Electrical Engineering and Computer Science and Research Laboratory of Electronics, Massachusetts Institute of Technology, Cambridge, MA*

MARK E. BREZINSKI • *Cardiac Unit and Department of Medicine, Massachusetts General Hospital and Harvard Medical School, Boston, MA*

ELIZABETH A. BUCHER • *Department of Cell and Developmental Biology, School of Medicine, University of Pennsylvania, Philadelphia, PA*

GANG CHENG • *Department of Cell Biology and Anatomy, Medical University of South Carolina, Charleston, SC*

CHENG-MING CHUONG • *Department of Pathology, School of Medicine, University of Southern California, Los Angeles, CA*

JAMES A. COFFMAN • *Division of Biology, Stowers Institute for Medical Research, California Institute of Technology, Pasadena, CA*

VICTOR G. CORCES • *Department of Biology, Johns Hopkins University, Baltimore, MD*

DIANA K. DARNELL • *Department of Neurobiology and Anatomy, University of Utah School of Medicine, Salt Lake City, UT*

TIMOTHY J. DAVIES • *Department of Zoology, University of Oxford, Oxford, UK*

ALAN R. DAVIS • *Shell Center for Cell and Gene Therapy, Baylor College of Medicine, Houston, TX*

FRANÇOISE DIETERLEN-LIÈVRE • *Institut d'Embryologie Cellulaire et Moleculaire du CNRS et du College de France, Nogent/Marnec Cedex, France*

ADAM S. DOHERTY • *Department of Biology, University of Pennsylvania, Philadelphia, PA*

CHRISTOPHER J. DRAKE • *Department of Cell Biology and Anatomy, Medical University of South Carolina, Charleston, SC*

ROBERT FINKELSTEIN • *Department of Neuroscience, University of Pennsylvania, Philadelphia, PA*

JOSIANE FONTAINE-PÉRUS • *Faculté des Sciences et des Techniques, Université de Nantes, Nantes Cedex, France*

JAMES G. FUJIMOTO • *Department of Electrical Engineering and Computer Science and Research Laboratory of Electronics, Massachusetts Institute of Technology, Cambridge, MA*

MASAHIKO FUJINAGA • *Department of Anesthesia, Stanford University School of Medicine, Palo Alto, CA*

ANDREA GAMBOTTO • *Department of Molecular Genetics and Biochemistry, University of Pittsburgh School of Medicine, Pittsburgh, PA*

VIRGINIO GARCIA-MARTINEZ • *Department of Anatomy and Neurobiology, University of Utah School of Medicine, Salt Lake City, UT*

RICHARD L. GARDNER • *Department of Zoology, University of Oxford, Oxford, UK*

LEANNE GODINHO • *Howard Florey Institute of Experimental Physiology and Medicine, University of Melbourne, Parkville, Victoria, Australia*

ROBERT G. GOURDIE • *Department of Cell Biology and Anatomy, Medical University of South Carolina, Charleston, SC*

C. MAY GRIFFITH • *Department of Cell Biology, Harvard Medical School, Boston, MA*

ABHA R. GUPTA • *Department of Ophthalmology, Scheie Eye Institute, University of Pennsylvania School of Medicine, Philadelphia, PA*

BRIAN K. HALL • *Department of Biology, Dalhousie University, Halifax, Nova Scotia, Canada*

ELIZABETH D. HAY • *Department of Cell Biology, Harvard Medical School, Boston, MA*

JEFF HARDIN • *Department of Zoology, University of Wisconsin, Madison, WI*

PAUL J. HEID • *Department of Zoology, University of Wisconsin, Madison, WI*

VLAD HERLEA • *Laboratory of Anatomy and Pathology, Bucharest, Romania*

JAMES C. HUHTA • *Department of Cell Biology, University of Medicine and Dentistry of New Jersey, Stratford, NJ*

E. SIDNEY HUNTER, III • *Reproductive Toxicology Division, National Health and Environmental Effects Research Laboratory, US Environmental Protection Agency, Research Triangle Park, NC*

GREG HUNTER • *Developmental Biology Program, Institute of Molecular Medicine and Genetics, Medical College of Georgia, Augusta, GA*

SEON HEE KIM • *Department of Department of Molecular Genetics and Biochemistry, University of Pittsburgh School of Medicine, Pittsburgh, PA, and Institute for Molecular Biology and Genetics, Seoul National University, Seoul, Korea*

SUNYOUNG KIM • *Institute for Molecular Biology and Genetics, Seoul National University, Seoul, Korea*

SIMON J. KINDER • *Embryology Unit, Children's Medical Research Institute, Wenworthville, New South Wales, Australia*

MARGARET L. KIRBY • *Developmental Biology Program, Institute of Molecular Medicine and Genetics, Medical College of Georgia, Augusta, GA*

SIMON J. KINDER • *Embryology Unit, Children's Medical Research Institute, Wenworthville, New South Wales, Australia*

MARGARET L. KIRBY • *Developmental Biology Program, Institute of Molecular Medicine and Genetics, Medical College of Georgia, Augusta, GA*

DAVID J. KOZLOWSKI • *Department of Biology, University of Pennsylvania, Philadelphia, PA*

DONNA KUMISKI • *Developmental Biology Program, Institute of Molecular Medicine and Genetics, Medical College of Georgia, Augusta, GA*

CAROLYN A. LARABELL • *Life Sciences Division, Lawrence Berkeley National Laboratory, University of California, Berkeley, CA*

PATRICK S. LEAHY • *Division of Biology, California Institute of Technology, Pasadena, CA*

ELIZABETH E. LECLAIR • *Department of Orthopaedic Surgery, Thomas Jefferson University, Philadelphia, PA*

NICOLE M. LE DOUARIN • *Institut d'Embryologie Cellulaire et Moleculaire du CNRS et du College de France, Nogent/Marnec Cedex, France*

KERSTI K. LINASK • *Department of Cell Biology, University of Medicine and Dentistry of New Jersey, Stratford, NJ*

CHARLES D. LITTLE • *Department of Cell Biology and Anatomy, Medical University of South Carolina, Charleston, SC*

CECILIA W. LO • *Department of Biology, University of Pennsylvania, Philadelphia, PA*

CARMEN LOPEZ-SANCHEZ • *Department of Neurobiology and Anatomy, University of Utah School of Medicine, Salt Lake City, UT*

CHRISTOPHER J. LOWE • *Department of Molecular and Cellular Biology, University of California–Berkeley, Berkeley, CA*

ALBERT M. MAGUIRE • *Department of Ophthalmology, Scheie Eye Institute, University of Pennsylvania Medical School, Philadelphia, PA*

GEORGE M. MALACINSKI • *Department of Biology, Indiana University, Bloomington, IN*

CRAIG S. MICKANIN • *Department of Pediatric Cardiology, Children's Hospital of Philadelphia, Philadelphia, PA*

TAKASHI MIKAWA • *Department of Cell Biology and Anatomy, Medical University of South Carolina, Charleston, SC*

TOM MIYAKE • *Department of Biology, Dalhousie University, Halifax, Canada*

SALLY A. MOODY • *Department of Anatomy and Cell Biology, The George Washington University Medical Center, Washington, DC*

KEN MUNEOKA • *Department of Cell and Molecular Biology, Tulane University, New Orleans, LA*

MICHAEL G. NAROTSKY • *Reproductive Toxicology Division, National Health and Environmental Effects Laboratory, US Department of Environmental Protection, Research Triangle Park, NC*

VALERIE NGO-MULLER • *Université Pierre et Marie Curie, Paris Cedex, France*

JOSEPH L. PALLADINO • *Institute for Human Gene Therapy, University of Pennsylvania, Philadelphia, PA*

JONATHON PINES • *Department of Zoology, Wellcome/CRC Institute, Cambridge, UK*

JEAN RICHA • *Department of Genetics, University of Pennsylvania, Philadelphia, PA*

PAUL D. ROBBINS • *Departments of Molecular Genetics and Biochemistry, University of Pittsburgh School of Medicine, Pittsburgh, PA*

JOHN M. ROGERS • *Reproductive Toxicology Division, National Health and Environmental Effects Laboratory, US Department of Environmental Protection, Research Triangle Park, NC*

GARY C. SCHOENWOLF • *Department of Neurobiology and Anatomy, University of Utah School of Medicine, Salt Lake City, UT*

RICHARD M. SCHULTZ • *Department of Biology, University of Pennsylvania, Philadelphia, PA*

DIANE C. SLUSARSKI • *Department of Biological Sciences, University of Iowa, Iowa City, IA*

BRADLEY R. SMITH • *Department of Radiology, Duke University Medical Center, Durham, NC*

JODI L. SMITH • *Department of Neurosurgery, University of Utah Medical Center, Salt Lake City, UT*

HARRIETT STADT • *Developmental Biology Program, Institute of Molecular Medicine and Genetics, Medical College of Georgia, Augusta, GA*

N. SUSAN STOTT • *Department of Surgery, Auckland Hospital, Auckland, NZ*

DAZHONG SUN • *Department of Cell Biology, Harvard Medical School, Boston, MA*

MEERA SUNDARAM • *Department of Genetics, School of Medicine, University of Pennsylvania, Philadelphia, PA*

KAZUHIRO TAKANO • *Department of Life Sciences (Biology), Graduate School of Arts and Sciences, The University of Tokyo, Tokyo, Japan*

PATRICK P. L. TAM • *Embryology Unit, Children's Medical Research Institute, Wenworthville, New South Wales, Australia*

SEONG-SENG TAN • *Howard Florey Institute of Experimental Physiology and Medicine, University of Melbourne, Parkville, Australia*

LUAN TAO • *Institute for Human Gene Therapy, University of Pennsylvania, Philadelphia, PA*

MARIE-AIMÉE TEILLET • *Institut d'Embryologie Cellulaire et Moleculaire du CNRS et du College de France, Nogent/Marnec Cedex, France*

CHARLES F. THOMAS • *Department of Integrated Microscopy Resource, University of Wisconsin–Madison, Madison, WI*

ROBERT P. THOMPSON • *Department of Cell Biology and Anatomy, Medical University of South Carolina, Charleston, SC*

TAKESHI TSUDA • *Department of Cell Biology, University of Medicine and Dentistry of New Jersey, Stratford, NJ*

ROCKY S. TUAN • *Department of Orthopaedic Surgery, Thomas Jefferson University, Philadelphia, PA*

DANIEL H. TURNBULL • *Skirball Institute of Biomolecular Medicine, New York University Medical Center, New York, NY*

B. RUSH WALLER, III • *Department of Cell Biology and Anatomy, Cardiovascular Developmental Biology Center, and Department of Pediatrics, Medical University of South Carolina, Charleston, SC*

ERIC S. WEINBERG • *Department of Biology, University of Pennsylvania, Philadelphia, PA*

ANDY WESSELS • *Department of Cell Biology and Anatomy, Medical University of South Carolina, Charleston, SC*

JOHN G. WHITE • *Department of Integrated Microscopy Resource, University of Wisconsin–Madison, Madison, WI*

JAMES M. WILSON • *Institute for Human Gene Therapy, University of Pennsylvania, Philadelphia, PA*

NELSON A. WIVEL • *Institute for Human Gene Therapy, University of Pennsylvania, Philadelphia, PA*

GREGORY A. WRAY • *Department of Medicine, State University of New York, Stony Brook, NY*

JOHN YOCHEM • *Department of Genetics and Cell Biology, University of Minnesota, St. Paul, MN*

SHIPENG YUAN • *Department of Neurobiology and Anatomy, University of Utah School of Medicine, Salt Lake City, UT*

YONG ZENG • *Department of Ophthalmology, Scheie Eye Institute, University of Pennsylvania School of Medicine, Philadelphia, PA*

MAGDALENA ZERNICKA-GOETZ • *Department of Genetics, Wellcome/CRC Institute, University of Cambridge, Cambridge, UK*

CATHERINE ZILLER • *Institut d'Embryologie, Cellulaire et Moleculaire du CNRS et du College de France, Nogent/Marnec Cedex, France*

ROBERT M. ZUCKER • *Reproductive Toxicology Division, National Health and Environmental Effects Research Laboratory, US Environmental Protection Agency, Research Triangle Park, NC*

Developmental Biology Protocols

I

INTRODUCTION

1

Developmental Biology Protocols

Overview I

Rocky S. Tuan and Cecilia W. Lo

1. Introduction

As the next millennium dawns, developmental biology, the study of the processes that give rise to cellular diversity and order within an organism and to the continuation from one generation to the next, has reached a most exciting stage as an experimental science. In particular, in the last two decades, the application of analytical and technical know-hows generated from the "molecular biology revolution" have critically advanced our understanding of development in a mechanistic way. There is every reason to believe that the study of development will be one of the most promising areas of the life sciences in the next millennium.

The goal of this three-volume set of *Developmental Biology Protocols* is to provide the reader with a richly annotated compendium of protocols representing current, state-of-the-art experimental approaches used in the study of development. The scope of the volumes is intentionally broad, as modern developmental biology is by necessity a wide-ranging discipline, involving multiple experimental systems, as well as using techniques generated from many fields. This chapter provides a brief overview of the protocols covered in this volume.

2. Systems: Production, Culture, and Storage

Beginning with Aristotle's elegant descriptive treatise on avian embryonic development (doubtlessly prompted by the incorporation of eggs as a food staple!), the use of animal model systems has been one of the most important aspects of the study of development. This volume has selected three model systems, echinoderm (sea urchin; Chapters 2 and 3), avian (chicken; Chapters 4, 5, and 6), and rodents (mouse; Chapters 7, 8, and 9), to illustrate the requirements and rationales for using particular model systems for the study of embryonic development. Readers are advised to consult other more specialized literature sources exclusively dedicated to a particular system for similar information on other experimental model systems of development, such as *Xenopus*, *Coenorhabditis elegans*, *Drosophila*, and zebrafish.

From: *Methods in Molecular Biology, Vol. 135: Developmental Biology Protocols, Vol. I*
Edited by: R. S. Tuan and C. W. Lo © Humana Press Inc., Totowa, NJ

3. Developmental Pattern and Morphogenesis

This section focuses on how the pattern and formation of specific organs and tissues may be experimentally examined. Examples include the analysis of inductive interactions (Chapter 11) and gastrulation and mesodermal patterning (Chapter 12), and the examination of head and brain (Chapter 10), craniofacial (Chapter 13) and axial skeletal development (Chapter 14), as well as cardiac morphogenesis (Chapter 15).

4. Embryo Structure and Function

The study of embryonic development depends on the precise analysis of structure and function in order to detect changes in form and shape as well as biological activities, particularly if experimental perturbations are performed. This section provides state-of-the-art methodologies for histological and immunohistochemical analyses (Chapters 16, 18, and 20), and high-resolution imaging using confocal laser scanning microscopy (Chapters 17, 19, and 20) and ultrasound backscatter microscopy (Chapter 23). Functional analyses include magnetic resonance imaging (Chapter 21), optical coherence tomography (Chapter 22), Doppler echocardiography (Chapter 24), and cellular calcium imaging (Chapter 25). The exciting application of information technology to imaging is highlighted in Chapter 26, which describes softwares developed for the acquisition, display and analysis of digital three-dimensional time-lapse data sets.

5. Cell Lineage Analysis

One of the ongoing challenges of developmental biology is to map the origin and the fate of progenitor cells in the course of tissue patterning and morphogenesis. This section presents examples of the many markers and microscopic imaging methods currently used. Cell labeling with fluorescent dyes is described (Chapters 30, 33, and 34). Gene markers, introduced recombinantly into specific cell populations, are powerful tools for cell lineage analysis (Chapters 27, 28, and 29). These approaches, coupled with new microscopic and digital computing instrumentations (e.g., Chapter 31), have provided exciting new information on cell lineage during development in many model systems.

6. Chimeras

Chimeras refer to individuals made up of the parts of more than one individual. Experimentally, by grafting cells or tissue from one embryo (donor) to another (host), transplantation chimeras can be produced in many species and often between species. Provided specific detection methods are available, such chimeras allow the investigator to follow a specific group of cells (the graft) through a period of development and to determine the fates and locations of their progeny. Chapters in this section cover multiple systems and approaches in using the chimera technology, both intra- and interspecific. Because of the oviparous nature of their development, avian embryos, specifically those of the chicken and quail, have long been used to generate transplantation chimeras (Chapters 35, 36, and 37). Recently, grafting technology has also been developed for mouse embryos (Chapter 39), as well as for interspecific chimeras, particularly in the analysis of neural crest cells (Chapter 40) and somites and neural tube (Chapter 41). For mouse embryos, the establishment of the embryonic stem cell (ES) technology has been one of the most important advances in transgenesis. The utilization of ES cells in the production of chimeras to permit developmental analysis is covered in Chapter 38. In the case of *C. elegans*, an animal whose nearly invariant cell lineage has been fully described, the use of genetic mosaics (i.e., individuals that harbor both genotypically mutant and genotypically wild-type cells), has been invaluable in determining the cells that need to inherit a functional copy of a gene in order to prevent a mutant phenotype (Chapter 42).

7. Experimental Manipulation of Embryos

A common theme in most of the chapters of this volume is the versatility of the developing embryo as an experimentally accessible system. In fact, it is the prospect of applying contemporary analytical tools to revisit "experimental embryology" that is creating the excitement among modern developmental biologists. This section describes some of the current methodologies in experimental embryology: (1) carrier-mediated delivery of growth factors (Chapter 43); (2) laser ablation and fate mapping (Chapter 44); (3) photoablation of cells expressing β-galactosidase (Chapter 45); and (4) *ex utero* surgery (Chapter 46). Given that many transgenic animals used in the study of development harbor the *LacZ* reporter gene under the regulation of promoters of putative importance, the ability to specifically ablate those cells that express β-galactosidase is of great potential application in assessing the functional importance of specific cell populations in development.

8. Application of Viral Vectors in the Analysis of Development

Retrovirus and adenovirus are the two most commonly used viral vectors for gene transduction in vertebrates. This section details the protocols in the construction and production of retroviral vectors (Chapter 47), the application of retroviral vectors in gene transduction in limb mesenchyme cultures (Chapter 48), and the construction of adenoviral vector (Chapter 49) and its application in the analysis of eye development and cardiovascular development (Chapters 50 and 51).

Volume I provides the reader with sophisticated and current information on issues of primary importance to experimental developmental biology. Practical details on the acquisition and setting up of the appropriate experimental model system, the means to analyze embryonic structure/function, the ways to perturb these processes both experimentally as well as taking advantage of current recombinant techniques, and the analysis of cell lineage, should all be of great utility to both the beginning and seasoned developmental biologists.

II

SYSTEMS: *PRODUCTION, CULTURE, AND STORAGE*

2

Rearing Larvae of Sea Urchins and Sea Stars for Developmental Studies

Christopher J. Lowe and Gregory A. Wray

1. Introduction

Sea urchins have long been used to study morphogenesis and cell fate specification and are an established model system in developmental biology *(1)*. Most contemporary studies have focused on early development, however, and few molecular genetic studies have examined larval development, or the formation of the highly derived radial body plan of the adult *(2)*. A better understanding of the molecular genetic basis of both the body plans of this phylum may contribute significantly to several fields of biology *(3,4)*.

Despite over a century of debate, the evolution of the chordate body plan from its invertebrate ancestors is still a contentious issue *(5–8)*. As a group closely related to the chordates *(8)*, echinoderms are in a crucial phylogenetic position for reconstructing the evolution of the chordate body plan *(7)*. The common ancestor of hemichordates, echinoderms, and chordates may have had a larva that resembled the early feeding larva of echinoderms *(5)*. Garstang proposed that the ciliated band of such a larva was modified by a dorsal fusion, resulting in the formation of structure that was further modified to become the chordate neural tube. A greater understanding of the molecular genetics of echinoderm larval development may provide critical insights into the evolution of key chordate innovations such as the neural tube and notochord *(8)*.

The orthologs of many body-patterning genes present throughout the bilateria have been isolated from echinoderms *(9,10)*. Understanding how these genes (seemingly so conserved in patterning the embryos of diverse metazoans), function to establish the echinoderm radial adult body secondarily from a bilateral larva should provide insights into the role of animal body-patterning genes in morphological evolution *(3,4)*. Preliminary studies have proposed that the evolution of many novel aspects of echinoderm morphology was associated with recruitment of body-patterning genes into several new developmental roles *(2,11)*.

Larval culturing techniques are described for three echinoids (*Lytechinus variegatus*, *Strongylocentrotus purpuratus*, and *Strongylocentrotus droebachiensis*) and one asteroid (*Pisaster ochraceus*). These species were chosen based primarily on practical considerations, adult availability and robustness, length of reproductive season, and ease

From: *Methods in Molecular Biology, Vol. 135: Developmental Biology Protocols, Vol. I*
Edited by: R. S. Tuan and C. W. Lo © Humana Press Inc., Totowa, NJ

of rearing. Sea urchin species were chosen because of the extensive developmental research already available from embryogenesis, and the asteroid, *P. ochraceus*, was chosen because the early bipinnarian larva may be ancestral to the echinoderms and therefore appropriate for testing hypotheses of chordate origins. Large amounts of larval material can be reliably reared using the protocols presented here. Much of the protocol described in the chapter is appropriate for rearing other echinoderm species and marine invertebrates.

2. Materials

1. Sterilized 0.53*M* KCl.
2. 1-Methyladenine (Sigma, St. Louis, MO) 100 μ*M* (1-MA) in seawater. Store at 4°C for up to 1 wk.
3. Seawater, 0.4-μm filtered (Millipore, Bedford, MA) or MBL artificial seawater *(12)* (composition per liter: 24.72 g NaCl, 0.67 g KCl, 1.36 g CaCl$_2$ · 2H$_2$O, 4.66 g MgCl$_2$ · 6H$_2$O, 6.29 g MgSO$_4$ · 7H$_2$O, 0.180 g NaHCO$_3$. Final salinity 31% pH approx 7.6 with 0.1*N* NaOH).
4. Sterilized algal culturing medium (*see* **Note 1**). Composition per liter: 1 L MBL artificial seawater and 129 μL of both A and B solutions of F/2 Algae Food (Fritz Industries, Dallas, TX) or 1 L MBL artificial seawater and 1 tube of Alga Gro® (Carolina Biological Supply Co., Burlington, NC).
5. Phytoplankton: *Rhodomonas lens* and *Dunaliella tertiolecta* (Algal Culture Collection, Botany Department, University of Texas, Austin, TX).
6. Embryological glass and labware (*see* **Note 3**).
7. Gravid adult echinoderms. Species and collection contacts: *Lytechinus variegatus* (Susan Decker, Davie, FL); *Strongylocentrotus droebachiensis* (Marine Biological Laboratory, Woods Hole, MA); *Strongylocentrotus purpuratus* and *Pisaster ochraceus* (Marinus, Long Beach, CA; Pacific Biomarine, Venice, CA).

3. Methods

For aquaria requirements and adult maintenance, refer to **Note 2**. All culturing glass and labware used for culturing should be clearly marked and maintained as separate stock (*see* **Note 3**). Refer to **Note 4** for a discussion of specific rearing requirements for each species. Procedures for rearing other echinoderm larvae are described in *(14)*.

1. Collection of sea urchin gametes: Invert urchin and inject approx 2 mL of 0.53*M* KCl into the coelomic cavity by directing the syringe needle through the peristomial membrane between the mouth and the perimeter of the test (*see* **ref.** *13* for description of adult anatomy). Repeat injection several times at different points around the peristomial membrane to ensure that each of the five gonads, lying on the inside of the test, are exposed to KCl. Place the inverted urchin onto the rim of a beaker filled with seawater (at the appropriate temperature, *see* **Note 4**), and wait for the urchin to spawn gametes through the gonopores at the apex of the test on the aboral side. Spawning typically begins approx 30 s after injection. Oocytes will fall in streams to the bottom of the beaker. Collect oocytes and rinse several times by allowing the eggs to settle, resuspending in seawater, and decanting. Sperm is white and will rapidly cloud the seawater. Once identified by release of sperm, males should be removed from the seawater and sperm collected "dry" by Pasteur pipet (**Note 5**).

2. Collection of asteroid gametes: Spawning of gametes is induced by injection of 100 μ*M* 1-methyladenine through the body wall and into the lumen of each arm close to the disk (*see* **ref. *13*** for description of anatomy). 1 mL of 1-MA should be injected for every 100 mL of body volume (*see* **Note 6**). Animals should be separated from each other, placed into buckets, covered with seawater at 12°C and left for approx 2 h. Spawning should begin shortly thereafter. Eggs and sperm are released through the gonopores on the aboral surface of each arm close to the disk. Sperm should be collected "dry" (*see* **Note 5**). Females are often very fecund. Collect oocytes, decant off seawater, and rinse several times in filtered seawater.

3. Fertilization: Fertilize eggs from stages 1 or 2 within 1 to 2 h (sooner if possible). For both asteroids and echinoids, dilute one drop of "dry" sperm in 100 mL of filtered seawater. Add 20 drops of this sperm suspension to each 200 mL of egg suspension and stir for 2 min. Allow eggs to settle, then resuspend in fresh filtered seawater *(14)*. For sea urchins, fertilization success can be determined by checking for raised vitelline envelopes under the microscope within minutes. For asteroids, the first unambiguous sign of successful fertilization will be cell division.

4. Culture of embryos: Zygotes should be transferred to large glass containers (1 gal pickling jars are ideal) and should not exceed a density of more than a monolayer on the bottom of the container. Embryos should be left to develop, at the appropriate temperature, until hatching (*see* **Note 7**). The seawater should be changed and cultures cleaned when the embryos begin swimming, to remove the shed vitelline envelopes, as these promote bacterial growth. The density of the hatched blastula should not exceed 1 individual/mL. Transfer excess embryos into new containers. Higher density cultures result in asynchrony and developmental abnormalities and increase risk of cultures crashing unpredictably (**Note 8**). After embryos have hatched, cultures should be stirred gently with paddles (**Note 9**).

5. Maintaining larval cultures: Seawater should be changed at least every other day, more often if there is evidence of algal or bacterial growth. Changing seawater requires care, as the larvae are easily damaged. *S. droebachiensis* is relatively robust, but both *L. variegatus* and *P. ochraceus* are particularly fragile. Several methods can be used for this purpose *(14)*. We prefer using a 200-mL plastic beaker whose bottom has been removed and covered with a Nytex® mesh, as this is the most efficient method for processing large numbers of cultures. The beaker is submerged in a shallow dish and placed in a sink (*see* **Note 10**), and cultures are gently poured into the beaker. The water flows through the Nytex mesh and overflows out of the shallow dish retaining the larvae behind the Nytex mesh. Fresh filtered seawater should be gently poured through to wash the larvae (several times). The culture container should be rinsed once with hot fresh water and once with filtered seawater before larvae are returned to it. To transfer larvae, gently pour the contents of the mesh-bottom beaker back into the clean container or pipet them using a turkey baster.

6. Culture feeding: Larvae should be fed every 2 d. From dense cultures of *R. lens* and *D. tertiolecta* (approx 10^5 cells/mL), spin down algal cells at 5000 rpm for 1 min. Pour off the supernatant, and resuspend the algal pellet in seawater to original starting volume. Failure to replace algal growth media will result in increased bacterial growth in cultures. Calculate the algal density with a hemacytometer and add an appropriate amount of each alga to reach a total of between 8000 and 10,000 cells/mL (*see* **Note 11**).

7. Algal culturing: Algae should be grown under full-spectrum fluorescent lights and aerated. A small aquarium pump can be used to aerate cultures. Syringe filters (Acrodisc®, 0.2 μm, Gelman Sciences, Ann Arbor, MI) can be inserted in the air lines to prevent culture contamination. For small cultures (100–200 mL), sterile Pasteur pipets can be used to aerate. Larger cultures require a larger-bore glass tube (such as 5- or 10-mL glass pipets)

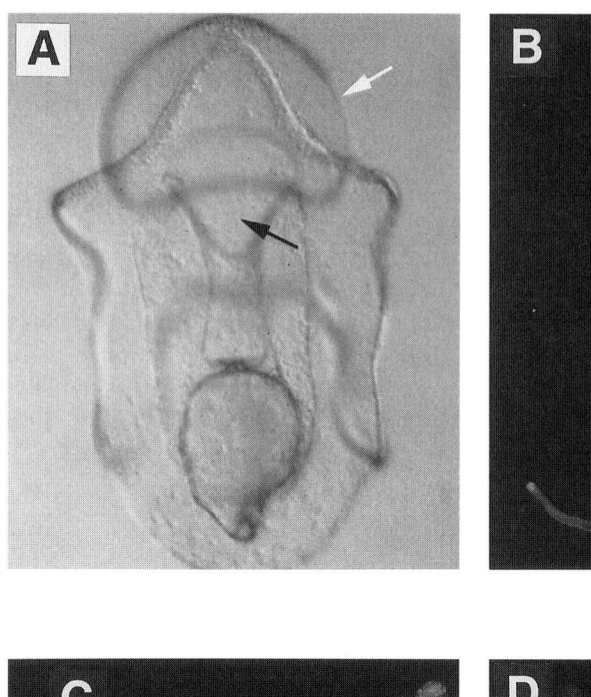
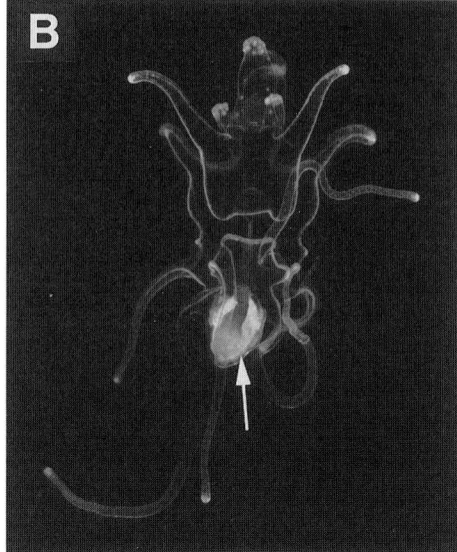
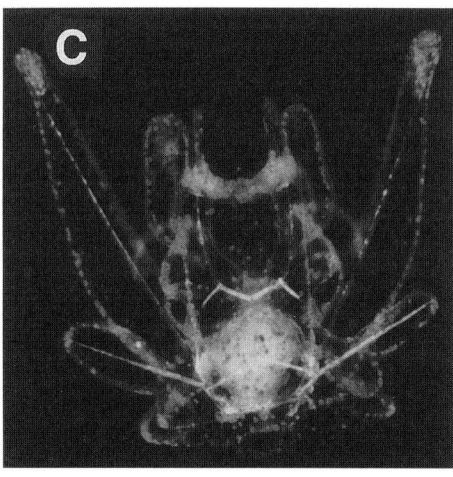
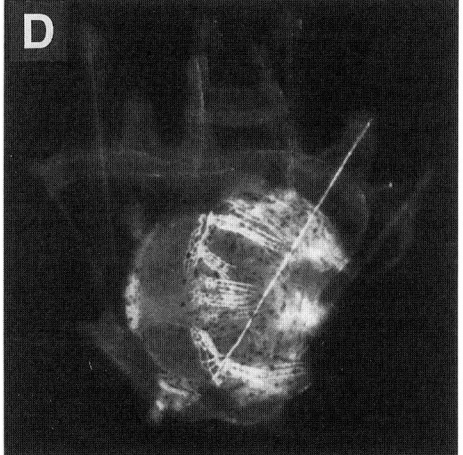

Fig. 1. (*See* color plate 1 appearing after p. 258.) Larvae of sea urchins and sea stars. All larvae are oriented with anterior up. (**A**) Bipinnaria larva of *Pisaster ochraceus*. A convoluted ciliated band is used for both locomotion and feeding (white arrow). The mouth is located in the mid anterior on the ventral surface (black arrow) and leads into the esophagus and stomach. (**B**) Brachiolarian larva of *Evasterias troschelii* with larval morphology very similar to that of *Pisaster ochraceus*. Later brachiolarian larvae develop large arms extending the length of the ciliated bands, and increase greatly in size. The development of the adult rudiments begins on the left hand side of the larva, spreading around the stomach to the right hand side (white arrow). (**C**) Pluteus larva of *Lytechinus variegatus*. Development of larval spicules support the extension of the ciliated band into arms, which are used for both feeding and locomotion. (**D**) Late larva of *Strongylocentrotus droebachiensis* close to metamorphosis. Development of the adult is clearly visible on the left hand side of the larva. The oral surface of the adult is against the larval ectoderm on the left-hand side of the larva. Larval structures such as the spicules are still clearly visible.

to maintain adequate culture mixing. Use only embryological labware (*see* **Note 3**), and maintain sterile technique when opening flasks. Foam bungs are convenient for sealing the flask. Starter culture innoculae arrive in small volumes. Add double the volume of sterilized algal media in a small sterile Erlenmeyer flask, and aerate with enough force to cause circulation of medium within the flask. Cultures should be doubled in volume every day, or every other day, by adding fresh medium, up to a final vol of 3 L. Every 2–3 d, dilute the cultures with at least an equal volume of fresh algal medium. *R. lens* will become a deep purple color with increasing density, and *D. teriolecta* will be bright green. Optimal density for algae is approx 10^5 cells/mL. Maintain several cultures of each alga at staggered densities to ensure a continuous supply of dense algae throughout rearing.

8. **Figure 1** shows the larval morphology of early and late feeding larva of sea urchin and sea star development. For complete description of the development of these species, refer to (*16,17*).

4. Notes

1. Autoclave seawater for 15 min on a liquid cycle. Extended autoclaving causes salts to precipitate out of solution. Seawater may be boiled briefly if an autoclave is not available.
2. Aquarium and adult requirements: Adult echinoderms should be kept in circulating marine aquaria. Ideally, several filter sets should be used: under gravel filter, activated charcoal filter, and particulate filter. Artificial seawater may be used (e.g., Instant Ocean®) but filtered natural seawater is preferable. Periodic checks on levels of nitrate, ammonia, nitrite, and pH should be carried out. Ask collectors to ship adult animals overnight to minimize losses in transit. Before placing animals into aquaria, remove any individuals that have spawned or are in the process of spawning. Check at regular intervals over the first 24-h period for spawning individuals and remove them from the aquaria. Presence of gametes in the seawater can induce mass spawning. Animals should be fed periodically to maintain ripe gonads. Echinoids can be fed sliced grapes or carrot shavings, and *P. ochraceus* should be fed live mussels.
3. Larval culturing requires laboratory space and glassware designated only for larval rearing. All materials and work areas should be kept free of detergent, toxins, fixatives, and heavy metals and should be washed only with fresh water. Label glass and plastic ware to avoid contamination.
4. The choice of species depends on a variety of factors, including:
    a. Available culture and aquaria facilities: Adult *S. purpuratus*, *S. droebachiensis*, and *P. ochraceus* require chilled seawater aquaria between 12 and 15°C. Ideally their larvae should also be reared between 12 and 15°C. Maintenance of adult *L. variegatus* requires heated seawater aquaria between 22 and 28°C, but their larvae can easily be reared at room temperature. *S. droebachiensis* and *L. variegatus* are usually easier to rear successfully than *S. purpuratus* and *P. ochraceus*.
    b. Experimental purpose of larval rearing: If the purpose of rearing larvae is for molecular genetic studies, then *S. purpuratus* or *L. variegatus* are recommended, as more molecular genetic information has been accumulated for these species. No asteroids are commonly used in developmental studies, but *P. ochraceus* is the most easily collected and maintained.
5. Once sperm are diluted in seawater, their motility declines rapidly. Collect sperm with as little seawater as possible (*14*). Undiluted sperm may be stored for up to 3 d at 5°C.
6. Asteroid oocytes are arrested in meiosis. Completion of meiosis and spawning is stimulated by exposure to 1-MA.

7.

Species	Optimal rearing temperature (°C)	Time to hatching (h)	Tolerance range (°C)
S. purpuratus	15	~18	10–18
S. droebachiensis	12	~24	10–15[a]
L. variegatus	28	~8	20–28
P. ochraceus	12	~29–32	10–15[a]

[a]We have had some success with rearing cold-water species at room temperature by allowing the embryos to first gastrulate at their optimal rearing temperature, then gradually increasing the culture temperature to room temperature. Larvae of individuals collected at lower latitudes in their range are probably more temperature tolerant than populations from more northern latitudes.

After *(14)* and *(15)*.

8. Generally larvae are less affected by density at early developmental stages and become increasingly sensitive as they grow. If the purpose of rearing is to obtain large quantities of larvae at a range of developmental stages, then earlier cultures may be maintained at higher densities and gradually thinned as each developmental time point is sampled. Late larvae of *P. ochraceus* are particularly sensitive to high densities. If there is a large amount of heterogeneity in developmental rate within the culture, then the density is probably too high.

9. Cultures should be stirred. Strathmann *(14)* describes several methods of culture stirring. We suggest the use of plastic paddles and a low rpm motor (Grainger Scientific, Lake Forest, IL). Large quantities of larvae may be reared using this apparatus. Use of a stir bar is not recommended. Sea urchins may be reared without stirring if the density is kept to approx 1 larva/10 mL.

10. A range of Nytex mesh sizes should be used. When cleaning early larval cultures, use a 30-μM mesh. Increase the size to 100 μM midway through larval development and use 200-μM mesh for late larvae. Larger mesh diameters allow for more effective larval rinsing and allow particulate culture contaminants to be washed out. Use a hot glue gun (not solvent-based glues) to attach the mesh to the base of the plastic beaker from which the bottom has been removed.

11. Reducing density and regular feeding will result in rapid and synchronous development. *S. droebachiensis* and *S. purpuratus* can reach metamorphosis in less than 25 d if fed at the recommended levels. *L. variegatus* can reach metamorphosis in 12 d if kept at lower densities but typically will take longer at a density of 1 larva/mL. *S. droebachiensis* seems less sensitive to density, and large numbers of larvae can regularly be reared at 12°C to metamorphosis in less than 30 d. High densities and lower algal densities will slow down the rate of development. *P. ochraceus* develops more slowly and can take up to 8 wk until metamorphosis. However, with low densities and high rates of feeding, metamorphosis can be reached in 5 wk.

References

1. Wray, G. A. (1997) Echinoderms, in *Embryology* (Gilberts, S. C. and Raunio, A. M., eds.), Sinaeur, Sunderland, MA, pp. 309–330.
2. Lowe, C. J. and Wray, G. A. (1997) Radical alterations in the roles of homeobox genes during echinoderm evolution. *Nature* **389,** 718–721.
3. Slack, J. M., Holland, P. W., and Graham, C. F. (1993) The zootype and the phylotypic stage. *Nature* **361,** 490–492.
4. Raff, R. A. (1996) *The Shape of Life.* University of Chicago Press, Chicago.

5. Garstang, W. (1928) The morphology of the Tunicata, and its bearings on the phylogeny of the Chordata. *Q. J. Microsc. Sci.* **72,** 51–187.

6. Berrill, N. J. (1955) *The Origin of Vertebrates*, Clarendon, Oxford.

7. Lake, J. A. (1990) Origin of the Metazoa. *Proc. Natl. Acad. Sci.* USA **87,** 763–766.

8. Gee, H. (1996) *Before the Backbone*. Chapman Hall, London.

9. Popodi, E., Kissinger, J. C., Andrews, M. E., and Raff, R. A. (1996) Sea urchin hox genes—Insights into the ancestral hox cluster. *Mol. Biol. Evol.* **13,** 1078–1086.

10. Wray, G. A. and Lowe, C. J. Developmental regulatory genes and echinoderm evolution. *Systematic Biology*, in press.

11. Lowe, C. J. and Wray, G. A. Gene recruitment in echinoderm early life history evolution, in preparation.

12. Cavanaugh, G. M., ed. (1975) *Formulae and Methods of the Marine Biological Laboratory Chemical Room*, 6th ed. Marine Biological Laboratory, Woods Hole, MA.

13. Brusca, R. C. and Brusca, G. J. (1990) *Invertebrates*, Sinauer, Sunderland, MA.

14. Strathmann, M. F. (1987) *Reproduction and Development of Marine Invertebrates of the Northern Pacific Coast*. University of Washington Press, Seattle.

15. Wray, G. A. and McClay (1989) Molecular heterochronies and heterotopies in early echinoid development. *Evolution* **43,** 803–813.

16. Kumé, M. and Dan, K. (1968) *Invertebrate Embryology*, Nolit Publishing House, Belgrade, Yugoslavia.

17. Fraser, A., Gomez, J., Hartwick, E. B., and Smith M. J. (1981) Observations on the reproduction and development of *Pisaster ochraceus* (Brandt). *Can. J. Zool.* **59,** 1700–1707.

3

Large-Scale Culture and Preparation of Sea Urchin Embryos for Isolation of Transcriptional Regulatory Proteins

James A. Coffman and Patrick S. Leahy

1. Introduction

The development of a complex multicellular organism from a single-celled zygote requires that the protein structures encoded in the DNA of the organism's genome be expressed in specified cells at specified levels, contingent on specific signals generated at specific stages of development. In fact, explicit "instructions" that direct gene expression during metazoan development are also genetically encoded, within regulatory domains of genes *(1,2)*. The binding of transcriptional regulatory proteins to specific DNA sequences within a gene's regulatory domain serves to modulate the transcriptional output of the gene, by intermolecular mechanisms of activation or repression. Genetic *cis*-regulatory domains therefore constitute information-processing systems that interpret information provided by the cell—that is, transcription factors active in the nucleus—in terms of transcriptional output. A fundamental approach to studying the developmental information flow that regulates gene expression is to characterize both the *cis* elements and *trans*-acting factors involved in the control of developmentally regulated genes *(2)*. In fact, analysis of *cis*-regulatory systems leads directly to the characterization of the *trans*-acting factors, since the latter are proteins that bind specific DNA sequences with relatively high affinity. This allows their direct isolation by affinity chromatography on columns bearing specific oligonucleotide target sites *(3–5)*.

A major technical difficulty of isolating transcription factors is presented by the fact that they are typically among the least abundant proteins in a cell *(4–6)*. Their purification therefore requires an abundance of raw material. A particularly good system in this regard is the sea urchin. Females of the species *Strongylocentrotus purpuratus* typically carry $>10^7$ eggs during the peak of their season, and it is thus possible to recover approx 10^{10} eggs by spawning 1000 female animals. This is well over the minimum amount of starting material required to purify even the least abundant transcription factors *(5,6)*. In addition, sea urchin eggs are easily fertilized in vitro and can be grown in culture, where they develop synchronously into larvae. While the culture of small (10^5–10^6) or moderate (10^7–10^8) numbers of sea urchin embryos is

From: *Methods in Molecular Biology, Vol. 135: Developmental Biology Protocols, Vol. I*
Edited by: R. S. Tuan and C. W. Lo © Humana Press Inc., Totowa, NJ

relatively straightforward and has been described in detail elsewhere *(7)*, the culture of huge numbers presents a number of unique technical challenges. Here we describe our two day procedure for the culture, harvest, freezing, and storage of $1–2 \times 10^{10}$ sea urchin embryos for the purpose of purifying transcriptional regulatory proteins.

2. Materials

1. Facilities (optimally at a marine lab):
   a. 4°C cold room with plenty of deck space.
   b. 16°C culture room big enough to fit 8–10 20-gal trash cans.
   c. Storage carboys containing at least 400 L of filtered seawater at 4°C.
   d. Aeration system large enough to saturate 200 gal of culture media.
   e. –70°C freezer with plenty of space.
2. Human resources: 12–18 able-bodied humans for day 1, 4–6 for day 2.
3. Sea urchins: 1000–2000 gravid *S. purpuratus*, supplied by various sources (available by request from authors).
4. Hardware:
   a. Syringes and needles: 20-cc syringes, 21-gauge needles (Becton-Dickinson, Rutherford, NJ).
   b. Basins for collecting eggs: rectangular plastic containers (16-cup Ultra Seal, Model #0228, Sterilite, Townsend, MA), each containing a plastic screen support resting on a plastic spiral of the same material (All Purpose Plastic Screening, 3/8 in. × 3/8 in. mesh size, #E-09403-30, Cole-Parmer Instrument Co., Vernon Hills, IL), raising the screen approx 1.25 in. off the bottom of the container.
   c. Nitex filters and apparatus: (i) For washing eggs (several): 150 μM Nitex mesh (Tetko Inc., Briarcliff Manor, NY), attached to the small end of a plastic 1-L tricorn beaker from which the bottom has been cut off. Either a heavy-duty rubber band or the top half of another tricorn beaker can be used to attach the Nitex. (ii) For collecting embryos (two): 51 μM Nitex mesh, lining plastic colanders (14 in. diameter × 5.5 in. deep), attached by plastic clips.
   d. 200 1-L plastic tricorn beakers for washing eggs and at least 4 8-L culture vessels (Nalgene multipurpose jars with lids, 8.3 L, Nalge # 5300-9910, Rochester, NY) for pooling washed eggs and harvesting embryos.
   e. Culture cans: 8–10 20-gal Rubbermaid refuse containers (19.5 in. × 23.5 in. depth) with modified lids (**Fig. 1**).
   f. Stir motors and paddles: For 8-L Nalgene jars (for pooling washed eggs): 3 in. × 4.5 in. paddles attached to nylon shaft via nylon screws, driven by a 20-rpm motor (Model #H1-11, H&R, Mt. Laurel, NJ). For large (20-gal) culture cans (for overnight culture; *see* **Fig. 1**): 10 in. × 12 in. nylon paddles attached to 21 in. nylon shaft with nylon screws, driven by a 60 rpm motor (Model #772RW9040, Bodine Electric, Chicago, IL).
   g. Tubing (3/16 in. ID, 5/16 in. OD) and aquarium airstones for aeration.
   h. Two 15-quart buckets for collecting embryos (Ultra Pail, #1124, Sterilite).
   i. 2 large and 2 small wash basins.
   j. 8 1-L centrifuge bottles (Nalgene).
   k. Ziploc heavy-duty freezer bags (gallon size).
   l. Heavy-duty aluminum foil.
   m. Liquid nitrogen cylindrical dewer (14 in. diameter × 24 in. deep).
5. Solutions and buffers:
   a. Millipore filtered seawater (MFSW) for sperm suspension, filtered seawater (FSW) for washing eggs and for cultures.

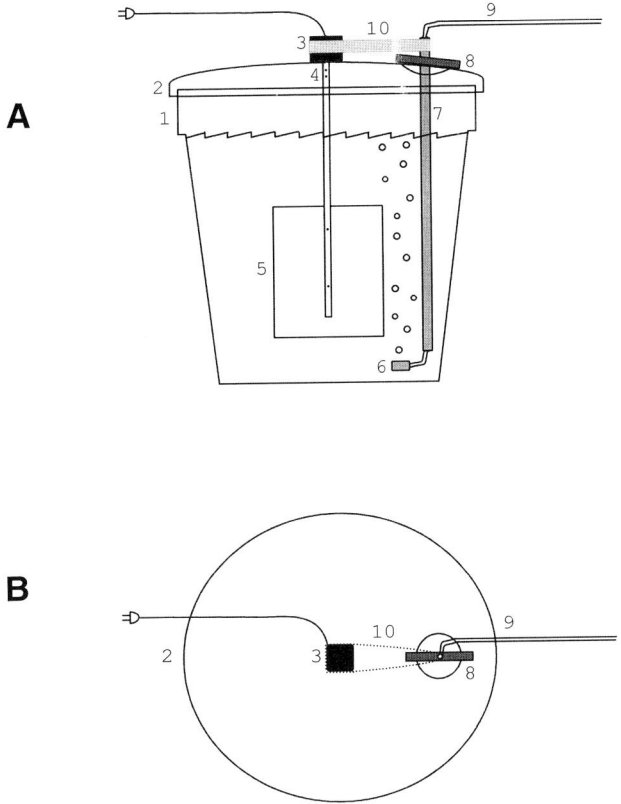

Fig. 1. Schematic of large-scale culture system for sea urchin embryos. **(A)** Side (cutaway) view showing the following: 1. 20-gal Rubbermaid trash can. 2. Trash can lid. 3. Electric (AC) motor that drives rotating shaft inserted through a hole in the trash can lid. The motor is mounted directly onto the lid. 4. Nylon paddle shaft (21 in. long), affixed to the motor shaft with two nylon screws. 5. Plastic paddle (10 in. × 12 in.), affixed to the paddle shaft with two nylon screws. 6. Standard aquarium air stone attached to the air line. 7. Stiff thin plastic cylinder (shower rod cover) used as a passageway for the air line and to keep the airstone away from the paddle ("air line assembly"). 8. Access hole cut in the trash can lid, bridged by a wooden block with a hole drilled in the center that serves as a support for the air line assembly. 9. Air line (5/16 in. OD tubing) connected to a regulated flow of compressed air. 10. Duct tape that wraps around the motor and air line assemblies, functioning to hold the air line assembly out of the way of the turning paddle. **(B)** Top (surface) view of modified trash can lid assembly. Numbers refer to the same features as in (A).

    b. 0.5 *M* KCl.
    c. 1 *M* glucose.
    d. Buffer A (TEESSD): 10 m*M* Tris-HCl, pH 7.4, 1 m*M* EDTA, 1 m*M* EGTA, 1 m*M* spermidine trihydrochloride, 1 m*M* DTT, 0.36 *M* sucrose.
    e. Liquid nitrogen (80-L bottle).

3. Methods

1. Spawning and collecting gametes: (**Note**: It is advisable to wear gloves during this procedure to avoid being pricked by spines.) It is not possible to discriminate between male and

female *S. purpuratus* without inspecting their gametes. However, a few gametes will often be shed if the animal is shaken vigorously, allowing selection of females before spawning. This is advantageous since sperm from only a few males is sufficient to fertilize billions of eggs—Selecting females for spawning thus saves both time and solutions. Spawning is induced by injection with 0.5 *M* KCl. In general, the minimum amount of 0.5 *M* KCl required to induce spawning should be used (usually ≤1 mL per sea urchin), as too much KCl can have deleterious effects on the gametes. The solution is injected into the coelomic cavity through the peristomial membrane surrounding the mouth (Aristotle's lantern) on the ventral side of the animal. The solution should be injected at several points around the circumference to ensure activation of all five gonopores. The animal is then shaken vigorously. The gametes will be shed from the gonopores on the dorsal surface of the animal. Eggs are pale yellow, sperm is milky white. Sperm is collected "dry" by placing the males upside down over a weighing dish, small beaker, or watch glass. Usually 10–15 mL of sperm (an amount typically shed by 20–30 males) is more than enough to fertilize $1–2 \times 10^{10}$ eggs. Eggs should be collected immediately after injection by placing the females, upside down, on the plastic screen support in plastic basins that have been prearranged on a table of ice and filled with enough seawater to cover the gonopores (*see* **Note 1**). The eggs will thus be shed into the seawater and settle on the bottom of the container. Alternatively, small jars (such as baby food jars) or beakers of seawater can be used to collect the eggs, but this is more cumbersome and takes up more space, major considerations when spawning 1000 animals at a time. Allow the animals to spawn to completion; generally, this takes 20–30 min. Sometimes they will appear to stop shedding but can be induced to begin again by vigorous shaking (*see* **Note 1**).

2. Washing eggs: The eggs are washed as follows. After the spawned eggs have settled in the containers, the sea urchins and screen support are removed, and as much seawater as possible (without losing eggs) is carefully decanted to waste. The settled eggs are then resuspended by swirling (giving an approx 30% slurry) and passed through a 150-μ*M* Nitex filter apparatus (*see* **Subheading 2.**) into plastic tricorn 1-L beakers, up to approx 1/3-full (approx 300 mL, containing approx 100 mL eggs). The beaker is then filled to the top with ice-cold FSW and placed in a 4°C cold room, where the eggs are allowed to settle (*see* **Note 2**). After the eggs have settled completely, as much of the seawater is removed as possible, and the eggs are again resuspended in fresh ice-cold FSW. This process is repeated until all the eggs have been washed four times (four settlings in FSW). After decanting the supernatant from the fourth wash, the eggs (approx 30% slurry) can be pooled in a larger container, such as an 8-L culture can, wherein they should be stirred continuously until distribution to the large culture cans. When all the eggs have been washed and pooled, a measured aliquot of the eggs should be counted.

3. Setting up the culture cans: The culture cans should be set up in the 16°C-culture room while the eggs are washing (i.e., concurrent with step 2). Typically, a culture of $1–2 \times 10^{10}$ embryos can be achieved in 8–10 20-gal trash cans. The cans are each filled with approx 60 L of filtered seawater and fitted with the modified lids in which motorized stir paddles are mounted (**Fig. 1**). Air lines fitted with stones are fed to the bottom of each can through a hole in the lid, and airflow is regulated to give a vigorous (but not violent) rate of bubbling.

4. Fertilization and culture of embryos: The eggs should be evenly distributed between the culture cans for fertilization. The maximum density for successful development to the blastula stage is approx 3×10^4/mL; a slightly lower density is better (e.g., 2×10^4/mL). Thus, for example, 1.5×10^{10} embryos will need to be evenly distributed between eight culture cans, each containing 60 L of seawater. Once the eggs are distributed into the cans,

the stir motors are turned on, and the eggs are fertilized by adding sperm that has been freshly diluted in MFSW (10 mL MFSW per milliliter "dry" sperm). To start, 10 mL of freshly diluted sperm are added to each 60-L culture can. Fertilization is monitored in an aliquot of eggs through a dissecting microscope; successfully fertilized eggs raise a distinctive fertilization envelope that appears as a halo surrounding the egg. If <90% of the eggs are fertilized, more sperm should be added. It is desirable to achieve at least 90% fertilization (*see* **Note 3**). A few (approx 5–10) drops of antifoam A emulsion (Sigma, No. A-5758, St. Louis, MO) are added to each culture can to prevent foaming, and the embryos are allowed to develop to the hatched blastula stage, which in *S. purpuratus* takes approx 24 h (*see* **Note 4**).

5. Harvesting, freezing, and storage of embryos: Once the embryos have hatched, they are harvested by straining through large plastic colanders lined with 50-μM Nitex mesh. The Nitex is affixed to the colander with plastic clips. This is a two-person job that is performed as follows. One person dips a bucket into the culture can to collect an approx 3-gal aliquot of embryos, then slowly pours the bucket's contents through a filter affixed to a colander that the second person holds and continuously rotates over a large wash basin. The rate of rotation of the colander should be such that the embryos are continuously in suspension and not clogging the filter. To achieve this, the first person should adjust their rate of pouring so that the filter does not get too full, allowing the second person to maintain a good "swirling" motion (i.e., the liquid retained in the filter should be low enough that most of it "hugs" the sides of the colander by centrifugal force). When a good amount of concentrated embryos (up to 1 L) have accumulated in the filter (they look like orange juice when they are concentrated enough) they are poured into a large container on ice (e.g., an 8-L beaker or culture vessel). This procedure will require a little practice. Typically, a single culture containing approx 2×10^9 embryos can be concentrated down to approx 4–6 L. Note that the seawater filtrate will fill the wash basin quickly and that the wash basin will thus need to be emptied several times during the filtration process. For this reason, this procedure should be performed in a location where several hundred liters of seawater can be conveniently disposed of. After 4–6 L of concentrated embryos have been accumulated, they are distributed to 1-L centrifuge bottles and centrifuged at 1500g for 5–10 min at 4°C. The supernatant is removed by aspiration, and the embryo pellet is resuspended in ice-cold 1 M glucose (approx 10 vol). It is best, at first, to only fill the centrifuge bottle with the glucose halfway, cap it, and shake it vigorously to break up the embryo pellet. Then the bottle is topped off with more 1 M glucose, and the embryos are recentrifuged at 2000g for 5–10 min at 4°C. The second pellet is generally much more compact, so the supernatant can be removed by careful decanting. (The pellet should stay in one piece.) The pellet is finally resuspended in approx 700 mL ice-cold Buffer A and shaken vigorously to resuspend the embryos. The embryos are transferred to 1-gal heavy-duty Ziploc bags, which are sealed and placed in a large dewer of liquid N_2 for approx 5 min to freeze the embryos. For storage, two bags of frozen embryos (approx 1×10^9 embryos) are wrapped in heavy-duty aluminum foil, forming a square "brick" that can be conveniently stacked in a –70°C chest freezer for storage.

6. Initial manipulation of frozen embryos for preparation of nuclear extracts: preparation of nuclear extracts from sea urchin embryos has been described in detail elsewhere (*4,6,8*) and will therefore not be described here. However, there are several practical considerations worth mentioning that affect the large-scale preparation of nuclear extract from frozen embryos. It is generally easier to prepare high-quality nuclear extract by limiting the number of embryos processed to approx 1×10^9 at a time (i.e., the amount typically frozen in two plastic bags wrapped together in an aluminum foil packet). The frozen

embryos must first be broken into small pieces. For this purpose, we generally use a hard rubber or plastic mallet and a large wash basin. The resultant chunks of frozen embryos are then transferred to a bucket, which is placed in a lukewarm water bath. The chunks are stirred vigorously with a stiff piece of PVC until they turn slushy. At this point, they can be stirred with a stainless steel mixing-propeller driven by a variable speed motor or other heavy-duty electric mixer such as those used in baking. In the process of thawing, the cells lyse. When the embryos are completely thawed (no ice chunks or slush remaining), they are transferred to 1-L centrifuge bottles, wherein the nuclei are recovered by centrifugation at 2500g at 4°C. The nuclei are washed several times by repeated resuspension in Buffer A followed by centrifugation at 2500g. Nuclear extract is then prepared exactly as described previously *(4,8)*.

4. Notes

1. When females begin spawning into the collection basin, clumps of eggs should visibly drop from their gonopores. Sometimes the eggs will stick to the dorsal surface of the animal at first. They can be dislodged by gently agitating the inverted animal, with the gonopores immersed in the seawater of the collection basin. The eggs will continue to stream out of the gonopores for some time. However, since time is a consideration in this procedure, spawning can be judged to be complete when, after shaking the animal, only a tiny trickle of eggs continues to be shed. Also, after awhile the animal might begin to shed coelomic contents other than gametes (i.e., coelomic fluid containing coelomocytes). This is evidenced by a reddish coloration in the eggs. As this material might cause problems with fertilization or contamination of the culture, it should be avoided if possible— that is, females that are shedding red should be removed from the spawning basin.

2. Washing the eggs is the most important part of the procedure, and the most laborious, because it must be done in the cold room. Unfortunately, inadequately washed eggs do not fertilize efficiently. After one or two washes, the rate of settling will increase and the supernatant will become clearer. One potential problem that can arise in the washing is that the eggs are allowed to sit too long after settling. This should be avoided, as it can lead to anoxia, which also inhibits fertilization and causes abnormal development or death.

3. Sometimes the requisite ≥90% fertilization cannot be achieved, no matter how much sperm is added. There are several possible reasons for this. In addition to problems with the washes (*see* **Note 2**), it is possible that the sperm is partially inactive. This is generally not a problem, however, when using fresh sperm from several individuals that has been kept "dry" (undiluted) on ice. In any event, care should be taken not to add too much sperm (greater than five times the amount recommended here), as it can serve as food for bacteria that might contaminate the culture and kill the embryos.

4. It is important that the embryos hatch before they are harvested. If they do not, their fertilization membranes will cause problems with the preparation of nuclear extracts (by coagulating with the nuclei, making clean isolation of nuclei difficult). After the embryos have hatched, the fertilization envelopes (and other contaminants present in the culture, such as unfertilized eggs) will wash through the Nitex filters during collection of the embryos. Hatched blastulae are obvious under a dissecting microscope: they no longer are surrounded by the "halo" of the fertilization membrane, and they are swimming (and thus also visible even to the naked eye). Embryos that are nearly ready to hatch can be seen to visibly rotate within their fertilization envelope. If embryos of a later stage are desired, they must be either grown at a lower density from the beginning (preferably) or diluted after hatching to a density of 5000–10,000/mL. It is therefore much less practical to culture 1–2 × 10^{10} embryos to gastrula stage or later.

Acknowledgments

The authors are indebted to Dr. Frank Calzone, who was instrumental in developing the protocol described here. We also thank members of the Davidson Lab, past and present, who have participated in the yearly large-scale harvest of sea urchin embryos at the Kerckhoff Marine Laboratory of the California Institute of Technology. Finally, we thank Dr. Eric Davidson, without whose vision and support this protocol would never have been developed. This work was supported by NIH Grant HD-05753, NSF Science and Technology Center Grant BIR 9214821, the Beckman Institute, and the Stowers Institute for Medical Research.

References

1. Britten, R. J. and Davidson, E. H. (1969) Gene regulation for higher cells: A theory. *Science* **165,** 349–358.
2. Arnone, M. I. and Davidson, E. H. (1997) The hardwiring of development: organization and function of genomic regulatory systems. *Development* **124,** 1851–1864.
3. Kadonaga, J. T. and Tjian, R. (1986) Affinity purification of sequence-specific DNA binding proteins. *Proc. Natl. Acad. Sci. USA* **83,** 5889–5893.
4. Calzone, F. J., Hoog, C., Teplow, D. B., Cutting, A. F., Zeller, R. W., Britten, R. J., and Davidson, E. H. (1991) Gene regulatory factors of the sea urchin embryo. I. Purification by affinity chromatography and cloning of P3A2, a novel DNA binding protein. *Development* **112,** 335–350.
5. Coffman, J. A., Moore, J. G., Calzone, F. J., Britten, R. J., Hood, L. E., and Davidson, E. H. (1992) Automated sequential affinity chromatography of sea urchin embryo DNA binding proteins. *Mol. Mar. Biol. Biotechnol.* **1,** 136–146.
6. Calzone, F. J., Thézé, N., Thiebaud, P., Hill, R. L., Britten, R. J., and Davidson, E. H. (1988) Developmental appearance of factors that bind specifically to *cis*-regulatory sequences of a gene expressed in the sea urchin embryo. *Genes Dev.* **2,** 1074–1088.
7. Leahy, P. S. (1986) Laboratory culture of *Strongylocentrotus purpuratus* adults, embryos, and larvae. *Methods Cell Biol.* **27,** 1–13.
8. Cameron, R. A., Zeller, R. W., Coffman, J. A., and Davidson, E. H. (1995) The analysis of lineage specific gene activity during sea urchin development, in *Molecular Zoology: Advances, Strategies & Protocols* (Ferraris, J. D. and Palumbi, S. R., eds.), Wiley-Liss, New York, pp. 221–243.

4

The Chick Embryo as a Model System for Analyzing Mechanisms of Development

Diana K. Darnell and Gary C. Schoenwolf

1. Introduction

The chick embryo provides an excellent model system for studying the development of higher vertebrates wherein growth accompanies morphogenesis. (Note: virtually all information given here for the chick embryo is applicable to the quail embryo, and much of it is applicable to embryos of other avians including domesticated and wild species.) There are many advantages to working with chick embryos. Chicken eggs are available year-round, they are inexpensive, and they can be purchased in any specified quantity (no excess or shortfall). If eggs are acquired and used within a week, unincubated eggs can be stored in any cool place, obviating the need for a special storage facility. Chicken eggs can be incubated to any stage of interest, simplifying experimental design and allowing the investigator to coordinate his or her schedule with the need to have embryos at the desired developmental stage for a particular experiment. At the time the egg is laid, the avian embryo consists of a flat, two-layered blastoderm that lies on the surface of the yolk and, therefore, is readily accessible. Subsequent development occurs with incubation at 38°C and is rapid. Within 2 to 3 d of laying, chick embryos gastrulate, neurulate, and fold into three-dimensional (3-D) animals with beating hearts, somites, and complex nervous systems. Such rapid development is an advantage for experimental design and timely data collection. During this period of early development when so much is occurring, chick embryos can be easily removed from the shell for culture, or they can be cultured *in ovo*. Embryos are semi-transparent, making viewing of internal tissues possible under the microscope, and they are of sufficient size to make several types of micromanipulation practical at these early stages. A large database exists on the descriptive aspects of normal and abnormal development of the early avian embryo, and numerous techniques for experimental manipulating of the avian embryo have been devised. Because of its many advantages, there is a long history of well-documented, experimental studies on the chick embryo, and detailed fate maps exist that show the locations of progenitor cells prior to gastrulation as well as at later stages, when many different organ rudiments are forming (e.g., the limb buds). The availability of this information greatly adds to the value of chick embryos as a model system for studying development.

From: *Methods in Molecular Biology, Vol. 135: Developmental Biology Protocols, Vol. I*
Edited by: R. S. Tuan and C. W. Lo © Humana Press Inc., Totowa, NJ

2. Materials

Acquiring fertile chicken eggs is usually straightforward. Local fertile egg suppliers may be found in the phone book or identified through health food stores that carry fertile eggs, although eggs for research should be purchased directly from the supplier, not from a store, to ensure freshness and fertility (as well as a better price). Also, eggs delivered to health food stores are washed, removing a coating from the surface of the egg that is necessary for successful long-term development of the embryo *in ovo*. In the event that fertile eggs are not available locally, they can be purchased from SPAFAS (Preston, CT). Arrangements can usually be made for local eggs to be delivered, either unincubated or incubated to a specified age. For experiments on early embryos, it is best to purchase fresh, unincubated eggs within a few days of the start of incubation to maximize embryo fertility, synchrony, and quality. For up to a few days, until incubation is initiated, eggs can be stored in their crates on the benchtop in a cool room or in a refrigerator (at 10°C or higher). When previously incubated eggs are purchased, these need to be placed directly into an incubator on arrival.

3. Methods

To develop normally, chicken embryos must be incubated in a humidified chamber that can be maintained at 38°C. If embryos will be incubated for more than a few days, then, in addition, they require turning every few days for optimum development and survival. Commercial incubators come in many shapes and sizes, with forced-air incubators being the most common. For maximum flexibility, we use several Marsh Automatic Incubators (Lyon Electric Co., Inc., Chula Vista, CA), each attached to a timer (e.g., Fisher*brand* Digital Outlet Controller, Fisher Scientific, Pittsburgh, PA) so that incubation of separate batches of eggs can be started independently. The reservoir in the bottom of the incubator should be filled with distilled water to provide a humidified atmosphere; distilled water is used in place of tap water to minimize mineral deposits. Prior to setting eggs (i.e., placing them in the incubator), the thermostat of the incubator should be adjusted to 38°C (temperature readings are only accurate when the water reservoir contains water; temperature should be periodically monitored to ensure temperature consistency), and eggs should be placed on wire racks above the water reservoir. Eggs should be placed on their sides (long axis horizontal to the horizon) if they will be windowed for *in ovo* experiments or either on their sides or large end up if embryos will be isolated for *ex ovo* culture. Eggs can be set with the incubator already running at 38°C, or the timer can be set to start the incubator at a later time after the thermostat has been fully tested. In the latter case, the incubator will reach 38°C in just a few minutes.

Accurate staging of avian embryos is important for the interpretation of experimental results, and several published stage series are available and commonly used. The most comprehensive and frequently cited is the Hamburger and Hamilton stage series (HH), which covers the entire 21-d period from laying to hatching and consists of descriptions and photographs of embryos at stages 1–45 *(1,2)*. Several other stage series are useful in that they deal with earlier stages or subdivide critical developmental periods into substages. The earliest stage of Hamburger and Hamilton (HH stage 1), along with the period of development occurring prior to laying (and therefore prior to HH stage 1), is divided into 14 stages (I–XIV), according to the criteria of Eyal-Giladi and Kochav *(3)*. HH stage 3, an important stage in gastrulation, has been subdivided by Vakaet *(4)* into four stages based on the length and structure of the primitive streak, and Vakaet's stages have in turn been modified by Schoenwolf and coworkers *(5)*, for clarity.

To determine the length of time for which batches of eggs must be incubated to reach a desired stage, the incubation times given by Hamburger and Hamilton *(1,2)* are used. There are many variables that affect the rate of embryonic development, including the strain of chickens producing the eggs, the exact temperature of incubation, and the season of the year. The times given by Hamburger and Hamilton, therefore, provide rough estimates that must be periodically refined empirically within each laboratory. When starting a new experiment with different stages than we used recently, we routinely incubate batches of eggs differing from each other by only a few hours to provide a range of stages around the desired stage. Then adjustments are made accordingly to maximize the number of embryos obtained at the desired stage. It is also important to realize that some manipulations require considerable time to do, especially when learning a new technique. Again, incubating batches of eggs helps ensure that embryos at the desired stage are available when the operation is actually done on them. For example, if one intends to graft three-dozen stage 4 (HH) eggs over a 4-h period, then one might set a dozen eggs each hour for 4 h (an extra dozen to allow for infertility, loss of embryos during manipulation, and embryos at incorrect stages) beginning 14 h before the experiment will commence and use them after incubation in the same order in which they were set. Alternatively, if the experimental period is only a few hours, then all the eggs can be set at the same time, and when the appropriate stage is reached, the incubator can be turned off and opened; development will slow substantially and several hours of experiments can be conducted while embryos remain at that stage. If these same eggs are reincubated they will resume development, although their development will be retarded compared with eggs incubated continuously.

Several culture methods can be used with avian embryos, each specifically adapted to a particular time during development and limited to a specific duration. *Ex ovo* culture *(6–9)* is ideal for experimental manipulation of whole embryos at pregastrula through neurula stages when culture for less than 2 d is desired. Several techniques also exist for tissue explant culture in collagen gel *(10)* or on a fibronectin matrix with serum-containing *(11)* or defined *(12)* medium. For whole embryos, if development beyond HH stage 16 (to hatching) is desired, *in ovo* culture is usually required *(13,14)*. Several of these culture techniques are discussed in detail in Chapter 5 of this volume.

Many different experimental strategies have been used to exploit the advantages of the chick embryo model, with each strategy lending itself to particular types of questions or issues. Strategies used routinely include

1. tissue grafting (**ref. *15***, discussed in detail in Chapter 35)—homotopic and isochronic grafting (for fate mapping using quail grafts in chick embryos or grafts of fluorescently labeled tissues in unlabeled embryos) and heterotopic and heterochronic grafting (to test a region's prospective potency and level of commitment as well as to study cell–cell inductive and suppressive interactions);
2. tissue ablation *(16)* and use of tissue explants, often to ask whether cell–cell inductive or suppressive interactions occur and whether tissues are committed or plastic. These experi-

ments, as well as heterotopic and heterochronic grafting, can often reveal whether tissue interactions are sufficient and/or necessary to regulate the development of interacting populations of cells;

3. use of implants of chromatography beads, coated with growth factors *(17)* or transfected cells *(18)*, both used for targeted overexpression of secreted signaling factors;

4. intracellular dye injection of single cells for lineage analysis and extracellular dye injection for fate mapping groups of cells (discussed in detail in Chapter 30); and

5. retroviral transformation, for lineage analysis, using replication-incompetent virus and a reporter gene, or misexpression, using replication-competent virus producing a wild-type or mutated gene of interest.

These techniques and the availability of a large database on normal and abnormal development make the chick embryo a powerful tool for studying important questions in developmental biology. For example, elucidating the cellular and molecular interactions involved in early induction and signaling events are possible because of the availability of detailed fate maps and experimental techniques such as those just described. These maps show the locations in gastrula-stage embryos of specific progenitor cell populations (e.g., prospective somites, heart, and forebrain; *see* **refs. *15*** and *19*). Thus specific progenitor cells can be selected, and their level of commitment, potency, and interactions can be defined.

In conclusion, using the chick embryo as a model system to study developmental events offers many advantages, including low cost, availability, ease of handling, well-established techniques, and a large database of developmental events. The single major advantage that this system currently lacks is direct genetics; that is, the ability to make transgenic animals and directly knock out (or in) specific genes. This deficiency will likely be overcome in future studies by using antisense RNA or retrovirally supplied dominant-negative receptors to block specific signaling proteins and by using transfected cell lines and retroviral infection for over- or misexpression of genes of interest. Therefore, it is clear that the chick model system will continue to be favored for answering many types of questions posed by scientists studying vertebrate development.

Acknowledgments

Original work described herein from the Schoenwolf laboratory was supported by Grants NS 18112 and HD 28845 from the National Institutes of Health. DKD was supported in part by NIH Developmental Biology Training Grant HD 07491.

References

1. Hamburger, V. and Hamilton, H. L. (1951) A series of normal stages in the development of the chick embryo. *J. Morphol.* **88,** 49–92.

2. Sanes, J. R. (1992) On the republication of the Hamburger-Hamilton stage series. *Dev. Dyn.* **195,** 227–275.

3. Eyal-Giladi, H. and Kochav, S. (1976) From cleavage to primitive streak formation: A complementary normal table and a new look at the first stages of the development of the chick. *Dev. Biol.* **49,** 321–337.

4. Vakaet, L. (1984) Early development of birds, in *Chimeras in Developmental Biology* (Le Douarin, N. and McLaren, A., eds.), Academic, London, pp. 71–88.

5. Schoenwolf, G. C., Garcia-Martinez, V., and Dias, M. S. (1992) Mesoderm movement and fate during avian gastrulation and neurulation. *Dev. Dyn.* **193,** 235–248.

6. New, D. A. T. (1955) A new technique for the cultivation of the chick embryo in vitro. *J. Embryol. Exp. Morphol.* **3,** 326–331.

7. Spratt, N. J. (1947) A simple method for explanting and cultivating early chick embryos *in vitro. Science* **106,** 452.

8. Packard, D. S., Jr. and Jacobson, A. G. (1976) The influence of axial structures on chick somite formation. *Dev. Biol.* **53,** 36–48.

9. Connolly, D., McNaughton, L. A., Krumlauf, R., and Cooke, J. (1995) Improved in vitro development of the chick embryo using roller-tube culture. *Trends Genet.* **11,** 259–260.

10. Placzek, M., Tessier-Lavigne, M., Jessell, T., and Dodd, J. (1990) Orientation of commissural axons in vitro in response to a floor-plate derived chemoattractant. *Development* **110,** 19–30.

11. Antin, P. B., Taylor, R. G., and Yatskievych, T. (1994) Specification of precardiac mesoderm occurs during gastrulation in quail. *Dev. Dyn.* **200,** 144–154.

12. Yatskievych, T. A., Ladd, A. N., and Antin, P. B. (1997) Induction of cardiac myogenesis in avian pregastrula epiblast: The role of the hypoblast and activin. *Development* **124,** 2561–2570.

13. Hamburger, V. (1960) *A Manual of Experimental Embryology*, rev. ed., University of Chicago Press, Chicago, IL.

14. Fisher, M. and Schoenwolf, G. C. (1983) The use of early chick embryos in experimental embryology and teratology: Improvements in standard procedures. *Teratology* **27,** 65–72.

15. Garcia-Martinez, V., Alvarez, I. S., and Schoenwolf, G. C. (1993) Locations of the ectodermal and nonectodermal subdivisions of the epiblast at stages 3 and 4 of avian gastrulation and neurulation. *J. Exp. Zool.* **267,** 431–446.

16. Yuan, S., Darnell, D. K., and Schoenwolf, G. C. (1995) Identification of inducing, responding, and suppressing regions in an experimental model of notochord formation in avian embryos. *Dev. Biol.* **172,** 567–584.

17. Crossley, P. H., Martinez, S., and Martin, G. R. (1996) Midbrain development induced by FGF8 in the chick embryo. *Nature* **380,** 66–68.

18. Fan, C. M. and Tessier-Lavigne, M. (1995) Patterning of mammalian somites by surface ectoderm and notochord: Evidence for sclerotome induction by a hedgehog homolog. *Cell* **79,** 1175–1186.

19. Hatada, Y. and Stern, C. D. (1994) A fate map of the epiblast of the early chick embryo. *Development* **120,** 2879–2889.

5

Culture of Avian Embryos

Diana K. Darnell and Gary C. Schoenwolf

1. Introduction

One of the virtues of using avian embryos for experimentally analyzing early developmental events is the ease with which they can be cultured and subsequently manipulated and observed. Gastrula- and neurula-stage avian embryos can be cultured for 24–48 h on their vitelline membranes (acellular membranes enclosing the embryo and the yolk), which are stretched over a glass ring placed on an egg-agar substrate. Their development occurs similarly to that of embryos developing *in ovo*, yet unlike *in ovo*, embryos are fully accessible for experimentation. This method of cultivation on the vitelline membranes/egg-agar substrate, called New culture, was developed by Denis New in 1955 *(1)*, and its usefulness has not been surpassed in the more than 40 yr since its inception *(2)*. New culture provides an ideal system for experiments involving grafting of tissues from one embryo to another, microinjection of dyes or drugs, time-lapse video recording, and where superior development is required from cultured embryos during the first 2 d of incubation. Blastoderm expansion occurs essentially normally in New culture until the blastoderm comes in contact with the encircling ring. Optimal stages for starting New cultures are between the pregastrula and 7-somite stage (HH [**refs.** *3* and *4*] stages 1–9), with the optimum time of development in culture without degeneration of some tissues being approx 24 h or development to about 22 somites (HH stage 14), whichever comes first. Differentiation of some tissues will continue well beyond this time. Whole-egg culture plates should be used for pregastrula stage embryos (HH stages 1–2), and agar-albumen culture plates should be used for later stages (HH stages 3–9) to optimize development (*see below* and **Note 1**).

Other culture techniques have been developed that are tailored to embryos at different stages or other experimental requirements *(5–11)*. For example, methods have been developed to allow culture in the absence of the vitelline membranes (Spratt culture), of embryo fragments (Spratt culture and Yuan culture), at earlier or later stages (Packard culture), or for longer culture periods (Connolly and McNaughton culture and *in ovo* culture). Spratt and Packard culture, which involves culturing embryos on egg-agar substrates without the vitelline membranes, are ideal for certain kinds of embryo manipulations, including extirpations and transections, and for short-term culture of donor embryos prior to collecting tissue for grafting into host embryos (except

From: *Methods in Molecular Biology, Vol. 135: Developmental Biology Protocols, Vol. I*
Edited by: R. S. Tuan and C. W. Lo © Humana Press Inc., Totowa, NJ

when donors are being labeled with fluorescent dyes, in which case New culture should be used for donors also; *see* Chapter 30). Embryos cultured using Spratt culture *(5)* develop relatively normally for 24 h, when the peripheral two thirds of the area opaca is removed prior to culture. However, their neuraxis does not extend to the normal lengths seen *in ovo* and in New culture. Cutting off the very caudal end of the area pellucida facilitates axis extension in Spratt culture (*see* **Subheading 3.3.**). Optimal stages for starting Spratt cultures are from mid-gastrula stage (HH stage 3) to 7 somites (HH stage 9); embryos can develop well for up to about 24 h or to about 22 somites (HH stage 14) in this culture system. In addition to whole-embryo culture, fragments of embryos can be cultured in isolation using the Spratt protocol, or a variation on the New culture method for use with embryo fragments and transections (*see* **Note 2**) can be found in recent papers by Yuan and coworkers *(6,7)*. For older embryos (up to 3 d or HH stage 18), a Spratt-like culture, Packard culture *(8)*, can be done using plates containing homogenized whole egg. Packard culture plates also work best for blastoderms obtained from unincubated eggs and cultured according to the method of New. Finally, if longer culture periods are desired, embryos can be cultured in one of two ways. First, using the Connolly and McNaughton technique *(9)*, the blastoderm is folded in half longitudinally, sealed along the edge of the area pellucida and cultured for a few days in liquid medium in roller bottles, much as mouse embryos are cultured. However, this technique is new, and its value has not yet been well established. The second and more common course for long-term experiments is the use of *in ovo* culture *(10)*, in which a window is cut into the egg shell, and the embryo is manipulated inside the shell and then returned to the incubator for culture (up to hatching stages). With advances in this technique that yield normal development in 95% of surviving embryos windowed on day 1 *(11)*, this versatile culture technique is the method of choice when development is required beyond day 3 of incubation (or beyond 1–2 d postmanipulation). Each of these culture techniques is to be described (*see* **Subheading 3.**).

2. Materials

2.1. Equipment, Supplies, and Tools

1. 35 × 10-mm culture dishes for making New, Spratt, or Packard culture plates.
2. 5-mL plastic tubes with loose fitting caps for the Connolly and McNaughton culture.
3. Sterile glassware and tools for making plates: For making agar-albumen plates: two 250-mL beakers, two 100-mL graduated cylinders, and a set of dull forceps (e.g., watchmaker's forceps that have been filed down). For making homogenized whole-egg plates: sterile blender container, 50-mL centrifuge tubes, and two 250-mL beakers (*see* **Note 1**).
4. Water bath.
5. Humidified chamber for storage of the plates in a refrigerator: A Tupperware-type plastic tub, with a tight-fitting lid, lined with damp paper towels to create humidity.
6. Forced-draft egg incubator (e.g., Marsh Automatic Incubator, Lyon Electric Co., Chula Vista, CA). Add distilled water to the water reservoir and set the thermostat at 38°C.
7. Humidified chamber for embryo culture: a large Petri dish (25 × 150 mm) with a wet piece of filter paper in the bottom.
8. Culture rings: We prefer square profile glass rings cut from glass or plastic tubing using a diamond saw. These may be made locally and are not available from a commercial supplier. We prefer rings with dimensions of approx 20-mm diameter for chick rings (15 mm for quail rings), 2.5 mm tall with a 2-mm wall thickness. Another laboratory *(2)* prefers

larger rings (30-mm diameter and 4–5 mm tall), claiming better survival over longer culture periods, as the blastoderms can expand for a longer period of time. Sterilize glass or plastic rings in 70% ethanol, then transfer them to a bowl containing sterile saline. Glass rings may be stored in 70% ethanol, but plastic rings will crack with similar storage.

9. General tools and glassware for embryo isolation: Two pairs of dull forceps, 1 pair of curved scissors, a spoon, dissection needles (small cactus needles [spines] or sharpened tungsten wires attached to wooden dowels or microinjection pipets pulled to a sharp point), sterile Pasteur pipets, sterile wide-mouth pipets (break off the tip of a sterile Pasteur pipet, cover the broken end with a bulb and use the wide end for pipeting), sterile Pasteur pipets with their tips pulled to a narrower diameter for fine work, a sterile conical evaporating dish, two sterile Petri dishes (one for embryos and one for rings) and a waste container (for temporary storage of yolk, etc., until discarded).

10. Vital dye/agar slides for *in ovo* staining: Coat a clean, sterile, glass slide on one side with 1% agar. Dry overnight, then soak agar slides for a week in 1% neutral red or Nile blue; rinse with sterile distilled water and allow to dry.

11. Tape to seal windowed eggs: For eggs windowed on day 1 that need to be inverted, Scotch brand Super 88 vinyl electrical tape works best and does not leak. For eggs windowed on day 2, any tape that will stick to the shell is fine.

12. Plate incubator: Add water to trays inside incubator to humidify, and set thermostat for 38°C.

13. Plastic dishes with tight-fitting lids for fixing embryos after culture.

2.2. Solutions and Culture Plates

1. Saline: 123 mM NaCl (7.19 g/L distilled water), autoclaved and cooled, with 1 mL penicillin/streptomycin (Gibco-BRL, Grand Island, NY) added at the time of use.

2. Agar-albumen culture plates: Adjust water bath to 49°C. Open the blunt end of an egg and remove and discard thick albumen and chalazae with dull forceps. Collect thin albumen in a sterile graduated cylinder (60 mL for 40 plates), then transfer to a sterile 250-mL beaker and place into the 49°C water bath. In another 250-mL beaker, make a 0.6% solution of Bacto-Agar (0140-01; Difco, Detroit, MI) in saline at room temperature (0.36 g in 60 mL for 40 plates). On a magnetic stirrer/heater, mix and bring to a slow boil to dissolve agar. Cool the beaker to 49°C in the water bath. When the temperatures of both agar/saline and albumen have equilibrated at 49°C, add albumen to sterile agar/saline and stir vigorously on a stir plate. Return the beaker to the water bath and pipet 2.5 mL agar-albumen into each 35 × 10 mm culture dish to produce a confluent layer on the bottom. Avoid transferring bubbles. Allow the plates to sit undisturbed at room temperature until the agar/albumen solidifies, then place plates into a humidified chamber for storage. Plates may be refrigerated for up to 2–3 wk but are best when used fresh.

3. Homogenized whole-egg culture plates: Adjust water bath to 47°C. Homogenize the contents of three cold, unincubated, fertile chicken eggs in the cooled, sterile container of a blender. Pour the foamy homogenate into sterile, 50-mL centrifuge tubes and spin at 14,900g for 30 min at 5°C. Pour supernatant from eggs into a sterile 250-mL glass beaker and warm to 47°C in the water bath. Prepare a 50-mL aqueous solution of 6% Bacto-Agar, autoclave for 8 min (120°C and 103.5 kPa), and place into the 47°C water bath. When the temperatures of both solutions have equilibrated at 47°C, add the agar slowly to the egg, swirling the beaker without forming bubbles, to a final ratio of 1:3 agar:egg. Pour 2–3 mL of this solution into the center of a warmed culture dish (60 × 15 mm), swirl gently to coat, and pour off excess liquid to leave a thin (1–1.5 mm) coating on the bottom of the dish. Allow the culture plates to sit undisturbed at room temperature until the culture medium

solidifies, then place the plates into a humidified chamber for storage. Plates may be refrigerated for up to 2–3 wk, but are best when used fresh.

4. Medium for Connolly and McNaughton culture: For culture day 1, use Leibovitz medium (L-5520, Sigma Chemical Co., St. Louis, MO) with 10% fetal bovine serum (FBS) and gentamicin at 50 µg/mL. For culture days 2–3, use Leibovitz medium with 50% FBS and gentamicin at 50 µg/mL.

3. Methods

3.1. New Culture: Culture on Vitelline Membranes/Egg-Agar Substrate

1. Incubation: Set the eggs into the egg incubator's rack, blunt end uppermost, and incubate until desired stages are reached (*see* Chapter 4). Use Hamburger and Hamilton *(3,4)* for approximate timing; exact timing may vary from their stated times depending on a number of factors (*see* Chapter 4).

2. Isolation of the embryo from the yolk: Select an egg from the incubator and open the blunt end with a pair of dull forceps; gently remove and discard albumen with forceps and by decanting. Try not to puncture the yolk. Gently transfer the yolk to a sterile dish (e.g., a conical evaporating dish) that is sufficiently full of sterile saline to submerge the yolk. Orient the yolk with the blastoderm uppermost and cut with scissors through the vitelline membranes around the *equator* of the yolk (or lower for bigger rings). Gently lift up on the cut edge of the vitelline membranes with forceps and, using scissors to hold the yolk down, peel the vitelline membranes with attached blastoderm away from the yolk.

3. Preparing the embryo for culture: With a spoon, gently transfer the blastoderm and vitelline membranes to a sterile glass Petri dish containing sterile saline, and place this dish on the microscope stage. Orient the membranes so the blastoderm is up (vitelline membranes down) and remove any remaining, adherent albumen using forceps to peel it away from the vitelline membranes. If yolk remains, remove it gently by puffing saline from a pipet onto the embryo and then aspirate the yolk from the saline, rather than aspirating the yolk directly off of the embryo. With practice, embryos can be collected virtually without adherent yolk. Stage the embryo accurately *(3,4,12,13)* and record this information.

4. Attaching the embryo to a culture ring: Make sure that the vitelline membranes are oriented so that the blastoderm is on top (i.e., the blastoderm will be viewed with its ventral or yolk side facing up). With forceps, gently transfer a ring from sterile saline onto the top of the vitelline membranes so that the embryo is centered in the ring. Remove saline from the dish until the top of the ring is exposed to the air (so the vitelline membranes will not float off the ring when you fold them over the edge). Use forceps to lift the part of the vitelline membranes remaining outside the ring up and over the top of the ring. Continue until the entire cut edge of the vitelline membranes is wrapped over the ring, making a small bowl containing the blastoderm. If necessary, gently pull the vitelline membranes so that they are taut and the blastoderm is roughly in the center of the ring. Then carefully remove saline from the interior of the ring with a pipet.

5. Transferring the embryo on the culture ring to the culture plate: Lift the ring with the attached vitelline membranes and blastoderm using forceps and transfer it to a culture plate (*see* **Note 3**). If the entire embryo will be used, as with most experiments, then graft, inject, or otherwise manipulate the embryo as appropriate and place the culture plate into a humidified chamber in a humidified plate incubator (or into a time-lapse setup) at 38°C for the duration of culture. If embryo fragments are desired as in the Yuan experiments, *see* **Note 2**.

3.2. Spratt or Packard Culture: Culture on Egg-Agar Substrates

1–3. Follow **steps 1–3** listed for **Subheading 3.1.**
   4. Preparing the blastoderm for culture: Remove the blastoderm from the vitelline membranes by locating the outer margin of the blastoderm and detaching it from the vitelline membranes by gently pulling the two layers apart with forceps or by cutting through the outer edge of the blastoderm with iridectomy scissors. Transfer the blastoderm via a wide-mouthed pipet to a culture plate with sufficient saline to facilitate orientation of the embryo on the plate. Use agar-albumen plates for Spratt culture with embryos between early gastrula and 7-somite stages (HH stages 2–9). Use homogenized whole-egg plates for Packard culture for older (HH stage 10–18) embryos. For Spratt and Packard culture, blastoderms can be oriented either dorsal-side or ventral-side up, depending on the type of experiment; development occurs in either orientation. The ventral/endodermal side of the embryo can determined by looking for attached yolk granules at early stages (the dorsal side, which was adjacent to the vitelline membranes, has no adherent yolk) and by differences in dorsoventral morphology of organs (heart, gut, etc.) at later stages. Pipet off excess saline to flatten the blastoderm onto the substrate. Trim away the outer two-thirds of the area opaca using a cactus needle-knife, glass micropipet, sharpened tungsten wire, or iris knife. At the caudal end of the blastoderm, remove the area opaca entirely and a little of the area pellucida to promote good extension of the axis. Add a few drops of saline to the culture dish and float the blastoderm to an undisturbed area of the substrate. Then pipet away excess saline (cultures should be fairly dry) and pieces of extirpated area opaca. At this point the embryo is ready to be transected, injected, grafted, or otherwise manipulated for the experiment. Following this, place the lid on the culture dish and place the dish into a humidified chamber in a 38°C, humidified, plate incubator.

3.3. Connolly and McNaughton Culture: Pita-Pocket Style Culture in Roller Bottles

1–3. Follow **steps 1–3** listed for New culture with embryos at the extended streak stage or older (HH stage 3+ or 4). Embryos that have been placed in New culture can be removed to Connolly and McNaughton culture after experimental manipulation and, for grafted embryos, after a short (1-h) recovery period for graft healing.
   4. Preparing the blastoderm for culture: Remove the blastoderm from the vitelline membranes by locating the outer margin of the blastoderm and detaching it from the vitelline membranes by gently pulling the two layers apart with forceps or by cutting through the outer edge of the blastoderm with iridectomy scissors. Transfer the blastoderms to a Petri dish containing Leibovitz air-buffered tissue-culture medium. Embryos may be stored at room temperature while additional blastoderms are collected. Place each blastoderm ventral-side (yolk-side) up, and remove any excess yolk by puffing the blastoderm gently with a stream of medium from a fine pipet. Fold the blastoderm along the longitudinal axis (i.e., left to right) so the endoderm is on the inside and the neural ectoderm is on the outside. The midline of the embryo should lie at the fold. Seal the free edges of the blastoderm by cutting with iridectomy scissors along a line passing just within the area opaca. Crimping the area opaca first with forceps may help keep the embryo positioned correctly.
   5. Culturing embryos: Up to five folded and sealed embryos may be transferred into each 5-mL plastic tube containing 500 µL of Leibovitz medium with 10% FBS and gentamicin. Lightly capped tubes are placed at a 10° angle in a 38°C roller-bottle incubator rotating at 30 rpm. No special gassing is required. Normal development can proceed for at least 48 h, but after the initial 24 h of culture, the medium should be replaced with Leibovitz medium with 50% FBS and gentamicin. Embryos can develop normally to the 28 somite stage (HH stage 16), and they establish an extensive extraembryonic vascular system.

3.4. In Ovo *Culture: Culture in the Egg for Long-Term Experiments*

1. Incubation: For *in ovo* culture, eggs should be placed in a humidified, forced-draft incubator with their long axis parallel to the horizon, and the uppermost spot on the egg should be marked with pencil. This will be the location of the window; the blastoderm will lie directly beneath this area. Eggs should be incubated for at least 24 h prior to windowing. Eggs for experiments at or beyond day 2 may be windowed on day 2, sealed, and returned to the incubator until needed (*see* **Notes 4** and **5**).

2. Windowing the egg: Eggs should be cleaned (but not soaked) with 70% ethanol on a damp paper towel. Puncture the air space at the blunt end of the egg by piercing the shell with a needle. After air has escaped, the hole will be sealed by the shell membranes. With the egg's long axis parallel to the horizon, window the egg shell by carefully removing, using a dental drill and forceps, approx 1 cm^2 of egg shell and the underlying shell membranes at the point marked on the side of the egg (i.e., the point that was uppermost during incubation).

3. Visualizing the embryo: *In ovo* embryos can be visualized for staging and manipulation by applying chips of agar impregnated with stain (*see* **Note 6**). Place a drop of sterile distilled water or saline on the agar-coated side of the slide and remove a tiny chip of agar from the moistened area using a scalpel blade. Transfer the chip with dull forceps to the vitelline membranes over the embryo, leave it for a brief period (less than 1 min), then remove it with forceps. This procedure minimizes the amount of vital stain that comes in contact with the embryo; do not overstain, as excess vital stain can be toxic.

4. *In ovo* manipulation: For manipulation, the egg should be cradled in a stable holder (such as a portion of an egg carton) with the window up and a fiber-optic lighting directed into the window. The vitelline membranes should be torn open with a tungsten needle over the area you wish to manipulate. Retroviruses or dyes can then be injected, or tissue can be removed and donor tissue grafted into the host.

5. Sealing the window: After surgical manipulation or injection, sterile saline should be added to bring the embryo up to the level of the window. Then, the window in the shell should be sealed well with tape and the egg should be rotated 180° along its long axis so that the embryo is subjacent to undisturbed shell; eggs are then returned to the incubator for the duration of the culture period, sealed-window-side down. This allows the embryo to develop adjacent to an undisturbed shell surface, preventing adherence to the disrupted shell and membranes and allowing unhampered respiration to occur. The addition of saline and rotation can be delayed for up to 3 h without having a negative impact on development. This is recommended in the case of retroviral labeling or the application of drugs so that the added substance is not diluted before its integration/diffusion into the embryo.

4. Notes

1. Although albumen is bacteriolytic, for the best results use sterile techniques throughout each procedure. All glassware should be sterilized by autoclaving; forceps and other tools may be sterilized with 70% ethanol. Working areas, eggs, and hands (or surgical gloves, if one prefers to wear these) should be cleaned frequently by wiping them with a cheesecloth or towel, damp but not saturated, with 70% ethanol. Instruments and so forth should be allowed to dry before proceeding because 70% ethanol can kill or damage avian embryos.

2. Normally, for optimum development to occur in New culture, it is necessary for the blastoderm to remain attached to the vitelline membranes around its circumference. However, for culture of embryo fragments on vitelline membranes (Yuan culture), the vitelline membranes should be cleaned of *all* yolk, the embryo should be transected or a fragment isolated, the undesired regions of the blastoderm should be moved away and removed with a

fine pipet, and finally, the edge of the blastoderm for the remaining fragment should be detached from the vitelline membranes and moved to the center of the ring. After 6–12 h, the cultures should be checked, and if the embryo fragment has moved toward the periphery (near the ring), then it should be detached from the vitelline membranes and moved gently back into the center of the culture. Also at this point, any fluid should be removed that has accumulated inside the ring. Embryo fragments can be cultured with this method for up to 36–48 h.

3. For New culture, embryos isolated at pregastrula stages (HH stages 1–2) should be cultured on homogenized whole-embryo culture plates, whereas gastrula stage embryos and later (HH stages 3–9) should be cultured on agar-albumen plates. Transfer between the Petri dish and culture plate is much easier with square-profile glass rings rather than circular profile rings.

4. Unless steps are taken to minimize the adverse effects of early windowing, as many as 70% of surviving embryos windowed on day 1 will have some abnormality, usually dysraphic (open) neural tubes. Even with precautions, mortality after windowing on day 1 is approx 50%. However, the protocol included here has been developed for windowing day-1 eggs, which results in more than 95% of the surviving embryos developing normally in culture *(11)*. Eggs for experiments beginning at later stages may be windowed on day 2 and returned to the incubator until needed. Windowing after day 2 becomes difficult because of the development of highly vascular extraembryonic membranes, which lie just beneath the shell and its membranes.

5. Before windowing, 1–2 mL of thin albumen can be removed from the pointed end of the egg with a syringe. (Seal the hole with tape after withdrawing the needle.) Puncturing the air space and/or removing some thin albumen allows the blastoderm to drop away from the shell and its membranes at the time of windowing, thereby reducing the likelihood that the embryo or its membranes will be damaged during windowing.

6. An alternative method for visualizing embryos *in ovo* is to inject 0.05% Nile blue in saline or full-strength India ink into the subgerminal cavity (between the embryo and the yolk). However, the exposure to vital dyes can be better limited using the agar-impregnated technique, and India ink is highly toxic to embryos prior to day 2 of incubation.

Acknowledgments

Original work described herein from the Schoenwolf laboratory was supported by Grants NS 18112 and HD 28845 from the National Institutes of Health. D. K. D. was supported in part by NIH Developmental Biology Training Grant HD 07491.

References

1. New, D. A. T. (1955) A new technique for the cultivation of the chick embryo in vitro. *J. Embryol. Exp. Morphol.* **3,** 326–331.
2. Stern, C. D. and Bachvarova, R. (1997) Early chick embryos in vitro. *Int. J. Dev. Biol.* **41,** 379–387.
3. Hamburger, V. and Hamilton, H. L. (1951) A series of normal stages in the development of the chick embryo. *J. Morphol.* **88,** 49–92.
4. Sanes, J. R. (1992) On the republication of the Hamburger-Hamilton stage series. *Dev. Dyn.* **195,** 227–275.
5. Spratt, N. J. (1947) A simple method for explanting and cultivating early chick embryos *in vitro. Science* **106,** 452.
6. Yuan, S., Darnell, D. K., and Schoenwolf, G. C. (1995a) Mesodermal patterning during avian gastrulation and neurulation: Experimental induction of notochord from non-notochordal precursor cells. *Dev. Genet.* **17,** 38–54.

7. Yuan, S., Darnell, D. K., and Schoenwolf, G. C. (1995b) Identification of inducing, responding and suppressing regions in an experimental model of notochord formation in avian embryos. *Dev. Biol.* **172,** 567–584.

8. Packard, D. S., Jr. and Jacobson, A. G. (1976) The influence of axial structures on chick somite formation. *Dev. Biol.* **53,** 36–48.

9. Connolly, D., McNaughton, L. A., Krumlauf, R., and Cooke, J. (1995) Improved in vitro development of the chick embryo using roller-tube culture. *Trends Genet.* **11,** 259–260.

10. Hamburger, V. (1960) *A Manual of Experimental Embryology*, rev. ed., University of Chicago Press, Chicago, IL.

11. Fisher, M. and Schoenwolf, G. C. (1983) The use of early chick embryos in experimental embryology and teratology: Improvements in standard procedures. *Teratology* **27,** 65–72.

12. Eyal-Giladi, H. and Kochav, S. (1976) From cleavage to primitive streak formation: A complementary normal table and a new look at the first stages of the development of the chick. *Dev. Biol.* **49,** 321–337.

13. Schoenwolf, G. C., Garcia-Martinez, V., and Dias, M. S. (1992) Mesoderm movement and fate during avian gastrulation and neurulation. *Dev. Dyn.* **193,** 235–248.

6

Exo Ovo Culture of Avian Embryos

Tamao Ono

1. Introduction

Exo ovo culture of avian embryos is a technique for long-term culturing of embryos outside of their own shell and shell membranes. The problem of how to gain access to the avian embryo while allowing it to grow normally has been the subject of many studies *(1)*. Avian embryos are subjected to various environmental conditions in the course of normal development. For example, in the chick embryo, in the first day, development takes place in the oviduct where egg formation is completed by deposition of thick and thin albumen, uterine fluid, chalaza, inner and outer shell membranes, and shell around the yolk; for the next 21 d, the enveloping layers act as a buffer between the embryo and the egg's environment *(2)*. Current technologies now permit the culture of avian embryos from the single-cell stage (which is normally in the oviduct) through hatching. We can take out just fertilized ovum, inject DNA and mRNA, and culture until hatching. The choice of culture method depends on the age of the embryo at the start of the experiment. In contrast to shell windowing techniques, the *exo ovo* culture allows easy access to the developing embryos and is thus useful for analysis of the developmental process of embryos and embryo manipulation. Injections and microsurgical operations can be made into a specific portion of the embryo, including transplantation of undifferentiated tissues and primordia and microsurgery of limb buds. Avian embryos, especially chick and quail embryos, have been widely used for studies of developmental and molecular biology. Fertile and embryonated eggs are accessible all over the world as well as throughout all seasons of the year. The key feature of the procedure described here is that embryos are initially taken out from the shell for ease of manipulation and then placed back in culture in addition to various operations midway during culture. Specifics of the culture systems may be modified depending on operations, treatments, species studied, and so forth. This chapter explains step-by-step protocols for chick and quail embryo cultures.

2. Materials
2.1. Culture of Chick Embryos

1. A laboratory egg incubator with an automatic turner (30°- and 90°-angle turnings every 30 min); for example, P-008-B special model for embryo culture (Showa Furanki Institute, Yono, Japan).

From: *Methods in Molecular Biology, Vol. 135: Developmental Biology Protocols, Vol. I*
Edited by: R. S. Tuan and C. W. Lo © Humana Press Inc., Totowa, NJ

2. An electric drill with a flexible shaft and a diamond disk; for example, Minitor M-17-BS and A5131 (Minitor, Tokyo, Japan).

3. Plastic cups: polystyrene, approx 70 mL (60-mm diameter and 35-mm height).

4. Gauze: 40×40 mm.

5. Scissors.

6. Fertilized ova: The number of artificially inseminated hens should be 3× that of the desired number of fertilized ova. Check the time of laying and save laid eggs for the second step culture (*see* **Note 1**).

7. Nembutal®: Pentobarbital sodium 50 mg/mL (Abbott Laboratories, North Chicago, IL).

8. 10- and 30-mL syringes.

9. Thick albumen of hen's eggs: Clean the shell with 70% ethanol and crack open into 90-mm Petri dish. Suck up the thick albumen with 30-mL syringe and transfer into a beaker. Add 10,000 IU penicillin and 50 mg streptomycin per liter of albumen (*see* **Note 2**). Adjust the pH to 7.2–7.4 by bubbling CO_2 through it (*see* **Note 3**). Remove foam from the albumen by centrifugation and breaking up with spatula. Then, warm in CO_2 incubator at 41.5°C and 20% CO_2 until use (*see* **Note 4**).

10. Thin albumen of hen's eggs: Crack open the cleaned egg into 90-mm Petri dish (*see* **Subheading 2.1.9.**). Dump out the yolk and the surrounding thick albumen capsule with spatula. Collect remaining thin albumen (*see* **Note 5**). Add penicillin and streptomycin, adjust pH to 7.2–7.4, and remove the foam (*see* **Subheading 2.1.9.** and **Note 2**). For the second step culture, adjustment of pH is not necessary, and keep at room temperature (RT) until use.

11. Surrogate shell for the second step culture: Prepare a similar-sized egg shell to approximate the size of the egg for the embryo if normal shell formation were permitted (e.g., the egg previously laid by the same hen).

12. Surrogate shell for the third step culture: Prepare egg shell from an approx 30-g heavier egg compared with the expected egg used for the culture (e.g., a double-yolk egg).

13. Rings: Prepare 7- and 15-mm-tall polyvinyl chloride rings (made from water pipe; 36 and 42 mm inner and outer diameters, respectively) with four projections attached to the outside (*see* **Note 6**).

14. Elastic bands: 16- and 32-mm diameter.

15. 100-mL glass beakers.

16. Spatula.

17. 70% ethanol.

18. Plastic wrap.

19. Circle template.

2.2. Culture of Quail Embryos

1. Plastic cups: 20-mL Polypropylene (32 mm tall, 30- and 35-mm lower and upper diameters, respectively).

2. Fertilized ova: The number of pair-mated quail should be 5× that of the number of fertilized ova (*see* **Note 7**). Check the time of laying and save laid eggs for the second step culture.

3. Surrogate shell for the second step culture: Prepare a similar sized eggshell to approximate the size of the egg for the embryo if normal shell formation were permitted (e.g., the egg previously laid by the same quail).

4. Surrogate shell for the third step culture: Prepare shell from small-size chicken egg.

5. Rings: Prepare 7- and 15-mm-tall polyvinyl chloride rings (made from water pipe; 20- and 26-mm inner and outer diameters, respectively) with four projections attached to the outside (*see* **Note 6**).

6. Elastic bands: 13-mm Diameter.
7. 50-mL glass beakers.
8. Others: Laboratory egg incubator with an automatic turner (30°- and 90°-angle turnings every 30 min), electric drill with a flexible shaft and a diamond disk, plastic wrap, Nembutal (Abbott), thick and thin albumen of hen's eggs, scissors, syringes, spatula, 70% ethanol and circle template (*see* **Subheading 2.1.**).

3. Methods

3.1. Culture of Chick Embryos

3.1.1. Principles

Chick development is divided into three periods for the purpose of *exo ovo* culture: fertilization to blastoderm formation lasts for 1 d, embryogenesis for 3 d, and embryonic growth for 18 d *(3)*. Cultures are divided into three steps, corresponding to these three periods, respectively. Fertilization takes place in the anterior oviduct, after which the yolk-laden ovum is encapsulated in albumen secreted by the magnum. Around the time of the first division of the zygote, some 4.5 h after ovulation, the shell membrane is deposited in the isthmus, and the albumen is doubled in volume by the absorption of uterine fluid. In the final 18 h of the oviductal phase, the shell is calcified. The second and third phases take place in the shelled egg. The three discrete culture steps meet the changing demands at successive stages of development, and the embryos are transferred step by step at appropriate times *(3)*. The protocol described here is basically according to Perry *(3)* and Naito et al. *(4,5)* with some modification.

3.1.2. Culture Step 1: Without Thick Albumen Capsule

This step deals with the culture from the single-cell stage ovum before the attachment of thick albumen capsule (or the capsule removed) to the blastoderm stage.

1. Kill hens by injection of Nembutal (2 mL) into ulnar vein of the wing 60–80 min after the preceding egg has been laid (approx 35–55 min after ovulation).
2. Remove abdominal feathers, laparotomize, and find the ovum in the magnum.
3. Cut out both ends of the magnum holding the ovum and transfer into a 70-mL plastic cup.
4. Place the cutout magnum on the palm of your hand, insert scissors between the magnum and the ovum, and carefully cut out the wall of the magnum.
5. Put a 100-mL glass beaker inside the wall of the magnum and slide the beaker gently into the ovum. Remove the thick albumen capsule by spatula if present.
6. Add 5 mL of the thick albumen to the ovum and to an empty 70-mL plastic cup. Transfer the ovum gently into the plastic cup. Add the thick albumen up to the ovum's equatorial level. Rotate the ovum with spatula so that the germinal disc is positioned at the top of the yolk.
7. Cover the top of the yolk with a sheet of gauze. The four corners of the gauze should be soaked in the surrounding thick albumen. Seal the plastic cup with plastic wrap (70×70 mm) and rubber bands (32-mm diameter; use two bands).
8. Incubate the culture-set (**Fig. 1A**) for 24 h in a humidified CO_2 incubator at 41.5°C under 20% CO_2 (*see* **Note 9**).

3.1.3. Alternative Culture Step 1: With Thick Albumen Capsule

This is an alternative method for the single-cell stage ovum after the attachment of thick albumen capsule to the blastoderm stage.

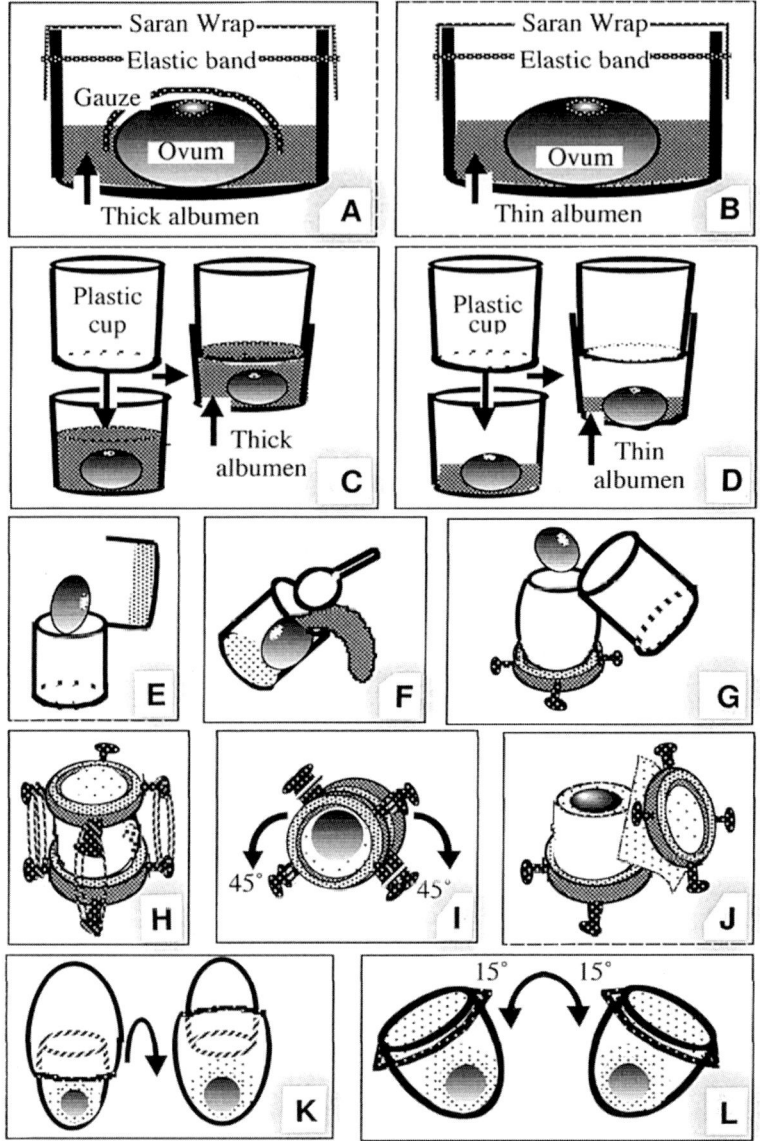

Fig. 1. A schematic drawing of *exo ovo* culture of chick and quail embryos. For the culture of chick embryo from single-cell ovum without its own albumen capsule, follow (**A**) and (**E**)–(**L**); chick embryo with its own albumen capsule, follow (**B**) and (E)–(L); quail embryo without its own albumen capsule, follow (**C**) and (E)–(L); and quail embryo without its own albumen capsule, follow (C) and (E)–(L). Detailed explanations are in the text.

1. Kill hens two hours and forty-five minutes after the preceding egg has been laid (*see* **Subheading 3.1.2., step 1**).
2. Follow **Subheading 3.1.2.**, **steps 2–7**, but culture the ovum with albumen capsule. Use thin albumen instead of thick albumen. The gauze cover is not necessary.
3. Incubate the culture-set (**Fig. 1B**) for 24 h in a humidified CO_2 incubator at 41.5°C under 100% air (*see* **Note 10**).

3.1.4. Culture Step 2

After culturing for 24 h in **step 1**, the embryos will have developed to the blastoderm stage, which is equivalent to that in a freshly laid egg. Fertilized ovum (germinal disc with yolk) in the natural condition forms shelled egg in the oviduct, which includes albumen, chalaza, shell membranes, and shell. During the first step culture, however, only the development of embryo occurs without the addition of these associated substances. In the second step, the cultured embryo is placed into a surrogate shell where embryogenesis can proceed. Alternatively, the culture can be started at this point from the blastoderm stage.

1. Place a circle template (34-mm diameter) to the narrow end of surrogate shell. Draw a line along with the inner circumference of the template by pencil. Wipe the lined area with 70% ethanol.
2. Cut the shell along with the line using a diamond disk attached to an electric drill. Cover with a Petri dish lid to prevent drying.
3. Dump out yolk and albumen. Use blunt-end half for the surrogate shell. Wipe the cut end by 70% ethanol.
4. Transfer the cultured embryo into a beaker (**Fig. 1E**). If the culture is started from here, wipe the shell with 70% ethanol and crack open the egg into a beaker. Remove the thick albumen capsule (if you cultured or used an albumen-capsuled embryo) using a spatula (**Fig. 1F**; *see* **Note 11**).
5. Remove the foam in the stock of thin albumen as much as possible, if present.
6. Place a surrogate shell on the ring (15 mm tall). Pour the thin albumen (about 20 mL) into the surrogate shell and then transfer the embryo with yolk (**Fig. 1G**).
7. Fill the thin albumen using a 10-mL syringe (without a needle) and then remove the foam with a spatula.
8. Cover with a sheet of plastic wrap (70×70 mm) and a ring (7 mm tall). The sheet cover is secured by elastic bands (16-mm diameter; two bands in each pair of screw projections).
9. Place the culture-set (**Fig. 1H**) in an incubator with the long axis of the shell held horizontally (cut-end vertically) (**Fig. 1I**), and culture the embryo for 3 d at 37.5°C and 70% relative humidity in an atmosphere of 100% air while being rocked round the long axis at a 90°C angle at 30-min intervals.

3.1.5. Culture Step 3

After culturing for 3 d in Culture Step 2, the embryos are transferred to the Culture Step 3 system for embryonic growth and hatching. If Culture Step 2 culture is continued, all the embryos die within several days in absence of an air space. Thus, the embryo is transferred into an extralarge shell with an artificial air space. Alternatively, the culture can be started newly from here.

1. Place a circle template (40–45 mm) to the blunt end of surrogate shell. Use narrow-end half for culture. Pour the thin albumen into 90-mm Petri dish and place the shell in the dish with the open face of the shell down until use. Follow **Subheading 3.1.4.**, **steps 1–3**.
2. Remove the rings and elastic bands of the second step culture-set (**Fig. 1J**). Wipe the second step shell with 70% ethanol.
3. Overlay the emptied large shell (open face down) to the Culture Step 2 setup (open face up). Gently turn the entire setup upside down and remove the smaller shell (**Fig. 1K**).
4. Cover with a sheet of plastic wrap (70×70 mm) and adhere the wrap to the cut end of the shell (*see* **Note 12**).

5. Place the culture-set thus made in an incubator with the cut end of the shell held upward, and culture the embryo at 37.5°C and 70% relative humidity in an atmosphere of 100% air while being rocked at a 30°C angle at 30-min intervals (**Fig. 1K**).

6. Stop rocking 1–2 d before expected day of hatch. Prick 5–10 holes in the wrap with a pin when the embryo has holed the chorioallantoic membrane with its beak. Replace the punctured wrap cover with a 60-mm plastic Petri dish lid when the chorioallantoic membrane has dried (*see* **Note 13**).

3.2. Culture of Quail Embryos

3.2.1. Principles

Culture period for quail embryos is shorter than that of chick embryos. Fertilization to blastoderm formation lasts for 1 d, embryogenesis for 2.5 d, and embryonic growth for 14 d. Culture protocol is basically according to Ono et al. *(6,7)*.

3.2.2. Culture Step 1: Without Thick Albumen Capsule

1. Kill quail by injection of Nembutal (0.2 mL) into ulnar vein of the wing or by decapitation 60–90 min after the preceding egg has been laid. Follow **Subheading 3.1.2., steps 2** and **3**.
2. Add 1 mL of the thick albumen to an empty 20-mm plastic cup.
3. Hold the cut end of the magnum above the cup with two pairs of forceps and tear the magnum until the ovum falls down into the cup.
4. Rotate the ovum with a spatula so that the germinal disc is positioned at the top of the yolk.
5. Add thick albumen to ovum until full. Seal the open surface of the cup by placing another cup inside it.
6. Make sure that any air pockets are eliminated and the ovum is submerged just below the surface.
7. Incubate the culture-set (**Fig. 1C**) for 24 h in a CO_2 incubator at 41.5°C under 20% CO_2.

3.2.3. Alternative Culture Step 1: With Thick Albumen Capsule

This step is the alternative culture protocol from the single-cell stage ovum after the attachment of thick albumen capsule to the blastoderm stage.

1. Kill quail two hours and thirty minutes to two hours and forty-five minutes after the preceding egg has been laid (*see* **Subheading 3.2.2., step 1**).
2. Follow **Subheading 3.2.2., steps 2–4**, but culture the ovum with albumen capsule and use the thin albumen instead of the thick albumen.
3. Add the thin albumen up to the equatorial level of the ovum. Make sure that the blastodisc is above the thin albumen.
4. Seal the open surface of the cup by placing another cup inside it. Make sure that air pocket is above the ovum.
5. Incubate the culture-set (**Fig. 1D**) for 24 h in a CO_2 incubator at 41.5°C under 100% air.

3.2.4. Culture Step 2

1. Place a circle template (18-mm diameter) to the narrow end of surrogate shell and cut out the shell (*see* **Subheading 3.1.4., steps 1–5**). Remove the thick albumen capsule (if you cultured or used an albumen-capsuled embryo) using forceps.
2. Follow **Subheading 3.1.4., step 6**, but pour 1 mL thin albumen instead of 20 mL.
3. Follow **Subheading 3.1.4., steps 7–9**. But use 30 × 30 mm plastic wrap cover and 13-mm-diameter elastic bands. The culture period is 2.5 d.

3.2.5. Culture Step 3

1. Cut the chicken egg shell along a line drawn around the equator level. Follow **Subheading 3.1.5.**, **steps 1–6**, but use 50×50 mm plastic wrap cover (*see* **Note 14**).

4. Notes

1. Keep hens in individual cages and artificially inseminate them 2 d before the culture. In our experiment for DNA injection, artificially inseminate approx 300 hens, check the time of laying during 7:00–10:00 AM and then collect approx 100 ova from the oviduct of hens. Under optimal lighting conditions (e.g., 14L/10D) hens usually lay eggs during early and middle hours of lighting period.
2. Antibiotics are not necessary when the experimental condition is clean, because lysozyme in the albumen is germicidal.
3. Use a CO_2 spray or an air gun attached to a pressure regulated CO_2 cylinder (for CO_2 incubator). Check pH using pH test papers (bromothymol blue). Under 20% CO_2, the albumen maintains its pH at 7.2–7.6.
4. Use freshly prepared albumen. The albumen has no buffering action, therefore its pH goes up to approx 8.8 in 100% air *(7)*. The pH of albumen in the oviduct is 7.0–7.4 *(7,8)*.
5. If you want to collect both thin and thick albumen at the same time, suck up the thick albumen first and then dump out the yolk.
6. Cut the water pipe using a high-speed saw (e.g., CC 12SA, Hitachi Koki USA Ltd., Chatsworth, CA). Drill four holes into each ring using a bench drill press and attach stainless steel screws with the screw tip cut off.
7. Artificial insemination of quail is not practical. Under 14L/10D photoperiod, approx 90% of ovipositions occur during the last 7 h of lighting period, and the mean oviposition time is 11.3 h after the onset of lighting *(9)*. In our experiment for DNA injection, lighting period is 2 AM–4 PM (14L/10D), check the time of laying at noon and during early afternoon, and finish culturing until late evening.
8. Once the ovum is taken out from the hen, keep it warm and culture it quickly.
9. Under 20% CO_2, the thick albumen maintains its pH at 7.2–7.6 during the 24-h period, which is similar to the natural condition.
10. With its own albumen capsule, the ovum can be cultured under 100% air.
11. With the albumen capsule present, movement of the yolk is obstructed inside the shell.
12. The thin albumen is applied around the cut end of the shell and it is used as an adhesive.
13. Make sure that the Petri dish lid is not adhered to the shell.
14. After culturing for 2.5 d in **Subheading 3.2.4.**, the embryos are transferred to **Subheading 3.2.5.** system for the embryonic growth. Hatchability is lower when the second step culture is continued *(7)*. Thus, the embryo is transferred into surrogate chicken shell with an artificial air space. Alternatively, the culture can be started a new from here.

Acknowledgment

The author thanks Dr. Rocky S. Tuan, editor of this book, for his helpful suggestions in the preparation of the manuscript, and Drs. G. Eguchi, K. Agata, M. Mochii, K. Kino, K. Noda, and H. Miyakawa for their participation in development of the protocol. This work was supported in part by a grant-in-aid from the Ministry of Education, Science, Sports and Culture, Japan, 09876079.

References

1. Selleck, M. A. (1996) Culture and microsurgical manipulation of the early avian embryo, in *Methods in Avian Embryology, Methods in Cell Biology, vol. 51* (Bronner-Fraser, M., ed.), Academic, San Diego, CA, pp. 1–21.
2. Perry, M. M. (1988) A complete culture system for the chick embryo. *Nature* **331,** 70–72.
3. Perry, M. M. and Mather, C. M. (1991) Satisfying the needs of the chick embryo in culture, with emphasis on the first week of development, in *Avian Incubation, Poultry Science Symposium, vol. 22* (Tullet, S. G., ed.), Butterworth-Heinemann, London, pp. 91–106.
4. Naito, M., Nirasawa, K., and Oishi, T. (1990) Development in culture of the chick embryo from fertilized ovum to hatching. *J. Exp. Zool.* **254,** 322–326.
5. Naito, M., Nirasawa, K., and Oishi, T. (1995) An *in vitro* culture method for chick embryos obtained from the anterior portion of the magnum of oviduct. *Br. Poult. Sci.* **36,** 161–164.
6. Ono, T., Murakami, T., Mochii, M., Agata, K., Kino, K., Otsuka, K., Ohta, M., Mizutani, M., Yoshida, M., and Eguchi, G. (1994) A complete culture system for avian transgenesis, supporting quail embryos from the single-cell stage to hatching. *Dev. Biol.* **161,** 126–130.
7. Ono, T., Murakami, T., Tanabe, Y., Mizutani, M., Mochii, M., and Eguchi, G. (1996) Culture of naked quail (coturnix coturnix japonica) ova in vitro for avian transgenesis: Culture from the single-cell stage to hatching with pH-adjusted chicken thick albumen. *Comp. Biochem. Physiol.* **113A,** 287–292.
8. Sauveur, B. and Mongin, P. (1971) Etude comparative du fluide uterine et de l'albumen de l'oeuf *in utero* chez la poule. *Ann. Biol. Anim. Biochem. Biophys.* **11,** 213–214.
9. Sonoda, Y., Kai, O., and Imai, K. (1997) Egg laying and ovarian follicular growth in Japanese quail under continuous lighting. *Jpn. Poult. Sci.* **34,** 308–317.

7

Culture of Preimplantation Mouse Embryos

Adam S. Doherty and Richard M. Schultz

1. Introduction

The preimplantation mammalian embryo develops as a free living entity within the mother. This internal development inherently precludes facile experimental manipulation necessary to study cellular and molecular mechanisms of preimplantation development. In turn, this has led to intense efforts over the course of decades to develop culture media that support the preimplantation development in vitro and, in particular, mouse preimplantation development. By the mid-1960s and early 1970s, these efforts led to the development of media such as Brinster's modified oocyte culture (BMOC) *(1)* and Whitten's medium *(2)*. Further research examined the effect of the composition of the gas phase and led to the general conclusion that 5% oxygen was better than 21% oxygen, which is present in air. In addition, an empirically driven approach led to the formulation of culture media that supported development in vitro of one-cell embryos to the blastocyst stage *(3,4)* and overcame the two-cell block, which is exhibited following the culture of one-cell embryos obtained from outbred or inbred mice; embryos obtained from F1 hybrid mice do not exhibit the two-cell block.

The culture media that are still widely used by the research community for preimplantation mouse development are based on culture media developed for somatic cells. During the late 1980s and through the mid-1990s, Biggers and his colleagues undertook a rational and systematic approach to the development of culture media for the preimplantation mouse embryo that was based on simplex optimization *(5–8)*. This work, which also defined positive roles for organic osmolytes in preimplantation development, resulted in the generation of the medium called potassium modified simplex optimized medium (KSOM) + amino acids (KSOM + AA) *(9)*. When compared to other commonly used culture media, KSOM + AA fosters rates of development in vitro that most closely approach those that occur in vivo of any medium used to date. Coupled with this improved rate of development is that the pattern of gene expression in these cultured embryos is virtually indistinguishable from that of embryos that develop in vivo. Such is not the case for embryos that develop in other culture media, such as Whitten's medium, where the expression of many genes is reduced *(9)* as well as the overall rate of protein and RNA synthesis.

From: *Methods in Molecular Biology, Vol. 135: Developmental Biology Protocols, Vol. I*
Edited by: R. S. Tuan and C. W. Lo © Humana Press Inc., Totowa, NJ

This chapter will briefly outline the culture of preimplantation mouse embryos and will focus exclusively on the use of medium KSOM + AA (*see* **Note 1**), because data obtained from culture experiments in which the embryos are cultured in media that clearly support suboptimal development may simply reflect the suboptimal culture conditions and not bear directly on regulatory mechanisms that operate in vivo. For these reasons, investigators are strongly encouraged to use this medium rather than many of the commonly used media (e.g., M2, M16, Whitten's medium).

2. Materials

1. Stereomicroscope, preferably with ground-glass stage, understage illumination, and up to at least 40X magnification, e.g., Wild M5A, Olympus SZH (Olympus, Melville, NY).
2. #5 Watchmaker's forceps and small, sharp scissors.
3. CO_2 incubator.
4. 35-mm and 60-mm plastic petri dishes; embryological watch glasses can also be used to collect the embryos.
5. Mouth-operated micropipet made from a drawn-out Pasteur pipet.
6. Flying saucer incubator (Modular Incubator Chamber, Billups-Rothenberg, Del Mar, CA).
7. 5% CO_2/5% O_2/90% N_2 gas mixture.
8. PMSG (Calbiochem, La Jolla, CA) and hCG (Sigma, St. Louis, MO); stock solutions contain 5–7.5 IU of either PMSG or hCG/0.1 mL of phosphate-buffered saline (PBS). The solutions are stored at –20°C and thawed immediately before use. The solutions should not be refrozen and hence should be stored in convenient aliquots; e.g., to inject 10 mice, make 1.1-mL aliquots.
9. Tuberculin syringe (1 mL) with 27.5-gage needle.
10. Lightweight mineral oil (Sigma, St. Louis, MO) or dimethylpolysiloxane (50 centristoke) silicone oil. To extract potential water-soluble contaminants from the oil, 900 mL of oil is stirred with a magnetic stirrer for 2 d at 4°C with 100 mL of the appropriate culture medium. After the phases are allowed to separate, the oil is sterile-filtered by passage through a 0.8-μm filter.
11. Hyaluronidase (Type I-S, Sigma).
12. KSOM + amino acids (KSOM + AA):
    a. The following are made up as 10X stock solutions: 950 mM NaCl, 25 mM KCl, 3.5 mM KH_2PO_4, 2 mM $MgSO_4$, 2 mM glucose, 2 mM sodium pyruvate, 250 mM $NaHCO_3$, and 17.1 mM $CaCl_2$. Glutamine is made as a 20 mM stock solution, and EDTA, pH 7.0, is made as a 100-mM stock solution. Each of these reagents is obtained from Sigma or any reliable supply house.
    b. To make up 100 mL of medium, 10 mL of each of the 10X stock solutions, except the $CaCl_2$ 10X stock, are added to a 200-mL beaker. Next 286 μL of a 60% solution of DL-sodium lactate (Sigma) is added, followed by the addition of 10 μL of the 100 mM EDTA stock solution. Bovine serum albumin (BSA; 100 mg of Fraction V, Sigma) is then carefully added so that it does not clump. Next 20 μL of a 50 mg/mL stock solution of gentamicin sulfate (Gibco-BRL, Gaithersburg, MD) is added. The solution is then stirred gently with a magnetic stir bar, and 10 mL of the $CaCl_2$ 10X stock solution is slowly added. To avoid the formation and precipitation of calcium phosphate, the $CaCl_2$ stock is added last. The solution is then transferred to a 100-mL graduate cylinder, and the volume is brought up to 100 mL with water. The medium is then sterilized by passage through a 0.2-μm filter using a Nalge Nunc (Rochester, NY) or other suitable disposable filtration apparatus. The medium can be stored at 4°C for up to 1 wk. This medium is called KSOM.

   c. To make KSOM + AA, the amino acids are added immediately prior to use. A MEM nonessential amino acid solution (Gibco-BRL, #11140-050) and MEM amino acids solution without glutamine (Gibco-BRL, #11130-051) are used. The nonessential amino acids come as a 100X stock, and the essential amino acids solution comes as a 50X stock. The correct amount of each of these amino acid solutions is then directly added to the desired amount of KSOM to be used.

13. Bicarbonate-free MEM/PVP: Minimal essential medium (containing Earle's salts and glutamine but not containing bicarbonate) is obtained from Sigma as a powder (#M-0268). The bicarbonate is replaced with 25 mM HEPES, pH 7.2, and is supplemented with pyruvate (100 µg/mL), gentamicin (10 µg/mL), and polyvinylpyrrolidone (PVP) (3 mg/mL); the PVP is of M_r = 40,000 (Sigma).

14. Water: Highly pure water is required for the preparation of the culture media, because preimplantation mouse embryo development is extremely sensitive to impurities present in the water. Highly purified water suitable for embryo culture can be obtained by first passing reverse-osmosis grade water through a set of ion-exchange and activated-charcoal filters. This water is then fed into a Milli-Q water purification system that generates 18.2 MΩ water. Initially passing the reverse-osmosis water through the ion-exchange and activated-charcoal filters greatly increases the life of the cartridges used in the Milli-Q system. Purified water can also be purchased from commercial sources such as Sigma.

3. Methods

3.1. Superovulation of Mice

Female mice at least 6 wk of age are superovulated by intraperitoneal injection of 5–7.5 IU of pregnant mare's serum gonadotropin (PMSG) followed by intraperitoneal injection of 5–7.5 IU of human chorionic gonadotrophin (hCG) 44–48 h later; use a 1-mL tuberculin syringe with a 27.5-gage needle. A single female mouse is then housed with a single male mouse of proven fertility. The next morning the females are removed and examined for the presence of a vaginal plug, which is indicative that mating has occurred and the female is assumed to be pregnant. Typically, the females are injected with gonadotropins between 2 and 4 PM and the plugs examined at approx 9 AM (*see* **Note 2**).

3.2. Embryo Collection

Subheading 3.3. describes the collection of one-cell embryos. A similar method is used to collect embryos at later stages of development (e.g., two-cell stage, except that the embryos are flushed from the oviducts/uteri [no hyaluronidase treatment is necessary]).

1. Following sacrifice of the mice by cervical dislocation, the entrance to the body cavity is gained and the uteri/oviducts are located. Removing as little as possible of adhering fat tissue and uterus, the oviducts from a single animal are excised and placed in a 100-µL drop of bicarbonate-free MEM/PVP containing 0.5 mg/mL of hyaluronidase. Typically, 6 drops can be placed on a 60-mm plastic Petri dish (either the top or the bottom), and thus embryos can be collected from six mice.

2. The ovulated and fertilized eggs that are embedded in a cumulus cell mass are readily detected by the distension of the oviduct that appears fairly translucent under the stereodissecting microscope. Holding the oviduct with a pair of #5 forceps, the fertilized eggs are released from the oviduct into the medium by nicking the distended portion of the oviduct with a 27.5–30 gage needle. The cumulus cell mass usually extrudes itself, but if

it does not, the needle can be used to tap on the oviduct to force the embryos out into the medium (*see* **Note 3**). Once the fertilized eggs are released, remove the oviduct and any attached tissue from the drop with the forceps.

3. Examine each drop carefully. As soon as the cumulus cells are uniformly dispersed, harvest the fertilized eggs from each drop with a mouth-operated micropipet and transfer them in a minimum volume (i.e., <5 µL) to a 100-µL drop of MEM/PVP-bicarbonate (*see* **Note 4**).

4. Wash the embryos through 5–6 100-µL drops of MEM/PVP-bicarbonate to remove residual hyaluronidase and cumulus cells that were initially collected with the embryos. The transfer volume should be <5 µL (*see* **Notes 5** and **6**).

5. The embryos are now ready to be transferred to the culture medium.

6. To obtain embryos at later stages of development, embryos from the 2-cell to 8-cell stage are flushed from the oviduct, and morula and blastocysts are flushed from the uteri using bicarbonate-free MEM/PVP; hyaluronidase is not present. Plastic Petri dishes (35 mm in diameter) are very convenient to use, because following flushing, the embryos are initially dispersed throughout the dish, can rapidly be gathered at the middle of the dish by a gentle swirling motion. Two-cell embryos, four-cell, eight-cell, morula, and early cavitating blastocysts can be collected 48, 60, 72, and 96 h post-hCG, respectively.

3.3. Embryo Culture

1. About 1–2 h prior to the start of culture, set up a series of 35- or 60-mm Petri dishes containing KSOM + AA. The volume of the drops can be between 10 and 100 µL (*see* **Note 7**). The drops are then overlaid with either silicon or light mineral oil that has been equilibrated with the culture medium (*see* **Note 8**). The dishes are then placed in a flying saucer incubator that is gassed for several minutes with 5% CO_2/5% O_2/90% N_2 prior to sealing the container, which is then placed in a 37°C incubator. The medium is allowed to equilibrate for at least 1 h.

2. The embryos present in the HEPES-buffered medium are first transferred in a vol <5 µL to a drop of equilibrated KSOM + AA prior to their transfer to the drop of medium in which they will be cultured. When using small culture volumes (e.g., 10–25 µL), the embryos should be transferred in 1–2 µL. Embryo washing and transfer should be done as rapidly as possible to minimize pH and temperature changes in the equilibrated drops of medium. The presence of the oil will overly minimize pH fluctuations, because of loss of CO_2 from the culture medium, as well as temperature decreases. This protective effect, however, is not robust and hence there is an urgency to conduct the transfers in as short as time as possible. Once the embryos are transferred, the culture dish is returned to the flying saucer incubator, which is then regassed with 5% CO_2/5% O_2/90% N_2 prior to returning it to the incubator.

3. Frequent removal of the embryos from the culture chamber to monitor the developmental progress of the embryos can retard their development. This is most likely caused by temperature fluctuations and changes in pH of the culture medium. To minimize this effect, when the embryos are not being examined, the culture dishes can be placed on a slide warmer set at 37°C and covered with a Plexiglas box that is connected to a 5% CO_2/5% O_2/90% N_2 gas tank; the gas mixture is first passed through a gas washing bottle.

4. Notes

1. KSOM fosters excellent development in vitro. The addition of amino acids has little, if any, effect on development of embryos to the blastocyst stage. Their inclusion, however, does stimulate the incidence of cavitation at early time points, although by 120 h follow-

ing fertilization the same percentage of embryos has reached the blastocyst stage in either KSOM or KSOM + AA. The inclusion of amino acids also increases the incidence of blastocyst hatching.

2. Because ovulation and fertilization occur some 12–14 h post hCG, the time of plug detection is routinely referred to as day 0.5. It should be noted, however, that some laboratories term this day 0 or day 1. What is a better way of monitoring time is to state the time following hCG administration.

3. If too long of a time elapses between hCG administration and collection of the one-cell embryos, hyaluronidase activity apparently present within the oviduct results in the dispersion of the embryos. This results in no longer being able to observe a visible distension in the oviduct and the embryos are no longer readily expelled. They can be recovered, however, by flushing the oviduct as one would do to collect two-cell embryos.

4. Work as quickly as possible, as many investigators note that the longer the cells are exposed to hyaluronidase, the poorer is their subsequent development.

5. Again, work as quickly as possible, because the longer the embryos are held in the wash medium prior to transfer to the culture medium, the poorer is the ensuing development. Although these steps are routinely conducted at room temperature—and this may contribute to the poorer development—it should be noted that early preimplantation mouse embryos do not possess certain ion channels that regulate intracellular pH *(10)*. The use of HEPES-buffered medium that either does not contain or contains lower concentrations of bicarbonate likely results in perturbations of intracellular pH that detrimentally effect development. The use of HEPES-buffered medium is required to minimize pH fluctuations that would occur during the course of embryo manipulation.

6. The diameter of the Pasteur pipets that are used should be about 1.5× the diameter of the embryos, i.e., about 120 μm. Too narrow of a pipet tip will result in embryo distortion and death. Too large of a pipet will result in unacceptably large transfer volumes. The tips of the pipets should be drawn out so that essentially all of the collected embryos are contained within the drawn out portion of the pipet. The embryos should do not enter the wider region just above the "needle" portion. If they do, they may be damaged during their expulsion from the pipet, since they may get damaged when too many try to enter the region of the pipet in which the diameter is of similar size to the embryos' diameter. In addition, even though the inclusion of PVP minimizes nonspecific adhesion to the glass pipet, such adhesion that results in embryo loss can still occur. Siliconizing the pipets prior to use can also minimize such nonspecific adhesion.

7. Embryo development is enhanced when the embryos are cultured at high density (e.g., 1 embryo/μL). The likely explanation for this effect is that the embryos synthesize and secrete growth factors into the culture medium that stimulate proliferation and reduce apoptosis *(11)*. When the embryos are cultured at a low density, their ability to condition the medium is reduced and this likely accounts for the poorer development that is observed.

8. This equilibration appears to remove contaminants from the oil that can inhibit development *(12)*.

Acknowledgment

This work was supported by a grant from the NIH (HD 22681) to R.M.S.

References

1. Brinster, R. L. (1965) Studies on the development of mouse embryos in vitro. IV. Interaction of energy sources. *J. Reprod. Fertil.* **10,** 227–240.
2. Whittne, W. K. (1971) Embryo medium. Nutrient requirements for the culture of preimplantation embryo in vitro. *Adv. Biosci.* **6,** 129–141.

3. Abramczuk, J., Solter, D., and Koprowski, H. (1977) The beneficial effect of EDTA on development of mouse one-cell embryos in chemically defined medium. *Dev. Biol.* **61,** 378–383.

4. Gardner, D. K. and Lane, M. (1996) Alleviation of the "2-cell block" and development to the blastocyst stage of CF1 mouse embryos: Role of amino acids, EDTA and physical parameters. *Hum. Biol.* **11,** 2703–2712.

5. Biggers, J. D., Lawitts, J. A., and Lechene, C. P. (1992) The protective action of betaine on the deleterious effects of NaCl on preimplantation mouse embryos in vitro. *Mol. Reprod. Dev.* **34,** 380–390.

6. Lawitts, J. A. and Biggers, J. D. (1992) Joint effects of sodium chloride, glutamine, and glucose in mouse preimplantation embryo culture media. *Mol. Reprod. Dev.* **31,** 189–194.

7. Erbach, G. T., Lawitts, J. A., Papaioannou, V. E., and Biggers, J. D. (1994) Differential growth of the mouse preimplantation embryo in chemically defined media. *Biol. Reprod.* **50,** 1027–1033.

8. Summers, J. C., Bhatnagar, P. R., Lawitts, J. A., and Biggers, J. D. (1995) Fertilization in vitro of mouse ova from inbred and outbred stains: Complete preimplantation embryo development in glucose-supplemented KSOM. *Biol. Reprod.* **53,** 431–437.

9. Ho, Y., Wigglesworth, K., Eppig, J. J., and Schultz, R. M. (1995) Preimplantation development of mouse embryos in KSOM: Augmentation by amino acids and analysis of gene expression. *Mol. Reprod. Dev.* **41,** 232 238.

10. Baltz, J. M., Biggers, J. D., and Lechene, C. (1993) A novel H+ permeability dominating intracellular pH in the early mouse embryo. *Development* **118,** 1353–1361.

11. Brison, D. R. and Schultz, R. M. (1997) Apoptosis during preimplantation mouse embryo formation: Evidence for a role for survival factors including TGF-α. *Biol. Reprod.* **56,** 1088–1096.

12. Erbach, G. T., Bhatnagar, P., Baltz, J. M., and Biggers, J. D. (1995) Zinc is a possible toxic contaminant of silicone oil in microdrop cultures of preimplantation mouse embryos. *Hum. Reprod.* **10,** 3248–3254.

8

In Vitro Culture of Rodent Embryos During the Early Postimplantation Period

Masahiko Fujinaga

1. Introduction

Mammalian embryos grow within the uterus. This inaccessibility makes mammalian embryo investigations more difficult than those using non mammalian embryos. Although attempts to grow mammalian embryos outside of the uterus have appeared in the literature for many decades (*see* **Note 1**), a satisfactory methodology for rodent embryos was first established by the efforts of Dr. Denis New and colleagues at Cambridge University during the 1960s and 1970s (*1*). One of the many technical breakthroughs made by these investigators was the introduction of a roller bottle system (*2*), which has allowed many investigators access to a simple, reliable system (*3*). Many investigators, especially teratologists, have since participated in improving and refining rodent whole embryo culture systems.

Rodent whole embryo culture systems provide several distinct benefits for mammalian embryo investigations. First, these systems permit direct observation of embryonic development during culture. In addition, embryos can be subjected to manipulations, for example, microinjection or microsurgery (*see* **Note 2**). Second, maternal factors are eliminated in these in vitro systems. Third, variations in development can be minimized in experimental settings by selecting embryos at the time of explantation (*see* **Note 3**). Furthermore, embryos already showing abnormal development, including those in the process of resorption, can be identified at the time of explantation and either excluded from or included in the experiment (*see* **Note 4**). Fourth, the time lag between maternal death and collecting embryos can be minimized and culture promptly initiated (*see* **Note 5**). One of the limitations of these systems is that embryos can be cultured only during the early postimplantation period (*see* **Note 6**). Nevertheless, various important developmental events are occurring during this period, which make rodent whole embryo culture systems extremely valuable tools for mammalian embryo investigations (*see* **Note 7**). (Many investigators have taken up the challenge of culturing embryos/fetuses at other stages of development with some success; however, it is beyond the scope of this chapter to cover such topics.)

At the present time, methodologies for rodent whole embryo culture systems have been well established, particularly those for the early postimplantation period. However,

From: *Methods in Molecular Biology, Vol. 135: Developmental Biology Protocols, Vol. I*
Edited by: R. S. Tuan and C. W. Lo © Humana Press Inc., Totowa, NJ

misunderstandings persist among some investigators, and one reads comments such as, "rodent whole embryo culture systems are difficult to perform," or "embryos do not grow as well as under in vivo conditions." Indeed, rodent embryos cannot be cultured in a similar fashion as nonmammalian embryos. Nevertheless, rodent whole embryo culture is not extremely difficult to perform, and almost all embryos should grow as well as under in vivo conditions during a certain periods of development, for example, from the so-called "early somite stage" through the "early limb bud stage" between GD 9 and GD 11 in the rat and GD 7 and GD 9 in the mouse (*see* **Note 8**). This chapter aims to help investigators who are not very familiar with mammalian embryos to culture rodent embryos during such developmental periods. After having succeeded in culturing embryos following a protocol in this chapter, readers are encouraged to apply or expand it for their own investigative purposes. Also, many other papers are available to compensate for material that could not be covered here (e.g., **refs. *4–7***).

2. Materials

1. Timed-pregnant animals: It is important to understand the exact stage of pregnancy (*see* **Note 9** for a terminology and staging system, *see* **Note 10** for breeding regimens, *see* **Note 11** for commercially available timed-pregnant animals, and *see* **Note 12** for suggestions for those who have never worked with rodent embryos).
2. Anesthesia machine: The ideal systems to provide anesthesia to the mouse or rat for the purpose of whole embryo culture and for obtaining serum (blood) for culture medium are those using surplus anesthesia machines and vaporizers for inhalational anesthetic agents for human use. (*See* **Note 13** for an example of such a system and *see* **Note 14** for alternative systems.)
3. Surgical tools to dissect the uterus: Large surgical scissors and toothed forceps to cut abdominal skin and muscle, and small surgical scissors and small toothed forceps to dissect the uterus (*see* **Note 15**).
4. Sterile applicators (so-called "cotton tips") to expose the abdominal aorta (*see* **Note 16**).
5. Sterile disposable syringes (10 mL) and needles (18-gage or 20-gage) to withdraw blood from the abdominal aorta and "alcohol prep pads" (*see* **Note 17**).
6. Sterile disposable glass tubes (10–15 mL) to centrifuge blood (*see* **Note 18**).
7. Clinical centrifuge for initial centrifugation of blood (*see* **Note 19**).
8. Refrigerated centrifuge for subsequent centrifugation of blood (*see* **Note 20**).
9. Sterile disposable polypropylene tubes (50 mL) to collect and store serum (*see* **Note 21**).
10. Sterile disposable Petri dishes (100×15 mm and 60×15 mm).
11. Hank's balanced salt solution (HBSS) as a dissecting medium and for supplementation of culture medium (*see* **Note 22**).
12. Antibiotics to prevent bacterial growth in culture medium (*see* **Note 23**).
13. Sterile glass dissecting dishes with a silicon bottom (so-called "silicon Petri dish") (*see* **Note 24**) and 1.25-in.-long dressmaker pins for dissection of deciduae.
14. Surgical tools to dissect embryos from the uterus: fine scissors and fine forceps to dissect deciduae from the uterus and to dissect embryos from deciduae (*see* **Note 25**).
15. Dissecting microscope (*see* **Note 26**).
16. Sterile transfer pipets (*see* **Note 27**).
17. Sterile culture bottles (*see* **Note 28**).
18. Culture medium: The most commonly used culture medium is a combination of rat serum (75–80%) and balanced salt buffer solution (25–20%) HBSS, for both rat and mouse systems (*see* **Note 29**).

19. Custom made mixed gases, for example, 5% O_2/5% CO_2/90% N_2, 20% O_2/5% CO_2/75% N_2, and 95% O_2/5% CO_2, and gas regulators (*see* **Note 30**).
20. Bottle rotator and incubator (*see* **Note 31**).

3. Methods

3.1. Explantation of Embryos

1. Anesthetize the animal (**Fig. 1A**, *see* **Note 32**).
2. Wet the abdomen with rubbing alcohol and blot with a piece of tissue paper (**Fig. 1B**, *see* **Note 33**).
3. Make a transverse abdominal skin incision with scissors, then make T-shaped muscle incisions. (Do not cut skin and muscle together, because this may cause an accidental incision of the intestine, which is a potential source of bacterial contamination.)
4. Expose the abdominal aorta using a pair of sterile cotton applicators for blunt dissection (**Fig. 1C,D**, *see* **Note 34**).
5. Put an alcohol prep pad over the urethra (to avoid contamination of the syringe and needle), and draw blood from the abdominal aorta using a disposable needle and syringe (**Fig. 1E**, *see* **Note 35**).
6. After collecting blood, remove the needle from the syringe, and immediately transfer the blood into a sterile 10–15-mL glass tube (*see* **Note 36**).
7. Immediately centrifuge the tube for 10–15 min (while dissecting embryos) at room temperature, and then keep the tubes in a refrigerator until all animals are killed (proceed to **Subheading 3.2.** for collection of serum for culture medium; *see* **Note 37**).
8. Grab the uterine cervix with a toothed forceps and cut the uterine cervix on the vaginal side with small scissors.
9. Pull the uterine cervix upward (hold tight while doing this, not to drop the uterus), remove fat attached to the uterus, and separate the uterus together with its ovaries (**Fig. 1F**, *see* **Note 38**). (The rat and mouse have a uterus with two horns, each with its ovary.)
10. Immediately rinse the uterus once in HBSS in a Petri dish (100×15 mm) to remove the adherent blood (**Fig. 2A**).
11. Open the thoracic cavity and incise the heart to insure death of the animal.
12. Transfer the dissected uterus to a silicon Petri dish containing HBSS, and pin through the uterine cervix and an ovary (**Fig. 2B**).
13. Make incisions on the opposite side of the mesenterium in the uterus using a fine pair of scissors, and dissect the deciduae from the uterus (**Figs. 2C** and **3**, *see* **Note 39**).
14. Remove the uterine horn from the dissected side, and repeat the procedures for the other side.
15. Transfer the deciduae into another Petri dish (60×15 mm) containing HBSS, and dissect the embryos from the decidua using two pairs of fine forceps (**Figs. 3** and **4**, *see* **Note 40**).
16. Remove Reichert's membrane (parietal yolk sac) using a pair of fine forceps (**Fig. 5**, *see* **Note 41**).
17. Transfer the embryos into the culture bottle using a sterile disposable transfer pipet together with minimum amount of HBSS (*see* **Note 42**).
18. Flush the inside of the culture bottle with an appropriate gas mixture depending on the developmental stage for 15–30 s (*see* **Note 43**), cap the bottle, and place it on a rotator at a minimum of 20 rpm (preferably 30 rpm or higher) in a 37–38°C incubator (*see* **Note 44**).

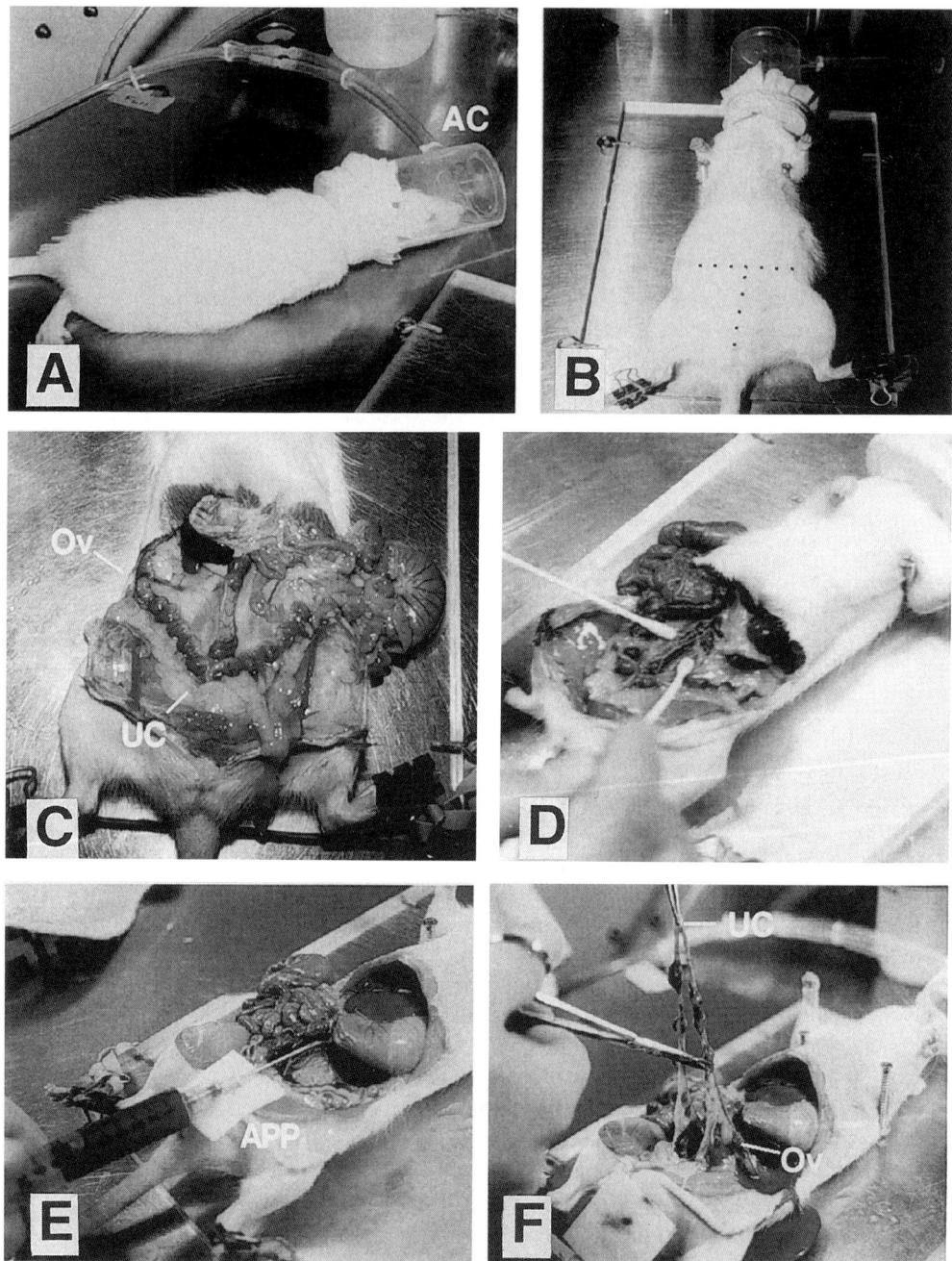

Fig. 1. **(A)** Anesthetizing a rat using isoflurane-N_2O-O_2. **(B)** The rat is placed on its back, the abdomen is made wet with rubbing alcohol and blotted with a piece of tissue paper. **(C)** The abdomen is incised in a T-shape as indicated in (B) by the dashed lines. This method of incision seems to cause minimal bleeding from incision sites. **(D)** The abdominal aorta is exposed using a pair of sterile cotton tips. **(E)** An alcohol prep pad is placed over the urethra, and blood is drawn from the abdominal aorta. **(F)** The uterine cervix is grabbed with a toothed forceps, and while pulling the uterine cervix upward, fat attached to the uterus is cut. Abbreviations: AC, anesthetic chamber; APP, alcohol prep pad; UC, uterine cervix; Ov, ovary.

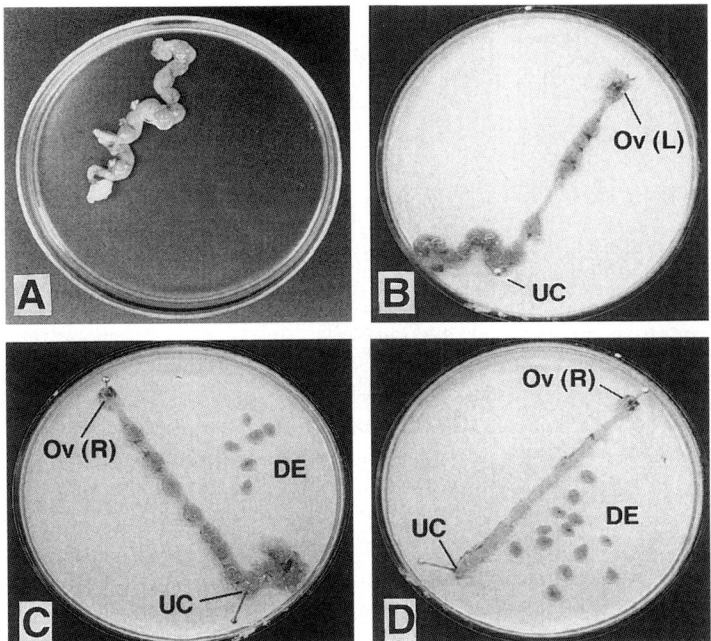

Fig. 2. **(A)** The dissected uterus in a Petri dish containing HBSS. **(B)** One uterine horn is pinned on a silicon Petri dish by the uterine cervix and an ovary. **(C)** The dissected deciduae from one uterine horn and the other intact uterine horn containing its deciduae. **(D)** Dissected deciduae from both uterine horns and the remaining uterine horn. Abbreviations: De, deciduae; UC, uterine cervix; Ov, ovary.

3.2. Collection of Serum for Culture Medium

1. Cover the precentrifuged tubes with Parafilm and centrifuge at 2000–4000 rpm (1000–2000g) at 4°C for 1–2 h or longer (*see* **Note 45**).
2. Collect and pool the serum in 50-mL sterile polypropylene centrifuge tubes (*see* **Note 46**).
3. Place a cap on the tube and incubate in water bath at 56°C for 30 min; this procedure is called "heat inactivation" (*see* **Note 47**, *see* **ref. 11**).
4. Cool the tube at room temperature.
5. Record the volume of serum, and store the tube in –80°C freezer (*see* **Note 48**).

3.3. Preparation of Culture Medium

1. When a large volume of serum has been accumulated, say, 100 mL or more (the larger, the better), thaw the collected serum (*see* **Note 49**).
2. Pool serum in a sterile container (*see* **Note 50**).
3. Add 0.25 vol HBSS; Final HBSS concentration = 20% and final serum concentration = 80% (*see* **Note 51**).
4. Add antibiotics as 1% of total volume; 10 mL/L of medium.
5. Make aliquots of appropriate volumes in 50-mL sterile polypropylene centrifuge tubes or cryotubes and store at –80°C (*see* **Note 52**).
6. Thaw the necessary amount of culture medium at 37°C, and transfer to culture bottles. Flush bottles with appropriate gas mixtures, cap with stoppers, and place in an incubator until used.

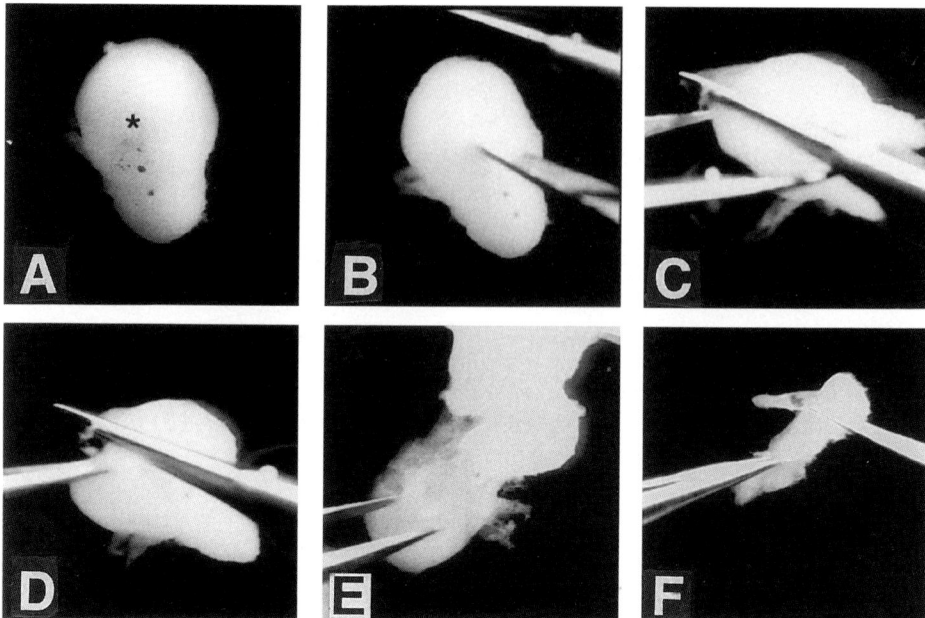

Fig. 3. A series of photographs showing how to dissect the embryo from the decidua using two pair of fine forceps on GD 9 in the rat (equivalent to GD 7 in the mouse). **(A)** There is a pair of pits on the decidua (one just beneath the asterisk mark, and another one on an opposite side). **(B)** First, stick one side of the forceps through those pits. **(C)** Close the forceps. **(D)** Along the side of this pair of forceps, cut off the top part of the decidua using another pair of forceps by closing the forceps. **(E)** Separate the decidua into two halves. Be careful not to injure the embryo during this procedure. Again, separate the piece of decidua containing the embryo into another two halves (not shown). **(F)** Gently detach the embryo from the decidua using forceps. Avoid injuring the embryo as much as possible. If an embryo is damaged or injured too much during these procedures, one should abandon this embryo.

3.4. Assessment of Embryos

1. Prepare at least two Petri dishes (60 × 15 mm) containing appropriate buffer solution, typically phosphate-buffered saline solution (PBS) or HBSS. (Use a sterile one if embryos are to be returned to culture.)
2. Carefully transfer embryos from a culture bottle into a Petri dish using a disposable transfer pipet, and rinse out the surrounding culture medium. (The yolk sac is a balloonlike structure and is easily ruptured if not handled with caution.)
3. Again, carefully transfer embryos into another Petri dish containing clean buffer solution.
4. Dissect the yolk sac from embryo using a pair of fine forceps.
5. Assess and process embryos as required (*see* **Note 53**).

4. Notes

1. A historical perspective on mammalian whole embryo culture is reviewed elsewhere *(12–14)*. The term "rodent" is used in this chapter to indicate mouse or rat to be consistent with the literature in this area of research, although it may not be precise.
2. Various types of embryo manipulation have been attempted using whole embryo culture systems and are reviewed elsewhere *(15–17)*.

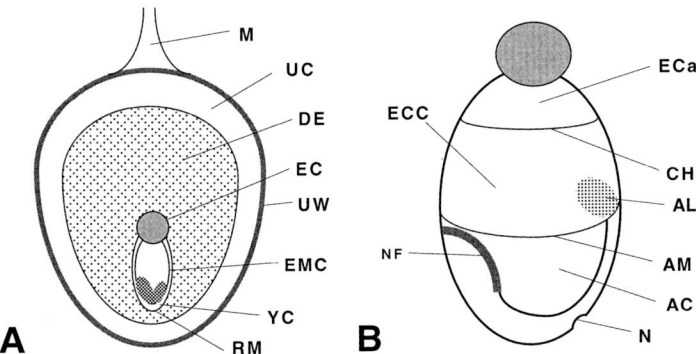

Fig. 4. Illustrations of embryos at stage 11b (on GD 7 in the mouse and GD 9 in the rat) indicating commonly used terminology. (**A**) Embryo within the uterus. (**B**) Embryo after dissection from its decidua. The terms "neural plate," "neural fold," and "head fold" are used inconsistently in the literature. The following sentences clearly distinguish these terms *(8)*: "As the notochord develops, the embryonic ectoderm over it thickens to form the neural plate. … [T]he neural plate invaginates along its central axis to form a neural groove that has neural folds on each side. … The head fold … the neural folds in the cranial region begin to thicken and project dorsally into the amniotic cavity." The term "head fold" is often misused to indicate "neural fold." (The term "head process" often used in the old literature indicates "notochordal process.") The term "archenteron" was commonly used to describe the indentation caused by development of notochordal plate *(9)*. The same areas of indentation is now more commonly called as "node." Abbreviations: AL, allantois; AM amnion; AC, amniotic cavity; CH, chorion; DE, decidua; EC, ectoplacental cone; ECa, ectoplacental cavity; ECC, extraembryonic coelomic cavity; EMC, embryonic cylinder; M, mesenterium; N, node; NF, neural fold; RM, Reichert's membrane (parietal yolk sac); UC, uterine cavity; UW, uterine wall; YC, yolk sac cavity.

3. Large variations in the development of rodent embryos are known to occur during gestation within litters (intralitter variability) and among litters (interlitter variability) *(18,19)*; examples are shown in **Figs. 6** and **7**. One advantage of the whole embryo culture system compared with in vivo systems is that embryos can be divided into different stages at the time of explantation, thus minimizing variations within and among experimental groups.

4. For those who have not worked with rodents before, the following are how intrauterine losses are usually expressed in rodent experiments. Preimplantation loss = (no. of corpora lutea – no. of implantation)/no. of corpora lutea). Postimplantation loss = (no. of implantation – no. of viable embryo/fetus)/no. of implantation. Postimplantation loss may be visually undetectable if it occurs many days in advance. If such a possibility is suspected, a chemical staining method is often used to detect implantation sites *(21)*. In brief:
   a. Trim excess fat from uterus and open; keep ovary attached.
   b. Pin the stretched uterus to a Pyrex tray, cover with a 10–12% ammonium sulfide solution (1 part of stock ammonium sulfide solution and 1 part of distilled water), and let stand for 10–15 min.
   c. Drain, and cover with distilled water for 1–2 min. Repeat several times.
   d. Implantation sites will be stained as small blue/green/black spots on the lumen surface.
   Steps b–d must be done in a fume hood. (Derived from a protocol in Dr. Thomas Shepard's laboratory at the University of Washington, Seattle.)

5. If an embryo is already in culture, the time lag between collection of the embryo and sample processing (for example, for fixation or freezing), will be greatly minimized com-

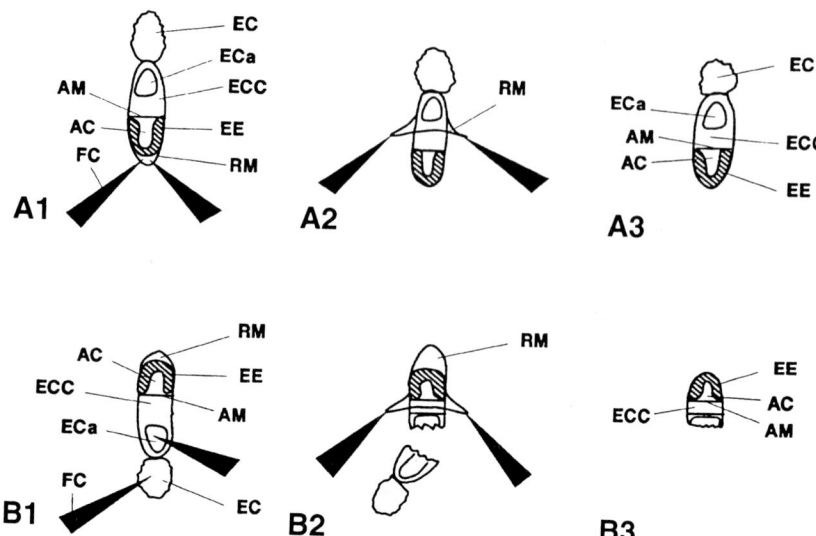

Fig. 5. Illustrations comparing two different ways of removing Reichert's membrane in the rat embryo at stage 11a, so-called "early neural plate stage"; derived from *(10)*. Embryos at this stage have three visible compartments: ectoplacental cavity, extraembryonic coelomic cavity, and amniotic cavity. The upper row **(A1)** to **(A3)** indicates the standard method of stripping Reichert's membrane away from the same side as the embryo. The lower row **(B1)** to **(B3)** indicates the method of removing Reichert's membrane from the side opposite the embryo after excising the ectoplacental cone and destroying part of other extraembryonic structures. The latter method is quicker and easier to perform and has a lesser chance of damaging the embryo. A similar method is used for younger embryos as early as stage 9, the so-called "primitive streak stage" *(10)*. However, this method is not applicable after the ectoplacental cavity closes, that is, after stage 12. Furthermore, this method is not recommended for mouse embryos because the chorion will protrude. Abbreviations: AC, amniotic cavity; AM, amnion; ECa, ectoplacental cavity; ECC, extraembryonic coelomic cavity; EC, ectoplacental cone; EE, embryo; FC, forceps; RM, Reichert's membrane.

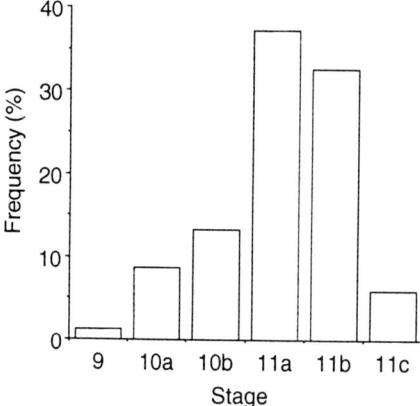

Fig. 6. The distribution of different stages of rat embryos explanted at 8–9 AM on GD 9 (203 embryos from 20 litters, unpublished data). Timed-pregnant animals were obtained by breeding for 2 h between 8 and 10 AM (B & K Universal, Fremont, CA).

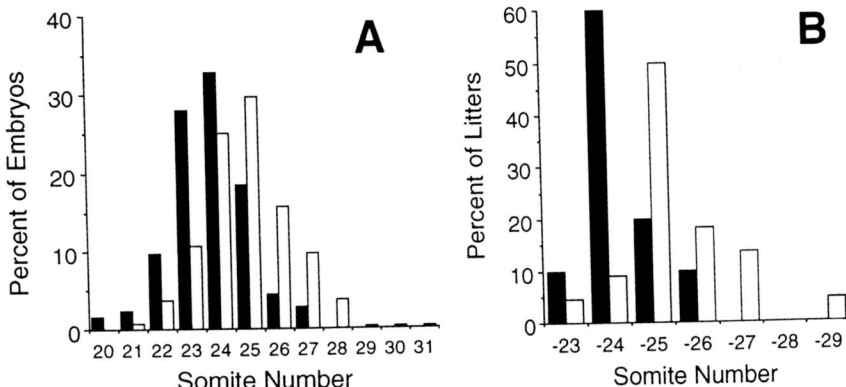

Fig. 7. The distribution of somite numbers in rat embryos examined at 11–12 AM on GD 11 (derived from **ref. *18***). Closed bars indicate embryos obtained from a morning short period breeding regimen, that is, 2 h between 8 and 10 AM (20 litters and 250 embryos). Open bars indicate embryos obtained from an overnight breeding regimen, i.e., 16 h between 5 PM and 9 AM (22 litters and 299 embryos). Mean somite numbers (± S.D.) were 23.7 ± 0.7 and 24.9 ± 1.1 (litter as a unit [**B**]) and 23.7 ± 1.3 and 24.9 ± 1.5 (embryo as a unit [**A**]), respectively. Based on the known linear increase of somites on GD 11 of approx 0.6 per hour *(20)*, the developmental difference between the two groups is approx 2 h.

pared with embryos that must be dissected from a mother. This is a particular advantage when the desired target signal is RNA.

6. **Figure 8** illustrates the development of rat embryos in culture explanted at different stages *(22)*. It has been demonstrated by many investigators that mouse or rat embryos grow as well in vitro using whole embryo culture systems as under in vivo conditions. For example, the increase in protein content between GD 9 and GD 11 in rats is almost identical in in vitro and in vivo conditions (**Fig. 9**). GD 11 coincides with the time when placental circulation is beginning to be established.

7. Representative developmental events occurring during the early postimplantation period are shown in **Fig. 10**; neurulation (which includes the process of neural tube closure), initiation of cardiogenesis, vasculogenesis, angiogenesis and hematopoiesis, initiation of heart beat and systemic circulation (**Fig. 11**), early development of optic and otic systems, development of branchial arches, early development of limbs, and early development of morphological asymmetry.

8. It is generally recognized that mouse embryos are more difficult to culture than rat embryos, particularly from earlier stages, before stage 12. Nevertheless, both embryos should be easily cultured after stage 12, the so-called "early somite stage." Thus, it is recommended for those without rodent embryo culture experience to first try culturing embryos from stage 12. Also, embryos at this stage are easier to handle than those of older stages because the exocoelomic cavity has not yet been expanded very much, thus there is less chance of rupturing the embryonic cylinder. In addition, embryos at stage 12 seem to be relatively more tolerant of surgical invasions, and they will accept damage relatively well that might occur during explantation procedures.

9. In nonmammals, the term "embryo" is used to refer to the organism at any time period between fertilization and birth. In mammals, however, the term "embryo" is defined as the developing organism during the period from the formation of the bilaminar embryonic disc (epiblast and hypoblast) until the end of organogenesis, when most of the major organs

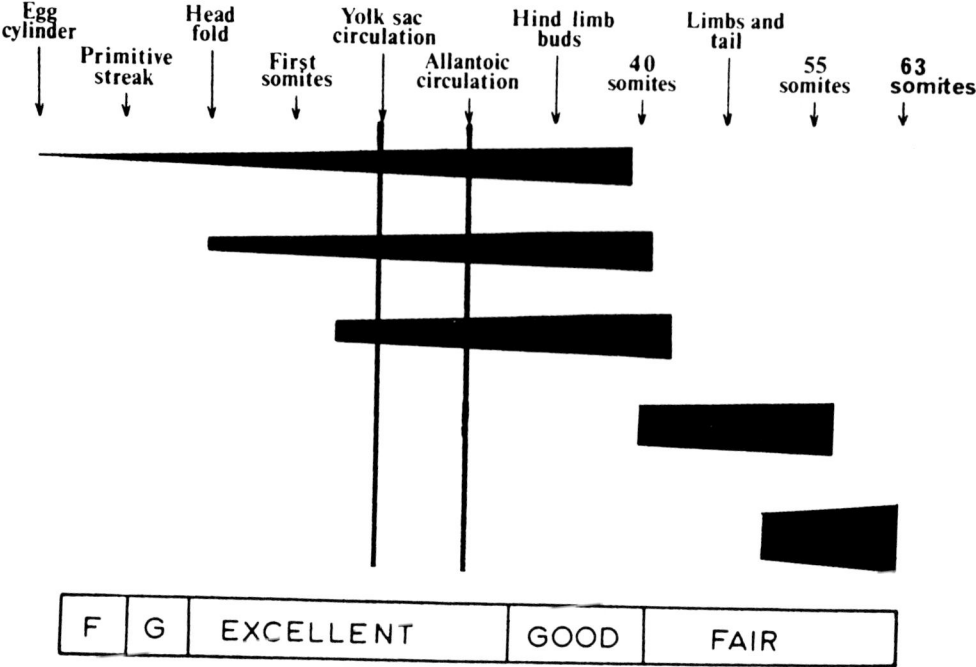

Fig. 8. Diagram derived from *(22)* showing the development obtainable from rat embryos in culture explanted at different stages of organogenesis. (The length of each black area indicates the extent of differentiation in culture. The increasing height of the black areas from left to right symbolizes the growth in size of the embryos.)

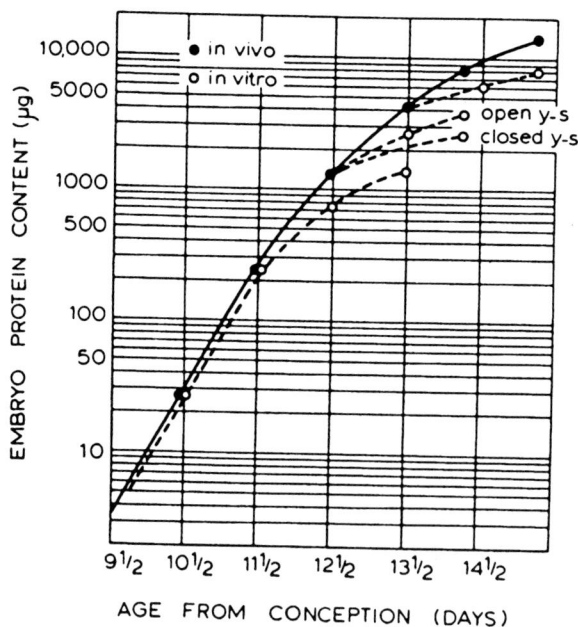

Fig. 9. Total protein content of rat embryos cultured from GD 9–GD 14 compared with growth in vivo; derived from *(22)*.

Approximate GD		Modified	
Mouse	Rat	Theiler's stage	Developmental events
5	7	9	Primitive streak presents
6	8	10a	Amniotic folds are visible
7	9	10b	Amniotic formation is complete
		11a	Neural groove is visible in the midline in frontal view
			Neural folds are not visible from lateral view
			Allantoic bud begins to develop
		11b	Neural folds are clearly visible from lateral view
		11c	Forgut pockets is visible (somites not visible)
			Cardiac primordial cells appear
		12/s1-2	First pair of somites is visible
		12/s3-4	Contraction of cardiac cells begins
8	10	12/s5-6	Bilateral heart tubes meet in the midline
			Primitive heart begins to beat
		12/s7-8	Cardiac looping begins
			Twisting of upper body begins
			Cephalic neural tube begins to close
			Hematopoiesis begins in the yolk sac
		13/s9-10	Allantois makes connection with chorion
			Yolk sac circulation becomes visible
		13/s11-12	Twisting of middle body begins
		14/s13-14	
		14/s15-16	Twisting of lower body begins
			Cephalic neural tube is almost closed
			Embryo detached from the yolk sac
9	11	14/s17-18	Axial rotation finishes
			Forelimb buds become visible

Fig. 10. Summary of the modified Theiler's staging system *(23)* and developmental events at each stage during the early postimplantation period. (Precise developmental events may vary in different species and strains.)

have been established *(8)*. In addition, the term "fetus" describes the organism from the end of organogenesis until birth. This discrepancy in definitions of embryos between nonmammals and mammals sometimes causes confusion among investigators. Furthermore, investigators use various staging systems for rodent embryos, which also causes confusion on some occasions *(23)*. In this manuscript, a modified Theiler's system proposed by this author is used (**Figs. 10** and **12**). In addition, gestational day (GD) is used to refer to the approximate time of development. GD 0 is defined as the day when a copulatory plug was observed. Although similar time-based staging systems are often used in the literature, such systems are of minimal value unless pictures or descriptions of the embryos or fetuses at each stage are presented, because the large variation in development among embryos and fetuses within a litter and among litters can vary by more than a half day in the mouse and rat over the whole gestation, including the early postimplantation period (*see* **Note 3**).

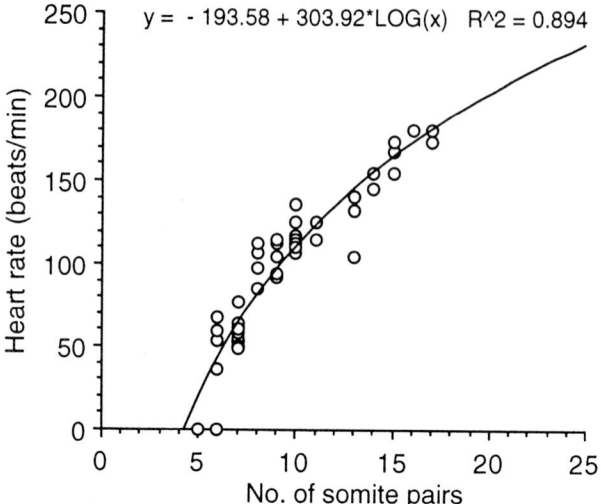

Fig. 11. Development of heart rate in rat embryos (unpublished data from a total of 90 embryos). Embryos were cultured from stage 11b/11c and were subjected to videotape recording under a dissecting microscope at different stages of development in the open perfusion microincubator system (Narishige Model PDMI-2, Tokyo, Japan) which was set at 37.7°C. Each embryo was used for heart rate determination only at a single developmental stage.

10. Two types of breeding regimens are commonly used, namely, overnight breeding and morning short-period breeding regimens. (*See* **ref. *18*** for a comparison of these regimens.) Although the short-period breeding regimen was originally developed to minimize developmental variations among litters by reducing variation in timing of copulation, it is now assumed that other factors such as the timing of implantation play more significant roles in developmental variation. Nevertheless, the short-period breeding regimen seems to be more advantageous than overnight breeding because copulation of animals from a previous "heat" cycle and the next "heat" cycle are eliminated. The time difference in development between the two breeding regimens is reported to be only 2 h in rats on GD 11 *(18)*.

11. Timed-pregnant animals are commercially available from most breeders. However, the following issues should be clarified before ordering animals:
    a. What pregnancy rate is guaranteed? Some breeders guarantee a 100% pregnancy rate, most 50%, but some none at all.
    b. What kind of breeding regimen is used? Although most breeders use a overnight breeding regimen, they sometime accept a request for a morning short-period breeding regimen.
    c. Age of female animals? Breeders tend to breed female animals as young as possible, perhaps for cost-effectiveness. However, according to this author's experience, younger females occasionally provide embryos almost 1 d younger than expected. It appears that younger female animals take a longer time to complete the implantation process, which results in younger than expected embryos. Indeed, this problem has been solved in this author's laboratory by asking the breeder to use older female rats (13–15 wk old) rather than younger female rats (9–11 wk old).
    d. Location of breeder? Try to obtain timed-pregnant animals locally, because shipment during the implantation period is generally known to result in a higher than normal resorption rate.

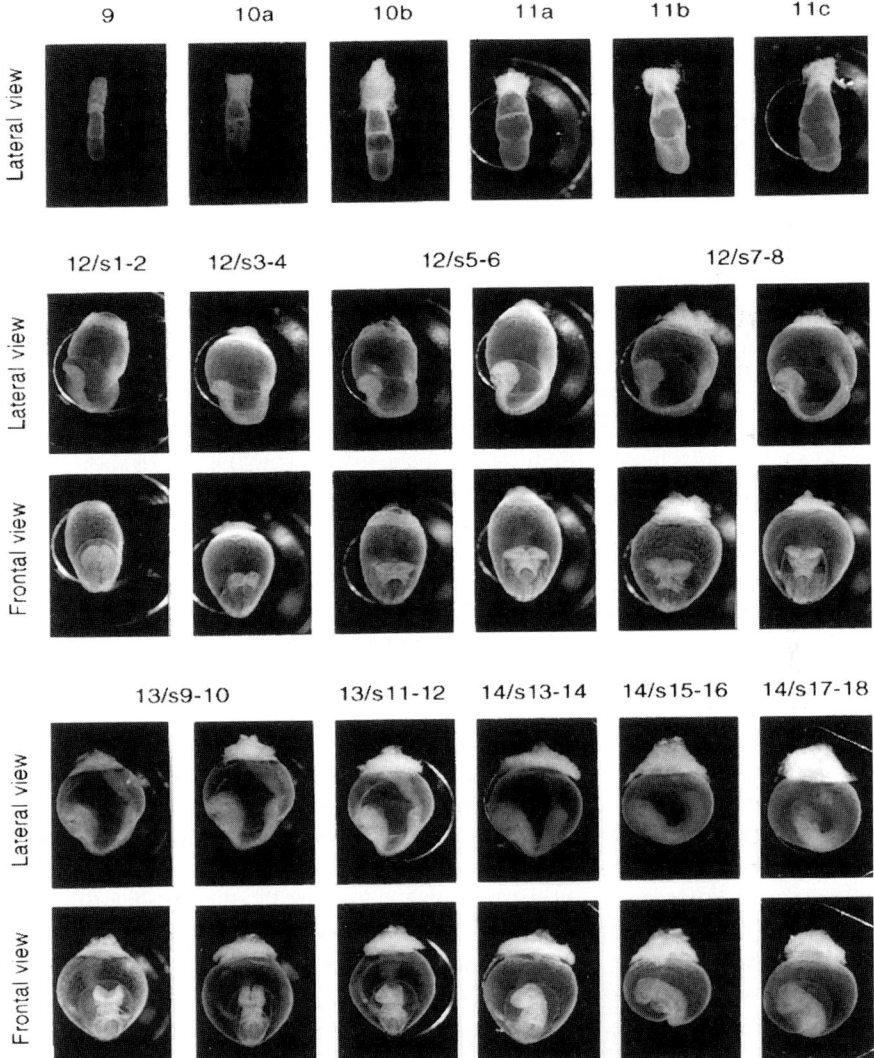

Fig. 12. A series of photographs of Sprague-Dawley rat embryos (B & K Universal) at different developmental stages, classified by a modified Theiler's staging system *(23)*.

12. For those who have never worked with rodent embryos, this author strongly recommends killing 5–10 animals at one time-point to create a frequency distribution figure such as shown in **Figs. 6** and **7**, which will help determine the timing of experiments. (Similar experiments at several other time-points will also be helpful.) After the first somite pair develops, at stage 12, the number of somite pairs is known to increase almost linearly *(20)*, 0.6 per hour, at least until GD 11 in rat and GD 9 in mouse. Thus, it is relatively easy to estimate the developmental stages of embryos to be harvested at a certain time-point. On the other hand, with embryos before stage 12, the so-called "presomite stage" and "primitive streak stage," it is more difficult not only to estimate the developmental stage but also to determine the embryonic stage itself. Inconsistency in terminology and staging systems among investigators makes this more difficult. Readers who need to work on those stages should first to read the referenced manuscripts *(19,23)*.

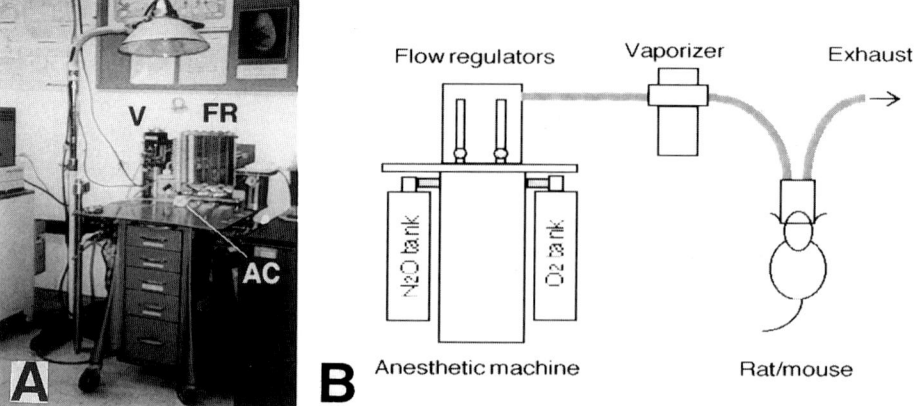

Fig. 13. **(A)** The anesthetic system (surplus anesthesia machine and vaporizer for human use) that is used in the author's laboratory. **(B)** A diagram of the system. The anesthesia chamber shown in this figure and **Fig. 1A** is made of a 120-mL plastic specimen container, whose mouth is covered with a disposable rubber globe using a rubber band; a cross-shaped slit is made by razor blade in the center. Abbreviations: AC, anesthetic chamber; FR, flow regulator; V, vaporizer.

13. This author believes that the ideal system to provide anesthesia is an anesthesia machine for human use (**Fig. 13**). This is ideal because anesthetic depth is easily controlled. Anesthesia machines for small animals are also commercially available, but they are relatively expensive, approx $5500 including a vaporizer (Stoelting, Wood Dale, IL). As anesthetic agents, inhalational anesthetics are ideal because the animals are anesthetized quickly and the effects of the agent on embryos and serum content are minimal. The most commonly used inhalational anesthetics at the present time are isoflurane and halothane, which are usually administered using flow- and temperature-independent vaporizers. Surplus anesthesia machines and vaporizers for human use are sometime available from operating rooms or research laboratories in anesthesia departments. Similarly, halothane and isoflurane for human use, but which are beyond their expiration dates, are sometimes available from operating rooms; such agents work fine with animals. A vaporizer specifically designed for use with halothane, enflurane, or isoflurane can be used with another agent. Although the concentration scale can be corrected to be precise, the differences are small enough to be ignored for animal usage. To clean the used vaporizer, run a low flow of compressed air through the vaporizer overnight in a fume hood.

14. If a vaporizer is available but not an anesthesia machine, the vaporizer can be used with either compressed air or 100% oxygen as a carrier gas, although the concentration of inhalational anesthetic has to be increased approx 2X to compensate for the lack of N_2O, and induction may take a little longer. Alternatively, a custom mixture of gases consisting of 75% N_2O/25% O_2, could be used as a carrier gas. If a vaporizer is not available, custom gas mixtures consisting of 2% isoflurane or 1% halothane/75% N_2O/25% O_2; 2% isoflurane or 1% halothane/100% O_2; or 2% isoflurane or 1% halothane in air can be used. These custom gas mixtures will cost approx $100–150 for a medical grade E-size tank. The most important issue is to establish a standard procedure to anesthetize the animals so that whole blood can be withdrawn at a reasonable speed (*see* **Note 35**) without under- or overanesthetizing the animal. Although methoxyflurane is often used by veterinarians, perhaps because it is easily used without an anesthetic machine and vaporizer, it is not

recommended for use because of potential renal toxicity in investigators. (Methoxyflurane is no longer used for clinical anesthesia.) If methoxyflurane or ether is used when collecting blood to obtain serum, make sure to leave the cap of the tube open during the heat inactivation procedure (*see* **Subheading 3.2.3.**) because these agents are extremely lipid soluble. They remain in serum even after short periods of administration. If not aiming to collect serum, an animal may be killed by cervical dislocation or by carbon dioxide inhalation, although this author does not prefer the latter method because it is now recognized to be extremely stressful for the animals unless given while gradually increasing concentration.

15. This author recommends storing surgical tools in a bactericidal solution such as zephiran chloride. To make 1 L of zephiran chloride solution, mix 7.8 mL zephiran chloride, 525.5 mL 95% ethanol and 466.7 mL autoclaved distilled water, and add 3–4 antirust tablets (#A-270, Winthrop-Breon Laboratories, New York, NY) after grinding to fine powder. Do not forget to rinse surgical tools with sterile water before use.

16. Individually wrapped sterile disposable cotton or polyester applicators are commercially available and are convenient.

17. Individually wrapped sterile disposable syringes and needles are commercially available. Syringes with a luer lock designed to prevent the piston from coming out when overdrawing are most convenient. For female rats, a 10-mL syringe is an appropriate size. For adult male rats, 20 mL is an appropriate size. For beginners, a 20-gage, 1.5-in. needle is recommended, but it takes longer to withdraw blood. Once accustomed to this procedure, a 18-gage 1.5-in. needle is recommended. Although mouse serum is not commonly used for culture medium, a 3-mL syringe and 25-gage needle are most appropriate in size if it is necessary to collect blood from mice.

18. Disposable borosilicate glass culture tubes are recommended; autoclave them before use. Polypropylene tubes are not appropriate for separating serum.

19. An inexpensive clinical centrifuge is good enough for this purpose. It will be convenient if two centrifuges are available, particularly when a large number of animals are to be sacrificed at one time to collect serum. Centrifugation speed is not a critical issue, but the blood has to be centrifuged before clotting begins. This is one of the key factors for successful culture, particularly when culturing from presomite stage, before stage 12 (*see* **Note 36**).

20. Although high-speed centrifugation is not necessary, refrigeration is needed to prevent evaporation, which may lead to higher serum osmolality.

21. Although various types of sterile disposable tubes are commercially available, a 50-mL polypropylene centrifuge tube is recommended. Make sure that the tubes tolerate ultra low temperatures, because some polypropylene tubes will crack. This author routinely uses Corning (Corning, NY) 50-mL centrifuge tubes (clear polypropylene, plug seal cap, sterile, #25330-50). 15-mL tubes are not recommended because they often crack. If storing a small volume, less than 3 mL, so-called "cryotubes" are appropriate.

22. Premade sterile HBSS is commercially available (#14025-092, Gibco-BRL, Gaithersburg, MD). Those with phenol red are not recommended because visibility is reduced by phenol red.

23. Premixed penicillin and streptomycin is commonly used and is commercially available, as 5000 IV penicillin and 5 mg streptomycin per milliliter in 0.9% sodium chloride solution, sterile-filtered (#P 0906, Sigma, St. Louis, MO). Adverse effects of these antibiotics on normal development have not been reported. When using antibiotics in the culture medium, autoclaving culture bottles, and storing surgical tools in bactericidal solution, bacterial contamination rarely occurs.

24. A glass dissecting dish (20 × 100 mm) with a silicon base is commercially available (#62-9016, Carolina Biological Supply, Burlington, NC); this is autoclavable. Alternatively, such dishes can be made by placing silicon sealer on the bottom of a regular glass Petri dish.

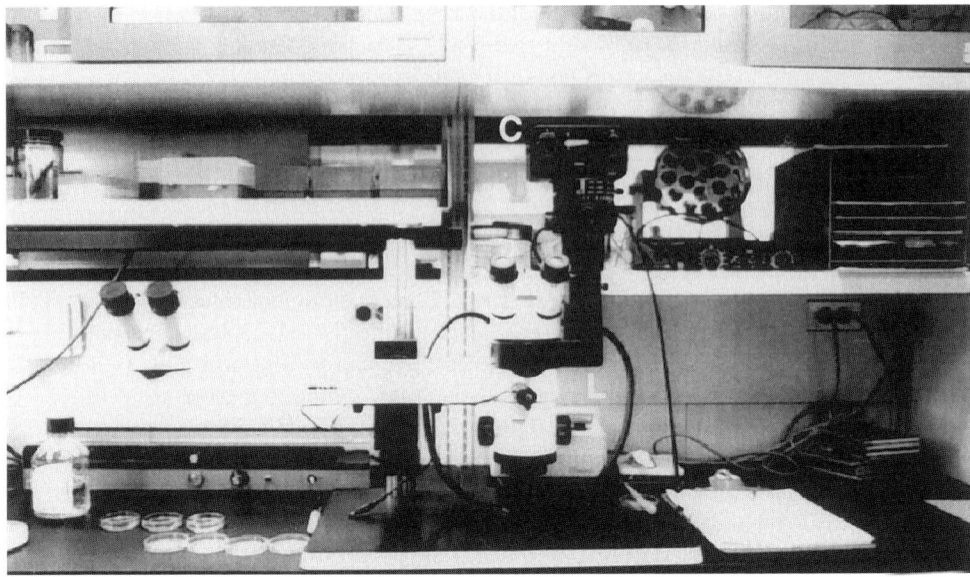

Fig. 14. The dissecting microscope used in this author's laboratory (Wild M3Z Stereo Dissecting Microscope). Abbreviations: C, camera; L, light source.

25. This author uses the following forceps (Genuine Dumont Forceps, Dumont No. 5, BIO INOX, 11 cm long, #11252-20) and fine scissors (Walton, slight curve, strong, narrow blades, very fine points, firm cutting action MORIA scissors, 10 cm long, #14077-10, Fine Science Tools, Foster City, CA).

26. Although any type of dissecting microscope that provides 10–40X magnification is appropriate, a light source coming from the side (usually by using a separate fiberoptic light source) rather than from the bottom of the stage works better. In addition, those with teaching scopes and photo systems would be more convenient. An example is shown in **Fig. 14**.

27. Individually wrapped sterile disposable plastic transfer pipets are commercially available and convenient. (Cut the tip depending on the size of embryo.)

28. There are several factors to be considered when choosing culture bottles (**Fig. 15**).

   a. Number of embryos to be cultured in a bottle has to be determined first, usually 3–5 embryos per bottle. Assuming that the culture period is 2 d from GD 9–GD 11 in rat (GD 7–GD 9 in mouse), approx 1.5 mL of culture medium per embryo is needed (**Fig. 16**). The size of culture bottle has to be at least 5–10X larger than the volume of culture medium for CO_2 in the gas phase to be able to maintain pH in the culture medium.

   b. The depth of culture medium in a bottle must be deeper than 5 mm for the embryo to grow normally because the diameter of the yolk sac is approx 3–4 mm on GD 11 in the rat (GD 9 in the mouse). Thus, a short bottle with a small mouth is ideal.

   c. An appropriate stopper must be available. A rubber stopper is most ideal.

   d. The cost of the bottle has to be reasonable. It is difficult to thoroughly clean a culture bottle after using it with a high concentration of serum; thus, a disposable one is preferable.

   e. The bottle must tolerate sterilization (autoclave or gas sterilization).

   f. Weight of bottle. Even though a heavy one (**Fig. 16A**), such a glass bottle, is preferable for a roller type rotator (**Fig. 17A**), a light one (**Fig. 16B**) is preferable for a wheel type rotator (**Fig. 17B**).

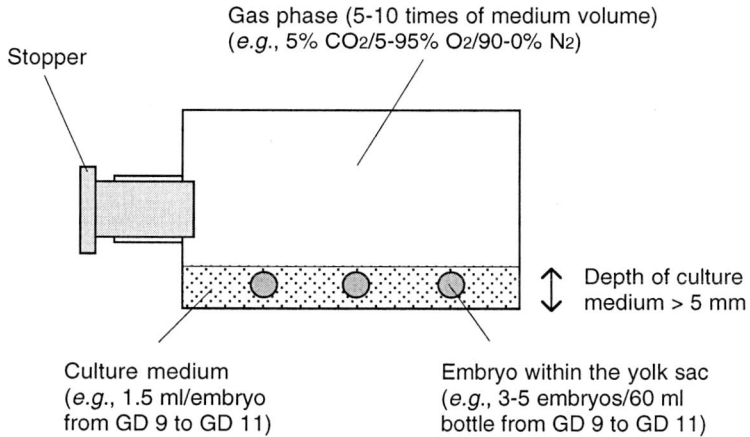

Fig. 15. Diagram of a culture bottle used for rat whole embryo culture.

Fig. 16. Culture bottles that this author uses. **Left:** Wheaton glass serum bottle (Lawson Mardon Inc., Mays Landing, NJ). 50-mL size (#223745); autoclave before use. Between three and five embryos can be cultured in a bottle. At least 4.5 mL of culture medium is needed to maintain enough depth, 4–5 mm. **Right:** Wheaton polyethylene terephtalate bottle, 45-mL size (Wheaton, #221133); gas sterilization is ideal, although rinsing the inside with alcohol seems to be good enough to prevent bacterial growth together with antibiotics in the culture medium. As many as three embryos can be cultured in a bottle. A rubber stopper of an appropriate size is recommended rather than the plastic screw cap that comes with the bottle, because it takes less time to close the bottle after gassing.

29. Many types of culture medium have been used in the past and are reviewed elsewhere *(24)*. At the present time, 75–80% rat serum supplemented with HBSS or Tyrode's solution is most commonly used for both rat and mouse cultures. Bovine serum is an attractive alternative, and its successful use has been reported *(25)*. However, it is not commonly used, because not all bovine serum provides for satisfactory outcomes, thus a screening experiment has to be conducted to find an appropriate serum. Human serum is another alternative source of serum and is used as a supplement to rat serum (for example, 25% of

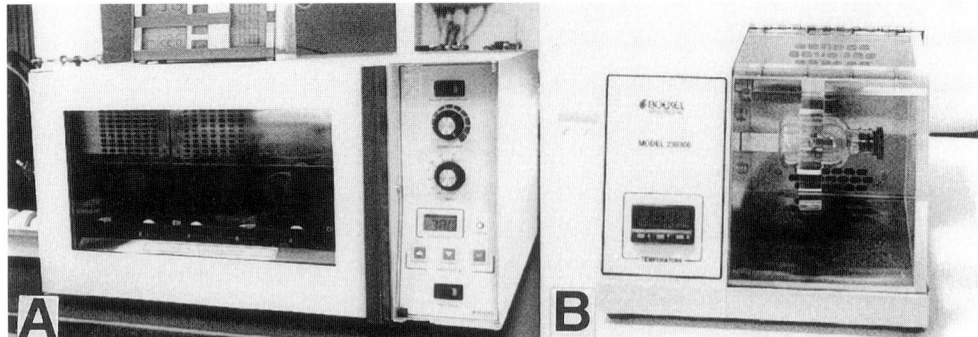

Fig. 17. Incubators that this author uses. (**A**) The Boekel Rock'N'Roll Rotisserie Incubator—Flat rolling action type (Model No. 131000, Boekel Scientific, Feasterville, PA); although rotation speed is adjustable, it is necessary to request a 2 times increased rotation speed capacity directly from the manufacturer when ordering.) A total of 15 culture bottles shown in (**A**) can be accommodated, that is, as many as 75 embryos can be cultured at one time when culturing rat embryos from GD 9–GD 11 or mouse embryos from GD 7–GD 9. (**B**) Boekel Bambino Hybridization Oven (Model 230300); no modification is necessary. This rotator/incubator is particularly useful when limited space is available in the laboratory. A total of four culture bottles shown in (**B**) can be accommodated, that is, as many as 12 embryos at one time when culturing from GD 9–GD 11 in the rat.

total volume). However, it is generally known that not all human sera work well. To date, an artificial culture medium capable of supporting normal development in rodent embryos has not been developed. Thus, practically speaking, rat serum is the ideal serum. Although many investigators use a mixture of serum from male, nonpregnant female, and pregnant female animals, it has not been well characterized how much of serum from a pregnant female animal is essential for a successful culture. As mentioned in **Note 49**, this author believes that consistency in culture media among experiments, pooling a large volume of serum from different animals, is most important. When culturing rat embryos from GD 9–GD 11 or mouse embryos from GD 7–GD 9, 1.5 mL per embryo of culture medium is the recommended amount to use (**Fig. 15**). Although there is no scientific basis, and it is not commonly described in the literature, this author usually changes culture medium at the midpoint when culturing mouse embryos. For example, start culturing embryos on GD 7 in 0.5 mL of culture medium per embryo. On GD 8, transfer embryos to another culture bottle containing 1.0 mL of culture medium per embryo. A similar method is used when culturing rat embryos from GD 8–GD 11. Changing culture medium on GD 9 seems to provide better outcomes.

30. Custom-made mixed gases of medical grade are commercially available. E-size tanks will be most convenient because of their small size and cost of approx \$100–150 per tank. At least the following three kinds of mixed gases are needed to culture embryos from GD 9–GD 11 in the rat (GD 7–GD 9 in the mouse): 5% O_2/5% CO_2/90% N_2 20%, O_2/5% CO_2/75% N_2 and 95% O_2/5% CO_2. Although specific gas regulators are commercially available, a surplus regulator for O_2, which is commonly used in hospitals, is more affordable and usable after the safety pins are destroyed.

31. Any type of apparatus that rotates culture bottles should work if rotation speed reaches 20 rpm or more, although most investigators recommend to use 30–60 rpm (*see* **Note 44**) (This author's first roller apparatus was a hot dog rotator modified by replacing an electric

motor and removing a heating unit.) Precise temperature control is not an essential issue for successful culture; that is, embryos grow well between 37.5°C and 38.5°C, although maintenance of consistent temperature is important to minimize experimental variations. When using a bottle rotator in a regular incubator, those with water jackets are not recommended because they tend to overheat *(4,6)*. **Figure 17** shows two types of rotator/incubators that this author has been using satisfactorily which are recommended to those who will be purchasing one.

32. When using isoflurane/N_2O/O_2 with an anesthetic machine and vaporizer, proceed as follows: Put the animal's head (rat) or whole animal (mouse) into a chamber while running 5% isoflurane/2 L/min N_2O/1 L/min O_2; induction of anesthesia usually takes only 10–15 s by this method. Putting a small piece of towel over the animal helps to introduce the head into the chamber. Massaging the animal's chest during induction helps calm the rat. Monitoring the respiration and be careful not to overanesthetize the animal, which makes blood drawing difficult. As soon as the animal is deeply anesthetized, turn down the isoflurane concentration to 1%. (If 5% isoflurane is maintained, an animal may die within a few minutes.) Immediately incise the abdominal skin and muscle as shown in **Fig. 1B** (5–10 s), then turn off isoflurane and N_2O but keep O_2 flow. The animal will remain anesthetized during the blood drawing procedures even if anesthesia is terminated at this stage. Blood drawing should take less than 30 s. If the anesthesia becomes too light, turn on the isoflurane and N_2O again to deepen it.

33. The use of alcohol is not to sterilize the animal surface but simply to prevent fur from sticking to the scissors.

34. Use two cotton applicators to "wipe" away overlying tissue and expose the abdominal aorta as much as possible; there are several layers of fascia surrounding the aorta. If the exposure is not adequate, the needle may go into fascia instead of into the aorta. If that happens, the blood cannot be withdrawn.

35. The bifurcation site of the abdominal aorta into the iliac arteries is a convenient place for puncture. Keep the needle bevel down at the time of puncture, and then turn the bevel up after inserting the needle into the aorta so that you can see blood coming into the tip of the needle. Do not spend too much time trying to get as much blood as possible. Cardiac massage by squeezing the closed chest or pausing the blood draw for a few seconds may help to get more blood. Approx 8–10 mL of blood should be obtained from a 250–300 g female rat.

36. Keep the tube and syringe vertical, and eject the blood slowly with a smooth action, aiming toward the bottom of the tube. Rapid ejection or contact of the blood stream with the tube wall may induce hemolysis, leading to a high potassium concentration in the serum, which might be harmful for embryos. When hemolysis is induced, the serum shows a reddish color.

37. Blood must be centrifuged immediately after sampling before clotting begins *(26,27)*. This is particularly important when culturing younger embryos, before stage 12, although the exact reasons are not yet fully understood.

38. Remove as much fat as possible. Fat remaining attached to the uterus will reduce the clarity of the solution.

39. Be careful not to damage the embryos located just beneath the incision site (**Fig. 4A**).

40. Rinse the deciduae well before transferring them to a new dish to keep the solution clean. Grab the large side (mesometrium side) of decidua with forceps for transport to avoid damaging the embryo.

41. Reichert's membrane must be removed for successful culture *(28,29)*. In mouse embryos, there is a relatively large gap between Reichert's membrane and embryo, and it is rela-

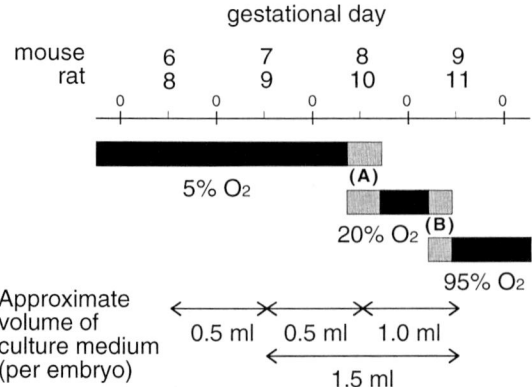

Fig. 18. Appropriate mixed gas O_2 concentrations at different stages of development and the approximate volumes of culture medium. The exact timing for changing O_2 concentrations has not been not well characterized but it does not seem to be critical for whole-embryo culture experiments. In addition, the developmental stages are indicated by gestational day in this figure. Thus, the gray areas are used for transitional periods between different O_2 concentrations; (**A**) at approx stage 13 and (**B**) at approx stage 15.

tively easy to remove the membrane. Contrarily, in rat embryos, there is sometimes no space; thus, it is more difficult to remove Reichert's membrane without damaging the rat embryo. (*See* **Fig. 5** for an easy way of removing Reichert's membrane in rat embryos.)

42. Transfer embryos into the culture medium as soon as possible with a minimal volume of HBSS. For a successful culture, one of the key factors seems to be not leaving embryos too long in HBSS.

43. Appropriate concentrations of mixed gases at different stages of development are shown in **Fig. 18**. At early stages of development, the embryo requires hypoxic conditions (*27*), because the embryo depends highly on anaerobic metabolism (*30*). Although the exact stage of development has not been determined when to switch from 5% to 20% O_2, it does not seem to be critical, that is, at approx stage 13 (in the afternoon on GD 10 in rat and GD 8 in mouse). It is more important to perform this procedure in a consistent manor in each experiment. Before this stage, however, it is well known that normoxic or hyperoxic conditions cause abnormal development. (Most cell lineage analysis studies using whole embryo culture systems during the gastrulation period, between stages 9 and 10, have been conducted under 5% CO_2 in air; thus, interpretation of those studies has to be done with caution.) Similarly, the exact stage of development has not been determined when to switch from 20% to 95% O_2, that is, at around stage 15 (in the morning on GD 11 in rat and GD 9 in mouse). Although some investigators use 45% O_2 during this transitional period, it does not seem to be necessary. **Figure 19** shows two examples of gassing systems, a sophisticated setup and a simple one. Gassing for 15–30 s should be enough to replace the gas phase inside of the bottle. The flow rate is not critical but it should be high enough to prevent the mixing of gases inside and outside of the bottle while closing the bottle, but it should not be too high to produce bubbles in the culture medium. Although some protocols suggest gassing for a longer period of time, this author does not recommend such long gassing because the temperature in the culture medium will decrease if the gassing period is prolonged. Close the cap quickly as the gassing needle is withdrawn from the bottle. Actual O_2 tensions in the culture medium after gassing with 5, 20, and 95% O_2 are reported to be 60 mm Hg, 150 mm Hg, and 680 mm Hg, respectively (*31*).

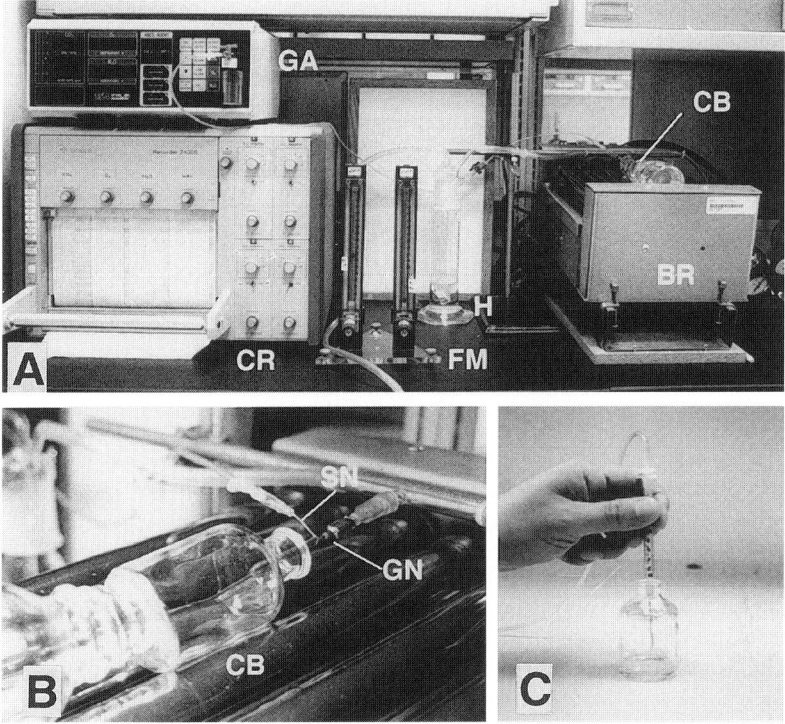

Fig. 19. Two examples of gassing systems. **(A)** An example of a sophisticated gassing system. The mixed gases come from a gas cylinder, go though a flow meter (FM), through a humidifier (H), and go into the culture bottle through a gassing needle (GN). During gassing, a portion of the gases is removed using a sampling needle (SN) to the gas analyzer (GA) to monitor gas concentrations: O_2, CO_2, N_2O, and inhalational anesthetic agents, and these are recorded on a chart recorder (CR). **(B)** A close-up picture of culture bottle during gassing. The bottle is rotated on a roller apparatus. The mixed gases go through a 16-gage needle attached to a plastic tuberculin syringe, which is connected to a gas cylinder. While pulling the gassing needle out of the bottle, quickly cap the bottle. If done reasonably quickly with high enough gas flow, only minimal mixing of gas in the bottle with outside gases occurs. This has been confirmed by this author using a system shown in **(A)**. **(C)** An example of a simple gassing system.

44. Rotation "promotes oxygenation of the medium by continuously exposing a fresh layer to the gas phase, and assists respiration by keeping the explants gently swirling about in the medium" *(22)*. Although 30–60 rpm is generally recommended, to this author's knowledge, there is no published data available for the effects of rotation speed on culture outcome. Vertical shakers seem to be too vigorous for embryos and are not recommended. Make sure that the embryos are not sticking on the inside wall of the culture bottle before placing the bottle on the rotator.
45. Cover the tube top with Parafilm to prevent contamination during centrifugation. Centrifugation speed does not seem to be very important.
46. To collect as much serum as possible but not to include any blood cells, this author leaves the bottom part of serum just above the blood clot (0.5 mL) when transferring serum to the storage tube. The remaining serum, together with some blood cells, is then transferred into a sterile microtube for additional centrifugation, 8000 rpm for 2 min, to obtain the remaining clear serum.

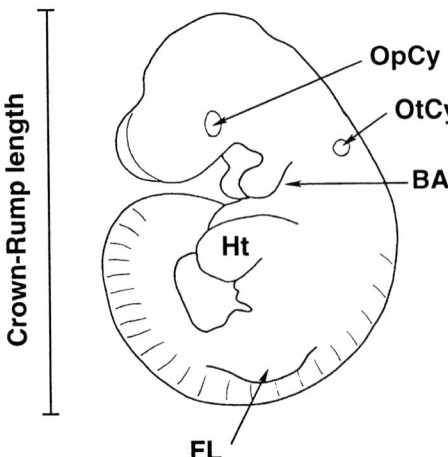

Fig. 20. An embryo at stage 15 on GD 11 in the rat or GD 9 in the mouse. Abbreviations: BA, branchial arch; FL, forelimb bud; OpCy, optic cyst; OtCy, otic cyst.

47. The original protocol by New recommends leaving the cap off to remove any dissolved ether from the serum when ether was used to provide anesthesia. However, if halothane or isoflurane is used for anesthesia, it is not necessary to leave the cap off because almost no anesthetic agent remains dissolved in the serum. This author recommends keeping the cap on during heat inactivation because evaporation during 30 min of incubation period will result in higher osmolality in the serum, which may be harmful to embryonic development.

48. Although there are no scientific data available, many investigators agree that keeping the serum at ultra low temperature, –80°C, is important. Serum kept in such a way would be usable for at least a year. Write the serum volume and date of collection on the tube and record in a book for future procedures that require pooling of serum.

49. Although it is not commonly described, this author believes that pooling a large volume of serum is one of the most important factors in obtaining consistent culture outcomes. There are many known factors (glucose, amino acids, and growth factors), and many yet unknown factors in the serum that are essential for normal development. By pooling the serum from many animals, those factors should be more standardized, thus resulting in an appropriate culture medium.

50. A sterile disposable filtration unit (Nalge Nunc, Rochester, NY), 500 mL or 1 L size, is convenient for this purpose.

51. If debris in the serum has to be removed (although it usually does not affect the performance of culture), filter the serum at this stage. A disposable vacuum filter unit would be convenient when used with prefiltration membrane.

52. This author makes aliquots of various volumes to accommodate the number of embryos for each experiment, 3 mL, 10 mL, and 20 mL. For example, at the time of explantation, pool the embryos in different stages of development in separate bottles. First, harvest all the embryos, then plan the experiment based on the number of embryos obtained. Only then prepare the culture bottles. This approach will help avoid repeated thawing of culture medium.

53. Several sophisticated scoring systems have been proposed (e.g., **refs.** *32* and *33*). However, it is not always necessary to use such systems. In most cases, measurement of crown–rump length (**Fig. 20**) and counting of somite numbers are sufficient to demonstrate normal growth. (Crown–rump length is usually determined using a scale attached to one of the eyepiece lenses.)

Acknowledgments

The author thanks Dr. James Scott for editing the manuscript and Dr. Cecilia Lo for encouraging me to prepare this manuscript. He also thanks Drs. Thomas Shepard, Alan Fantel, and Phil Mirkes at the Central Laboratory for Human Embryology (University of Washington, Seattle), who kindly provided me with basic knowledge and the techniques of whole embryo culture systems in 1987. In addition, I would like to dedicate this manuscript to Dr. Denis New, the father of whole embryo culture systems, who recently retired from research activity *(1)*. A visit to his laboratory and home in Cambridge, UK in 1988 inspired me to pursue research using whole embryo culture systems.

References

1. Cockroft, D. L. (1997) Vertebrate development in vitro. *Int. J. Dev. Biol.* **41,** 123–423.
2. New, D. A. T., Coppola, P. T., and Terry, S. (1973) Culture of explanted rat embryos in rotating tubes. *J. Reprod. Fertil.* **35,** 135–138.
3. Shepard, T. H., Fantel, A. G., and Mirkes, P. E. (1987) Somite-stage mammalian embryo culture—Use in study of normal physiology and mechanisms of teratogenesis, in *Developmental Toxicology: Mechanisms and Risk, Banbury Report No. 26.* Cold Spring Harbor Laboratory, Cold Spring Harbor, NY, pp. 29–44.
4. Cockroft, D. L. (1990) Dissection and culture of postimplantation, in *Postimplantation Mammalian Embryos—A Practical Approach,* IRL, Oxford, pp. 15–40.
5. Beddington, R. (1987) Isolation, culture and manipulation of post-implantation mouse embryos, in *Mammalian Development—A Practical Approach* (Monk, M., ed.), IRL, Oxford, pp. 43–69.
6. Freeman, S. J., Coakley, M. E., and Brown, N. A. (1987) Post-implantation embryo culture for studies of teratogenesis, in *Biochemical Toxicology—A Practical Approach* (Snell, K. and Mullock, B., eds.), IRL, Oxford, pp. 83–107.
7. Morriss-Kay, G. M. (1993) Postimplantation mammalian embryos, in *Essential Developmental Biology—A Practical Approach* (Stern, C. D. and Holland, P. W. H., eds.), IRL, Oxford, pp. 55–66.
8. Moore, K. L. (1988) *The Developing Human. Clinically Oriented Embryology,* 4th ed. W. B. Saunders, Philadelphia, PA.
9. Theiler, K. (1989) *The House Mouse. Atlas of Embryonic Development.* Springer-Verlag, New York, p. 29.
10. Fujinaga, M. and Baden, J. M. (1991) A new method for explanting early postimplantation rat embryos for culture. *Teratology* **43,** 95–100.
11. Steel, C. E. and New, D. A. T. (1974) Serum variants causing the formation of double hearts and other abnormalities on the growth in vitro of explanted rat embryos. *J. Embryol. Exp. Morph.* **31,** 709–719.
12. New, D. A. T. (1990) Introduction, in *Postimplantation Mammalian Embryos—A Practical Approach,* IRL, Oxford, pp. 1–14.
13. Cockroft, D. L. (1997) A comparative and historical review of culture methods for vertebrates. *Int. J. Dev. Biol.* **41,** 127–137.
14. Arechaga, J. (1997) Technique as the basis of experiment in developmental biology. An interview with Denis A. T. New. *Int. J. Dev. Biol.* **41,** 139–152.
15. Copp, A. J. and Cockroft, D. L. (1990) *Postimplantation Mammalian Embryos. A Practical Approach,* IRL, Oxford.
16. Eto, K. and Osumi-Yamashita, N. (1995) Whole embryo culture and the study of postimplantation mammalian development. *Dev. Growth Differ.* **37,** 123–132.

17. Naruse, I., Keino, H., and Taniguchi, M. (1997) Surgical manipulation of mammalian embryos *in vitro*. *Int. J. Dev. Biol.* **41,** 195–198.
18. Fujinaga, M., Jackson, E. C., and Baden, J. M. (1990) Interlitter variability and developmental stage of day 11 rat embryos produced by overnight and morning short-period breeding regimens. *Teratology* **42,** 535–540.
19. Fujinaga, M. and Baden, J. M. (1992) Variation in development of rat embryos at the presomite period. *Teratology* **45,** 661–670.
20. Butcher, E. O. (1929) The development of the somites in the white rat (Mus Norvegius Albinus) and the fate of the myotomes, neural tube, and gut in the tail. *Am. J. Anat.* **44,** 381–439.
21. Wislocki, G. B., Deane, H. W., Dempsey, E. W. (1946) The histochemistry of the rodent placenta. *Am. J. Anat.* **78,** 281–345.
22. New, D. A. T. (1978) Whole embryo culture and the study of mammalian embryos during organogenesis. *Biol. Rev.* **53,** 81–122.
23. Fujinaga, M., Brown, N. A., and Baden, J. M. (1992) Comparison of staging systems for the gastrulation and early neurulation period in rodents: A proposed new system. *Teratology* **46,** 183–190.
24. Cockroft, D. L. (1991) Culture medium for postimplantation embryos. *Reprod. Toxicol.* **5,** 223–228.
25. Klug, S., Lewandowski, C., Wildi, L., and Neubert, D. (1990) Bovine serum: An alternative to rat serum as a culture medium for the rat whole embryo culture. *Toxicol. In Vitro* **4,** 598–601.
26. Steele, C. E. (1972) Improved development of "rat egg-cylinders" *in vitro* as a result of fusion of the heart primordia. *Nat. New Biol.* **237,** 150–151.
27. New, D. A. T., Coppola, P. T., and Cockroft, D. L. (1976) Improved development of head-fold rat embryos in culture resulting from low oxygen and modifications of the culture serum. *J. Reprod. Fertil.* **48,** 219–222.
28. Nicholas, J. S. and Rudnick, D. (1934) The development of rat embryos in tissue culture. *Proc. Natl. Acad. Sci. USA* **20,** 656–658.
29. Jolly, P. J. and Lieure, C. (1938) Recherches sur la culture des oeufs des mammifères. *Arch. D'Anat. Microsc.* **34,** 307–374.
30. Shepard, T. H., Tanimura, T., and Robkin, M. A. (1970) Energy metabolism in early mammalian embryos. *Dev. Biol. Suppl.* **4,** 42–58.
31. Ohsaki, T., Fujimoto, E., and Miki, A. (1987) Angiogenesis and haemopoiesis in the rat visceral yolk sac under different oxygen concentrations in vitro. *Kobe J. Med. Sci.* **33,** 125–141.
32. Brown, N. A. and Fabro, S. (1981) Quantitation of rat embryonic development in vitro: A morphological scoring system. *Teratology* **24,** 65–78.
33. Van Maele-Fabry, G., Delhaise, F., and Picard, J. J. (1992) Evolution of the developmental scores of sixteen morphological features in mouse embryos displaying 0 to 30 somites. *Int. J. Dev. Biol.* **36,** 161–167.

9

Cryopreservation of Mouse Embryos

Jean Richa

1. Introduction

For almost two decades, introduction of new genes, transgenic technology, has been the source of many new strains of mice that have become valuable tools in various fields of research. More recently, knockout technology has generated a large number of mutant strains. To maintain all these lines and make them available to the scientific community, a strict breeding schedule must be followed. This can quickly become a financial burden, even for large institutions with adequate funding. Alternatively, cryopreservation offers the means to maintain mouse lines in the smallest space available, thereby reducing the requirements of time, energy, and funds. Such practices would involve freezing either mouse gametes (1–8), where the lines would be revived through in vitro fertilization, or freezing mouse embryos where the line would be reestablished after recovering the embryos.

Since it became a reality almost 25 yr ago (9), mouse embryo freezing has offered an excellent tool to preserve many mouse lines that otherwise would have vanished through breeding difficulties or health hazards. Several cryopreservation methods have been described in the literature; some relying on different cryoprotectants (10–12), to minimize damage from ice crystals, whereas others follow different cooling procedures (13,14). This chapter will cover a particular method, the "one-step" procedure, using one particular cryoprotectant, propylene glycol, and following a specific cooling process, the equilibrium freezing. Such a method was adapted from **ref. 15**, with minor modifications, and has been used extensively in our facility (Transgenic & Chimeric Mouse Facility, University of Pennsylvania, Philadelphia, www.med.upenn.edu/tcmf/) in order to preserve 25 mouse lines (to date). The success ratio of these practices (80–90% survival after thawing) has provided us with a high confidence level that has allowed us to continue with this process.

2. Materials
2.1. Cryoprotectant Solution (PG)

1. To a 50-mL tube, add 20 mL of Dulbecco's phosphate-buffered saline (PBS) (cat. #14280-010 or 036, Gibco-BRL, Gaithersburg, MD). This PBS contains 1000 mg/L of D-glucose, 36 mg/L of sodium pyruvate, and 5 mg/L of phenol red.

From: *Methods in Molecular Biology, Vol. 135: Developmental Biology Protocols, Vol. I*
Edited by: R. S. Tuan and C. W. Lo © Humana Press Inc., Totowa, NJ

2. Add 0.5 mL of penicillin/streptomycin (pen/strep) solution (cat. #15140-031, Gibco-BRL). This solution contains 10,000 µ/mL of penicillin G and 10,000 mg/mL of streptomycin sulfate in 0.85% saline.

3. Add 5.7 g of propylene glycol (cat. #P-1009, Sigma, St. Louis, MO). To measure propylene glycol (PPG) accurately, weigh the amount in a weighing boat, add it to the tube, and rinse the boat into the tube with extra PBS. Swirling the tube and warming it in a 37°C water bath will help dissolve the PPG faster.

4. Add 0.15 g of bovine serum albumin (BSA) and let it dissolve without agitation (we use Pentex™ from Bayer, Kankakee, IL, but other grades and sources can be used).

5. Fill the tube up to the 50-mL mark with PBS and invert several times to mix. Make sure pH is at 7.2–7.3 (adjust with 1 *N* NaOH solution).

6. Filter sterilize through 0.2-mm filter into a sterile tube. Store at 4°C.

7. Label and date for expiration: 6 mo.

2.2. Sucrose Diluent (SD)

1. To a 50-mL tube, add 20 mL of Dulbecco's PBS (Gibco-BRL), 0.5 mL of pen/strep solution, and 0.15 g of BSA.

2. Add 17.12 g of D(+)sucrose (cat. #25070-012, Gibco-BRL) and allow to dissolve completely.

3. Fill to the 50-mL mark with PBS, and mix well by inverting the tube several times.

4. Filter sterilize through 0.2-mm filter into a sterile tube. Store at 4°C.

5. Label and date for expiration: 6 mo.

2.3. Phosphate Buffer Nutrient (PB)

1. To a 50-mL tube, add 30 mL of Dulbecco's PBS, 0.5 mL of pen/strep solution, and 0.15 g of BSA.

2. Bring vol up to 50 mL with PBS and adjust pH to 7.2 if necessary.

3. Filter sterilize through 0.2-mm filter into a sterile tube. Store at 4°C.

4. Label and date for expiration: 6 mo.

Miscellaneous items ordered from IMV International (Minneapolis, MN): 1/2-cc straw (cat. #AAA101), 1/4-cc straw (cat. #AAA201), 13-mm goblets (cat. #PA003), and 13-mm canes (cat. #XC055). The liquid nitrogen (LN2) dewar is supplied by Cryomed (Model 35VHC, Forma Scientific, Marietta, OH).

2.4. Programming Bio-Cool (BC-III-80) Freezer (FTS Systems)-Prog1

1. SP°C	20.0	10. R3/m	6.0
2. R1/m	6.0	11. H3°C	22.0
3. H1°C	–6.0	12. T3m	10.0
4. T1m	300.0	13. E3	0.0
5. E1	0.0	14. R4/m	3.0
6. R2/m	0.6	15. H4°C	20.0
7. H2°C	–35.0	16. T4m	5.0
8. T2m	20.0	17. E4	0.0
9. E2	0.0		

3. Method: Freezing and Thawing Embryos—"One-Step" Procedure

3.1. Using the Bio-Cool Freezer

1. Turn on COOL and PROGRAMMER, and turn magnetic stirrer all the way (*see* **Notes 1** and **2**).

2. Wait until the temperature comes up and green light is on. Hit PROG, and the number 1 appears.
3. Hit RUN; the temperature will go to –6°C in about 20 min and hold there until straws are ready (for up to 5 h).
4. Proceed to prepare the straws and the embryos.
5. When freezing is finished, hit ADV and the temperature will return to 20°C in about 30 min.
6. Hit RUN and turn off COOL, PROGRAMMER, and turn back magnetic stirrer.

3.2. Preparing the Straws

1. Mark the 1/4-cc straw at 1 cm from the unplugged end.
2. Push the plug toward the open end, leaving 1 cm at the plugged end for sealing.
3. Label the straw with I.D. # and date. Also label a colored 1/2-cc straw with the I.D. #. Use methanol-resistant markers (cat. #1341, Precision Detectors, Franklin, MA).
4. Connect the straw to a 1-cc syringe (cat. #9623, Becton-Dickinson, Franklin Lakes, NJ) and aspirate 1 cm of PG solution (cryoprotectant) into the 1/4-cc straw, followed by 1-cm of air followed by another 1 cm of PG (*see* **Notes 3** and **4**).
5. Keep the straws in a horizontal position and start preparing the embryos.

3.3. Preparing the Embryos

1. After completing the embryo harvest (check **ref. *16*** for details), incubate the embryos in HEPES-buffered embryo culture medium, under oil at 4°C for 30 min. Examine embryos to eliminate any embryo of a questionable health status. The remaining procedures are performed at room temperature.
2. Separate the embryos into batches of 15–20, depending on the total number harvested.
3. Using the top of a 100-mm culture dish, dispense a row of 100-mL droplets of phosphate buffer (PB) and a row of 100-mL droplets of PG to accommodate the number of embryo batches to be processed.
4. Transfer embryos into each of the PB droplets and let them incubate for 5 min.
5. Transfer embryos into the PG droplets and start timing for 10 min.
6. After approx 5 min (the embryos will have shrunk), pick up each batch using a flame-pulled pipet connected to a mouthpiece and deposit them in the second PG column aspirated in the straw.
7. Once embryo-loading is completed, aspirate another 1 cm of air into each straw followed by a fill with sucrose diluent (SD), leaving 1 cm of air at the end to serve as sealing area.
8. When all the straws are ready (should now be at the end of 10 min), both ends are sealed and a 1/2-cc straw of a chosen color is connected to the unplugged end (**Fig. 1A**) and the assembly is dipped into the methanol bath through the rack of the Bio-Cool with the plugged end down.
9. Within a few minutes pull each straw out enough to "seed" it by touching with a pair of metal forceps previously dipped, end down, in liquid nitrogen (LN_2). **Figure 1B** indicate the spots where seeding should take place: Touch the top of the SD column and the top of the PG column (not containing the embryos—*see* **Note 5**).
10. Allow the straws to sit at –6°C for 5 min then hit ADV; this will start the slow cooling to the –35°C setpoint (about 50 min—*see* **Note 6**).
11. Prepare a cane with a goblet to hold the straws, label it and precool it by dipping into the LN_2 storage tank (*see* **Note 7**).
12. Once at –35°C, wait for 5 min and pull straws out and plunge them into the dewar containing LN_2. Now they can be transferred to the designated cane in the LN_2 storage vessel (*see* **Note 8**).

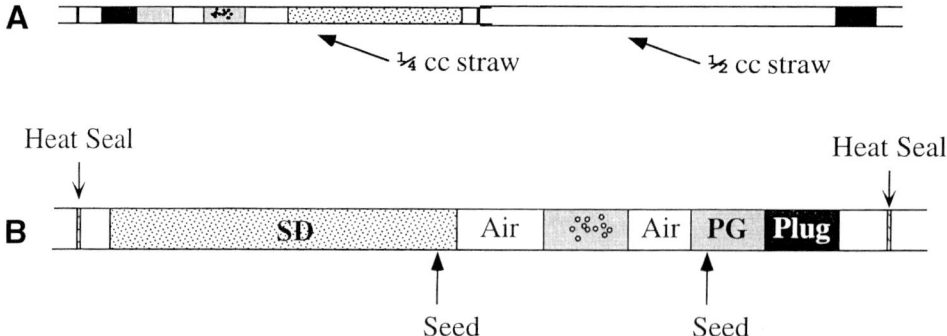

Fig. 1. Schematic diagram of straws used in cryopreservation of mouse embryos. (**A**) The assembly of 1/4 cc straw to a 1/2 cc holder straw that will be placed in liquid nitrogen by positioning the 1/4 cc straw toward the bottom of the storage container. (**B**) 1/4 cc straw filled with the required fluids. Arrows indicate the position of heat-sealing and/or freeze-seeding.

3.4. Thawing the Embryos

1. Prepare a water bath at 37°C and a test tube filled with water at room temperature. Also have ready a rod to push the plug out of the freezing straw (*see* **Note 9**).
2. Pull the desired straw out of storage and place it horizontally in air for 2–3 min (the ends of the straws should sit on hard surfaces, e.g., racks, but the rest of the straw is in air). Do not thaw embryos near an air current (heater, fan, etc—*see* **Note 10**).
3. Pull off the colored holder straw and wipe off the straw.
4. Hold unplugged end and shake the straw down 3–4X (as if it were a fever thermometer). This will mix the embryo column with the SD column. There should be one air space at the unplugged end and no air space in the rest of the straw after mixing.
5. Place straw in 37°C water bath, with plugged end down, for 3 min.
6. Place straw in room temperature water, with plugged end up for 2 min.
7. Prepare a drop or a small well of PB medium.
8. Wipe off straw and, with sharp scissors, cut the unplugged end.
9. Cut plugged end as the opposite end is dipped in the PB medium.
10. With the help of the rod, push the plug and rest of the fluid out of the straw.
11. Collect embryos and transfer to a fresh drop of PB medium for 10 min (*see* **Note 11**).
12. Transfer embryos to culture medium and place in incubator. Embryos that have less than 50% survival of the blastomeres should not be transferred.

4. Notes

1. It is necessary to replace the methanol bath as often as possible to maintain high cooling efficiency.
2. The well of the Bio-Cool should be filled with 100% methanol (cat. #9076-03, J. T. Baker, Phillipsburg, NJ) up to 1 in. from the top. Make sure not to wet the black rubber gasket; it will deteriorate, leak, and destroy the insulation in the freezer.
3. It is recommended that a sample of each solution used be passed through an acrodisc filter (cat. no. 4192, Gelman, Ann Arbor, MI) just before use. Let the samples warm up to room temperature before loading into the straws; this will prohibit the formation of minute air bubbles in the straw.
4. Avoid creating air bubbles in the PG section containing the embryos at all cost, as it will result in killing the embryos during the freezing process.

5. It is of utmost importance to seed the straws properly. Failure to do that will yield dead embryos.
6. It is important to monitor the performance of the Bio-Cool during the slow cooling process; any accidental warming caused by malfunction will result in dead embryos later.
7. Make it routine to include a couple of extra samples of frozen embryos (one at the beginning and another at the end of a batch of freezing); these are used as a means of quality control by thawing and transferring into foster females to evaluate development to term. Additional samples may be used to culture embryos in vitro to the blastocyst stage. In the event that you encounter a high incidence of developmental failure, technical problems should be considered.
8. It is recommended to freeze a minimum of 300 embryos per strain and provide duplicate storage locations to minimize the incidence of accidental loss caused by power or equipment failure.
9. Thaw out one straw at a time to have better control over the stages of the warming process.
10. Because of the small surface area of the freezing straw, any fluctuations in surrounding temperature will influence the embryo microenvironment. Therefore, it is critical to maintain a stable cold environment for frozen embryos; when retrieving a sample, make sure not to expose the rest of the straws to room temperature and then return them to LN_2 storage.
11. After thawing, 85–90% of the embryos should be live and in good condition. A very low percentage indicates a problem in the freezing procedure.

Acknowledgment

The author gratefully acknowledges the helpful comments of Susan Heyner and William Garside and the technical support of Bob Monks in the initial phase of establishing the procedure in our facility. Thanks are also due to Craig R. Street for preparing the figure.

References

1. Leibo, S. P., DeMayo, F. J., and O'Malley, B. (1991) Production of transgenic mice from cryopreserved ova. *Mol. Reprod. Dev.* **30,** 313–319.
2. Leibo, S. P. (1992) Techniques for preservation of mammalian germ plasm. *Anim. Biotech.* **3,** 139–153.
3. Whittingham, D. G. (1977) Fertilization in vitro and development to term of unfertilized mouse oocytes previously stored at –196°C. *J. Reprod. Fertil.* **49,** 89–94.
4. Fuller, B. J. and Bernard, A. (1984) Successful in vitro fertilization of mouse oocytes after cryopreservation using glycerol. *Cryo-Letters* **5,** 307–312.
5. Candy, C. J., Wood, M. J., Whittingham, D. G., Merriman, J. A., and Choudhury, N. (1994) Cryopreservation of immature mouse oocytes. *Hum. Reprod.* **9,** 1738–1742.
6. Songsasen, N., Betteridge, K. J., and Leibo, S. P. (1997) Birth of live mice from oocytes fertilized in vitro with cryopreserved spermatozoa. *Biol. Reprod.* **56,** 143–152.
7. Tao, J., Du, J., Kleinhans, F. W., Critser, E. S., Mazur, P., and Critser, J. K. (1995) The effect of collection temperature, cooling rate and warming rate on chilling injury and cryopreservation of mouse spermatozoa. *J. Reprod. Fertil.* **104,** 231–236.
8. Storey, B. T., Thompson, K. A., Richa, J., and Liebhaber, S. A. (1999) Mouse sperm cryopreservation: dialysis addition of cryoprotectant and IVF with direct insemination post-thaw provides promising route to transgene transmission. American Society and Andrology Annual Meeting, Louisville, KY, April 11–13, 1999.

9. Whittingham, D. G., Leibo, S. P., and Mazur, P. (1972) Survival of mouse embryos frozen to –196°C and –296°C. *Science* **178,** 411–414.

10. Liehman, P., Tepla, O., and Fulka, J. (1990) Vitrification of mouse 8-cell embryos with glycerol as a cryoprotectant. *Folia Biol.* **36,** 240–251.

11. Kasai, M., Komi, J. H., Takakomo, A., Tsudera, H., Sakurai, T., and Machida, T. (1990) A simple method for mouse embryo cryopreservation in a low toxicity vitrification solution, without appreciable loss of viability. *J. Reprod. Fertil.* **89,** 91–97.

12. Scheffen, B., Van Der Zwalen, P., and Massip, A. (1986) A simple and efficient procedure for preservation of mouse embryos by vitrification. *Cryo-Letters* **7,** 260–269.

13. Rall, W. F. (1987) Factors affecting the survival of mouse embryos cryopreserved by vitrification. *Cryobiology* **24,** 387–402.

14. Pomeroy, K. O. (1991) Cryopreservation of transgenic mice. *Genet. Anal. Tech. Applicat.* **8,** 95–101.

15. *Advanced Hands-On Workshop on Cryopreservation of Spermatozoa and Embryos.* June 11–13, 1993, Indianapolis, IN.

16. Hogan, B., Beddington, R., Costantini, F., and Lacy, E. (1994) *Manipulating the Mouse Embryo: A Laboratory Manual.* Cold Spring Harbor Laboratory Press, Cold Spring Harbor, NY.

III

Developmental Pattern and Morphogenesis

10

Studying Head and Brain Development in *Drosophila*

Robert Finkelstein

1. Introduction

Perhaps the most remarkable recent discovery in developmental biology is that the molecular mechanisms that pattern the animal embryo have been conserved throughout evolution. Many of the genes that specify embryonic pattern were first identified in the fruitfly *Drosophila melanogaster* and later studied in vertebrates. In this chapter, I focus on the anterior region of the fly embryo, which includes the head and brain. I describe the morphological and molecular tools that have been used to identify and analyze the genes that specify the head and brain.

2. Morphological Analysis

The fly embryo is divided along its anteroposterior axis into a series of metameric units or segments. Although the nature of anterior segmentation has been controversial, the head and brain are generally considered to consist of either six or seven segments (reviewed in **refs.** *1* and *2*). The three posterior head segments (referred to as the gnathal segments) are easy to detect morphologically. The gnathal segments give rise to the subesophageal ganglion (the posterior portion of the brain) as well as other head structures.

The more controversial portion of the fly head consists of the cephalic segments. These anterior head segments have been highly modified throughout evolution and are difficult to identify morphologically. From anterior to posterior, the cephalic metameres include the labral, ocular, antennal, and intercalary subdivisions. A recent study suggests that the labral segment may not be a true metamere (*3*).

The focus of this chapter will be the cephalic segments. These segments are particularly interesting because they give rise to the supraesophageal ganglion, the most anterior and largest region of the brain. How are the cephalic segments recognized? By definition, each segment should include a neuromere, a pair of appendages, a pair of mesodermal somites, and a pair of apodemes (which give rise to muscle attachment sites). In the case of the cephalic segments, not all these characteristics are always present. In recent analyses of the fly head, the cephalic segments have been recognized either by the analysis of brain neuromeres, specific sensory organs, or characteristic cuticular structures. In the next two sections, I discuss these segment-specific markers.

From: *Methods in Molecular Biology, Vol. 135. Developmental Biology Protocols, Vol. I*
Edited by: R. S. Tuan and C. W. Lo © Humana Press Inc., Totowa, NJ

Table 1
Subset of the Markers Used
to Identify Each of the Three Anterior (Cephalic) Segments

Cephalic segment	Cuticular structures	Sensory organs (22C10 antibody)[a]	Brain neuromere (anti-HRP)
Ocular	DMP	DMP, HPO, BO	Protocerebrum
Antennal	ANSO	ANSO, CH1	Deutocerebrum
Intercalary	DLP	DLP, AO	Tritocerebrum

Note that the DMP, ANSO, and DLP include both cuticular and neural portions.

[a]DMP, dorsomedial papilla; HPO, hypopharyngeal organ; BO, Bolwig's organ; ANSO, antennal sense organ; CH1, chordotonal organ; DLP, dorsolateral papilla; AO, associated organ.

2.1. Brain Neuromeres

The supraesophageal ganglion consists of three neuromeres. The most anterior of the three is the protocerebrum, which is thought to derive from the ocular segment. The remaining two neuromeres are the deutocerebrum (from the antennal segment) and the tritocerebrum (from the intercalary segment). These neuromeres are easily discernible by labeling with antibodies to horseradish peroxidase (HRP). Anti-HRP antibodies have been shown to recognize the surfaces of all insect neurons *(4)*. Use of these antibodies, in conjunction with confocal microscopy, has proven a powerful tool for observing brain development (*see* **ref. 5**).

A second method for recognizing specific brain regions is the use of enhancer trap strains *(6)*. In each of these strains, a *lacZ* reporter gene is inserted near a genomic regulatory region that drives expression in a tissue-specific pattern. In the strain A6-2-45, for example, the *lacZ* gene is specifically expressed in the optic lobes, which are part of the protocerebrum *(7)*. Labeling embryos from this strain with antibodies to β-galactosidase or X-gal permits the easy visualization of these structures during development.

2.2. Anterior Sensory Organs

Outside the CNS, specific sensory organs provide easily visible markers for specific cephalic segments. In a recent analysis, a large number of sensory organs were cataloged and mapped to each head segment (**ref. 8**; *see* **Table 1**). These neural structures were observed by labeling embryos with the monoclonal antibody 22C10, which recognizes the membranes of sensory neurons *(9)*. Using this antibody, Schmidt-Ott and colleagues were able to determine which head segments are disrupted by a series of mutations that affect head/brain development.

2.3. Cuticular Structures

In addition to the neural structures previously discussed, many investigators have studied segmental development by analyzing cuticular structures, which are secreted by the embryo late in embryogenesis. In cuticular preparations, the soft tissues of the fly embryo are eliminated, leaving the characteristic structures of the embryonic cuticle *(10)*. Every segment of the embryo exhibits specific cuticular structures that can easily be observed by phase contrast microscopy. In the anterior segments, these structures include elements of the "head skeleton" and the epidermal portions of specific sensory organs (**Table 1**).

3. Molecular Analysis

During embryogenesis, a complex series of movements called head involution bring most external head structures inside the anterior end of the fly embryo. Because of this, strictly morphological studies of head development can be extremely difficult. This is particularly true in mutant embryos, in which head involution often fails and specific structures become mislocalized and impossible to identify unambiguously.

To overcome this problem, many investigators have turned to gene expression as an alternative method for analyzing head segmentation. This approach was hinted at in the previous section, in which I described antibodies that label neurons in the anterior embryo. Many fly genes have been identified that are expressed in particular regions of the head and brain (reviewed in **ref. 2**). Among these are representatives of the different classes of segmentation genes. In this section, I will focus only on the segment polarity genes, which have provided powerful markers for understanding cephalic development.

In the trunk segments, specific segment polarity genes are expressed within subdomains of every segment of the embryo. Two examples are *engrailed (en)* and *wingless (wg)*, which encode a homeodomain protein and a secreted protein, respectively (for a review of segment polarity gene expression and function, *see* **ref. 11**). *en* is expressed in the posteriormost cells of each trunk segment, whereas the *wg* gene product is found in immediately adjacent anterior cells.

wg and *en* are also expressed in each of the head segments *(12)*. Their expression can easily be detected using antibodies, RNA probes, or *lacZ* marker strains (*see*, e.g., **ref. 13**). The expression of these two genes has made it possible to study the development of these segments and to determine what happens to them in mutant embryos. The use of segment polarity genes as markers led to rapid progress in understanding head/brain development.

4. Combining Morphological and Molecular Analysis: An Example

An important example of the power of the methods I have described is the analysis of the cephalic gap genes. Classic mutagenesis screens resulted in the identification of a group of mutations that disrupt head development. However, the mechanisms by which these mutations act remained obscure. Recently, however, the use of the techniques just discussed have led to important insights into how the head and brain are specified.

Three mutations, *orthodenticle*, *empty spiracles*, and *buttonhead* were among those shown to affect head formation in the early mutant screens. More recently, a series of investigators studied the phenotypes of embryos mutant for each of these genes using morphological and molecular markers (reviewed in **ref. 1**). Their analyses revealed that each mutation led to the deletion of a group of segments in the head/brain. As a result, the genes they affect are generally referred to as "gap" genes. Interestingly, the domains of expression and function of the three cephalic gap genes overlap during embryonic development *(14)*. The discovery of this overlap led to models of how the anterior head is subdivided. It has been proposed, for example, that these genes both create head/brain segments and specify their individual identities *(15)*. This mechanism is quite different from the molecular paradigm that governs trunk formation.

These genes, and others involved in head/brain formation in the fly, have vertebrate homologs. *Orthodenticle* and *empty spiracles*, for example, each have multiple homologs in higher animals expressed in the developing forebrain and in sensory organs (reviewed in **refs.** *16* and *17*). Targeted inactivation of these genes in the mouse shows that they play critical roles in regional specification of the rostral brain. Continued analysis of invertebrates and vertebrates is sure to produce further insights into how the head and brain are specified.

References

1. Finkelstein, R. and Perrimon, N. (1991) The molecular genetics of head development in *Drosophila melanogaster*. *Development* **112,** 899–912.
2. Jurgens, G. and Hartenstein, V. (1993) The terminal regions of the body pattern, in *The Development of Drosophila melanogaster*, Cold Spring Harbor Laboratory Press, Cold Spring Harbor, NY, pp. 687–746.
3. Rogers, B. T. and Kaufman, T. C. (1996) Structure of the insect head as revealed by the EN protein pattern in developing embryos. *Development* **122,** 3419–3432.
4. Jan, L. Y. and Jan, Y. N. (1982) Antibodies to horseradish peroxidase as specific neuronal markers in *Drosophila* and grasshopper embryos. *Proc. Natl. Acad. Sci. USA* **79,** 2700–2704.
5. Hirth, F., Therianos, S., Loop, T., Gehring, W. J., Reichert, H., and Furukubo-Tokunaga, K. (1995) Developmental defects in brain segmentation caused by mutations of the homeobox genes *orthodenticle* and *empty spiracles* in Drosophila. *Neuron* **15,** 769–778.
6. O'Kane, C. and Gehring, W. (1987) Detection *in situ* of genomic regulatory elements in *Drosophila*. *Proc. Natl. Acad. Sci. USA* **84,** 9123–9127.
7. Green, P., Younossi-Hartenstein, A., and Hartenstein, V. (1993) Embryonic development of the *Drosophila* visual system. *Cell Tissue Res.* **273,** 583–598.
8. Schmidt-Ott, U., Gonzalez-Gaitan, M., Jackle, H., and Technau, G. M. (1994) Number, identity, and sequence of the *Drosophila* head segments as revealed by neural elements and their deletion patterns in mutants. *Proc. Natl. Acad. Sci. USA* **91,** 8363–8367.
9. Fujita, S. C., Zipursky, S. L., Benzer, S., Ferrus, A., and Shotwell, S. L. (1982) Monoclonal antibodies against the *Drosophila* nervous system. *Proc. Natl. Acad. Sci. USA* **79,** 7929–7933.
10. Van Der Meer, J. (1977) Optical clean and permanent whole mount preparations for phase contrast microscopy of cuticular structures of insect larvae. *Drosophila Inf. Serv.* **52,** 160.
11. Martinez Arias, A. (1993) Development and patterning of the larval epidermis of *Drosophila*, in *The Development of Drosophila melanogaster*, Cold Spring Harbor Laboratory Press, Cold Spring Harbor, NY, pp. 517–608.
12. Schmidt-Ott, U., Sander, K., and Technau, G. M. (1994) Expression of *engrailed* in embryos of a beetle and five dipteran species with special reference to the terminal regions. *Roux's Arch. Dev. Biol.* **203,** 298–303.
13. Finkelstein, R. and Perrimon, N. (1990) The *orthodenticle* gene is regulated by *bicoid* and *torso* and specifies *Drosophila* head development. *Nature* **346,** 485–488.
14. Cohen, S. M. and Jurgens, G. (1990) Gap-like segmentation genes that mediate *Drosophila* head development. *Nature* **346,** 482–485.
15. Cohen, S. M. and Jurgens, G. (1991) *Drosophila* headlines. *Trends Genet.* **7,** 267–272.
16. Finkelstein, R. and Boncinelli, E. (1994) From fly head to mammalian forebrain: The story of *otd* and *Otx*. *Trends Genet.* **10,** 310–315.
17. Rubenstein, J. L. R., Martinez, S., Shimamura, K., and Puelles, L. (1994) The embryonic vertebrate forebrain: The prosomeric model. *Science* **266,** 578–580.

11

Bioassays of Inductive Interactions in Amphibian Development

Takashi Ariizumi, Kazuhiro Takano, Makoto Asashima, and George M. Malacinski

1. Introduction

Amphibian embryos provide excellent material for understanding the establishment of the vertebrate body plan during early development. Fertilized eggs are readily obtained by hormone-induced spawning, and their developmental rate can be adjusted by ambient temperature regulation. Eggs and early embryos are large enough in size for surgical manipulations, especially when compared with eggs of other vertebrates. Also, embryos and isolated embryonic tissues can be easily cultured for a minimum of several weeks in a simple salt solution containing antibiotics. These advantageous features of amphibian eggs and embryos have led to the discovery of the primary embryonic organizer, various inducing factors, and a variety of cellular events involved in inductive interactions.

In the original "induction experiments," the primary embryonic organizer (blastopore lip) was transplanted into the ventral side of a host embryo. The neural tissues of the induced secondary embryo were almost entirely derived from the host ventral ectoderm, normally fated to become epidermal tissue (1). Another set of classical "induction experiments," Nieuwkoop's "tissue recombinations," involved using endodermal (vegetal) cells combined with ectodermal (animal) cells to induce the formation of mesodermal tissue (2). The original primary organizer inductive phenomenon is called "neural induction," and the phenomenon generated by the recombination manipulation is referred to as "mesoderm induction." The basic body plan of vertebrate embryos is generally considered to be established as a result of these two major inductive interactions.

The principles underlying those classical induction assays remain valid today, and updated versions of them will be described herein. A conceptually simpler induction assay system is, however, now available. It is based on the pluripotency of blastula or gastrula ectoderm. The presumptive ectoderm (animal cap) forms an undifferentiated epidermal cell mass, the so-called "atypical epidermis," when cultured in isolation. It can, however, be induced to differentiate into neural tissue, mesoderm, and also endoderm by exposure to an appropriate inducing agent.

From: *Methods in Molecular Biology, Vol. 135: Developmental Biology Protocols, Vol. I*
Edited by: R. S. Tuan and C. W. Lo © Humana Press Inc., Totowa, NJ

This simple "animal cap assay," which uses an isolated ectoderm as a reacting tissue, has generated remarkable advances in the identification of inducing factors in recent years. Several peptide growth factors (PGFs) belonging to the fibroblast growth factor (FGF) and transforming growth factor-β (TGF-β) families have been identified as strong candidates for the natural (in vivo) mesoderm-inducing factor (reviewed in **ref.** *3*). Moreover, during the past decade several genes activated by the PGFs and/or responsible for the formation of the embryonic axis have been cloned and studied with the aid of the animal cap assay. The ease with which amphibian eggs can be microinjected is often exploited for testing the function of such genes.

In this chapter, we outline several bioassay systems, review microinjection procedures for studying inductive interactions during early amphibian development, and provide practical notes (cautions) for the uninitiated.

2. Materials

2.1. Animals and Production of Embryos

Amphibians, whose embryos are commonly used for the bioassays of inductive interactions, are listed in **Table 1**. They can be commercially purchased, are easily maintained, and can be coaxed to breed in the laboratory. Urodela, such as *Cynops* and *Ambystoma*, lay relatively large eggs (approx 2-mm diameter), and their embryos develop relatively slowly, which are advantageous features for performing surgical manipulation. Consequently, urodele embryos have proved to be the most useful material for traditional (i.e., classical) experimental embryology *(4)*. Urodele populations are, however, currently decreasing throughout the world. The amphibian most widely used today is the African clawed frog, *Xenopus laevis*. It is commercially available, produces large numbers (1000–2000) of eggs at one spawning in response to hormone stimulation, and its embryos develop relatively rapidly (hatch in 37 h at room temperature). Those features of *Xenopus* provide significant advantages for biochemical and molecular biological studies *(5)* when compared with urodeles.

2.1.1. Ambystoma mexicanum

Eggs of *Ambystoma mexicanum* (axolotl) are inseminated as they pass through the cloaca of the female. (After a ritual courtship, she picks up sperm packets [spermatophores] fastened to the substrate [e.g., rocks] by a male.) Fertilized eggs are usually obtained by natural spawnings in the laboratory. Artificial insemination is also possible *(6)*, but rather difficult when compared with the *Xenopus* artificial insemination procedure. To obtain natural axolotl spawnings, a male is placed together with a female in the mating container (e.g., plastic dishpan lined with pebbles) and left in the dark. Enhanced spawning frequencies can be obtained by injecting a female intramuscularly with 250 IU of human chorionic gonadotropin *(6)*. The male is removed the next day after spermatophores have been deposited on the pebbles. The container with the female is then returned to the dark. The female usually starts to shed fertilized eggs about 24 h after she has picked up spermatophores. Staging series published by Schreckenberg and Jacobson *(7)* and by Bordzilovskaya and Dettlaff *(8,9)* are available for the staging of *A. mexicanum* embryos.

Table 1
Amphibians Commonly Used in Experimental Embryology

Amphibia	Popular name	Number of eggs	Egg size (mm)	Natural breeding period	Locality
Urodela					
Ambystoma mexicanum	Axolotl	200–1000	2.0	October to April	Central Mexico
Cynops pyrrhogaster	Japanese fire-salamander	50–250	2.0	March to June	Japan (except Hokkaido)
Anura					
Xenopus laevis	African clawed-toad	1500–2000	1.2	April to September	South Africa

2.1.2. Cynops pyrrhogaster

Although adult *Cynops* females can be commercially purchased, Japanese laboratories usually collect them from the ditches around rice fields in early spring and late autumn. The majority of females collected in those seasons already carry spermatophores. The females are stored in water at 4–6°C and need not be fed. They retain the spermatophores for many months. To obtain fertilized eggs, females are injected with 50–100 IU of human chorionic gonadotropin (HCG; e.g., GESTRON®, Denka Seiyaku, Kawasaki, Japan) on alternate days for two or three injections *(10)*, then placed in water at 20°C together with a polyethylene tape on which they can lay eggs. A robust female will lay as many as 250 eggs over 1 wk (**Fig. 1A**). A staging series for *Cynops pyrrhogaster* has been published by Okada and Ichikawa *(11)*, based on external morphology and on features visible in dissected embryos.

2.1.3. Xenopus laevis

Xenopus can be easily induced to spawn by HCG injection every 3 mo. To obtain fertilized eggs, the dorsal lymph sacs of both male and female are injected with 600–800 IU of HCG. The animals are then placed in a container filled with water to approx 5 cm in depth and kept at 20°C overnight (**Fig. 1B**). Fertilized eggs laid on the bottom of the container are scraped off and collected with a wide-bore pipet (5 mm in diameter). For artificial insemination, HCG-injected females are kept at 20°C overnight to induce ovulation (**Fig. 1C**). Approximately 100 eggs are squeezed into a 60-mm plastic Petri dish and fertilized with 0.2–0.4-mL sperm suspension, which is prepared by suspending finely macerated testis in 2 mL modified De Boer's saline (MDB; 110.00 mM NaCl, 1.30 mM KCl, 0.44 mM CaCl$_2$, 3.00 mM HEPES, pH 7.3) *(12)*. Artificial insemination is advantageous, especially in the microinjection study, to simultaneously bring embryos to exactly the same developmental stage. A normal table published by Nieuwkoop and Faber *(13)* is available for staging *Xenopus* embryos.

2.2. Thermal Regulation of Development

The early stages of development are influenced by external circumstances, such as water temperature, because all amphibia lay their eggs in water. The effect of temperature on the developmental rate of popular laboratory amphibian embryos is given in **Fig. 2**. Within the normal temperature tolerance range it is possible to retard or accelerate their developmental rate without altering embryonic processes. This feature facilitates bringing embryos from different batches or spawnings to exactly the same developmental stage in preparation for experimentation. It should be noted, however, that the optimum range for amphibian embryos is 14°C to 25°C, though most species of urodele embryos seem to display better low-temperature tolerance compared with anuran eggs (especially *Xenopus*). Most laboratories are kept between 20 and 23°C, which is satisfactory for the experimental manipulation of virtually all amphibian embryos.

2.3. Equipment and Instruments

Equipment and instruments required for microinjection and microsurgery are illustrated in **Figs. 3** and **4**, respectively. Rugh *(14)* provides detailed descriptions for instrument making. In this section, we will describe the production and sterilization of the major instruments.

Fig. 1. Obtaining eggs. (**A**) Obtaining *Cynops* fertilized eggs: *Cynops* females, which have already received spermatophore from males, are induced for oviposition by the injection of human chorionic gonadotropin (HCG) into the body cavity. After two or three injections of 100 IU of HCG at 1-d intervals, females continue laying fertilized eggs for a week. 1) Female, 2) fertilized egg adhering to a polyethylene tape (10 × 200 mm), and 3) polypropylene case (100 × 70 × 40 mm). (**B**) Obtaining *Xenopus* fertilized eggs: Fertilized eggs are obtained by the injection of HCG into the dorsal lymph sacs of male and female (600 IU each). 1) Male, 2) female, 3) fertilized egg adhering to the bottom of container, and 4) polypropylene container (530 × 350 × 120 mm). (**C**) HCG-induced ovulation of *Xenopus* female: Eggs for artificial insemination can be obtained by the injection of 600 IU of HCG into the dorsal lymph sac. After 12 h (depending on the temperature), the female will lay eggs with her cloaca tinged with red (arrow).

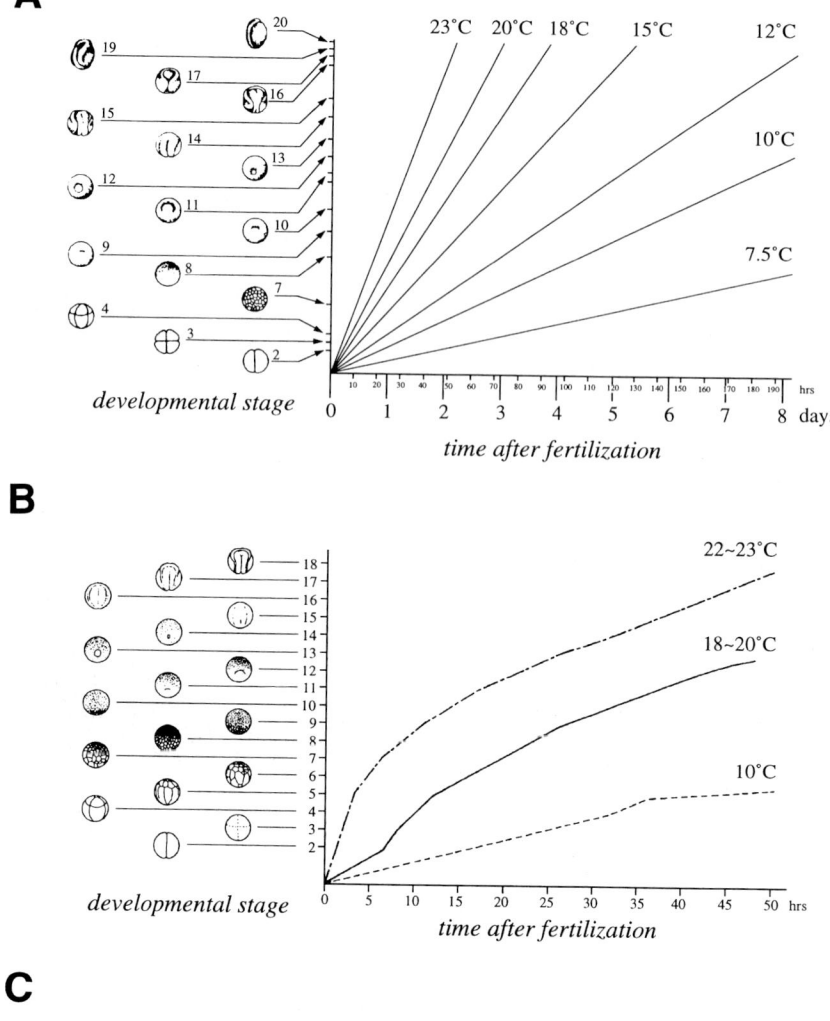

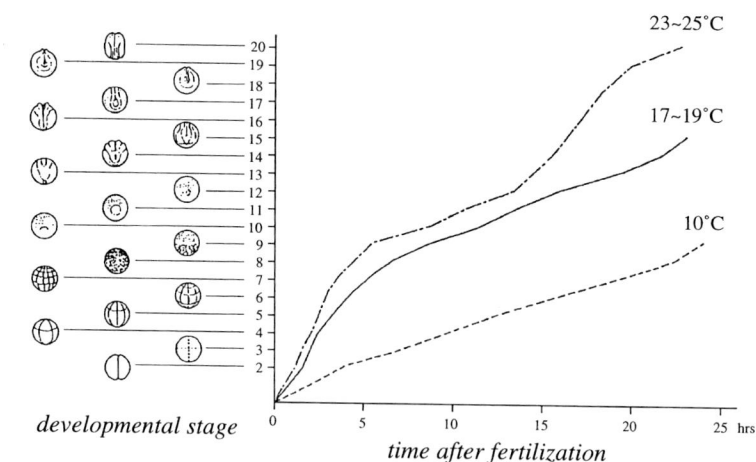

Fig. 2. Temperature-dependent development of amphibian embryos. (**A**) *Ambystoma mexicanum.* (**B**) *Cynops pyrrhogaster.* (**C**) *Xenopus laevis.*

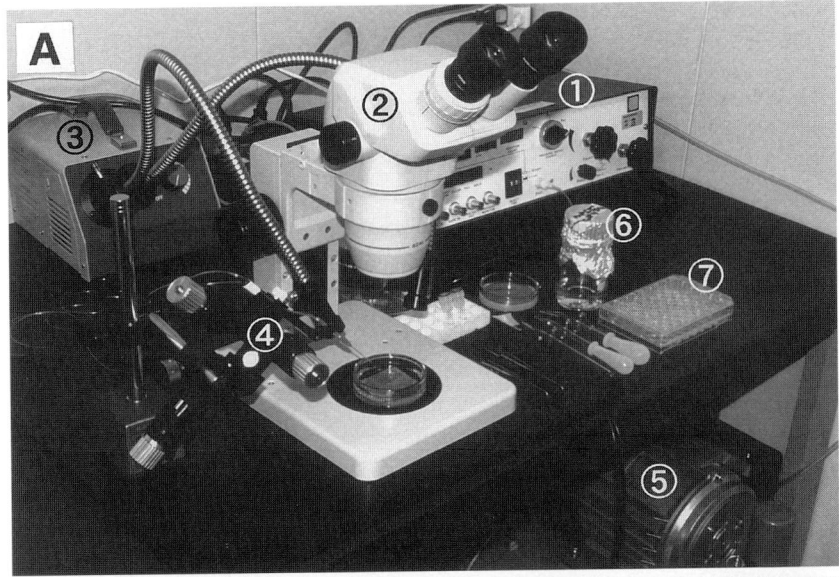

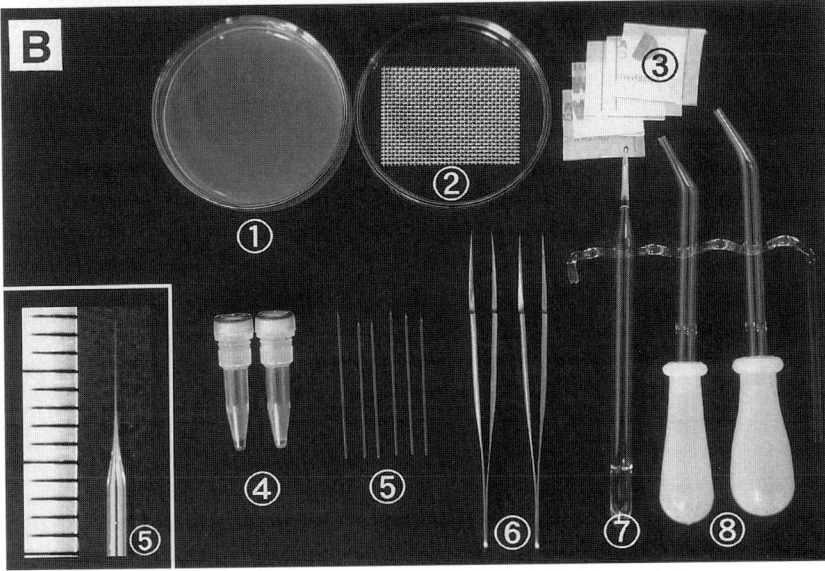

Fig. 3. Equipment and instruments for microinjection. (**A**) Arrangement of equipment: 1, microinjector; 2, binocular microscope; 3, illuminator; 4, manipulator; 5, air compressor; 6, Ficoll solution; and 7, 48-well plate. (**B**) Operating instruments: 1, 2% agar-coated Petri dish; 2, stainless mesh; 3, Parafilm; 4, samples; 5, micropipets (scale, mm); 6, watchmaker's forceps; 7, hair-loop; and 8, transfer pipets.

2.3.1. Clean Bench and Optical Equipment

Operations are best performed on a clean bench with airflow hood whenever possible to avoid infection by airborne microorganisms. A binocular microscope with ×10 oculars and ×1–4 objectives and an illuminator (fiber-optic light is desirable) are required. This equipment should be swabbed with 70% ethanol before use.

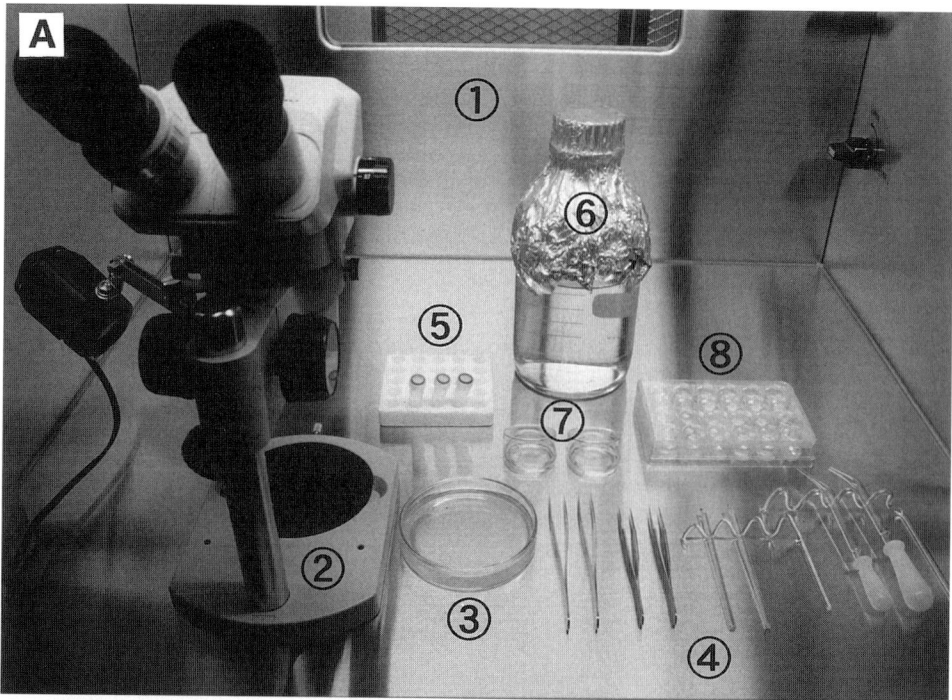

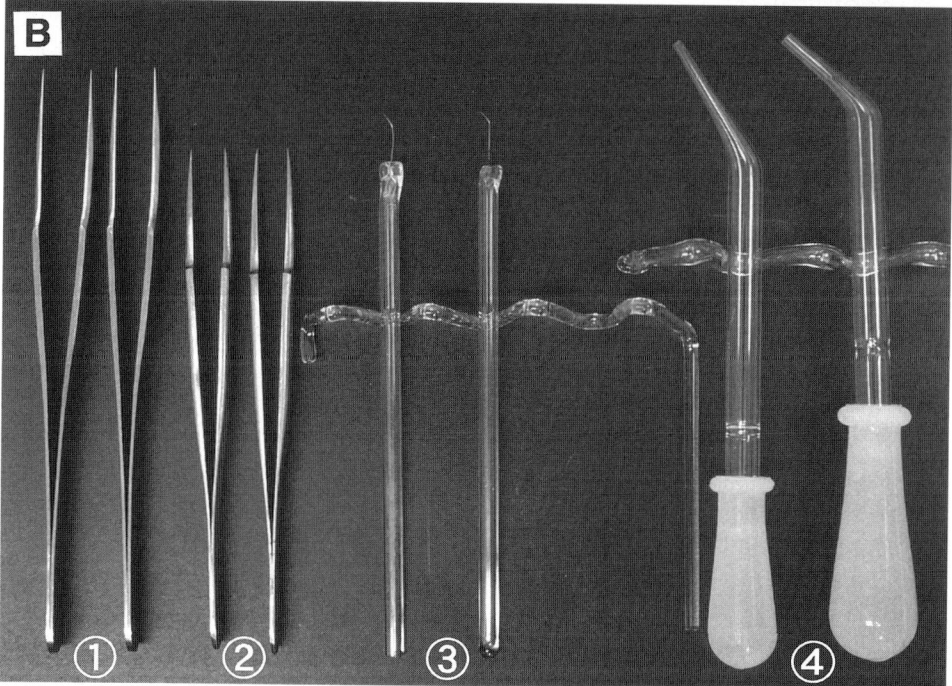

Fig. 4. Equipment and instruments for microsurgery. (**A**) Arrangement of equipment: 1, clean bench; 2, binocular microscope and illuminator; 3, 2% agar-coated Petri dish; 4, operating instruments; 5, samples; 6, culture medium; 7, Petri dishes; and 8, 24-well plate. (**B**) Operating instruments: 1, regular forceps (for dejelling); 2, watchmaker's forceps; 3, tungsten needles; and 4, transfer pipets.

2.3.2. Metal Instruments

Two pairs of watchmaker's forceps (e.g., Fontax No. 5 [Fontax Technologies, SA, Prilly, Switzerland]) are required to remove the vitelline membrane, which acts as a "corset" by tightly surrounding the egg. Some workers also use such forceps for the dissection of animal caps. Two pairs of thick-waisted forceps (e.g., Fontax No. 2 or KFI type-GG) are required for the removal of the urodele jellycoat. Forceps are sharpened on an oilstone and heat-sterilized at 180°C for 2 h.

For the dissection of tissues (e.g., animal caps), electrolytically sharpened tungsten needles have been used in our laboratory. Tungsten needles are superior to ordinary glass needles in that they are easy to make, are durable, and they can be resharpened and heat-sterilized. The 0.2-mm tungsten wire (Nilaco, Tokyo, Japan) is cut into 2-cm pieces using pliers and mounted on a 10-cm length of 3-mm soft glass tubing in a flame. A right angle is bent in the wire at approx 3–5 mm above its end. Sharpening of the wire end is done in 1–10 N NaOH solution or saturated sodium nitrite solution using a 3–12 V adjustable direct current system. The negative pole is connected to a carbon electrode in the solution and the positive pole is attached to the tungsten wire. A repeated dipping motion into the solution will sharpen the wire to a fine point. Tungsten needles are also heat-sterilized before use.

2.3.3. Glass Instruments

Pasteur pipets are used for making the transfer pipets, glass-ball tips, and handles for hair loops. Transfer pipets are made by flaming Pasteur pipets at their center and drawing them out at a 45° angle. For transferring whole embryos, it is best to cut the pipets at 1.5 mm (for *Xenopus*) or 2.5 mm (for urodele embryos) in diameter with an ampule cutter and smoothing the cut edge in a small flame. Similarly, small transfer pipets for pieces of tissue (e.g., animal caps) are made by cutting the tapered Pasteur pipets at 0.5–1 mm in diameter. These transfer pipets should be heat-sterilized and used with an ordinary rubber bulb that has been sterilized in 70% ethanol.

Glass-ball tips are used for moving the embryos and also for making depressions in operating dishes to hold embryos and tissues. They are simply made by holding the pointed end of a Pasteur pipet in a flame while rotating constantly. Hair loops are also used for handling embryos and isolated tissues. The handles for loops are made from Pasteur pipets, tapered in the manner used to make small transfer pipets. Both ends of a 1-cm length of baby's hair or superfine nylon line are inserted into the tube opening using forceps and held in place with a small amount of paraffin. Hair loops are sterilized in 70% ethanol.

2.3.4. Operating Dishes

Operations are usually carried out in 90-mm glass Petri dishes. The base of the dishes should be lined with 2–3% agar to prevent the embryonic tissues from adhering to glass surfaces. Some workers prefer to use modeling clay/Paraplast® (Sherwood Medical Co., Plano, TX) mixture for the lining. To make approx 10 operating dishes, dissolve 2–3 g agar in 100 mL of distilled water or culture medium using a microwave oven, thinly pour the solution over the base of each dish and let harden by cooling. Dishes are then wrapped in aluminum foil, autoclaved at 120°C for 20 min, and allowed to reharden by cooling. They can be kept in the refrigerator for a few months. Because they are disposable, plastic Petri dishes lined with autoclaved agar solution are more suitable when toxic chemicals are used in the operation.

2.4. Culture Media and Antibiotics (see Note 1)

Exogenous nutrients are unnecessary for the culture of embryos or isolated tissues, because embryonic cells have an abundance of yolk that provides sufficient nutrition for survival for a considerable period of development. A variety of media have been developed for operating on and culturing amphibian embryos, and the choice depends on the species, developmental stage, and condition of the embryo and/or tissue. The most common ones are listed in **Table 2**. These are simple saline solutions with 60–110 mM NaCl concentrations at pH 7.4–7.6. The salt concentration and pH of the media are the most important factors. The pH should be stable. Many workers have found that HEPES is an excellent buffer for culture media. Full-strength modified Holtfreter's solution (MHS) is the usual choice for urodele operation and culture media, whereas full-strength modified Steinberg's solution (MSS) and half-strengths of the other media (MMR, MBS, NAM) are suitable for *Xenopus* embryos.

The animal cap explants often curl up soon after the dissection, which may reduce the binding of inducers (e.g., PGFs) to their receptors on the inner blastocoelic cell surface. To prevent this curling, the NaCl concentration of the medium is increased to 90–100 mM so that the caps stay flat for at least a few hours *(15)*. It should be noted, however, that hypertonic media often cause the neuralization of animal caps. Therefore, hypertonic medium is replaced with a standard one when treatment with inducers is complete. Concentrations of more than 100 mM NaCl should be avoided for this purpose. For the culture of whole embryos, full-strength media (e.g., MHS for urodeles; MSS for *Xenopus*) are sufficient up to the blastula stage. For the gastrula stages onward, however, embryos should be cultured in lower strength media (10–20% MHS or MSS) to prevent exogastrulation and abnormal morphogenesis.

For the culture of whole embryos and embryonic tissues, 12, 24, and 48 wells of non-surface-treated polystyrene plates (e.g., Sumilon®, Sumitomo Bakelite, Osaka, Japan) are used in our laboratory. Embryonic tissues often adhere to plastic surfaces during the culture period. To prevent this, 0.1% bovine serum albumin (BSA; A-7888, Sigma, St. Louis, MO) is dissolved in the culture medium for *Xenopus* tissues, and 2–3% agar (or a piece of filter paper) is laid on the bottom of each well for urodele tissues.

Antibiotics are added as supplements to culture media to reduce the possibility of bacterial contamination. The choice of antibiotic depends on the requirements of the experiment. We usually add 100 mg of kanamycin sulfate (e.g., Banyu, Tokyo, Japan; Meiji Seika, Tokyo, Japan) per liter of medium. This medium can be autoclaved (120°C, 20 min), because kanamycin sulfate is heat stable. Some workers find that 50 mg of gentamicin per liter is a satisfactory antibiotic. Alternatively, other researchers add up to 400 mg penicillin-G and 400 mg streptomycin together with 25 mg gentamicin to 1 L of medium. These media with antibiotics may be filter-sterilized (or autoclaved before the addition of antibiotics).

Table 2
Culture Media for Amphibian Embryos and Embryonic Tissues

Components (mM)	Modified Holtfreter's solution (MHS)	Marc's modified Ringer's solution (MMR)	Modified Barth's solution (MBS)	Normal amphibian medium (NAM)	Modified Steinberg's solution (MSS)
NaCl	60.00	100.00	88.00	110.00	58.00
KCl	0.67	2.00	1.00	2.00	0.67
CaCl$_2$	0.90	2.00	0.41	—	—
Ca(NO$_3$)$_2$	—	—	0.33	1.00	0.34
MgCl$_2$	—	1.00	—	—	—
MgSO$_4$	—	—	0.82	1.00	0.83
Disodium EDTA	—	—	—	0.10	—
Sodium phosphate	—	—	—	2.00	—
NaHCO$_3$	—	—	2.40	1.00	—
HEPES	4.60	5.00	10.00	—	3.00
pH	7.60	7.40	7.40	7.40	7.40

3. Methods

3.1. Preoperative Treatment of Embryos (see Note 2)

All amphibian embryos are surrounded by jelly membranes (jellycoat) and a tight-fitting vitelline membrane. Any operation on the embryos requires that these membranes be removed as the first step.

Most urodele jellycoats consist of two or three membranes and form rather tough capsules. Before dejelling, embryos should be sterilized by bathing in 70% ethanol for 30 s and then quickly washed at least five times with modified Holtfreter's solution containing antibiotics (MHS, pH 7.6). Urodele jelly can be manually removed with two pairs of sharp pointed forceps. Dejelling is performed in the Petri dish filled with MHS: Pierce the capsule with one prong of each pair of forceps and then quickly tear the capsule, producing a large split through which the embryo is extruded (**Fig. 5A**). A chemical method, originally devised for removing *Xenopus* jelly and vitelline membranes *(16)*, can be used for the removal of the *Cynops* jelly capsule (**Fig. 5B**). After being sterilized in 70% ethanol and washed with MHS, embryos are transferred to MHS containing 1% sodium thioglycolate (pH 8–10). The jelly capsule will be shaken off within a few minutes. The jelly-free embryos are then *quickly* transferred to two or three washes in MHS, otherwise the vitelline membrane will also be removed.

Many anuran eggs are surrounded by a relatively loose jellycoat that is adherent to the vitelline membrane. *Xenopus* jellycoats are usually removed by chemical methods (**Fig. 5B**). Embryos are washed with sterile medium containing antibiotics, such as modified Steinberg's solution (MSS, pH 7.4). Then the medium is decanted and the dejelling solution added. Culture media containing 1% sodium thioglycolate (pH 8–10) or 4.5% cysteine hydrochloride (pH 7.8) are normally used as dejelling solutions.

Fig. 5. *(opposite page)* Removal of jellycoat and vitelline membrane. **(A)** Manual dejelling procedure. Urodele (*Cynops* and *Ambystoma*) jelly capsules should be removed in a Petri dish filled with modified Holtfreter's solution. (1) Hold the capsule with the left-hand forceps and pierce one tip of the right-hand forceps into the fluid-filled space. (2) Touch the tip to the bottom of dish and detach the left-hand forceps from the capsule. (3) Pierce one tip of the left-hand forceps into the fluid-filled space firmly and tear the capsule quickly with the right-hand forceps. The embryo with its vitelline membrane intact will pop out. **(B)** Chemical dejelling procedure. 1% sodium thioglycolate in a culture medium (pH 8–10) is sufficient to remove both *Xenopus* and *Cynops* jellycoat, whereas 4.5% cysteine hydrochloride (pH 7.8) can be used (only) for *Xenopus*. (1) Place embryos with jellycoats in the dejelling solution. The jelly can be shaken off within a few minutes. (2) Spill the dejelling solution and then quickly rinse at least 10X in the sterile culture medium with gentle swirling. (3) The jelly-free embryos (with vitelline membrane) will lie close to one another at the bottom of beaker. **(C)** Removal of vitelline membrane. All amphibian embryos are surrounded by a tight-fitting vitelline membrane. Microsurgery requires that it be manually removed in a 2% agar-coated Petri dish filled with culture media. (1) Keep the embryo upside down and then quickly grasp and tear the membrane with two pairs of watchmaker's forceps. (2) Place the naked embryo with the animal pole facing up with the aid of forceps or a hair-loop.

a, jelly capsule; b, fluid-filled space; c, embryo; d, vitelline membrane; e, Petri dish; f, dejelling solution; g, embryo with jelly; h, jelly-free embryo; i, culture medium; j, animal pole; k, 2% agar.

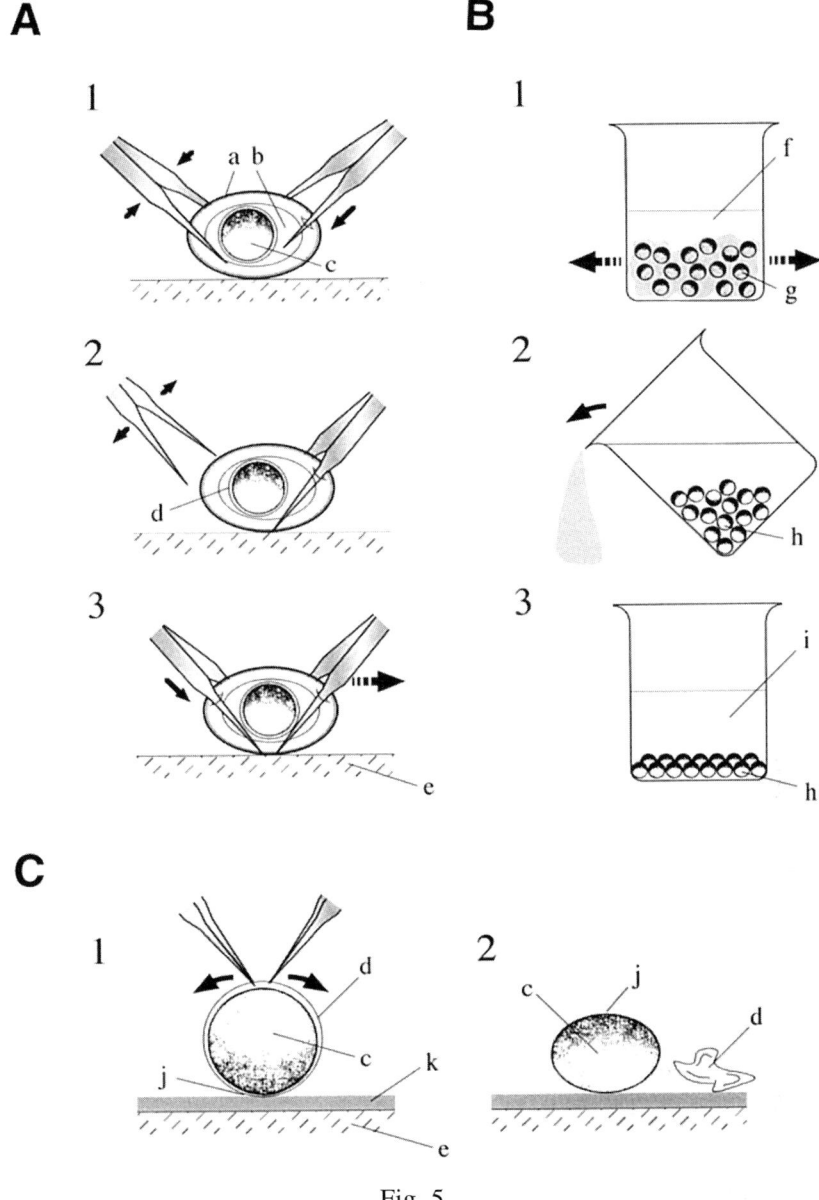

Fig. 5.

By gentle agitating, the jellycoats dissolve within a few minutes, and the jelly-free embryos settle close to one another in the bottom of the dish. Then the dejelling solution is immediately decanted and the eggs quickly rinsed at least 10X in sterile culture medium with gentle swirling.

The vitelline membrane, which lies close to the egg surface, can be manually removed with two pairs of watchmaker's forceps (**Fig. 5C**). For microsurgery on blastula-stage embryos the embryo is held upside down and then quickly grasped to tear the membrane with the two pair of forceps. No matter if a few vegetal cells are injured when the membrane is grasped. Some workers prefer to immerse embryos in a sterile culture

Samples   **Methods**

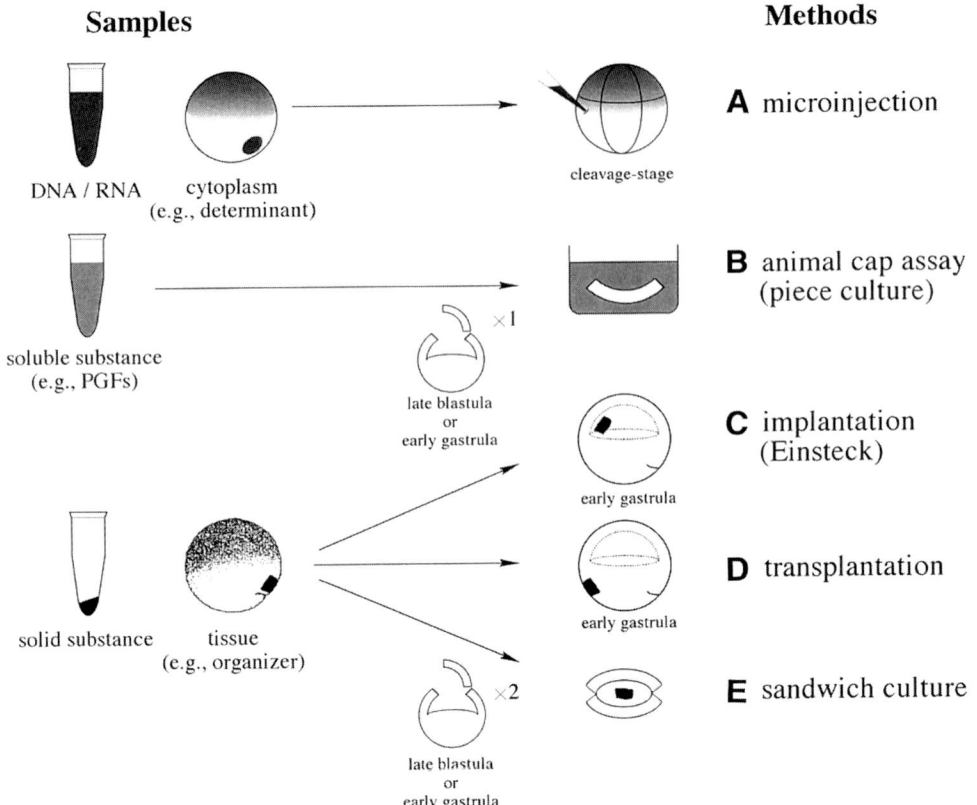

Fig. 6. Bioassays for testing inducing activity. The functions of nucleic acids and cytoplasmic determinants can be examined employing the "microinjection" technique on early cleavage-stage embryos. Embryos will show various phenotypes, such as twinning, axial deficiency, and hyperdorsalization, depending on the activity of factor injected. The soluble factors (e.g., peptide growth factors [PGF]) are often tested for their inducing activities by adding them to blastula/early gastrula ectoderm (animal cap) in a culture medium ("animal cap assay" or "piece culture" method). Activities of tissues (e.g., the organizer) or coagulated proteins are examined by grafting them into the blastocoel or ventral marginal zone of an early gastrula embryo. As a "positive" result, they will induce secondary embryos on the ventral side of host embryos (the "implantation" and "transplantation" methods). Such substances also induce head or trunk-tail structures in the "sandwich culture" method, in which inducers are sandwiched between two sheets of early gastrula ectoderm.

medium containing 10% Ficoll® (F-4375, Sigma) to dehydrate the perivitelline space. This treatment facilitates the grasping and tearing of the membrane without injury to the embryos, especially for early pre-blastula-stage embryos.

3.2. Microinjection

The "microinjection" technique is employed for the introduction of genes, radioactive precursors, cytoplasmic components, and cell lineage tracers into uncleaved eggs and early embryos (**Fig. 6A**). A variety of specialized instruments and media are required in addition to the conventional operating tools (**Fig. 3**).

3.2.1. Injection System and Micropipets

The injection system consists of a micromanipulator and an injector (e.g., PLI-100, Medical Systems Co., Greenvale, NY). The volume of material injected is controlled by the pressure/time of the compressed gas going through the injector. Micropipets are prepared from glass capillary tubes (e.g., GD-1, Narishige, Tokyo, Japan) using a micropipet puller (PC-10 or PN-30, Narishige). The pipet is held by a micromanipulator at a 45° (approx) angle, and its larger open end is connected by stiff plastic tubing to the injector. The pipet tip is broken with a pair of watchmaker's forceps and beveled at an angle with diamond paste and a grinding wheel (e.g., EG-3, Narishige) if necessary. Prior to microinjection, the inner surface of the pipet should be coated with SIGMACOTE® (SL-2, Sigma) to avoid adsorption of the injection material to the glass surface. Micropipets are air-dried and sterilized by autoclaving or UV irradiation.

3.2.2. Procedures

First, the micropipet is filled with the injection material from a drop placed on a piece of Parafilm (American National Can Co., Greenwich, CT). Then that sample is expelled for a predetermined unit of time. The diameter of the droplet at the pipet tip is measured using a micrometer and the volume calculated. The volume can be adjusted by the combination of pressure and time. The amount of the material injected, of course, depends on the end results desired. It should be noted, however, that larger amounts of sample (e.g., more than 200 pg of DNA) frequently cause abnormal development or death of the injected embryos.

After removing the jellycoat, eggs/embryos are placed in a 60-mm glass Petri dish that has a piece of stainless mesh (grid size: 0.7-mm square for *Xenopus*, 1.2-mm square for urodele embryos) on the bottom. The dish is filled with a full-strength culture medium. Ficoll is added to the medium (to 5%) to reduce the pressure that fluid in the perivitelline space (between the membrane and the egg surface) exerts on the egg (*see* **Note 3**). For the introduction of cytoplasmic components, the vitelline membrane is usually removed and membrane-free embryos are placed on the 2–3% agar lining instead of the stainless mesh. The injected embryos are allowed to develop in the full-strength medium until they reach the late blastula stage, and then they are transferred to the 10–20% medium.

3.3. Microsurgery

3.3.1. Animal Cap Assay (Piece Culture Method) (see **Note 4**)

Animal caps (presumptive ectoderm region) of blastula and early gastrula are competent to respond to an inducing stimulus and can be induced to form a variety of neural, mesodermal, and endodermal tissues. Recognizing the pluripotency of animal caps, a simple but excellent in vitro assay system, the so-called "animal cap assay," has been devised *(17,18)* (**Fig. 6B**). It is quite easy to perform and has many advantages over other methods. Investigators can test large numbers of fractions of the factors in solution and can estimate their inducing activity both qualitatively and quantitatively. Synergistic effects of two or more factors can be examined by combining them in the culture medium (e.g., **ref. 19**). It is also possible to know the competence of reacting tissues when animal caps of different age or size are treated with various concentra-

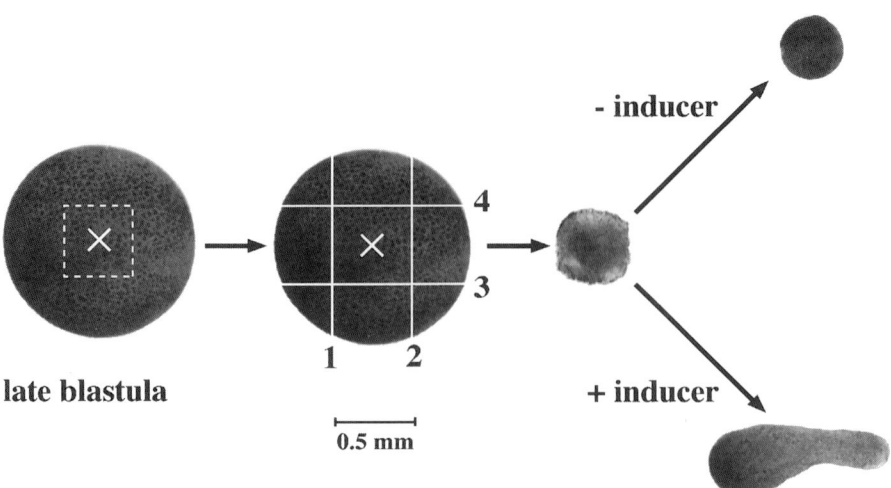

Fig. 7. Procedure for the animal cap assay using the *Xenopus* late blastula tissue. Place the membrane-free embryos with the animal pole (X) facing up. Next, dissect the animal caps (consisting of presumptive neural and epidermal regions) squarely using a pair of tungsten needles (1–4). Transfer the caps into test solutions with the inner surface facing up. Exposure to a dorsal mesoderm inducer (e.g., 10 ng/mL of activin) leads to elongation of caps after 6–12 h.

tions of the inducer for a specific period (e.g., **ref. 20**). Use of this assay system has enabled remarkable advances in the discovery of mesoderm-inducing factors such as bFGF *(21)* and activin A *(22)* during the past decade.

Figure 7 outlines an animal cap assay that tests the effect of activin on the *Xenopus* blastula (stage 9) animal caps. In this example, 10 ng/mL of activin A is employed as an inducer for dorsal mesoderm. After the removal of jellycoat and vitelline membrane, embryos are placed with the animal pole side up in the operating dish filled with a culture medium (e.g., MSS). The animal caps are dissected squarely using a pair of tungsten needles and then transferred to the test solutions with the superficial cell side down. Test solutions are usually prepared in a 24-well plate (e.g., Sumilon®, Sumitomo Bakelite). BSA at a final concentration of 1 mg/mL is present in the solutions to avoid adsorption of activin to the plastic surfaces. After exposure to the test solutions for several hours, the animal caps are washed in a culture medium by gentle pipeting and cultured in a fresh medium at 20–22°C. One may expose the explants in the test solution for the entire culture period (2–3 d). In the absence of an inducer, animal caps remain spherical and form irregular-shaped epidermis referred to as "atypical epidermis." The activin-treated caps begin to show the elongation movement that mimics the convergent extension of dorsal mesoderm during gastrulation *(23)* after approx 6 h. A large mass of well-differentiated muscle cells can be observed in the explants at the end of culture (2–3 d).

3.3.2. Implantation and Transplantation Methods (see **Note 5**)

For testing the activity of tissues and solid factors, the following two methods have been devised. The factor to be tested is mixed in different ratios with a noninducing protein, and a pellet is prepared after precipitation of the mixture. In the "implantation method" (Einsteck method, **ref. 25**), a piece of tissue or pellet is pushed into the blasto-coel of an advanced blastula or early gastrula through a slit made at the animal pole using a tungsten needle (**Fig. 6C**). As gastrulation proceeds, the implant becomes pressed against the ventral ectoderm of the host several hours after implantation. If the implant has the ability to induce axial structures (axial mesoderm and central nervous system), a secondary embryo is eventually formed on the ventral side of the host.

The "transplantation method" was originally devised for testing the inducing ability of prospective organizer tissue (**Fig. 6D**). The natural organizer (blastopore lip) dissected from a gastrula is transplanted into the ventral marginal zone of another early gastrula. The transplant will itself involute away from influence of the host's organizer and induce a secondary embryo on the ventral side of the host. This method can be adapted for other living tissues that have the ability to invaginate (e.g., activin-treated animal caps).

3.3.3. Sandwich Culture Methods

The explantation procedure called the "sandwich culture method" was improved by Holtfreter to eliminate any possible influence by the host embryo on interpretation of induction data *(26)*. In this method, prospective inducers are placed between two pieces of animal cap, and the inducers interact with the tissue from the inside of the implanted inducer-containing vesicle (**Fig. 6E**). The dissection procedures for preparing animal caps and for the culture of the explants are the same as described in **Subheading 3.3.1.** The dissected animal cap is placed with its superficial cell side down in the operating dish. Then, the inducer to be tested is placed on the first cap and covered with a second. The explants will combine and heal quickly (within 15 min). **Figure 8C–E** illustrates histological sections of 2-wk-old sandwich cultures that test the inducing ability of activin-treated animal caps in *Cynops (27)*. The control explants that are not exposed to inducer form atypical epidermis (**Fig. 8C**). In the explants subjected to archencephalic induction, forebrain with eyes will be induced (**Fig. 8D**), and axial structures are induced by a spinocaudal induction (**Fig. 8E**).

3.4. Scoring of Results (see Note 6)

3.4.1. External Morphology of the Operated Embryos

Embryos often exhibit hyperdorsalization, axial deficiency, and axial duplication (**Fig. 8A,B**) in the microinjection and implantation studies. To describe a continuous range of such phenotypes, several "indices," "grading schemes," and numerical scoring systems have been devised. **Figure 9** shows some such indices for scoring results in *Xenopus*. The dorsoanterior index (DAI) was originally devised to describe the degree of axial deficiencies caused by UV irradiation (grades 0–4) and hyperdorsoanterior enhancements caused by lithium or D_2O treatment (grades 6–10) *(28)*. For describing the degree of completeness of a secondary axis, the index devised by Cooke *(29)* is available. Most phenotypes can be classified by these indices.

3.4.2. Histological Examination of Embryos and Explants (see Note 7)

For interpreting results from microinjection and microsurgical studies, it is essential to prepare histological sections of the embryos and explants. **Figure 8C–E** shows some examples of sections of explants *(27)*. Standard protocols include Bouin's fluid fixa-

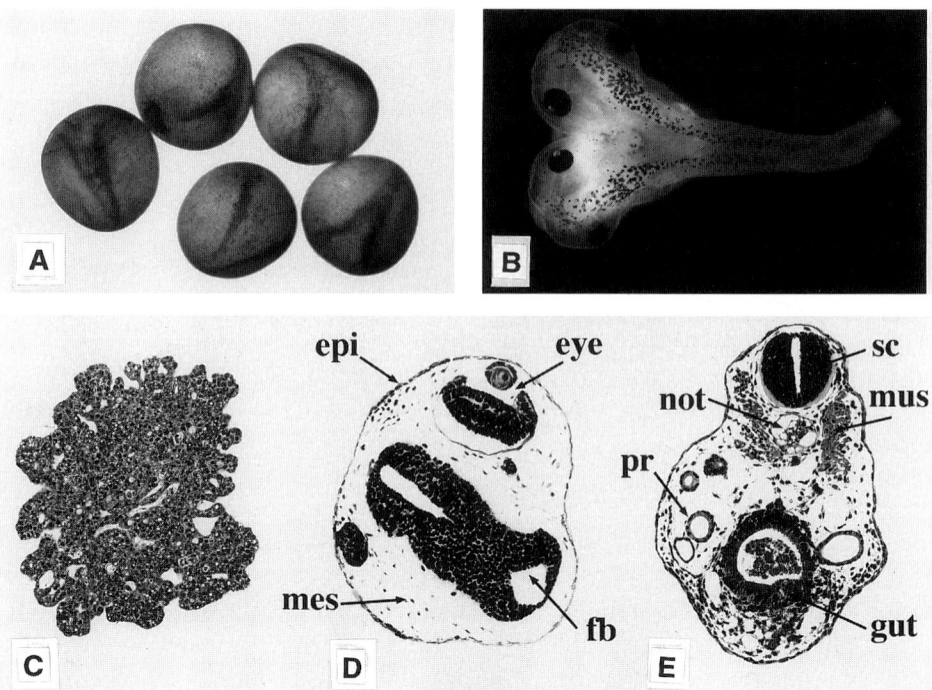

Fig. 8. (*See* color plate 2 appearing after p. 258.) Examples of phenotypes and histological sections from various induction assays. Complete twins induced by *Xwnt-8*. (**A**) Microinjection of *Xwnt-8* mRNA into ventro-vegetal blastomeres of *Xenopus* 8 cell-stage embryo leads to the duplication of neural tubes. (**B**) These embryos develop into Siamese twins. Sections of *Cynops* sandwich cultures. (**C**) Control, cultured without inducer, forms atypical epidermis. (**D**) Explant subjected to archencephalic induction contains forebrain with an eye, whereas (**E**) axial structures are formed by the spinocaudal induction. epi, epidermis; fb, forebrain; mes, mesenchyme; mus, muscle; not, notochord; pr, pronephros; sc, spinal cord (from **ref. 27**).

tion, paraffin embedding and sectioning, and hematoxylin/eosin staining. Early embryonic cells are rich in yolk platelets, which represent a major impediment to histological sectioning and staining procedures.

We outline here the techniques used in our laboratory for routine light microscopy. For more detailed descriptions of histological preparations, including possibilities for modifications, consult a traditional embryology or histology textbook (e.g., *14*).

The choice of fixatives depends on the developmental stages or the size of embryos (explants), and the end results desired. Fixatives routinely used in our laboratory are listed in **Table 3**. Bouin's fluid is the most satisfactory fixative. Fixation may be 1–12 h for the explants; 24 h for whole embryos. Specimens are then washed for 12–24 h in several changes of 70% ethanol to bleach the yellow picric color. It may be speeded up by adding 2% ammonium hydroxide or a few drops of saturated lithium carbonate to the 70% ethanol. Smith's fluid is good for early (yolk-rich) embryos. Fixation may be 12–24 h, followed by washing in running water for 12–24 h and dichromate bleaching (10 mL of 2% sodium disulphite with a few drops of conc. HCl) for 6–12 h. Although $HgCl_2$-containing fixatives such as Helly's, Zenker's, and Susa are also excellent for

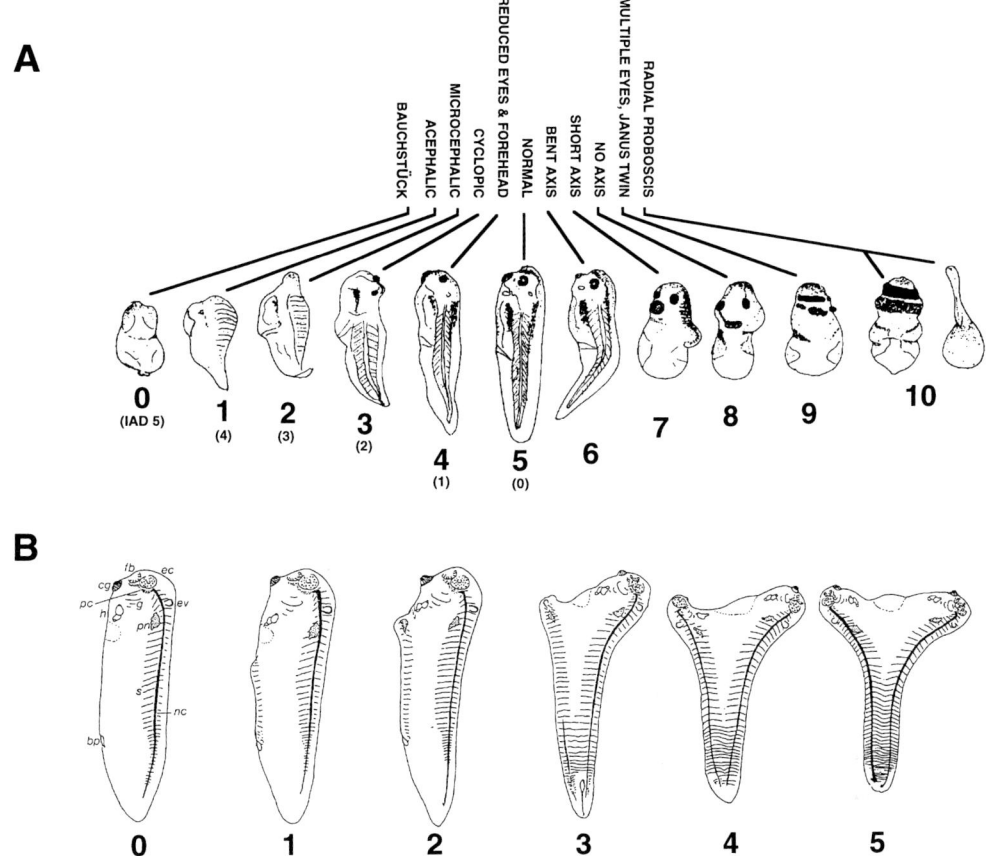

Fig. 9. Scoring of phenotypes. (**A**) Dorsoanterior index (DAI): grade 0, severe axial deficient; grade 5, normal; grade 10, hyperdorsalized embryos (from **ref. *28***). (**B**) Classification
of degree of completeness in secondary embryos: grade 0, normal; grade 5, Siamese twins
(from **ref. *29***).

both embryos and explants, excess corrosive sublimate needs to be removed by alco-
holic iodine.

Specimens are dehydrated through a graded series of ethanols (70, 90, 99.5, and
100%, with 1-h shifts) and cleared by replacing 100% ethanol with xylene (3X with
15-min shifts). Then, the specimens are transferred to a glass Petri dish containing
molten paraffin (e.g., Histprep 568, Wako, Osaka, Japan; m.p. 56–58°C) and gradually
infiltrated (2X with 15-min shifts). A special basket consisting of a glass tube
(1-cm length and diameter) with a nylon mesh (148-µm grids) on the bottom is used
for handling small explants. The explants are placed in the basket so that solution
changes (ethanols, xylene, and paraffin) can be done by moving the basket to a new
solution. After the infiltration, a ceramic dish (lined with glycerin) is filled with fresh
molten paraffin. The specimens are transferred and positioned appropriately in the dish
using a transfer pipet. The dish is left in water to harden the paraffin. Sections are
generally made at 6–10-µm thickness using a rotary microtome. For sectioning yolk-
rich embryos, cut them from the vegetal toward animal pole very slowly so that cracks

Table 3
Fixatives for General Histology[a]

Fixative	Bouin	Smith	Helly	Zenker	Susa (Heidenhain)
Formalin (37% formaldehyde)	25 mL	10 mL	5 mL	—	20 mL
Mercury (II) chloride	—	—	5 g	5 g	4.5 g
Saturated (aqueous) picric acid	75 mL	—	—	—	—
Potassium dichromate	—	0.5 g	2.5 g	2.5 g	—
Glacial acetic acid	5 mL	2.5 mL	—	5 mL	4 mL
Trichloroacetic acid	—	—	—	—	2 g
Sodium chloride	—	—	—	—	0.5 g
Sodium sulfate	—	—	1 g	1 g	—
Distilled water	—	87.5 mL	100 mL	100 mL	80 mL

[a]To be made up just before use.

will not develop. Then mount the ribbons of sections on a slide coated with Para-Tissuer (Daido Sangyo, Japan), add distilled water to the slide, and heat at 45°C over a slide warmer to allow sections to spread. Dry excess water with blotting paper, and leave the slide for overnight at room temperature.

After deparaffinization with xylene and hydration through a graded series of ethanols (100, 99.5, 90, 70%, and distilled water, with 5-min shifts), sections are stained with Delafield's hematoxylin for 1 min and rinsed in running tap water until nuclei stain blue and the cytoplasm stains pink. For counterstaining, sections are placed in 1% eosin Y solution for 1 min and rinsed briefly in water. Sections are dehydrated with ethanols (70, 90, 99.5, 100%, with 5-min shifts), cleared in xylene (3X with 5-min shifts), and sealed with a cover slip and a drop of Canada balsam or Permount (VWR Scientific, Atlanta, GA).

3.5. Concluding Remarks

We have presented several simple methods that can be used for the assay of inductive interactions in amphibian development. It is possible to combine two or more of the methods described herein (e.g., "microinjection" and "animal cap assay," "animal cap assay" and "sandwich culture method," and so forth). It has been demonstrated by the sandwich culture method that ectoderm treated with activin using the animal cap assay has a complete organizer activity (**Fig. 8C–E, ref.** *27*). This will serve as a suitable test system for analyzing how basic morphogenetic features of the body plan of amphibian embryos can be established in vitro. Gross morphological and histological criteria are emphasized as a starting point for studying inductive interactions. Now, however, several specific antibodies are available for the identification of induced tissues, which can make the histological analyses more informative. In addition, many practical methods that employ molecular markers have also been developed for *Xenopus* embryos in recent years *(5)*. These will make the analysis of inductive interactions much simpler and more direct. However, it should be emphasized that assay protocols based on both morphological criteria (e.g., **Figs. 8** and **9**) will need to be supplemented with molecular markers to achieve an enhanced understanding of amphibian inductive events.

4. Notes

1. The extent to which attention is devoted to maintaining sterile operating conditions and the use of antibiotics in culture media depends on the length of the culture period. The duration of the culture period in turn depends to some extent on the source of eggs/embryos. *Xenopus* embryos often develop to the experiment's end point before bacterial infection becomes a problem. More slowly developing axolotl eggs, however, are usually cultured longer; hence, sterile conditions and antibiotics are often more important for those more slowly developing species.

2. Exposure to chemical dejelling agents for too long will soften the vitelline membrane, causing normally spherical egg to "sag." As well, permanent damage to the egg surface likely occurs during prolonged exposure to those dejellying agents. Thus, quick rinsing is highly recommended.

   Mechanical removal of the vitelline membrane from blastula/gastrula stage embryos is a relatively straightforward procedure, but removing that membrane from an uncleaved egg requires substantial skill, because although a wound to the surface of a blastula will quickly heal, a puncture of the egg's surface is unlikely to heal. Excessive leakage of cytoplasm usually results.

3. The addition of Ficoll to the injection dish markedly reduces the amount of leakage of egg cytoplasm by counterbalancing the turgor pressure of the egg. Batches of eggs from various spawnings will often display different amounts of cytoplasmic leakage. Special attention should therefore be paid to not over-treating eggs with chemical dejellying agents.

4. Investigators should pay close attention to the following features in order to obtain reliable results with animal cap tissues. For *Xenopus*, animal caps are competent between stages 7 (middle blastula) and 10 (early gastrula) in general, but their responses to various exogenous inducers can be slightly different. The late blastula (stage 9) of *Xenopus* and very early gastrula (stage 11) of *Cynops* are used for animal cap assays in our laboratory. Although the choice of the stage of caps to use depends on the end results desired, the staging of embryos must be done very accurately.

   A second consideration is the size of the dissected animal caps. The size is acceptable as long as the control animal cap forms atypical epidermis without added exogenous inducer. It should be noted, however, that the differentiation pattern of caps is influenced by the number of cells in the animal cap and the presumptive region used. When animal caps are treated with activin, a large animal cap containing cells close to the marginal zone often differentiates into neural tissue and cement gland in addition to axial mesoderm *(24)*. The neural tissues should be considered to have been induced by the mesoderm as a secondary induction. In our experience, the most reliable sizes of animal caps are 0.5-mm square in *Xenopus* (stage 9) and 1.2-mm square in *Cynops* (stage 11).

   Third, the duration of exposure of animal caps to the inducer also influences their differentiation patterns. For example, a brief exposure (5 min) to 10 ng/mL of activin causes the differentiation of ventral mesoderm such as mesenchyme and mesothelium, whereas a long exposure (3 h) to the same dose leads to the muscle differentiation in the animal caps *(20)*.

5. The embryos operated on should be cultured in 10–20%-strength culture medium to prevent them from exogastrulating. It is also desirable that the bottoms of the culture dishes be coated with 2–3% agar with an appropriate depression for holding the embryos.

6. These numerical grading or scoring systems represent categorizations of gross phenotypic features which are easily recognized by the researcher. They do not represent direct, linear estimates of extent of development. Thus, a score of "6" should not imply that those effects are twice the magnitude of a score of "3." In reality, if biochemical or molecular indices were employed, concentrations of specific macromolecules might differ by 2, 10, or even 100 between embryos scored as "3" and "6." Nevertheless, these grading systems provide quick and convenient ways to assess the effects of inducing agents.

7. It is likely that some trial-and-error modifications at each step will be required to obtain satisfactorily stained histological sections. Because the yolk content of cells varies according to stage and region, protocols that accommodate specific target cells or tissues often need to be optimized.

Acknowledgments

We are grateful to members of our laboratory, particularly Chan Te-chuan, for helpful comments on the manuscript. Most of this work was supported by Grants-in-Aid for Scientific Research from the Ministry of Education, Science, Sports and Culture of Japan, and by CREST (Core Research for Evolutional Science and Technology) of the Japan Science and Technology Corporation. GMM's research is funded by the National Science Foundation.

References

1. Spemann, H. and Mangold, H. (1924) Über Induktion von Embryonalanlagen durch Implantation artfremder Organisatoren. *Arch. Mikrosk. Anat. Entwicklungsmech.* **100,** 599–638.
2. Nieuwkoop, P. D. (1969) The formation of mesoderm in urodelan amphibians, Pt. 1: Induction by the endoderm. *Wilhelm Roux' Arch. Entwicklungsmech. Org.* **162,** 341–373.
3. Asashima, M. (1994) Mesoderm induction during early amphibian development. *Dev. Growth Differ.* **36,** 343–355.
4. Hamburger, V. (1988) *The Heritage of Experimental Embryology. Hans Spemann and the Organizer.* Oxford University Press, New York.
5. Kay, B. K. and Benjamin Peng, H., eds. (1991) *Methods in Cell Biology.* Xenopus laevis: *Practical Uses in Cell and Molecular Biology.* Academic, San Diego, CA.
6. Armstrong, J. B. and Duhon, S. T. (1989) Induced spawnings, artificial insemination, and other genetic manipulations, in *Developmental Biology of the Axolotl* (Armstrong, J. B. and Malacinski, G. M., eds.), Oxford University Press, New York, pp. 228–235.
7. Schreckenberg, G. M. and Jacobson, A. G. (1975) Normal stages of development of the axolotl, *Ambystoma mexicanum. Dev. Biol.* **42,** 391–400.
8. Bordzilovskaya, N. P. and Dettlaff, T. A. (1975) Axolotl: *Ambystoma mexicanum* (Cope), in *Objects of the Biology of Development* (Dettlaff, T. A., Geycinovich, A. E., and Brodsky, V. Y., eds.), Series on the Problems of the Biology of Growth, Academy of Sciences, USSR, Moscow, pp. 370–389.
9. Bordzilovskaya, N. P. and Dettlaff, T. A. (1979) Table of stages of the normal development of axolotl embryos and the prognostication of timing of successive developmental stages at various temperatures. *Axolotl Newsletter* **7,** 2–22.
10. Matsuda, M. and Oya, T. (1977) Induced oviposition in the newt *Cynops pyrrhogaster* by subcutaneous injection of human chorionic gonadotropin. *Zool. Mag.* **86,** 44–47.
11. Okada, Y. K. and Ichikawa, M. (1947) Normal stages of development of the Japanese newt, *Triturus pyrrhogaster* (Boie). *Jpn. J. Exp. Morphol.* **3,** 1–6 (in Japanese).
12. Takano, K., Kikkawa, M., and Shinagawa, A. (1996) Production of hyperdorsal larvae by exposing uncleaved *Xenopus* eggs to a centrifugal force directed from the animal pole to the vegetal pole. *Dev. Growth Differ.* **38,** 537–547.
13. Nieuwkoop, P. D. and Faber, J. (1956) *Normal Table of* Xenopus laevis *(Daudin),* North-Holland, Amsterdam.
14. Rugh, R. (1956) *Experimental Embryology.* Burgess, Minneapolis, MN.
15. Komazaki, S. (1993) Movement of an epithelial layer isolated from early embryos of the newt, *Cynops pyrrhogaster.* I. Development of folding movement of the blastocoelic wall isolated from embryos before and during gastrulation. *Dev. Growth Differ.* **35,** 461–470.
16. Okamoto, M. (1972) A method for the removal of the jelly and vitelline membrane from the embryos of *Xenopus laevis. Dev. Growth Differ.* **14,** 37–41.
17. Becker, U., Tiedemann, H., and Tiedemann, H. (1959) Versuche zur Determination von embryonalem Amphibiengewebe durch Induktionsstoffe in Lösung. *Z. Naturforsch.* **14b,** 608–609.
18. Yamada, T. and Takata, K. (1961) A technique for testing macromolecular samples in solution for morphogenetic effects on the isolated ectoderm of the amphibian gastrula. *Dev. Biol.* **3,** 411–423.
19. Moriya, N., Uchiyama, H., and Asashima, M. (1993) Induction of pronephric tubules by activin and retinoic acid in presumptive ectoderm of *Xenopus laevis. Dev. Growth Differ.* **35,** 123–128.

20. Ariizumi, T., Sawamura, K., Uchiyama, H., and Asashima, M. (1991) Dose and time-dependent mesoderm induction and outgrowth formation by activin A in *Xenopus laevis*. *Int. J. Dev. Biol.* **35,** 407–414.

21. Slack, J. M. W., Darlington, B. G., Heath, J. K., and Godsave, S. F. (1987) Mesoderm induction in early *Xenopus* embryos by heparin-binding growth factors. *Nature* **326,** 197–200.

22. Asashima, M., Nakano, H., Shimada, K., Kinoshita, K., Ishii, K., Shibai, H., and Ueno, N. (1990) Mesodermal induction in early amphibian embryos by activin A (erythroid differentiation factor). *Roux's Arch. Dev. Biol.* **198,** 330–335.

23. Keller, R. E., Danilchik, J., Gimlich, R., and Shih, J. (1985) The function and mechanism of convergent extension during gastrulation of *Xenopus laevis*. *J. Embryol. Exp. Morphol.* **89 (Suppl),** 185–209.

24. Sokol, S., Wong, G. G., and Melton, D. A. (1990) A mouse macrophage factor induces head structures and organizes a body axis in *Xenopus*. *Science* **249,** 561–564.

25. Mangold, O. (1923) Transplantationsversuche zur Frage der Spezifität und der Bildung der Keimblätter. *Arch. Mikrosk. Anat. Entwicklungsmech.* **100,** 193–301.

26. Holtfreter, J. (1933) Nachweis der Induktionsfähigkeit abgetöteter Keimteile. Isolations- und Transplantationsversuche. *Wilhelm Roux' Arch. Entwicklungsmech. Org.* **128,** 584–633.

27. Ariizumi, T. and Asashima, M. (1995) Control of the embryonic body plan by activin during amphibian development. *Zool. Sci.* **12,** 509–521.

28. Kao, K. R. and Elinson, R. P. (1988) The entire mesodermal mantle behaves as Spemann's organizer in dorsal enhanced *Xenopus laevis* embryos. *Dev. Biol.* **127,** 64–77.

29. Cooke, J. (1989) Mesoderm-inducing factors and Spemann's organizer phenomenon in amphibian development. *Development* **107,** 229–241.

12

Gastrulation and Early Mesodermal Patterning in Vertebrates

Gary C. Schoenwolf and Jodi L. Smith

1. Introduction

The process of gastrulation occurs during the early embryogenesis of vertebrates, immediately following the phase of cleavage and when the embryo is at the blastula stage. Gastrulation results in the formation of the primitive gut or archenteron. With the initial formation of the archenteron, the embryo reaches the gastrula stage. As more broadly defined, gastrulation is the phase of development in which the three primary germ layers are established: the inner endoderm (i.e., the layer closest to the yolk and the layer that forms the archenteron), the outer ectoderm (i.e., the layer on the surface of the egg), and the mesoderm (i.e., the layer interposed between the endoderm and ectoderm).

The goal of gastrulation is to reorder cells within the early embryo, bringing various groups of cells into close proximity to one another where they can interact, leading to the induction or suppression of developmental fate within the interacting populations. Thus, to fully understand gastrulation, it is necessary to construct prospective fate maps—that is, maps that show the locations of cells prior to gastrulation—and to track cells during their movements. Studies of the timing of cell commitment within the early embryo are also important.

Gastrulation involves a series of form-shaping movements, called morphogenetic movements, in which the overall shape of the embryo is altered as cell positions are reordered. These morphogenetic movements are under the control of a specialized region of the embryo known as the organizer. The organizer, also called the embryonic shield in fish (1,2), the dorsal lip of the blastopore in amphibians (3), the lip of the blastopore in reptiles (4), Hensen's node in birds (5), and the node in mammals (6), not only coordinates or choreographs the morphogenetic movements of gastrulation, but additionally choreographs a series of morphogenetic movements that results in the formation of the vertebrate body plan. Next, we describe the vertebrate body plan; the morphogenetic movements that characterize gastrulation in five classes of vertebrates; the origin, displacement, and commitment of cells during gastrulation; the morphogenetic movements that lead to the formation of the vertebrate body plan; and the role of the organizer in choreographing morphogenetic movements.

From: *Methods in Molecular Biology, Vol. 135: Developmental Biology Protocols, Vol. I*
Edited by: R. S. Tuan and C. W. Lo © Humana Press Inc., Totowa, NJ

2. The Vertebrate Body Plan

The vertebrate body plan consists of a tube-within-a-tube organization, which differs somewhat at the three major rostrocaudal levels of the body axis: the head, the trunk, and the tail (**Fig. 1**). One of the two tubes of the tube-within-a-tube organization of the vertebrate body plan is an outer tube. This tube consists of epidermal ectoderm and forms the skin of the embryo as well as its derivative appendages (e.g., feathers and scales in birds, hair and nails in mammals). In addition, this outer layer of the embryo gives rise to placodes—platelike thickenings that form, for example, the inner ear vesicle and the lens of the eye. Moreover, the outer ectodermal tube forms the neural plate, the earliest rudiment of the adult central nervous system (*see* last paragraph of this section and Vol. II, Chapter 15). The outer tube of the tube-within-a-tube organization of the vertebrate body plan forms at all three major rostrocaudal levels of the body axis.

The second tube of the tube-within-a-tube organization of the vertebrate body plan is an inner tube. This tube consists of endoderm and forms the archenteron. This later tube forms at all three major rostrocaudal levels of the developing embryo, but its structure differs at each level. In the early embryo, the inner tube consists of the foregut, within the head region, a blind tube that opens caudally; the midgut, within the trunk region, an area that opens broadly into the yolk sac prior to formation of the umbilical cord; and the hindgut/tail gut, within the caudal trunk and tail regions, a blind tube that opens rostrally. These three areas of the archenteron form through the action of three types of folds called the body folds. The foregut is formed through the action of the single head fold of the body, the midgut is formed through the action of the paired lateral body folds, and the hindgut is formed through the action of the single tail fold of the body.

The space between the ectodermal and endodermal tubes of the vertebrate body plan is filled with mesoderm, which differs in its structure at the head, trunk, and tail levels. At the head level (**Fig. 1A,B**), the mesoderm consists mainly of loosely packed cells called head mesenchyme. The head mesenchyme has three origins: two from the mesoderm and one from the ectoderm. The two mesodermal contributions are from a rostral midline population called the prechordal plate mesoderm and bilateral, partially segmented areas called somitomeres. In the chick embryo, there are seven pairs of somitomeres contributing to the head mesenchyme. A specialized population of ectodermal cells also contributes to the head mesenchyme: neural crest cells. These cells are left behind during the process of neurulation (*see below*). Besides contributing to the head mesenchyme, the mesoderm within the head region also forms a midline, longitudinal rod of cells lying beneath the neural tube; this rodlike structure is the notochord (the notochord is absent beneath the most rostral level of the neural tube—the future forebrain level). Moreover, the ventral mesoderm of the head contributes to the paired heart-forming regions (**Fig. 1A**), which fuse ventrally beneath the foregut to form the heart tube (**Fig. 1B**; *see below*).

At the trunk level (**Fig. 1C**), the mesoderm also consists of the midline notochord as it does at the head level, but head mesenchyme and heart mesoderm are absent. Instead, the mesoderm is organized into the following subdivisions on each side of the midline (listed in medial-to-lateral order): somites (the source of the dermis of the skin, the skeletal muscles, and the skeletal elements of the vertebral column), intermediate mesoderm (the source of the urogenital ducts and organs), and lateral plate mesoderm. The lateral plate mesoderm is split frontally into a dorsal somatic mesoderm (adjacent to the epidermal ectoderm) and a ventral splanchnic mesoderm (adjacent to the endoderm). The space between the somatic and splanchnic mesoderm is the coelom or body cavity. Collectively, the ectoderm and somatic mesoderm of the lateral plate constitute the somatopleure (the source of the body wall and the parietal layer of the coelomic epithelium), and the endoderm and splanchnic mesoderm of the lateral plate constitute the splanchnopleure (the source of the gut wall and the visceral layer of the coelomic epithelium). The coelom becomes partitioned during subsequent development into the three main types of body cavities: the paired pleural cavities, the pericardial cavity, and the peritoneal cavity.

At the tail level, the mesoderm also consists of the midline notochord as it does at the head and trunk levels. The tail notochord, however, arises within the trunk and grows into the tail from the caudal trunk region. In addition to the notochord, the tail mesoderm also consists of loosely packed mesenchymal cells (**Fig. 1D**). These cells are derived from the tail bud, which arises near the end of gastrulation from persisting remnants of the rostral part of the primitive streak. The tail bud mesenchyme forms two main structures during subsequent development—a period called secondary body development in contrast to primary body development or development of the body from the germ layers: the caudal neural tube (which, therefore, has a very different origin and mechanism of development than the more rostral neural tube; *see below* and Vol. II, Chapter 15) and the paired somites (**Fig. 1E**).

In addition to the two tubes of the tube-within-a-tube organization of the vertebrate body plan just described, two other tubes form within the early embryonic body. One of these tubes is an ectodermal tube, the neural tube. The neural tube forms at all levels of the developing embryo during the process of neurulation; it then undergoes regional specialization. The neural tube forms from the neural plate, throughout the head and trunk regions (its method of formation is different in the tail region where it forms from the tail bud), as the neural plate narrows mediolaterally and lengthens rostrocaudally to form the neural groove, a gutterlike space, which is flanked laterally by ectodermal neural folds (**Fig. 2**). During neurulation, the neural folds are brought into the dorsal midline where they fuse to establish the neural tube. The neural crest cells are ectodermal cells derived from the neural folds. These cells detach from the neural folds (or from the roof of the incipient neural tube, depending on the exact rostrocaudal level and particular organism) and migrate laterally where they contribute to the head mesenchyme as well as to a variety of other structures. In addition to the neural tube and the two tubes of the tube-within-a-tube organization of the vertebrate body plan, one other tube forms in the early embryonic body: the heart tube. The heart tube arises from paired tubes, which fuse together in the ventral midline of the head owing to the action of the lateral body folds (**Fig. 1B**). The heart tube is initially a straight structure lying in the ventral midline just beneath the foregut. Two changes occur in the position of the heart tube during subsequent development. First, owing to the action of the head fold of the body, the straight-heart tube is displaced caudally from the head region to the trunk region. Second, but concomitant with the first change, the heart begins to loop on itself, projecting from the midline to the embryo's right side. With looping of the heart, the embryo overtly exhibits sidedness or right-left asymmetry (**Fig. 1C**).

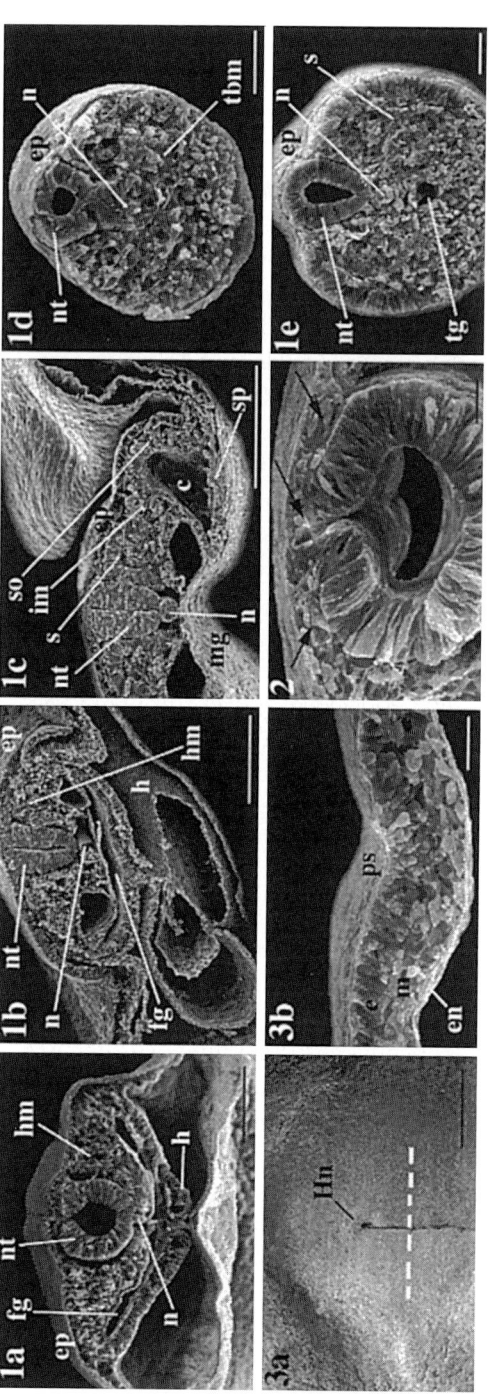

Fig. 1. Scanning electron micrographs showing differences in the organization of the mesoderm in transverse sections at the three major rostrocaudal levels of the body axis: Head, trunk, and tail. (**A**) Chick embryo, head. Within the head region immediately after closure of the neural groove, the mesoderm is organized into a midline notochord (n), beneath the future brain level of the neural tube; loosely packed cells of the head mesenchyme (hm); and mesoderm of the heart-forming region (h). ep, epidermal ectoderm; fg, foregut; nt, neural tube. (**B**) Chick embryo, heads. At a later stage, the same three mesodermal subdivisions are still present within the head; however, the mesoderm of the heart-forming region has now formed a heart tube (h), which has begun to loop. ep, epidermal ectoderm; fg, foregut; hm, head mesenchyme; n, notochord; nt, neural tube. (**C**) Chick embryo, trunk. Within the trunk region, the mesoderm is organized into a midline notochord (n), beneath the future spinal cord level of the neural tube; somites (s); intermediate mesoderm (im); and lateral plate mesoderm, which is split frontally into a dorsal somatic mesoderm (so) and a ventral splanchnic mesoderm (sp). c, coelom; ep, epidermal ectoderm; mg, midgut; nt, neural tube. (**D**) Mouse embryo, tail. Within the tip of the tail region, the incipient notochord (n) has formed and the remainder of the mesoderm consists of loosely packed mesenchymal cells derived from the tail bud (tbm). ep, epidermal ectoderm; nt, neural tube. (**E**) Mouse embryo, tail. Nearer the base of the tail region, the notochord (n) is well defined, as are the paired somites (s), which are derived from the tail bud mesenchyme. ep, epidermal ectoderm; nt, neural tube; tg, tail gut. (A)–(C) Bar = 100 μm; (D) and (E) Bar-400 μm.

Fig. 2. Scanning electron micrograph of a transverse section through the incipient neural tube of a chick embryo at the future midbrain level. Arrows indicate neural crest cells. Bar = 200 μm.

Fig. 3. Scanning electron micrographs. (**A**) Whole mount: The rostral end of the primitive streak (ps) is capped by Hensen's node (Hn), the organizer of the avian embryo. The dashed line indicates the level of the section shown in (**B**). (B) Transverse section through the primitive streak in the early chick embryo at the level shown by the dashed line in (A). e, epiblast; m, mesoderm; en, endoderm. (A) Bar = 100 μm; (B) Bar = 200 μm.

116

3. Morphogenetic Movements Characterizing Gastrulation in Five Classes of Vertebrates

Gastrulation involves a series of morphogenetic movements that vary from species to species. The main reason for the variation is probably the structure of the egg, particularly the amount of yolk contained within it, which necessitates that embryos devise differing strategies to displace cells and to move them into the interior of the egg. Here, we will briefly discuss the morphogenetic movements of gastrulation in five classes of vertebrates: fish, amphibians, reptiles, birds, and mammals.

Gastrulation involves three types of morphogenetic movements: epiboly, or spreading of cells over the surface of the yolk; emboly, or movements of cells into the interior of the embryo; and convergent extension, the concomitant narrowing and lengthening of an area of the gastrula. In addition, different terms are used to describe these movements in different species. Because so little is known about the cellular and molecular control of these movements, the traditional use of different terms in different species may or may not be justified. In this brief overview of gastrulation of five classes of vertebrates, we will stick with tradition and use those terms commonly used in each example of the class we discuss.

3.1. Gastrulation in Fish

The fish egg at the end of cleavage consists of a large spherical yolk cell capped at the animal pole with a cytoplasmic blastoderm composed of hundreds of cells. The egg at this stage is called a blastula. Gastrulation involves three principal movements: epiboly, that is, the spreading of the blastoderm vegetally over the yolk, eventually completely enclosing it; ingression or involution, two terms for emboly (which term is used depends on the species of fish embryo being described); and convergent extension. The blastoderm consists of two layers during epiboly: a superficial or enveloping layer and a deep layer. During epiboly, cells within the deep layer rearrange radially so that this layer thins from four to five cells thick to one to two cells thick. The enveloping layer is one-cell thick at the onset of epiboly and remains one-cell thick throughout epiboly. The cells within the marginal zone of the deep-cell layer undergo extensive movement as epiboly is underway, and they pile up at the future caudal end as the embryonic shield. Within the embryonic shield, marginal zone cells intercalate across the now-defined caudal midline, extending the embryonic shield rostrally as the primitive embryonic axis. The narrowing and concomitant lengthening of the embryonic shield during cell intercalation is called convergent extension. During convergent extension of the embryonic shield, cells migrate through its caudal part (i.e., they involute or ingress), moving into the interior of the embryo to establish the endoderm and mesoderm; cells remaining on the surface of the egg constitute the ectoderm. For additional details, *see* **ref. 7.**

3.2. Gastrulation in Amphibians

The egg of the amphibian at the blastula stage consists of an asymmetric structure. Unlike the fish egg, all yolk is contained within cells distributed throughout the animal-vegetal extent of the blastula. Three areas can be identified within the blastula based on the size of the blastomeres (and, consequently, the amount of yolk they contain). Cells at the vegetal pole represent the primitive endoderm and are large cells containing copious yolk. Cells at the animal pole form a cytoplasmic cap, called the animal cap, and are much smaller cells than are vegetal cells, containing little yolk. Animal cap cells lie above a blastocoel, a fluid-filled space displaced toward the animal pole. Blastomeres between the vegetal cells and the animal cap are intermediate in size and encircle the blastocoel; these cells are called marginal cells.

Gastrulation in amphibians, as in fish, also involves epiboly, emboly, and convergent extension. During epiboly, marginal cells and the animal cap spread downward over the vegetal cells. This spreading occurs more rapidly at the future dorsal side of the embryo than at its lateral or ventral sides, and it creates a fold on the future dorsal side of the embryo called the dorsal lip of the blastopore. Just below the dorsal lip, the large vegetal cells form a structure called the yolk plug. The yolk plug eventually becomes surrounded by lateral lips and ventral lips and forms a plug that fills the blastopore; the latter is the opening of the archenteron to the outside and it eventually forms the anus. The lips of the blastopore form in two ways: by convergent extension of the marginal zone toward the vegetal region and by formation of bottle cells, flask-shaped cells that undergo emboly (i.e., invaginate) into the interior of the gastrula during formation of the archenteron. As bottle cells form, cells turn inward or involute (i.e., undergo emboly) at the lips of the blastopore, forming the endodermal archenteron and the mesoderm. Cells remaining on the surface of the gastrula form the ectoderm. For additional details, *see* **ref. 8**.

3.3. Gastrulation in Reptiles and Birds

The eggs of reptiles and birds consist of a huge mass of yolk on which floats a small cytoplasmic cap consisting of thousands of blastomeres collectively forming the blastoderm. We discuss gastrulation in reptiles and birds together because gastrulation occurs similarly in these two classes of vertebrates. Gastrulation in reptiles and birds also involves epiboly, emboly, and convergent extension. Epiboly occurs as the extraembryonic, future yolk sac portion of the blastoderm grows downward and ultimately completely encloses the yolk. Although the movements of epiboly are more extensive than those of emboly that occur during the phase of gastrulation proper, they are often considered more a part of extraembryonic membrane formation than of gastrulation, because they occur throughout the first half of the incubation period (21 d in the chick), far beyond the period of gastrulation proper. Emboly consists of two movements in birds and reptiles: delamination and ingression or involution. In birds, the process of hypoblast formation is initiated slightly before the egg is laid. During formation of the hypoblast, cells delaminate individually from the blastoderm (also said to polyingress), moving toward the yolk where they form a loosely arranged, thin, sheet-like layer, the hypoblast. Hypoblast formation occurs more vigorously near the future caudal margin of the blastoderm (near a fold called Koller's sickle) as cells delaminate in mass and join the sheet. Formation of the hypoblast continues during the first few hours of incubation after the egg is laid. With formation of the hypoblast, the blastoderm becomes bilaminar.

Formation of the hypoblast is followed in birds by formation and progression (i.e., rostral elongation) of the primitive streak and ingression of cells into the interior of the blastoderm and in reptiles by formation of a blastopore and involution of cells into the interior. In birds, the primitive streak forms near the caudal margin of the blastoderm as cells migrate toward the caudal midline and intercalate with one another. This intercalation results in convergent extension of the primitive streak. As the primitive streak is progressing, cells move from the epiblast into the primitive streak and ingress into the interior of the embryo, where they (1) enter the hypoblast layer as the endoderm, displacing the former to an extraembryonic site (called the germ cell crescent) and (2) enter the space between the epiblast and endoderm, where they form the mesoderm (**Fig. 3**). Ingression of cells through the primitive streak occurs for a number of hours in bird embryos. As the cells derived from the three germ layers assemble into an embryo, the primitive streak moves caudally or regresses. It eventually forms the tail bud, which contributes cells to the organ rudiments in the caudal trunk and tail regions of the embryo (previously described). In reptiles, a blastopore forms at the caudal margin of the blastoderm as cells undergo convergent extension toward the caudal midline. Cells then move inward or involute, defining a blastopore and its lip. For additional details, *see* **ref. *9***.

3.4. Gastrulation in Mammals

Despite the major differences in the structures of the reptilian, avian, and mammalian eggs, the process of gastrulation occurs similarly in these organisms. The mammalian egg contains little yolk. At the end of cleavage, the egg is called a blastocyst. It consists of a hollow sphere of cells containing a blastocoel surrounded by a layer called the trophoblast. A cluster of cells forms on one side of the blastocoel; this cluster is called the inner cell mass. The inner cell mass will form the blastoderm during subsequent development. The blastocyst invades the uterine wall during implantation, and subsequent *in utero* development occurs until the time of birth or parturition.

The inner cell mass undergoes a number of changes during gastrulation. First, a layer delaminates from it to form the primitive endoderm or hypoblast. With formation of the hypoblast, the inner cell mass becomes a bilaminar blastoderm consisting of an upper (i.e., toward the trophoblast) epiblast and a lower (i.e., toward the blastocoel) hypoblast. As delamination is occurring, an extraembryonic cavity begins to form between the epiblast and trophoblast; formation of this cavity invaginates the blastoderm into a cuplike structure. This cuplike configuration is the main difference between the structure of the avian (or reptilian) and mammalian blastoderms—the former is flat. At the caudal margin of the mammalian blastoderm, cells migrate toward the midline and pile up as a primitive streak, which like the avian primitive streak undergoes convergent extension as cells intercalate across the midline. Ingression of cells through the mammalian primitive streak occurs similarly to that in birds, but the primitive streak apparently does not undergo regression. Instead, the early embryonic axis grows forward as primitive-streak cells are largely depleted. Remnants of the primitive streak form a tail bud as they do in bird embryos; the tail bud contributes cells to the structures of the tail region (previously described). For additional details, *see* **ref. *10***.

4. Origin, Displacement, and Commitment of Cells During Gastrulation

The locations of groups of cells with various prospective fates can be identified at stages prior to and during gastrulation by the process of fate mapping. During fate mapping, a localized region of the embryo is labeled, and such labeled cells are subsequently followed to track their movements and fates. Fate mapping results in the con-

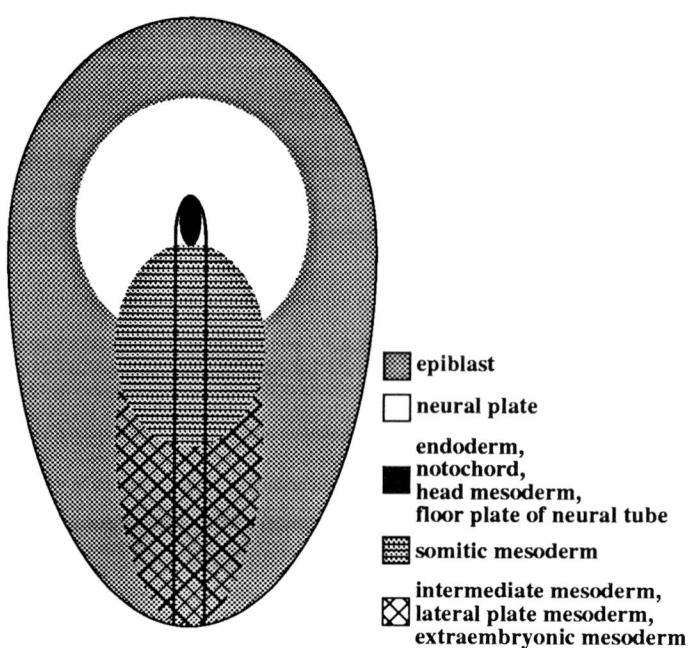

epiblast

neural plate

endoderm,
notochord,
head mesoderm,
floor plate of neural tube

somitic mesoderm

intermediate mesoderm,
lateral plate mesoderm,
extraembryonic mesoderm

Fig. 4. Prospective fate map of the bird blastoderm at the full primitive-streak stage.

struction of prospective fate maps (e.g., **Fig. 4**; **ref. *11***), which show the locations of cells within a snapshot of time. Various techniques are used (*see* Chapter 35), including in birds, the transplantation of quail donor cells to chick host embryos. For fate mapping, such transplantation is done homotopically (i.e., from one site of the donor to the same site of the host) and isochronically (i.e., from one stage of the donor to the same stage of the host). Prospective fate maps tell us what a group of cells will form during normal development, and a series of fate maps constructed for progressively advancing stages shows us the displacement of cells over time. However, prospective fate maps fail to provide information on the state of commitment of cells to their prospective fates. To obtain such information as well as to test cell potency, cells are grafted heterotopically (and/or sometimes heterochronically). Cells that adapt to their new location are said to be plastic, totipotent, or pluipotent and are uncommitted to their fate. Cells that self-differentiate at a heterotopic site are said to be unipotent and are committed to their fate.

Prospective fate maps reveal that cells of the gastrula are highly organized. For example, in the chick gastrula at the full primitive-streak stage (**Fig. 4**), cells of the epiblast consist of prospective ectoderm, mesoderm, and endoderm. Cells within the perimeter of the epiblast consist of prospective epidermal ectoderm. More centrally located epiblast cells surrounding Hensen's node, near the rostral end of the primitive streak, consist of prospective neural plate. Hensen's node contains prospective ectodermal cells (i.e., prospective floor plate of the neural tube), prospective mesodermal cells (i.e., prospective notochord and head mesoderm), and prospective endoderm (i.e., of the prospective foregut). More caudally, cells within the primitive streak and flanking epiblast consist of prospective somitic mesoderm, and more caudally, of prospective intermediate mesoderm, lateral plate mesoderm, and extraembryonic mesoderm. Prospective heart mesoderm ingresses through the primitive streak at an earlier stage *(12)*.

Heterotopic grafting studies reveal that most cells of the early embryo are plastic during gastrulation stages, with one exception: The cells of Hensen's node that form the notochord become committed prior to or during early gastrulation *(13,14)*. Other populations of cells, such as prospective heart and other mesodermal cells as well as ectoderm cells are labile *(15,16)*. Plasticity also exists between germ layers such that at the full primitive-streak stage, prospective mesoderm can form ectodermal derivatives and vice versa *(17)*. However, after mesodermal cells undergo ingression, their fate becomes fixed, but commitment to cell fate is apparently independent of ingression because ectodermal cells remaining on the surface of the epiblast become committed at a similar time in development *(18)*.

Various other experiments, particularly cell grafting and cell labeling, have begun to reveal the cellular mechanisms underlying morphogenetic movements. For example, during amphibian gastrulation, cells undergo either radial intercalation (during epiboly) or medial-lateral intercalation (during involution of cells over the lips of the blastopore) as well as cell-shape changes during epiboly (i.e., cell flattening) and invagination of the archenteron (cell wedging during formation of bottle cells). In birds and mammals, similar cell behaviors occur during morphogenetic movements, but additionally, cell division and growth and oriented mitoses play important roles in generating such movements *(19–21)*.

5. Other Forming-Shaping Movements Involved in Formation of the Vertebrate Body Plan

After formation of the three primary germ layers, the vertebrate body plan is sculptured from the germ layers by four major morphogenetic events: cardiogenesis, body folding, somitogenesis and neurulation. The first three of these events are briefly discussed here. Neurulation was briefly discussed previously, and more details can be found in Vol. II, Chapter 15. Additionally, the body plan within the caudal part of the body is established through secondary body development, which unlike primary body development, occurs independently of gastrulation and formation of the definitive germ layers. Secondary body development was briefly discussed above.

5.1. Cardiogenesis and Body Folding

Cardiogenesis and body folding are two of the earliest morphogenetic movements involved in formation of the body plan, with both occurring while gastrulation is underway. Recall that the heart forms from bilateral heart-forming regions. The head fold of the body defines the rostral end of the head and establishes the foregut. It displaces the heart rudiments caudally from the future head to the future thoracic region. Concomitantly, two events occur:

1. Each heart-forming region buds off an inner tubular structure—the future endocardium of the heart—and assembles into an enveloping epithelial layer: the future myocardium of the heart.
2. The lateral body folds bring the two endocardial tubes together in the ventral midline where they fuse as a single straight-heart tube consists of two cellular and one extracellular layer: the endocardium and myocardium, separated from one another by the cardiac jelly.

In addition to the head fold of the body and the lateral body folds, a third type of body fold forms: the tail fold of the body. This fold establishes the caudal gut and defines the tail end of the embryo. For additional details on cardiogenesis and body folding, *see* **ref. 22**.

5.2. Somitogenesis

Formation of the somites, paired blocks of mesodermal tissue flanking the notochord and overlying neural tube, provides an early example of the segmentation of the vertebrate embryo. Shortly after ingression through the primitive streak, the somitic mesoderm consists of loosely packed mesenchymal cells. During somitogenesis, this mesenchyme undergoes epithelialization on each side of the notochord/neural tube, forming a pair of longitudinal bands called the segmental plates. Interactions between the segmental plates and the neural tube, notochord, epidermal ectoderm, and intervening extracellular matrix result in segmentation of the segmental plates into the somites, which extend throughout the trunk and tail regions. Recently, a chick homolog of the *Drosophila* segmentation gene, *hairy*, has been cloned. This homolog is expressed in the segmental plates in a cyclical fashion at the same rate as that of somite formation (i.e., one pair every 90 min), suggesting that *hairy* might function as a molecular clock for segmentation *(23)*.

Shortly after their formation, the somites become compartmentalized. Three major subdivisions form in each somite: the dermatome (source of the dermis of the skin), the myotome (source of skeletal muscle), and the sclerotome (source of the bones of the vertebral column). For additional details on somitogenesis, *see* **ref. 24**.

5.3. Patterns of Gene Expression

The vertebrate body plan is patterned both structurally and molecularly. We just described the structural or anatomical patterning of the body plan. Here we briefly describe some of the molecules expressed in highly organized temporal and spatial patterns within the forming body plan. Space constraints preclude a detailed discussion of this immense topic, so only a general scenario will be outlined with a few key examples.

Three types of molecules are expressed during formation of the vertebrate body plan, which experimental studies (such as gene knockouts, targeted overexpression of protein, or expression of dominant-negative receptors) suggest may play a causal role in generation of the body plan: secreted signaling molecules/growth factors, signaling intermediaries (such as kinases), and transcription factors.

Formation of the mesoderm during early gastrulation seems to involve growth factors of the FGF, TGFβ, and Wnt families. In amphibian embryos, it is well documented that vegetal cells induce the mesoderm by secreting growth factors that act on the marginal cells. An early response to such induction is the expression of the transcription factor, *Brachyury*. Similar growth factors are present in avian and mammalian embryos, but details of mesodermal induction are far less clear in higher vertebrates. *Brachyury* expression also occurs in higher vertebrates and is correlated with mesoderm formation. In amphibian embryos, a localized region of the embryo, called the Nieuwkoop Center, induces the organizer. Again, less is known about the Nieuwkoop Center in higher vertebrates, but recently an experimental paradigm has been developed using avian lateral blastoderm isolates that shows promise for elucidating cell interactions underlying induction of the organizer in higher vertebrates *(25)*.

Once the organizer is established through the action of the Nieuwkoop Center, it secretes *FRZ-B*, a soluble receptor for Wnt protein. By doing so, it is believed that the organizer dorsalizes the embryo by antagonizing the function of Wnt, a secreted ventralizing protein. A similar antagonism seems to act in neural induction: Secretion of chordin and noggin, two "neural inducers," by the organizer inhibits the action of the secreted growth factor, Bmp4. Bmp4 is necessary for differentiation of the epidermal ectoderm. Thus neural induction seems to involve the inhibition of formation of epidermis and the resulting "self-differentiation" of the ectoderm into neural plate, which, according to this scenario, is the default state of the ectoderm. For additional details, *see* **refs. *26*** and ***27***).

The expression pattern of combinations of genes can be used to molecularly define the various rudiments of the vertebrate body. For example, the organizer can be defined by the localized expression of several genes, including the transcription factors *goosecoid, Hnf3β*, and *cNot*, and the secreted signaling molecule, Sonic hedgehog. Similarly, the mediolateral subdivisions of both the ectoderm and mesoderm can be defined by their patterns of gene expression. Within the ectoderm, the epidermal ectoderm expresses the secreted growth factor Bmp4, whereas the neural plate expresses the transcription factors *Sox-2* and *Sox-3* as well as a cell adhesion molecule, L5. The floor plate of the neural tube expresses both *Hnf3β* and Sonic hedgehog. Within the mesoderm, the notochord expresses Sonic hedgehog, *Hnf3β*, and *cNot*, as well as an unknown antigen recognized by the Not-1 antibody. The somites express various *Pax* genes (paired-box transcription factors) and the transcription factor, *Paraxis*, the intermediate mesoderm expresses *Pax2*; and the lateral plate mesoderm expresses a cytoskeletal antigen, cytokeratin. The early heart mesoderm expresses the transcription factors, *tinman, eHand*, and *dHand*, as well as the cytoskeletal antigens, actin and myosin. Finally, rostrocaudal (and mediolateral) regional markers are expressed in early rudiments of the vertebrate body. For example, within the early neural tube, the transcription factors, *BF1, Six3*, and *Otx2*, are expressed at the forebrain level as are the growth factors Bmp6 and fgf8. At the midbrain/hindbrain juncture, the transcription factors *Pax2, Otx2*, and *En-2*, are expressed as is fgf8. Within various rhombomeres of the hindbrain, numerous *Hox* genes (transcription factors) are expressed as well as a variety of other genes. Within the spinal cord *Pax3* is expressed. Finally, within the tail bud fgf8 and *Brachyury* are expressed. Presumably, the expression of multiple genes within individual units of the future central nervous system establishes a combinatorial code that ultimately leads to a commitment in cell fate that is regionally specific.

6. The Role of the Organizer in Choreographing Morphogenetic Movements

Recent studies in amphibian embryos suggest that the morphogenetic movements involved in formation of the vertebrate body plan are controlled by the organizer *(28,29)*. This control includes initiating these complex movements, maintaining them, and establishing their directionality. How the organizer accomplishes these tasks is currently unknown, but the fact that it does so highlights the importance of the organizer in establishing the vertebrate body plan.

Acknowledgment

Original work described herein from the Schoenwolf Laboratory was supported by Grants NS 18112 and HD 28845 from the National Institutes of Health.

References

1. Oppenheimer, J. M. (1959) Extraembryonic transplantation of sections of the *Fundulus* embryonic shield. *J. Exp. Zool.* **140,** 247–268.
2. Ho, R. K. (1992) Axis formation in the embryo of the Zebrafish, *Brachydanio rerio. Semin. Dev. Biol.* **3,** 53–64.
3. Spemann, H. and Mangold, H. (1924) Über induktion von Embryonalanlagen durch Implantation artfremder Organisatoren. *Roux Arch. Entwicklungsmech. Org.* **100,** 599–638.
4. Yntema, C. L. (1968) A series of stages in the embryonic development of *Chelydra serpentina. J. Morphol.* **125,** 219–252.
5. Waddington, C. H. (1932) Experiments on the development of chick and duck embryos, cultivated in vitro. *Phil. Trans. R. Soc. Lond. (Biol.)* **221,** 179–230.
6. Beddington, R. S. P. (1994) Induction of a second neural axis by the mouse node. *Development* **120,** 613–620.
7. Warga, R. M. and Kimmel, C. B. (1990) Cell movements during epiboly and gastrulation in zebrafish. *Development* **108,** 569–580.
8. Keller, R. E. (1986) The cellular basis of amphibian gastrulation, in *Developmental Biology: A Comprehensive Synthesis* (Browder, L., ed.), Plenum, New York, pp. 241–327.
9. Lemaire, L. and Kessel, M. (1997) Gastrulation and homeobox genes in chick embryos. *Mech. Dev.* **67,** 3–16.
10. Tam, P. P. L. and Behringer, R. R. (1997) Mouse gastrulation: The formation of a mammalian body plan. *Mech. Dev.* **68,** 3–26.
11. Garcia-Martinez, V., Alvarez, I. S., and Schoenwolf, G. C. (1993) Locations of the ectodermal and non-ectodermal subdivisions of the epiblast at stages 3 and 4 of avian gastrulation and neurulation. *J. Exp. Zool.* **267,** 431–446.
12. Garcia-Martinez, V. and Schoenwolf, G. C. (1993) Primitive-streak origin of the cardiovascular system in avian embryos. *Dev. Biol.* **159,** 706–719.
13. Schoenwolf, G. C., Garcia-Martinez, V., and Dias, M. S. (1992) Mesoderm movement and fate during avian gastrulation and neurulation. *Dev. Dyn.* **193,** 235–248.
14. Selleck, M. A. J. and Stern, C. D. (1992) Commitment of mesoderm cells in Hensen's node of the chick embryo to notochord and somite. *Development* **114,** 403–415.
15. Inagaki, T., Garcia-Martinez, V., and Schoenwolf, G. C. (1993) Regulative ability of the prospective cardiogenic and vasculogenic areas of the primitive streak during avian gastrulation. *Dev. Dyn.* **197,** 57–68.
16. Schoenwolf, G. C. and Alvarez, I. S. (1991) Specification of neuroepithelium and surface epithelium in avian transplantation chimeras. *Development* **112,** 713–722.
17. Garcia-Martinez, V., Darnell, D. K., Lopez-Sanchez, C., Sosic, D., Olson, E. N., and Schoenwolf, G. C. (1997) State of commitment of prospective neural plate and prospective mesoderm in late gastrula/early neurula stages of avian embryos. *Dev. Biol.* **181,** 102–115.
18. Darnell, D. K. and Schoenwolf, G. C. (1997) Vertical induction of Engrailed-2 and other region-specific markers in the early chick embryo. *Dev. Dyn.* **209,** 45–58.
19. Sausedo, R. A. and Schoenwolf, G. C. (1993) Cell behaviors underlying notochord formation and extension in avian embryos: Quantitative and immunocytochemical studies. *Anat. Rec.* **237,** 58–70.

20. Sausedo, R. A. and Schoenwolf, G. C. (1994) Quantitative analyses of cell behaviors underlying notochord formation and extension in mouse embryos. *Anat. Rec.* **239,** 103–112.

21. Sausedo, R. A., Smith, J. L., and Schoenwolf, G. C. (1997) Role of nonrandomly oriented cell division in shaping and bending of the neural plate. *J. Comp. Neurol.* **381,** 473–488.

22. Harvey, R. and Rosenthal, N. (1998) *Heart Development*. Academic, New York.

23. Palmeirim, I., Henrique, D., Ish-Horowicz, D., and Pourquie, O. (1997) Avian *hairy* gene expression identifies a molecular clock linked to vertebrate segmentation and somitogenesis. *Cell* **91,** 639–648.

24. Christ, B. and Ordahl, C. P. (1995) Early stages of chick somite development. *Anat. Embryol.* **191,** 381–396.

25. Yuan, S. and Schoenwolf, G. C. (1998) De novo induction of the organizer and formation of the primitive streak in an experimental model of notochord reconstitution in avian embryos. *Development* **125,** 201–213.

26. Slack, J. M. W. (1993) Embryonic induction. *Mech. Dev.* **41,** 91–107.

27. Lemaire, P. and Kodjabachian, L. (1996) The vertebrate organizer: Structure and molecules. *Trends Genet.* **12,** 525–531.

28. Keller, R., Shih, J., Sater, A. K., and Moreno, C. (1992) Planar induction of convergence and extension of the neural plate by the organizer of *Xenopus. Dev. Dyn.* **193,** 218–234.

29. Elul, T., Koehl, M. A. R., and Keller, R. (1997) Cellular mechanism underlying neural convergent extension in *Xenopus laevis* embryos. *Dev. Biol.* **191,** 243–258.

13

Craniofacial Development of Avian and Rodent Embryos

Brian K. Hall and Tom Miyake

1. Introduction

In this overview, we examine approaches to the embryological development of the craniofacial region of avian and rodent embryos. By craniofacial region we mean: externally, the face and head, excluding the caudal pharyngeal arches; internally, the brain, skull, jaws, and facial skeleton, teeth (in rodents), and associated soft tissues, including muscular, vascular, and connective tissues. The terms "avian" and "rodent" are often equated with the domestic fowl *(Gallus domesticus)*, mouse *(Mus musculus)*, and rat *(Rattus* sp.); most studies of craniofacial development have been performed on these three species, mostly on chick and mouse. Such a model organism approach is unfortunate. It is typical of the type approach of pre-Darwinian biology and negates the richness afforded by comparative and phylogenetic analyses *(see* **refs.** *1* and *2)*, but is perhaps unavoidable, given both the convenience of working with chick and mouse embryos and the breadth of knowledge that has accumulated on these two species. There is a rich literature on aspects of the development of different strains of mice but much of the literature on avian development assumes that all strains of chickens are alike. The major staging tables for the chick *(3)* does not identify the strain; that for the mouse *(4)* is based on hybrids between C57BL/6 females and CBA males. We have recently expanded staging tables for murine craniofacial development for several inbred strains *(5–8)*, but comparative analyses are few.

Some classic works on comparative avian or rodent development that include craniofacial development are: for birds, Romanoff *(9)* and Landauer *(10,11)*; for rodents, Rugh *(12)* and Johnson *(13)*. Recent overviews on chick and/or mouse development include an atlas of mouse development *(14)*, three edited books *(15–17)* and three laboratory manuals *(18–20)*. Although it may appear unlikely that craniofacial development of avian and rodent embryos could be discussed together, in fact, early stages of craniofacial development in both these vertebrate classes involve similar populations of cells and similar (often identical), developmental processes. We examine cellular origins and craniofacial tissues (**Subheadings 2.** and **3.**) and 12 approaches to the analysis of craniofacial development (**Subheading 4.**).

From: *Methods in Molecular Biology, Vol. 135: Developmental Biology Protocols, Vol. I*
Edited by: R. S. Tuan and C. W. Lo © Humana Press Inc., Totowa, NJ

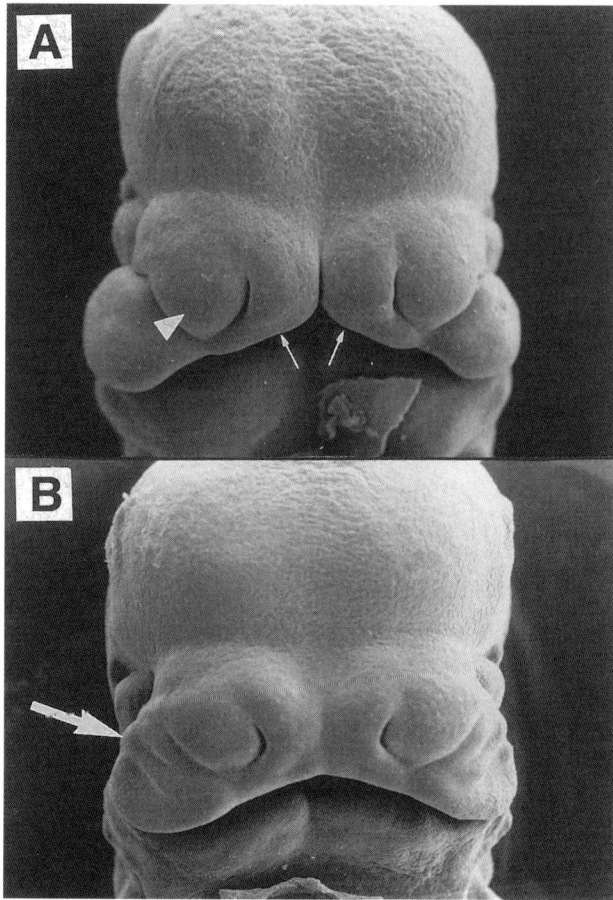

Fig. 1. Scanning electron micrographs of frontal views of the developing faces of mouse embryos (C57BL/6 strain) of stages 20.2 and 21.12 (as defined by Miyake et al. in **ref. 5**). **(A)** The medial nasal prominence (arrows) and lateral median prominence (arrow head) are marked. **(B)** Developing vibrissae are marked by the arrow.

2. Cellular Origins

Craniofacial development is characterized externally by the formation of masses of mesenchymal cells that form the craniofacial processes (maxillary and mandibular processes from which the upper and lower jaws develop; frontonasal [nasomedial] processes from which the median portion of the face [including the nose in rodents] will develop, and epithelial placodes [thickenings] and pits that mark the location of the major sense organs [nasal placode and pits, etc.] *see* **ref. 21**). These structures are seen in the greatest detail when studied with scanning electron microscopy, as depicted in **Fig. 1**.

Internally, we see that the craniofacial processes are composed of mesenchymal cells, most of which arose in the neural tube as neural crest cells. The migration of neural crest cells away from the neural tube creates the facial processes (*see* **ref. 22** and Vol. II, Chapters 6 and 7 and Vol. III, Chapter 21). The brain is developing at the same time; there is now a vast literature on brain development.

Neural crest cells emigrate from the neural tube along the length of the brain, except for the most rostral portion of the forebrain and individual rhombomeres of the hindbrain (*see* **ref.** *23* and references therein for rhombomere patterning, and **ref.** *24* for patterning of the neural tube). In the chick, rostral forebrain neural epithelium provides much of the ectoderm covering the head and is therefore the source of, or intimately associated with, the placodes *(25,26)*. The transcription factor AP-2 is essential for closure of the neural tube (there is excessive apoptosis in the neural tubes and branchial arches of AP-2$^{-/-}$ mice) and therefore plays an important role in neural crest cell emigration *(27)*.

Migrating neural crest cells are mesenchymal in appearance (initiation of migration involves a cellular transformation from epithelial → mesenchymal organization and concomitant reorganization of the cytoskeleton and modification of extracellular matrix products *[28]*), and consequently are very difficult to distinguish from mesenchyme of mesodermal origin. Craniofacial muscles, indeed, all the muscles of the head, are products of head mesoderm but are patterned by the neural crest cells that form the connective tissue sheaths of the muscles *(29–31)*. Neural crest cells also play an important role in angiogenesis *(32)*. Indeed, it is not an overstatement to claim that neural crest cells pattern craniofacial development *(31,33)*.

A number of molecular markers aid investigation of neural crest cells. The antibody HNK-1 identifies migrating neural crest cells, but does not label all neural crest cells, and also labels mesodermally derived mesenchyme. *Hox-2* and *Hox-11* label distinct subpopulations of neural crest cells emigrating from the murine hindbrain *(34,35)*; indeed, a *Hox* code for the head and branchial arches has now been identified. A *lacZ* reporter, specific to migrating neural crest cells, has been used to investigate the mode of action of the *Splotch* (sp) mutation in mice (which affects neural crest cells indirectly by disrupting cell interactions *(36)*. Two receptor tyrosine kinases (EphA4 and EphB1) and their ligands maintain discrete populations of migrating neural crest cells *(37)*. The recent development of cDNA libraries against subpopulations of quail neural crest cells *(38)* opens up a new avenue for identification and exploration of the behavior and function of these all-important cell populations.

3. Tissues

Neural-crest-derived mesenchymal cells undergo extensive proliferation to form condensations of cells from which individual cartilages, bones, teeth, ligaments, and connective tissues will develop. Establishment of a minimal cell size for a condensation is crucial for initiation of cytodifferentiation *(6,8,39–41)*. Many mutants in birds and rodents that affect craniofacial development have condensations as their initial site of action *(13,39,41)*. Tissue interactions between mesodermal or neural-crest-derived mesenchyme and epithelia are critical for initiation of differentiation and also regulate basic shape and size of tissues formed through their effect on condensations *(2,42–44*; and *see* Chapter 24 in Vol. III).

4. Approaches to Craniofacial Development

Our aim is to provide an overview of 12 approaches to craniofacial development that may be taken. It would be impossible to be encyclopedic, and, in any event, a substantial number of the number of the chapters in this volume—including virtually all those in Parts V and VI—provide methods of choice for analysis of craniofacial development. They should be consulted.

We begin with the resolution obtained when viewing the entire embryos or craniofacial region at the level of detail provided by a dissecting microscope and then proceed to the increasing resolution provided by finer scale analyses.

1. Observations of live embryos: Although it may appear old-fashioned, or too obvious even to mention, much can be learned from a close observation of live embryos as dissected either from the shell or from the uterus. All that is required is a good-quality dissecting microscopy with ×10 or ×15 eyepieces; forceps, camel hair brush or glass probes with which to *gently* manipulate the embryo; and close observation. Embryos are best examined in clean saline or phosphate-buffered saline, after removing any yolk or blood by suction with a pipet. This is the time for maximal observation and the best time to stage embryos. Embryos will begin to turn opaque and loose much detail and relief as soon as fixative is applied. A wide, shallow dish, such as a 10-cm diameter glass Petri dish, containing a shallow depression slide, both holds the embryo and offers ready access to the embryo.

2. Staging: Knowing the age and stage of development of the embryo under study sounds easy; you know how long the embryos have been in the incubator or when the mice were mated. But avian development is highly temperature dependent (*see* **refs. *10*** and ***11***), and if mating of rodents is carried out overnight, there may be as much as a 12-h difference between embryos of females mated "at the same time." We find that mating for only 2 h, although it decreases the success rate and requires that more females be mated (typically we set up 10 females to obtain one pregnancy), allows very precise aging of embryos (*see* **refs. *5–8***).

   Nevertheless, morphological stage of development can vary greatly among avian embryos in eggs incubated under identical conditions and opened at the same time or among rodents of the same litter. Consequently, placing embryos into morphological stages is critical. Basic staging tables are available for chick and mouse (*3,4,14*). Tables for more precise staging of embryos at particular developmental periods are also available: Eyal-Giladi and Kochav (*45*) for the first day of chick development, and Miyake et al. (*5*) for the major stages of murine development when the craniofacial region is forming. Staging is best done on fresh embryos; fine details will be obscured after fixation (*see* Chapter 23 in **refs. *2*** and ***38*** for further details on the importance of fine-grained staging of embryos).

3. Scanning electron microscopy: Perhaps the next most informative level of analysis for resolution of external and internal features is scanning electron microscopy (SEM). Many studies of craniofacial development have been undertaken using this method. The atlas edited by Schoenwolf (*46*) is an excellent source of methods and analyses of embryos. **Reference *5*** contains the use of SEM analysis for staging mouse embryos and a discussion of past literature. The studies by Sulik on normal, alcohol- and vitamin A-treated mice (*47,48*) are among the best applications of SEM to craniofacial development.

4. Whole-mount embryos: Visualization of whole embryos as complete specimens, especially when coupled with specific staining of individual tissues or organs, offers a powerful way to analyze timing, patterning and morphogenesis. "Clearing" specimens to render them transparent, and differentially staining the skeleton is a long-established method that allows very early stages of chondrogenesis and osteogenesis to be examined (*49*). More recently, methods have been developed to visualize individual tissues or even cell types that contain specific molecules to which antibodies may be attached. In this way, antibodies to type II collagen can be used to visualize the very earliest stages of cartilage or

notochordal sheath development; antibodies to myosin or actin to visualize myogenesis; antibodies to neurofilament proteins to visualize nerve cells; antibodies to neural crest cells to visualize their migration and patterning and so forth (*see* **ref. 50** for an excellent analysis of the methods). Whole-mount preparation can be combined with use of a fluorochrome to visualize the antibody and visualized, either with immunofluorescence or with confocal microscopy (*see* approach 6).

5. Histology: Light microscopic histology after either paraffin or plastic embedding is a mainstay of any analysis of craniofacial development and forms the technical basis for histochemical, immunohistochemical, and immunocytochemical analyses, as well as for *in situ* hybridization. While whole-mount visualization provides the resolution required for analysis of patterns, pathways, and morphogenesis, knowledge of the behavior of individual cells or small groups of cells almost always requires histological analysis, often of serially sectioned embryos in which structures can be traced from section to section. (Serially sectioned embryos can be reconstructed in three dimensions (3-D) *[6]* or used for histochemical, immunohistochemical, or *in situ* hybridization *[7–9]*).

    Patterns seen in whole-mount embryos can be misleading, especially when different cell populations are contiguous, as they are during much of craniofacial development: neural, epidermal, neural crest, and placodal ectoderm are juxtaposed early in development; neural crest- and mesodermally derived mesenchyme are contiguous in facial processes and in those portions of skull or facial skeleton that contain bones of mixed (mesoderm and neural crest) cellular origins. There is no substitute, and there are no short cuts, to acquiring the necessary skills and equipment to perform histological analysis as a routine in studies of craniofacial development. Although such manuals as the latest edition of Humason's *Animal Tissue Techniques* (*51*) are invaluable guides, in histology, doing is learning.

    Transmission electron microscopic analysis offers the highest resolution for histological analyses and may be coupled with histochemistry or autoradiography to localize the sites of synthesis or storage of individual cellular constituents (*52*).

6. 3-D reconstruction: Reconstructing specimens in 3-D from serial sections used to be an absolute pain. No longer must pain be endured to obtain rapid and accurate reconstructions. We (*53*) have used 3-D reconstruction to identify a previously unknown connection between the neural-crest-derived mesenchymal cell populations that comprise the mandibular and maxillary arches in chick embryos; cells for both arches accumulate at the base of the arches (**Fig. 2**). Knowledge of this common condensation provided an explanation for the similar timing of the epithelial-mesenchymal interactions that regulate maxillary and mandibular mesenchyme (the cells are induced as a single population in a single interaction) and an explanation for the "piling up" of cells at the base of the arches in vitamin-A-treated embryos—movement into the individual arches is inhibited (*see* **ref. 53** for discussion).

    We have also used 3-D reconstruction to visualize development of such complex craniofacial regions as the middle ear and first arch cartilages (*6*) and the distribution of the earliest osteogenic precursors (**Fig. 3, ref. 7**) using Image 1.38, a 3-D image software package developed by Wayne Rasband, NIH and reconstruction in an ICAR 80.8 Workstation with Silicon Graphics Personal Iris 4D/25 Server and UNIX V Operating System (*6*). The recent refinement of methods to align adjacent sections (*54*) will enhance 3-D reconstruction even further.

    Confocal microscopy provides a further means of visualizing and reconstructing tissues with a resolution of the order of 0.1 μm. Because of its high resolution and the need to optically "section" the tissues, it is most applicable to specimens considerably smaller than those normally analyzed in studies of craniofacial development, several hundred

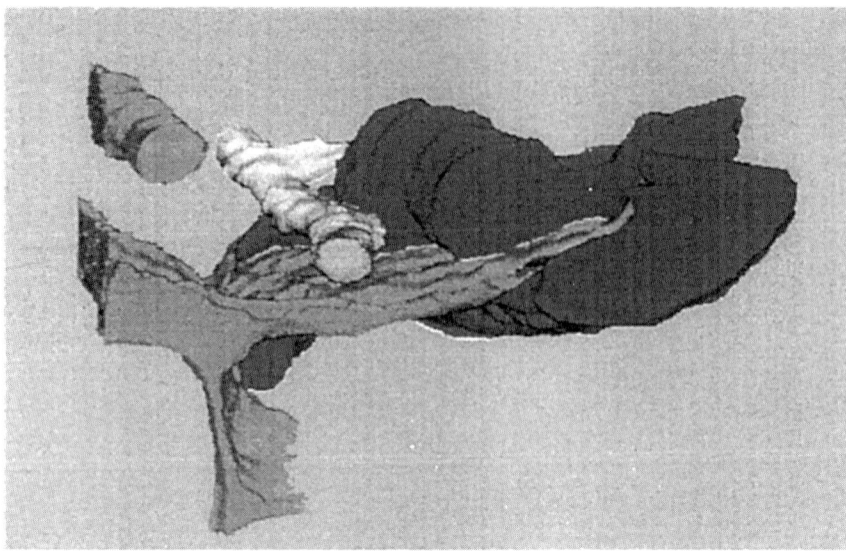

Fig. 2. A 3-D reconstruction (in frontal view) of the common osteogenic condensation for the maxilla and mandible (darkest color) above and below the buccal cavity. The white vessel is the maxillary vein (*see* **ref.** *53* for further details).

microns thickness being the realistic upper limit for confocal microscopy-based 3-D reconstructions; *see* Chapters 17, 19, and 20 in this volume for methods.

7. Histochemistry: The reconstruction of early osteogenic precursors of the murine man-dibular arch noted earlier (shown in **Figs. 4** and **5**) and in **ref.** *7* was made possible because the cells were visualized using a histochemical procedure for alkaline phosphatase. This is but one of many procedures available for the visualization of cellular constituents (*see* **ref.** *55* for methods).

8. Immunohistochemistry and immunocytochemistry: Similarly, the ability to bind antibod-ies against specific molecules or portions of molecules, and to visualize that antibody-ligand complex with a histochemical stain or fluorescent marker, is a class of techniques that has had wide application in craniofacial development. HNK-1 to label neural-crest cells, or antibodies against the nerve cell adhesion molecule N-CAM to separate chondro-from osteogenic precursor cells (*56*) are but two examples of techniques that are limited only by generation of the appropriate probe; *see* **refs.** *18*, *57*, and *58* for methodology.

9. *In situ* hybridization: *In situ* hybridization with RNA probes provides a means to localize the mRNA products of individual genes. It is, of course, important to remember that pres-ence of the mRNA need not mean that the message is translated into protein. It is therefore important to combine *in situ* hybridization with visualization of the gene product, usually with immunohistochemical approaches, Northern blot analysis or RT-PCR. Even then, posttranslational modification may provide a further level of regulation not evident from *in situ* or antibody-based methods. **References** *18*, *20*, and *57* provide detailed protocols; see also the chapters in Part III in Vol. III of this series.

10. Genetic manipulations: The use of mutant strains of mice (and to a lesser extent, of birds) has a long history in craniofacial development (*10,13*); much of our knowledge of mecha-nisms of normal development has come from analysis of mutants; *see* **ref.** *36* for one example and **ref.** *13* for many more. More recently, the ability to knock out individual genes from rodent embryos has opened a window to craniofacial development previously

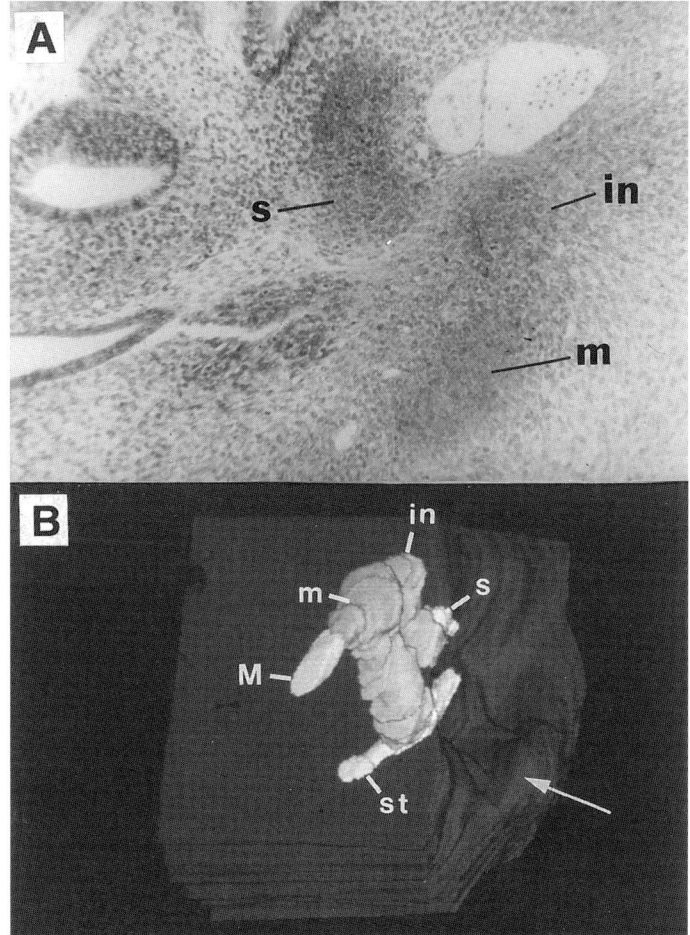

Fig. 3. **(A)** A horizontal section of the developing middle ear region of a mouse embryo (stage 23.31 as defined in **ref. 5**) stained with toluidine blue to show the common condensation for the incus (in), malleus (m), and stapes (s) of the developing middle ear. **(B)** A 3-D reconstruction of the condensation for the middle ear (lateral to the right, medial to the left, dorsal to the top), showing the incus (in), malleus (m), caudal end of Meckel's cartilage (M), stapes (s), and stylohyal cartilage (st). The arrow marks the auricular hillock of the developing external ear.

restricted to analysis of mutant strains, random mutations, or mutations affecting many genes or entire chromosomes. **References *18* and *58*** and the chapters in Part V of Volume II of this series, provide detailed methodologies for generating knock-out mice. The chapters in Part VI, Vol. III in this series (especially Chapter 45) provide discussions of individual genes in disease states affecting craniofacial development. The chapters in Part V provide methods for generating chimeras of avian and rodent embryos.

11. Cell labeling and surgical intervention: The availability of lipophilic dyes that can be injected or electrophoresed into individual cells, or into small numbers of cells, has enormously increased our understanding of the cell populations that comprise the craniofacial region; *see* **refs. *20*** and *24* for examples; Chapters 10 and 11 in **ref. *15*** for methods for injecting cells in avian and rodent embryos, respectively; and Part V for methods of cell lineage analysis.

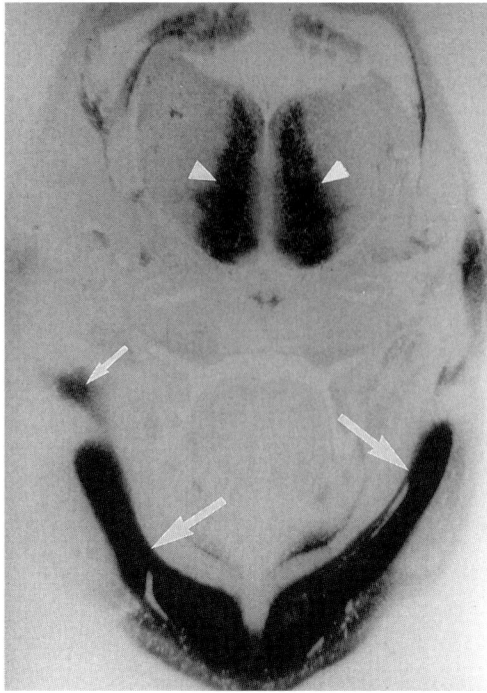

Fig. 4. A horizontal section through the head of a mouse embryo of stage 21.32 in which alkaline phosphatase can be visualized (*see* **ref. 7** for methods) as parallel centers within the midbrain (arrow heads) and condensation for the dentary bones (arrows).

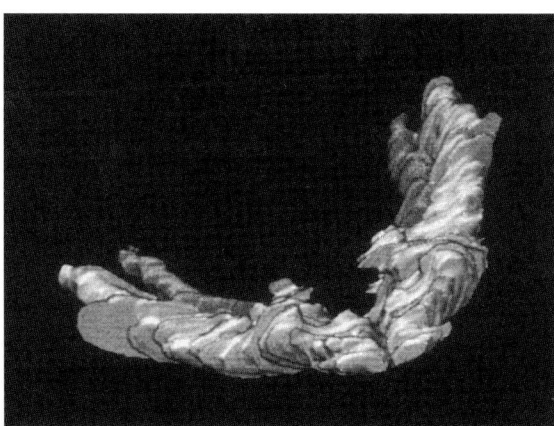

Fig. 5. A 3-D reconstruction of the developing dentary bones of a mouse embryo, in which the dentary bones are visualized by their positive staining for alkaline phosphatase (*see* **ref. 7** for details).

Much has also been learned from surgical removal and/or transplantation of tissues involved in craniofacial development. Again, *see* Chapters 10 and 11 in **refs. *15*** and ***20*** for basic methods. Also *see* Chapters 44 and 46 in this volume, and Chapter 7, Vol. III in this series for details of individual procedures.

12. Culture techniques: Many of the techniques available for the culture of avian and rodent embryos are detailed in other chapters. Those that are relevant to craniofacial development are Chapter 6 for avian embryos and Chapters 7 and 8 for rodents; *see also* **refs.** *15* and *20* for microsurgery on avian embryos, and **refs.** *15*, *18*, and *58* for rodents.

Acknowledgment

We thank Alma Cameron for her technical expertise in the studies on murine craniofacial development, much of which was supported by a grant from the National Institutes of Health (Bethesda, MD). We thank Bill Atchley (North Carolina State University, Raleigh, NC) for collaboration in that work. Ongoing work is supported by the Natural Sciences and Engineering Research Council of Canada (Grant A5056), NATO, and by the Killam Trust of Dalhousie University, Halifax, Nova Scotia, Canada.

References

1. Bolker, J. A. (1995) Model systems in developmental biology. *BioEssays* **17,** 451–455.
2. Hall, B. K. (1998) *Evolutionary Developmental Biology.* 2nd ed. Kluwer Academic Publishers, Dordrecht, The Netherlands.
3. Hamburger, V. and Hamilton, H. L. (1951) A series of normal stages in the development of the chick embryo. *J. Morphol.* **88,** 49–92. [Reprinted in (Hamilton, H. L., ed.), *Lillie's Development of the Chick: An Introduction to Embryology*, 3rd ed. (1965). Holt, Rinehart and Winston, New York.]
4. Theiler, K. (1972) *The House Mouse. Development and Normal Stages from Fertilization to 4 Weeks of Age.* Springer-Verlag, Berlin.
5. Miyake, T., Cameron, A. M., and Hall, B. K. (1996a) Detailed staging of inbred C57BL/6 mice between Theiler's [1972] stages 18 and 21 (11–13 days of gestation) based on craniofacial development. *J. Craniofac. Genet. Dev. Biol.* **16,** 1–31.
6. Miyake, T., Cameron, A. M., and Hall, B. K. (1996b) Stage-specific onset of condensation and matrix deposition for Meckel's and other first arch cartilages in inbred C57BL/6 mice. *J. Craniofac. Genet. Dev. Biol.* **16,** 32–47.
7. Miyake, T., Cameron, A. M., and Hall, B. K. (1997a) Stage-specific expression patterns of alkaline phosphatase during development of the first arch skeleton in inbred C57BL/6 mouse embryos. *J. Anat.* **190,** 239–260.
8. Miyake, T., Cameron, A. M., and Hall, B. K. (1997b) Variability and constancy of embryonic development in and among three inbred strains of mice. *Growth Dev. Aging* **61,** 141–155.
9. Romanoff, A. L. (1965) *The Avian Embryo. Structural and Functional Development.* Macmillan, New York.
10. Landauer, W. (1967) *The hatchability of chicken eggs as influenced by environment and temperature.* Monograph 1 (revised). 315 pp, University of Connecticut, Storrs Agricultural Station, Storrs, CT.
11. Landauer, W. (1973) *The hatchability of chicken eggs as influenced by environment and heredity (A supplement to Monograph Number One and Number One Revised).* Monograph 1 (revised). 54 pp, University of Connecticut, Storrs Agricultural Station, Storrs, CT.
12. Rugh, R. (1968) *The Mouse: Its Reproduction and Development.* Burgess, Minneapolis, MN.
13. Johnson, D. R. (1986) *The Genetics of the Skeleton. Animal Models of Skeletal Development.* Oxford University Press, Oxford, UK.
14. Kaufman, M. H. (1992) *The Atlas of Mouse Development.* Academic, London.
15. Bard, J. (ed.) (1994) *Embryos: Color Atlas of Development.* Wolfe, London.

16. Gilbert, S. F. and Raunio, A. M. (eds.) (1997) *Embryology: Constructing the Organism.* Sinauer Associates, Sunderland, MA.

17. Thorogood, P. (ed.) (1997) *Embryo, Genes and Birth Defects.* Wiley, Chichester, UK.

18. Hogan, B., Beddington, R., Constantini, F., and Lacy, E. (1994) *Manipulating the Mouse Embryo: A Laboratory Manual*, 2nd ed. Cold Spring Harbor Laboratory Press, Cold Spring Harbor, NY.

19. Rossant, J. and Pedersen, R. A. (eds.) (1986) *Experimental Approaches to Mammalian Embryonic Development.* Cambridge University Press, Cambridge, UK.

20. Bronner Fraser, M. (ed.) (1996) *Methods in Avian Embryology. Methods in Cell Biology,* Vol. 51. Academic, San Diego, CA.

21. Webb, J. F. and Noden, D. M. (1993) Ectodermal placodes: Contributions to the development of the vertebrate head. *Am. Zool.* **33,** 434–447.

22. Hall, B. K. (1999) *The Neural Crest in Development and Evolution*, Springer-Verlag, New York.

23 Saldivar, J. R., Krull, C. E., Krumlauf, R., et al. (1996) Rhombomere of origin determines autonomous versus environmentally regulated expression of *Hoxa3* in the avian embryo. *Development* **122,** 895–904.

24. Lumsden, A. and Krumlauf, R. (1996) Patterning the vertebrate neuraxis. *Science* **274,** 1109–1115.

25. Couly, G. F. and Le Douarin, N. M. (1990) Head morphogenesis in embryonic avian chimeras: Evidence for a segmental pattern in the ectoderm corresponding to the neuromeres. *Development* **108,** 543–558.

26. Northcutt, R. G. (1996) The origin of craniates—Neural crest, neurogenic placodes, and homeobox genes. *Israel J. Zool.* **42,** S273–S313.

27. Schorle, H., Meier, P., Buchert, M., et al. (1996) Transcription factor AP-2 essential for cranial closure and craniofacial development. *Nature* **381,** 235–238.

28 Duband, J. L., Monier, F., Delannet, M., and Newgreen, D. (1995) Epithelium-mesenchyme transition during neural crest development. *Acta Anat.* **154,** 63–78.

29. Noden, D. M. (1983) The embryonic origins of avian cephalic and cervical muscles and associated connective tissues. *Am. J. Anat.* **168,** 257–276.

30. Noden, D. M. (1987) Interactions between cephalic neural crest and mesodermal populations, in *Developmental and Evolutionary Aspects of the Neural Crest* (Maderson, P. F. A., ed.), Wiley, New York, pp. 89–110.

31. Köntges, G. and Lumsden, A. (1996) Rhombencephalic neural crest segmentation is preserved throughout craniofacial ontogeny. *Development* **122,** 3229–3242.

32. Noden, D. M. (1989) Embryonic origins and assembly of blood vessels. *Am. Rev. Respir. Dis.* **140,** 1097–1103.

33. Hall, B. K. (1988) Patterning of connective tissues in the head: Discussion report. *Development* **103(Suppl),** 171–174.

34. Hunt, P., Wilkinson, D., and Krumlauf, R. (1991) Patterning the vertebrate head: murine *Hox 2* genes mark distinct subpopulations of premigratory and migrating cranial neural crest. *Development* **112,** 43–50.

35. Hatano, M., Iitsuka, Y., Yamamoto, H., et al. (1997) *Ncx*, a *Hox11* related gene, is expressed in a variety of tissues derived from neural crest cells. *Anat. Embryol.* **195,** 419–425.

36. Serbedzija, G. N. and McMahon, A. P. (1997) Analysis of neural crest cell migration in *Splotch* mice using a neural crest-specific LacZ reporter. *Dev. Biol.* **185,** 139–147.

37. Smith, A., Robinson, V., Patel, K., and Wilkinson, D. G. (1997) The EphA4 and EphB1 receptor tyrosine kinases and ephrin-B2 ligand regulate targeted migration of branchial neural crest cells. *Curr. Biol.* **7,** 561–570.

38. Bevan, S. G., Southey, M. C., Armes, J. E., et al. (1996) Spatiotemporally exact cDNA libraries from quail embryos: A resource for studying neural crest development and neurocristopathies. *Genomics* **38,** 206–214.

39. Hall, B. K. and Miyake, T. (1992) The membranous skeleton: The role of cell condensations in vertebrate skeletogenesis. *Anat. Embryol.* **186,** 107–124.

40. Hall, B. K. and Miyake, T. (1995a) Divide, accumulate, differentiate: Cell condensation in skeletal development revisited. *Int. J. Dev. Biol.* **39,** 881–893.

41. Hall, B. K. and Miyake, T. (1995b) How do embryos measure time?, in *Evolutionary Change and Heterochrony* (McNamara, K. J., ed.), Wiley, New York, pp. 3–20.

42. Dunlop, L.-L. T. and Hall, B. K. (1995) Relationships between cellular condensation, preosteoblast formation and epithelial-mesenchymal interactions in initiation of osteogenesis. *Int. J. Dev. Biol.* **39,** 357–371.

43. Hall, B. K. (1982) How is mandibular growth controlled during development and evolution? *J. Craniofac. Genet. Dev. Biol.* **2,** 45–49.

44. Hall, B. K. (1988) Mechanisms of craniofacial development, in *Craniofacial Morphogenesis and Dysmorphogenesis* (Vig, K. W. and Burdi, A. R., eds.), The University of Michigan Press, Ann Arbor, pp. 1–21.

45. Eyal-Giladi, H. and Kochav, S. (1976) From cleavage to primitive streak formation: A complementary normal table and a new look at the first stages of the development of the chick. *Dev. Biol.* **49,** 321–337.

46. Schoenwolf, G. C. (ed.) (1986) *Scanning Electron Microscopy Studies of Embryogenesis*. Scanning Electron Microscopy, Inc., AMF O'Hare, IL.

47. Sulik, K. K. and Johnston, M. C. (1983) Sequence of developmental alterations following acute ethanol exposure in mice: Craniofacial features of the fetal alcohol syndrome. *Am. J. Anat.* **166,** 257–270.

48. Sulik, K. K. and Schoenwolf, G. C. (1985) Highlights of craniofacial morphogenesis in mammalian embryos, as revealed by scanning electron microscopy. *Scan. Electron Microsc.* **1985,** 1735–1752.

49. Hanken, J. and Wassersug, R. (1981) The visible skeleton. *Funct. Photogr.* **16(4),** 22–26, 44.

50. Klymkowsky, M. W. and Hanken, J. (1991) Whole-mount staining of *Xenopus* and other vertebrates. *Methods Cell Biol.* **36,** 419–441.

51. Presnell, J. K. and Schreibman, M. P. (1997) *Humason's Animal Tissue Techniques*, 5th ed., The Johns Hopkins University Press, Baltimore.

52. Erickson, C. A. (1993) Morphogenesis of the avian trunk neural crest—Use of morphological techniques in elucidating the process. *Microsc. Res. Tech.* **26,** 329–351.

53. Dunlop, L.-L. T. and Hall, B. K. (1995) Relationships between cellular condensation, preosteoblast formation and epithelial-mesenchymal interactions in initiation of osteogenesis. *Int. J. Dev. Biol.* **39,** 357–371.

54. Streicher, J., Weninger, W. J., and Müller, G. B. (1997) External marker-based automatic congruencing: A new method of 3D reconstruction from serial sections. *Anat. Rec.* **248,** 583–602.

55. Pearse, A. G. E. (1985) *Histochemistry—Theoretical and Applied*. Churchill-Livingstone, Secaucus, NJ.

56. Fang, J. and Hall, B. K. (1995) Differential expression of neural cell adhesion molecule (NCAM) during osteogenesis and secondary chondrogenesis in the embryonic chick. *Int. J. Dev. Biol.* **39,** 519–528.

57. Johnstone, A. and Thorpe, R. (1987) *Immunochemistry in Practice*, 2nd ed., Blackwell, Oxford, UK.

58. Wassarman, P. M. and DePamphilis, M. L. (eds.) (1993) *Guide to Techniques in Mouse Development. Methods in Enzymology Vol. 225*. Academic, San Diego, CA.

14

Examination of the Axial Skeleton of Fetal Rodents*

Michael G. Narotsky and John M. Rogers

1. Introduction

The axial skeleton represents one product of the metameric segregation of the mesoderm in the developing embryo. The mechanisms underlying this pattern formation remain poorly understood. Genetic alterations, either resulting from spontaneous mutation or as a result of xenobiotic exposure, may disrupt this patterning and lead to a variety of skeletal alterations. Chemical agents including valproic acid *(1)*, retinoic acid *(2)*, salicylate *(3)*, and acetazolamide *(4)* have been shown to cause supernumerary ribs in rodents. Fewer agents cause a reduction in the number of ribs or vertebrae. Agents or conditions in the latter category include boric acid *(5,6)*, arsenate *(7)*, methanol *(8)*, 2-chlorodeoxyadenosine *(9)*, and hyperthermia *(10)*. Effects on axial development have also been associated with changes in homeotic gene expression *(11–13)* as well as deletion of the *bmi-1* proto-oncogene *(14)*. Posteriorization of *Hoxa10* expression has been associated with lumbar ribs in mice following prenatal exposure to salicylate *(15)* as well as retinoic acid *(2)*. Careful characterization of the morphology of the axial skeleton is critical in identifying homeotic shifts and other changes produced by xenobiotics or altered gene expression. Here we present our methods for preparing and examining rodent skeletons, including anatomical landmarks and a brief discussion of methods for analyzing and presenting these data.

2. Materials

We use 18-cell polystyrene trays (Flambeau Products Corp., Roanoke Rapids, NC) to process rat and mouse fetuses for skeletal examination. We process one fetus per cell (and one litter per tray), thus maintaining the identity of each individual fetus. A flexible screen is used to keep the fetuses in their respective cells when draining solutions from the trays. These trays, however, are incompatible with acetone; thus, when specimens require acetone for dehydration and fat removal, we transfer the specimens to 30-mL polypropylene jars (Nalge Nunc, Rochester, NY, #2118-0001) that fit well into 12-cell trays. As an alternative to polystyrene trays, some laboratories process entire litters together in glass or polypropylene jars and use tags to identify individual fetuses.

---

*This document has been reviewed in accordance with U.S. Environmental Protection Agency policy and approved for publication. Mention of trade names or commercial products does not constitute endorsement or recommendation for use.

From: *Methods in Molecular Biology, Vol. 135: Developmental Biology Protocols, Vol. I*
Edited by: R. S. Tuan and C. W. Lo © Humana Press Inc., Totowa, NJ

The formulations for solutions used in preparing skeletal specimens are as follows.

1. Alcian blue solution (150 mg/L): 3 g alcian blue 8GX, 16 L 95% ethanol, 4 L glacial acetic acid.
2. 1% potassium hydroxide: 200 g potassium hydroxide (KOH), 20 L water.
3. Alizarin red S solution (25 mg/L): 500 mg alizarin red S, 20 L 1% potassium hydroxide.
4. EtOH:glycerin:benzyl alcohol (2:2:1): 2 parts 70% ethanol, 2 parts glycerin, 1 part benzyl alcohol.
5. EtOH:glycerin (1:1): 1 part 70% ethanol, 1 part glycerin.

In addition, as mentioned previously, acetone may be used for dehydration and fat removal. For safety, a fumehood should be used when working with acetone, the alcohols, and acetic acid. Also, ethanol solutions at concentrations greater than 70% are flammable; thus appropriate precautions (e.g., metal containers or flammable safety cabinets) should be used when fixing specimens in 95% ethanol as well as for storage of ethanol (*16*).

All of the solutions can be prepared in advance and stored at room temperature. However, we recommend discarding solutions in which alcian blue has precipitated. The alizarin red S solution is light sensitive; accordingly, this solution should be protected from light. The ethanol-glycerin solutions (with and without benzyl alcohol) may be reused provided they are not too dark. If they have drawn up too much stain, their destaining properties may be limited. Also, the benzyl alcohol solution may become too alkaline after several uses (*17*).

3. Method
3.1. Preparation of the Skeletal Specimen

Following cesarean section (gestation day 20 or 21 for rats, day 17 or 18 for mice), rodent fetuses are examined for gross external alterations, weighed, and euthanized. Specimens may be either single or double stained. In both staining procedures, alizarin red S is used to visualize bone; in double-stained specimens, alcian blue is also used to visualize cartilage. We generally prefer double staining, as the alcian blue facilitates visualization of important cartilaginous landmarks that would otherwise be difficult or impossible to examine. Several procedures for preparing single- (e.g., **refs.** *18–20*) and double-stained specimens (e.g., **refs.** *11* and *21–24*) have been published. Recipes that work well for us are as follows.

3.1.1. Procedure for Preparation of Single-Stained Rodent Fetuses

1. After the fetuses are removed from the uterus, weighed, examined for external alterations, and euthanized, make an abdominal incision in each fetus selected for skeletal examination. This enhances penetration of the fixative.
2. Place the selected fetuses from each litter in 70% EtOH. Use at least approx 10 mL solution for each gram of tissue.
3. If the sex of the fetus has not already been recorded, this may be done after fixation, but should be done prior to any further processing.
4. Remove the intrathoracic and intraabdominal viscera, the skin, and the subcutaneous fat between the scapulae and along the back of the neck. The eyes may also be removed. Be careful not to damage the skeleton. The fetuses may be returned to the 70% EtOH solution until further processing.

5. Transfer the specimens to glass jars, or other acetone-resistant containers, and add acetone; work in a fume hood. Allow specimens to remain in acetone for at least 1 d.

6. Place the specimens in staining solution containing alizarin red S (25 mg/L) and 1% KOH for approx 2 d for rats or 1 d for mice until they are sufficiently macerated. Because overexposure to KOH can damage the specimens, monitor them periodically so that they do not overmacerate.

7. If they are overstained, transfer the specimens to a solution of 2 parts glycerin, 2 parts 70% ethanol, and 1 part benzyl alcohol for 1–3 d. This step can be skipped if the soft tissue is only minimally overstained.

8. Transfer the specimens to a solution of 1 part of 70% EtOH and 1 part of 100% glycerin.

9. Examine the specimens in a glass dish. We recommend submerging the specimen in ethanol:glycerin to minimize glare.

10. When ready for final storage, transfer the specimens to glycerin (with a few crystals of thymol to prevent mold growth).

3.1.2. Procedure for Preparation of Double-Stained Fetal Rodent Skeletons

1. After the fetuses are removed from the uterus, weighed, sexed, examined for external alterations, and euthanized, eviscerate and completely remove the skin. Dipping the specimen in warm water (e.g., 30°C) for approx 30 s may facilitate removal of skin from the extremities. Remove the intrathoracic and intraabdominal viscera and the subcutaneous fat between the scapulae and along the back of the neck. The eyes may also be removed. Be careful not to damage the skeleton.

2. Place the fetuses in 95% EtOH. Use at least approx 10-mL solution for each gram of tissue.

3. Stain according to the following schedule.
   Day 1: *Afternoon*: Transfer the specimens to alcian blue solution. Leave overnight.
   Day 2: *Morning*: Transfer the specimens to fresh 95% EtOH for destaining/rinsing. *Afternoon*: Transfer to 1% KOH. Leave overnight.
   Day 3: *Morning*: Transfer to solution of alizarin red S (25 mg/L) in 1% KOH.
   Day 4: *Morning*: Transfer to solution of 1 part of 70% ethanol and 1 part glycerin. (If the specimens are overstained on the morning of day 4, they should be placed in a solution of 2 parts of 70% ethanol, 2 parts glycerin, and 1 part benzyl alcohol for 1 d. Then transfer the specimens to a 1:1 solution of glycerin and 70% ethanol.)

4. Examine the specimens in a glass dish. We recommend submerging the specimen in ethanol:glycerin to minimize glare during the examination.

5. When ready for final storage, transfer the specimens to glycerin (with a few crystals of thymol to prevent mold growth).

When preparing double-stained specimens, note that the overlying skin of the specimen must be removed to allow alcian blue to penetrate and stain the cartilage. Because the torso's skin is easily removed after fixation, evaluation of the *axial* skeleton will not be compromised if the euthanized specimen is placed directly into fixative (95% ethanol); the fixed specimen can be skinned days or weeks later. However, the skin is typically difficult to remove from the extremities of fixed specimens; thus, if cartilage formation in the digits is of interest, it is important that the specimen *not* be placed in fixative until the skin has been removed. In this case, we either skin the specimen immediately after euthanasia or, if a limited work force is available, we refrigerate the specimen or place it in 4% saline, and skin it later that same day.

Table 1
Key Morphological Criteria Used
in Assessing Normal Vertebral Regions of Full-Term Rodent Fetuses

Region, vertebrae	Number of vertebrae	Key features
Cervical		
C1 (atlas)	1	Wide transverse process, dorsolateral foramina
C2 (axis)	1	Ventrolateral foramina, spinous process
C3–C5	3	Cartilaginous lateral tubercula (CLT), ventrolateral foramina
C6	1	Tubercula anterior, CLT, ventrolateral foramina
C7	1	CLT, no peripheral foramina
Thoracic		
T1	1	First with rib and costal cartilage, cephalad diapophysis
T2	1	Prominent spinous process, cephalad diapophysis
T3–T10	8	High ossified arch, CLT
T11	1	Blunted ossified arch, reduced CLT
T12–T13	2	Virtually no CLT
Lumbar		
L1–L3	3	Reduced/absent pleurapophysis
L4–L6	3	Pleurapophysis projecting cephalad
Sacral		
S1–S2	2	CLT fused with other sacral vertebrae, pleurapophysis angled caudad
S3–S4	2	CLT fused with other sacral vertebrae, pleurapophysis projecting cephalad

3.2. Practical Anatomy of the Vertebrae

The normal rat skeleton has 7 cervical (C), 13 thoracic (T), 6 lumbar (L), 4 sacral (S), and 27–30 caudal vertebrae *(25)*. In the full-term fetus, only a few of these caudal vertebrae are ossified. Also, there are normally six sternebrae and, on each side, seven costal cartilages connecting the first (i.e., cephalad-most) seven ribs to the sternum. In the adult, the sacral vertebrae are fused and ossified to form the sacrum, the sternebrae are fused to form the sternum, and the costal cartilages are ossified (thus, technically, there are no longer any costal "cartilages").

The key criteria for distinguishing the vertebral regions are simply the presence of ribs in the T region and the cartilaginous fusion of the lateral tubercula in the S region. Within the C, T, L, and S regions, smaller subregions—or even individual vertebrae—can be differentiated by their morphological characteristics; these features are outlined in **Table 1**.

The cervical vertebrae can normally be differentiated into five subregions: C1, C2, C3–C5, C6, and C7 (**Fig. 1**). The atlas (C1) and axis (C2) are the two largest vertebrae and can readily be distinguished from the other vertebrae as well as from each other. C1 and C2 are also different from each other in that C1 has dorsolateral foramina and an anterior arch, whereas C2 has ventrolateral foramina and a middorsal spinous process. C3–C5 are essentially indistinguishable from one another, with all three of these vertebrae having ventrolateral foramina and cartilaginous lateral tubercula (CLT). In a nondisarticulated specimen, the CLT, per se, will not be readily visible; however, their presence can be discerned from the dark-textured bone-cartilage interface that is visible when viewing the specimen from the side. Although C6 has similar features to C3–C5, it is the only vertebra with tubercula anterior. The tuberculum anterior (TA), also called the ventral lamina *(26)* or Chassaignac's tubercle *(25)*, is a cartilaginous caudad projection from the ventral aspect of the vertebra (**Figs. 1–3**); it is an attachment point of the longus colli muscle. C7 is distinguishable from the other cervical vertebrae in that it has neither ventrolateral nor dorsolateral foramina. The five cervical subregions can generally be identified in an intact double-stained specimen, primarily by recognizing the unique features of C1, C2, and C6 (**Fig. 1**). It should be noted, however, that disarticulation of the cervical vertebrae may be necessary to adequately visualize the peripheral foramina *(8)* (**Fig. 3**).

Based on their morphology, the thoracic vertebrae can be differentiated into five subregions: T1, T2, T3–T10, T11, and T12–T13. However, T3–T10 may be further subdivided, because only the first seven thoracic vertebrae have costal cartilages that articulate with the sternum. The features distinguishing the thoracic subregions are more subtle than those demarcating the cervical subregions. T1 normally differs from C7 by the presence of a rib and by the lack of a CLT. Unlike the remaining thoracic vertebrae, T1 and T2 both have cephalad projecting diapophyses, but only T2 has a prominent spinous process (**Figs. 1** and **4**). The next eight vertebrae, T3–T10, have high ossified arches as well as CLT. There are gradual changes between adjacent vertebrae in this subregion; thus, T3 is clearly different from T10. However, as there is no clear landmark (other than articulation with the sternum) to morphologically distinguish them, we classify T3–T10 as one subregion. T11, sometimes referred to as the "transitional" vertebra, has characteristics of the T3–T10 region as well as the T12–T13 subregion (**Fig. 4**). Whereas T3–T10 have high arches and CLT, T11 has blunted arches and reduced CLT. In contrast, T12–T13 have relatively low arches and virtually no CLT. Although the identification of T11 is, in our experience, the most subjective in nature, it is a useful landmark for characterizing homeotic shifts.

The lumbar vertebrae can be differentiated into two subregions: L1–L3, and L4–L6. The former three vertebrae have reduced or absent pleurapophyses; whereas the latter three have cephalad-projecting pleurapophyses (**Figs. 5** and **6**).

The defining characteristic of the sacral region is the fusion of the CLT (which later ossify to form the sacrum). Similar to the lumbar region, the sacral vertebrae can be differentiated into two subregions: S1–S2, and S3–S4 (**Fig. 5**). The pleurapophyses of the former two vertebrae are angled caudad; whereas for the latter two, the pleurapophyses project cephalad. At the lumbosacral junction, two important criteria help identify S1 and distinguish it from the lumbar region. The first, and defining, feature is that the CLT is fused with that of the other sacral vertebrae; this may be readily visible only in double-stained fetuses. The second feature, visible in both double- and single-stained fetuses, is the direction that the pleurapophysis projects. For L6, the pleurapophysis projects cephalad, but for S1 it is angled caudad.

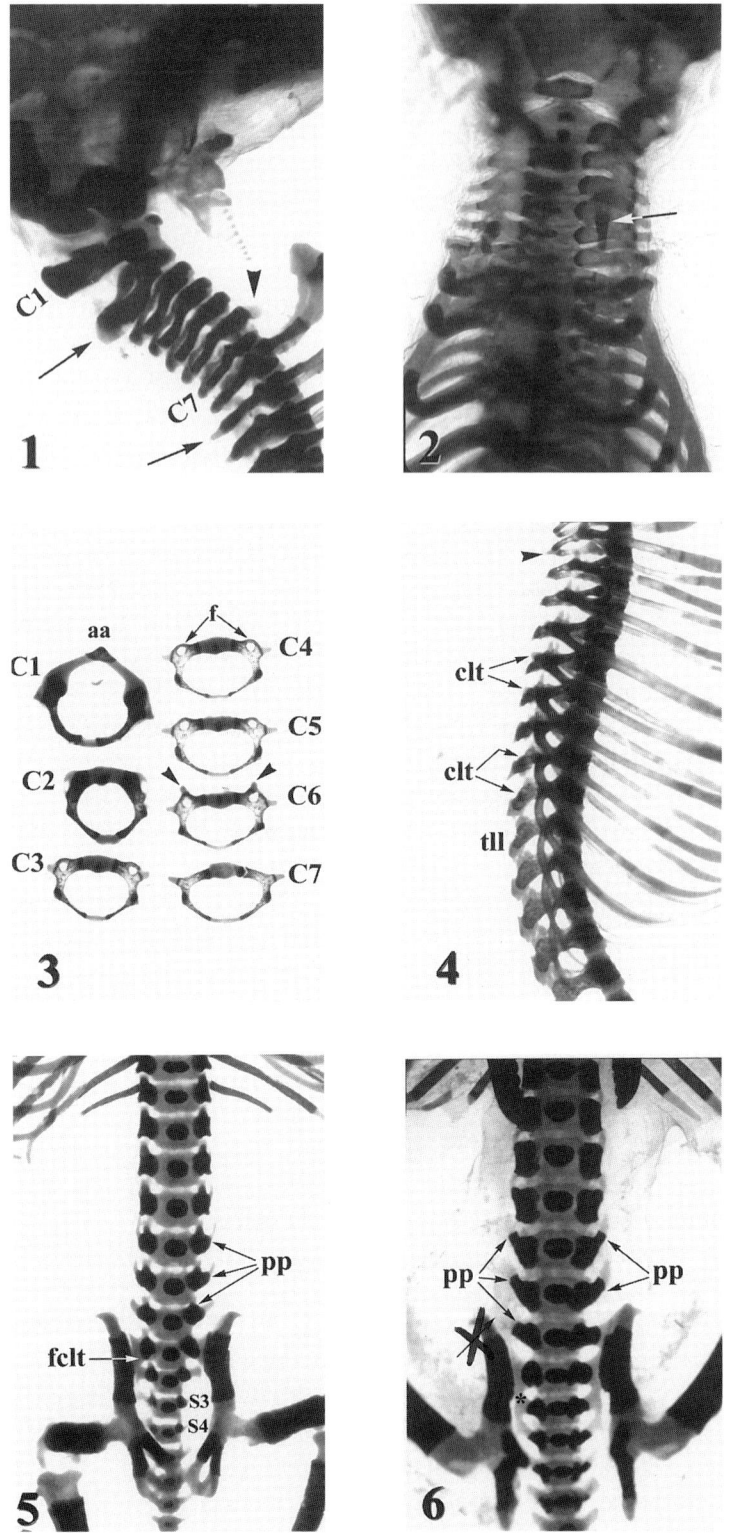

Figs. 1–6. (*See* color plate 3 appearing after p. 258.)

3.3. Boundaries

We make special mention of the landmarks and ambiguities involved at the boundaries of the different axial regions:

3.3.1. Craniocervical

The exoccipital and the atlas are readily distinguished, but note whether or not the two bones are fused. The presence of an extra vertebra, the proatlas, has been reported at the craniocervical border *(11)*.

3.3.2. Cervicothoracic

C7 normally has a well-developed CLT and lacks a rib; whereas T1 lacks a CLT and (by definition) bears a rib. By our working definition, T1 is the first vertebra bearing a rib with a prominent costal cartilage. For most intents and purposes, this definition can be interpreted as "bearing a rib that articulates with the sternum"; however, our working definition also includes fetuses with T1-costal-cartilage defects (e.g., interrupted, or attached to an adjacent rib rather than the sternum). Conversely, we describe vertebrae at the C-T border that have ribs without costal cartilage as cervical vertebrae bearing "cervical ribs."

Fig. 1. *(previous page)* Lateral view of the cervical vertebrae (C1–C7) and cervicothoracic border in a rat fetus on gestation day 21. Note the cartilaginous tuberculum anterior (arrowhead) on C6 and the dorsal spinous processes (arrows) of the axis (C2) and T2.

Fig. 2. *(previous page)* Ventral view of a gestation day 18 mouse fetus from a dam exposed to 5 g/kg methanol on gestation day 7. Note the anterior homeotic shift of the left tuberculum anterior (arrow), whereas the right tuberculum anterior is positioned normally on C6.

Fig. 3. *(previous page)* Cranial view of disarticulated cervical vertebrae (ventral aspect at top) of a mouse fetus on gestation day 18. Prominent distinguishing features include the anterior arch (aa) of the atlas (C1), tubercula anterior (arrowheads) on C6, and ventrolateral foramina (f) present on C2–C6, but absent on C1 and C7. C3–C5 are similar in appearance.

Fig. 4. *(previous page)* Lateral view of the thoracic region of a gestation day 20 rat fetus. The spinous process (arrowhead) can be seen on T2. Note the gradual shift in morphology of the vertebral arches in this region, from high arches with prominent cartilaginous lateral tubercula (clt) present on T3–T10 to blunted arches and reduced clt on T11 (t11) to low arches and minimal clt on T12–T13.

Fig. 5. *(previous page)* Ventral view of the lumbosacral region of a gestation day 20 rat fetus. Prominent pleurapophyses (pp) are present on L4–L6 but absent or greatly reduced on L1–L3. Fusion of the cartilaginous lateral tubercula (fclt) can be seen for S1 and S2. The pleurapophyses on S1 and S2 project caudad in contrast to the more cephalad projections on S3–S4.

Fig. 6. *(previous page)* Ventral view of the lumbar region of a gestation day 21 rat fetus. The lumbosacral junction is asymmetrical in that the caudal-most lumbar vertebra on the right side is sacral in character on the left side. The right side of the fetus has three lumbar vertebrae with cephalad-projecting prominent pleurapophyses (pp), whereas the left side has only two such vertebrae; these features are typical of L4–L6. The right side has three sacral vertebrae with fusion of the cartilaginous lateral tubercula (asterisk); the left side, however, has the normal complement of four.

3.3.3. Thoracolumbar

Except for the presence or absence of a rib, there are no salient features to define the T-L border. This can be problematic when there are more or less than the usual number of ribs or the last rib(s) are shorter than usual or both. Are these situations best described as a lumbar vertebrae bearing a rib or as an altered number of thoracic vertebrae? We have successfully used different approaches in different studies. In studies where only mild effects on the axial skeleton are observed, we described any supernumerary ribs at the T-L border as "lumbar ribs" (e.g., **ref. *1***). However, in studies where more extensive and varied effects were seen, we considered any rib (of any size) at the T-L border to be born by a thoracic vertebra *(6)*. In addition, we classified (longer than one half the length of the adjacent rib, less than one half the length of the adjacent rib, or focal) or measured (with a micrometer) the length of the last rib. This classification system is a modification of the rib-length classifications proposed by Kimmel and Wilson *(27)*.

3.3.4. Lumbosacral

Two important criteria help define the L-S border. The defining feature is that the CLT is fused with that of the other sacral vertebrae; however, this may be readily visible only in double-stained fetuses (**Figs. 5** and **6**). The second feature, visible in single- as well as double-stained fetuses, is the direction that the pleurapophysis projects. For L6, the pleurapophysis projects cephalad, but for S1 it is angled caudad. An alternative, albeit inferior, method is to define the L-S border using the iliac crests; any vertebra at least one half of which is cephalad to the iliac crests would be considered presacral.

3.3.5. Sacrocaudal

Except for the fused CLT, there is no clear feature to distinguish S4 from the first caudal vertebra. Unfortunately, the CLT around S4 sometimes stains very weakly with alcian blue, making it difficult to positively determine the number of sacral and caudal vertebrae.

3.4. Evaluation of the Axial Skeleton in Rodent Fetuses

Alterations in ossification, morphology, and number of the skeletal elements should be recorded for the entire, not just axial, skeleton. Focusing on the axial skeleton, however, we will address issues regarding examination of the sternum, ribs, and vertebrae. There is a wide variety of acceptable approaches for recording these findings. In most routine studies, elaborate notation systems may be unnecessary for adequate data recording. Recording the number of vertebrae in each subregion may also be dispensable; however, recognizing important morphological features along the spinal column may be important in assessing fetal developmental.

In our evaluations of several compounds with notable affects on the axial skeleton, we have developed a system for recording (and presenting) our findings. We record the number of vertebrae in each region (or subregion); because effects may be unilateral (e.g., **Figs. 2** and **6**), we record our findings for each side independently. We also record sternebral patterns (fused, bipartite, asymmetric), the number of intact costal cartilages articulating with the sternum (normally seven on each side), and aberrant patterns of attachment to the sternum.

In addition to counting the vertebrae, we also routinely count the ribs, as the numbers of thoracic vertebrae and ribs may differ. We record the number of ribs at the proximal (vertebral) end; rib counts at the proximal and distal ends may also differ because of a variety of rib abnormalities (e.g., fused, bifurcated). When examining the ribs, we not only record the type of abnormality (e.g., fused, bifurcated, absent, short, detached, discontinuous) *(28)*, we also, when possible, identify the affected rib(s). This information may be important, as the relative location of dysmorphologies may reflect the developmental stage of insult to the conceptus *(6)*.

3.4.1. Ambiguities

The boundaries of the different regions may sometimes be ambiguous, even in control specimens. For example, the rib field may be extended craniad or caudad resulting in a rudimentary (or larger) rib at the vertebra normally designated as C7 or L1. Generally, we describe such findings as "cervical ribs" or "lumbar ribs"; however, more severe effects may warrant different terminologies to adequately describe a toxicant's effects. To illustrate, in separate studies, we have demonstrated marked effects on the cervical/thoracic vertebrae in rats exposed to boric acid *(6,29)* and in mice exposed to methanol *(8)*. In the boric acid study, T1 was defined as the most cephalad vertebra bearing a rib with a prominent costal cartilage. In contrast, in the methanol study, seventh-vertebra ribs that were attached to the sternum and/or fused to the second rib were tabulated as ribs on C7. Although mutually exclusive descriptors were used in these two studies, each system was appropriate for the study in which it was applied.

A single vertebra may have both cervical and thoracic characteristics. For example, a vertebra at the C-T junction may bear a rib with a costal cartilage articulating with the sternum (thus meeting our criteria to be designated thoracic), but be more similar to C7 in that it lacks a CLT. Sometimes the extent of the CLT is intermediate between the typical C and T vertebra, thus further complicating the examiner's description (and interpretation) of the specimen.

Severe and frequent dysmorphologies (e.g., fusions, duplications, hemivertebrae) may sometimes render vertebrae unidentifiable. Nonetheless, as we mentioned, the relative location of dysmorphologies of the ribs or vertebrae may reflect the developmental stage of insult to the conceptus *(6)*.

3.4.2. Statistics

In typical developmental toxicity studies, it is appropriate to use the litter, rather than the fetus, as the basic unit of analysis *(30)*; however, in some gene knockout experiments, a fetus-based analysis may be appropriate. Generally, we use analyses of variance on the litter incidences (i.e., percentage affected per litter) of findings such as lumbar ribs, cervical ribs, <27 presacral vertebrae, >27 presacral vertebrae, <14 intact sternocostal connections, <7 cervical vertebrae, etc.

In addition to knockout experiments, some analyses of vertebral-count distributions (e.g., 7-13-6 cervical-thoracic-lumbar vertebrae) may not be conducive to litter-based statistical analyses. However, depending on one's experimental objectives, vertebral-count distributions can be calculated for each side independently and presented graphically. With this approach, one can concisely present a great deal of information in a manner that facilitates interpretation of the data. For example, Narotsky et al. *(6)* used bar-graph presentations (**Figs. 7** and **8**) to show the percentage of fetuses with each

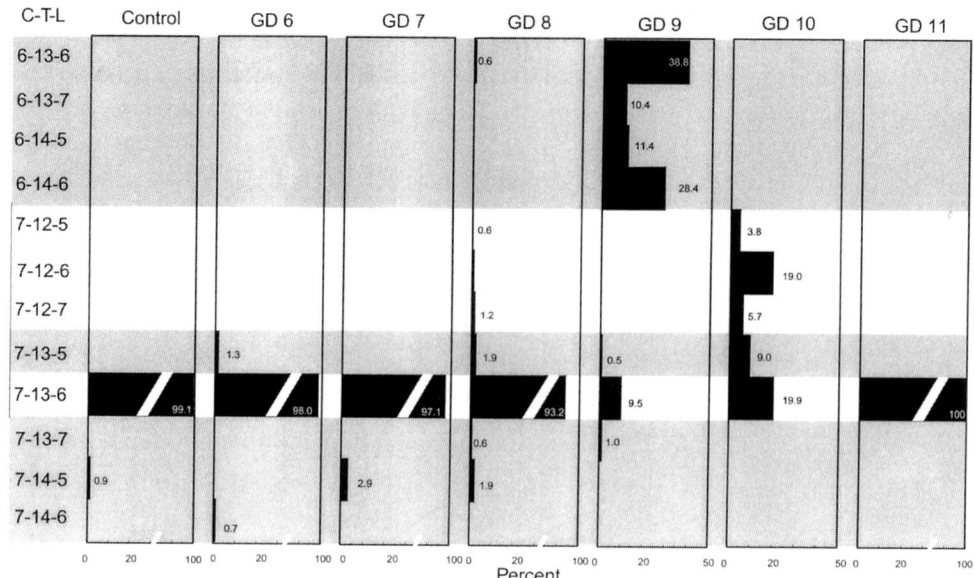

Fig. 7. Graphical presentation of incidences of fetuses with different counts of cervical (C), thoracic (T), and lumbar (L) vertebrae following single-day treatment with 500 mg boric acid/kg b.i.d. Data are for the right side of the fetus only. Although the right and left sides of any given fetus were not necessarily symmetrical, the left-side profiles were very similar to the right-side profiles shown.

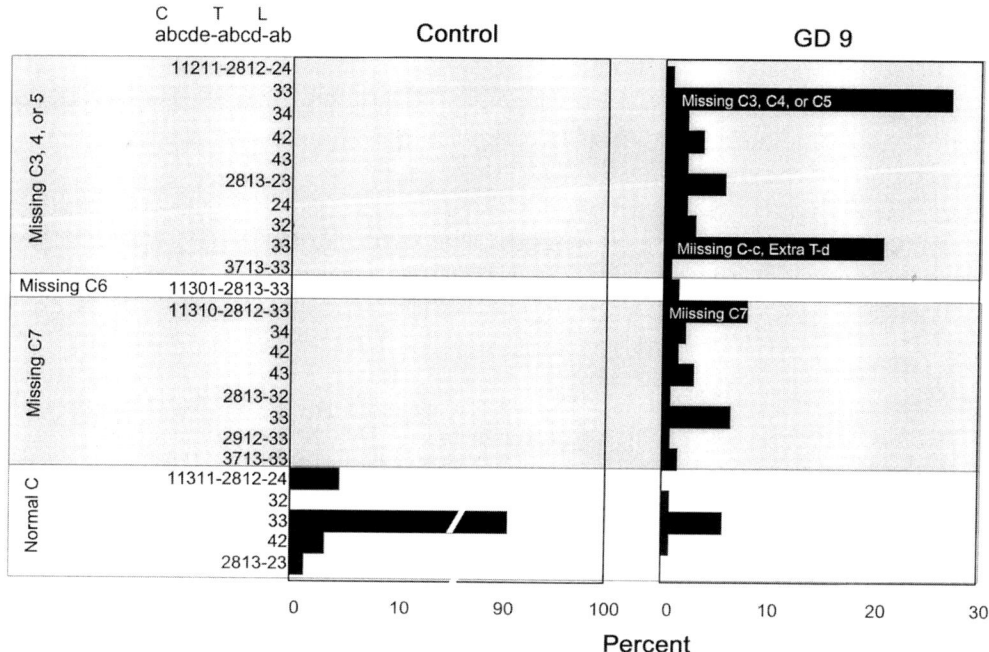

Fig. 8. Graphical presentation of incidences of fetuses with different counts for each vertebral (sub)region (right side) for controls and fetuses exposed to boric acid on gestation day 9. The cervical subregions a–e correspond to C1, C2, C3–C5, C6, and C7; the thoracic subregions a–d correspond to T1–T2, T3–T10, T11, and T12–T13; and lumbar subregions a–b correspond to L1–L3 and L4–L6, respectively.

unique vertebral configuration. As shown in **Fig. 7**, virtually all control fetuses had the normal 7-13-6 profile, but fetuses exposed to boric acid on gestation day 9 or 10 had reduced numbers of cervical or lumbar vertebrae, respectively. Furthermore, although day-9-exposed fetuses had an increased number of thoracic vertebrae (i.e., supernumerary ribs), it is evident in **Fig. 7** that all of the affected fetuses had only six cervical vertebrae.

4. Notes

We have focused on the use of full-term fetuses for the preparation of skeletal specimens; however, the procedures described here also work well for neonatal rats and mice. Earlier developmental stages may also be assessed by these methods, but the alcian blue staining becomes more critical as one processes earlier, less-ossified stages. Selby *(17)* has described a procedure for preparing single-stained skeletons of adult mice.

Although we have treated rats and mice as being identical in their anatomy, it is worth noting that some of the features that are fairly obvious in the rat (e.g., spinous process on T2, differences in pleurapophyses in lumbar subregions) are less so in the mouse, whereas the anterior arch of the atlas and the TA on C6 are more readily observed in mice.

Acknowledgments

The authors thank Bonnie Hamby, Deborah Best, Ellen Rogers, and Lynn Connelly for their expertise in preparing and examining skeletal specimens.

References

1. Narotsky, M. G., Francis, E. Z., and Kavlock, R. J. (1994) Developmental toxicity and structure-activity relationships of aliphatic acids, including dose-response assessment of valproic acid in mice and rats. *Fundam. Appl. Toxicol.* **22,** 251–265.
2. Kessel, M. (1992) Respecification of vertebral identities by retinoic acid. *Development* **115,** 487–501.
3. Wickramaratne, G. A. de S. (1988) The postnatal fate of supernumerary ribs in rat teratogenicity studies. *J. Appl. Toxicol.* **8,** 91–94.
4. Beck, S. L. (1983) Assessment of adult skeletons to detect prenatal exposure to acetazolamide in mice. *Teratology* **28,** 45–66.
5. Heindel, J. J., Price, C. J., and Schwetz, B. A. (1994) The developmental toxicity of boric acid in mice, rats, and rabbits. *Environ. Health Perspect.* **102(Suppl 7),** 107–112.
6. Narotsky, M. G., Schmid, J. E., Andrews, J. E., and Kavlock, R. J. (1998) Effects of boric acid on axial skeletal development in rats. *Biol. Trace Element Res.* **66,** 373–394.
7. Beaudoin, A. R. (1974) Teratogenicity of sodium arsenate in rats. *Teratology* **10,** 153–158.
8. Connelly, L. E. and Rogers, J. M. (1997) Methanol causes posteriorization of cervical vertebrae in mice. *Teratology* **55,** 138–144.
9. Narotsky, M. G., Best, D. S., Guidici, D. L., Hamby, B. T., Knudsen, T. B., Kavlock, R. J., et al. (1998) 2-Chlorodeoxyadenosine-induced lumbar hernia in rats. *Teratology* **57,** 245 [abstract].
10. Kimmel, C. A., Cuff, J. M., Kimmel, G. L., Heredia, D. J., Tudor, N., Silverman, P. M., et al. (1993) Skeletal development following heat exposure in the rat. *Teratology* **47,** 229–242.
11. Kessel, M., Balling, R., and Gruss, P. (1990) Variations of cervical vertebrae after expression of a *Hox-1.1* transgene in mice. *Cell* **61,** 301–308.

12. Small, K. M. and Potter, S. S. (1993) Homeotic transformations and limb defects in *Hox A11* mutant mice. *Genes Dev.* **7**, 2318–2328.

13. Charité, J. W., de Graaff, W., Shen, S., and Deschamps, J. (1994) Ectopic expression of *Hoxb-8* causes duplication of the ZPA in the forelimb and homeotic transformation of axial structures. *Cell* **78**, 589–601.

14. van der Lugt, N. M., Domen, J., Linders, K., van Roon, M., Robanus-Maandag, E., te Riele, H., et al. (1994) Posterior transformation, neurological abnormalities, and severe hemato-poietic defects in mice with a targeted deletion of the *bmi-1* proto-oncogene. *Genes Dev.* **8**, 757–769.

15. Daston, G. P. and Overmann, G. J. (1996) Lumbar ribs associated with posteriorization of Hox a10 expression in salicylate-treated mouse embryos. *Teratology* **53**, 85 [abstract].

16. National Fire Protection Association (1996) *NFPA 30: Flammable & Combustible Liquids Code*, Quincy, MA.

17. Selby, P. B. (1987) A rapid method for preparing high quality alizarin stained skeletons of adult mice. *Stain Technol.* **62**, 143–146.

18. Dawson, A. B. (1926) A note on the staining of the skeleton of cleared specimens with alizarin red S. *Stain Technol.* **1**, 123–124.

19. Kawamura, S., Hirohashi, A., Kato, T., and Yasuda, M. (1990) Bone-staining technique for fetal rat specimens without skinning and removing adipose tissue. *Cong. Anom.* **30**, 93–95.

20. Staples, R. E. and Schnell, V. L. (1964) Refinements in rapid clearing technic on the KOH-alizarin red S method for fetal bone. *Stain Technol.* **39**, 61–63.

21. Inouye, M. (1976) Differential staining of cartilage and bone in fetal mouse skeleton by alcian blue and alizarin red S. *Cong. Anom.* **16**, 171–173.

22. Kimmel, C. A. and Trammell, C. (1981) A rapid procedure for routine double staining of cartilage and bone in fetal and adult animals. *Stain Technol.* **56**, 271–273.

23. Whitaker, J. and Dix, K. M. (1979) Double staining technique for rat foetus skeletons in teratological studies. *Lab. Anim.* **13**, 309–310.

24. Miller, D. M. and Tarpley, J. (1996) An automated double staining procedure for bone and cartilage. *Biotech. Histochem.* **71**, 79–83.

25. Greene, E. C. (1949) Gross anatomy, in *The Rat in Laboratory Investigation* (Farris, E. J. and Griffith, J. Q., Jr., eds.), Hafner, New York, pp. 24–50.

26. Yasuda, M. and Tsunetsugu, Y. (1996) *Color Atlas of Fetal Skeleton of the Mouse, Rat and Rabbit*. Ace Art Co., Ltd., Osaka, Japan, pp. 10, 26.

27. Kimmel, C. A. and Wilson, J. G. (1973) Skeletal deviations in rats: Malformations or variations. *Teratology* **8**, 309–316.

28. Wise, L. D., Beck, S. L., Beltrame, D., Beyer, B. K., Chahoud, I., Clark, R. L., et al. (1997) Terminology of developmental abnormalities in common laboratory mammals (version 1). *Teratology* **55**, 249–292.

29. Narotsky, M. G., Hamby, B. T., Best, D. S., and Kavlock, R. J. (1996) Effects of single-day boric acid treatment on axial skeletal development in rats. *Teratology* **53**, 101 [abstract].

30. Haseman, J. K. and Kupper, L. L. (1979) Analysis of dichotomous response data from certain toxicological experiments. *Biometrics* **35**, 281–293.

15

Cardiac Morphogenesis and Dysmorphogenesis

An Immunohistochemical Approach

B. Rush Waller, III and Andy Wessels

1. Introduction
1.1. Cardiovascular Anomalies

   Cardiovascular anomalies are the most common birth defects in man, accounting for 0.5–1% of live births. They are typically classified according to the affected segment of the heart. The complexity of an anomaly, its relationship to other cardiac structures, and its physiologic consequences, determine its clinical significance. Ventricular and atrial septal defects are the two anomalies most commonly seen and can produce a spectrum of problems, including congestive heart failure and pulmonary hypertension *(1)*. Abnormal atrioventricular valves, such as found in tricuspid atresia or congenital mitral stenosis, are associated with underdevelopment of either the right or left ventricle, respectively, and have more severe clinical implications. Maldevelopment of either the aorta or the pulmonary artery leads to conditions such as tetralogy of Fallot, double-outlet right ventricle, and transposition of the great arteries. All of these congenital anomalies require surgical intervention for long-term survival. The events involved in normal cardiac development are described in Chapter 5 in Vol. II in this series. This chapter addresses a method that enhances the study of cardiac dysmorphogenesis.

1.2. The Mouse

   The mouse as a model for human disease has become more important because, in part, of the availability of genetically modified mouse mutants *(2)*. Targeted and insertional mutations through transgenic manipulation have resulted in several developmental defects of the cardiovascular system *(3)*. **Table 1** lists several murine genetic mutations that produce cardiac malformations homologous to congenital defects seen in humans. Several trisomic mice models producing cardiac anomalies are listed as well.

   Ideally, one would like to examine the development of these cardiovascular anomalies by comparing the morphological features of the developing abnormal heart with those of hearts of carefully staged normal embryos. The ability to discern between tissues such as ventricular myocardium, atrial myocardium, and endocardial cushion

From: *Methods in Molecular Biology, Vol. 135: Developmental Biology Protocols, Vol. I*
Edited by: R. S. Tuan and C. W. Lo © Humana Press Inc., Totowa, NJ

Table 1
Murine Cardiovascular Anomalies

Name	Genetic mutation	Phenotype
Spontaneous or induced mutants		
Splotch	Pax3	PTA *(4)*
Patch	α-subunit PDGF-receptor	ASD, VSD, DORV, tricuspid atresia *(5)*
inv/inv		Situs inversus, L-loop, pulmonary atresia *(6)*
iv/iv		L-loop, atrial isomerism, DORV, TGA, tetralogy of Fallot, situs inversus
Knockout mice	Sox 4	PTA *(7)*
	Connexin 43	Abnormal looping *(8)*
	Endothelin 1	VSD, DORV *(9)*
	Neurofibromin 1	VSD, PTA *(10)*
Trisomic mice	Chromosome 10	VSD *(11,12)*
	Chromosome 12	VSD
	Chromosome 13	Pulmonary stenosis, VSD, DORV
	Chromosome 14	VSD, DORV, TGA
	Chromosome 16	VSD, DORV, PTA *(11,12)*
	Chromosome 19	VSD, DORV *(13)*

Abbreviations: VSD, ventricular septal defect; ASD, atrial septal defect; PTA, persistent truncus arteriosus; DORV, double outlet right ventricle; TGA, transposition of the great arteries.

tissue is important in tracing the developmental fates of each region. The subpopulations of myocardial cells are characterized by specific phenotypes that may transform throughout development. These phenotypes are determined by the expression of a distinct set of genes that lead to the existence of a panel of corresponding mRNAs and proteins. On sections, mRNAs can be detected (localized) using *in situ* hybridization. Proteins can be identified using techniques such as immunohistochemistry.

1.3. Immunohistochemistry

Immunohistochemistry is a histological technique that allows the unambiguous detection of specific proteins, thereby locating the subpopulations of cells that are characterized by the expression of these "tissue-specific" markers in serial sections of embryonic hearts at different stages of development *(14–16)*.

This chapter describes a method of processing, embedding, and staining a series of embryos for immunohistochemical evaluation. The specific antibodies that are used depend on the purpose of the experiment, the tissue investigated, and the developmental process studied.

2. Materials

1. Phosphate-buffered saline (PBS), pH 7.4; yields 3 L of 10X solution: 500 mL of 0.5 M Na_2HPO_4 (71 g/L); add 100 mL of 0.5 M NaH_2PO_4 (60 g/L) while adjusting pH to 7.4. Dilute with 2400 mL of distilled water. Add 262.98 g of NaCl (1.5 M/L). Store at room temperature.

2. Amsterdam fixative: 350 mL Methanol, 350 mL acetone, 50 mL acetic acid, 250 mL distilled H$_2$O. Store at room temperature.

3. Ethanol solutions: 50, 70, 80, 95, 100% in distilled water.

4. Molecular sieves (cat. no. M-2635; Sigma Chemical Co., St. Louis, MO).

5. Paraffin embedding medium: Polyfin (Electron Microscopy Sciences, Ft. Washington, PA).

6. Xylene.

7. Toluene.

8. 3% Hydrogen peroxide in PBS: Mix 25 mL 30% hydrogen peroxide (w/w) solution (cat. no. H-1009; Sigma) in 225 mL 1X PBS.

9. 1% PBS-BSA: Add 1 g bovine serum albumin (BSA) fraction V (cat. no. A-4503; Sigma) to 100 mL 1X PBS. Store at 4°C.

10. 2.5% Rabbit serum in 1% PBS-BSA: Add 100 µL rabbit serum (cat. no. R-9133; Sigma) to 4 mL 1% PBS-BSA.

11. TENG-T (5X) solution, pH 8.0: 5 mL of 1 *M* TRIZMA base (cat. no. T-6791; Sigma), 5 mL of 0.5 *M* EDTA (cat. no E-4884; Sigma), 4.38 g NaCl, 1.25 g gelatin, 250 µL Tween-20 (cat. no. P-7949; Sigma).

12. Antibodies (used in the studies presented in this chapter):
    a. Monoclonal antibodies:
       i. MF20 (hybridoma bank, D. Bader, Vanderbilt University, Nashville, TN).
       ii. Anti-α smooth muscle actin (clone 1A4, cat. no. A-2547; Sigma), store at –20°C.
    b. Secondary antibody: Rabbit antimouse peroxidase (RAM-Po; Sigma).

13. DAB: Immunopure® Metal Enhanced DAB Substrate Kit (Pierce, Rockford, IL). Store at or below –20°C; do not bring solution to room temperature. Includes stable peroxide buffer, to be stored at 4°C. Contents: Methanol 70%, 3,3' diaminobenzidine tetra-hydrochloride 0.5%, cobalt chloride 1%. Carcinogenic, wear gloves.

14. Mounting medium: Cytoseal™ 60 (Stephens Scientific, Riverdale, NJ), contains toluene (CAS-108-88-3, acrylic resin).

15. Mounting slides: Silane-Prep™ slides, (3 in. × 1 in.) coated with aminoalkylsilane, precleaned frosted ends (Sigma), store at 18–26˚C.

16. Microscope cover glass, 22 × 60 mm (e.g., Fisher*brand*®, Fisher Scientific, Pittsburgh, PA).

17. Embedding mold dish.

18. Plastic embedding ring (Histoprep, Fisher).

19. Embedding mold release spray (Histoprep, Fisher).

20. Spatula (e.g., Fisher*brand* Spoonula, Fisher).

21. Orbital shaker (e.g., The Belly Dancer, Stovall Life Science Inc., Greensboro, NC).

22. Forceps.

23. Humidification chamber (e.g., a 10 in. × 10 in. × 1 in. covered tray; line bottom with moist paper towels covered by thin plastic mesh).

24. Microtome (e.g., Leitz, Germany).

25. Microtome blades, disposable (Reicher-Jung Stainless Steel Blades, low profile, Fisher).

26. Polyethylene disposable transfer pipets (e.g., Fisher*brand*®, Fisher).

27. Glass scintillation vials, 20 mL (VWRbrand, cat. no. 66022-081; VWR Scientific Products, West Chester, PA).

28. Paraffin oven.

29. Vacuum oven (NAPCO®, Precision Scientific, Chicago, IL) with vacuum pump (Sargent-Welch DirecTorr®, Sargent-Welch Scientific Co., Skokie, IL).

30. Adjustable pipetters (e.g., Pipetman, Rainin, Woburn, MA) with disposable plastic pipet tips (e.g., Fisher*brand*®, Fisher).

31. Embedding station (e.g., Shandon Lipshaw™, Pittsburgh, PA).

3. Methods

3.1. Embryo Fixation and Preparation

Male and female mice are placed together in cages. The females are examined each morning for a vaginal plug. When a plug is seen, that morning is considered day 0.5.

On the desired embryonic day, the females are sacrificed using a CO_2 chamber followed by cervical dislocation. The abdominal surface is cleaned with 70% ethanol. Using sharp scissors, the abdomen is opened in the midline, and the uterine horns are removed, explanting the conceptuses into Petri dishes containing PBS. Under a binocular dissecting microscope, the extraembryonic membranes are removed. The thorax of each specimen is left intact. In larger specimens, approx 13 embryonic days or older, the limbs and the head may be removed both to make sectioning easier and to improve penetration of the fixative into the tissue. The yolk sacs or livers may be preserved for karyotyping.

The embryos are staged according to Theiler *(17)* using external features and crown–rump length and then placed into individually labeled glass vials containing fixative. The nature of the fixative depends on the antibodies that will be used for immunohistochemistry. Each specimen is given an individual reference number, written on labeling tape affixed to the vial and protected with clear cellophane tape to prevent "smearing" of writing by organic solvents during the processing stages (*see* **Note 1**).

The embryos are fixed in Amsterdam fixative overnight, followed by dehydration in a graded series of ethanol (70, 80, 90, 95, 100, 100% [containing molecular sieves]; 1 h per step) (*see* **Note 2**). If necessary, specimens can be stored in 70% ethanol before processing through the entire series (*see* **Note 3**). After dehydration, the tissues are "cleared" in toluene for a total of 2 h. For paraffin penetration, an equal volume of paraffin is added to the toluene, and each vial is placed in a paraffin oven at 60°C for at least 1 h (The specimens can be kept overnight at this step.) The toluene/paraffin is then replaced with paraffin only and placed back in the oven for 1 h. After a final paraffin change, the vials, without caps, are placed in a vacuum oven at –18 atm for 1 h to remove any air bubbles that could eventually impair sectioning.

3.2. Embedding

Metal embedding molds are sprayed with mold release spray and preheated in the oven at 60°C. The molds are then filled with melted paraffin and kept on the hot plate of the embedding station. One at a time they are placed on the cold plate to allow the deepest portion of the mold to begin to harden; they are then removed from the cold plate to a flat surface at room temperature. Heat lamps are placed immediately above the mold to keep the upper portion of the paraffin in its liquid state until the embryo can be properly positioned. The vials containing embryos are removed from the oven and placed on the hot plate. The embryos are then individually transferred from their vials to the molds. The embryos are hard yet very fragile at this point and must be handled carefully. All tools used in this process should be kept hot. For transferring embryos, we recommend the use of plastic transfer pipets.

Orientation of the embryo is critical because many cardiac defects constitute anomalies of right–left sidedness (**Fig. 1**). Standard orientation planes for sections include sagittal, transverse, and frontal. The deepest aspect of the mold will be the front of the paraffin block once it is placed onto the microtome and will therefore be the first aspect of the embedded specimen to be sectioned.

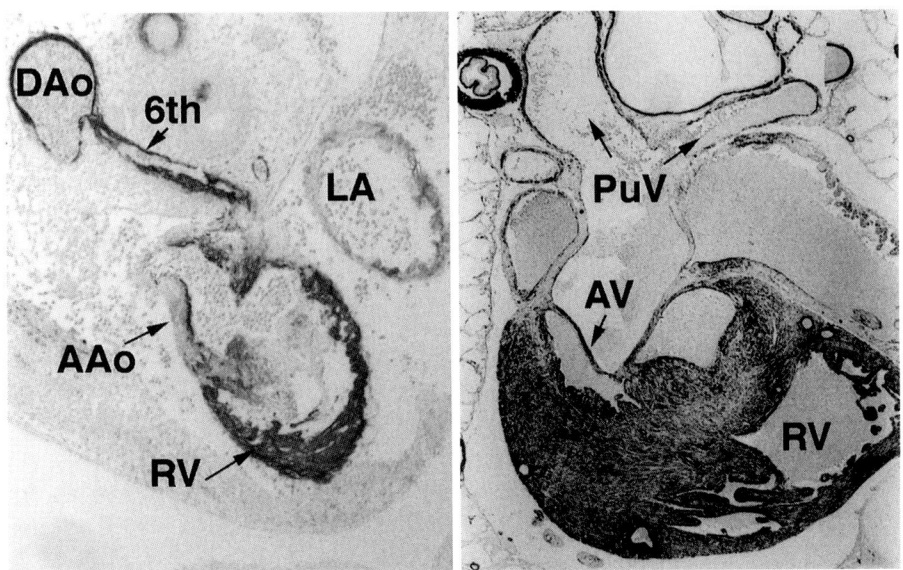

Fig. 1. This figure shows two examples of mouse models with congenital heart malformations. The left panel shows a section from a trisomy 16 heart at 13 embryo day (ED). It shows a hypoplastic right sixth-sided aortic arch connecting with a dominant right dorsal aorta. The right panel shows a section of an *inv/inv* embryo with complete situs inversus in which the morphological right ventricle is located at the left side of the embryo, and the pulmonary veins enter the morphological left atrium situated at the right side of the embryo. The left atrium connects to the morphological left ventricle via the mitral atrioventricular valve. AV = atrioventricular valve, 6th = 6th aortic arch, LA = left atrium, PuV = pulmonary vein, RV = morphological right ventricle, DAo = (right sided) dorsal aorta, AAo = ascending aorta.

3.2.1. Sagittal Orientation

The right side of the embryo should be facing down into the deepest part of the mold. The caudal portion should be directed toward the label. The right side will be sectioned first (**Fig. 2A**).

3.2.2. Transverse Orientation

The cephalic portion of the embryo should be placed into the deepest portion of the mold with the dorsum of the embryo facing away from the label. The head will be sectioned first (**Fig. 2B**).

3.2.3. Frontal Orientation

The ventral aspect of the embryo should be facing down into the deepest part of the mold. The caudal portion should be directed toward the label. The front of the embryo will be sectioned first (**Fig. 2C**).

Once the embedded specimen is in the desired orientation, a labeled plastic ring holder is fitted onto the metal mold (**Fig. 2D**). Melted paraffin is poured slowly into the mold until it reaches the top of the ring holder. The paraffin is allowed to harden at room temperature, and then the block is gently removed from the mold and may be stored at room temperature (*see* **Note 5**).

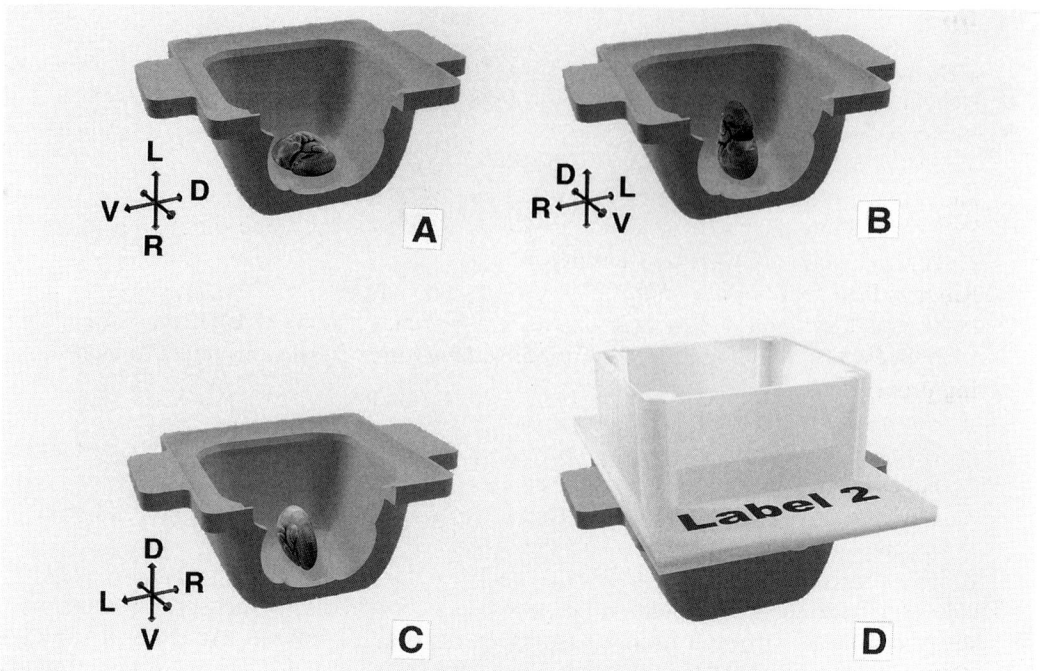

Fig. 2. This figure shows, when following the instructions in the text, how an embryo should be embedded for standard sectioning in a sagittal (**A**), transverse (**B**), and frontal (**C**) plane, using metal-embedding molds and plastic-embedding rings. Important note: in panel **B** the axis for left and right is incorrectly labeled, R should be L, and L should be R. L = left, R = right, D = dorsal, V = ventral.

3.3. Sectioning and Mounting

The specimen block is fixed onto the microtome with the label pointing to the right. The block is then trimmed down on the edges to reduce excess paraffin; the amount of excess paraffin remaining around the specimen determines the number of sections that can be placed on each glass slide. One top corner should be notched to aid in maintaining orientation once the sections have been shaved from the block.

The embedded specimen is then cut into 5 μm-thick sections with the sections coming off of the microtome in ribbons of paraffin. The ribbons can be placed on construction paper in trays and numbered according to depth of tissue by microns (*see* **Note 6**).

Glass slides are prepared by marking a rectangular area with a PAP pen and pipeting previously boiled distilled water into the middle of the marked area. From the ribboned sections, every tenth section (every 50 μm) can be mounted for H&E staining as a guide series. (For very small embryos, the use of a guide series is not recommended.) Selected sections for immunohistochemistry are mounted onto Silane-prep glass slides. The frosted label end of the glass slide is to the right, and the sections are mounted from left to right. The slides are then placed on a slide warmer at 45°C, which allows the sections to "stretch." Once the specimen and paraffin are smooth, the water is gently aspirated off the slides. The slides are then placed in racks in an oven at 37°C overnight for drying. After drying, they are stored in slide boxes at 4°C.

3.4. Immunohistochemistry

1. Deparaffinate the selected slides in xylene (2 X 5–7.5 min).
2. Rehydrate sections by washing for 2 min per step in 100, 95, 80, 70, and 50% ethanol.
3. Rinse in distilled water for 1 min.
4. Rinse in PBS (1X) for a minimum of 1 min.
5. An optional pronase pretreatment may be used at this step (*see* **Note 4**).
6. Pretreat sections in 3% hydrogen peroxide in PBS for 30 min on the orbital shaker to reduce endogenous peroxidase activity.
7. Rinse in PBS for at least 1 min.
8. Individual specimens are then incubated at room temperature with TENG-T for 15 min to reduce nonspecific binding of antibodies. *Incubation* in all steps refers to the following process:
   a. Place slides on a flat and dark surface.
   b. Place 5–10 μL of antibody or staining solution on each section (make sure that the edges of the sections are also covered by solution).
   c. Place incubated slides in humidified incubation chamber (to prevent sections from drying out).
   d. Place incubation chamber in a quiet corner of the lab (to prevent solutions on neighboring sections from running through).
9. Rinse in PBS for at least 1 min.
10. The specimens are then incubated overnight at room temperature with the primary monoclonal antibodies diluted in 2.5% rabbit serum with 1% PBS-BSA.
11. Wash slides in PBS 3X for 5 min each on orbital shaker.
12. Incubate with secondary antibody RAM-Po (1:400 dilution in 2.5% rabbit serum with 1% PBS-BSA) for 2 h at room temperature.
13. Wash slides 3X for 5 min each in PBS on orbital shaker.
14. Form immunocomplex by incubating with DAB solution. These usually take approx 2 min. Observe under light microscope to know when staining is adequate.
15. Terminate this last reaction by washing in distilled water.
16. Dehydrate slides through graded ethanols: 70, 80, 90, 100, 100% (containing molecular sieves), and xylene 2X.
17. Finally, cover slip using Cytoseal™ 60 (Stephens Scientific, Riverdale, NJ) with a cover slip.

3.5. Examples

The identification of abnormal cardiovascular phenotypes requires a knowledge of the normal anatomy of the species being studied, especially of the changes and progression expected with development. **Table 2** lists several of the major events in the development of the normal mouse heart. The description of human cardiac anatomy has been enhanced over the past several decades by the use of a "segmental" approach *(18)*. This approach describes the morphology of the cardiac chambers with specific emphasis given to their spatial relationships and connections to each other and to the great vessels *(19,20)*. Extracardiac structures such as the aortic arch, caval veins, and pulmonary veins are also described. This methodical "segmental" approach can also be applied to the analysis of embryos.

Although the general morphology of the embryos can certainly be studied using standard histologic techniques such as hematoxylin and eosin staining, the cardiac embryologist can discern much greater cardiac-specific information using immunohistochemistry. For instance, the ability to discern the distal boundary of the myocardium

Table 2
Major Developmental Events of the Normal Mouse Heart

Embryonic day	Observations *(3,28)*
D 8	Looping being established
D 9.5	Trabeculation of primitive ventricle more prominent
D 10	Early aortopulmonary septation
D 10.5–11	First sign of primary atrial septum
D 11.5	Pulmonary arteries first noted
D 12	Semilunar valves forming
	Left-sided 4th and 6th aortic arches and dorsal aorta are dominant
	Primary atrial septum fusing with AV cushion tissue
D 13	Complete separation of OFT into two separate channels
	Myocardialization of conal septum begins
D 14	VSD closure at outlet foramen
	Coronary arteries seen
	All valves distinguishable
D 15	Chambers and great vessels with definitive connections

of the distal outflow tract has aided in the description of the asymmetric development of the outflow tract myocardium of the rat heart. Very early in development, this asymmetry accounts in part for the ultimate difference in the level of the aortic and pulmonary valves *(21)*. This distinction between myocardial and mesenchymal tissue in the outflow tract of the rat has also allowed further description of the establishment of fibrous continuity between the aortic and mitral valves *(22)*. Antibodies to myosin heavy-chain proteins specific to atrial and ventricular myocardium have assisted detailed analysis of the development of the atrioventricular junction *(23–25)*. Experiments in connexin43-deficient mice have shown the development of intertrabecular pouches and excessive delamination at the ventricular base of both outflow tract ridges, leading to shelves of dense myocardium in the right ventricular outflow tract *(26)*.

In our experiments in mice, we have used antibodies both to smooth muscle actin, as a marker for neural-crest derived cells *(27)*, and to myosin heavy chain as a marker for myocardial tissue. These markers have helped to demonstrate the first stages of myocardialization in the normal embryonic mouse at embryonic day 13 (**Fig. 3**). In experiments with Trisomy-16 mice designed to evaluate the development of the outflow tract, abnormal orientation of the aortopulmonary septum in its earliest stages has been identified. A critical part of the defective development of some of these mice, especially those with double-outlet right ventricle and persistent truncus arteriosus, is the absence of myocardialization in the conal septum of the developing outflow tract (**Fig. 4**).

4. Notes

1. It is best not to label the vial caps as the only method of labeling.
2. Length of time in each step of ethanol and fixative depends on the size of the specimen. Larger specimens may require more time.
3. If specimens are to be transported from one lab to another, the 70% ethanol solution is the best step at which to stop and to use for transport.

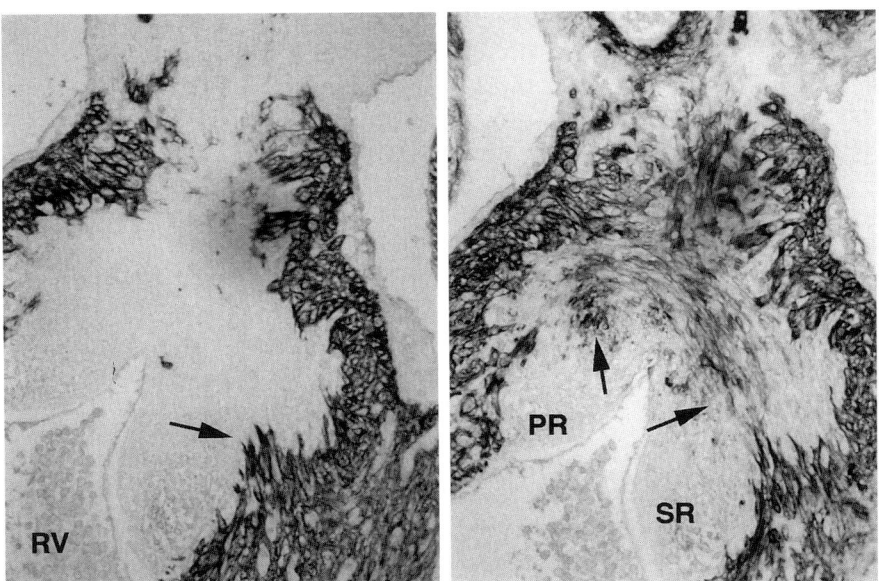

Fig. 3. This figure shows two sister sections from a mouse embryo at 13 ED at the level of the conal outlet septum. The left panel shows a staining for myosin heavy chain and shows the myocardial nature of the cellular projections (arrow) into the mesenchymal cushions; the right panel was stained for the presence of smooth-muscle actin and demonstrates a band of, presumably neural crest-derived, cells reaching toward the myocardial projections (arrows). PR = parietal endocardial conal ridge, SR = septal endocardial conal ridge.

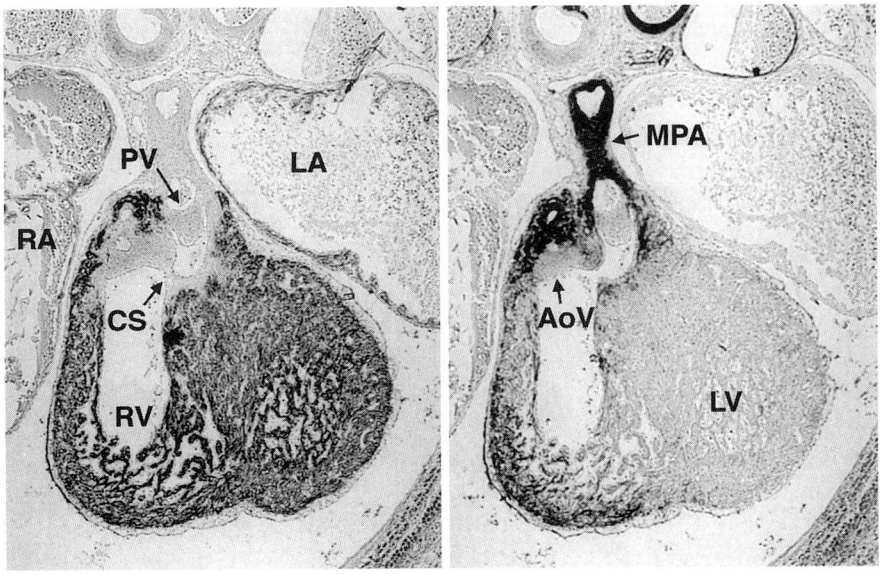

Fig. 4. Double outlet right ventricle in a trisomy 16 embryo at 14.5 ED. In this specimen the semilunar outlet valves (PV and AoV) are located side by side, and both arise from the right ventricle. The right ventricle (RV) communicates with the main pulmonary artery (MPA) through a small communication in a primitive malaligned mesenchymal, nonmyocardialized, conal septum (CS). AoV = aortic valve, CS = conal septum, LA = left atrium, LV = left ventricle, MPA = main pulmonary artery, PV = pulmonary valve, RA = right atrium, RV = right ventricle.

4. Pronase pretreatment: Although most antigens are easily detected using this protocol, the detection of some antigens can benefit significantly from a pretreatment with a proteolytic enzyme. Pronase (Sigma type XIV protease) in a concentration of 0.1 mg/mL distilled water for 30 min followed by rinsing for 5 min in PBS between steps 4 and 6. However, no general protocol can be given, as optimal results of this pretreatment depend strongly on the nature of the antigen and the nature of the tissue in which this antigen should be detected. Furthermore, it should be stressed that some antigens can be destroyed by proteolytic digestion.

5. If paraffin block is difficult to remove from the mold, then the mold and block may be cooled briefly for a few minutes. Excess cooling may crack the block.

6. The ribboned sections that are stored on trays should be covered to avoid losing the order of the sections.

Acknowledgments

The authors acknowledge the support from NIH Grants 5 T 32 HL07710 and HL 52813. Computer-generated images were provided by Tom Trusk. Many thanks to Aimee Phelps and Rossi Bennington for their assistance.

References

1. Rosenthal, G. (1997) Prevalence of congenital heart disease, in *The Science and Practice of Pediatric Cardiology* (Garson, Bricker, Fisher, and Neish, eds.), Williams & Wilkins, Baltimore, pp. 1083–1107.

2. Rossant, J. (1996) Mouse mutants and cardiac development: New molecular insights into cardiogenesis. *Circ. Res.* **78,** 349–353.

3. Webb, S., Brown, N. A., and Anderson, R. H. (1996) The structure of the mouse heart in late fetal stages. *Anat. Embryol. (Berl.)* **194,** 37–47.

4. Franz, T. (1989) Persistent truncus arteriosus in the Splotch mutant mouse. *Anat. Embryol.* **180,** 457–464.

5. Leveen, P., Pekny, M., Gebre-Medhin, S., Swolin, B., Larsson, E., and Betsholtz, C. (1994) Mice deficient for PDGFB show renal, cardiovascular and hematological abnormalities. *Genes Dev.* **8,** 1875–1887.

6. Yokoyama, T., Copeland, N. G., Jenkins, N. A., Montgomery, C. A., Elder, F. F., and Overbeek, P. A. (1993) Reversal of left-right asymmetry: A situs inversus mutation. *Science* **260,** 679–682.

7. Schilham, M. W., Oosterwegel, M. A., Moerer, P., et al. (1996) Defects in cardiac outflow tract formation and pro-B-lymphocyte expansion in mice lacking Sox-4. *Nature* **380,** 711–714.

8. Reaume, A. G., de Sousa, P. A., Kulkarni, S., Langille, B. L., Zhu, D., Davies, T. C., Juneja, S. C., Kidder, G. M., and Rossant, J. (1995) Cardiac malformation in neonatal mice lacking connexin43. *Science* **267,** 1831–1834.

9. Kurihara, Y., Kurihara, H., Oda, H., Maemura, K., Nagai, R., Ishikawa, T., and Yazaki, Y. (1995) Aortic arch malformations and ventricular septal defect in mice deficient in endothelin-1. *J. Clin. Invest.* **96,** 293–300.

10. Brannan, C. I., Perkins, A. S., Vogel, K. S., et al. (1994) Targeted disruption of the neurofibromatosis type-1 gene leads to developmental abnormalities in heart and various neural crest-derived tissues. *Genes Dev.* **8,** 1019–1029.

11. Miyabara, S. (1990) Cardiovascular malformations of mouse trisomy 16: pathogenetic evaluation as an animal model for human trisomy 21, in *Developmental Cardiology: Morphogenesis and Function* (Clark, E. B. and Takao, A., eds.), Futura, Mount Kisco, NY, pp. 409–430.

12. Pexieder, T., Miyabara, S., and Gropp, A. (1981) Congenital heart disease in experimental (fetal) mouse trisomies: Incidence, in *Perspectives in Cardiovascular Research: Mechanisms of Cardiac Morphogenesis and Teratogenesis* (Pexieder, T., ed.), Raven, New York, pp. 389–399.

13. Bacchus, C., Sterz, H., Buselmaier, W., Sahai, S., and Winking, H. (1987) Genesis and systematization of cardiovascular anomalies and analysis of skeletal malformations in murine trisomy 16 and 19: Two animal models for human trisomies. *Hum. Genet.* **77,** 12–22.

14. Wessels, A., Vermeulen, J. L., Viragh, S., Kalman, F., Morris, G. E., Man, N. T., Lamers, W. H., and Moorman, A. F. (1990) Spatial distribution of "tissue-specific" antigens in the developing human heart and skeletal muscle. I. An immunohistochemical analysis of creatine kinase isoenzyme expression patterns. *Anat. Rec.* **228,** 163–176.

15. Wessels, A., Vermeulen, J. L., Viragh, S., Kalman, F., Lamers, W. H., and Moorman, A. F. (1991) Spatial distribution of "tissue-specific" antigens in the developing human heart and skeletal muscle. II. An immunohistochemical analysis of myosin heavy chain isoform expression patterns in the embryonic heart. *Anat. Rec.* **229,** 355–368.

16. Wessels, A., Vermeulen, J. L., Verbeek, F. J., Viragh, S., Kalman, F., Lamers, W. H., and Moorman, A. F. (1992) Spatial distribution of "tissue-specific" antigens in the developing human heart and skeletal muscle. III. An immunohistochemical analysis of the distribution of the neural tissue antigen G1N2 in the embryonic heart; implications for the development of the atrioventricular conduction system. *Anat. Rec.* **232,** 97–111.

17. Theiler, K. (1989) *The House Mouse: Atlas of Embryonic Development.* 2nd ed., Springer-Verlag, New York, p. 178.

18. Anderson, R. H. (1991) Simplifying the understanding of congenital malformations of the heart. *Int. J. Cardiol.* **32,** 131–142.

19. Anderson, R. H. and Becker, A. E. (1984) Sequential segmental analysis of congenital heart disease. *Pediat. Cardiol.* **5,** 281–288.

20. van Praagh, R. and van Praagh, S. (1982) Embryology and anatomy: Keys to the understanding of complex congenital heart disease. *Coeur* **13,** 315–337.

21. Ya, J. (1997) Normal and abnormal development of the outflow tract of the embryonic heart, in *Anatomy & Embryology*, University of Amsterdam, Amsterdam, p. 156.

22. Jackson, M., Connell, M. G., Smith, A., and Anderson, R. H. (1995) Immunohistochemical evaluation of the developing outflow tract in the rat: Achieving aortic to mitral fibrous continuity. *Cardiovasc. Res.* **30,** 262–269.

23. Lamers, W. H., Viragh, S. S., Wessels, A., Moorman, A. F. M., and Anderson, R. H. (1995) Formation of the tricuspid valve in the human heart. *Circ. Res.* **91,** 111–121.

24. Wessels, A., Vermeulen, J. L., Verbeek, F. J., Viragh, S., Kalman, F., Lamers, W. H., and Moorman, A. F. (1992) Spatial distribution of "tissue-specific" antigens in the developing human heart and skeletal muscle. III. An immunohistochemical analysis of the distribution of the neural tissue antigen G1N2 in the embryonic heart; implications for the development of the atrioventricular conduction system. *Anat. Rec.* **232,** 97–111.

25. Wessels, A., Markman, M. W., Vermeulen, J. L., Anderson, R. H., Moorman, A. F., and Lamers, W. H. (1996) The development of the atrioventricular junction in the human heart. *Circ. Res.* **78,** 110–117.

26. Ya, J., Erdtsieck-Ernste, E. B. H. W., De Boer, P. A. J., Van Kempen, M. J. A., Jongsma, H., Gros, D., Moorman, A. F. M., and Lamers, W. H. (1998) Heart defects in connexin-43 deficient mice. *Circ. Res.* **82,** 360–366.

27. Rosenquist, T. H. and Beall, A. C. (1990) Elastogenic cells in the developing cardiovascular system. Smooth muscle, nonmuscle, and cardiac neural crest. *Ann. NY Acad. Sci.* **588,** 106–119.

28. Kaufman, M. H. (1992) *Atlas of Mouse Development*, 1st ed., Harcourt Brace, New York.

IV

Embryo Structure and Function

16

Application of Plastic Embedding for Sectioning Whole-Mount Immunostained Early Vertebrate Embryos

Kersti K. Linask and Takeshi Tsuda

1. Introduction

Development of tissues and organs relies on the constant interplay of intracellular events and molecules in the microenvironment of the cells. Dynamics of patterning of molecules in a spatiotemporal manner become important in understanding any developing system. Immunohistochemistry has become an important technique for the study of specific proteins in the early embryo. Ability to localize molecules in the embryo with the greatest resolution possible provides much critical information regarding intracellular and extracellular localization of specific molecules and changes in their patterning over time. For example, an apical or basal cellular localization of a molecule may be critical in relation to its function. Thus, the definition and correlation of the spatiotemporal patterns of protein expression in the developing embryo with events of morphogenesis may give useful information on the possible roles of these molecules. Such analyses will provide a more precise basis for further experimentation.

To obtain the highest resolution possible in antigen localization, a pre-embedding immunohistochemical technique followed by whole-mount plastic embedding and sectioning at 1 μm has been modified to make it useful for analyzing events of early chick and mouse morphogenesis in specific regions of the embryo. The description of the pre-embedding immunohistochemistry method and embedding procedure provided is an adaptation of the technique of Franklin and Martin (1) and others (2,3). This technique has been used with good results for analyses of whole chick embryos up to 48 h of development and for whole mouse embryos up to 9.5 d of gestation and isolated organs (i.e., heart) up to day 15 of gestation.

For presentation purpose, this chapter is divided into two parts, dealing with chick embryos and mouse embryos, respectively.

2. Sectioning Whole-Mount Immunostained Chick

In terms of the requirements of the procedures, there are no differences in immunohistochemistry between using polyclonal or monoclonal antibodies. It is important that the primary antibodies are of well-established specificity, that a suitable detection system is available, either fluorochrome-based or enzyme-linked, and that the embryonic

From: *Methods in Molecular Biology, Vol. 135: Developmental Biology Protocols, Vol. I*
Edited by: R. S. Tuan and C. W. Lo © Humana Press Inc., Totowa, NJ

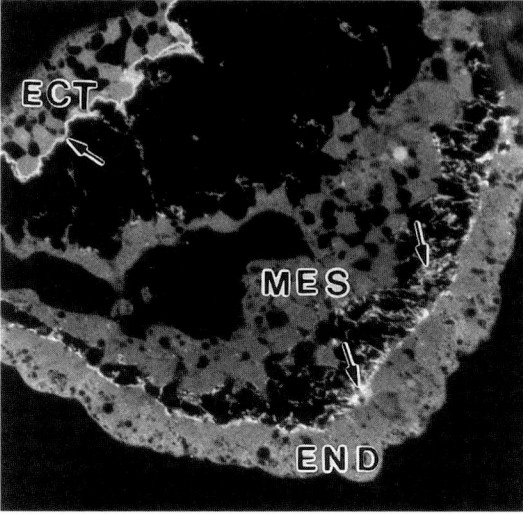

Fig. 1. Whole-mount immunostained stage 7+ chick embryo is shown sectioned through heart-forming region. Cellular fibronectin (white fluorescent pattern; see arrows) is observed associated with fibrils at cardiac mesoderm (mes)-endoderm (end) interface and underneath ectoderm (ect). This embryo was embedded in araldite, sectioned at 1 μm and observed with a Nikon Optiphot microscope equipped with epifluorescence attachment.

specimen is processed optimally for antibody penetration and recognition of its antigen for subsequent observation. The methodology described provides optimal resolution of antigen within the embryo that is sectioned at 1–2 μm (*see* **Fig. 1**). Control embryos should always be processed according to the same protocol as the experimental embryos. A negative control should also be processed (i.e., an embryo to which no primary antibody has been applied). This procedure takes 1 wk to carry out.

2.1. Materials

1. Phosphate-buffered saline solution (PBS): NaCl, 8 g/L; KCl, 0.2 g/L; $CaCl_2$, 0.1 g/L; MgCl $6H_2O$, 0.1 g/L; Na_2HPO_4, 0.92 g/L; KH_2PO_4, 0.2 g/L. pH 7.4. Add the $CaCl_2$ to the solution last, after addition of water; otherwise, it may precipitate.
2. 3% paraformaldehyde fixative: Dissolve 3.0 g paraformaldehyde in 50 mL of water. The paraformaldehyde is converted to formaldehyde by adding approx 3 drops of 1*N* NaOH and heating the mixture to 70°C in a laboratory ventilated hood. The solution is allowed to cool and added to 50 mL of double-strength PBS. Alternatively a 9% paraform-aldehyde solution can be made and frozen as aliquots of 5-mL or 10-mL tubes. For use, dilute down to 3% paraformaldehyde with PBS. Initially, the solution will show floccu-lence, but on standing to room temperature (RT) will become clear.
3. Permeabilization of embryos: 100% methanol kept at –20°C.
4. Rehydration: Use ethanol.
5. PBS/BSA: 1% bovine serum albumin (BSA) in PBS.
6. Primary antibodies: Dilute to necessary concentrations with PBS/BSA.
7. Secondary antibodies: Dilute to necessary concentrations with PBS/BSA.
8. Plastic embedding: Araldite embedding medium is made by mixing 5 mL araldite (araldite 502 resin, cat. no. 18060), 5 mL dodecenyl succinic anhydride (DDSA; cat. no. 18022),

and 200 μL benzyldimethylamine (cat no. 18241). All embedding reagents are available from Ted Pella, Inc., Redding, CA. If too many bubbles have formed while mixing reagents, deaerate this mixture with a vacuum pump.

9. Nitex filters: 5-μm nylon netting (Sefar America Inc., Kansas City, MO). Cut netting into approx 1.5-cm squares.

10. Nuclepore filters: Nuclepore® Track-Etch Membrane, PC MB 19 × 42 mm, 8 μm (cat. no. 113314, Corning Separations, Acton, MA). Cut into 1-cm squares. Store in 70% ethanol. Rinse with PBS before use.

11. Petri dishes: 100- and 35-mm dishes.

12. 18-well glass slides (5 mm) with hydrophobic coating (cat. no. 10-943, Cel-Line Associates, Inc., Newfield, NJ).

13. Flat embedding mold: See any catalog of electron microscopy supplies. A good one to use is one made of blue silicone rubber.

14. Toluidine blue: Make up toluidine blue stain as with Na borate, 2 g and 0.2 g toluidine blue stain (Gibco-BRL, Grand Island, NY). Bring up to a vol of 200 mL deionized, distilled H_2O. filter. When staining, heat sections on hot plate and then rinse with water. Prevent any running of stain into fluorescent sections.

15. Mounting medium: Vectashield®. A mounting medium for fluorescence (Vector Lab, Burlingame, CA). If not using fluorescent-conjugated secondary antibodies, use any mounting medium as, for example, Cytoseal (Stephens Scientific, Riverdale, NJ).

2.2. Methods

1. Embryo collection: Incubate eggs until desired stage of development. Remove from incubator and allow to cool to RT. Let eggs sit on side in egg carton before opening. This will allow the blastoderm to come to lie on top of the yolk. Keep egg in this position when cracking eggshell and allow egg yolk to flow gently into a 100-mm Petri dish. Blastoderm should be visible as a white circle on top of yolk. Cut a square around the blastoderm through the vitelline membrane (*see* **Note 2.3.**, **step 1**). With a forcep lift a PBS-rinsed Nitex filter and gently slip into the yolk underneath the blastoderm. Then lift the blastoderm away from the yolk with the Nitex and transfer into a 35-mm Petri dish containing PBS. With forceps, peel the blastoderm and vitelline membrane away from the Nitex, and rinse blastoderm with PBS to remove as much of the yolk as possible. Remove Nitex from the dish. Remove blastoderm from the vitelline membrane, if still attached, and now place on a Nuclepore filter with the embryo's ventral (yolky) side up. Keep forcep tips away from the embryo. Lift up the Nuclepore and embryo with forceps, holding the area opaca only, to place in fresh PBS in another 35-mm Petri dish. Lift up at an angle when removing from the PBS. This allows the surface tension to remove any yolk still clinging to the embryo as well as to flatten the embryo for fixation. Rinse one more time in PBS. While still on Nuclepore, transfer now into paraformaldehyde fixative in a small glass dish. Can transfer many embryos in this manner into the fixative.

2. Fixation (*see* **Note 2.3.**, **step 2**): Fix embryos in 3% paraformaldehyde/PBS for 1 h at RT or overnight at 4°C. Embryos are harder with paraformaldehyde fixation. After fixation, trim away area opaca and discard. Transfer embryo into PBS in a small glass dish for further processing.

3. Permeabilization and rehydration: Remove the PBS with a Pasteur pipet and add cold (from the freezer at –20°C) 100% methanol. The glass dish is covered and placed in the freezer. The cold methanol is changed 3X in the course of 1 h. Thus the embryos remain in methanol 1 h. Using a Pasteur pipet, the embryos are brought back to PBS through a cold (about 4°C) ethanol series of 100, 90, 70, 50% and PBS. The embryos should remain at each concentration for 10 min (*see* **Note 2.3.**, **step 3**).

4. Rinses: The embryos are rinsed in 3 changes of PBS, 10–15 min for each change (*see* **Note 2.3.**, **step 4**).

5. Blocking step for nonspecific binding of antibody: Add 20–30 µL of preimmune serum or 1% BSA/PBS into separate wells of glass slides with hydrophobic coating. Number of wells to be used depends on number of embryos to be immunostained. Embryos are transferred on the tip of a scalpel blade into the preimmune serum or BSA/PBS. Embryos tend to flatten out, but may need some adjustment with forceps. Be sure that each embryo is completely immersed in solution and not on the surface of the solution. Slide is placed in a moistened chamber and then placed into a 4°C refrigerator overnight and for approx 1 h at RT the next day (*see* **Note 2.3.**, **step 5**) (**Fig. 2**).

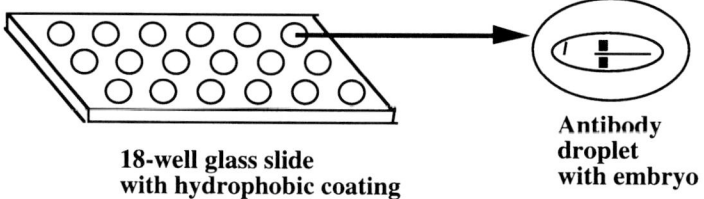

**18-well glass slide
with hydrophobic coating**

**Antibody
droplet
with embryo**

Fig. 2. Glass slide and antibody droplet.

6. Rinses and primary antibody incubation step (*see* **Note 2.3.**, **step 6**): Embryos are removed from the well slides and placed in a glass dish filled with PBS at RT. Embryos are rinsed in 3 changes of PBS, 30 min for each change. Embryos are transferred into approx 20 µL of primary antibody placed in wells of an 18-well slide, transferred on the tip of a scalpel blade. Control embryos are placed in preimmune serum or PBS/BSA. Slides are returned to moistened chamber and placed at 4°C overnight.

7. Rinses and secondary antibody incubation step: Next morning, embryos are removed from the well slides and placed in separate wells of 24-well plates filled with PBS. Keep at RT. Embryos are rinsed in 3 changes of PBS, 30 min for each change. Each embryo is transferred into 20 µL of secondary antibody (fluorescent conjugate or enzyme conjugate) that has been placed in wells of an 18-well slide as earlier. Slides are returned to moistened chamber and placed at 4°C overnight.

8. Rinses and dehydration: Next morning, as above, embryos are removed from the well slides, and each embryo is placed in separate wells of 24-well plates filled with PBS and kept at RT. Embryos are rinsed in 3 changes of PBS, 30 min for each change. At this point, the embryos may be set up for whole-mount observation and localization of specific antigen (*see* **Note 2.3.**, **step 7**). Embryos for sectioning are dehydrated to 100% ethanol by reversing the ethanol alcohol series of steps described in step 3 (i.e., 50, 70, 90, and 100%). While in 70% alcohol, mark region of interest on embryo with powdered or granular carbon (*see* **Note 2.3.**, **step 8**). Also mark anterior end of embryo.

9. Plastic embedding: Embryos are transferred on the tip of a scalpel dedicated for plastic embedding from the 100% ethanol to araldite embedding medium in a small watch glass (approx 2–3 mL of embedding medium). Embryos are kept in the embedding medium for 2–3 h (*see* **Note 2.3.**, **step 9**). Embryos become almost transparent in the araldite, and the carbon dots become the only visible marker to orient the embryos in the molds after this step. Transfer embryos to fresh embedding medium, into a mold, onto araldite rafts (*see* **Note 2.3.**, **step 10**) using the end of a scalpel blade. With a needle, the embryos are positioned in the mold with the head of the embryo toward the tapered end of the mold. The mold is placed overnight in a 60°C oven to allow the araldite to harden.

10. Sectioning: Before sectioning, the tapered end of the plastic block that contains the embryo should be trimmed down to a small face on the block. This will allow one to get better and flat sections. A razor blade or microtome knife can be used to trim the face of the block. Embryos are sectioned through desired region at 1–2 μm using a glass knife and an ultra-microtome. Sections pass automatically into water in small rafts underneath the blade. The plastic sections are transferred from the water onto a slide with a small loop and allowed to dry for approx 5–10 min on the surface of a hot plate. The heat helps to flatten the sections. The first section and the last section of a slide are placed on one side, toluidine blue stained, and can be used for orientation of sections seen on that slide. Underneath the slide (*see* **Note 2.3.**, **step 11**), circle with a black marker where the sections are placed. Mount slides with Vectashield and place a cover slip over sections. The slide can be observed as soon as the mounting medium is set (**Fig. 3**).

Two toluidine blue stained sections 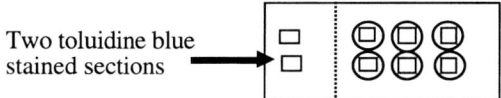   Do not get toluidine blue on the rest of the sections.

Fig. 3. Toluidine blue stained sections.

2.3. Notes

1. If there is a problem with the blastoderm slipping away and becoming lost in the yolk, try cutting membrane on one side of the embryo only. Then insert square of Nitex (plastic mesh square for supporting embryo) underneath the embryo and cut the remaining membrane around the embryo.
2. Sometimes paraformaldehyde may cause a loss of antigenicity. A good fixative for use with monoclonal antibodies is Histochoice® (Amresco, Solon, OH). Place embryo into Histochoice for at least 1 h at RT or overnight. Remove Histochoice, and rinse embryo 2X in 70% ethanol for 10 min at RT. Then proceed to permeabilizing with 100% methanol.
3. Through all of the preceding steps after placing embryos into PBS after fixation, the embryos have remained in the same dish. Solutions have been removed and added with a Pasteur pipet.
4. If there are problems with air bubbles sticking to embryos, deaerate PBS using a vacuum pump.
5. A moistened chamber can be made with a 100-mm Petri dish with a piece of filter paper moistened with H_2O placed on the bottom of the dish. Two paper clips are placed on top of the filter paper and the 18-well slide is then placed on the two paper clips. This keeps the preimmune serum/antibody from drying out and also keeps the slide from sitting directly in any excess water. Seal chamber with Parafilm and place at 4°C overnight.
6. Dilute 1° antibody to its specific concentration in PBS/BSA. Place 20 to 30 μL of antibody onto an 18-well slide, using every other well. Submerge embryo in the antibody. Make sure embryo is flat and antibody is completely covering it. Long incubations with a low concentration of antibody at 4°C will give lower background staining than short incubations with high concentration of antibody. Do not contaminate well with antibody when transferring embryos with scalpel. Use a clean scalpel for each different antibody.
7. If one wishes to observe antigen localization in a whole-mount preparation, one can now place the embryo in 70% glycerol/PBS and cover slip on a support to prevent excessive flattening of the embryo; observe with microscope.
8. Carbon marks can be placed on embryo with the bevel of a syringe needle. When doing this, it is best to remove most of the alcohol and leave enough so that the embryo does not become dry. Never let the embryos dry at any point in this procedure.

9. When transferring embryos from the alcohol into araldite, do this underneath a dissecting microscope. Then, with the scalpel, gently mix the plastic around the embryo. Swirls of alcohol can be seen in the araldite coming off the embryo. After mixing well, but being sure that embryo remains intact, let stand for the specified 2–3 h at RT to allow sufficient penetration of the plastic into the embryo. The embryo turns transparent in araldite. Can only see embryos by the carbon markings during this period. Importantly, if sufficient time is not given for penetration, inferior sections will result.

10. Mold preparation: Place first on the bottom of each well, a thin, flat "raft" of previously hardened araldite (as shown): One may have to use sandpaper to get a nice flat surface on the araldite raft. On this raft place fresh araldite. Then transfer and position the embryo for cross or sagittal sectioning in each mold. Avoid air bubbles. The purpose of the raft is to prevent the embryo in the liquid araldite from sinking to the bottom of the mold before the araldite hardens. By doing this, the embryo will be positioned in the middle of each mold and will be easier to trim and section. With leftover araldite, make new "rafts" in the unused wells for future use (**Fig. 4**).

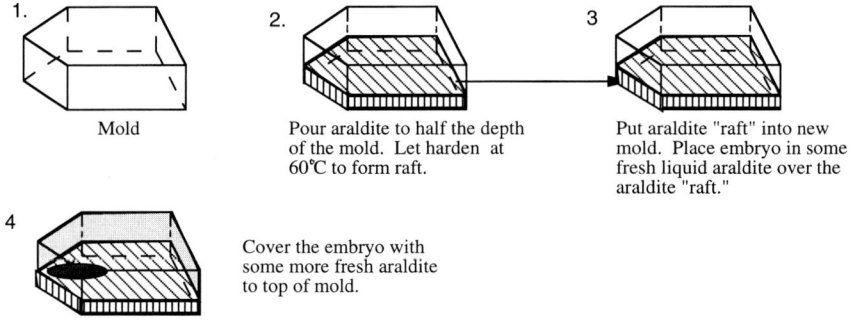

Fig. 4. Preparing molds.

11. Mark the *underside* of the slide by circling the small area around each plastic section. After mounting, the sections are barely visible and are hard to find for observation. Marker pen ink fluoresces and can ruin sections if they come in contact with the ink. Red markers are to be avoided because the ink gives a very bright fluorescence.

3. Sectioning Whole-Mount Immunostained Mouse

This technique works well for obtaining an overall, three-dimensional image of the localization of a specific protein in mouse embryos at early stages of development, up to approximately embryonic day (ED) 9.5 of gestation. Embryos must be fixed and permeabilized so that antibodies can penetrate into the tissues of the embryo. As for chick embryos, as the embryos become larger, different regions probably will need to be dissected out and immunostained. To obtain better rinsing of cavity regions in slightly older embryos, make a small hole with a needle, as into the apex of the heart, to avoid trapping of antibody in the cavities. Mouse antibodies can be used to immunostain for specific proteins without much problem of background. Degree of background must be tested for each antibody. Use of blocking solution and number and length of rinses, however, are critical when using antibodies made in mouse. These same embryos set up for whole-mount observation can then be dehydrated and set up for embedding in plastic and sectioning at 1–2 μm as with chick embryos.

3.1. Materials

1. PBSMT: 50 mL PBS (Ca^{2+}/Mg^{2+}-free); 1 g 2% nonfat instant skim milk (Carnation); 50 µL 0.1% Tween-20 (or 0.5% Triton X-100, 250 µL); 0.5 g 1% BSA.
2. PBT: 50 mL PBS; 50 µL 0.1% Tween-20.

3.2. Method

During the entire whole-mount procedure, it is necessary that reagents fully penetrate the sample and react. It is critical that during the rinse steps, excess reagents are completely removed. To facilitate both, the embryos should be gently rocked throughout the procedure (*see* **Note 3.3., step 1**).

1. Embryo removal: Open up killed pregnant female and expose uterine horns. Remove uterine horns quickly into a Petri dish with PBS. Rinse well. Transfer to a fresh dish of PBS. Conduct all subsequent steps under the dissecting microscope in a sufficient depth of PBS to completely immerse the tissue. Using forceps, tease the decidual masses free of the uterine wall. Transfer all decidual masses to fresh PBS. Gently open up each decidua to expose the embryo. Remove the embryos and transfer with siliconized Pasteur pipets to fresh PBS. One should try to remove the embryos within the shortest time possible, within 30 min at most. The embryo can be flattened if, before fixing, the amnion cap is dissected off and two cuts are made in the lateral amnion. Remove all extraembryonic membranes. Keep the embryos on ice as much as possible.
2. Fixation: 4% paraformaldehyde/PBS overnight at 4°C. Make up fixative as for chick embryos except at 4%.
3. Wash embryos with PBS 2X for 15 min at RT with gentle agitation or rocking (*see* **Note 3.3., step 2**).
4. Dehydration and permeabilization: Dehydrate using an ascending methanol series of 30, 50, and 70%, each for 15 min (all methanol diluted with PBS). The older the embryos, the longer the time for each step (≤ED 8.5, 15 min; >ED 8.5, 30 min). Permeabilization/blocking endogenous peroxidase activity step: Use 100% methanol:DMSO:30% H_2O_2 solution made up in a ratio of 4:1:1, respectively. Leave in this solution for 5 h or overnight at 4°C.
5. Wash in 70% methanol for 30 min at RT with rocking. Transfer to 100% methanol for long-term storage (–20°C).
6. Rehydrate the embryos at RT in microcentrifuge tube in following series: 70% methanol/PBS, 30 min with rocking; 50% methanol/PBS, 30 min with rocking; 1 mL PBS, 30 min with rocking; 1 mL PBSMT (*see* **Subheading 2.1.**); 2X for 1 h with rocking (30 min each).
7. Primary antibody step (*see* **Note 3.3., step 3**): Incubate with 1 mL of primary antibody diluted in PBSMT with rocking in microcentrifuge tubes at 4°C overnight. Antibody must completely cover embryos. Correct dilution must be determined empirically for each antibody. It is recommended to use lower concentrations, lower temperature (4°C), longer exposure time, and with rocking for the best final localization of protein (overnight incubations and up to 2 d).
8. Rinses: Wash 6X with rocking in PBSMT: 2X with 1 mL for 1 h at 4°C; 4X in 1 mL for 1 h each at room temperature (total of 4 h).
9. Secondary antibody incubation: Incubate embryos in 1 mL of secondary antibody (i.e., goat antimouse Cy3-conjugated or rhodamine-conjugated goat antimouse antibody) diluted in PBSMT with rocking at 4°C.
10. Wash as in **step 7**.

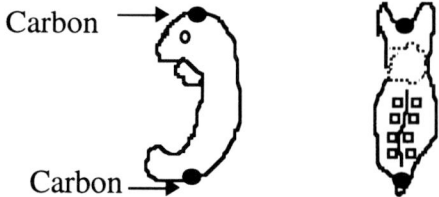

Fig. 5. Positioning embryo in araldite.

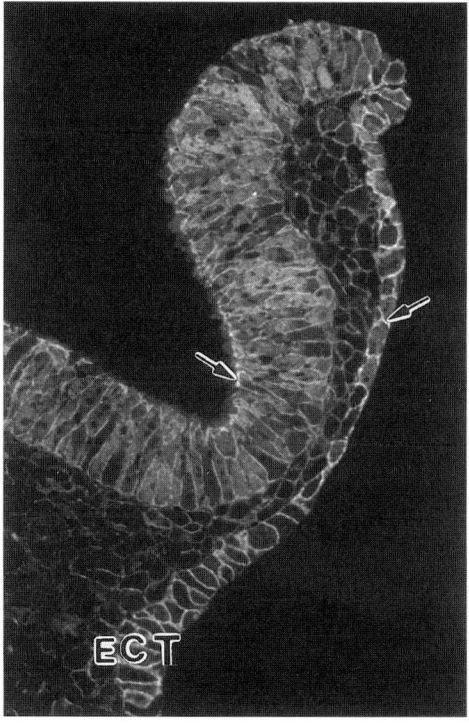

Fig. 6. Section through neural fold and hindbrain region of ED8.5 mouse embryo. Whole-mouse embryo was immunostained with a monoclonal antibody against β-catenin. After immunostaining, the embryo was embedded in araldite and sectioned at 1 μm through specific regions marked by carbon particles. As expected, all cells seen here stain with β-catenin. Prominent localization, however, is seen at apical cell junctions in neuroepithelium and in association with cell membranes of ectoderm (*see* arrows).

11. Rinse in PBT (as indicated in **Subheading 2.1.**). Wash in 1 mL of PBT with rocking for 20 min, 3X.
12. Postfixation: Postfix in 4% paraformaldehyde in PBS at 4°C (a few hours to overnight).
13. Dehydration: 1 mL PBT quick rinse; 1 mL PBT 30 min at RT with rocking; 1 mL 50% methanol 30 min at RT; 1 mL 70% methanol 30 min at RT; 1 mL 100% methanol 30 min at RT (2X). While the embryo is in 70% alcohol, mark anterior and posterior end of the embryo with powdered or granular carbon for better positioning in araldite, as shown in **Fig. 5**.

14. Plastic embedding with araldite: *See* **Subheading 2.2., step 9**. Follow same procedure after dehydration step. Duration of permeabilization with araldite depends on the size of the embryos. The following are recommended as guidelines:

ED (d)	Duration (h)
8.5	4–5
8.5–9	6–8
9.5	8 (to overnight)

15. *See* **Fig. 6** for a representative section of a mouse embryo immunostained for β-catenin.

3.3. Notes

1. The whole procedure can be carried out in any kind of tube. We commonly use 1.5-mL microcentrifuge tubes (especially when using genotyped embryos). Try not to transfer embryos from tube to tube, but rather keep embryo(s) in the same tube and change solutions.
2. When changing solutions, do it under using a dissecting microscope with light coming from below so that you may not lose your embryos while rinsing (especially when dealing with younger embryos).
3. If you are using a primary antibody not made in mouse, you may shorten blocking step. This blocking step is used when using antimouse antibody as a primary antibody for mouse embryos.

References

1. Franklin, R. M. and Martin, M. T. (1981) Pre-embedding immunohistochemistry as an approach to high resolution antigen localization at the light microscope level. *Histochemistry* **72,** 173–190.
2. Davis, C. A. (1993) Whole-mount immunohistochemistry, in *Methods in Enzymology, Guide to Techniques in Mouse Development* vol. 225 (Wassarman, P. M. and DePamphilis, M. L., eds.), Academic, San Diego, CA, pp. 502–516.
3. Hogan, B., Beddington, R., Costantini, F., and Lacy, E. (1994) Immunohistochemistry of whole mount embryos, in *Manipulating the Mouse Embryo*, 2nd ed. Cold Spring Harbor Laboratory Press, Cold Spring Harbor, NY, pp. 340–343.

17

Confocal Microscopy of Live
Xenopus Oocytes, Eggs, and Embryos

Carolyn A. Larabell

1. Introduction

The use of the confocal microscope to study living *Xenopus* eggs affords the opportunity to obtain four-dimensional (4-D) data (three-dimensional [3-D] data over time) throughout early development of these large cells. Microscopy of living cells often reveals important information about dynamic cellular events that cannot be gleaned from analyses of fixed cells; however, certain compromises must be made to assure cell viability. Several approaches are typically used for examining living cells. One approach is to collect successive images at fixed time intervals over extended periods of time (e.g., time-lapse data collection), examining either the same optical plane or multiple different planes (3-D data sets, referred to as a z-series). Another approach is to collect, as rapidly as possible, multiple images over relatively short periods of time, again from a single optical plane or many different optical planes. In either case, it is imperative that the data be collected without altering normal cell function. Although multiphoton microscopy is the preferred approach for studying living cells, as it results in less damage to the cell, it is difficult with *Xenopus* oocytes and embryos because the energy of the longer wavelengths used is absorbed by the pigment granules, causing intense local heating (Larabell, personal observations). We have found, however, that standard confocal microscopy using a krypton–argon laser can be used to examine *Xenopus* eggs and embryos with undetectable adverse effects as determined by the ability of the eggs to develop into swimming tadpoles *(1,2)*.

A critical parameter in assuring cell viability when examining living cells is the extent of laser exposure the specimen receives. Laser exposure is determined by the intensity of the laser beam utilized and the dwell time of the laser beam. Therefore, when examining living cells with a scanning laser confocal microscope, it is best to use the lowest possible laser power, which typically requires very bright signals or longer dwell times at each region of the specimen. However, when examining rapidly occurring events, obviously it is also important to collect the data as rapidly as possible, which requires using short dwell times. As a result, it is often necessary to compromise the quality of the image. All of these considerations are critical when recording the rapidly occurring events that occur during cortical rotation in the first cell cycle of

From: *Methods in Molecular Biology, Vol. 135: Developmental Biology Protocols, Vol. I*
Edited by: R. S. Tuan and C. W. Lo © Humana Press Inc., Totowa, NJ

Xenopus eggs. Another important consideration when studying living cells is the choice of fluorescent markers, because fluorescently tagged antibodies, which are typically used for fixed cells, can interfere with cell function. In recent years, numerous fluorescent probes that can be used for live cell studies, such as fluorescently tagged proteins and RNAs, have been developed and provide a myriad of opportunities for studying cell dynamics. In this chapter, I have outlined several approaches we use for studying dynamic events in living *Xenopus* oocytes, eggs, and embryos.

2. Materials
2.1. Buffers

1. Modified amphibian Ringer (MR): 100 mM NaCl, 2 mM KCl, 1 mM MgCl$_2$, 2 mM CaCl$_2$, and 5 mM HEPES, adjusted to pH 7.2 with NaOH.
2. Injection buffer: 91 mM Potassium gluconate, 19 mM KCl, 19 mM NaCl, 0.5 mM MgCl, 10 mM HEPES, pH 7.3.
3. Modified Barth's saline: 88 mM NaCl, 1.0 mM KCl, 2.4 mM NaHCO$_3$, 0.82 mM MgSO$_4$ · 7$_2$O, 0.33 mM Ca(NO$_3$)$_2$ · 4H$_2$O, 0.41 mM CaCl$_2$ · 6H$_2$O, 10 mM HEPES, pH 7.5.
4. Dejellying solution: 2.5% cysteine hydrochloride in 1/3 MR, adjusted to pH 8.0 with NaOH.
5. Agarose: Low-gelling-temperature agarose (Sigma Chemical Co., St. Louis, MO) in 1/3 strength MR.
6. 5% Ficoll (Sigma) in 1/3 MR.
7. Oocyte culture medium (from **ref. 3**): 50% Liebowitz L-15 medium containing 5 mg/mL vitellogenin, 1 mM L-glutamine, 15 mM HEPES, 1 µg/mL insulin, gentamycin (100 µg/mL), tetracycline (50 µg/mL), nystatin (50 U/mL).
8. Oocyte storage buffer (defolliculated, stage VI oocytes): MR plus 1 mg/mL bovine serum albumin (BSA) (to coat the Petri dish) and 50 µg/mL gentamycin at pH 7.0–7.8 (pH 7.0 was used for long-term storage of defolliculated oocytes).
9. Maturation buffer: MR plus 50 µg/mL gentamycin, pH 7.0–7.8 MR (up to pH 7.8 was used for faster maturation).
10. Progesterone solution: Stock was made at 10 mg/mL in 100% ethanol.

2.2. Vital Dyes

1. Nile Red (PolySciences Inc., Warrington, PA). Make a stock solution at a concentration of 1 mg/mL in 100% ethanol; use at a final concentration of 1 µg/mL in 1/3 MR.
2. DiOC$_6$(3): Make a stock solution at a concentration of 1 mg/mL in 100% methanol; use at a final concentration of 1–2 µg/mL in 1/3 MR.

2.3. Fluorescently Labeled RNAs

1. Fluorescently labeled RNAs were made in vitro by incorporation of fluorescein (BMB)- or Texas Red (Molecular Probes, Eugene, OR)-derivatized UTP by Dr. Malgosia Kloc (*see* **ref. 6**).

2.4. Fluorescently Tagged Proteins

1. Actin: Fluorescent (rhodamine) nonmuscle actin was obtained from Cytoskeleton (Denver, CO) at a concentration of 10 mg/mL in 2 mM Tris-HCl, pH 8.0, 0.2 mM CaCl$_2$, 0.2 mM ATP, 0.5 mM DTT.
2. Tubulin: Fluorescent (rhodamine or fluorescein) tubulin was obtained from Cytoskeleton at a concentration of 10 mg/mL in G-PEM buffer (80 mM piperazine-N,N-*bis*[2-ethane-sulfonic acid sequisodium salt; 1 mM magnesium chloride; 1 mM ethylene glycol-*bis*

(b-amino-ethyl ether) N,N,N-,N-tetra-acetic acid; 1 m*M* guanosine 5'-triphosphate, pH 6.8]) minus glycerol. Although this company also provides tubulin in a solution that contains glycerol, we had better results generating microtubule arrays in living eggs when using the buffer lacking glycerol.

2.5. Fluorescent, Carboxylated Beads

1. Yellow-green FluoSpheres, 0.2 μm in diameter and coated with a hydrophilic polymer containing multiple carboxylic acids were obtained from Molecular Probes (Eugene, OR). They were suspended 1:1 in distilled water, centrifuged, and resuspended to remove the sodium azide immediately prior to injection.

2.6. Viewing Dishes for Use with an Inverted Microscope (see Note 1)

1. A circular hole (approx 20 mm in diameter) was cut in the bottom of a 35-mm plastic Petri dish. A No. 1 glass cover slip was placed over this hole and attached to the bottom of the dish using silicon-based glue and cured overnight. It is important to use a thin layer of glue so that the outer rim of the Petri dish (rather than the cover slip) rests flat on the microscope stage.

3. Methods

We use a Nikon Diaphot 200 (inverted microscope) attached to the Bio-Rad (Richmond, CA) MRC 1024 confocal laser scanning microscope for most of our studies of living *Xenopus* oocytes, eggs, and embryos. For low-magnification global views of the eggs and embryos we use a Nikon 20X Fluor lens with a 0.75 numerical aperture, and for high-magnification views of organelles, microtubules, and GFP-constructs we use a Nikon 60X PlanApo oil immersion lens with a 1.4 numerical aperture (*see* **Note 2**).

3.1. Labeling Cytoplasmic Constituents

1. Yolk platelets: Incubate dejellied eggs/embryos in a solution of 1 μg/mL Nile Red in 1/3 MR for 5 min, then rinse in 1/3 MR. Nile Red labels virtually all yolk platelets and can be detected using standard rhodamine filters.
2. Mitochondria, endoplasmic reticulum, and germ plasm: Incubate dejellied eggs/embryos in a solution of 1–2 μg/mL $DiOC_6(3)$ in 1/3 MR for 5 min, then rinse in 1/3 MR. Organelles labeled with $DiOC_6(3)$ can be visualized using standard fluorescein isothiocyanate (FITC) filters. Simultaneous incubation of eggs/embryos in Nile Red and $DiOC_6(3)$ works quite well and permits visualization of the vast majority of organelles in the egg/embryo periphery.
3. Fluorescently labeled tubulin: For analyses of microtubule polymerization and dynamics during cortical rotation, eggs were fertilized (add sperm to eggs in 1/3 MR for 5 min, rinse, dejelly with cysteine, then rinse well in 1/3 MR to remove all cysteine). After the final rinse, eggs were placed in 1/3 MR for microinjection of the tubulin monomers. The tip of the microinjection needle was inserted into the animal hemisphere, typically midway between the animal pole and the equator, and brought to rest at the center of the egg, just beneath the nucleus. Approximately 20 nL of a 10 mg/mL solution of either fluorescein- or rhodamine-labeled tubulin (Cytoskeleton) was delivered to each egg using a glass injection needle with a 5 μm-diameter tip within 30 min after fertilization. It is important that the tubulin monomers are injected as soon as possible after fertilization to allow time for their diffusion to the egg periphery prior to the onset of cortical rotation. Immediately after injection, the eggs were positioned in the viewing dish (described in **Subheading 3.3.**), with the vegetal hemisphere downward (animal hemisphere upward), for examination.

4. Fluorescently labeled actin: Approximately 20 nL of fluorescent (rhodamine) nonmuscle actin was injected into the center of manually defolliculated, meiotically immature oocytes. To do this, the tip of the microinjection needle was inserted in the animal hemisphere, typically midway between the animal pole and the equator, and brought to rest just beneath the nucleus (*see* **Note 3**). After 8–10 h, oocytes were matured by adding 1 µL progesterone solution to the oocytes in Barth's medium and overnight incubation at 16–17°C. The next morning, oocytes were examined for the appearance of the white spot at the apex of the animal pole, signifying germinal vesicle breakdown. Several hours after the appearance of the white spot, the oocytes were placed in viewing dishes (described in **Subheading 3.3.**), then pricked in the animal hemisphere with a finely drawn glass needle (gravity-drawn and hand-tapered Drummond microcapillary tubes, Drummond Scientific, Broomall, PA) held in a micromanipulator. The eggs were then immediately transferred to the microscope stage for data collection.

5. Carboxylated, fluorescein-filled beads: A glass microneedle was inserted into the ventral equatorial region of dejellied, fertilized eggs. The needle tip was positioned at the vegetal pole, as close as possible to the cell surface, where approx 2–4 nL of washed FluoSpheres (Molecular Probes) were deposited in the cytoplasm. We monitored movement of the fluorescent beads along microtubules in live embryos during cortical rotation (and throughout embryogenesis) using FITC filters. Alternatively, embryos can be fixed at various time-points throughout embryogenesis and then examined, because there was minimal detectable quenching of the fluorescent signal after fixation.

6. Green fluorescent protein (GFP) constructs: To examine movement of fluorescently tagged proteins during the first cell cycle, GFP-tagged mRNA molecules (kindly provided by, and used in collaboration with, Dr. Randall T. Moon) were injected into meiotically immature oocytes. Oocytes were then matured in vitro by addition of progesterone. Several hours after the appearance of the white spot at the apex of the animal pole, the oocytes were mounted in viewing dishes (described in **Subheading 3.3.**) and activated by pricking with a sharp glass needle (*see* in **Subheading 3.1.4.**). The GFP constructs were monitored during the first cell cycle using FITC filters. To examine movement of fluorescently tagged proteins during embryogenesis, the GFP-tagged mRNA molecules were injected at the one- or two-cell stage. Embryos can then be examined at successive stages throughout embryogenesis.

3.2. Visualization of Exogenous mRNAs

Fluorescently labeled RNAs were made in vitro by incorporation of fluorescein (BMB)- or Texas Red (Molecular Probes)-derivatized UTP. These RNAs, alone or in pairs, were injected into manually defolliculated, or collagenased, oocytes of various stages. Although the RNAs were targeted for the nucleus, some leakage into the cytoplasm from the pipet was unavoidable. After injection, oocytes were cultured in 50% Liebowitz L-15 medium plus antibiotics at 18°C for up to 3 d. At periodic time intervals, the oocytes were placed in 35-mm viewing dishes (for these studies the oocytes were not immobilized in agarose wells) for examination in the confocal microscope (*3*).

3.3. Immobilizing Embryos for Observation
Using an Inverted Microscope

Immediately prior to the onset of the experiment, pour a thin layer of low-gelling-temperature agarose (Sigma) in 1/3 MR into a viewing dish. Quickly place nine steel balls (1.6-mm diameter), in an evenly spaced 3 × 3 grid, in the molten agarose solution.

It is important that the steel balls directly contact the cover slip so the well that is formed allows the egg/embryo to rest directly on the cover slip. Remove the steel balls after the agarose has cooled, then fill the dish with a solution of 5% Ficoll (Sigma) in 1/3 MR. Position one egg or dejellied embryo in each well with the region to be examined facing downward, in contact with the cover slip.

3.4. Collecting Data from Live Cells
Using a Confocal Laser-Scanning Microscope (see Note 4)

1. Yolk platelet (cytoplasmic) displacements: Live embryos that had been incubated in Nile Red were placed in viewing dishes on the microscope stage between 15 and 20 min postfertilization in order to detect the earliest possible displacements of organelles. Because we did not use antifade reagents or other methods for reducing photobleaching, we used the fastest possible scan speed to establish the appropriate settings for data collection, e.g., gray-scale range, box size, and beginning and ending optical sections for z-series. Once all parameters were set, we began collecting consecutive z-series with no lapse of time between z-series. Once rotation achieves maximal velocity, organelle movements occur too rapidly to do any image averaging during data collection. Therefore, with the Bio-Rad 1024 we used small boxes (128×128 pixels or smaller) and the "normal" scan speed to detect fastest movements. After data collection was completed, we imported the images into ImageSpace (Molecular Dynamics, Mountain View, CA). In this program we selected a single optical plane from each of the z-series and made a video of the organelle movements in that optical plane over time.

2. Microtubule-mediated organelle transport: Fertilized eggs were incubated in $DiOC_6(3)$ alone or $DiOC_6(3)$ plus Nile Red for 5 min. After rinsing, they were placed in a viewing dish and positioned on the microscope stage by approx 20–30 min postfertilization (prior to the onset of cortical rotation). To image movements of these organelles, we used a 60X objective lens and began collecting data in the optical plane believed to be the "shear zone" *(2)*. This plane was the outermost section in which we could detect $DiOC_6(3)$-labeled organelles. To collect data as rapidly as possible, we chose a single optical section prior to the onset of rotation (typically that region approx 4–8 µm from the cell surface that contains the parallel array of microtubules) and collected images of that same plane throughout cortical rotation. These images can be played back using ImageSpace or other imaging-processing programs.

3. Microtubule polymerization: Fluorescently labeled tubulin monomers were injected into dejellied, fertilized eggs prior to the onset of cortical rotation, as described in **Subheading 3.1.3.** The eggs were then positioned in viewing wells and placed on the microscope stage. When FITC-labeled tubulin was used, we also labeled yolk platelets with Nile Red in order to assist with finding the appropriate region to monitor for microtubule polymerization (which we now know to be 4–8 µm inside the cell surface). We then monitored this region over time using either successive z-series, or by collecting data from a single optical section over time, as described in items 1 and 2. When using Nile Red, the "transport zone" is identified as that region of the egg containing the smallest labeled yolk platelets. When using rhodamine-labeled tubulin, we used $DiOC_6(3)$-labeled organelles to aid in finding the transport zone, identified as that region approx 4 µm deeper than the outermost $DiOC_6(3)$-labeled organelles and approx 4–5 µm shallower than the germ plasm (seen as clusters of $DiOC_6(3)$-labeled mitochondria). The use of either Nile Red or $DiOC_6(3)$ seems to mask the subtleties of microtubule polymerization, and our most informative data were collected without using either Nile Red or $DiOC_6(3)$ for visualization of organelles. It is much more difficult to locate the transport zone using this approach,

but the slight auto-fluorescence of the yolk platelets (which is minimal in living eggs) can aid in finding the appropriate region. A cell surface label could also be used to aid in locating the egg cortex but becomes unnecessary with experience examining labeled organelles in the egg periphery.

4. Actin polymerization: Rhodamine-labeled actin monomers were injected into meiotically immature (stage VI) oocytes. Approx 12 h later, after progesterone-induced meiotic maturation (distinct white spot visible at the apex of the animal pole), oocytes were placed in viewing dishes. Oocytes were positioned on the microscope stage and a series of images (z-series) were collected at 0.2-μm intervals to determine the organization of cortical actin. An oocyte was then activated (by pricking with a sharp glass electrode), and actin dynamics were monitored throughout the first cell cycle by collecting either consecutive z-series over time or by examining the same optical section over time (as described in items 1 and 2).

4. Notes

1. Viewing dishes: It is important when making the viewing dishes that the steel balls directly contact the cover slip so that the well being formed will allow the egg/embryo to rest directly on the cover slip. Any agarose between the egg/embryo and the cover slip will decrease the ability to focus and image deeper regions of the specimen.

2. Lenses and cover slips: Objective lenses are corrected for cover slips of a known composition and thickness and it is important, particularly when using high numerical aperture lenses, that the refractive index of the cover slip match that of the lens being used for data collection. For example, 60X and 63X, 1.4 NA oil-immersion lenses must be used with cover slips of approx 0.17-mm thickness (we always use a No. 1 glass cover slip with our viewing dishes to allow us to use high magnification lenses, with greater NA, with every specimen preparation). For lower magnification work, it is possible to use a lens with a correction collar that can be rotated to match the internal lens elements with the cover slip thickness and immersion medium. This is convenient when examining tissue culture cells. The best approach for viewing living cells in an aqueous medium, however, is to use a water-immersion objective lens, (which can be used with or without a cover slip, when using an upright microscope) to reduce spherical aberrations. It is also important to avoid attempting to focus deeper into the sample than is allowed by the working distance of the lens.

3. Microinjections: It is very important that the eggs are not tipped and that the animal hemisphere remain upward during the microinjections to avoid inducing developmental abnormalities associated with inversion of the eggs during the first half of the first cell cycle.

4. Collecting data from live cells using a confocal laser-scanning microscope: The simplest approach for studying living cells is to collect data from the same optical plane over time, generating a time-lapse video sequence of images. Collecting 4-D data (3-D data over time) is somewhat more complex. Special temperature control and environmental chambers are not necessary with *Xenopus* eggs/embryos, unlike somatic cells, since they develop quite normally at room temperature. There are other conditions typically associated with imaging living cells, however, that must be considered. In order to achieve good temporal resolution, each experiment requires optimization of the following parameters: laser intensity, rate of data collection (scan speed and box size), detector (photomultiplier) sensitivity, and aperture size. The increased temporal resolution achieved, unfortunately, is often accompanied by a decrease in the image quality; but the information acquired compensates for this loss.

   a. Laser intensity: When examining the first cell-cycle events, our goal is to rapidly collect multiple images of the specimen over extended time periods (3-D and 4-D imaging). Therefore, we use the lowest possible laser intensity (highest neutral density filters) to avoid photobleaching of the labeled probes and reduce the possibility of significant phototoxicity. With the Bio-Rad MRC 1024, equipped with a mixed gas krypton/argon laser, we use neutral density filters to provide 10% laser power if possible; in some cases, however, it has been necessary to use filters yielding 30% laser power. Moderate photobleaching of most fluorescent reagents occurs over time when we using 10% laser power, whereas 30% laser power typically results in significant photobleaching and reduced observation times. Green fluorescent protein constructs, however, are much more resistant to photo damage and demonstrate only minimal detectable photobleaching at either 10% or 30% laser power. We did not observe phototoxicity in *Xenopus* eggs/embryos using either 10% or 30% laser power with up to 45 min continuous exposure during the first cell cycle, as determined by the ability of the embryos to develop into swimming tadpoles.

b. Rate of data collection (scan speed/box size): When examining the dynamics of cortical rotation, during which time organelles are moving along the microtubules at velocities up to 60 μm/min, it is important to collect images as rapidly as possible. Adjusting the size of the image data set (the pixel by pixel box size) and/or the speed of the scan directly affects the rate at which data are collected. When collecting data from the same optical plane over time, generating a time-lapse video sequence of images, we can use a box size of 512×512 pixels for monitoring slower events. When using the Bio-Rad MRC 1024, with a box size of 512×512 pixels and the scan speed set at "slow," images are collected at a rate of about one frame every 3 s (additional time is required to store the data to the hard drive). Using the slower scan speeds results in loss of information between scans and yields discontinuous, jerky movements in the time-lapse video playback. By using the "normal" setting, the rate of image collection can be increased to one frame every second; this is accomplished by decreasing the pixel dwell time and yields a noisier image due to the decreased signal-to-noise ratio. When monitoring rapid movements, it is not possible to use image averaging or smoothing techniques (e.g., kalman filtering) during data collection because there is detectable organelle displacements between scans. To detect the most rapid organelle movements during cortical rotation, therefore, we typically used small box sizes (e.g., 128×128 pixels), the normal scan rate setting, and no image averaging or filtering. Although this compromises the quality of the image, we can successfully detect the rapid organelle transport associated with cortical rotation. When using the Leitz DM-IRB (Leica Microsystems, Inc., Deerfield, IL) to examine cortical rotation, we were able to obtain high-quality data sets at a rate of one frame every 2 s. The NORAN OZ CLSM (NORAN Instruments Inc., Middleton, WI), which has the ability to collect images as rapidly as 480 frames per second with a box size of 256×60 or 512×30 pixels, generated excellent videos of continuous organelle movements throughout cortical rotation. To collect 4-D data (3-D data over time), we collect successive z-series, each z-series consisting of between 6 and 8 optical sections that are approx 0.5–1.0 μm apart using box sizes of between 128×128 pixels and 512×512 pixels. Each Z-series starts as close to the egg periphery as possible, and the final optical plane is about 20 μm inside the cell surface. For very fast movements, we collect optical sections every 0.2–0.5 μm, (with a 60X lens) using box sizes of 128×128 pixels or smaller. We collect successive z-series; in other words, as soon as one z-series ends, we immediately begin the next z-series in that same region of the egg and do not allow any time to lapse between the end of one z-series and the onset of the next. This approach allows us to monitor movements in the entire egg periphery over extended time periods. We then import the data sets into ImageSpace (Molecular Dynamics) for data analysis and image processing.

c. Aperture (iris) size: The magnification and numerical aperture of the lens being used determine the pinhole size. For example, it is recommended that the iris be set at 2 mm when using the Nikon 60X, 1.4 NA lens. Increasing the size of the iris allows more light to enter the photomultiplier tube but reduces the z-resolution, or confocal nature, of the data. When working with very-low-intensity signals in living cells, it is often better to use larger iris settings rather than increase the laser intensity. This decreases photobleaching and phototoxicity that can plague live cell imaging, but does compromise the resolution and optical sectioning ability.

d. Detector sensitivity (gain and black level): The photomultiplier tube (PMT) settings control the intensity of the image. Proper gain settings ensure that the brightest areas of the image are white, whereas proper black level settings assure that low-intensity data are not lost. These settings must be optimized for the data set to utilize the full gray scale. Increasing the gain setting produces a brighter image and, therefore, permits the use of lower intensity laser settings. Although this also results in noisier images, it is the preferred option for live cell imaging. A proper black level setting assures that those parts of the image that are not emitting any photons are displayed as true black.

References

1. Larabell, C. A., Rowning, B. A., Wells, J., Wu, M., and Gerhart, J. C. (1996) Confocal microscopy analysis of living *Xenopus* eggs and the mechanism of cortical rotation. *Development* **122,** 1281–1289.
2. Rowning, B. A., Wells, J., Wu, M., Gerhart, J. C., Moon, R. T., and Larabell, C. A. (1997) Microtubule-mediated transport of organelles and localization of β-catenin to the future dorsal side of *Xenopus* eggs. *Proc. Natl. Acad. Sci. USA* **94,** 1224–1229.
3. Kloc, M., Larabell, C., and Etkin, L. D. (1996) Elaboration of the messenger transport organizer pathway for localization of RNA to the vegetal cortex of *Xenopus* oocytes. *Dev. Biol.* **180,** 119–130.

18

Whole-Mount Immunolabeling of Embryos by Microinjection

Increased Detection Levels
of Extracellular and Cell Surface Epitopes

Charles D. Little and Christopher J. Drake

1. Introduction

A major conceptual breakthrough has occurred in the field of developmental morphogenesis in the past decade. It is now clear that molecular mechanisms at the external cell surface and within the surrounding extracellular matrix are fundamental to embryogenesis. Adhesion receptors, receptor tyrosine kinases, cell-cell recognition machinery, adhesion substrate molecules, and matrix-bound growth factors are all part of an instructive molecular language. Together these various kinds of molecules encode information that is displayed in a topographical manner. Understanding the physical and temporal distribution of these molecules is necessary if the morphogenetic code is to be deciphered.

Enormous amounts of excellent immunolabeling data have been generated using fixed embryonic specimens; however, prefixation immunolabeling has received much less attention. Here we describe an in vivo microinjection method for immunolabeling embryos (1–3). This protocol employs immunofluorescence labeling and microscopy; however, other detection methods should be equally efficacious. Examples will be described in which attempts at conventional immunofluorescence labeling failed to label cells or extracellular matrix, whereas the prefixation microinjection approach was entirely successful. In other cases, we noted dramatically improved signal-to-noise ratios compared with conventional postfixation immunolabeling. This chapter outlines a protocol for avian embryos; however, the method should be applicable to fish, amphibian, and mammalian embryos. Similarly, it is reasonable to assume that this approach could be used on excised embryonic rudiments, such as the eye, heart, and limb buds.

2. Materials

1. Buffers and solutions:
   a. 10X Phosphate-buffered saline (PBS), pH 7.2: 2 L distilled H_2O, 160 g NaCl, 4 g KCl, 23 g Na_2HPO_4 (dibasic anhydrous), 4 g KH_2PO_4 (monobasic anhydrous).

From: Methods in Molecular Biology, Vol. 135: Developmental Biology Protocols, Vol. I
Edited by: R. S. Tuan and C. W. Lo © Humana Press Inc., Totowa, NJ

b. Embryonic PBS (EPBS), pH 7.4: Composition per 2 L, use double-distilled water (DDH$_2$O), 16 g NaCl, 0.4 g KCl, 2.3 g Na$_2$HPO$_4$ (dibasic anhydrous), 0.4 g KH$_2$PO$_4$ (monobasic anhydrous), 0.12 g CaCl$_2$ (dihydrate), 0.2 g MgCl$_2$ (hexahydrate). CaCl$_2$ and MgCl$_2$ should be added after other chemicals are dissolved. Sterile filter the EPBS into 1-L flasks and store at 4°C.

c. PBS with 0.01% sodium azide (NaAz): To 1 L of 1X PBS, add 1 mL of 10% NaAz.

2. Plasticware and paper rings: Plastic Petri dishes 150×25 mm, 100×15 mm, and 35×10 mm; paper rings cut from Whatman 52 filter paper with a cork borer (Whatman International Ltd., Maidstone, UK); the ring has an inside diameter of 13 mm and an outside diameter of 23 mm.

3. Agarose bed for injection: A 5% solution of Bacto-Agar (Difco Laboratories, Detroit, MI) agarose is autoclaved and while warm, a thin layer is poured into 35 × 10-mm plastic Petri dishes.

4. Micropipets for injection: Sigmacote-treated (Sigma, St. Louis, MO) glass tubes (150 mm long, outer diameter of 0.9 mm, inner diameter of 0.6 mm) are pulled using a Narishige PB-7 pipet puller (Narishige, Tokyo, Japan) to obtain a micropipet with a 18-μm diameter.

5. Incubation chamber: Injected embryos are incubated in a humidified chamber made from a 150 × 25-mm plastic Petri dish. Moist paper towels are placed on the bottom of the Petri dish, and the dish is then placed into a 37°C incubator.

6. Fixatives: Paraformaldehyde fixative, pH 7.4. In a fume hood, add 1 drop of 1 *M* NaOH and 3 g of paraformaldehyde to 70 mL DDH$_2$O. Stir using medium heat until dissolved (20 min). Let cool and then add 10 mL of 10X PBS and then bring to 100 mL with DDH$_2$O. Store at –20°C for up to 1 mo. Storage at 4°C should be no longer than 1 d.

7. Blocking solution: 3% Bovine serum albumin (BSA): Add 3 g BSA to 100 mL 1X PBS + NaAz; store at –20°C.

8. Antibodies: Both monoclonal and polyclonal antibodies are used as primaries. Appropriate fluorochrome-labeled antibodies, such as FITC, rhodamine, and Cy5 conjugates are used as secondaries.

9. Fluorochrome conjugation of immunoglobulins: Fluorescent probes were conjugated by amine linkage to primary antibody using Amersham FluoroLink-Ab Labeling Kit (Amersham Life Science Inc., Arlington Heights, IL). Primary antibody is isolated using protein-G affinity chromatography. The IgG fraction is then resuspended in 1 *M* sodium carbonate buffer, pH 9.3, and mixed with the desired fluorescent dye that is supplied as a bifunctional NHS-ester. After incubation, the antibody labeling mixture is applied to a gel filtration column for separation of the labeled protein from the unconjugated dye. The labeled protein is eluted with neutral-pH PBS containing 0.1% NaAz. Protein concentration is determined and the solution is stored at 4°C in a light-tight container.

10. Antibleaching mounting solution for fluoroprobes: To 45 mL glycerol, add 5% *n*-propyl gallate, 0.25% 1,4-diazabicyclo-[2,2,2] octane, 0.0025% p-phenylenediamine. Bring to vol of 50 mL with glycerol and cover with aluminum foil. Rotate for 24 h and stand for 24 additional hours to remove air bubbles *(4)*.

3. Methods

1. Embryo preparation: Quail and chicken embryos at Hamburger and Hamilton stages 6–10 are mounted on paper rings as a means of manipulating the embryos. The egg is gently broken into a dish. The albumen covering the embryo is removed using a transfer pipet exposing the vitelline membrane. A Whatman 52 (Whatman International) paper ring, 23 mm in diameter with a 13-mm hole punched in the center, is carefully centered over the

embryo and positioned on the vitelline membrane. A period of 2–3 min is allowed for the paper to adhere to the vitelline membrane (*see* **Notes 1** and **2**). Using sharp iridectomy scissors, the embryo-ring assembly is cut away from the yolk. The paper ring is grasped by forceps and lifted from the yolk (*see* **Note 3**). Embryos are then immersed in sterile EPBS and gently washed to free loosely adherent yolk particles. It is not necessary to completely clean the embryo, and attempts to do so will damage it. The washed embryos are placed on a solid agarose bed cast into a plastic 35-mm bacteriology dish. Unless very extended incubation periods are planned, these steps do not require sterile conditions. The procedures for preparing embryos have been previously described (*1,2*).

2. Primary antibodies and controls: The concentration of antibody required to generate an acceptable immunofluorescent signal varies. In the past, antibodies at concentrations from 0.5–4.0 mg/mL of immunoglobulin have proven successful. Purified, high-affinity antibodies are best. However, one of the advantages of this technique is that even antibodies that gave "marginal" results in conventional whole-mount immunolabeling have proven to be surprisingly effective when microinjected (*3*). The volume that can be injected is dictated by the stage/size of the embryo. Commonly, at stages 7–9, a single bolus injection is delivered with a vol of 25 nL. A 25-nL injection of a sample at a concentration of 1 mg/mL (1 ng/nL) represents 25 ng of delivered immunoglobulin. The delivery of more reagent is achieved by injecting both sides of the embryo.

   Most experiments to date have involved injection of unconjugated "primary" antibodies followed by fixation, permeabilization, and immunolabeling with fluorochrome-conjugated secondary antibodies. Recently, however, excellent results have been obtained by injection of fluorochrome-conjugated primary antibodies (*see* **Note 4**).

   Controls for nonspecific labeling of two types have been performed repeatedly with no evidence of such problems. The first control method is the conventional practice of using preimmune or nonspecific immunoglobulins; in this case, however, these reagents are injected. A second control method, used during initial development of this method, was to compare immunofluorescence labeling patterns obtained using whole-mount postfixation methods (*5*) to the pattern obtained using in vivo microinjection (*3*). For example, when the staining patterns of fibronectin or the QH1 antigen are analyzed using both methods, the pattern observed is identical. To date, multiple antigens have been detected using microinjection; for example, $\alpha v \beta 3$ integrin, vitronectin, elastin, fibrillin-2, fibronectin, fibulin-1, vascular endothelial growth factor, matrix metalloproteinase-2, and the quail endothelial marker epitope QH1 (**refs. 3** and **6** and unpublished observations). Of these, elastin and vascular endothelial growth factor were not detectable using conventional methods, even after multiple trials with various fixatives. Often, as was the case for elastin, the inability to detect protein contradicted other data (i.e., immunoblotting data) that suggested the presence of the protein. Indeed, this in vivo immunolabeling technique was devised in response to such results. When elastin antibodies were microinjected, brightly fluorescent extracellular fibrils were observed (unpublished observations).

3. Microinjection and culture: Washed embryos are placed on agar plates and moved to the injection apparatus, which consists of a dissecting microscope with a variable-angle mirrored transmitted light base, a micromanipulator, and a micropump (*see* **Note 5**). High-quality optics with an upper magnification of 50X provides both sufficient resolution and the necessary working distance to allow for micromanipulation. There is considerable flexibility regarding the nature of the micromanipulator and the pump. The simplest suitable micromanipulators are mechanical three-axis models. More sophisticated models having hydraulic controls are helpful but are not required for basic injections. There are a variety of pumps that are suitable to drive the microinjection.

Examples of pumps that have been used successfully are hydraulic pumps, such as the Sage syringe pump (Orion Research, Beverly, MA), or pneumatic pumps, such as the Pico-Spritzer 2, (General Valve Corp., Fairfield, NJ) and the Pico-Injector (Medical Systems Corp., Greenvale, NY).

Micropipets for microinjection are prepared by breaking the tip of a micropipet using a pair of forceps such that an 18-μm diameter opening is formed. The injection volume is calibrated by filling the tip with EPBS and injecting 10 vol onto the lid of a plastic Petri dish, forming a single droplet. The droplet volume is determined using a glass Microcap pipet (Drummond Scientific, Broomall, PA). The volume of a single bolus is calculated by dividing by the droplet volume by 10 (droplet vol/10).

The actual microinjection is the most difficult aspect of the in vivo immunolabeling protocol. In early-stage avian embryos, the extracellular space of the somatopleure or splanchnopleure (i.e., the space between the mesoderm and the ectoderm/endoderm) has proven to be a suitable site for the delivery of reagent. Critical elements to the delivery of reagents to this space are the angle of the micropipet and the lighting. Micropipet angles of 45–70° are generally used. Because of the variability in lighting provided by the gimbal-mounted mirror in the brightfield base, and the preference of the user, it is not possible to provide a standard lighting setting. Empirical manipulation, however, will quickly establish a preferred setting. Microinjected embryos are incubated in a humidified atmosphere at 37°C for as little as 15 min to as long as 60 min after injection (*see* **Notes 6** and **7**). For most antigens, 30 min appears to be adequate. Tissue culture incubators are convenient for this purpose.

4. Fixation and immunolabeling: Embryos still mounted on the paper ring are washed 3X in PBS and fixed in freshly prepared 3% paraformaldehyde/PBS for 30 min. After fixation, the vitelline membrane is removed and the embryo is made permeable by two baths in ice-cold absolute methanol (45 min each) (*see* **Notes 8** and **9**). Embryos are rehydrated in a graded ethanol (100–30%) series and then placed in PBS plus azide (AZ). Embryos are removed from the ring using a microcutter to facilitate the later step of mounting. The cuts freeing the embryo are made just outside the boundary of the area pellucida/opaca such that a rectangle is formed. The embryos are transferred to a 24-well cell culture plate and placed in 3% BSA + AZ for 12 h to overnight at 4°C. This step and all subsequent labeling and washing steps are done using a rocker capable of gentle agitation. After BSA blocking steps, the embryos are washed once in PBS + Az and immunolabeled using fluorochrome-conjugated secondary antibodies at a concentration of 10–15 μg/mL in PBS + NaAz, 12 h to overnight at 4°C. The secondary antibody solution is removed by pipeting and replacing with PBS + NaAz. Embryos are washed a total of 3X with the PBS + NaAz, remaining in the final wash for 12 h to overnight at 4°C. The procedures for fixation and immunolabeling embryos have also been previously described *(1,2)* (*see* **Note 2**) (*see* **Fig. 1**).

5. Mounting: Using a shortened spatula, the washed embryo is transferred from the 24-well plate to a glass microscope slide, where a drop of mounting medium has been placed. To transfer the embryo, it is first placed into a dish of PBS + NaAz and gently coaxed onto the spatula with a pair of forceps. It is then lifted and moved to the slide, which is viewed on a dissecting scope. The tip of the spatula is placed into the mounting media and the embryo is lowered into the media. After the embryo slides off of the spatula and is positioned as desired, spacers cut from a #1 glass cover slip (previously scribed with a diamond point) are placed on both sides of the embryo, and a #0 glass cover slip glass is positioned over the embryo. The cover slip is then sealed with clear nail polish to prevent slipping.

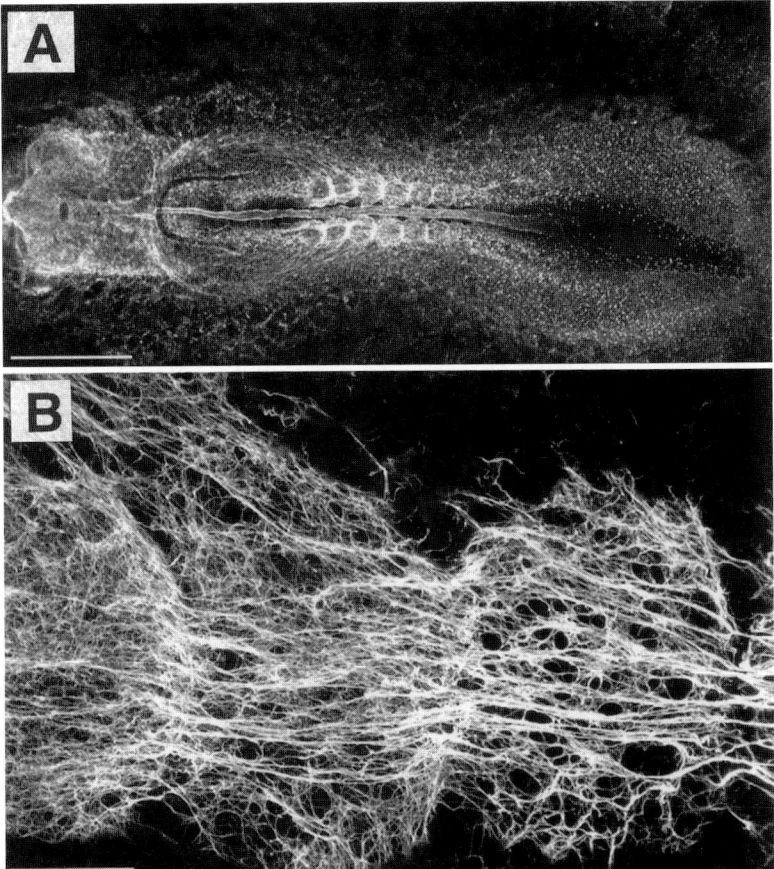

Fig. 1. One of the most successful uses of the in vivo labeling technique is the results obtained using the JB3 monoclonal antibody. This antibody recognizes avian fibrillin-2. **(A)** A low magnification image of a 6-somite quail embryo that was microinjected with immunopurified JB3 (25 nL at 1.3 ng/nL), cultured for 30 min, fixed, permeabilized, and immunolabeled. Fibrillin-2 immunostaining is evident as bright filamentous strands that are most conspicuous along the somitic field. **(B)** A higher magnification view of fibrillin-2 filaments associated with the somites. Bar: a-500 μm; b-50 μm. For details *see* **ref. 5**.

4. Notes

1. The paper ring assembly can be positioned with either the ventral or dorsal surface facing upward, depending on which areas of the embryo are to be microinjected. Even confined compartments such as the somitocoele, endocardial cushions, and the neural tube lumen can be injected successfully.
2. *After stage 10, it is not practical to ring-mount the embryos*, because of the presence of blood vessels, which would be compromised. Thus, older embryos must be injected *in ovo* or in shell-less culture.
3. When removing the embryo from the yolk, it is helpful to cut circumferentially through paper-ring/vitelline composite. This last step crimps the paper edge to the vitelline membrane. It is the vitelline membrane that supports the embryo during subsequent washing and transfer steps.

4. One significant advantage of prefixation immunolabeling is the strikingly low degree of background noise. This is partially attributable to the fact that there is no "intracellular" trapping of secondary antibodies. It is important to point out that *this method does not label intercellular proteins.*

5. After transferring the embryo to the agar plate for microinjection, a small volume of EPBS should be placed directly on the embryo. This facilitates microinjection and prevents desiccation.

6. Postinjection incubation times must be determined empirically. If a large excess of antigen is present in the extracellular space, a radial diffusion pattern of fluorescence is observed. Hypothetically, in order to label all available antigens, a condition of antibody-excess must be achieved. While no definitive data are available on immunoglobulin diffusion rates, unpublished empirical data suggest that control antibodies diffuse throughout the mesoderm on one side of the notochordal axis within 10–15 min. The diffusion across the midline appears to be constrained somewhat by the notochord/neural tube; eventually, antibodies can be detected in the side opposite the injection site (unpublished observations). For most antigens tested to date, the conditions described herein have provided adequate labeling. It is important to note that if the antibody perturbs the function of a given antigen, the incubation times must be kept to a minimum, as prolonged incubation falls into the realm of a perturbation study.

7. Preliminary studies using simultaneous, indirect, double-immunofluorescence show that respective antibodies segregate in the correct manner (unpublished observations) provided the appropriate host-specific secondary antibodies are used. Thus, the primary antibodies must be from two different hosts (e.g., mouse antiprotein X and rabbit antiprotein Y) and the fluorochrome-conjugated secondary antibodies must be from the same host (e.g., goat antimouse IgG-fluorochrome A and goat antirabbit IgG-fluorochrome B).

8. *The vitelline membrane must be removed prior to methanol permeabilization.* This semitransparent membrane is a difficult to see. If visualization of the membrane is difficult, cutting the embryo off the ring will sometimes help identify the membrane for removal.

9. When adding ice-cold methanol to the cut embryos, it is important that they are flat on the bottom of the dish. To ensure this, remove the PBS and put one small drop of methanol onto the embryo for several seconds, then add enough methanol to cover. This prevents folding of the embryo, thus facilitating later imaging.

Acknowledgments

This work is supported by the March of Dimes Birth Defects Foundation, grants FY95-0453 and RO1 HL57645 to C. D. L. and grant RO1 HL57375 to C. J. D.

References

1. Drake, C. J., Davis, L. A., and Little, C. D. (1992) Antibodies to β_1 integrins cause alterations of aortic vasculogenesis, in vivo. *Dev. Dyn.* **193,** 83–91.
2. Drake, C. J., Davis, L. A., Hungerford, J. E., and Little, C. D. (1992) Perturbation of Beta-1 integrin-mediated adhesions results in altered somite cell shape and behavior. *Dev. Biol.* **149,** 327–338.
3. Drake, C. J., Cheresh, D. A., and Little, C. D. (1995) An antagonist of integrin avb3 prevents maturation of blood vessels during embryonic neovascularization. *J. Cell Sci.* **108,** 2655–2661.

4. Giloh, H. and Sedat, J. W. (1982) Fluorescence microscopy: Reduced photobleaching of rhodamine and fluorescein protein conjugates by *n*-propylgallate. *Science* **217,** 1252–1255.

5. Little, C. D., Piquet, D. M., Davis, L. A., Walters, L., and Drake, C. J. (1989) The distribution of laminin, collagen type IV, collagen type 1 and fibronectin in the cardiac jelly basement membrane. *Anat. Rec.* **224,** 417–425.

6. Rongish, B. J., Drake, C. J., Argraves, W. S., and Little, C. D. (1998) Identification of the developmental marker, JB3-antigen, as fibrillin-2 and its de novo organization into embryonic microfibrous arrays. *Dev. Dyn.* **212,** 461–471.

19

Confocal Laser Scanning Microscopy of Morphology and Apoptosis in Organogenesis-Stage Mouse Embryos*

Robert M. Zucker, E. Sidney Hunter III, and John M. Rogers

1. Introduction

In our efforts to use confocal laser scanning microscopy for study of organogenesis-stage rodent embryos, we have developed fixation and clearing methods to allow optical sectioning through embryos with thickness approaching 1 mm (z-axis). We have combined fixation and clearing methods with fluorochrome staining for several purposes. In this chapter we present two methods; first, clearing with methyl salicylate (oil of wintergreen) and staining with Nile blue sulfate (NBS) (not used as a vital dye for this protocol) for general morphological assessment, and second, staining live embryos with the vital stain LysoTracker® Red (LT), followed by fixation and clearing with benzyl alcohol:benzyl benzoate (BABB) to visualize areas of apoptosis (*see* **Note 1**, **ref. *1***). With both protocols, an entire organogenesis-stage rodent embryo can be optically sectioned and reconstructed in three dimensions (3-D) to reveal areas of dye staining.

In the morphological staining procedure, embryos are fixed in 4% paraformaldehyde overnight, dehydrated in a graded methanol series with NBS added to the 95% methanol solution, and cleared in methyl salicylate. Fluorescence appears to come from both the methyl salicylate and the NBS, and staining is uniform and nonspecific, giving good morphological detail. For analysis of apoptosis, embryos were incubated in the LT stain, fixed in 4% paraformaldehyde overnight, dehydrated in a graded methanol series, and cleared in BABB. LT is an aldehyde-fixable lysosomotropic dye that stains large phagolysosomes and apoptotic bodies in a manner that appears to be very similar to several other vital dyes used for this purpose, including acridine orange, NBS, and neutral red. However, LT has the great advantage of being aldehyde fixable, is also quite bright and shows little evidence of bleaching with repeated scans. To test our staining protocol, apoptosis was induced by the chemotherapeutic hydroxyurea or arsenic by adding these compounds to the medium of gestation day (GD) 8 mouse

---

*The research described in this article has been reviewed and approved for publication as an EPA document. Approval does not necessarily signify that the contents reflect the views and policies of the Agency, nor does mention of trade names or commercial products constitute endorsement or recommendation for use.

From: *Methods in Molecular Biology, Vol. 135: Developmental Biology Protocols, Vol. I*
Edited by: R. S. Tuan and C. W. Lo © Humana Press Inc., Totowa, NJ

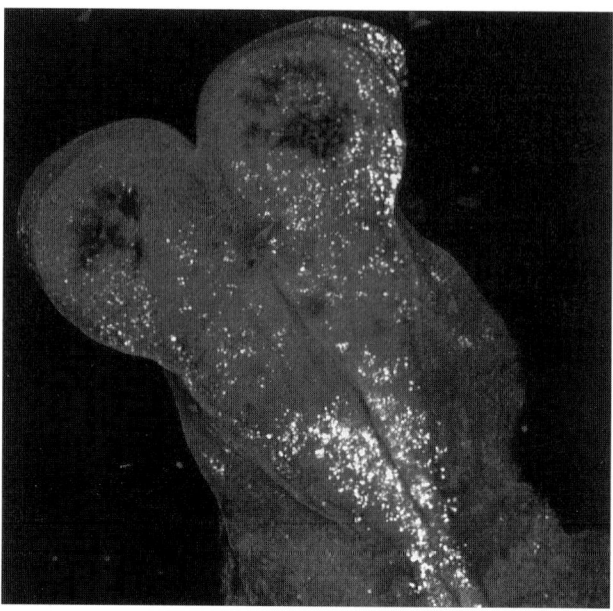

Fig. 1. 3-D reconstruction of an unfixed, uncleared GD 8 mouse embryo treated in vitro with arsenic for 6 h and vitally-stained with 5 μm LT for 30 min. The small size and flatness of the embryo at this stage allows for the imaging of this tissue without clearing. However, even here most of the detectable fluorescence originates near the surface. LT is staining regions of lysosomal activity in the embryo (original magnification ×125).

embryos developing in vitro (**Fig. 1**). Alternatively, pregnant mice were given a teratogenic dose of methanol on GD 7; on GD 9, embryos were removed and processed for imaging. For the figures presented in this chapter, mouse embryos were harvested on GD 8 or 9. However, we have used these techniques on mouse embryos from GD 7 through GD 10, as well as on isolated limb buds from GD 14 rat embryos.

2. Materials

1. Animals and teratogenic chemicals: CD-1 mice were obtained from Charles River Laboratories (Raleigh, NC). Animals were kept on a 14-h light:10-h dark lighting schedule and provided water and Prolab Rat/Mouse/Hamster 3000 formula (PMI Feeds, St. Louis, MO) ad libitum. Females were housed with males for the last 2 h of the light period, and females with copulatory plugs were considered to be at GD 0. Teratogenic chemicals used for this presentation include hydroxyurea (HU; Sigma, St. Louis, MO) and methanol (Optima grade; Fisher, Pittsburgh, PA).
2. Stains: Morphology: NBS (Fluka, Ronkonkoma, NY). Apoptosis: 1 mM LT, in DMSO (Molecular Probes, Eugene, OR) was obtained in 20 vials (50 μL each). A 50-μL vial of LT was diluted with 50 μL DMSO (Sigma) to make a 500-μM stock solution. The LT red stock solution was diluted 100-fold with Dulbecco's phosphate-buffered saline (PBS) (Gibco-BRL, Gaithersburg, MD) to yield a working solution with a final concentration of 5 μM.
3. Fixative: Paraformaldehyde (20%; Electron Microscopy Sciences, Fort Washington, PA) was diluted to 4% with PBS and stored frozen at −20°C in 5-mL aliquots. Use only elec-

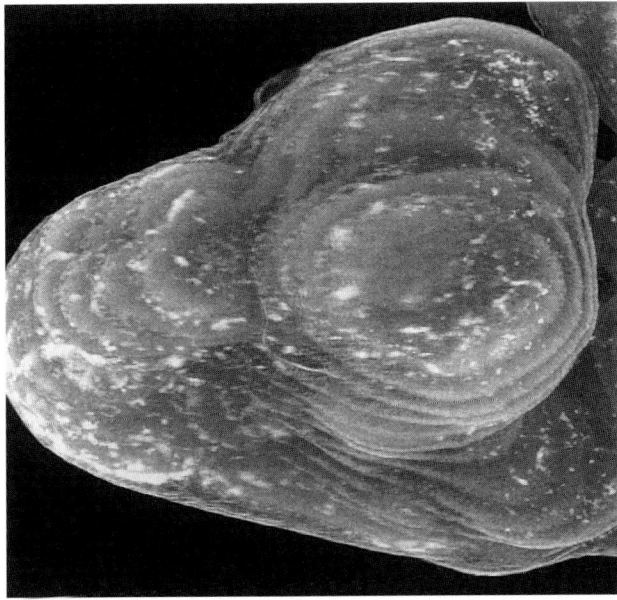

Fig. 2. 3-D reconstruction of 21 sections of a control GD 9 mouse embryo stained with NBS and cleared with methyl salicylate. The depth of the embryo head is greater than 600 µm. The edges of the optical sections can be seen, revealing the stacking of these individual scanned sections (original magnification ×100).

tron microscopy-grade paraformaldehyde that is freshly made or stored frozen at –20°C. Higher concentrations of paraformaldehyde will yield more autofluorescence. It is advisable to use lower concentrations for probes that excite with 488-nm light and emit in the green range.

4. Methanol dehydration series: After fixation, embryos were washed extensively with PBS to remove the fixative. The embryos were then dehydrated in methanol (#A-412-500; Fisher) with successive 15-min washes of 50, 70, and 95% (v/v) methanol in distilled water and then 2X with 100% methanol. Methanol dehydration is superior to acetone or ethanol dehydration. For NBS staining, NBS powder (1:20,000 w/v) was dissolved in 100% ethanol, and this solution was added dropwise to the 95% methanol dehydrating solution.

5. Clearing agents: A clearing solution consisting of 1:2 (v/v) BABB (Sigma) is used in conjunction with LT staining for apoptosis. The refractive index of the BABB solution is similar to the refractive index of the embryo, yielding an embryo that is nearly transparent *(2–4)*. For Nile blue staining of morphology, methyl salicylate (Fisher or Mallinkrodt [J. T. Baker, Phillipsburg, NJ]) is used for clearing instead of the BABB (**Figs. 2–4**).

6. Slides, cover slips, sealants: Cleared embryos are transferred to 2-mm thick depression slides (#12-265a; Fisher) and fresh BABB or methyl salicylate is added. Each depression slide is covered with a 20 × 30-mm cover slip (1.5 size), which is sealed with 3–4 successive coats of clear nail polish. It is essential that the depression slides be sealed well, as the BABB and methyl salicylate are potentially damaging to various components of the microscope. The slides can be carefully cleaned with acetone. Acetone will dissolve nail polish so it must be used sparingly and with care. After use, slides can be soaked in acetone to remove cover slips, and depression slides can then be cleaned and reused.

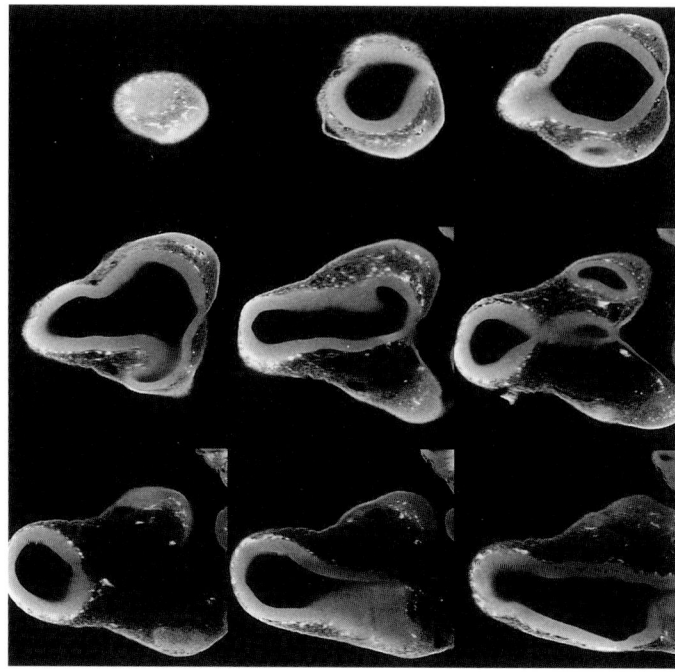

Fig. 3. Optical sections from the GD 9 mouse embryo 3-D reconstruction shown in **Fig. 2**. The individual sections were obtained using 568-nm excitation light, and the emitted light was collected through a 605/32 barrier filter (Chroma). Note the internal morphology of the embryo revealed in the individual sections. For clarity every second section is represented in the figure (original magnification ×100).

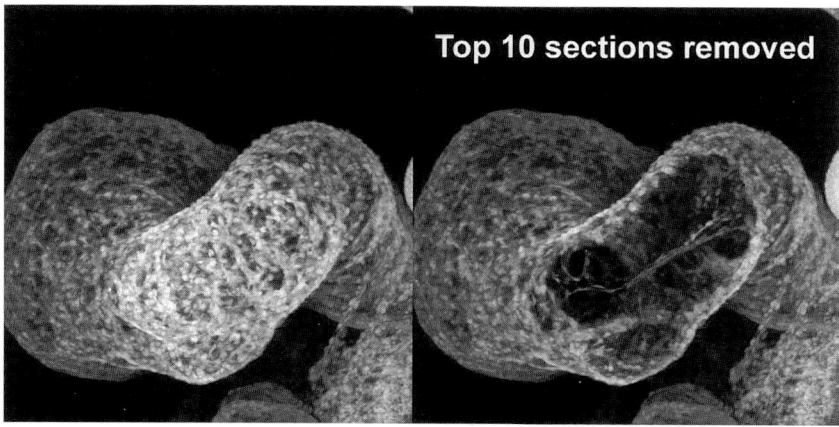

Fig. 4. **(A)** 3-D reconstruction of 30 sections of a heart from a GD 9 mouse embryo stained with NBS and cleared with methyl salicylate. The section images were obtained using 568 excitation light and emitted light was collected through a 605/32 barrier filter. Here, by choosing a higher number of sections, the individual sections are not apparent and a smooth 3-D image is obtained (i.e., compare to **Fig. 2**). **(B)** The top 10 sections are deleted from the 3-D reconstruction by the Leica imaging software and the remaining 22 serial sections are converted into a 3-D reconstruction, revealing internal structure (original magnification ×200).

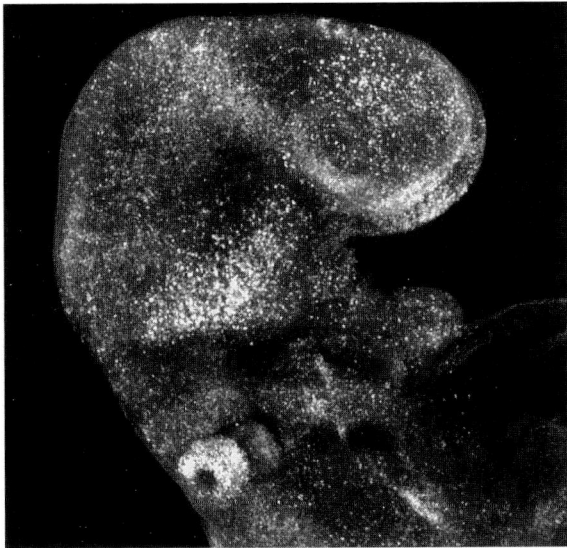

Fig. 5. 3-D reconstruction of 25 sections of a GD 9 embryo stained with LT and cleared with benzyl alcohol/benzyl benzoate to visualize regions of apoptosis after maternal treatment with methanol on GD 7. The fluorescence occurs throughout the embryo with the individual sections showing specificity of staining. The individual section images were obtained using 568 excitation light and the emitted light was collected through a 598/40 barrier filter (Chroma) (original magnification ×100).

7. Confocal microscope: This work requires a point-scanning confocal microscope that has good optical efficiency (*see* **Note 2**). For these studies, the Leica TCS4D with a Leica inverted DMIRB microscope (Leica Microsystems Inc., Deerfield, IL), and an argon-krypton laser (Omnichrome, Chino, CA) emitting three wavelengths (488, 568, and 647 nm) were used. A ×5 or ×10 objective with a high numerical aperture (NA) is desirable. Lenses that we have found to be acceptable include the following: Zeiss ×5 fluor (NA 0.25); Zeiss ×10 fluor (NA 0.5); Leica ×10 Plan APO (NA 0.5) and Leica multi-immersion ×10 (NA 0.4). A Zeiss lens (Carl Zeiss Inc., Thornwood, NY) will fit on a Leica microscope, but the magnification will be increased by 20%. The working distances of the air lens are in millimeters, whereas the Leica multi-immersion lens is approx 350 μm. Other lenses may also work adequately (*see* **Notes 6** and **7**).

3. Methods

1. Animals and treatments: Mice were mated as described above. For embryo culture, 3–6 somite embryos (GD 8) were removed from the gravid uteri and dissected free of decidua, Reichert's membrane and the parietal yolk sac, leaving the visceral yolk sac intact. Embryos were cultured for 24 h in 75% rat serum:25% Tyrode's by standard procedures. Hydroxyurea was added to the culture medium of treated embryos to give a final concentration of 250 μM for the entire culture period. At the end of the culture period, embryos were divested of the yolk sac and amnion and further processed as detailed below. Animals dosed with methanol were weighed on the morning of GD 7 and given two intraperitoneal injections of methanol 4 h apart, for a total dosage of 4.9 g methanol per kg maternal body weight. On GD 9, dams were killed and embryos removed from the gravid uteri and divested of their surrounding membranes (**Figs. 5** and **6**).

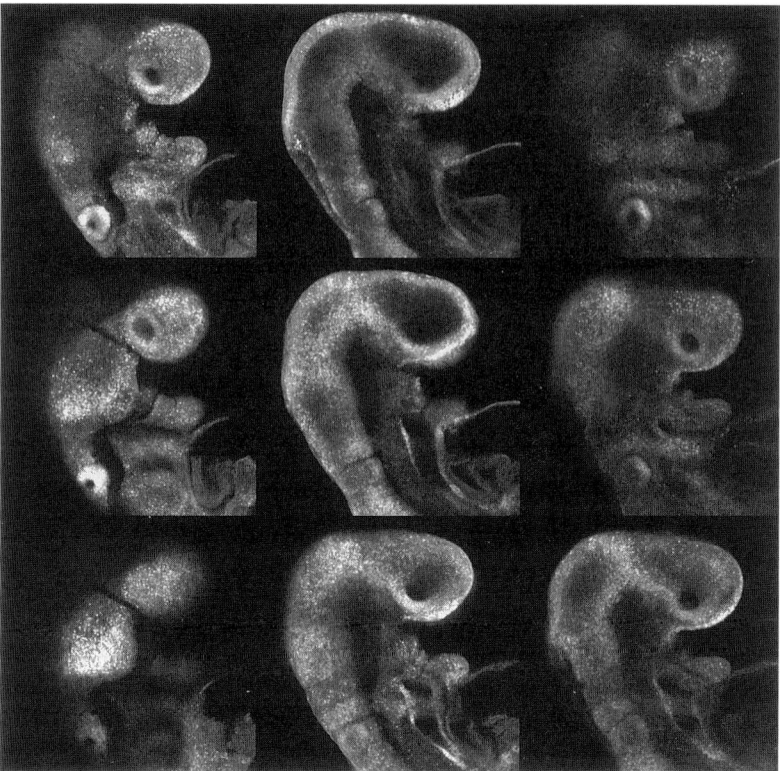

Fig. 6. Optical sections used to derive **Fig. 6**. The neuroepithelium shows increased cell death in the individual tissues. To illustrate the depth of laser penetration the circular structure (otic pit) at the top of the picture becomes bright and then disappears and then reappears as the sections go deeper into the embryo. This represents the future ears of the embryo. The sections above and below the middle sections were removed and every second section in the middle of the embryo is represented as a nine-section display. The sections are approx 20 μm apart with the embryo having a thickness of 500 μm.

2. Staining:
    a. Morphological staining with NBS: NBS was used to stain embryos to aid in general morphological assessment. NBS staining is accomplished in fixed embryos at the 95% methanol dehydration stage. NBS staining conditions were derived empirically. NBS powder is dissolved in 100% alcohol (approx 1:20,000 w/v) and was added dropwise to the 95% dehydrating solution to yield a light blue color. The embryo should become dark blue after a 1–2 h staining time. The final intensity of stain in the embryo is decreased during the clearing procedure, resulting in a lightly stained embryo that facilitates handling and observation on the microscope stage and enhances fluorescence.
    b. Vital staining of apoptosis with LysoTracker red: LT was used at a final concentration of 5 μ*M* in Hank's Balanced Salt Solution (HBSS). It is advisable to use a nonbuffered salt solution for vital staining with LT, as specific staining depends on a pH gradient across membrane-bound compartments. Between 1 and 3 embryos were incubated in this medium for 30 min at 37°C. The dye concentration used was decided on after preliminary experiments were carried out using 1, 2, or 5 μ*M* LT for either 30-min or

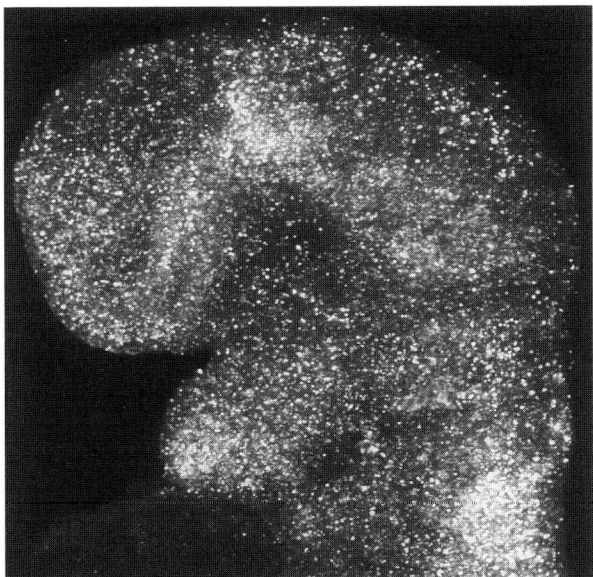

Fig. 7. 3-D reconstruction of 25 sections of a GD 9 mouse embryo that was treated with 250 μm hydroxyurea for 6 h in vitro. The bright spots represent LT staining and indicate regions of high lysosomal activity. The areas containing bright spots are located throughout the embryo but concentrations can be seen in the neural tube and otic pit (ear) regions (original magnification ×100).

      1-h incubation. Consistent staining was achieved with all combinations of LT concentration and time, but the best resolution with the highest signal-to-noise ratio, and best tissue conditions were achieved using 5 μ*M* LT for 30 min (**Figs. 7** and **8**). Embryo heartbeats were usually maintained through the end of the 30-min incubations.

3. Fixation: Paraformaldehyde (20%; Electron Microscopy Sciences) was diluted to 4% with PBS and stored frozen at -20°C. Fresh embryos were fixed immediately with ice-cold 4% paraformaldehyde for 24 h. Embryos stained with LT were washed with PBS after the 30-min stain period and fixed in the same manner. Embryos fixed overnight at 4°C were then stored at the 70% methanol dehydration stage (*see* #4) until further processing.

4. Dehydration: Embryos were dehydrated in methanol with successive 15-min washes of 50, 70, and 95% (v/v) methanol in distilled water and then 100% methanol. For NBS staining, the stain is added to the 95% methanol dehydrating solution, and embryos are left in this solution for 1–2 h (*see* 2a). Two additional 100% methanol washes were made to ensure that the embryo was completely dehydrated. When the dehydrating solutions were changed, the solutions were removed carefully with a transfer pipet and new solutions were added to the same glass tube containing the embryos rather than transferring the embryo into a different tube. In this way, potential pipetting damage to the embryo was minimized.

5. Clearing (*see* **Note 3**): For uniform morphological staining, NBS and methyl salicylate were used. Embryos were successively placed in 50% and 75% methyl salicylate in methanol, followed by two changes of 100% methyl salicylate, each step for 15 min. For staining of apoptosis with LT, the BABB clearing protocol was used. The BABB protocol produced an embryo that was more transparent than the one produced by the methyl salicylate protocol, allowing better discrimination of specific staining for apoptosis. Care

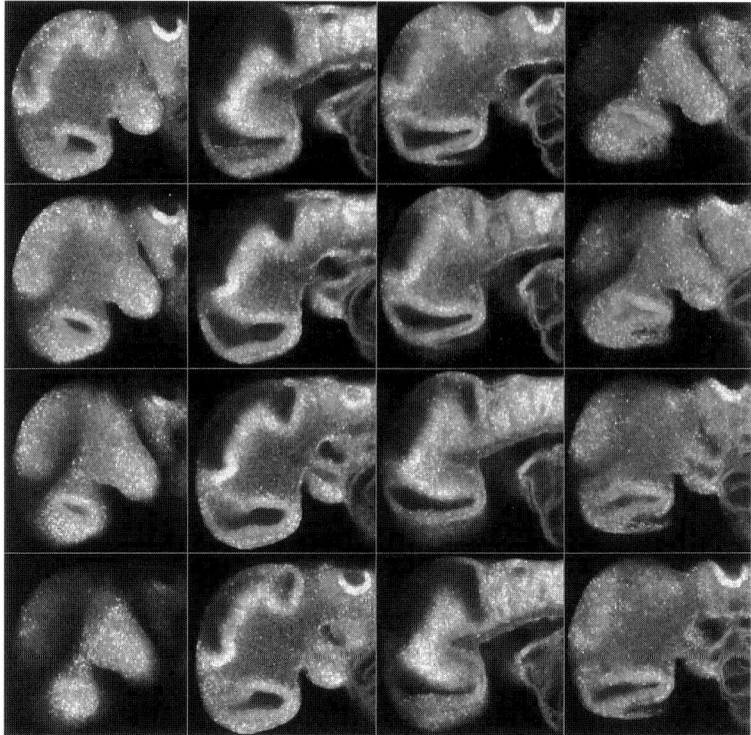

Fig. 8. Individual optical sections of the GD 9 hydroxyurea-treated embryo represented in **Fig. 7**. The sections are separated by about 20 μm. The sections above and below these sections were removed to present a 16-section display. There is an increased amount of fluorescence throughout the embryo compared to control embryos, but individual sections show specificity of label. Both the left and right otic pits can be observed, demonstrating the depth of scanning with resolution and brightness of fluorescence maintained. The neuroepithelium shows dramatic hydroxyurea-induced cell death in the individual sections.

must be taken during the clearing procedure because the embryos become totally transparent and could be damaged or lost. The refractive index of the BABB solution is similar to the refractive index of the embryo, yielding an embryo that is transparent *(1,5,6)*. Embryos were first put into a solution of 1:1 MEOH/BABB for 30 min and then the solution was removed and 100% BABB was added for at least 30 min. After clearing with either methyl salicylate or BABB, most of the clearing agent was carefully decanted and the embryo at the bottom of the tube was transferred into a large glass depression slide with some of the clearing agent.

6. Mounting and cover-slipping of embryos (*see* **Notes 4** and **5**): Methyl salicylate- or BABB-cleared embryos: After carefully removing the methyl salicylate or BABB from the large glass depression slide with a transfer pipet, the transparent embryos were visualized by surface light reflection. The embryos were then transferred to 2-mm-thick depression slides (Fisher 12-265a) and fresh methyl salicylate or BABB was added. The depression slide was then covered with a 20 × 30-mm cover slip (1.5 size) and was sealed with 3–4 successive coats of "Sally Hard as Nails" nail polish. It is essential that the depression slide be sealed, as the methyl salicylate and BABB are potentially damaging to various components of the microscope.

7. Confocal microscopy (*see* **Notes 2** and **8–10**): The embryo contained in the sealed depression slide was imaged using a Leica laser scanning confocal microscope (TCS4D). The 568 line excited the LT dye and a BP-TRITC filter was used to measure the emitted light. The sample was line-averaged at 16 or 32 scans per line, and a total of 20–30 sections per embryo were found to be sufficient to reconstruct the entire thickness of the embryo in 3-D. Embryos as thick as 700 μM have been completely sectioned optically by this technique with good resolution of internal staining both in individual optical sections and in 3-D reconstructions. The stack of confocal images was combined using a Leica 3-D program and the resultant composite, 256 gray-level TIFF file was transferred to a Pentium PC operating under Windows NT 4.0. Thumbs Plus (Cerrious Software, Charlotte, NC) was used to display and categorize the images obtained from the confocal microscope. Image Pro Plus (Media Cybernetics LP, Silver Spring, MD) was used to label the images prior to digital printing. Digital printing was done using a NP1600 (Codonics, Middleburg Heights, OH) dye sublimation printer (gamma = 2.0, contrast = 15). In order to represent a composite image form a stack of images, a few of the top and bottom images having minimal information were eliminated. The resultant image consisted of 16 individual TIFF images instead of the entire stack of approx 25 images.

4. Notes

1. A fixable vital stain and subsequent clearing procedure has been developed to study cell death in murine embryos at mid-gestation by confocal laser scanning microscopy, enabling the visualization of structures inside the embryo that are several hundreds of micrometers below the surface. The fluorescent dye, LT Red, allowed visualization of known areas of programmed cell death in normal embryos and increased staining in embryos treated with chemicals (hydroxyurea, MEOH) that correlated to lysosomal activity and apoptosis. It has been reported previously in embryos that dyes that stain lysosomes (Nile blue, neutral red, or acridine orange) were correlated to lysosomal activity, phagocytosis, and apoptosis *(6–16)*. LT Red was used to observe these regions of high lysosomal activity, which related directly to engulfment of apoptotic bodies and indirectly to cell death.

2. There are many technical difficulties associated with imaging of an object greater than 50 μm *(1,5,17,18)*. One approach used to solve this imaging problem is to cut serial sections of the object, image a thin slice, and then reconstruct the object with software. Serial section analysis is a time-consuming, tedious process in which errors can be introduced by sectioning the embryo and then realigning the measured sections. Maintaining the embryo as an intact structure can eliminate many problems associated with the reconstruction of serial sections. The whole mount provides a 3-D view of stained components located inside the embryos. However, imaging whole mounts also has technical difficulties. By using a confocal laser microscope instead of a normal fluorescent microscope, the laser light can penetrate deeper into the tissue for better visualization. However, because the tissue is dense and opaque (not transparent), the light becomes scattered, reflected, refracted, and absorbed when entering the tissue and then when passing back through the tissue into the objective *(17,18)*. This results in a reduction in the intensity of laser light and decreased optical quality as increased penetration into the tissue is attempted. Thus, depending on the light gathering characteristics of the objective and the power of the laser light, imaging of an embryo can be severely affected.

3. To overcome the opaqueness of the tissue, a clearing procedure previously applied to histological sections and amphibian eggs was used. The clearing procedure involves extracting water from the tissue and then replacing it with a solution that has a similar refractive index as the tissue has *(2–4)*. Previous clearing agents used include the follow-

ing: potassium hydroxide/glycerol, methyl salicylate (artificial oil of wintergreen), carbon disulfide, glycerol, and xylene. The clearing solution [1:2 mixture of benzyl alcohol (refractive index = 1.54035) and benzyl benzoate (refractive index = 1.5681)] (BABB) used in this study was derived by Murray and Kirschner *(2)*. Although the BABB mixture was designed for amphibian eggs and embryos, it appears to work equally well with mammalian embryos. Experiments varying the percentages of the components did not seem to increase the image quality of the mammalian embryo image. In choosing a clearing agent to render the tissue transparent, one must consider its compatibility with the staining reagents used, compatibility with aqueous solutions, and effects on the fixed tissue.

4. In the BABB protocol, the embryos were put first into a solution of 1:1 MEOH and BABB and then into a solution of 100% BABB. The refractive index of this solution is similar to the refractive index of the embryo, yielding an embryo that is nearly transparent *(2,3)*. After the final step, the medium was carefully decanted and the embryo contained in the bottom of the tube was transferred into a large glass depression slide. After carefully removing the medium, the embryo could be observed by light refraction. The embryos from both procedures were then transferred into a thick depression slide and fresh methyl salicylate or BABB was added and covered with a 20×30-mm cover slip. The cover slip was sealed with 3–4 successive coats of clear nail polish. It is essential that the depression slide be sealed, as the BABB cocktail is toxic to microscopes. Although the embryos are stable for a long time, the clearing agents will gradually dissolve the seal. It is sometimes useful to reseal the slides prior to using them on the confocal microscope to ensure that no spills occur.

5. The BABB clearing procedure has some disadvantages. Because the clearing agent is not compatible with aqueous solutions, the water in the tissue sample must be totally displaced by alcohol prior to the addition of the clearing agent. We have been successful in using fixable dyes like ethidium bromide or LT Red, but have been unsuccessful in using dyes like Hoechst 33342, which are removed by the alcohol dehydration procedure. Another disadvantage with this clearing agent is that the embryos become very brittle during the procedure. Care must be used in sealing the slide to immobilize the embryo. In addition, care must be taken to remove excess clearing agent to eliminate the possibility that the solution come into contact with the microscope or its objectives.

6. The proper choice of objectives is critical for this work. Most confocal microscopy applications use lenses with high power and high numerical apertures (NA) to image cells *(5,17,18)*. However, to image parts of embryos, it is necessary to use low-power objectives. In our applications it was important to observe large regions of the embryo to determine the cell death patterns. To obtain these images, it was necessary to use low-power objectives ($\times 5$ and $\times 10$) with relatively high NAs to allow sufficient visible fluorescent wavelengths to be transmitted.

7. In our experiments, the use of depression slides yielded the best results with dry objectives having high NAs. Both sides of the chamber can be scanned with almost equal resolution. Although an oil lens will give better resolution than a dry lens, the oil lens will introduce a series of problems that include shallow depth of focus (usually less than 300 µm), possible contamination with clearing agent residues on the cover glass surface, and possible exertion of mechanical stress on the cover slip and thus the seal. Removal of the oil from the cover glass may also break the nail polish seal on the depression slide. We have added a $\times 5$ Zeiss Fluor objective (.25 NA) to the Leica microscope, based on its higher NA and its ability to effectively transmit more fluorescence at lower power. The use of the Zeiss lens on the Leica infinity-corrected microscope results in a 20% increase in magnification. Thus, the effective magnification of the $\times 5$ Zeiss lens is $\times 6.2$ on the Leica sys-

tem. Leica has just released a ×10 multi-immersion lens (0.4 NA) and a ×10 dry (0.4 NA) that have depth of field of 350 μm making them extremely useful in observing entire small embryos or limbs. Although other objectives may be used for these applications, it is important to consider the NA, magnification, and the working distance of the lens.

8. Image quality can be affected by a number of controls on the confocal microscope. These include the selection of dichroic bandpass filter, averaging, laser power, and bleaching of the dye. It was observed that a narrower bandpass instead of a long-pass filter decreased the reflections. Methyl salicylate also yielded a clear uniform embryo that did not contain reflections observed with the BABB clearing procedure. It is important that the embryo not have saturation pixels. For the current applications, image quality was superior when bandpass filters were used instead of conventional Schott filters.

9. This procedure is rapid and flexible, although sample preparation variables including staining, fixation, dehydration, and clearing must be carefully controlled for optimum resolution. Image quality will also be affected by the choice of laser line, specific filter selections, and specific confocal microscope controls *(1)*. It appears from our studies using LT Red that sufficient fluorescence occurs to allow the pinhole size to be decreased, the laser power increased, and the photomultiplier (PMT) to be operated below their saturation points. These factors combined with the minimal bleaching observed with the LT Red dye have produced clear images and a straightforward way to monitor apoptosis and morphology in the embryo. In fact, if enough sections are obtained (60) and the z-distance between sections is reduced to approx 2X the x/y distance, the embryo can be visualized by rotating it 360° using 3-D software (Voxblast, Vaytec, Fairfield, IA).

10. Although this assay has been done primarily on embryos, it should be applicable to a broad range of tissues and other fixable dyes. In order to achieve fluorescence with less-efficient systems than the Leica TCS4D point scanner, it may be necessary to incorporate other more efficient dyes into the embryo. We have successfully used nucleic acid stains (ethidium bromide, YO-PRO, and BO-PRO) prior to MEOH dehydration. Final concentrations of the nucleic acid stains in PBS were: ethidium homodimer, 25 μg/mL; YO-PRO, 5 μ*M*; and BO-PRO, 5 μ*M*. Ethidium bromide, YO-PRO, or BO-PRO staining is done after fixation. We stained the samples at 37°C for up to 24 h to ensure that the stain penetrated the entire tissue. The excess stain is then washed out of the sample with PBS and dehydration is initiated with the 50% MEOH stage. The YO-PRO dye was excited with a 488 line using a BP-FITC bandpass filter. Ethidium bromide was excited with the 488-nm line and a BP-TRITC filter was used in the detection. BO-PRO was excited with a 568 line and a BP-TRITC or BP 598/40 filter (Chroma, Brattleboro, VT) was used in the detection. These different filters may be used to increase the amount of transmitted light.

Acknowledgment

We wish to thank Owen Price for creating the composite figures.

References

1. Zucker, R. M, Hunter, S., and Rogers, J. M. (1998) Confocal laser scanning microscopy of apoptosis in organogenesis-stage mouse embryos B. *Cytometry* **33,** 348–354.
2. Gard, D. L. (1993) Confocal immunofluorescence microscopy of microtubules in amphibian oocytes and eggs. *Methods Cell Biol.* **38,** 231–264.
3. Klymkowsky, M. W. and Hanken, J. (1991) Whole mount staining of Xenopus and other vertebrates. *Methods Cell Biol.* **36,** 419–441.

4. Maziere, A. M., Hage, W. J., and Ubbels, G. A. (1996) A method for staining of cell nuclei in Xenopus laevis embryos with cyanine dyes for whole-mount confocal laser scanning microscopy. *J. Histochem. Cytochem.* **44,** 399–402.

5. Paddock, S. W. (1994) To boldly glow ... Applications of laser scanning confocal microscopy in developmental biology. *Bioessays* **16,** 357–365.

6. Rogers, J. M., Francis, B. M., Sulik, K. K., Alles, A. J., Massaro, E. J., Zucker, R. M., et al. (1994) Cell death and cell cycle perturbation in the developmental toxicity of the demethylating agent, 5-aza-2'-deoxycytidine. *Teratology* **50,** 332–339.

7. Abrams, J. M., White, K., Fessler, L. I., and Steller, H. (1993) Programmed death during drosophila embryogenesis. *Development* **117,** 29–43.

8. Alles, A. J. and Sulik, K. K. (1993) A review of caudal dysgenesis and its pathogenesis as illustrated in an animal model. *Birth Defects* **29,** 83–102.

9. Chernoff, N., Rogers, J. M., Alles, A. J., Zucker, R. M., Elstein, K. H., Massaro, E. J., et al. (1989) Cell cycle alterations and cell death in cyclophosphamide teratogenesis. *Teratogenesis Carcinog. Mutagen* **9,** 199–209.

10. Hurle, J. M. (1988) Cell death in developing systems. *Methods Achievements Exp. Pathol.* **13,** 55–86.

11. Knudsen, T. (1990) In vitro approaches to the study of embryonic cell death in developmental toxicity, in *In Vitro Methods in Developmental Toxicology: Use in Defining Mechanisms and Risk Parameters* (Kimmel, G. L. and Kochhar, D. M., eds.), CRC, Boca Raton, FL, pp. 129 142.

12. Menkes, B., Prelipceanu, O., and Capalnasan, I. (1979) Vital fluorochroming as a tool for embryonic cell death research, in *Advances in the Study of Birth Defects, vol. 3: Abnormal Embryogenesis* (Persaud, T. V. N., ed.), University Park Press, Baltimore, MD, pp. 219–241.

13. Philips, F. S., Sternberg, S. S., Schwartz, H. S., Cronin, A. P., Sodergrenae, A. E., and Vidal, P. M. (1967) Hydroxyurea acute cell death in proliferating tissues in rats. *Cancer Res.* **27,** 61–74.

14. Rogers, J. M., Taubeneck, M. W., Dalston, G. P., Sulik, K. K., Zucker, R. M., Elstein, K. H., et al. (1995) Zinc deficiency causes apoptosis but not cell cycle alterations in organogenesis-stage rat embryos: Effect of varying duration of deficiency. *Teratology* **52,** 149–159.

15. Saunders, J. W., Jr. and Gasseling, M. T. (1962) Cellular death in morphogenesis of the avian wing. *Dev. Biol.* **5,** 147–178.

16. Saunders, J. W., Jr. (1966) Death in embryonic systems. *Science* **154,** 604–612.

17. Pawley, J. B. (1995) *Handbook of Biological Confocal Microscopy.* Plenum, New York.

18. White, N. S., Errington, R. J., Fricker, M. D., and Wood, J. L. (1996) Aberration control in quantitative imaging of botanical specimens by multi dimensional fluorescence microscopy. *J. Microsc.* **181,** 99–116.

20

Embryo/Fetal Topographical Analysis by Fluorescence Microscopy and Confocal Laser Scanning Microscopy*

Robert M. Zucker and John M. Rogers

1. Introduction

For topographical (surface) analysis of developing embryos, investigators typically rely on scanning electron microscopy (SEM) to provide the surface detail not attainable with light microscopy. Although it provides beautiful surface detail, SEM is an expensive and time-consuming technique, and the preparation procedure may alter the morphology of the specimen. SEM sample preparation techniques of fixation, dehydration, critical point drying, mounting, and coating may not only introduce artifacts but also make the tissue friable and difficult to position. As a rapid, simple, and economical alternative, it is possible to use fluorescent staining with acridine orange (AO) (*see* **Notes 1** and **2**) of fresh or fixed rodent embryonic tissues, combined with either a confocal or standard fluorescence microscope to visualize the surface images of embryonic/fetal structures *(1)*. In addition, the use of confocal microscopy provides better resolution and depth of field. However, inexpensive low-power objectives can be used, as they yield a good depth of field of the topographical surface. With both the epifluorescence and the confocal microscopes, we have used the fluorochrome, AO, to obtain surface images of fresh or fixed rodent embryos. For low-magnification applications, this technique provides images rivaling SEM at a great reduction in cost, time, and effort. Many of the difficulties introduced by the SEM technique are eliminated by this AO procedure (*see* **Note 3**).

2. Materials

1. Fixation: Paraformaldehyde (20%, Electron Microscopy Sciences, Fort Washington, PA) is diluted to 4% in Dulbecco's phosphate-buffered saline (PBS; Gibco-BRL, Gaithersburg, MD) and frozen at −20°C until use. Glutaraldehyde (25%, Electron Microscopy Sciences) was diluted in PBS to 2% and frozen at −20°C until use or kept at 4°C. For this procedure, it does not appear that the exact method of fixation is critical, as AO coats the surface of the embryo. However, glutaraldehyde is preferable to paraformaldehyde in this procedure.

*The research described in this article has been reviewed and approved for publication as an EPA document. Approval does not necessarily signify that the contents reflect the views and policies of the Agency, nor does mention of trade names or commercial products constitute endorsement or recommendation for use.

From: *Methods in Molecular Biology, Vol. 135: Developmental Biology Protocols, Vol. I*
Edited by: R. S. Tuan and C. W. Lo © Humana Press Inc., Totowa, NJ

2. AO staining: The embryos are stained with AO (Molecular Probes, Eugene, OR) by diluting a stock solution of 1 mg/mL AO with PBS to get a working concentration of 5 µg/mL for epifluorescence microscopy or 0.1 µg/mL for confocal applications (*see* **Notes 1** and **2**). A row of plastic 35-mm culture dishes or large glass depressions slides is used to sequentially stain the embryo and wash off the excess AO from the embryos. After staining, the embryos are transferred into a 2-mm-thick depression slide (Fisher Scientific, Pittsburgh, PA, #12-265a), slide chamber or 35-mm tissue culture dish containing PBS.

3. Standard fluorescence microscope (*see* **Note 4**): An epifluorescence microscope with low-power objectives can be used. The quality of the objectives does not appear to be very important. Depending on the size of the embryo, ×1, ×2, ×2.5, and ×4 objectives can be chosen to visualize the embryos. The resultant magnification will range between 12.5 and ×60 because of the magnification factor if using a finite optics microscope. The newer infinity-corrected microscopes do not have an extra 1.25 magnification factor and are preferred for the lower magnification factor with greater field of view.

4. Confocal microscope (*see* **Notes 5** and **6**): Any point-scanning or slit-scanning confocal microscope with any optical efficiency is acceptable. In our studies, a Leica TCS4D confocal with a Leica inverted DMIRB microscope was used. Depending on the size of the embryo or fetal tissue we can choose either a ×1.6, ×2.5, or ×5 lens. The numerical aperture (NA) is not important for this assay and the lower NA will yield a greater depth of focus. Less-expensive lenses are acceptable and will yield adequate resolution.

3. Methods

Animals: For all timed-mated animals, presence of a copulatory plug on the morning after breeding was considered gestation day (GD) 0. On the day of sacrifice, pregnant animals were killed by carbon dioxide asphyxiation. Timed pregnant C57BL6J mice were killed on GD 8 or 9 (**Fig. 1**) Rats (Sprague-Dawley) were killed between GD 9.5 and 13 in **Figs. 2**, **3**, and **4**.

1. Fixation: Embryos can be stained unfixed or fixed with paraformaldehyde or glutaraldehyde. Paraformaldehyde (20%, Electron Microscopy Sciences) is diluted to 4% in PBS and frozen at –20°C until use. Glutaraldehyde (25%, Electron Microscopy Sciences) is diluted in PBS to 2%. The stock solution appears to be stable and can be stored for a year at either 4°C or –20°C. All embryos are fixed at 4°C. Embryos can be stored at 4°C in the fixative until use or dehydrated to 70% MEOH or ETOH and stored at 4°C until use. Glutaraldehyde is a superior fixative for this application, as its autofluorescence appears to enhance the image and actually be beneficial for image quality.

Fig. 1. *(opposite page)* The confocal laser scanning microscopy (CSLM) technology has been used in our laboratory to measure surface features. By coating the embryo with a dilute solution of AO, the embryo becomes opaque and reflects most of the light striking the embryo. The resultant topographical image is one of surface fluorescence and resembles that obtained with a SEM. Here we show a 3-D reconstruction of 40 sections from a GD 8 mouse embryo imaged with a ×10 Leica objective on the confocal microscope (original magnification ×100). By coating the embryo with AO, the surface appears to act as a barrier to laser light penetration. Since the fluorescence efficiency of the surface AO was far greater than internal stain, the image is derived almost entirely from surface fluorescence.

Fig. 2. *(opposite page)* Composite epifluorescence photomicrograph illustrating representative stages of rat development that can be imaged *in toto* by the described technique. Right to left: GD 9.5, 10.5, 11.5, 12.5. Nikon Optiphote-2 microscope (objective ×1, total magnification ×12.5).

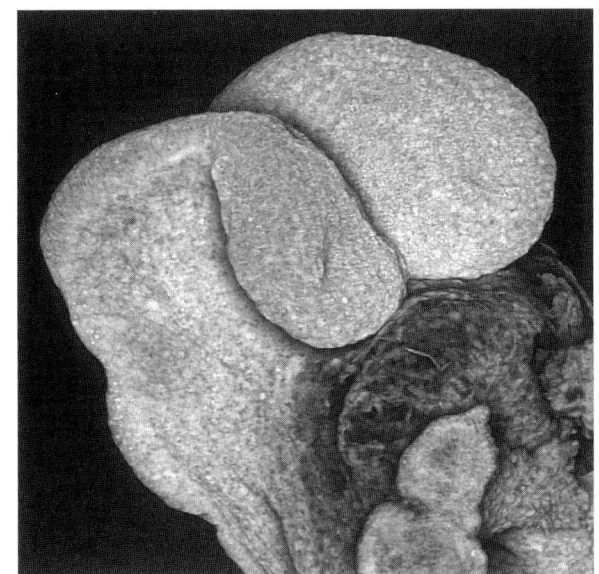

Fig. 1

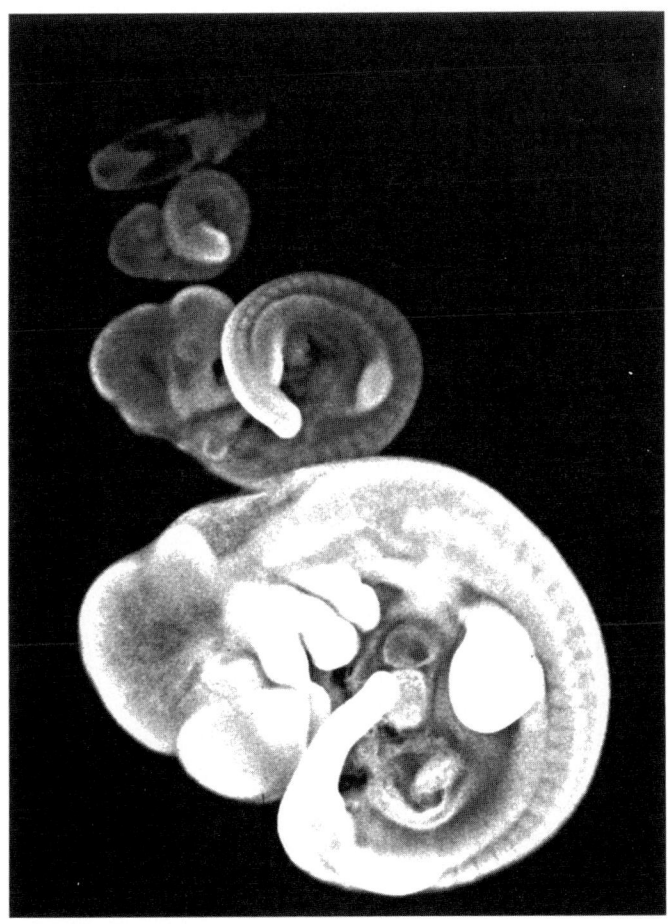

Fig. 2

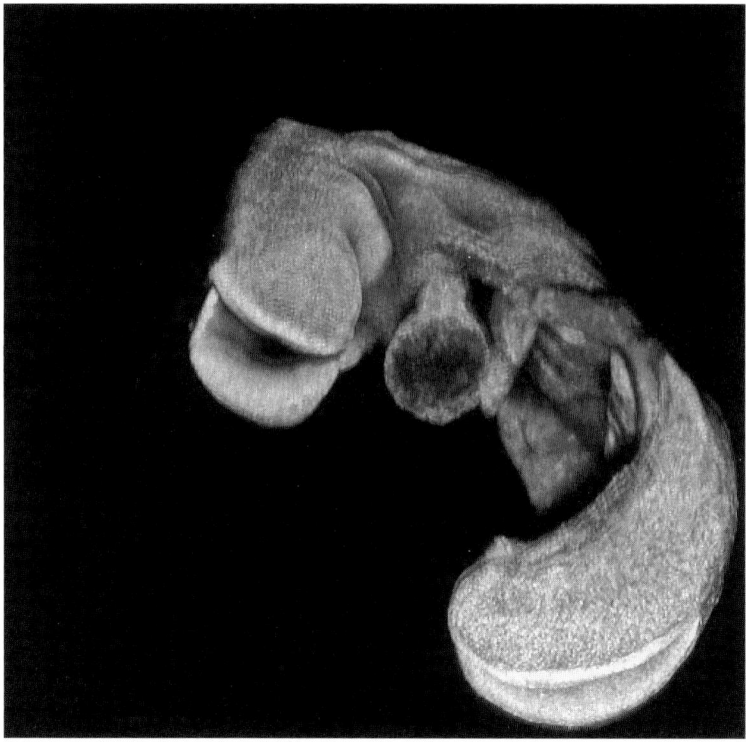

Fig. 3. CLSM-reconstructed image included 45 sections of a 10.5-d rat embryo showing an incomplete neural tube closure. Because of the size of this embryo, it would be impossible to have the whole embryo in focus using the epifluorescence microscope (microscope objective ×2.5; total magnification ×25).

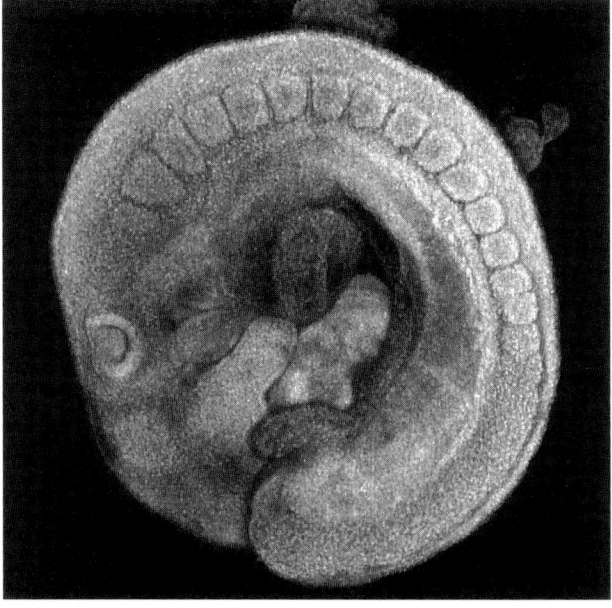

Fig. 4. CLSM-reconstructed image of a GD 11 rat embryo stained with AO and made up of 40 individual sections. The embryo is lying flat, making the image obtainable with either an epifluorescence microscope or confocal microscope.

2. Acridine orange staining: Fresh embryos or embryos fixed with either 4% paraformaldehyde or 2% glutaraldehyde are stained with AO by diluting a stock solution of 1 mg/mL AO
with PBS to get a working concentration of 5 µg/mL for epifluorescence or 0.1 µg/mL for
confocal scanning laser microscope (CLSM). The greater efficiency of the laser based
confocal microscope using the 488 laser line allows for the embryos to be stained at
1/50th the concentration of that used with a conventional epifluoresence microscope.
Higher concentrations of AO are not beneficial with CLSM and should be avoided as they
may yield nonspecific fluorescence. The embryos are stained for approx 5 min in a
35-mm plastic dish or large depression slide and then briefly washed 2X in PBS to remove
the excess dye. After washing, it is important that the medium in which the embryo is
suspended be free of AO dye. After staining, the embryos can be observed by suspending
them in a hanging-drop depression slide and sealing the top with a cover glass. Both sides
of the embryo can be analyzed on an inverted or upright microscope. The preferable orientation is the one with the cover glass closest to the objective. On an upright microscope,
the samples can be examined in a depression slide with or without a cover slip. We have
found, however, that the cover slip tends to stabilize the embryo and minimize movements
in the slide chamber. This staining procedure was determined empirically. Staining times
and concentrations may have to be adjusted for different specimens or in different laboratories to optimize intensity and/or resolution of the sample. A sufficient amount of AO
should be used to acquire an adequate signal with either optical system. Understaining
will yield a reduced signal, and overstaining will yield both red and green fluorescence.
Overall, we have found the staining procedure to be quite forgiving and flexible.

3. Conventional fluorescence microcopy (*see* **Notes 4, 7, 8**): Our conventional fluorescence
microscopy was accomplished with a Nikon Optiphot-2 epifluorescence microscope
equipped with an Olympus S Plan ×2 (0.08 NA) and a Leica Pl ×1 (0.04 NA) objective, a
100-W mercury arc lamp, and a B-2A filter cube (510-nm dichroic with a 450–490 excitation and a 520-nm emission filter for visible excitation). The fluorescent image was transmitted to a Nikon microflex UFX-DX 35-mm camera system either through a ×2.0 or ×2.5
projection lens. Fluorescence photomicrographs were taken with film between ASA 100
and 400 dependent on the brightness of the image. Exposure times should be not more
than a few seconds. Images can also be obtained using video microscopy cameras like the
Dage 330 or Dage 300 (Dage-MTI; Michigan City IN, cooled CCD) or a digital camera
such as (Hamamatsu, Bridgewater, NJ) (C4742-95, cooled CCD). Other configurations
should also be acceptable.

4. Confocal microscopy (*see* **Notes 5, 6, 9**): AO-stained embryos contained in a sealed
depression slide were imaged using a Leica confocal microscope (TCS4D) with an
Omnichrome laser emitting 488-nm lines with an DMIRB inverted microscope (Leica
Microsystems, Deerfield, IL). Imaging from both sides of the glass depression slide
seemed to work well with the low-power objectives. There was a slight image improvement with the cover glass face down, closest to the objective. Depending on the size of the
embryo, the following objectives were used: Leica ×1.6 (0.04 NA), ×2.5 (0.06 NA), Leica
×5 (0.12 NA). The emitted light was obtained through a 522/22 filter (Chroma,
Brattleboro, VT). Other bandpass or Schott longpass filters (Schott Corp., Mt. Laurel, NJ)
will work equally well. The sample was averaged at 4 frames per second using a linear
look-up table (LUT) and usually 20–50 sections were obtained. The thickness of the sections generally ranged between 10–20 µm. If the individual sections representing the
scanned focal planes are visible in the reconstructed image, more sections can be obtained
to eliminate this effect by essentially yielding a smaller separation between adjacent
images. The Leica TCS4D confocal system has the capacity to obtain up to 90 sequential

images. Newer systems can obtain larger stacks of images allowing for thicker specimens to be studied. Embryos as thick as a few millimeters have been imaged by this technique. The stack of confocal images was combined using a Leica 3-D program and the resultant composite, 256 gray-level TIFF file was transferred to a Pentium PC operating under Windows NT 4.0. Thumbs Plus (Curious Software, Charlotte, NC) was used to display and categorize the images obtained from the confocal microscope. Image Pro Plus (Media Cybernetics, Silver Spring, MD) or Adobe Photo Shop (Adobe Systems Inc., San Jose, CA) was used to label the images prior to digital printing. The digital printing was done using a NP1600 (Codonics, Middleburg Heights, OH) dye sublimation printer (gamma = 2.0, contrast = 15). In order to present a composite image from a stack of images, a few of the top and bottom images having minimal information were eliminated. The resultant image consisted of 16 individual TIFF images instead of the entire stack of approx 25 composite images.

4. Notes

1. Acridine orange is a fluorescent compound that has been used to detect nuclear structure, lysosomal activity, and apoptosis. We have used this compound to coat the surface of the embryo to increase the amount of light scattered and reflected off the embryo surface. The resulting image represents surface fluorescence of the object. Acridine orange is a suspected carcinogen and should be handled with caution.

2. Acridine orange is an inexpensive, visible-light-excited dye that has been used at low concentration for DNA and RNA analysis, lysosomal staining, and spatial detection of cell death. In these applications, protocols must be adhered to meticulously, as the staining reaction is unforgiving to minor variation in AO concentrations or ionic strength of pH of the buffer. However, when AO is used at as high a concentration as was done in this application, it binds to other macromolecules, including proteins, polysaccharides, and glycosamines, resulting in a green fluorescence topographical image. It is important to use a sufficiently high concentration to provide a surface rendered with minimal light penetration.

3. SEM sample preparation techniques of fixation, dehydration, critical point drying, mounting, and coating may not only introduce artifacts but also make the tissue friable and difficult to position. This AO procedure eliminates many of these difficulties.

4. For fluorescence microscopy, it is important to understand the lens limitations and the relationship between NA and depth of field. Higher NAs yield greater light-absorbing efficiency with less depth of field. For this technique to work, it is important to have a large depth of field to observe surface features from the embryo that are located on different planes. Increased intensity can be obtained by increasing the staining time and staining concentration. It appears that the best resolution is obtained from the less-expensive lenses having greater depth of field and low NAs. The embryo should be positioned so that it lies flat and not at an angle. This will increase the ability to observe surface features and retard the out-of-focus light. If the embryo is too large to image in its entirety, it is important to focus on the region closest to the objective and have the background out of focus to create the best possible picture. The objectives of the fluorescence microscope should be focused to produce the fluorescence intensity of the area of interest.

5. Confocal microscope: Using low-power objectives ($\times 1.6$, $\times 2.5$, and $\times 5$), the image plane is not completely in focus and contains planes from below and above the plane of focus. However, information above and below the plane of focus will not interfere with the image, as adjacent planes of focus will just add this information to the overall image. This is not a procedure to quantify fluorescence but a procedure to observe the surface structure of the embryo. The pinhole does not have to be closed down to minimal values. The number

of sections should be adequate to produce a reconstructed image that does not display steps between adjacent optical sections. The out-of-focus light will also tend to reduce step observation. As a practical rule, we ran with a larger pinhole, more sections (40–50), and minimal noise averaging to save time. The embryo can be stained with a much lower concentration of AO than with the epifluoresence microscope because of the optical efficiency of the confocal microscope compared with the epifluoresence microscope.

6. The depth of field limits normal lenses. CLSM does not have the same restrictions, as out-of-focus light can be recombined to yield a 3-D image of an object. This procedure uses reflected light and is the direct opposite to a clearing procedure (used to observe interior structures) of the embryo in which the amount of light scattered reflected and refracted is minimized *(2)*.

7. With noninfinity microscopes, a ×1.25 power has to be added to the total power as a result of the fluorescence module. Because the viewing of the embryo is dependent on the field, it is advisable to use the newer infinity-corrected microscopes, which do not have this 25% magnification factor. If there are other modules on the microscope, such as interference contrast or Nomarski contrast, these optics will also increase the magnification factor by an additional 25%, and they should be removed.

8. The two factors to consider are the depth of field and the fluorescence throughput of the system. The NA should not be too high, as the depth of field decreases with increasing NA. It appears that inexpensive lenses with lower NAs but greater depth of field are better for this application than are more-expensive lenses that accumulate more fluorescent light.

9. The application of confocal microscopy to whole-mount mammalian embryos has been limited by a number of optical factors. To see a significant portion of the embryo, it is necessary to use low-power objectives, which contain low NAs. This creates a situation in which the laser cannot penetrate deep into the tissues, because the light is reflected, absorbed, and scattered prior to penetrating deep into the tissue and returning to the objective. The AO coating of the embryo essentially prohibits the laser light from penetrating into the embryo, resulting in an image of only the surface fluorescence.

References

1. Zucker, R. M., Elstein, K. H., Shuey, D. L., Embron-McCoy, M., and Rogers, J. M. (1995) Utility of fluorescence microscopy in embryonic/fetal topographical analysis. *Teratology* **51,** 430–434.
2. Zucker, R. M., Hunter, E. S., and Rogers, J. M. (1998) Confocal laser scanning microscopy of apoptosis in organogenesis-stage mouse embryos. *Cytometry* **33,** 348–354.

21

Magnetic Resonance Imaging Analysis of Embryos

Bradley R. Smith

1. Introduction

Analyzing phenotypic effects in embryos produced with directed gene manipulation is a major challenge for developmental biologists who are trained in molecular techniques but who are not well-versed in embryology. Analysis techniques that preserve the three-dimensional (3-D) integrity of the specimen while also demonstrating internal structural detail provide an advantage for researchers who are not prepared to study hundreds of optical histological sections and mentally reconstruct the embryo. Magnetic resonance (MR) imaging nondestructively generates 3-D and cross-sectional views of embryos to provide macroscopic structural detail while preserving the spatial information.

MR imaging as described in this chapter generates images of water in the tissues of the embryo (**Fig. 1**). Image contrast can result from tissue-to-tissue differences in water density, water diffusion, bulk movement of water, water-binding properties of the tissues, extraneous contrast agents, and the impact of all these characteristics on the magnetization relaxation rates of the hydrogen protons in that water. The brightness or intensity of each pixel in the image is proportional to the amount of MR signal detected at the corresponding position within the embryo.

This chapter details embryo preparation methods (including a method for contrast enhancement of vascular spaces) and MR imaging protocols. This embryo perfusion technique is a modification of the microangiography procedure developed by Eric Effmann *(1)* for MR microscopy *(2)*.

2. Materials

2.1. Embryo Preparation

1. Sorenson's phosphate buffer: Combine 7.13 g sodium phosphate dibasic anhydrous (Mallinckrodt-Baker, Chesterfield, MO) in 250 mL distilled water and label as solution A and 5.52 g sodium phosphate monobasic (Mallinckrodt) in 200 mL distilled water and label as solution B. Solutions A and B can be stored in separate stoppered flasks for approximately several weeks at 4°C.
2. Perfusion fixative: 2% glutaraldehyde and 1% formalin in 300 mosM phosphate buffer *(3)*. The perfusion fixative should be mixed fresh each day by combining: 45 mL Sorenson's buffer solution A, 22.5 mL Sorenson's buffer solution B, 20 mL of 25% glutaraldehyde

From: *Methods in Molecular Biology, Vol. 135: Developmental Biology Protocols, Vol. I*
Edited by: R. S. Tuan and C. W. Lo © Humana Press Inc., Totowa, NJ

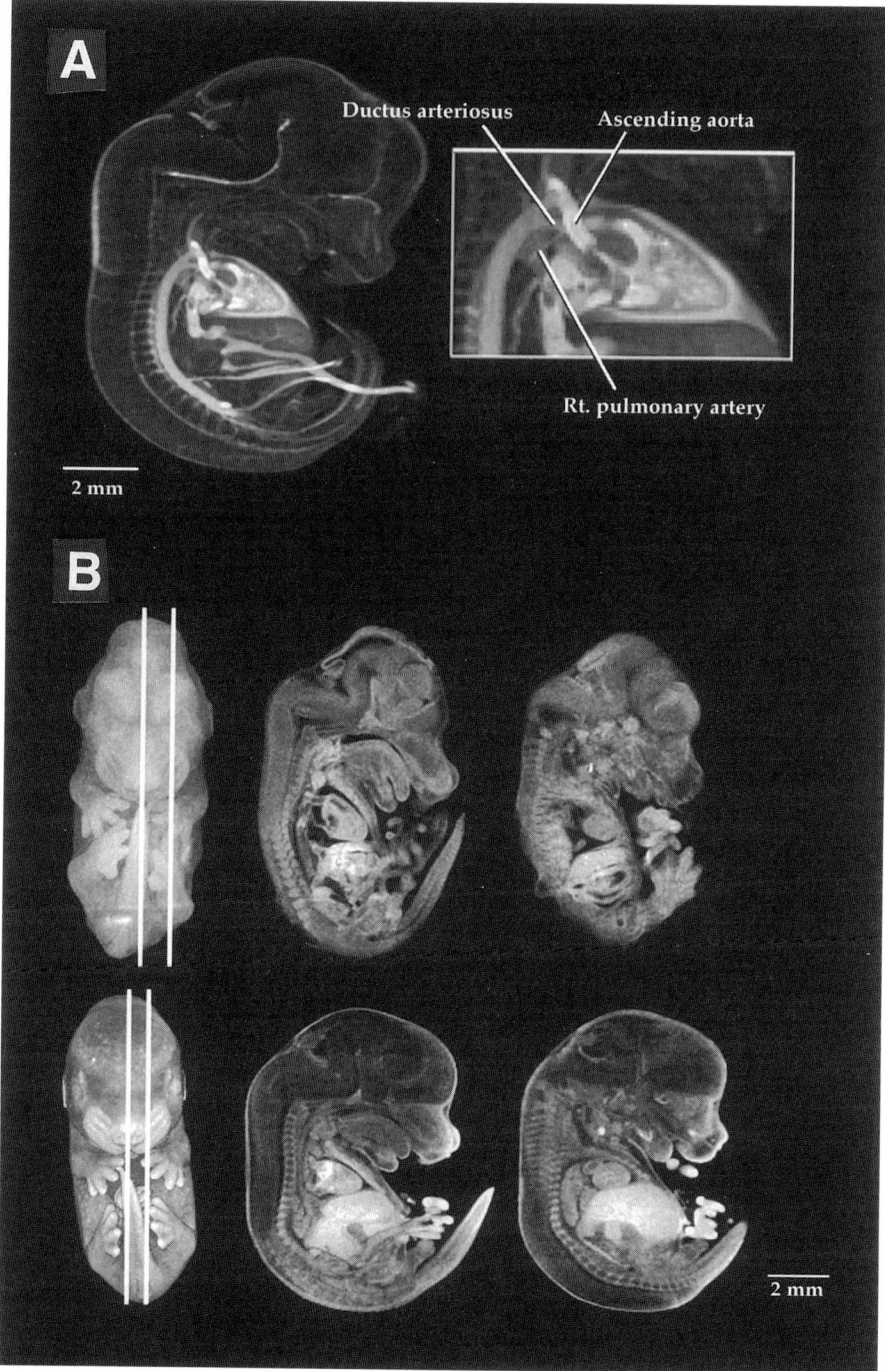

Fig. 1. **(A)** Details of the cardiovascular system of a normal day-14.5 mouse embryo are depicted in this MR image. The MR contrast agent BSA-DTPA-Gd provides high signal in the lumen of blood vessels and chambers of the heart. **(B)** Anatomic detail is also available without the use of contrast agents as shown in this comparison of a 14.5-day, LH-2 knockout mouse embryo and a wild-type control (LH-2 knockout courtesy of Forbes D. Porter, Duke University, Durham, NC).

(Grade I, Sigma, St. Louis, MO), 6.75 mL of 37% formaldehyde (i.e., formalin, Sigma), 1.57 g NaCl, and 160.75 mL distilled water for a total of 250 mL fixative. Glutaraldehyde is toxic and carcinogenic.

3. 0.2 *M* cacodylate buffer stock: Combine 42.8 g cacodylic acid with 1000 mL distilled water and store at 4°C. Avoid contact with skin: contains arsenic.

4. Verapamil stock: For a dosage of 0.3145 mg verapamil/kg embryo prepare the following stock: 0.817 mg verapamil in 100 mL cacodylate buffer stock. This assumes approx 10 μL of solution is delivered to a 14.5-d mouse embryo weighing approx 0.25 g. Use the same day.

5. Gadolinium contrast material (BSA-DTPA-Gd):
   a. Obtain the following: 5 g bovine serum albumin (cat. no. A 8022; Sigma, St. Louis, MO), 2 g DTPA (cat. no. 28,402-5; Aldrich, Milwaukee, WI), 1 g gadolinium chloride (cat. no. 27,825-1; Aldrich), SpectraPor 2 dialysis membrane (#08-670-3B, 63.7 mm × 15 m; VWR Scientific, West Chester, PA or Fisher Scientific, Pittsburgh, PA), 500 g type A, porcine skin gelatin (cat. no. G-2500; Sigma).
   b. Prepare the following: 500 mL sodium carbonate (0.1 *M*, adjust to pH 9.5 with HCl), approx 20 mL hydrochloric acid (HCl, 37%), 500 mL sodium hydroxide (1 *M*), 1 L sodium chloride (0.1 *M*).
   c. Coupling of bovine albumin to DTPA dianhydride:
      Step 1. Dissolve 5 g (2.5×10^{-4} mol) albumin in 100 mL sodium carbonate and adjust pH to 9.5 with 1 *M* sodium hydroxide.
      Step 2. Add 1.78 g (5×10^{-3} mol) DTPA dianhydride to the albumin solution from step 1. After complete dissolution of the DTPA dianhydride, adjust the pH to 9.5 with 1 *M* sodium hydroxide. Readjust the pH to 9.5 after 15 and 30 min of reaction. After 60 min of reaction, adjust the pH to 7.0 (with HCl) and dialyze the reaction mixture exhaustively against water (use approx 9-in. length of dialysis membrane with a clip at each end). After dialysis is complete, the volume in the dialysis bag is approx 200 mL, and the pH is approx 8.0.
      Step 3. Add $GdCl_3$ in the following way to bring the solution to approx 3.3 m*M* in $GdCl_3$ (the final mixture will contain approx 35 mg of gadolinium chloride): Dissolve 742 mg of gadolinium chloride in 10 mL of distilled water. Adjust pH to 6.0 with 1 *M* sodium hydroxide. (Keep a record of how much sodium hydroxide is used in this step to ensure that roughly 35 mg of gadolinium is added to the final mixture during this next step.) Add 3.5 mL (or some adjusted volume) of this gadolinium chloride solution to the solution from step 2 so that approx 35 mg of gadolinium chloride is transferred. Addition of gadolinium chloride causes a white precipitate that clears after the first milliliter is added. The pH drops to approx 6.5. Readjust the pH to 8.0 with 1 *M* sodium hydroxide to remove all of the precipitate. Let the reaction mixture sit for 10 min and dialyze exhaustively against 0.1 *M* sodium chloride.
      Step 4. Concentrate the conjugate solution to about 100 mg protein/mL. Aliquot the conjugate solution (5–10 mL) and lyophilize. Store the contrast material in a tightly sealed container at 4°C up to 12 mo. Rehydrate immediately prior to use.
      Step 5. Add 0.7 g gelatin to 10 mL BSA-DTPA-Gd at room temperature. Warm in a double beaker to 40–50°C. The mixture will turn clear. Keep warm at 40°C until used or store at 4°C and rewarm to 40°C immediately before use. (The solution will congeal near room temperature.)

6. Glass cannulas: Glass micropipets (0.75 mm OD, Sutter Instrument Co., San Francisco, CA) are drawn to a fine point with a pipet puller (Brown-Flaming Micropipette Puller, Sutter) to create cannulas, and are connected to a microperistaltic pump (Ismatec, Cole-Parmer, Vernon Hills, IL) via Tygone tubing (Cole-Parmer, 0.51-mm ID, 2.33 mm-OD).

Immediately prior to use, the tip of the cannulas are broken off to the appropriate diameter. This is done under the microscope where the umbilical vessel size can be compared to the cannula tip size. The tip should be approx 80% the diameter of the lumen to ensure a snug fit when run up to the shoulder of the flair in the cannula.

7. Embryo embedding medium: Fomblin LC08 (Ausimont USA, Inc., Thorofare, NJ) is a dense, hydrogen-free, inert, and hydrophobic liquid, well suited as an embedding agent for MR imaging. It provides a signal void around the specimen, and its hydrophobic nature prevents it from exchanging with the water in the specimen which would lead to dilution of the MR signal in the embryo. Fomblin is highly inert and nontoxic.

2.2. MR Imaging System

A 9.4-Tesla imaging magnet (Bruker Instruments Inc., Billerica, MA) with shielded gradients producing 80 G/cm and rise times of approx 110 μs. 1-cm-diameter radio frequency imaging coil constructed from a single sheet of dielectric microwave substrate *(4)*.

2.3. Data Visualization

Indigo[2] Maximum Impact workstation (Silicon Graphics, Mountain View, CA) running VoxelView_ULTRA 2.5 volume-rendering software (VitalImages, Inc., Fairfield, IA).

3. Methods
3.1. Harvesting the Embryos

Harvest embryos of the desired age as follows. Euthanize the pregnant female by cervical dislocation (*see* **Note 1**). Use a midline incision through the abdominal skin and peritoneum (from the lower margin of the ribs to the vagina) to exposes the uterine horns. Cut the uterine horns free from their mesentery and uterine vessels by starting at one end of the uterus, working toward the bifurcation of the uterus, and then cutting toward the end of the opposite uterine horn. Place the intact uterine horns into cold PBS, and keep the PBS on ice (*see* **Note 2**).

3.2. Perfusion Fixation

Isolate a single conceptus (embryo, membranes and placenta) by carefully tearing it free from the uterine musculature using two pairs of forceps. Remove Reichert's membrane (a small dark cap of tissue usually positioned opposite the placenta on the surface of the yolk sac), and then tear open the yolk sac 2/3 the way around its junction with the placenta, taking care to avoid the larger vitelline blood vessels. Pull the embryo through the opening taking care not to stretch or tear the umbilical vessels that still attach the embryo to the placenta. Place the embryo on its left side (right side up) to position the umbilical vessels properly. Remove the amnion, a very thin membrane covering the entire embryo. Identify the umbilical artery and vein (*see* **Note 3**). With fine forceps, make a small tear in the umbilical vein and then in the artery (*see* **Note 4**). Place the tip of a glass cannula attached to a line with PBS into the open umbilical vein directed toward the embryo. Activate the microperistaltic pump to approx 10% power (*see* **Note 5**) to infuse the saline until the majority of the blood in the embryo is replaced with PBS (the blood will escape through the torn umbilical artery). Replace the PBS cannula with a cannula delivering the fixative, and infuse until the PBS has been replaced with fixative (*see* **Note 6**). Replace the cannula with a third cannula containing the contrast material, and infuse either the umbilical artery or vein until the contrast material has filled the desired vessels (*see* **Note 7**). Tie off the umbilical vessels proximal to the point of infusion with 3-0 silk suture to prevent backflow of the contrast material. Rinse spilled contrast off the surface of the embryo with warmed PBS and place the embryo (still connected by the umbilical vessels to the placenta) into cold fixative for immersion fixation (the same fixative used for perfusion). The embryos should be stored in the fixative at 4°C until MR imaging. It is best to image contrast-perfused specimens within 1 wk of infusion.

3.3. Embedding Embryos for MR Imaging

Place the embryo in an MR-imaging-compatible container that can be sealed and that fits tightly into the MR imaging coil. Fill the container with Fomblin, taking care to release all air bubbles, and seal the container. Small wedges of Tygone tubing can be used to keep the embryo well centered inside its holder.

3.4. MR Imaging

MR imaging is performed at 9.4 T with a 1-cm solenoid imaging coil using a 3-D spin-echo pulse sequence with the following parameters: repetition time (TR) = 198 ms, echo time (TE) = 7 ms, number of excitations per view (nex) = 2, flip angle (a) = 90°, field of view (FOV) in the *x*- and *y*-axes for 14.5-d mouse embryos = 14 mm, number of samples in the *x*- and *y*-axes = 256, FOV in the slice direction = 7 mm, number of image slices = 128, resulting in isotropic *voxels* with a resolution of 54.7 μm in each dimension and a total scan time of 3 h 36 min.

4. Notes

1. Cervical dislocation of the female is the most appropriate euthanasia for maintaining viable embryos because this precludes introducing drugs into the embryos which usually compromises their heart rate and reduces the time they remain viable.
2. The embryos will remain viable for as long as 2 h when removed intact with the uterus and submerged in PBS on ice. It is important to warm each embryo individually with PBS at 37°C immediately prior to isolating the umbilical vessels. The warmed PBS will reconstitute a vigorous heartbeat and plump the umbilical vessels to their natural diameter, making them much more accessible.
3. The umbilical artery and vein can be distinguished according to the following criteria:
   a. The umbilical veins pass over the top of (superficial to) the umbilical arteries as they enter the placenta.
   b. The umbilical artery has a more purple color (the better oxygenated blood in the umbilical vein is more orange).
   c. The blood in the umbilical artery is more pulsatile and flows toward the placenta, the blood in the umbilical vein is less pulsatile and flows toward the embryo. Proper identification of these vessels is critical for infusing contrast material into the desired system (arterial vs venous).
4. Before the umbilical vessels can be individually manipulated it is helpful to use two pair of forceps to tease away some of the very thin membranous material that surrounds the cord. The size of the umbilical vessels and their sensitivity to touch (expressed by vessel constriction) varies dramatically with embryonic age. For vessels that constrict when

touched, dab a small amount of the fixative from the tip of the glass cannula onto the surface of the umbilical artery and vein prior to opening them for cannulation. Both umbilical vessels should be opened prior to infusion of any material in order to provide a vent for the excess fluid that would otherwise cause the embryo to balloon.

5. The perfusion should proceed with low pressure and high volume in order to maintain physiologic dimensions of the blood vessels and heart. Where heart dimensions are critical, the initial PBS perfusion solution should be replaced by the verapamil stock solution to arrest the heart in diastole. The very small size of the cannula tip will limit the pressure in the embryo as long as there is a vent in one of the blood vessels for excess fluid to escape. Accordingly, the pump can be set to deliver the solutions at a rate that will replace the blood volume approx 2X per minute. The pump's rate will vary according to the embryo's age.

6. A dye can be added to the fixative to visualize the extent of the infusion. Two drops of food coloring to 20-mL fixative serves well. The same concentration with a second color is helpful to visualize the MR contrast solution.

7. The following steps help minimize clogging of MR contrast solution in the cannula:
   a. Keep the embryo and the cannula tip warm with a heat lamp.
   b. Keep PBS flowing through the line used to deliver contrast material at all times.
   c. Immediately prior to infusion of the contrast material, stop the pump, insert the tip of the cannula into the warm contrast, reverse the pump to draw up approx 50 µL of contrast, set the pump to forward again, and do not remove the cannula from the contrast material until the solution is flowing forward. The cannula can now be inserted into the umbilical vessel.

Acknowledgment

The author thanks G. Allan Johnson, Elwood Linney, Gary Cofer, Steve Suddarth, Sally Gewalt, and Eric Effmann for their contributions to the development of this technique. This work is supported in part by grants from the NIH (P41 RR0 5959 and NO1-HD-6-3257) and the North Carolina Biotechnology Center (9210-IDG-1016).

References

1. Effmann, E. L. (1982) Development of the right and left pulmonary arteries: A micro-angiographic study in the mouse. *Invest. Radiol.* **17,** 529–538.
2. Smith, B. R., Johnson, G. A., Groman, E. V., and Linney, E. (1994) Magnetic resonance microscopy of mouse embryos. *Proc. Nat. Acad. Sci. USA* **91,** 3530–3533.
3. Pexieder, T. (1978) Development of the outflow tract of the embryonic heart, in *Morphogenesis and Malformation of the Cardiovascular System* (Rosenquist, G. C. and Bergsma, D., eds.), Birth Defects: Original Article Series, **15,** 19–68.
4. Suddarth, S. A. and Johnson, G. A. (1991) Three-dimensional MR microscopy with large arrays. *Magn. Reson. Med.* **18,** 132–141.

22

Optical Coherence Tomography Imaging in Developmental Biology

Stephen A. Boppart, Mark E. Brezinski, and James G. Fujimoto

1. Introduction

Optical coherence tomography (OCT) is an attractive imaging technique for developmental biology because it permits the imaging of tissue microstructure *in situ*, yielding micron-scale image resolution without the need for excision of a specimen and tissue processing. OCT enables repeated imaging studies to be performed on the same specimen in order to track developmental changes. OCT is analogous to ultrasound B mode imaging except that it uses low-coherence light rather than sound and performs cross-sectional imaging by measuring the backscattered intensity of light from structures in tissue *(1)*. The principles of OCT imaging are shown schematically in **Fig. 1**. The OCT image is a gray-scale or false-color two-dimensional (2-D) representation of backscattered light intensity in a cross-sectional plane. The OCT image represents the differential backscattering contrast between different tissue types on a micron scale. Because OCT performs imaging using light, it has a one- to two-order-of-magnitude higher spatial resolution than ultrasound and does not require specimen contact.

OCT was originally developed and demonstrated in ophthalmology for high-resolution tomographic imaging of the retina and anterior eye *(2–4)*. Because the eye is transparent and is easily optically accessible, it is well suited for diagnostic OCT imaging. OCT is promising for the diagnosis of retinal disease because it can provide images of retinal pathology with 10-μm resolution, almost one order of magnitude higher than previously possible using ultrasound. Clinical studies have been performed to assess the application of OCT for a number of macular diseases *(3,4)*. OCT is especially promising for the diagnosis and monitoring of glaucoma and macular edema associated with diabetic retinopathy because it permits the quantitative measurement of changes in the retinal or retinal-nerve fiber layer thickness. Because morphological changes often occur before the onset of physical symptoms, OCT can provide a powerful approach for the early detection of these diseases.

Recently, OCT has been applied for imaging in a wide range of nontransparent tissues *(5–9)*. In tissues other than the eye, the imaging depth is limited by optical attenuation resulting from scattering and absorption. Ophthalmic imaging is typically performed at 800-nm wavelengths. However, because optical scattering decreases with

From: *Methods in Molecular Biology, Vol. 135: Developmental Biology Protocols, Vol. I*
Edited by: R. S. Tuan and C. W. Lo © Humana Press Inc., Totowa, NJ

Incident Light   **Delayed Reflected Light**

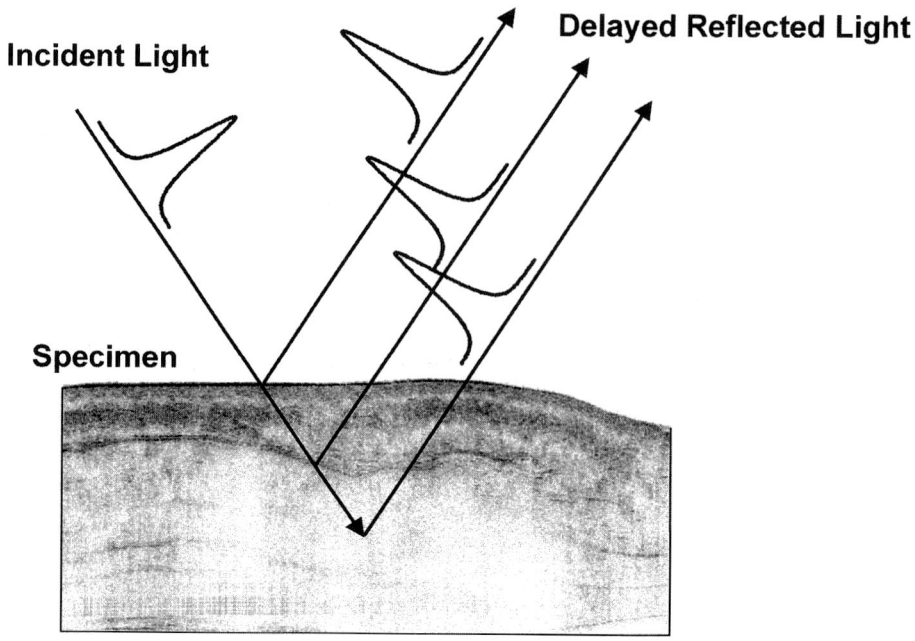

Specimen

Fig. 1. OCT imaging is performed by directing an optical beam at the object to be imaged, and the echo delay of backscattered light is measured.

increasing wavelength, OCT imaging in nontransparent tissues is possible using 1.3 μm or longer wavelengths. In most tissues, imaging depths of 2–3 mm can be achieved using a system detection sensitivity of 100–110 dB. OCT has been applied to image arterial pathology in vitro and has been shown to differentiate plaque morphology with superior resolution to ultrasound *(10–12)*. Imaging studies have also been performed to investigate applications in gastroenterology, urology, and neurosurgery *(13–15)*. High-resolution OCT using short-coherence-length, short-pulsed light sources has also been demonstrated and axial resolutions of <5 μm achieved *(16,17)*. High-speed OCT at image acquisition rates of 4–8 frames/s for a 250- to 500-square pixel images has been achieved *(18,19)*. OCT has been extended to perform Doppler imaging of blood flow and birefringence imaging to investigate laser intervention *(20–22)*. Different imaging delivery systems, including transverse imaging catheter/endoscopes and forward-imaging devices, have been developed to enable internal body OCT imaging *(23,24)*. Most recently, OCT has been combined with catheter/endoscope-based delivery to perform in vivo imaging in animal models *(25)*.

2. Principles of Operation

OCT is based on optical ranging, the high-resolution, high-dynamic-range detection of backscattered light as a function of delay. In contrast to ultrasound, because the velocity of light is extremely high, the echo time delay of reflected light cannot be measured directly and interferometric detection techniques must be used. One method for measuring echo time delay is to use low-coherence interferometry or optical-coherence domain reflectometry *(26)*. Low-coherence interferometry was first developed for

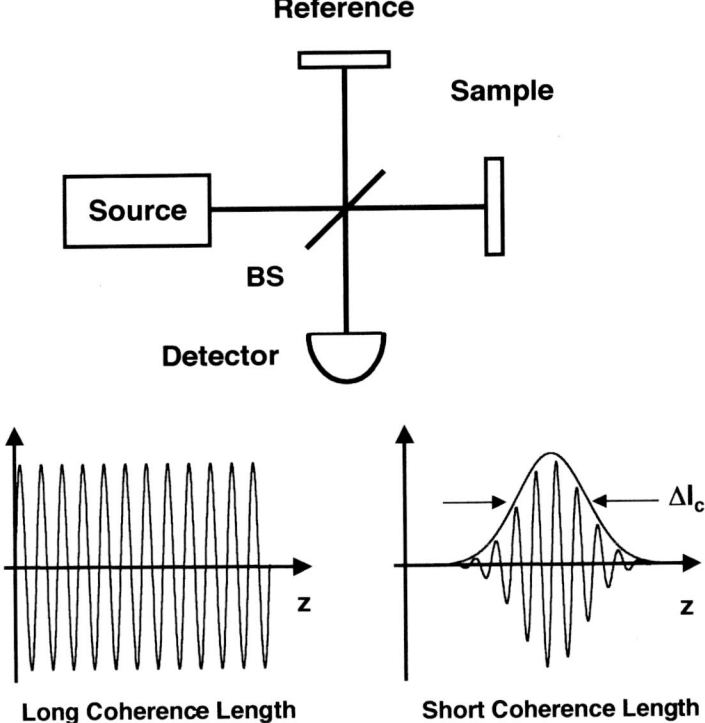

Fig. 2. Schematic showing the concept of low-coherence interferometry. Using a short-coherence length light source and a Michelson-type interferometer, interference fringes are observed only when the path lengths of the two interferometer arms are matched to within a coherence length. Abbreviations: BS, beam splitter; Δlc, coherence length.

measuring reflections in fiber optics and optoelectronic devices and was demonstrated in ophthalmology for measurements of axial eye length and corneal thickness *(27,28)*.

The echo time delay of reflected light is measured by using a Michelson-type interferometer (**Fig. 2**). The interferometer can be implemented using a fiber optic coupler in order to yield a compact and robust system (**Fig. 3**). The light reflected from the specimen or sample is interfered with light, which is reflected from a reference path of known path length. Interference of the light reflected from the sample arm and reference arm of the interferometer can occur only when the optical path lengths of the two arms match to within the coherence length of the optical source. As the reference-arm optical-path length is scanned, different echo delays of backscattered light from within the sample are measured. The interference signal is detected at the output port of the interferometer, electronically bandpass filtered, demodulated, digitized, and stored on a computer. The position of the incident beam on the specimen is scanned in the transverse direction, and multiple axial measurements are performed. This generates a 2-D data array, which represents the optical backscattering through a cross-sectional plane in the specimen. The logarithm of the backscatter intensity is then mapped to a false color or gray scale and displayed as an OCT image (**Fig. 4**).

In contrast to conventional microscopy, the axial resolution in OCT images is determined by the coherence length of the light source. The axial point-spread function of

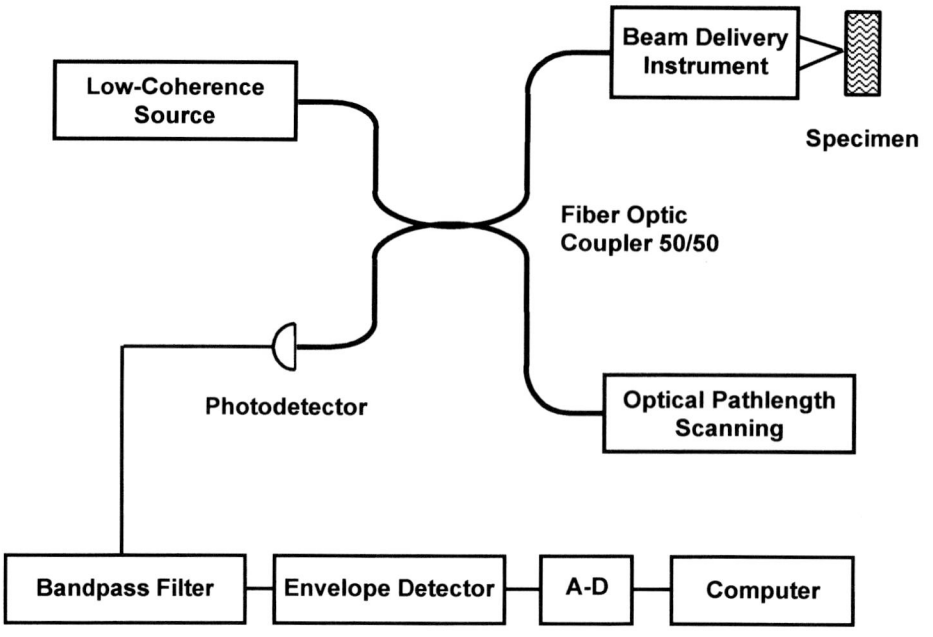

Fig. 3. Schematic representation of OCT system implemented using fiber optics. The Michelson interferometer is implemented using a 3-dB (50/50) fiber coupler. The output of the interferometer is detected with a photodiode, demodulated, and processed by a computer.

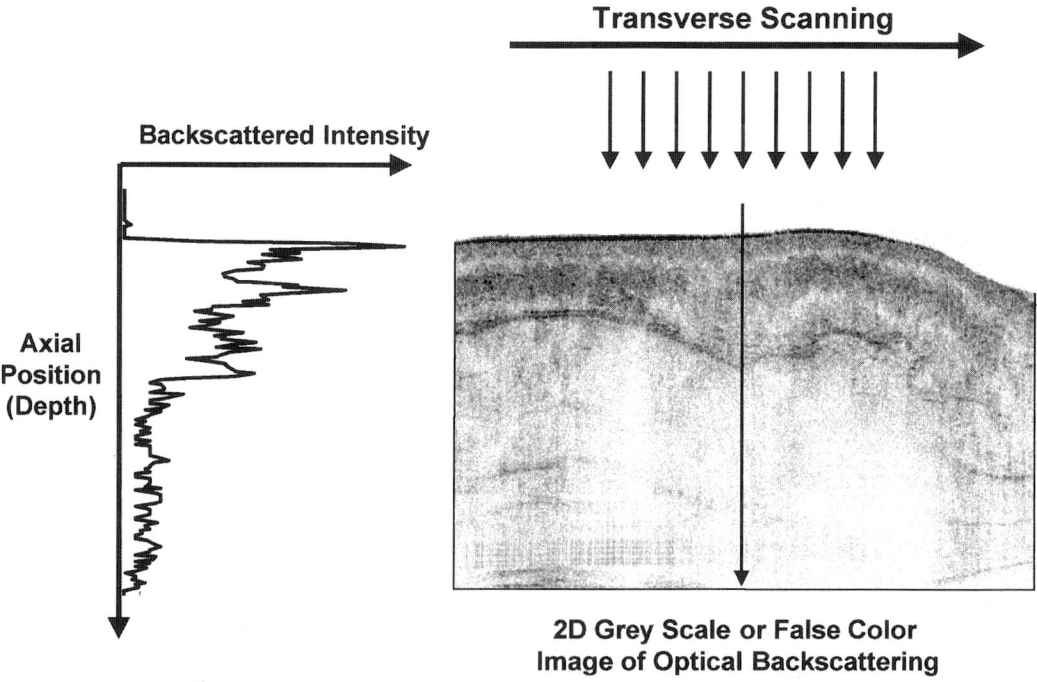

Fig. 4. The OCT image is acquired by performing axial measurements of optical backscatter at different transverse positions on the specimen and displaying the resulting 2-D data set as a gray-scale or false-color image.

the OCT measurement as defined by the signal detected at the output of the interfero-meter is the electric-field autocorrelation of the source. The coherence length of the light is the spatial width of the field autocorrelation, and the envelope of the field autocorrelation is equivalent to the Fourier transform of its power spectrum. Thus, the width of the autocorrelation function, or the axial resolution, is inversely proportional to the width of the power spectrum. For a source with a Gaussian spectral distribution, the axial resolution Δz is given by

$$\Delta z = \frac{2\ln 2}{\pi} \cdot \frac{\lambda^2}{\Delta \lambda}$$

where Δz and $\Delta \lambda$ are the full-widths-at-half-maximum of the autocorrelation function and power spectrum, respectively, and λ is the source central wavelength. This means that high-axial resolution requires broad bandwidth optical sources.

The transverse resolution in an OCT imaging system is determined by the focused spot size in analogy with conventional microscopy and is given by

$$\Delta x = \frac{4\lambda}{\pi} \cdot \frac{f}{d}$$

where d is the spot size on the objective lens and f is its focal length. High-transverse resolution can be obtained by using a large numerical aperture and focusing the beam to a small spot size. The transverse resolution is also related to the depth of focus or the confocal parameter $2z_R$ (two times the Raleigh range):

$$2z_R = \frac{\pi \Delta x^2}{2\lambda}$$

Thus, increasing the transverse resolution produces a reduced depth of field. Typically, the confocal parameter or depth of focus is chosen to match the desired depth of imag-ing. Increased resolution may also be obtained by spatially tracking the focus.

Finally, the detection signal-to-noise ratio (SNR) is given by the optical power backscattered from the sample divided by the noise equivalent bandwidth (NEB):

$$SNR = 10 \log \left(\frac{\eta}{2\hbar\omega} \frac{P_{SAM}}{NEB} \right)$$

Depending on the desired SNR performance, incident powers of 5–10 mW are typi-cally required for OCT imaging of 250–500-square pixel images at several frames per second. If lower data-acquisition speeds or SNR can be tolerated, power requirements can be reduced accordingly.

The majority of OCT imaging systems to date have used superluminescent diodes (SLDs) as low-coherence light sources. SLDs are commercially available at a range of wavelengths including 800 nm, 1.3 μm, and 1.5 μm and are attractive because they are compact and have high efficiency and low noise. However, output powers are typically limited to hundreds of microwatts and the available bandwidths are relatively narrow, permitting imaging with 10–15-μm resolution. Recent advances in short-pulse solid-state laser technology make these sources attractive for OCT imaging in research applications. Femtosecond solid-state lasers can generate tunable, low-coherence light at powers sufficient to permit high-speed OCT imaging. Short-pulse generation has been achieved across the full wavelength range in $Ti:Al_2O_3$ from 0.7–1.1 μm and over more limited tuning ranges near 1.3 and 1.5 μm in $Cr^{4+}:Mg_2SiO_4$ and $Cr^{4+}:YAG$ lasers. OCT imaging with resolutions of 4 and 6 μm has been demonstrated at 800 nm and 1.3 μm, respectively, using $Ti:Al_2O_3$ and $Cr^{4+}:Mg_2SiO_4$ sources *(16,17)*. More-compact and more-convenient sources such as superluminescent fiber sources are currently under investigation *(29–32)*. Recently, amplified SLD sources have become commercially available that have approx 15-μm resolutions at 1.3 μm and sufficient powers (>10 mW) to permit real-time OCT imaging (AFC Technologies Inc., Hull, Quebec, Canada).

3. Methods: Developmental Biology Applications

3.1. Morphological Imaging

Imaging studies were performed on several standard biological animal models commonly employed in developmental biology investigations *(33,34)*. OCT imaging was performed in *Rana pipiens* tadpoles (in vitro), *Brachydanio rerio* embryos and eggs (in vivo), and *Xenopus laevis* tadpoles (in vivo). Tadpoles were anesthetized by immersion in 0.05% tricaine until they no longer responded to touch. Specimens were oriented for imaging with the optical beam incident from either the dorsal or ventral sides. After imaging, specimens for histology were euthanized in 0.05% benzocaine for 30 min until no cardiac activity was observed. Specimens were then fixed in 10% buffered formalin for 24 h, embedded in paraffin, sectioned, and stained with hematoxylin and eosin. In order to facilitate the registration between OCT images and corresponding histology, numerous OCT images were first acquired at desired anatomical locations in 25–50-μm intervals between cross-sectional planes. Serial sectioning at 20-μm intervals was performed during histological processing. Following light microscopic observations of the histology, OCT images from the same transverse plane in the specimen were selected based on correspondence to the histological sections.

To illustrate the ability of OCT to image developing internal morphology in optically opaque specimens, a series of cross-sectional images were acquired in vitro from the dorsal and ventral sides of a Stage 49 (12-day) *R. pipiens* tadpole. The plane of the OCT image was perpendicular to the anteroposterior axis. **Figure 5** shows representative OCT images displayed in gray scale. The gray scale indicates the logarithm of the intensity of optical backscattering and spans a range of approx –60 to –110 dB of the incident optical intensity. These images (7×3 mm, 500×250 pixels, 12-bit) were each acquired in 40 s.

Features of internal architectural morphology are clearly visible in the images. The image of the eye (**Fig. 5A**) differentiates structures corresponding to the cornea, lens, and iris. The corneal thickness is on the order of 10 μm and can be resolved because of the differences in index of refraction between the water and the cornea. By imaging through the transparent lens, the incident OCT beam images several of the posterior ocular layers; including the ganglion cell layer, retinal neuroblasts, and choroid. The thicknesses of these layers were measured from the corresponding histology using a microscope with a calibrated reticule. The thicknesses of the ganglion cell layer, retinal neuroblasts, and choroid were 10, 80, and 26 μm, respectively, and demonstrate the imaging resolution of the OCT system. Identifiable structures in **Fig. 5E** include

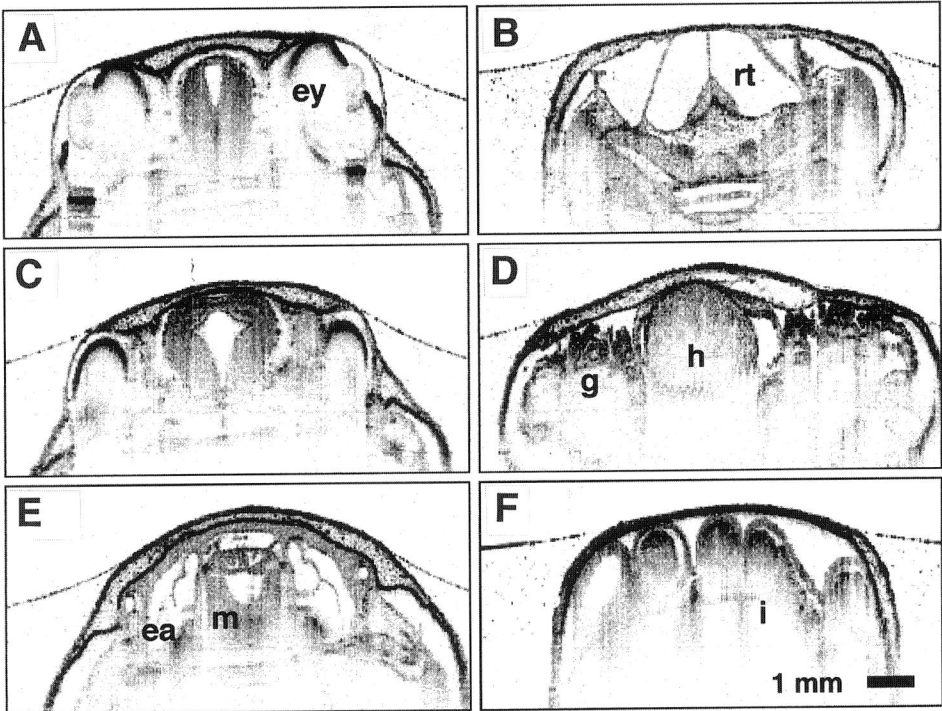

Fig. 5. OCT images of Stage 49 (12-d) *Rana pipiens* tadpole. Images in the right column, **(B)**, **(D)**, and **(F)**, were acquired with the OCT beam incident from the ventral side to image the respiratory tract, ventricle of the heart, internal gills, and gastrointestinal tract. Abbreviations: ea, ear; ey, eye; g, gills; h, heart; i, intestine; m, medulla.

the medulla and the ear vesicle. The horizontal semicircular canal and developing labyrinths are observed. Internal morphology not accessible in one orientation because of the specimen size or shadowing effects can be imaged by reorienting the specimen and scanning in the same cross-sectional image plane. The images in **Fig. 5B,D,F** were acquired with the OCT beam incident from the ventral side to image the respiratory tract, ventricle of the heart, internal gills, and gastrointestinal tract.

Figure 6 demonstrates the application of OCT for sequential imaging of development in vivo. Repeated images (3 × 3 mm, 300 × 300 pixels) were made of a developing zebrafish embryo within its egg beginning immediately after fertilization, every 15 min, for 24 h, at the same cross-sectional plane within the specimen. Developmental changes illustrate early cleavage beginning at the two-cell stage. Because of the high mitotic rate of the embryo, results of cellular division and migration are observed as well as the establishment of the anteroposterior axis. In this example, the zebrafish egg and embryo were semitransparent, and the use of OCT significantly complemented observations made using light microscopy. By imaging subtle differences in backscattering intensity, interfacial structural layers millimeters deep within specimens can be clearly delineated. In **Fig. 6A,B**, images of the zebrafish embryo are shown at the two- and four-cell stage, just after fertilization. Images of a zebrafish embryo prior to and immediately after hatching are shown in **Fig. 6C,D**.

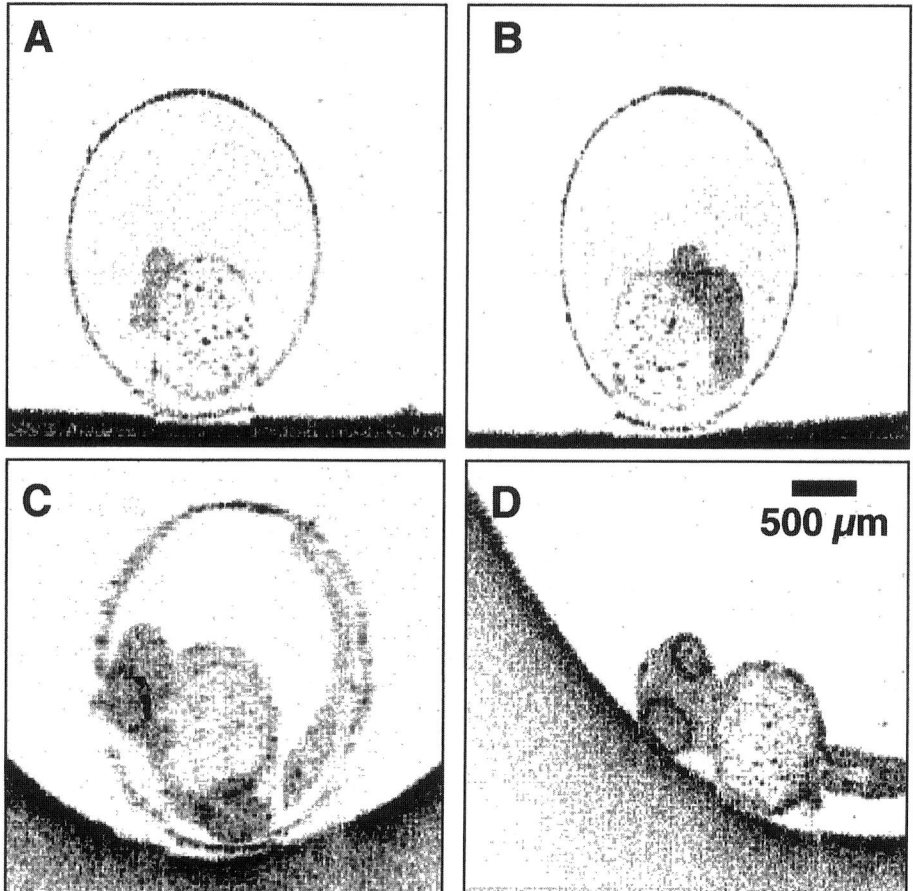

Fig. 6. Sequential OCT imaging of in vivo development in zebrafish embryo. Although embryo was semitransparent, the following demonstrates imaging in single specimens for extended periods of time. Images were acquired at (**A**) 1 h, (**B**) 1.75 h, (**C**) 24 h, and (**D**) 48 h.

3.2. Functional Imaging

Previous OCT images have characterized morphological features within biological specimens. These structures are static even though they may have been acquired from in vivo specimens. In vivo imaging in living specimens, particularly in larger organisms and for medical diagnostic applications, must be performed at high speeds to eliminate motion artifacts within the images. Functional imaging is the quantification of in vivo images, which yields information characterizing the functional properties of the organ system or organism. High-speed OCT permits both the positioning and manipulation of specimens as well as imaging in real time and is a powerful technology for functional imaging in developmental biology animal models.

Studies investigating normal and abnormal cardiac development are frequently limited by an inability to access cardiovascular function within the intact organism. OCT has been demonstrated for the high-resolution assessment of structure and function in the developing *Xenopus laevis* cardiovascular system *(35,36)*. OCT, unlike tech-

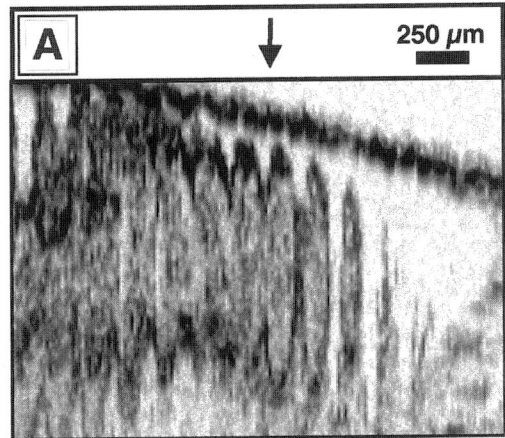

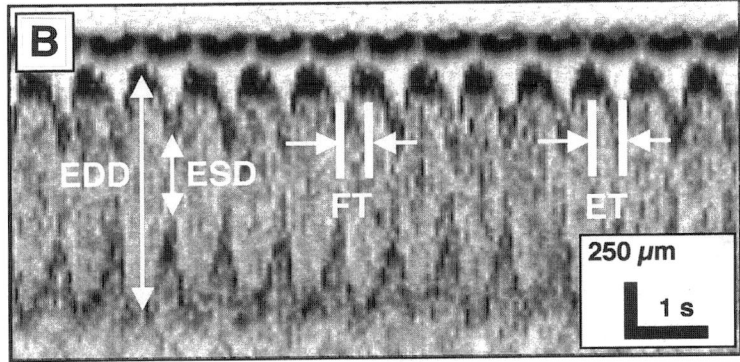

Fig. 7. **(A)** OCT image acquired with the diode-based system of the beating *Xenopus* ventricle. **(B)** OCT optical cardiogram of a normal functioning anesthetized heart. Abbreviations: EDD, end diastolic dimension; ESD, end systolic dimension; ET, ejection time; FT, filling time.

nologies such as computed tomography and magnetic resonance imaging, provides high-speed in vivo imaging, allowing quantitative dynamic activity, such as ventricular ejection fraction, to be assessed. This is analogous to both ultrasound M-mode and 2-D echocardiography.

The super-luminescent diode-based OCT system is capable of acquiring information of in vivo cardiovascular dynamics. Because a single axial scan can be acquired in approx 100 ms with the diode-based system, an OCT optical cardiogram, analogous to a M-mode echocardiogram, can be obtained with this slower system. **Figure 7A** is an OCT image acquired with the diode-based system of the beating *Xenopus* ventricle. The periodic bands within the image correspond to the movement of the cardiac chamber walls during the cardiac cycle. Both the ventral and dorsal walls of the ventricle, in addition to the ventricular lumen, can be identified, as well as the oscillatory nature inherent during the cardiac cycle. Shown in **Fig. 7A**, the OCT beam was positioned and held stationary over the ventricle at the site corresponding to the center of the image (arrow). Axial (depth) scans from this location were acquired over time and are shown in **Fig. 7B**, which is an OCT optical cardiogram of a normal functioning anesthetized heart.

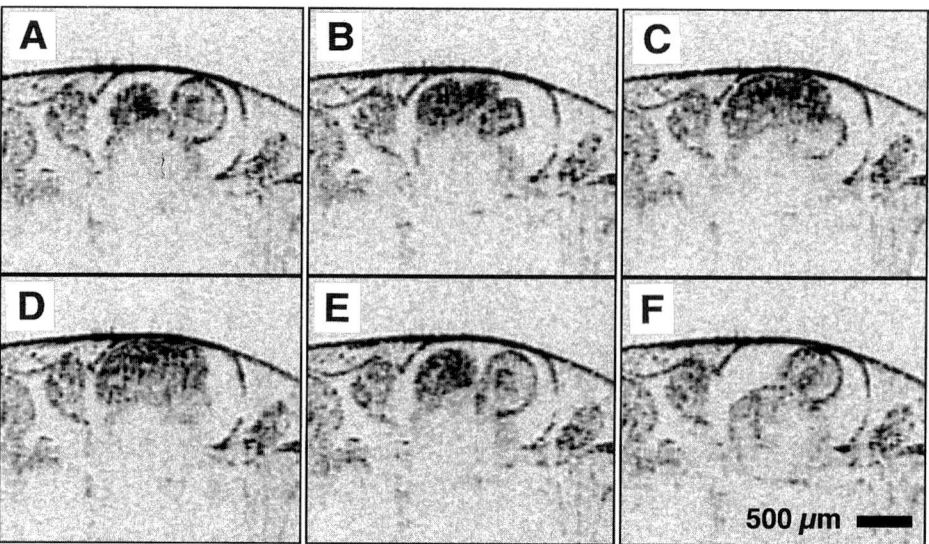

Fig. 8. High-speed OCT images of the beating *Xenopus* heart. Acquisition rates of 8 frames/s permit imaging through a single cardiac cycle.

OCT optical cardiograms permit quantitative measurements of chamber function and allow assessment of changes over time or resulting from the administration of pharmacological agents. Measurements included heart rate, end-diastolic and end-systolic dimensions, and filling and emptying times, as illustrated in **Fig. 7B**. From these measured parameters, and by using established heart volume models, ventral wall velocity, fractional shortening, and ejection fraction can be calculated.

Though the relatively slow data acquisition rate of the diode-based system (30 s/image) is adequate for in vitro imaging of microstructure and obtaining optical cardiograms from a single location, 2-D in vivo imaging of the rapidly beating heart requires considerably faster imaging speeds. The Cr^{4+}:forsterite laser was used as a light source to enable high-speed OCT imaging. Using a power of 2 mW incident on the specimen, the resulting SNR is 110 dB for high-speed imaging. Other modifications to the OCT system include the incorporation of a new optical delay line in place of the mechanical galvanometer reference arm scanner. This enabled faster axial scanning and an image acquisition rate of 4–8 images (256 × 256 pixels) per second. Both the axial and transverse resolutions were comparable to those for the super-luminescent diode source.

In contrast to the motion artifacts present in the image in **Fig. 7A**, the images in **Fig. 8** were acquired with the high-speed OCT system and are free of artifacts. The morphology of the in vivo cardiac chambers is clearly delineated at this faster imaging speed. Image acquisition is fast enough to capture the cardiac chambers in mid-cycle. With this capability, images were acquired at various times during the cardiac cycle. A sequence of six images is shown in **Fig. 8** that comprise a complete beat beginning with the initiation of diastole, the filling of the ventricle. These frames can be displayed in real time to produce a movie illustrating the dynamic, functional behavior of the developing heart.

3.3. High-Resolution Imaging

Although previous studies have demonstrated in vivo OCT imaging of tissue morphology, most have imaged tissue at approx 10–15 µm resolutions, which does not allow differentiation of cellular structure. The *Xenopus laevis* (African frog) tadpole was used to demonstrate the feasibility of OCT for high-resolution in vivo cellular and subcellular imaging *(37)*. The ability of OCT to identify the mitotic activity, the nuclear-to-cytoplasmic ratio, and the migration of cells was evaluated. OCT images were compared to corresponding histology to verify identified structures.

A short-pulse, solid-state, Cr^{4+}:forsterite laser operating near 1.3 µm was used for high-resolution imaging. The axial resolution was determined by the bandwidth of the laser source and was 5 µm. The transverse resolution was determined by the beam focusing parameters and was 9 µm. Cells as small as 15 µm in diameter could be imaged. *Xenopus* specimens ranged in age from 14 to 28 d (Stage 25–30). The specimens were irrigated at 5-min intervals to prevent dehydration during imaging. Multiple 2-D cross sections (900×600 µm, 300×300 pixels) were acquired perpendicular to the anteroposterior axis along the entire length of the specimen. To follow cellular mitosis and migration processes, three-dimensional (3-D) OCT volumes were acquired over time. Fifteen 2-D cross sections acquired at 5-µm intervals were assembled to produce a 3-D dataset. Three-dimensional volumes were acquired every 10 min from the region posterior to the eyes and lateral to the neural tube of the specimen.

Immediately following image acquisition, the location of the image planes was marked with India ink for registration between OCT images and histology. Specimens were euthanized by immersion in 0.05% Benzocaine for 1 h and then placed in a 10% buffered solution of formaldehyde for standard histological preparation. Histological sections, 5 µm thick, were sectioned and stained with hematoxylin and eosin for comparison with acquired OCT images. Correspondence was determined by the best match between OCT images and light microscopy observations of the histology. Images were processed using IP Lab 3.0 (Signal Analytics Corp., Vienna, VA) on an Apple Computer Power Macintosh 9500/200. The 3-D volumes were analyzed to identify individual cells, image mitotic activity, and track migrating cells within the acquired volumes over time. Cell position was determined by measuring distances from two internal reference points, the neural tube and the outer membrane.

Representative high-resolution images of *Xenopus* architectural and cellular microstructure are shown in **Fig. 9**. The arrow in **Fig. 9A** identifies the intricate gill structure within the respiratory tract. A superficial artery less than 100 µm in diameter is illustrated by the arrow in **Fig. 9B**. The lower right corner of this image reveals mesenchymal cells with distinct membranes and cell nuclei. A region near the nasal placode is shown in **Fig. 9C**. The arrows indicate pairs of daughter cells immediately following cell division. The arrows in **Fig. 9D** indicate melanocytes believed to have originated from the neural crest. Melanin has a high index of refraction ($n = 1.7$) compared with surrounding tissue structures ($n = 1.35$) and therefore results in higher optical backscattering in OCT images. This facilitates tracking of neural crest melanocyte migration deep within scattering specimens.

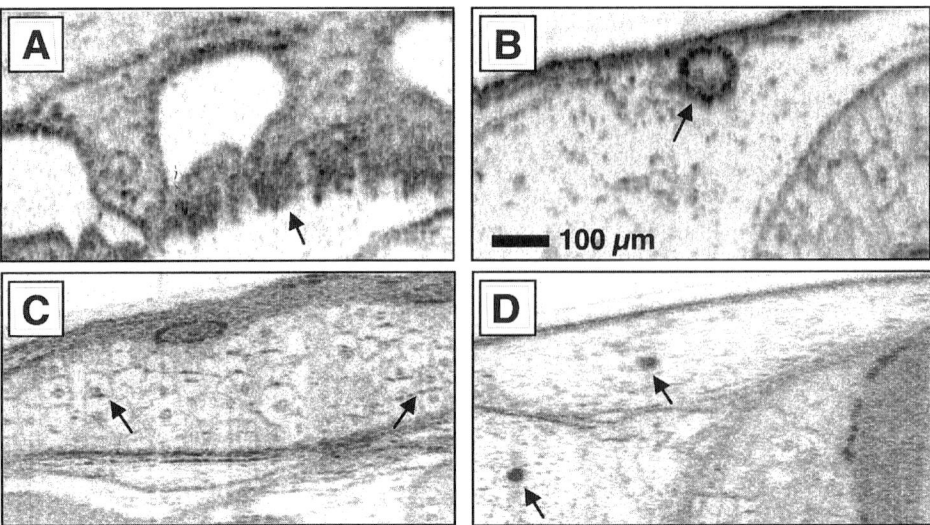

Fig. 9. Cellular-level OCT in vivo imaging in a developing *Xenopus laevis* tadpole performed with a high-resolution OCT system using a short-pulse laser as a low-coherence light source. The image resolution was 5 μm in the axial direction and 9 μm in the transverse direction. Arrows indicate **(A)** gill structures within respiratory tract, **(B)** superficial artery, **(C)** daughter cells immediately after cell division, and **(D)** melanocytes migrating from the neural crest.

4. Notes

1. The capabilities of OCT offer a unique and informative means of imaging specimens in developmental biology. The noncontact nature of OCT and the use of low-power, near-infrared radiation for imaging causes few harmful effects on living cells. OCT imaging does not require the addition of fluorophores, dyes, or stains in order to improve contrast in images. Instead, OCT relies on the inherent optical contrast generated from variations in optical scattering and index of refraction. These factors permit the use of OCT for extended imaging of development over the course of hours, days, or weeks. As shown in **Fig. 6**, images can be acquired over extended periods of time, in vivo, without complications from toxic byproducts, which can adversely affect the viability of the developing specimen. OCT permits the cross-sectional imaging of tissue and organ morphology and enables developing in vivo structure to be visualized in opaque specimens or in specimens too large for high-resolution confocal or light microscopy.

2. High-quality histology, especially serial sectioning, is often difficult, time consuming, and costly. It is especially difficult to histologically prepare the large numbers of specimens typically needed for genetic and developmental studies. OCT technology offers a promising alternative for rapidly accessing changes in architectural morphology and functional parameters. Because the position of the OCT optical beam on the specimen is precisely controlled by micron-stepping-motor stages and galvanometric scanners, the registration of the OCT imaging beam on the specimen is precisely established. Repeated serial OCT cross-sectional images can easily be acquired to construct a 3-D representation of specimen morphology. In contrast, in histology, alignment of sectioned planes is often difficult and not repeatable. The major discrepancy in registration between our OCT images and histology occurs as the result of small discrepancies in the tilt angle of the image planes rather than axial (posterior–anterior) registration.

3. OCT functions as a type of optical biopsy that permits real-time, cross-sectional in vivo imaging of architectural morphology and functional changes. Additionally, OCT may complement the results obtained with histological preparations by reducing the uncertainty associated with artifacts often attributed to histological processing. Finally, with the ability to detect morphological changes on the order of 5–10 μm, OCT offers an opportunity to identify small variations from normal development.

4. OCT image contrast results from the different optical backscattering properties between different structures. Tissue structures are differentiated according to their varying degrees of optical backscattering, whereas fluid-filled cavities within the specimen have low backscattering. The cartilaginous skeletal system of the tadpole appears highly scattered and is clearly identified in **Fig. 5A**. As light propagates deeper through the specimen, a larger percentage of the incident beam is either scattered or absorbed. Hence, less signal is available from deeper structures, and in a gray-scale image, the image becomes lighter as the SNR is reduced. Morphology located directly below a highly backscattering structure can be shadowed by the structure above it. For example, in **Fig. 5**, retinal layers are not imaged throughout the entire globe because of shadowing effects from the highly backscattering iris and sclera, which attenuate the transmission of light to deeper structures directly below. A sharp vertical boundary demarcates the regions where light is transmitted through the lens and where light is shadowed. Variation of the specimen orientation will vary the shadow orientation and permit the imaging of different internal structures. These effects are analogous to attenuation and shadowing observed in ultrasound. If the biological specimen is relatively homogeneous, the signal attenuation with depth can be compensated by simple image-processing techniques. However, because the morphology of the specimens used in this study is complex, these techniques were not applied here.

5. A number of comparisons and contrasts can be drawn between the OCT images and histological preparations. There is a strong correlation of the tissue architectural morphology between OCT images and the histology. However, it is important to note that OCT images tissue properties in a completely different manner than does histology. Histology relies on the differences in the transmission of light through stained tissue while OCT relies on differentials in optical backscattering. Histology with light and electron microscopy offers unprecedented resolution on the cellular and subcellular level. OCT does not have comparable resolution but has the ability to rapidly and repeatedly perform imaging in vivo. Histological images have artifacts caused by tissue dehydration, shrinkage, and stretching during processing. OCT images have artifacts that arise from optical attenuation, with depth, shadowing, and refractive index effects. The axial distances measured in OCT images represent the echo time delay of light, and thus, in order to convert this information into physical dimensions, it is necessary to know the index of refraction of the tissue. The index of most tissues varies between 1.35 and 1.45; thus, possible errors in longitudinal range can be of the order of only 5–10% if the index is unknown.

6. In addition to axial scale changes, the index of refraction also produces refraction of light rays when they traverse boundaries with different indexes. This effect is most significant at the proximal boundary of the specimen where the OCT beam is incident on the specimen from air. This refraction effect can cause internal features to appear as if they are angularly displaced. Refraction depends on the mismatch of the index across a boundary and is negligible if the light enters the tissue perpendicular to the boundary and becomes larger when the light ray is more oblique. These errors can be minimized by either partially or fully submerging the specimens in liquid, which produces refractive index matching. It is important to note that these same scale uncertainties and refractive effects are also present in ultrasound imaging. However, if measurements are performed in a consistent manner, these effects are considered part of the baseline. The diagnostic power of

the imaging technique is not compromised, because it relies on detecting deviations from the baseline.

7. The image scale and distortion effects are actually present in both ultrasound and MRI images, although they arise from slightly different physics principles. Histology, to some extent, suffers from similar problems in obtaining measurements from preparations resulting from tissue preservation, dehydration, and sectioning. Tissue configuration following preparation may not always reflect the in vivo orientation, making quantification difficult. Despite the artifacts in both techniques, these artifacts are reproducible and can be perceived as the baseline. These differences in calibrating histopathology against real-tissue dimensions usually do not compromise diagnostic utility.

8. Imaging at cellular and subcellular resolutions with OCT is an important area of ongoing research. The *Xenopus* developmental animal model was selected because its care and handling are relatively simple, at the same time allowing cells with a high mitotic index to be assessed. Many of the cells observed were as large as 100 μm in diameter, but ranged in size down to a few microns, below the resolution of this OCT system. Precise image registration with histology is more problematic at the cellular level. In addition to the problem of registering the OCT image with histology, cells may divide, grow, and move between the time the cells are imaged and the time the specimen is euthanized and fixed for processing. However, high-speed imaging at cellular resolutions should permit real-time tracking of a cell. The ability to position the OCT imaging plane at arbitrary angles is advantageous, especially when cell division is to be observed. The division of a cell into two daughter cells may not always occur in a predicted plane.

9. With further advances in OCT technology, improved discrimination and imaging of more-detailed morphology should be possible. New laser sources at other wavelengths in the near-infrared can enhance tissue contrast as well as potentially provide functional information, because tissue scattering and absorbance properties in specimens are wavelength dependent. Short-coherence-length short-pulse laser sources have been used to achieve higher axial resolutions on the order of 2–4 μm. Unfortunately, unlike superluminescent diode sources, these high-speed and high-resolution systems utilize femtoscond lasers, which are relatively complex and costly. As noted previously, simpler and more economical sources such as superluminescent fibers are currently under development. OCT can be performed using high numerical aperture lenses to achieve high transverse resolutions, as in confocal microscopy, at the expense of reducing the depth of field. Real-time image acquisition has been demonstrated. High-speed imaging reduces specimen motion artifacts and should allow images to be obtained without the need for anesthesia. Imaging acquisition speeds may be further increased by increasing source power and scanning speed. In combination with specially designed catheters or endoscopes, in utero imaging of embryonic and fetal development in live-bearing species may be possible.

10. Imaging developing embryonic morphology with OCT offers many new possibilities for studies in developmental biology as well as for microscopy in general. Optical coherence tomography provides high-resolution morphological, functional, and cellular information on developing biology in vivo. These results demonstrate the feasibility of gaining further insight into the morphological and functional expression of the genetic program. High-resolution, in vivo optical imaging using OCT permits the identification of morphological variations during embryogenesis. By tracking the results of cellular differentiation and rapidly identifying normal and abnormal organo- and morphogenesis in embryos, OCT represents a multifunctional investigative tool that should not only complement many of the existing imaging technologies available today, but will also reveal previously unseen dynamic changes during development.

Acknowledgments

We would like to thank Dr. Hazel Sive and her associates at the Massachusetts Institute of Technology for invaluable scientific advice and assistance. We also acknowledge the contributions of Drs. Brett Bouma and Gary Tearney. This research was supported in part by NIH contracts R01-EY11289, R01-CA75289, and the Office of Naval Research contract N00014-97-1-1066.

References

1. Huang, D., Swanson, E. A., Lin, C. P., Schuman, J. S., Stinson, W. G., Chang, W., et al. (1991) Optical coherence tomography. *Science* **254,** 1178–1181.
2. Hee, M. R., Izatt, J. A., Swanson, E. A., Huang, D., Lin, C. P., Schuman, J. S., et al. (1995) Optical coherence tomography of the human retina. *Arch. Ophthalmol.* **113,** 325–332.
3. Puliafito, C. A., Hee, M. R., Lin, C. P., Reichel, E., Schuman, J. S., Duker, J. S., et al. (1995) Imaging of macular disease with optical coherence tomography (OCT). *Ophthalmology* **102,** 217–229.
4. Puliafito, C. A., Hee, M. R., Schuman, J. S., and Fujimoto, J. G. (1995) *Optical Coherence Tomography of Ocular Diseases.* Slack, Thorofare, NJ.
5. Schmitt, J. M., Knuttel, A., and Bonner, R. F. (1993). Measurement of the optical properties of biological tissue using low-coherence reflectometry. *Appl. Opt.* **32,** 6032–6042.
6. Schmitt, J. M., Knuttel, A., Yadlowsky, M., and Eckhaus, A. A. (1994) Optical coherence tomography of a dense tissue: Statistics of attenuation and backscattering. *Phys. Med. Biol.* **39,** 1705–1720.
7. Fujimoto, J. G., Brezinski, M. E., Tearney, G. J., Boppart, S. A., Bouma, B. E., et al. (1995) Biomedical imaging and optical biopsy using optical coherence tomography. *Nat. Med.* **1,** 970–972.
8. Schmitt, J. M., Yadlowsky, M. J., and Bonner, R. F. (1995) Subsurface imaging of living skin with optical coherence microscopy. *Dermatology* **191,** 93–98.
9. Sergeev, A., Gelikonov, B., Gelikonov, G., Feldchetin, F., Pravdenki, K., Kuranov, R., et al. (1995) High-spatial resolution optical-coherence tomography of human skin and mucus membranes in *Conference on Lasers and Electro Optics '95*, Vol. 15 of 1995 OSA Technical Digest Series (Optical Society of America, Washington, DC), paper CThN4.
10. Brezinski, M. E., Tearney, G. J., Bouma, B. E., Izatt, J. A., Hee, M. R., Swanson, E. A., et al. (1996) Optical coherence tomography for optical biopsy: Properties and demonstration of vascular pathology. *Circulation* **93,** 1206–1213.
11. Tearney, G. J., Brezinski, M. E., Boppart, S. A., Bouma, B. E., Weissman, N., Southern, J. F., et al. (1996) Catheter-based optical imaging of a human coronary artery. *Circulation* **94,** 3013.
12. Brezinski, M. E., Tearney, G. J., Weissman, N. J., Boppart, S. A., Bouma, B. E., Hee, M. R., et al. (1997) Assessing atherosclerotic plaque morphology: Comparison of optical coherence tomography and high frequency intravascular ultrasound. *Br. Heart J.* **77,** 397–404.
13. Brezinski, M. E., Tearney, G. J., Boppart, S. A., Swanson, E. A., Southern, J. F., and Fujimoto, J. G. (1997) Optical biopsy with optical coherence tomography, feasibility for surgical diagnostics. *J. Surg. Res.* **71,** 32–40.
14. Tearney, G. J., Brezinski, M. E., Southern, J. F., Bouma, B. E., Boppart, S. A., and Fujimoto, J. G. (1997) Optical biopsy in human gastrointestinal tissue using optical coherence tomography. *Am. J. Gastroenterol.* **92,** 1800–1804.
15. Tearney, G. J., Brezinski, M. E., Southern, J. F., Bouma, B. E., Boppart, S. A., and Fujimoto, J. G. (1997) Optical biopsy in human urologic tissue using optical coherence tomography. *J. Urol.* **157,** 1915–1919.

16. Bouma, B., Tearney, G. J., Boppart, S. A., Hee, M. R., Brezinski, M. E., and Fujimoto, J. G. (1995) High-resolution optical coherence tomographic imaging using a mode-locked Ti:Al$_2$O$_3$ laser source. *Opt. Lett.* **20,** 1486–1489.

17. Bouma, B. E., Tearney, G. J., Biliinski, I. P., Golubovic, B., and Fujimoto, J. G. (1996) Self-phase-modulated Kerr-lens mode-locked Cr:forsterite laser source for optical coherence tomography. *Opt. Lett.* **21,** 1839–1842.

18. Tearney, G. J., Bouma, B. E., Boppart, B. E., Golubovic, B., Swanson, E. A., and Fujimoto, J. G. (1996) Rapid acquisition of *in vivo* biological images using optical coherence tomography. *Opt. Lett.* **21,** 1408–1410.

19. Tearney, G. J., Bouma, B. E., and Fujimoto, J. G. (1997) High-speed phase- and group-delay scanning with a grating-based phase control delay line. *Opt. Lett.* **22,** 1811–1813.

20. de Boer, J. F., Milner, T. E., van Germert, M. J. C., and Stuart Nelson, J. (1997) Two dimensional birefringence imaging in biological tissue by polarization sensitive optical coherence tomography. *Opt. Lett.* **22,** 934–936.

21. Chen, Z., Milner, T. E., Srinivas, S., Wang, X., Malekafzali, A., van Germert, M. J. C., et al. (1997) Noninvasive imaging of *in vivo* blood flow velocity using optical Doppler tomography. *Opt. Lett.* **22,** 1119–1120.

22. Izatt, J. A., Kulkarni, M. D., Yazdanfar, S., Barton, J. K., and Welch, A. J. (1997) *In vivo* bidirectional color doppler flow imaging of picoliter blood volumes using optical coherence tomography. *Opt. Lett.* **22,** 1439–1441.

23. Tearney, G. J., Boppart, S. A., Bouma, B. E., Brezinski, M. F., Weissman, N. J., Southern, J. F., et al. (1996) Scanning single-mode fiber optic catheter-endoscope for optical coherence tomography. *Opt. Lett.* **21,** 543–545.

24. Boppart, S. A., Bouma, B. E., Pitris, C., Tearney, G. J., Fujimoto, J. G., and Brezinski, M. E. (1997) Forward-scanning instruments for optical coherence tomographic imaging. *Opt. Lett.* **22,** 1618–1620.

25. Tearney, G. J., Brezinski, M. E., Bouma, B. E., Boppart, S. A., Pitris, C., Southern, J. F., et al. (1997) *In vivo* endoscopic optical biopsy with optical coherence tomography. *Science* **276,** 2037–2039.

26. Takada, K., Yokohama, I., Chida, K., and Noda, J. (1987) New measurement system for fault location in optical waveguide devices based on an interferometric technique. *Appl. Opt.* **26,** 1603–1606.

27. Fercher, A. F., Mengedoht, K., and Werner, W. (1988) Eye-length measurement by interferometry with partially coherent light. *Opt. Lett.* **13,** 186–190.

28. Hitzenberger, C. K. (1991) Measurement of the axial eye length by laser Doppler interferometry. *Invest. Ophthalmol. Vis. Sci.* **32,** 616–624.

29. Chernikov, S. V., Zhu, Y., Taylor, J. R., Platonov, N. S., Samartsev, I. E., and Gapontsev, V. P. (1996) 1.08–2.2 μm supercontinuum generation from Yb3+ doped fiber laser. *Conference on Lasers and Electro Optics CLEO 96,* Vol. 9 of 1996 OSA Technical Digest Series (Optical Society of America, Washington, DC) paper CTuU4.

30. Swanson, E. A., Chinn, S. R., Hodgson, C. W., Bouma, B. E., Tearney, G. J., and Fujimoto, J. G. (1996) Spectrally shaped rare-earth doped fiber ASE sources for use in optical coherence tomography. *Conference on Lasers and Electro Optics CLEO 96,* Vol. 9 of OSA 1996 Technical Digest, (Optical Society of America, Washington, DC) paper CTuU5.

31. Chernikov, V., Taylor, J. R., Gapontsev, V. P., Bouma, B. E., and Fujimoto, J. G. (1997) A 75 nm, 30 mW superfluorescent ytterbium fiber source operation around 1.06 μm. *Conference on Lasers and Electro Optics CLEO 97,* Vol. 11 of OSA 1997 Technical Digest, (Optical Society of America, Washington, DC) paper CTuG8.

32. Bouma, B. E., Nelson, L. E., Tearney, G. J., Jones, D. J., Brezinski, M. E., and Fujimoto, J. G. (1998) Optical coherence tomographic imaging of human tissue at 1.55 μm and 1.8 μm using Er- and Tm-doped fiber sources. *J. Biomed. Opt.* **3,** 76–79.

33. Boppart, S. A., Brezinski, M. E., Bouma, B. E., Tearney, G. J., and Fujimoto, J. G. (1996) Investigation of developing embryonic morphology using optical coherence tomography. *Dev. Biol.* **177,** 54–63.

34. Boppart, S. A., Brezinski, M. E., Tearney, G. J., Bouma, B. E., and Fujimoto, J. G. (1996) Imaging developing neural morphology using optical coherence tomography. *J. Neurosci. Methods* **2112,** 65–72.

35. Nieuwkoop, P. D. and Faber, J. (1994) *Normal Table of Xenopus Laevis.* Garland, New York.

36. Boppart, S. A., Tearney, G. J., Bouma, B. E., Southern, J. F., Brezinski, M. E., and Fujimoto, J. G. (1997) Noninvasive assessment of the developing xenopus cardiovascular system using optical coherence tomography. *Proc. Natl. Acad. Sci.* **94,** 4256–4261.

37. Boppart, S. A., Bouma, B. E., Pitris, C., Southern, J. F., Brezinski, M. E., and Fujimoto, J. G. (1998) In vivo cellular optical coherence tomography imaging. *Nat. Med.* **4,** 861–865.

23

Ultrasound Backscatter Microscopy of Mouse Embryos

Daniel H. Turnbull

1. Introduction

Extensive genetic information and transgenic techniques available in the mouse have led to its wide use in studies of mammalian development and models of human disease. A basic limitation of analyzing dynamic developmental processes in mouse embryos is their inaccessibility, because they are encased in the maternal uterus. In particular, cardiovascular development has been difficult to study and an understanding of the underlying mechanisms regulating heart development is incomplete, in part because of the lack of methods to measure cardio- and hemodynamics in live mouse embryos. Also, in contrast to lower vertebrate species such as frog, chick, and zebrafish, direct *in utero* manipulation of mouse embryos, through injection of dyes, cells, or retroviruses are difficult or impossible at most embryonic stages. High-frequency (40–50 MHz) ultrasound imaging and Doppler scanners, also called ultrasound backscatter microscopes (UBM), have recently been developed that allow noninvasive measurements of cardiovascular structure and function in live mouse embryos and that make possible direct *in utero* manipulations through high-resolution ultrasound-guided injections. This protocol describes methods that have been employed to image and measure blood flow in live mouse embryos and to inject cells, retroviruses and other agents into specific embryonic target sites using UBM imaging as a guidance system.

2. Materials

1. UBM: Custom-built UBMs have been developed to perform the procedures described in this protocol *(1,2)*. Alternatively, it is possible to modify commercial high-frequency ultrasound scanners (e.g., Ultrasound Biomicroscope, Paradigm Medical Industries, Salt Lake City, UT) to perform *in utero* mouse embryo imaging (**Fig. 1A**). In either case, the imaging system consists of a high-frequency (40–50 MHz) focused ultrasound transducer that is scanned mechanically to produce two-dimensional (2-D) images at real-time frame rates (≥8 images per second). The scanning probe is mounted on a three-axis motorized micropositioning system (e.g., UMR8-25 translation stages with 860A-1 motorized drives and 860-c2 motor controller, Newport Corp., Irvine, CA) with a joystick controller (PMC200-J; Newport) for precise, reproducible positioning of the UBM image plane.
2. UBM-guided Doppler system: A high-frequency (40–50 MHz) ultrasound Doppler system has been implemented on a UBM to perform image-guided measurements of blood

From: *Methods in Molecular Biology, Vol. 135: Developmental Biology Protocols, Vol. I*
Edited by: R. S. Tuan and C. W. Lo © Humana Press Inc., Totowa, NJ

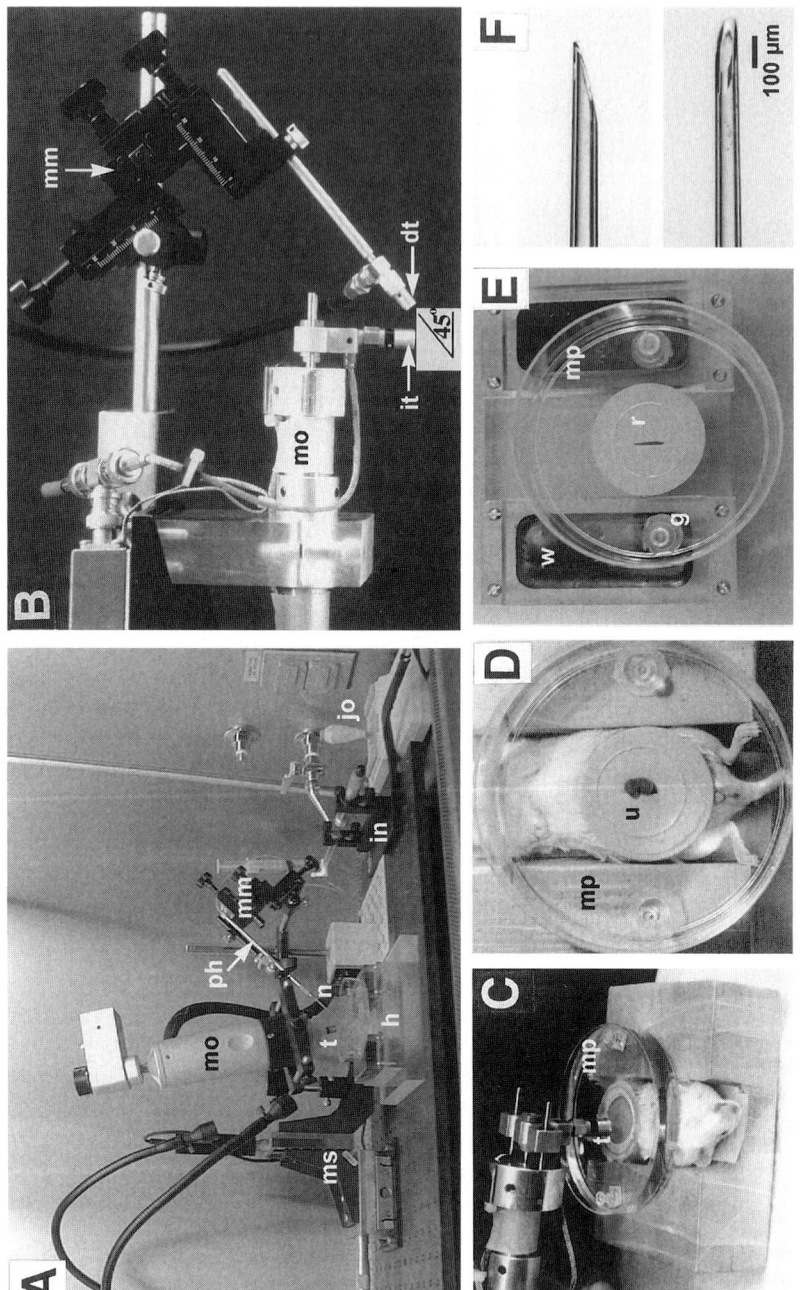

Fig. 1.

velocity in live mouse embryos *(2)*. The custom-built Doppler transducers are mounted on a 3-D micromanipulator (M-152, Narishige, Tokyo, Japan), attached to the motion stage housing the imaging transducer (**Fig. 1B**), allowing calibration of the Doppler sample volume at a predetermined position in the UBM image plane *(2)*.

3. Mouse holding stage: A two-level stage is used to hold the anesthetized mice during these procedures. This stage can be made from wood (**Fig. 1C,D**) or Plexiglas, in which case the top section has two wells filled with dissection wax to facilitate pinning a modified Petri dish over the mouse (**Fig. 1E**). Rubber strips (e.g., O-LRS-50 latex sheets; Small Parts Inc., Miami Lakes FL) or other spacer material should be cut to fit under the mouse to raise her to the top level of the holding stage.

4. Modified Petri dishes: Standard 100-mm diameter plastic Petri dishes are modified by punching a 25-mm diameter hole in the center and drilling two small holes at opposite sides of the dish for pinning it over the anesthetized mouse (**Fig. 1D,E**). The small holes are covered with silicone grease (High vacuum grease; Dow Corning, Midland, MI) to minimize leakage of water or phosphate-buffered saline (PBS) during the UBM procedures. For UBM-guided injections, a thin rubber membrane (Silastic L room temperature vulcanization (RTV) silicone rubber; Dow Corning), cast on a standard 35-mm plastic Petri dish, is placed over the central hole and an opening cut to allow the exposed uterus to be pulled into the Petri dish (*see* **Subheading 3.**).

5. Injection needles: Injection needles are made from glass microcapillary pipets (outer diameter = 1 mm, inner diameter = 0.5 mm, length = 100 mm: B100-50-10; Sutter Instrument Co., Novato, CA). The needles (**Fig. 1F**) are pulled to produce a long taper (Flaming/Brown Micropipette Puller, Sutter Instruments), broken under a dissection microscope and sharpened with a micropipet beveler (EG-40; Narishige) (*see* **Subheading 3.**).

6. Guided-injection apparatus: Needles are held in a micropipet holder (HI-7 pipet holder and CI-1 connector; Narishige) connected with 1-mm (inner diameter) tubing to an oil-filled (M-8410 Mineral Oil; Sigma Chemical Co., St. Louis, MO) microinjector (Manual Microsyringe pump; Stoelting Research Instruments, Wood Dale, IL). The micropipet holder is held in a 3-D micromanipulator (M-152; Narishige), which is used to position the injection needle (**Fig. 1A**).

7. Cell suspensions: *In utero* UBM-guided injections of single-cell suspensions have been made into mouse embryos staged between 9.5 *(3)* and 13.5 days *(4)* post coitus.

Fig. 1. *(previous page)* UBM systems and ancillary equipment for noninvasive imaging, Doppler blood flow measurements, and UBM-guided *in utero* injections. (**A**) A modified commercial UBM setup for guided injections. The mechanical scan-head (mo) is mounted on a motorized micropositioning system (ms) with joystick (jo) controller. Injections are performed using a manual microinjector (in) with the micropipet holder (ph) and injection needle (n) mounted on a 3-D micromanipulator (mm). The mouse holding stage (h) is shown below the UBM transducer (t). (**B**) The front-end of a prototype UBM scanner set up for blood-flow measurements, with the Doppler beam aligned at 45° to the image plane. Labels: mo, motorized scan-head; mm, micromanipulator; it, imaging transducer; dt, Doppler transducer. (**C**) Mouse positioned under the UBM transducer (t) for noninvasive imaging of embryos. Labels: mp, modified Petri dish. (**D**) Mouse prepared, with a piece of the uterus (u) containing two E9.5 embryos exposed in PBS, for *in utero* injection. (**E**) Two-level mouse holding stage, with modified Petri dish (mp) and rubber membrane (r) prepared for E9.5 embryos. Labels: g, silicone grease; w, wax. (**F**) Tips of beveled glass injection microcapillaries showing side and front views. (**Figure 1B** reprinted with permission from **ref. 2**; **Figure 1C** reprinted with permission from **ref. 6**).

The preparation of the cells depends on the type of cell being injected. Success has been obtained injecting both cultured *(3)* and primary *(4)* neural cells suspended in Dulbecco's modified Eagle's medium (DMEM; Sigma) with 0.05% DNase at a concentration of approx 5×10^4 cells/μL.

8. Anesthetic: Timed pregnant mice are anesthetized with sodium pentobarbital (0.6 mg/10 g body weight, injected intraperitoneally) mixed with magnesium sulfate ($MgSO_4 \cdot 7H_2O$, 1 mg/10 g body weight) as a mild muscle relaxant to decrease the incidence of spontaneous uterine contractions that interfere with image acquisition (*see* **Note 1**).

9. Miscellaneous materials: During noninvasive (transabdominal) imaging of mouse embryos, the Petri dish is filled with distilled water (*see* **Note 2**) to couple the UBM transducer to the skin. For procedures involving surgical exposure of the uterus, sterile PBS (e.g., PBS with calcium- and magnesium chloride, Sigma) should be used as the coupling medium. A set of sterile surgical instruments (two blunt forceps, small scissors, scalpel, hemostatic forceps), razor blades, suture (5-0 silk), skin clips (9-mm Autoclip wound clips; Benton Dickenson, Sparks MD), syringes and needles (1-cc disposable insulin syringe with 28-gage needle), and cotton-tipped applicators are also required.

3. Methods

1. Noninvasive mouse embryo imaging: Timed pregnant mice (Note: 0.5 d post coitus [dpc], is defined as noon of the day a vaginal plug is detected after overnight mating) are anesthetized, and the lower abdomen and/or back are wet shaved (*see* **Note 3**) to provide a clear acoustic window to the embryos *(5,6)*. The mouse is placed in the lower level of the holding stage, and a modified Petri dish is pinned over the shaved skin, raising the mouse until a leak-proof seal is formed between the skin and Petri dish (**Fig. 1C**; *see* **Note 4**). The Petri dish is filled with distilled water to allow scanning through the skin to the underlying embryos (**Fig. 2A**). Before scanning, check the front surface of the UBM transducer for air bubbles, and brush them off with a cotton-tipped applicator if necessary (*see* **Note 5**).

   With practice, it is possible to image most or all of the embryos in a litter, starting with the embryo in the upper lateral position (closest to the ovary) on the right side and systematically imaging down the uterine horn, past the vagina and up the left side of the horn. Imaging the whole litter requires repositioning the Petri dish 4–5X to visualize all the embryos. Some embryos, especially those close to the vagina, may be too deep to obtain high-quality UBM images, and gas in sections of bowel overlying the uterus will form an impenetrable barrier to the ultrasound beam, showing up as a dark shadow in the UBM image. Gentle manipulation through the skin of the anesthetized mouse can sometimes be used to position the uterus above obstructing bowel and fat layers. The most obvious anatomical features in early stage mouse embryos are the fluid-filled (dark) neural tube cavity and echogenic (bright) beating heart *(5)* (**Fig. 2A**). With experience, it is possible to

---

Fig. 2. *(opposite page)* Noninvasive *in utero* UBM imaging and UBM-guided Doppler measurements in mouse embryos. (**A**) Sagittal UBM image of an E10.5 embryo, demonstrating the definition of neural tube (n), heart (ht), and dorsal aorta (da). (**B**) *In utero* image of E13.5 mouse heart, showing the definition of the developing left (la) and right (ra) atria and left (lv) and right (rv) ventricles. Smallest scale markers on right (A, B) = 100 μm. (**C**) E10.5 mouse embryo with Doppler sample volume (white box) positioned over the primitive ventricle (v) and atria (a). Labels: n, neural tube cavity. (**D**) Doppler measurements of ventricular inflow and outflow blood velocity, where initial ventricular filling (E), followed by filling resulting from atrial contraction (A), were identified on the inflow waveform. (**Figure 2A,B** reprinted with permission from **ref. 6**; **Figure 2C,D** reprinted with permission from **ref. 2**).

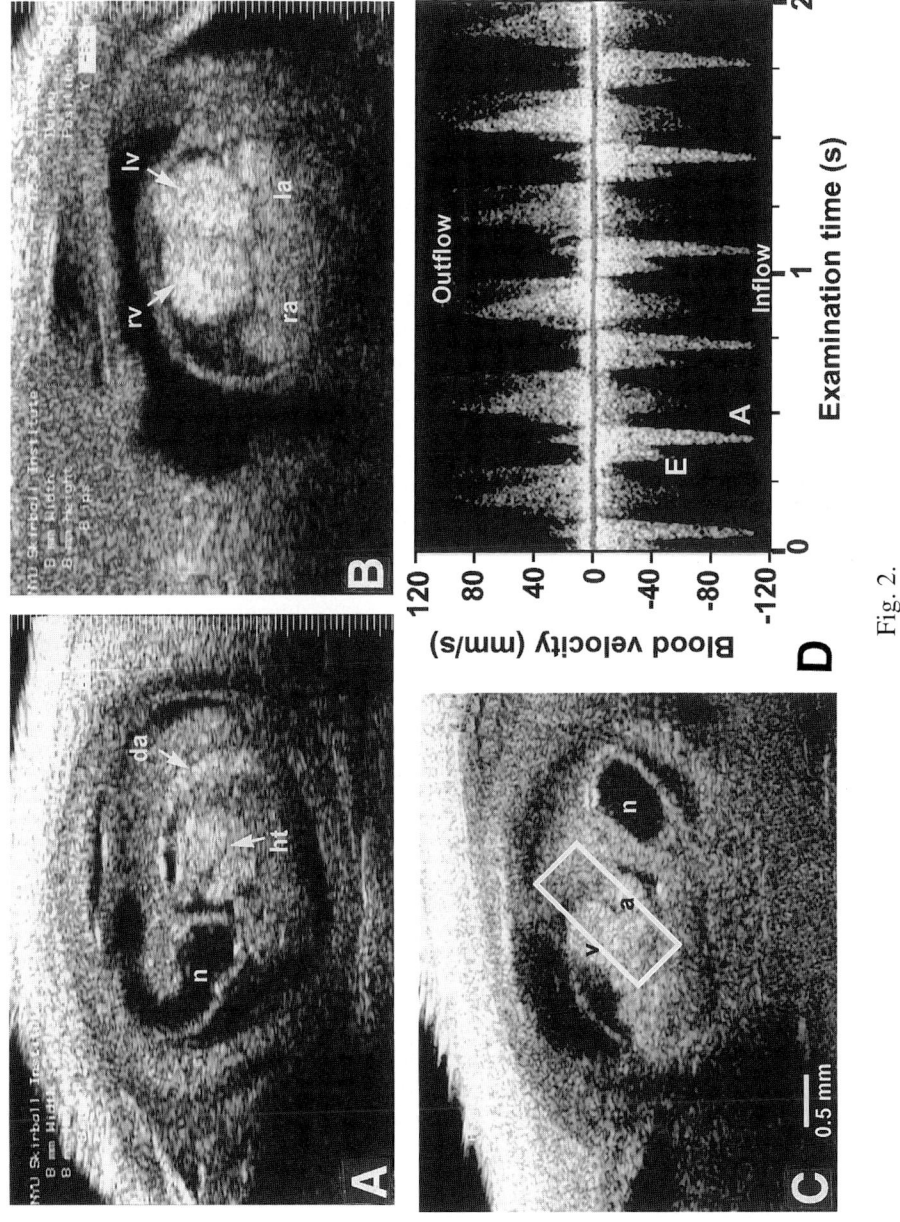

Fig. 2.

239

identify numerous features in developing mouse embryos, including limb buds, large blood vessels, cardiac chambers, brain, spinal cord, and internal organs. This has been especially useful for noninvasive measurements of cardiac chamber dimensions and contractility *(6)* (**Fig. 2B**). As in human fetal ultrasound imaging, *in utero* UBM imaging of mouse embryos in ideal transverse, coronal or sagittal planes is often impossible, and embryos must be viewed in oblique planes approximating the desired orientations.

2. UBM-guided blood flow measurements: Ultrasound Doppler blood velocity measurements are made using a second ultrasound transducer whose sampling volume is centered at a predetermined position in the UBM image *(2)*. Large blood vessels (e.g., umbilical vessels, vitelline vessels, aorta, cerebral vessels) are identifiable on real-time UBM images from the moving speckle pattern produced by moving blood (**Fig. 2A**). Blood velocity waveforms can be measured in specific vessels and in the hearts of developing mouse embryos (**Fig. 2C,D**). In particular, the umbilical vessels are easily identified, forming the connection between the embryo and the maternal placenta, and Doppler waveforms have been obtained in embryos staged as early as 9.5 dpc, soon after the chorio-allantoic connection is formed *(2)*.

3. Making injection needles: Glass microcapillary pipets are pulled to form a long taper (10 mm or more) and then broken with fine forceps under a dissection microscope. A calibrated length scale in one eyepiece of the microscope is useful to ensure reproducible-sized needles. For injection of cell suspensions, the micropipet should be broken with an outer diameter of 70–80 μm (corresponding inner diameter should be approx 40–50 μm). For injection of dyes and retroviruses, the needle tips can be smaller, but it is not recommended to use needles with an outer diameter less than 50 μm (inner diameter approx 30 μm), as these will be too fragile and tend to break during the injection procedure. A very sharp needle is required to pierce the uterine wall. The method used to sharpen the needles is to bevel the tips in three stages, keeping the grinding surface of the beveler wet at all times (*see* **Note 6**): Start with a 30° beveling angle, then progress to 25° and finish with a 20° angle. This procedure is time-consuming (10–15 min per needle) and is best performed with a stereo microscope mounted over the beveler to visually monitor the needle as it is sharpened. The result of the beveling procedure is a needle with a very sharp tip with an opening on the side (**Fig. 1F**), which helps avoid needle clogging during injections. Several extra needles should be prepared before starting a set of injections, because broken and clogged tips are inevitable.

4. Making rubber membranes: Silastic RTV rubber is mixed in a 10:1 (rubber: curing agent) ratio until the mixture is a uniform color (*see* **Note 7**). The result is a thick paste that can be spread thinly over the outside surfaces (bottoms and lids) of 35-mm dishes with a flat spatula and left to set overnight. Avoid introducing bubbles in the rubber, as these will produce holes in the resulting membranes. Mixing 10-g rubber with 1-g curing agent should produce enough rubber to make 16 membranes with very little extra material. After setting, the rubber membranes are pressed over the hole in the modified Petri dishes (**Fig. 1E**); they will stick to the clean, dry plastic with no additional adhesives required. An opening should be made in the rubber membrane to fit the embryonic stage being injected. **Table 1** indicates approximate dimensions for the openings used for embryos staged between 8.5 and 13.5 dpc.

5. Aligning injection needle in UBM image plane: Prior to injecting mouse embryos, the needle should be aligned in the UBM image plane (*see* **Note 8**). This is done by trial and error, adjusting the angle between the axis of the needle and the image plane by adjusting the orientation of the 3-D micromanipulator and imaging an injection needle in a small water bath (e.g., a standard 100-mm plastic Petri dish). The tip of the needle appears as a bright spot on the UBM image, larger than the true physical size of the needle tip *(3,4)*.

Table 1
Guidelines for Size of Openings in Rubber Membranes

Embryonic stage (dpc)	8.5	9.5	10.5	11.5	12.5	13.5
Dimensions of opening (mm)	Slit	Slot	Slot	Oval	Oval	Oval
	12	12×1	12×2	15×2	15×3	15×5

When the needle is properly aligned, the brightness from the tip will remain constant as the tip is advanced through the image plane. If alignment is perfect, the entire length of the injection needle will be visualized in the UBM image (*see* **Note 9**).

6. *In utero* UBM-guided injections: After anesthetizing the pregnant mouse, wet shave the abdomen with a razor blade (*see* **Note 10**), and make a 2-cm midline incision, first through the skin and then through the peritoneal muscle. Place the mouse in the lower level of the holding stage, raising the mouse until the level of her abdomen is slightly above the upper level of the stage (*see* **Note 11**). Carefully locate and pull out the uterus, noting the number of embryos on each side of the uterine horn (*see* **Note 12**). Select one side to inject, and pull the first (closest to ovary) one (11.5 dpc or more) or two (10.5 dpc or less) embryos through the opening in the rubber membrane. Pin the modified Petri dish to the top level of the mouse holding stage and fill the Petri dish with sterile PBS: The rubber membrane should press snugly against the mouse's shaved abdomen, holding the uterus in place and making a leak-proof seal (**Fig. 1D**). Using the real-time UBM images, locate the specific region of the embryo for injection. Monitor the tip of the injection needle as it is pushed through the uterine wall and into the target region (**Fig. 3**). After injecting the exposed embryos, gently pull one or two new embryos through the rubber membrane, and gently push the injected embryos back into the abdominal cavity. In this way, work from the ovary to the vagina, injecting each embryo in the selected target site. If both sides of the uterine horn are to be injected, pour off the PBS and reposition the modified Petri dish, starting with the embryo closest to the opposite ovary and working down to the embryo closest to the vagina, as on the first side of the horn (*see* **Note 13**). After injecting the desired number of embryos, gently replace the uterus as closely as possible to its original position, suture the peritoneal muscle, and use clips to close the skin.

7. Animal recovery: In all procedures (noninvasive or surgical) involving anesthetizing pregnant mice, keep the animals warm during recovery from anesthesia. This can be done using a warming pad, slide warmer, or a recovery chamber specifically designed for small animal surgery. After recovery, the mouse is housed in a separate cage until giving birth or until being killed if the embryos are being collected before birth (*see* **Note 14**).

4. Notes

1. Other anesthetics (e.g., ketamine/xylazine mixture) can be substituted for sodium pentobarbital. The anesthetic known as avertin is not recommended for mid-gestation pregnant mice, because it causes arrythmia in the embryonic hearts.
2. If distilled water is obtained from a laboratory faucet, allow it to stand in a container overnight to remove excess dissolved air that can form bubbles on the skin surface and block or reduce the penetration of the ultrasound beam.
3. The skin must be shaved very clean, as any remaining hair will trap tiny air bubbles and reduce the penetration of the high-frequency ultrasound beam.
4. Leakage problems can be minimized through the use of a rubber ring fitted around the opening in the modified Petri dish. Use the same membranes described for UBM-guided injections, but cut out a central hole with a diameter of approx 20 mm to form a ring that

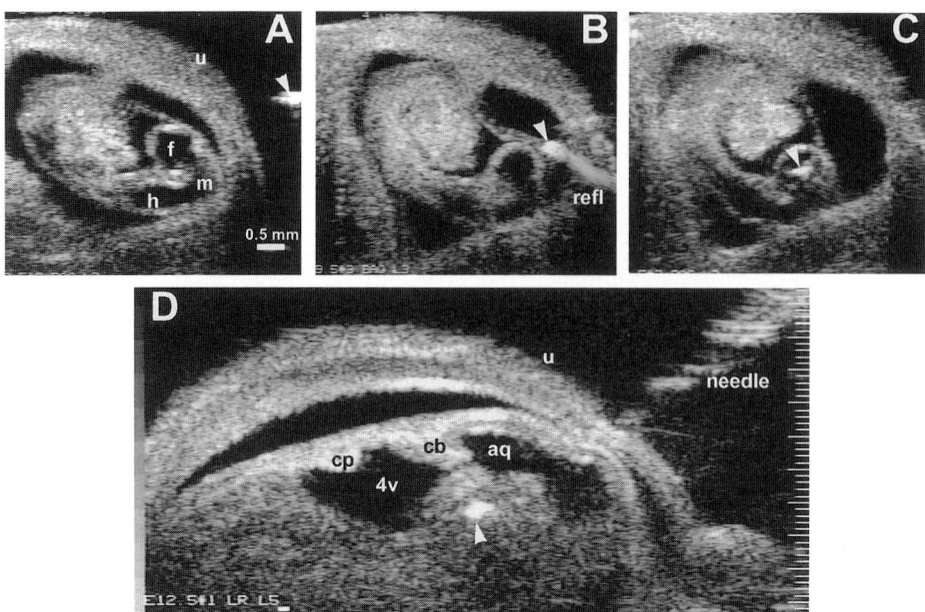

Fig. 3. *In utero* UBM-guided injections into mouse embryos. **(A)** The needle tip (arrowhead) is monitored on real-time UBM images as it advances through the uterine wall **(B)** and into the forebrain (f) ventricle **(C)** of an E9.5 mouse embryo. Note the reflection artifact (refl) from the glass injection needle. Labels: m, midbrain; h, hindbrain; u, uterus. **(D)** Injection needle tip (arrowhead) positioned in the mid-hindbrain isthmus region of an E12.5 mouse embryo shown in sagittal orientation. Smallest scale markers on right = 100 μm. Labels: aq, midbrain aqueduct; cb, cerebellum; cp, choroid plexus; 4v, fourth ventricle; u, uterus. (**Figure 3A,C** reprinted with permission from **ref. 3**).

     is placed around the 25-mm opening in the plastic dish. Carefully shave the region of skin that contacts the rubber, because any remaining hair will interrupt the seal between rubber and skin, leading to leakage of water during the imaging procedure.

5. Air bubbles on the transducer face are easy to remove by gently brushing the submerged face of the transducer with a cotton-tipped applicator. Usually, this will only be necessary the first time a dry transducer is placed in fluid. For the rest of the imaging session, a water film will remain on the transducer, making it impossible for new air bubbles to form.

6. If the beveling surface becomes dry during needle sharpening, small particles of grit from the grinding surface can become lodged in the needle tip. These will inevitably cause the needle to clog during the injection procedure. Inspect the needles after sharpening and discard any that have grit particles in the tip.

7. Stirring the thick two-component rubber with a spatula introduces air into the mixture. If a low-vacuum system (e.g., vacuum oven) is available, it is helpful to put the rubber mixture under vacuum (1 atm for 10 min is sufficient) to draw out the trapped air. Otherwise, examine the rubber surface after spreading onto the 35-mm plastic dishes, and break any bubbles that are apparent with a spatula before the rubber sets.

8. The procedure to align the injection needle in the image plane need only be performed once if the setup is not altered. It is best to clamp or screw the base of the micromanipulator onto the base plate housing the motorized micropositioning system. In this way, the orientation between the UBM image plane and microinjection system will be maintained after the alignment procedure.

9. Reflection artifacts from the injection needle are unavoidable (**Fig. 3B**). The echo closest to the transducer (top) indicates the true position of the glass needle, whereas lower echos are caused by multiple reflections inside the glass and make the injection needle appear larger than its real physical size.

10. The region of shaved skin should be large enough to provide a good seal between skin and rubber. A large shaved region on the lower abdomen, 30–40 mm in diameter, will allow ample shaved skin to form the seal around the 2-cm-long incision.

11. The abdomen should be high enough to make good contact with the rubber membrane, but not so high as to restrict breathing or put too much pressure on the uterus. Lower the level of the mouse if there are any signs of difficult breathing or gasping, or if there are puffs of blood into the PBS from the uterus.

12. Handle the uterus as gently as possible, with blunt forceps or cotton-tipped applicators. Use the forceps to gently pull the uterus, holding the tissue between the decidual swellings, always taking care to avoid breaking or even touching the blood vessels feeding into the uterus.

13. The embryos closest to (and tethered to) the vagina are the deepest, and they are the most difficult to pull out for injection. Too much pulling on the uterus at the vagina should be avoided. If thc entire litter is to be injected, reposition the setup for the second side of the horn, rather that attempting to go across the vagina.

14. Embryonic survival using this method is approx 50% *(3,4)*, with survival after a given injection session ranging from 0 to 100%.

Acknowledgment

The author thanks Orlando Aristizabal, Scott Baldwin, Kenneth Campbell, Gordon Fishell, Nicholas Gaiano, Alexandra Joyner, Robin Kimmel, Aimin Liu, Maria McCarthy, Martin Olsson, Colin Phoon, and Shardha Srinivasan for their participation in the development of these methods. This work was supported by grants from the NIH (GM057467, HL62334), NSF (IBN9728287), and the Whitaker Foundation. The author is an Investigator of the American Heart Association/New York City Affiliate.

References

1. Turnbull, D. H., Starkoski, B. G., Harasiewicz, K. A., Semple, J. L., From, L., Gupta, A. K., Sauder, D. N., and Foster, F. S. (1995) A 40–100 MHz B-scan ultrasound backscatter microscope for skin imaging. *Ultrasound Med. Biol.* **21,** 79–88.

2. Aristizabal, O., Christopher, D. A., Foster, F. S., and Turnbull, D. H. (1998) 40 MHz echocardiography scanner for cardiovascular assessment of mouse embryos. *Ultrasound Med. Biol.* **24,** 1407–1417.

3. Liu, A., Joyner, A. L., and Turnbull, D. H. (1998) Alteration of limb and brain patterning in early mouse embryos by ultrasound-guided injection of *Shh*-expressing cells. *Mech. Dev.* **75,** 107–115.

4. Olsson, M., Campbell, K., and Turnbull, D. H. (1997) Specification of mouse telencephalic and mid-hindbrain progenitors following heterotopic ultrasound-guided embryonic transplantation. *Neuron* **19,** 761–772.

5. Turnbull, D. H., Bloomfield, T. S., Baldwin, H. S., Foster, F. S., and Joyner, A. L. (1995) Ultrasound backscatter microscope analysis of early mouse embryonic brain development. *Proc. Natl. Acad. Sci. USA* **92,** 2239–2243.

6. Srinivasan, S., Baldwin, H. S., Aristizabal, O., Kwee, L., Labow, M., Artman, M., and Turnbull, D. H. (1998) Noninvasive *in utero* imaging of mouse embryonic heart development using 40 MHz echocardiography. *Circulation* **98,** 912–918.

24

Use of Doppler Echocardiography to Monitor Embryonic Mouse Heart Function

Kersti K. Linask and James C. Huhta

1. Introduction

Current understanding of organ-level vertebrate heart development has been gained primarily from studies of avian and amphibian heart development from whole-mount culture studies and subsequent dissections and serial sections of the heart during different stages of development of the organism. Chick embryos have been particularly accessible for experimental manipulation and functional analyses as compared with normal mouse embryonic hearts, which are poorly accessible without disturbing the placental blood flow. Specifically, in the study of vertebrate heart development, there has been a lack of noninvasive techniques for physiologic assessment of mouse cardiac function. There have been several studies describing hemodynamic changes in the developing chick embryonic heart using Doppler techniques (1–4). Most of these studies have been invasive in nature such that the hemodynamic status may be disturbed.

A noninvasive method of cardiac assessment of normal embryonic blood velocities using Doppler ultrasonography has been described in the chick embryo (5) and more recently in the mouse by our laboratory (6–8). The methodology to assess embryonic mouse heart function is described in detail here. The noninvasive methodology and patterns of heart function provided will be useful in the analyses of transgenic mouse models to analyze litters where heart defects may be suspected. Improved assessment of the mouse embryo may be possible in the future as new imaging technology is perfected (e.g., see **ref. 9**).

1.1. Echocardiography

This methodology was introduced into clinical medicine in the 1960s, became popular in the 1980s, and is widely used in present-day pediatric cardiology (10). The resolving power of this method is a function of the frequency of the transmitted ultrasound. The higher the frequency, the better the resolution, but the poorer the penetration. The signal-to-noise ratio deteriorates with increasing distance. Thus, relatively good signals can be obtained in small animals. Furthermore, the patterns obtained during early mouse gestation very closely resemble the wave forms of the inflow and outflow of normal human fetal Doppler echocardiography results (11).

From: *Methods in Molecular Biology, Vol. 135: Developmental Biology Protocols, Vol. I*
Edited by: R. S. Tuan and C. W. Lo © Humana Press Inc., Totowa, NJ

Using a clinical transducer, this methodology has been applied by our laboratory to analyze mouse embryonic heart function in normal embryos, trisomic embryos, and transgenic animals. Velocities detected in small embryonic hearts result from the instantaneous changes in the energy differences of the blood cells during acceleration and deceleration. The maximal velocities originate at the smallest area of the inflow or outflow regions of the heart. It is not possible, however, to locate the precise location of the velocity using this technique. Other anatomical or pathological studies of the heart need to be carried out on the same hearts that are interrogated by echocardiography to obtain more precise areas that may be responsible for the waveform patterns. Blood waveform patterns as early as day 10 of gestation can be picked up. The earliest day of gestation for the best imaging results on the embryonic mouse heart is approximately day 13 of gestation.

2. Materials

2.1. Sedatives

Either of two sedatives (*see* **Note 1**) have been used in our lab to immobilize the pregnant female mouse: tribromoethanol and pentobarbital.

2.1.1. Tribromoethanol (TBE)

1. Stock solution: Dissolve 1 g of 2,2,2-tribromoethanol (Aldrich Chemical Co., Milwaukee, WI) in 1 mL of 2-methyl-2-butanol (Aldrich Chemical Co.). Aliquot and store at –20°C in the dark.
2. Working solution: Thaw TBE and dilute with sterile distilled water to make a working concentration of 25 mg/mL. TBE is stable at pH 7.0 for an extended time period.

2.1.2. Pentobarbital Sodium

The commercial stock (Abbott Laboratories, North Chicago, IL) is diluted with 0.9% sodium chloride inj (Abbott) to the concentration of 5 mg/mL.

2.1.3. Administration of Sedative

The sedative is administered intraperitoneally (ip). Dosage is 0.1 mL of working solution per 10 g of adult mouse weight. The pregnant female is sedated and immobilized usually within 1 min of administration of sedative. Sedation duration is 30–60 min. The doses used of these sedatives are comparable in their abilities to sedate the mice. Tribromoethanol mice, however, become immobilized faster and the animals recover quicker from sedation than mice in the pentobarbital group.

2.2. Instrumentation

Any Doppler ultrasound instrument with a ultrasonic transducer of 7.5 MHz used for clinical purposes can be used for interrogating mouse embryonic heart function. However, many clinical departments may not allow ultrasound instruments used for human sonography for studies on animals. Therefore, one must have access to an ultrasound instrument dedicated for use on animals.

The instrument used in our studies was the Interspec XL (Conshohocken, PA) with a 7.5 MHz ultrasonic transducer. The heart inflow and outflow velocity waveforms are

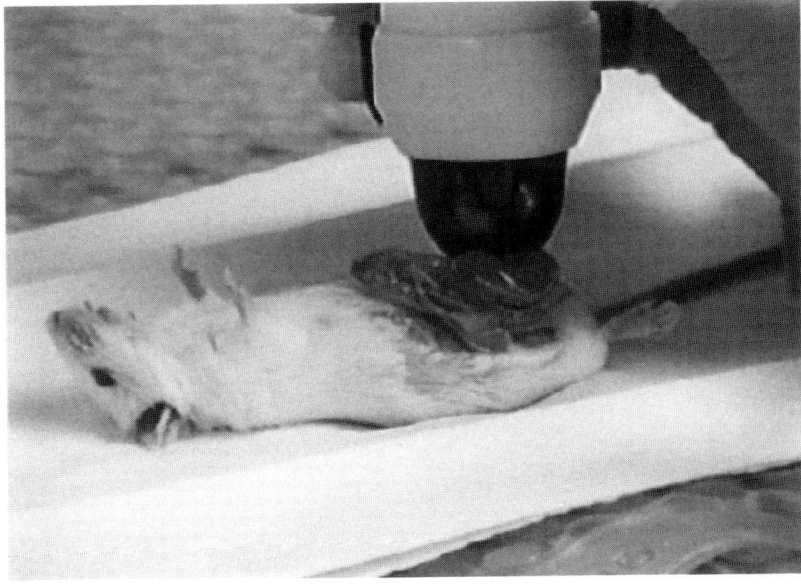

Fig. 1. Acoustic gel is spread relatively thick over abdominal area of pregnant female for noninvasive Doppler evaluation of embryonic mouse heart function using a clinical transducer (*see* Chapter 24). The thick gel allows placement of the embryonic heart at the necessary focal depth of the transducer.

recorded using a sweep speed of 100 mm/s on videotape, using a Panasonic video cassette recorder AG6300 (Tokyo, Japan).

2.3. Coupling Gel

Doppler velocities are obtained from the embryonic hearts by coupling the ultrasonic energy from the transducer with an acoustic gel (Aquasonic 100™ Ultrasound Transmission Gel, Parker Laboratories, Orange, NJ) that is placed on the female's abdomen.

2.4. Analysis and Software

An echocardiography analysis system from Digisonics, Inc. (Houston, TX) was used for analyses of peak inflow velocities, the systolic ejection, and diastolic filling time intervals. It is advised to consult with a pediatric cardiologist or, if possible, a fetal echocardiologist in interpreting waveform patterns that are obtained.

3. Method

1. Pregnant female is sedated with ip administration of the sedative. Thicker fur over the abdomen is cut short or can be shaved after sedation. 70% alcohol is used to wet down any remaining fur. Animal is placed on a warm pad and is interrogated while under a heat lamp to maintain the temperature of the dam.
2. Acoustic gel is spread relatively thick over abdominal area of pregnant female (*see* **Fig. 1**). Alternatively, an ultrasonic standoff made from a gel-filled latex glove is placed over the abdomen of the female. This is to allow placement of the embryonic heart at the necessary focal depth of the transducer (**Fig. 2A**).

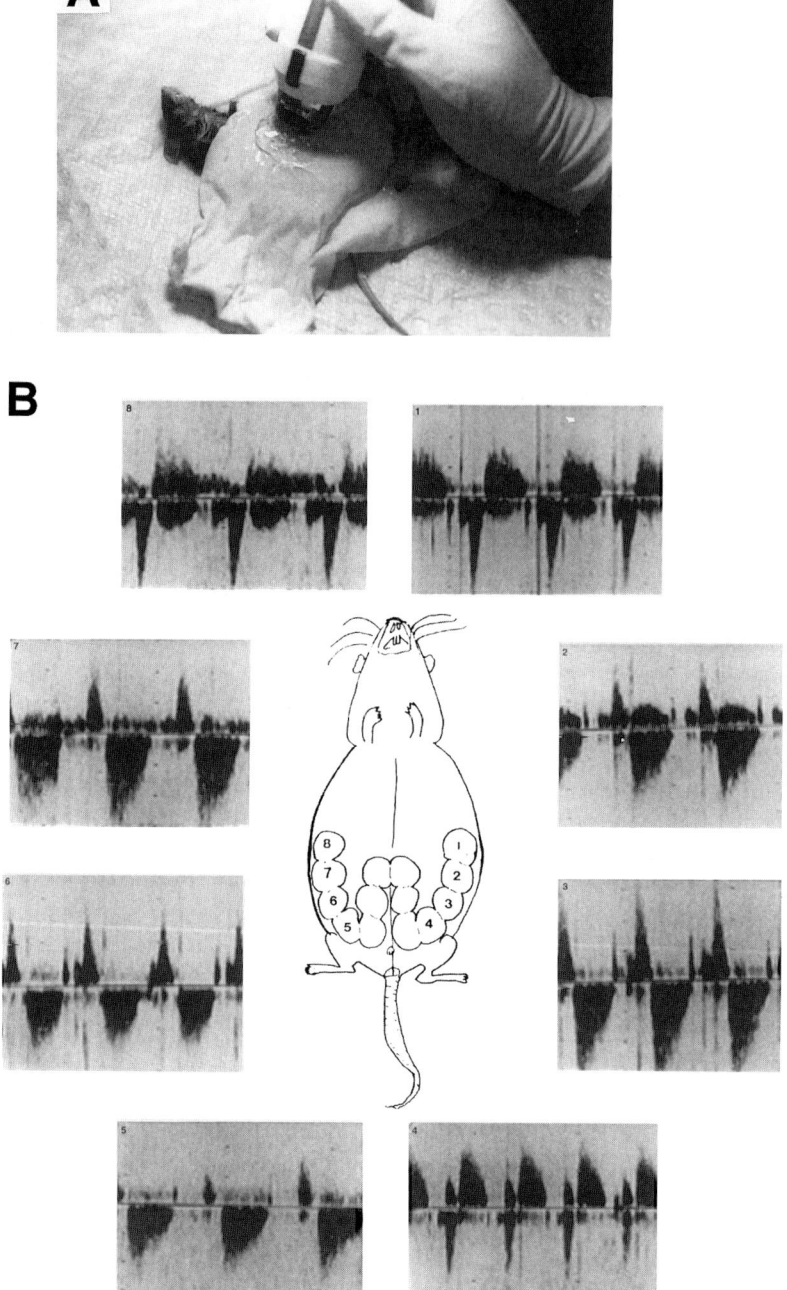

Fig. 2. **(A)** Technique of mouse embryo Doppler examination. Note that the ultrasound transducer is placed here on a gel-filled latex glove in order to position the mouse abdomen in the focal depth of the transducer. After the individual embryonic hearts are visualized beating, the pulsed Doppler sample volume is placed over the heart and **(B)** the blood velocity patterns of each are obtained for the *n* number of embryos that are present (eight different embryos evaluated here). The angle of Doppler interrogation is changed for each to optimally obtain the maximal velocity signal. Note the short atrial contraction wave followed by the outflow velocity. The orientation of the waveform will change depending on the orientation of the mouse embryo. Used with permission from **ref. 7**.

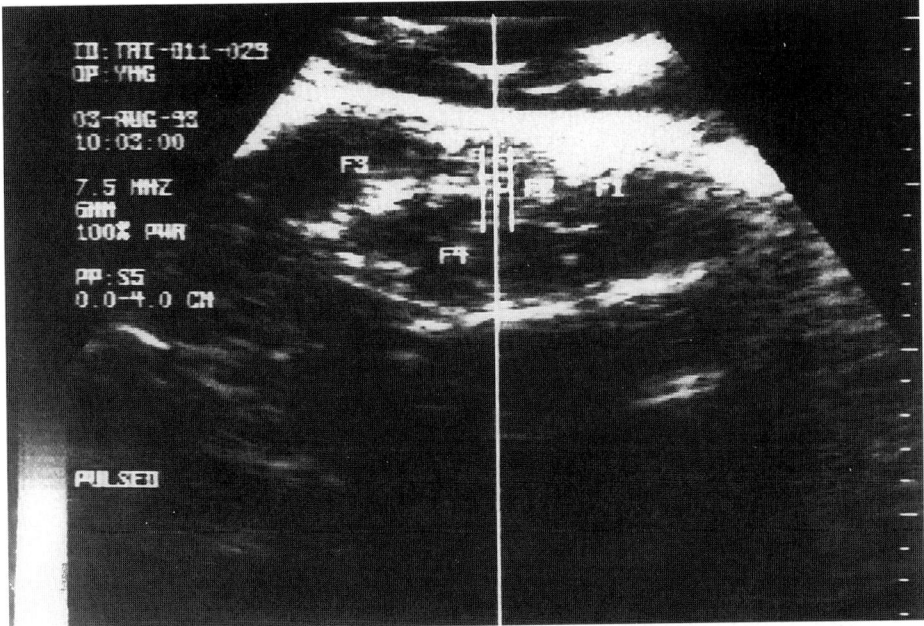

Fig. 3. A two-dimensional echocardiography image of a section of the pregnant mouse abdomen with the individual mouse embryos marked F1, F2, F3, and F4. The pulsed Doppler sample volume marker is placed over the F2 embryo at 11 d of gestation.

3. The abdominal area is partitioned into imaginary quarters; that is, upper left and right; lower left and right. Each quarter is quickly scanned (*see* **Note 2**) to obtain possible numbers and relative location of embryos. Then each embryo is returned to systematically for evaluation (*see* **Fig. 1B**).

4. After identifying the position of the beating embryonic heart, the sample volume (*see* **Note 3**) of the pulsed Doppler is placed over the entire heart to obtain blood velocities (*see* **Fig. 3**).

5. The same embryo is evaluated from several angles of insonation to obtain the maximal velocities. Sequential inflow A, inflow E, and outflow waveforms (*see* **Note 4**) are obtained together. *See* **Fig. 4** for typical blood flow waveform pattern on day 13 of gestation (*see* **Note 5**).

6. At the end of the noninvasive evaluation and at the time of sacrifice of the dam, the abdomen is quickly opened to scan the embryos with the Doppler one last time and to check the location of the embryos with respect to the Doppler patterns that were obtained (*see* **Note 6**).

7. Relationship between the Doppler measurements and gestational age in the normal mouse embryo is evaluated as follows: Mean values are obtained for each litter and at one gestational age. By obtained mean values per litter, samples are obtained that contain independent observations. This allows valid analysis using correlational and linear regression techniques. Pearson correlation coefficients are computed for each measure. Subsequently, Spearman rank correlations are computed to assess whether the Pearson values are unduly affected by extreme values. Simple linear regressions are used to estimate the average changes per gestational day using data from normal embryos for parameters with evidence of a significant correlation. Mean values for each gestational age are measured for graphical presentation. *t*-test is used to compare abnormal velocities and intervals for ejection and to analyze parametric variable for normal littermates (*see* **Note 7**).

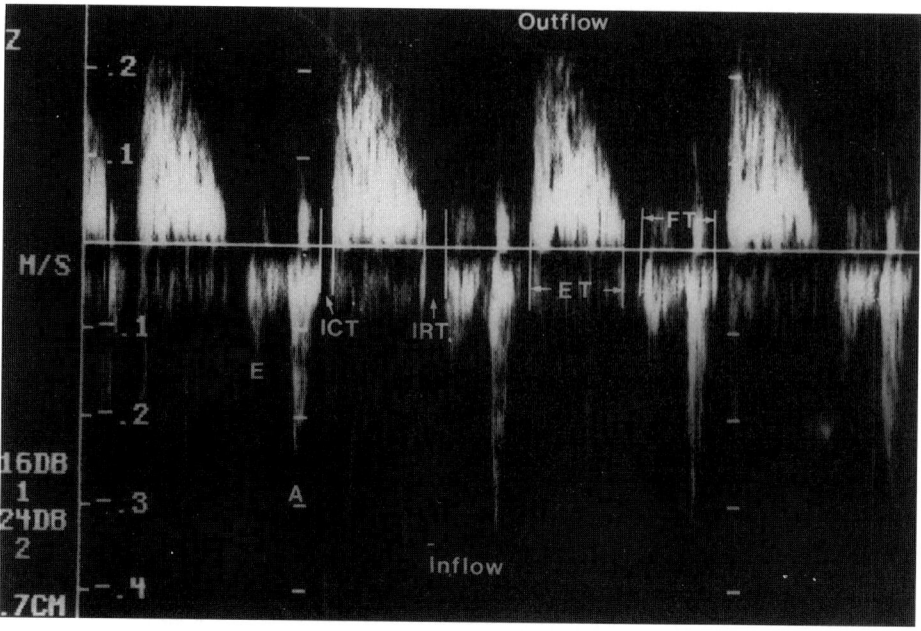

Fig. 4. The Doppler waveform is calibrated from 0 to 0.4 m/s (0–40 cm/s). The outflow in this example is above the zero line with a peak velocity of 22 cm/s, and the peak inflow signal (A-wave) is below the line at 30 cm/s. Note the biphasic inflow pattern consisting of the E (passive filling) wave and the A (atrial contraction) wave. The time interval measurements are shown, including the filling time (FT), the ejection time (ET), the isovolemic contraction time (ICT), and the isovolemic relaxation time (IRT). Used with permission from **ref. 7**.

4. Notes

1. It has been found that sedatives may have different effects on embryonic mouse heart function during specific developmental time periods (*see* **ref. 8**).
2. Imaging at 7.5 MHz is sufficient to locate the embryo and the beating heart and to orient the sagittal plane of the embryo. This frequency is not adequate to directly image the internal anatomy of the heart.
3. The sample volume length is adjusted to 2–4 mm to completely insonate the contracting embryonic heart. The high-pass filter is at its lowest setting of 50 Hz. It is critical to minimize the angle of insonation to obtain the maximal blood flow velocity because of the relation between the angle of insonation and the actual velocity *(5)*.
4. The E and A velocities and embryonic heart rate are measured on the inflow velocity waveform. The peak systolic ejection velocity is measured on the outflow waveform. In addition, the isovolemic contraction time (ICT; interval from the end of filling to the onset of ejection) and the isovolemic relaxation time (IRT; interval from the end of ejection to the onset of filling) are obtained. All time intervals are measured in milliseconds and are expressed as a percentage of the beat-to-beat cardiac cycle interval. The data are plotted in relation to the day of gestation that the animal is examined. *See* **Fig. 5** (used with permission from **ref. 7**) for normal changes in the patterns of mouse embryo Doppler between days 11 and 17 of gestation.
5. Because of the small size of the embryonic mouse heart, it is not possible to separately image the left and right sides of the heart, but rather the peak ejection velocities represent

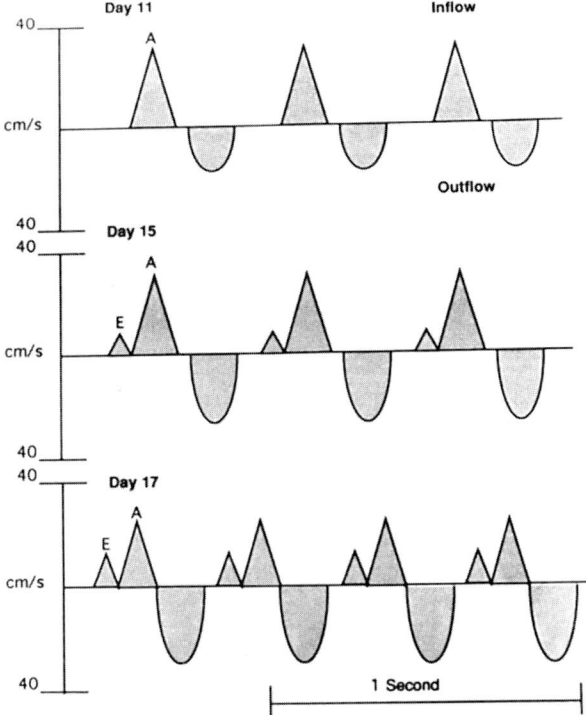

Fig. 5. Normal changes in the patterns of mouse embryo Doppler. At day 11, the inflow signal is a pure A wave with a long ICT and slower heart rate. At day 15, a biphasic pattern is seen indicating improved myocardial compliance, and the ICT is shorter. The mature pattern of inflow/outflow is present at day 17 with biphasic inflow (still A dominant) and no measureable ICT. The heart-rate-corrected IRT decreases over this time period. Used with permission from **ref. 7**.

velocities for both sides of the heart. Also, the Doppler hemodynamic measurements are not absolute values, even for a particular strain.

6. Embryonic heart rates drop with opening of the maternal abdomen, so valid quantitative comparisons with noninvasive data are not possible. Applicability of Doppler echocardiography to litter sizes greater than 10 is not known. It is expected that with increased number of embryos, there is an increased possibility that not all embryos may be detected for assessment.

7. Interobserver variability in obtaining measurements is a possibility. Thus, in evaluations, observer variation may be computed by using two-way analyses of variance with random effects.

Acknowledgments

The authors thank three visiting pediatric cardiologists for the invaluable input in the development of the protocol: Drs. Pongsak Khowsathit, Bangkok, Thailand; Yong-Hao Gui, Shanghai, China; and Guoying Huang, Shanghai, China. This work was supported in part by grants from the March of Dimes, W.W. Smith Charitable Trust, and NIDR (DE11467) to KKL.

References

1. Hu, N. and Clark, E. B. (1989) Hemodynamics of the stage 12 to stage 29 chick embryo. *Circ. Res.* **65,** 1665–1670.
2. Hu, N., Connuck, D. M., Keller, B. B., and Clark, E. B. (1991) Diastolic filling characteristics in the stage 12 to 27 chick embryo ventricle. *Pediatr. Res.* **29,** 334–337.
3. Nakazawa, M. F., Ikeda, K. K., and Takao, A. (1990) Effect of atrial natriuretic peptide on hemodynamics of the stage 21 chick embryo. *Pediatr. Res.* **27,** 557–560.
4. Wagman, A. J., Hu, N., and Clark, E. B. (1990) Effect of changes in circulating blood volume on cardiac output and arterial and ventricular blood pressure in the stage 18–24 and 29 chick embryo. *Circ. Res.* **67,** 187–192.
5. Huhta, J. C., Borges, A., Yoon, G. Y., Murdison, K. A., and Wood, D. C. (1990) Noninvasive ultrasonic assessment of chick embryo cardiac function. *Ann. NY Acad. Sci.* **588,** 383–386.
6. Linask, K. K., Gui, Y. H., Huhta, J. C., and Khowsathit, P. (1993) Evaluation of heart defects in early TS 16 mouse embryos using echocardiography. *Mol. Biol. Cell.* **4,** 144a.
7. Gui, Y. H., Linask, K. K., Khowsathit, P., and Huhta, J. C. (1996) Doppler echocardiography of normal and abnormal embryonic mouse heart. *Ped. Res.* **40,** 633–642.
8. Huang, G. Y. and Linask, K. K. (1998) Doppler echocardiographic analysis of effects of tribromoethanol anesthesia on cardiac function in the mouse embryo: A comparison with pentobarbital. *Lab. Anim. Sci.* **48,** 206–209.
9. Turnbull, D. H., Bloomfield, T. S., Baldwin, H. S., Foster, F. S., and Joyner, A. L. (1995) Ultrasound backscatter microscope analysis of early mouse embryonic brain development. *Proc. Natl. Acad. Sci. USA* **92,** 2239–2243.
10. Sanders, S. P. (1990) Echocardiography, in *Fetal and Neonatal Cardiology* (Long, W. A., ed.), W.B. Saunders, Philadelphia, pp. 301–329.
11. Tulzer, G., Khowsathit, P., Gudmundsson, S., Wood, D. C., Tian, Z. Y., Schmitt, K., and Huhta, J. C. (1994) Diastolic function of the fetal heart during second and third trimester: A prospective longitudinal Doppler-echocardiographic study. *Eur. J. Pediatr.* **153,** 151–154.

25

Calcium Imaging in Cell–Cell Signaling

Diane C. Slusarski and Victor G. Corces

1. Introduction

The development of fluorescent calcium (Ca^{2+}) indicators has been a very powerful tool for looking at the role of this second messenger in signal transduction. The ability to load these indicators into cells in a nondisruptive manner allow for the visualization of intracellular-free Ca^{2+} ion ($[Ca^{2+}]_i$) in a living developing embryo (reviewed in **refs. 1** and **2**). The vertebrate zebrafish (*Danio rerio*) is ideally suited for fluorescent studies by virtue of its transparent embryos. The zebrafish has two phases of $[Ca^{2+}]_i$ release in the first few hours of development. The first is a phase of dramatic long-lived $[Ca^{2+}]_i$ elevations associated with the forming cleavage furrows during the first few cell divisions *(3)*. After the 16–32-cell stage, a subset of cells displays rapid aperiodic $[Ca^{2+}]_i$ increases, localized to the enveloping region *(4)*, and persist until the 1000–2000-cell stage *(4,5)* (*see* **Fig. 1**).

The focus of this chapter is calcium imaging and cell–cell interaction in development; in particular, early embryonic development in the zebrafish. In both frogs (*Xenopus*) and zebrafish, overexpression of Wnts, a family of signaling proteins involved in cell–cell communication, has developmental defects similar to the effects of agents which modulate the phosphatidylinositol (PI) cycle *(5,6)*. Intracellular $[Ca^{2+}]_i$ release is one of the products of the phosphatidylinositol (PI) cycle signal transduction pathway *(7,8)*. In using intracellular $[Ca^{2+}]_i$ release as a read out, a member of the *Wnt* gene family has been shown to modulate this signal transduction pathway *(5)*. A member of the frizzled gene family, the putative Wnt receptor, modulates $[Ca^{2+}]_i$ release in a G-protein-dependent manner *(9)*. Four methods are used in this application; microinjection to deliver reagents to the cytoplasm, expression of exogenous proteins from micro-injected mRNAs, fluorescence ratio imaging, and whole-mount *in situ* hybridization. The following procedure will focus on the use of fluorescence ratio imaging of $[Ca^{2+}]_i$ release as an assay for early signal transduction events in zebrafish development.

2. Materials

2.1. Buffers

1. Stock salts: 280 g Instant Ocean Sea Salts (Aquarium Systems, Mentor, OH) in 2 L dH_2O. Recipe from **ref. 10**.

From: *Methods in Molecular Biology, Vol. 135: Developmental Biology Protocols, Vol. I*
Edited by: R. S. Tuan and C. W. Lo © Humana Press Inc., Totowa, NJ

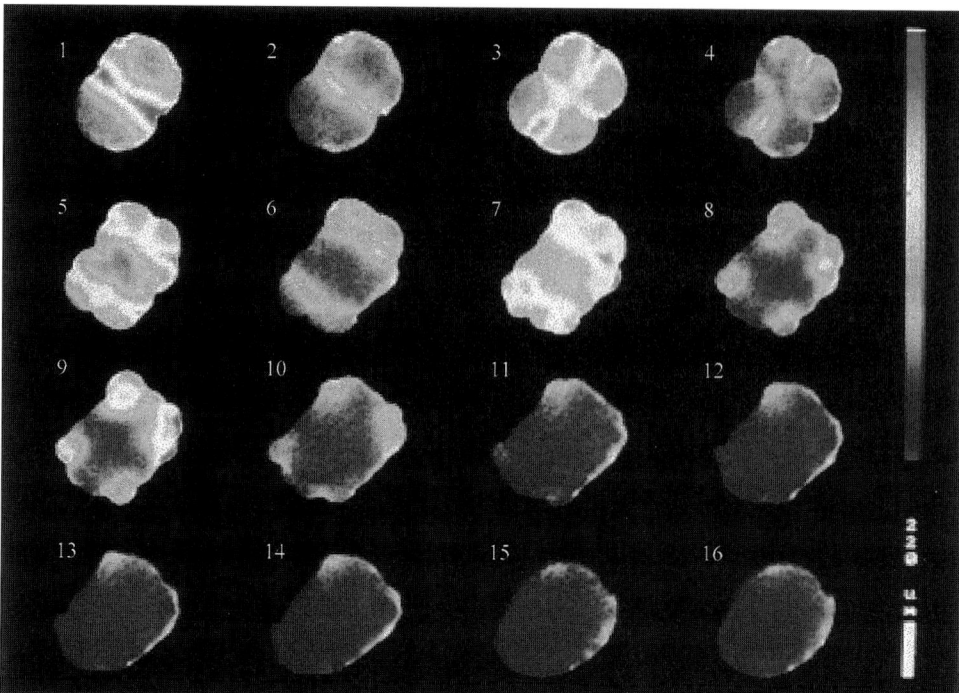

Fig. 1. (*See* color plate 4 appearing after p. 258.) A sequence of pseudocolor ratio images of one zebrafish embryo as it proceeds through the earlier stages of development (late 2-cell [#1] to >1000-cell [#16]). A zebrafish egg was microinjected with fura-2 prior to the first cleavage (final concentration approx 20 μM) and retained in an inverted position (blastodisc down). The resulting ratio images were rendered in pseudocolor to produce semiquantitative maps of $[Ca^{2+}]_i$ distribution, where violet = low ratio (low $[Ca^{2+}]_i$) and red = high ratio (high $[Ca^{2+}]_i$) (*see* color bar). The first cleavages characteristically generate extremely high $[Ca^{2+}]_i$ transients coincident with the forming cleavage furrow, as illustrated by the yellow/red lines through the embryo (2-cell [#1], 4-cell [#3], 8-cell [#5], 16-cell [#7], and 32-cell [#9]). The cleavage-associated transients generally quiet down after approximately the 32-cell stage (#11–16). The second wave of $[Ca^{2+}]_i$ transients are not as apparent in the figure because of the contrast set to accommodate the high intensity of the cleavage-associated $[Ca^{2+}]_i$ transients.

  2. Egg medium: 1.5 mL Stock salts per liter of dH_2O, final concentration, 60 μg/mL. Egg medium is the same as egg water recipe from **ref. *10***.
  3. 5-Hydroxytryptamine (5-HT) (Sigma Co., St. Louis, MO): Final concentration of $10^{-8}M$ in egg medium. Prepare immediately before use from frozen stock of $10^{-4}M$; protect from light and keep on ice.
  4. Mianserin (Sigma): Make a stock solution of $10^{-4}M$ in DMSO (store at 4°C). Use at a final concentration of $10^{-8}M$ in egg medium (avoid contact with skin).

2.2. Microinjection Reagents

  1. Fura-2 dextran-conjugated (fura-2): 10,000 M_r (Molecular Probes, Eugene, OR). Working solution of 5 mM in sterile ddH_2O.
  2. Texas Red dextran-conjugated (TxR): 10,000 M_r (Molecular Probes). Stock solution 5 mg/mL in sterile ddH_2O.

3. Synthetic mRNA (5-HT receptor, 5-HT$_{1c}$R) *(11)*. Capped mRNA concentration of 100–200 ng/µL, mix 1:1 with TxR for a final concentration of 50–100 ng/µL of mRNA in the injection cocktail.

2.3. Zebrafish

1. Adult zebrafish were maintained at 27.5°C in aquarium (60 mg Instant Ocean/liter of water) with biological filters. Day/night cycle of 14 h light/10 h dark Rearing, staging, and embryo collection are detailed in **ref. 10**.
2. Eggs were collected from mating tanks from natural spawnings at "dawn" (when the lights come on) and washed with egg medium.
3. Zebrafish were microinjected with a pressure injector (Narishige-IM-200, Narishige USA, Inc., East Meadow, NY) with approx 200 pL at the 1-cell stage and approx 100 pL at the 8-cell stage.

2.4. Instrumentation (Hardware/Software)

1. Inverted epifluorescence microscope (Zeiss axiovert 135, Carl Zeiss, Jena, Germany) equipped with: cover-slip-bottomed heating chamber (Biophysica Technologies, Inc.), filters for epifluorescence (Chroma Technology Corp., Brattleboro, VT) and ×10 Plan-Neofluar objective (Zeiss, N.A.: 0.3).
2. PTI Deltascan dual monochromator illumination system with Ludl Electronic Products, Ltd. (LEP), Hawthorne, NY, controller.
3. Xenon arc lamp (75 W).
4. Slow-scan CCD camera (CH250 Photometrics Ltd., Tucson, AZ) (1.3 k × 1 k by 12 b).
5. Uniblitz shutter drive (Vincent Associates, Rochester, NY).
6. Computer (Sun SPARC20 workstation [Sun Microsystems, Palo Alto, CA]) and Image analysis software (IC-300 and Ratiotool programs from Inovision Corp., Raleigh, NC) to digitize and manipulate the images.

3. Methods
3.1. Embryo Collection

Set up a mating tank, bottom lined with marbles, the evening prior to the experiment with equal numbers of male and female zebrafish. Adult zebrafish can be obtained from a local pet store (*see* **Note 1**). The number of adult fish used depends on the size of the mating tank. A small "mouse cage" tank can easily house approx 3 adults of each sex and should yield sufficient eggs. Shortly after the lights come on, collect eggs by siphoning the bottom of the tank. Rinse the eggs with egg medium.

3.2. Microinjection

1. Preparation of synthetic mRNA for microinjection: Linearize the plasmid of interest. In this example, a plasmid containing the gene for the serotonin receptor, 5-HT$_{1c}$R, is linearized with NotI and used as template *(11)*. Set up the in vitro transcription reaction using T7 RNA polymerase and cap analog (m^{7}G[5']ppp[5']G, Pharmacia & Upjohn, Bridgewater, NJ). After DNAse treatment, run the reaction over a Chromaspin (Clontech, Palo Alto, CA) column, precipitate with ethanol, and air-dry. Resuspend RNA pellet in sterile water (*see* **Note 2**). Just prior to microinjection, mix 0.5 µL of RNA (from 100 ng/µL working stock) with 0.5 µL of TxR. Store on ice, protect from light.
2. Preparation and calibration of injection micropipet: Pull a tapered needle with a sharp tip from a micropipet (25 lambda, Drummond, Broomall, PA) on a standard needle puller

(vertical pipet puller, Kopf Instruments, Tujunga, CA). Backfill the needle with 0.5–1 μL of fura-2. Mount the needle onto the micromanipulator or some other support. Break off the needle tip with fine forceps under a dissecting microscope (Wild M7) and lower into a Petri dish with silicon oil. Inject a drop of fura-2 into the oil, and measure the diameter with a calibrated eyepiece micrometer. Adjust the injection pressure/time to yield an injection volume of approx 100–200 pL (*see* **Note 3**).

3. Microinjection into the 1-cell embryo: As soon as the fertilized eggs are collected and rinsed, transfer into an injection dish (*see* **Note 4**). In order to obtain even dye distribution, inject the embryos in the 1-cell stage before the next cell division (*see* **Note 5**). Gently orient the embryos using fine forceps. Position the needle to penetrate the chorion (either by moving the needle with a micromanipulator to the embryo or moving the injection plate to bring the embryo to the needle). Once the chorion is penetrated, gently insert the needle through the yolk into the blastodisc (which is clearer than the yolk region) and inject the fura-2. Slowly withdraw the needle, using forceps to support the embryo if needed. Allow the injected embryos to develop to the 8-cell stage (approx 1 h, depending on the temperature).

4. Microinjection into the 8-cell embryo: Load the needle with 5-HT_{1C}R/TxR cocktail and calibrate for 50–100 pL injection volume. It is best to do this when the fura-2-injected embryos are at the 4-cell stage. When the fura-2-injected embryos enter the 8-cell stage, inject 1–2 blastomeres with the RNA cocktail as described in the previous step. Maintain the embryos in egg medium with mianserin (*see* **Note 6**).

3.3. Microscopy and Image Analysis

To image intracellular calcium $[Ca^{2+}]_i$ in living cells, the ratiometric dye fura-2 will be used. It is a fluorescent derivative of the Ca^{2+} chelator EGTA (*12*). The excitation spectra are different between the Ca^{2+}-bound (340-nm) and Ca^{2+}-free (380-nm) forms. By taking the ratio of the fluorescence intensity at these two wavelengths, an estimate of intracellular-free Ca^{2+} can be derived, to some degree, independent of cell thickness and distribution of the fluorescent indicator (which can vary in living cells). There is a wide range of hardware and software that can be used for fluorescence ratio imaging. Our system is briefly outlined below; consult manufacturer instructions for details on other setups.

1. Set up the microscopy workstation: Turn on the hardware by first firing the Xenon arc lamp. After the arc lamp is powered, the other components can be switched on: the computer to run the software (SPARC workstation with data cube), shutter drive (pti LPS-220), controllers for the Deltascan (Optical Chopper OC-4000, Shutter Controller SC-500), Photometrics CCD Camera (CH250), and the motorized stage controller. Install fura-2 and TxR filter cubes in a slider.

2. Select embryos that are cleaving properly and do not show any obvious signs of injection damage. Orient embryo as desired on the cover-slip floor in the thermostated chamber of the inverted microscope; set at 28.5°C. Analysis is possible at room temperature. Two orientations are useful for these analyses. One orientation places the blastodisc face down such that the top of the embryo is facing the objective (*see* **Note 7**; **Fig. 1**). The other orientation places the embryo such that a side view of the blastodisc (one side of the long axis) is facing the objective (**Fig. 2**). Select the 340-nm excitation wavelength (on the Deltascan or filter wheel) and have the fura-2 filter cube in place. Looking through the eyepieces of the inverted microscope, select an embryo with good fluorescence intensity

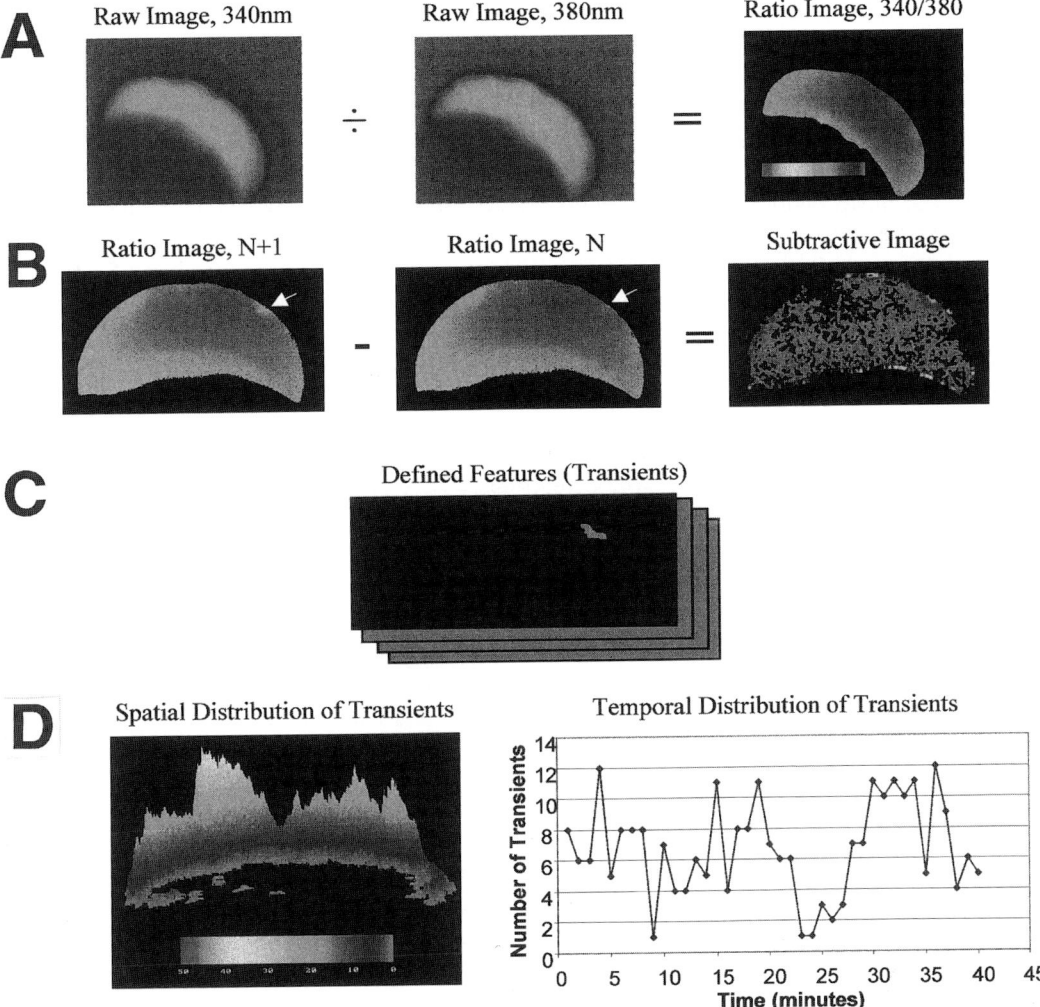

Fig. 2. (*See* color plate 5 appearing after p. 258.) Example of the subtraction analysis algorithm to detect $[Ca^{2+}]_i$ transients. (**A**) An example of the intensity of image pairs collected at 340- and 380-nm excitation wavelengths, digitized and ratioed by computer. (**B**) Two sequential ratio images acquired from a zebrafish embryo microinjected with fura-2 and 5-$HT_{1c}R$ mRNA and subjected to image analysis. The resulting ratio images are centered in the frame-store. Successive images are subtracted from the previous image to generate the subtractive image. The digital analysis detects changes in the original 12-bit image, but all the changes in intensity are not apparent in this print (8-bit for display). (**C**) Regions of increase in fluorescence intensity of the appropriate size are defined as features. (*See* white arrows in (**B**) for an example of a transient.) The number of transients are compiled over the duration of the time course. (**D**) Complied transients are represented topographically to demonstrate the distribution of the total number of transients over the region of the blastodisc. The higher the number of transients, the higher the peaks and the color range; in this example, the color bar ranges from 1 (purple) to the low 40s (yellow). The other form of representation is graphically to demonstrate the number of transients (y-axis) as a function of time (x-axis). Data were plotted in Microsoft Excel.

(*see* **Note 8**; **Fig. 2A**). Check the same embryos at the 380-nm excitation wavelength (510-nm emission). After selecting embryos with adequate fura-2 intensity, change the excitation wavelength to 540 nm and move the Texas Red cube in place. Select embryos with the desired TxR distribution (*see* **Notes 8** and **9**).

3. After selecting an embryo and the appropriate exposure times, collect a background exposure by moving the microscope stage such that the embryo is out of the field of view. Collect background images at 340 and 380 nm; archive to store raw data. The background will be subtracted from all further images. Move the embryo back into the field of view and center. Collect a sample image pair at 340 and 380 nm, and review a ratio calculated by the computer software (e.g., Ratiotool 1.22, Inovision). The ratio image is a pixel-by-pixel match of the two excitation wavelengths (340-nm image/380-nm image). Use the sample ratio to determine the proper threshold. Set at a level to exclude areas outside of the embryo, such as yolk or chorion, from the ratio image (**Fig. 2A,B**). Set to desired contrast. Determine the TxR distribution by collecting a reference exposure at 540-nm excitation.

4. Reset the excitation wavelengths to 340/380 nm. Determine the stage of the embryo (i.e., 8- or 32-cell). Initiate the time course by collecting image pairs at 15-s intervals (*see* **Note 10**). The first phase of calcium release will be very obvious. There will be long-lived calcium elevations associated with the forming cleavage furrows. These will persist until the 32–64-cell stage (**Fig. 1**). The second phase of calcium release will appear as localized calcium elevations, roughly the size of single cells (**Fig. 2B**, white arrows).

5. After a baseline of samples has been collected (4–10 min), add 5-HT to activate the receptors. Reorient the embryo if it has shifted during the solution addition (if the embryo has moved, collect another TxR reference). Collect 340/380-nm image pairs at 15-s intervals. The intracellular calcium transients last until approx 1000-cell stage (approx 2–2.5 h). After the data collection is completed, return the embryo to egg medium plus mianserin.

6. The manipulations described result in alterations in embryonic development. There are several molecular markers that can be used to assess the impact of manipulations on axis formation. Determine how the embryo develops by looking at either the morphology at a later stage with DIC optics or marker gene expression via whole-mount *in situ* hybridization (*see* **Note 11**).

7. Data analysis. The archived raw data is then processed to generate a sequence of ratio images using Ratiotool (Inovision) (**Fig. 2A**). The ratio files are then processed in the ISEE, Inovision Network editor. In short, the region of interest (ROI) is defined as the blastodisc. The ROI is then positioned in the image frame-store such that the series of images are aligned relative to each other (**Fig. 2B**). The images are low-pass-filtered, and the transients are determined by a subtractive analog (patterned after **refs.** *15* and *16*; *see* **Note 12**). Sequential ratio images are subtracted from each other to generate a subtractive image (**Fig. 2B**). In the subtractive image, a transient is defined as a feature approximately the size of a cell with an increase in fluorescence intensity (*see* **Note 13**; **Fig. 2B**, arrows). The number of transients are compiled over the duration of the time course (**Fig. 2C**). The output of the compiled features can be demonstrated in two manners. A topographical image showing the total number of transients located along the blastodisc for the spatial distribution or the temporal distribution plotted graphically (*see* **Note 14**; **Fig. 2D**).

Many proteins have been identified that modulate mesoderm induction, but the details of intracellular signaling are less known. The technique just described serves as a powerful tool to dissect the steps of the intracellular signaling pathway modulated in axis specification.

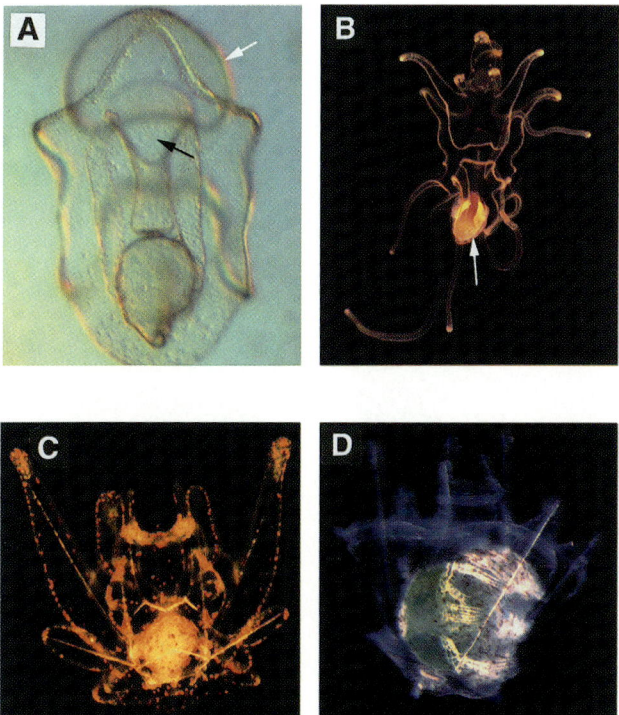

Plate 1, Fig. 1 (*see* discussion in Chapter 2, and full caption on p. 12). Larval morphology of sea urchin and sea star development. All larvae are oriented with anterior up.

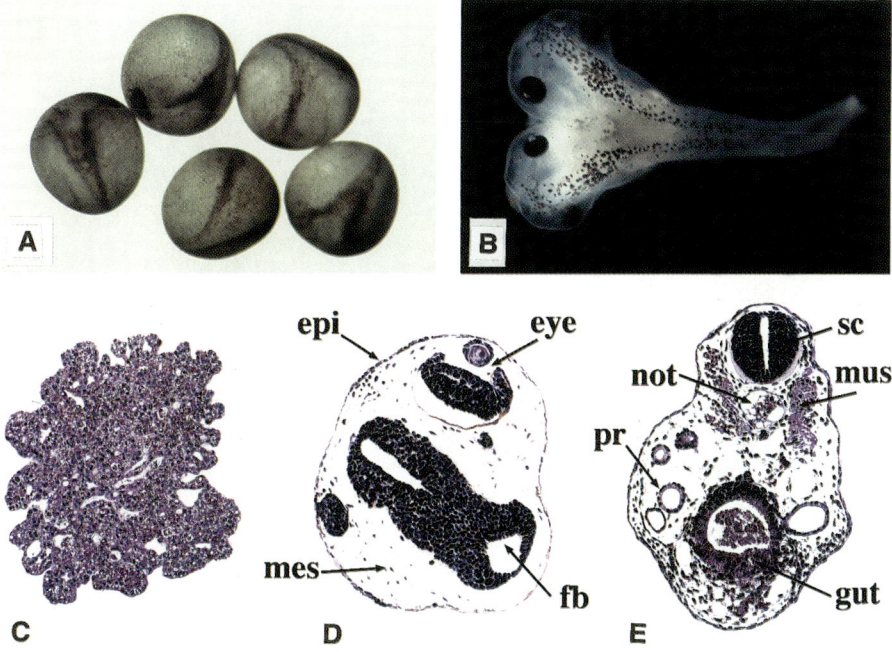

Plate 2, Fig. 8 (*see* discussion in Chapter 11, and full caption on p. 106). Examples of phenotypes and histological sections from various induction assays. Complete twins induced by *Xwnt-8*.

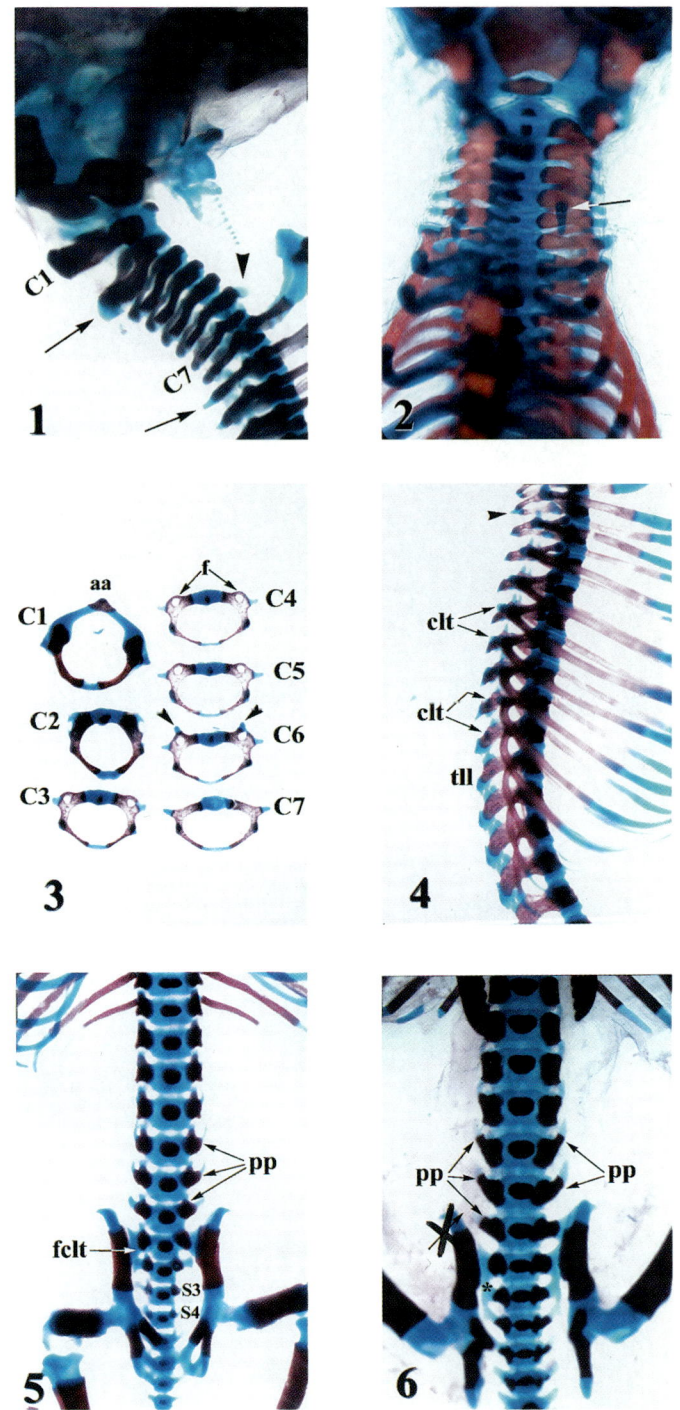

Plate 3, Figs. 1–6 (*see* discussion in Chapter 14, and full captions on p. 145). **Fig. 1.** Lateral view of the cervical vertebrae and cervicothoracic border in rat fetus. **Fig. 2.** Ventral view of a gestational d 18 mouse fetus. **Fig. 3.** Cranial view of disarticulated cervical vertebrae. **Fig. 4.** Lateral view of a thoracic region of gestation d 20 rat fetus. **Fig. 5.** Ventral view of lumbosacral region of a gestation d 20 rat fetus. **Fig. 6.** Ventral view of the lumbar region of a gestation d 21 rat fetus.

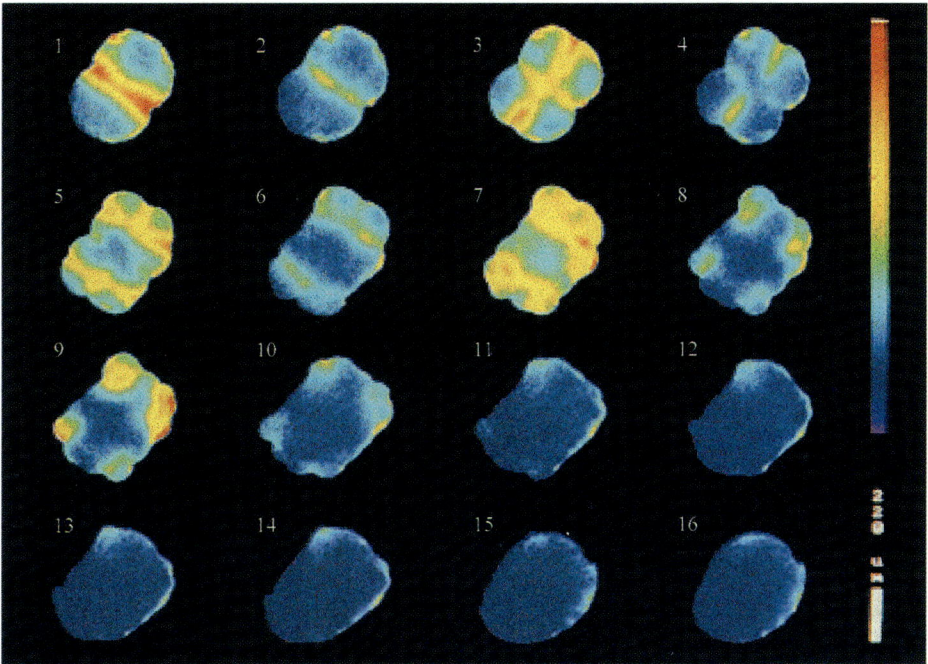

Plate 4, Fig. 1 (*see* discussion in Chapter 25, and full caption on p. 254). Sequence of pseudocolor ratio images of zebrafish embryo in early stages of development.

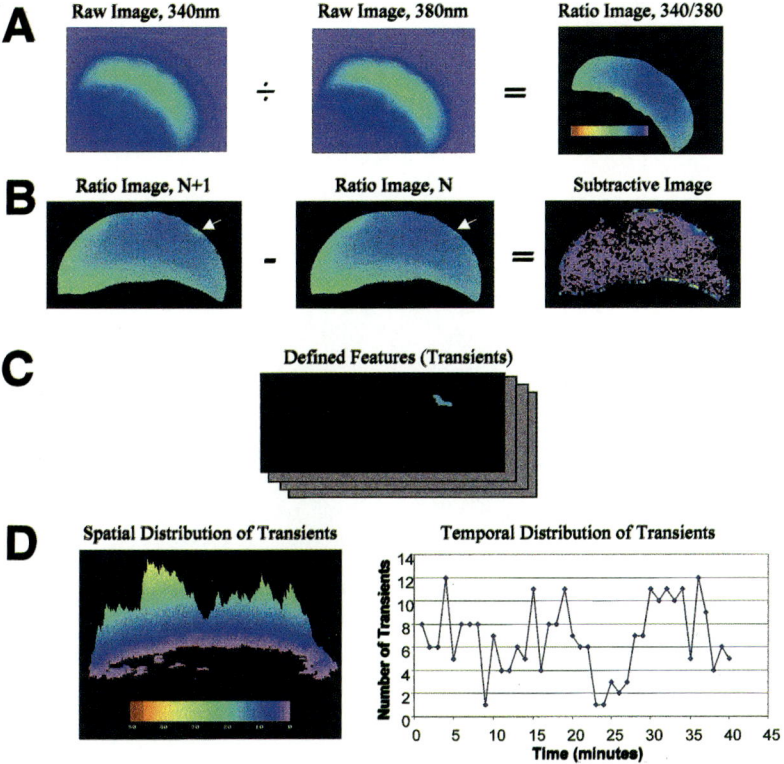

Plate 5, Fig. 2 (*see* discussion in Chapter 25, and full caption on p. 257). Example of subtraction analysis algorithm to detect [Ca^{2+}] transients.

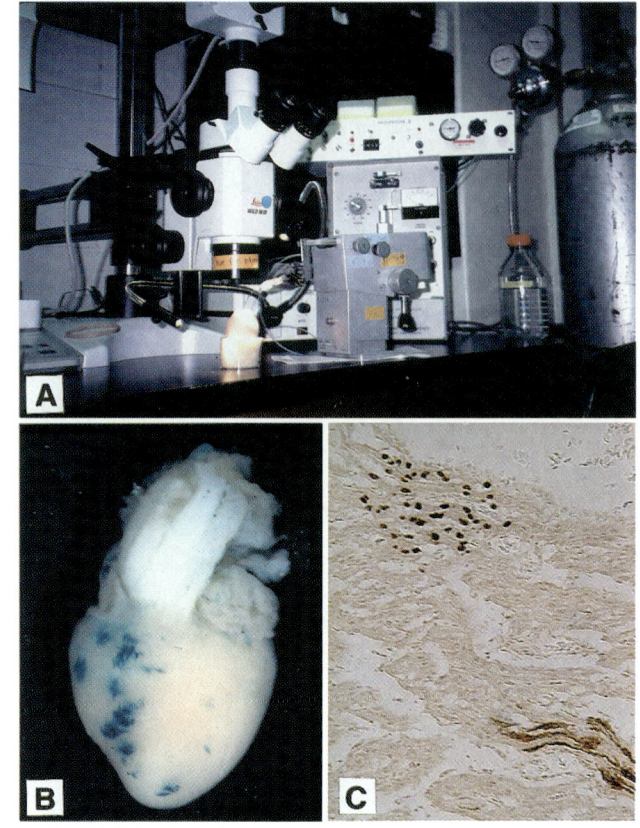

Plate 6, Fig. 1 (*see* discussion in Chapter 29 and full caption on p. 300). **(A)** Basic micro-injection equipment. **(B)** Whole mount of chick embryo heart at d 14 of embryonic incubation. **(C)** Two independent group of myocytes infected with different replication-defective constructs.

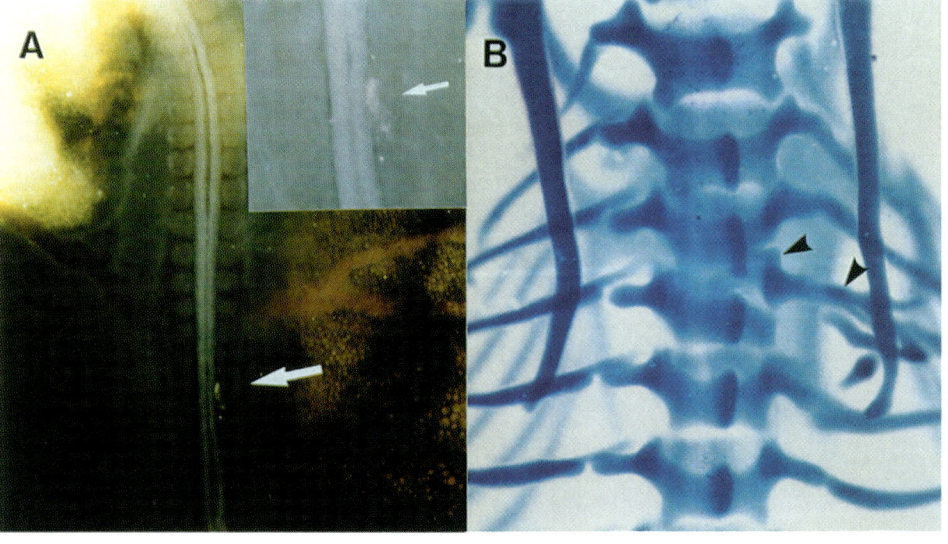

Plate 7, Fig. 2 (*see* discussion in Chapter 43 and full caption on p. 468). Implantation of FGM containing BMP2 protein into E2 chick embryos.

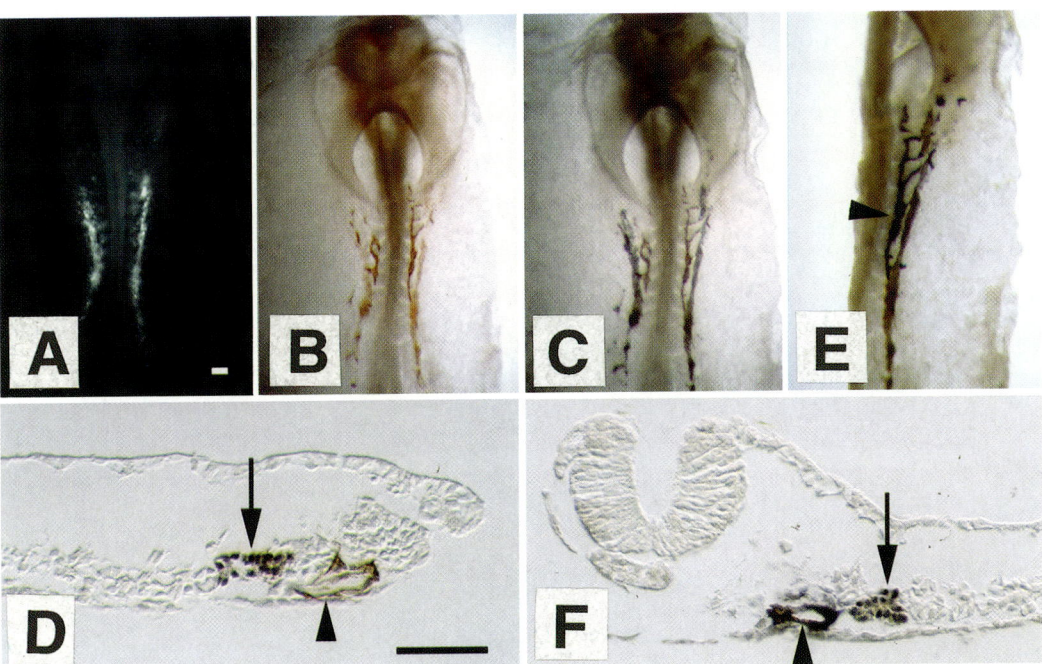

Plate 8, Fig. 5 (*see* discussion in Chapter 30 and full caption on p. 312). Creativity in the use of dynamic labeling techniques. **(A)** Whole mount 24 h after grafting. **(B)** Same embryo with anti-rhodamine antibody/peroxidase without cobalt-nickel intensification. **(C)** Same embryo after QCPN antibody/peroxidase labeling with cobalt-nickel intensification. **(D)** Transverse section of half of the embryo shown in **Fig. 5C**. **(E)** Other half of the embryo shown in **Fig. 5C**.

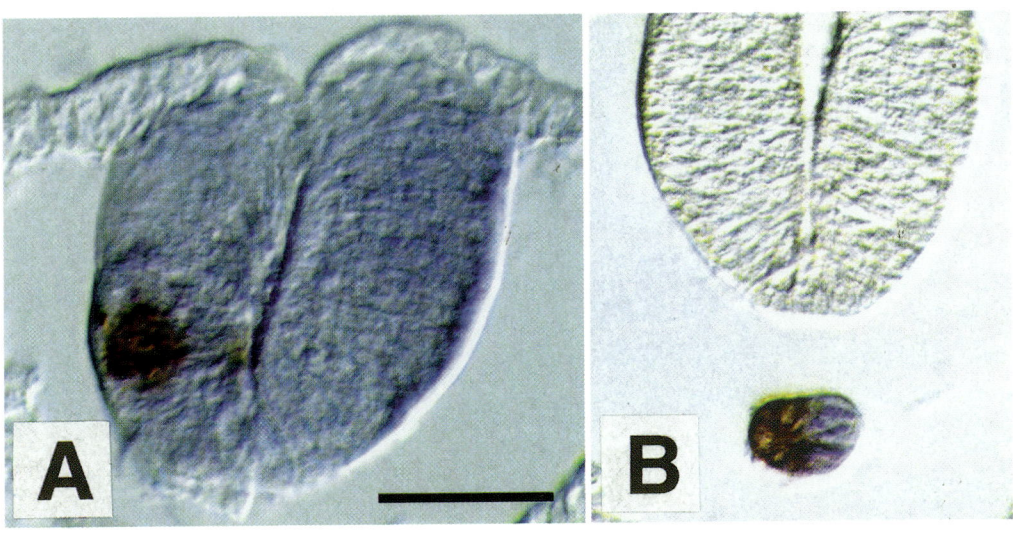

Plate 9, Fig. 7 (*see* discussion in Chapter 30 and full caption on p. 314). Combination of *in situ* hybridization with immunocytochemical detection of graft cells. **(A)** Transverse section through the neural tube of an embryo showing quail graft cells within the neural tube of an embryo. **(B)** Transverse section through the neural tube and notochord of an embryo showing graft cells within the notochord of an embryo.

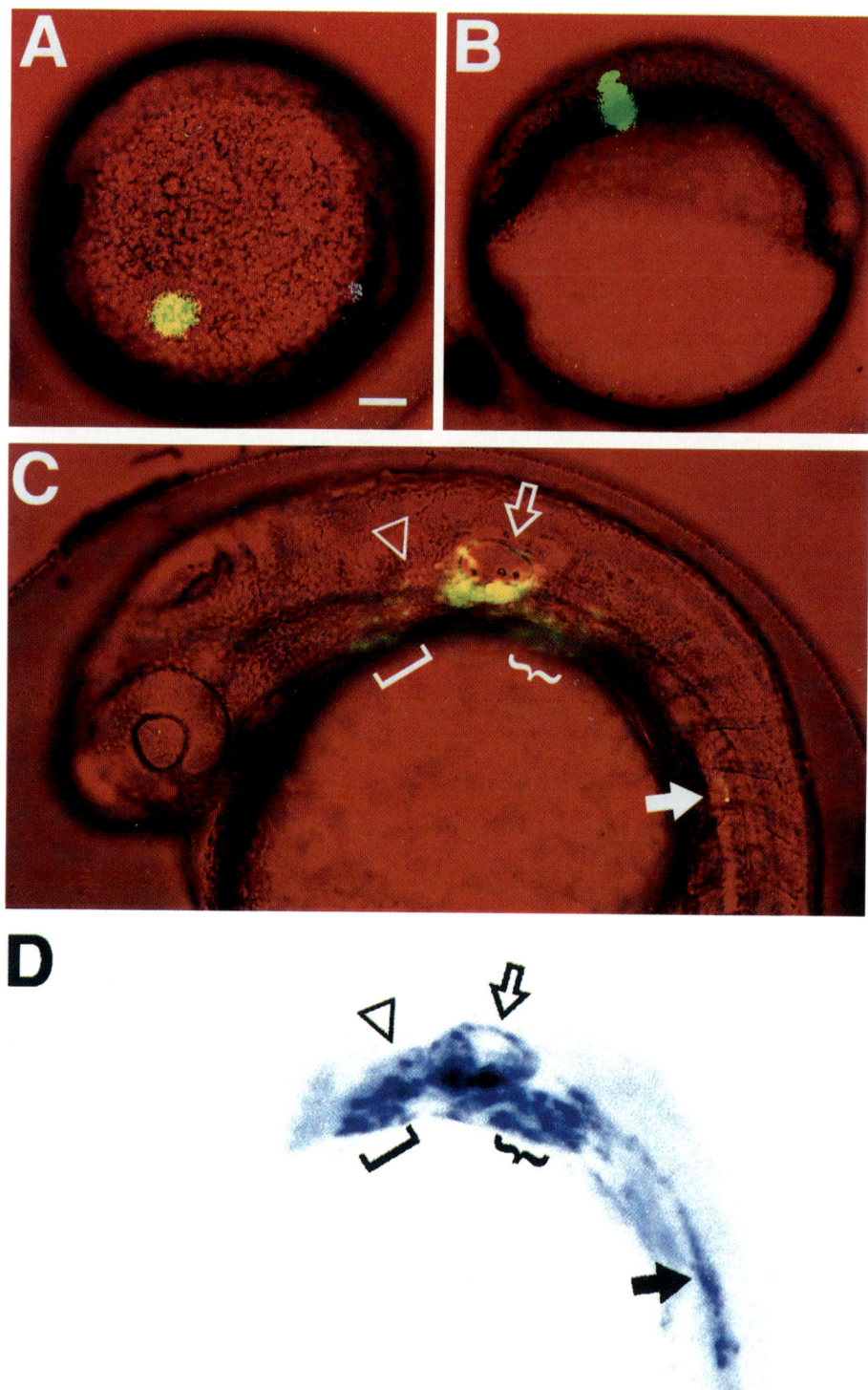

Plate 10, Fig. 1 (*see* discussion in Chapter 33 and full caption on p. 352). Detection of uncaged fluorescein (lysine-fixable) in zebrafish embryos. **(A)-(C)** Live embryos viewed with combined I3 epifluorescent and brightfield illumination. **(D)** Embryo from **(C)** fixed and stained with an alkaline phosphatase-conjugated antifluorescein antibody to detect cells containing uncaged fluorescein.

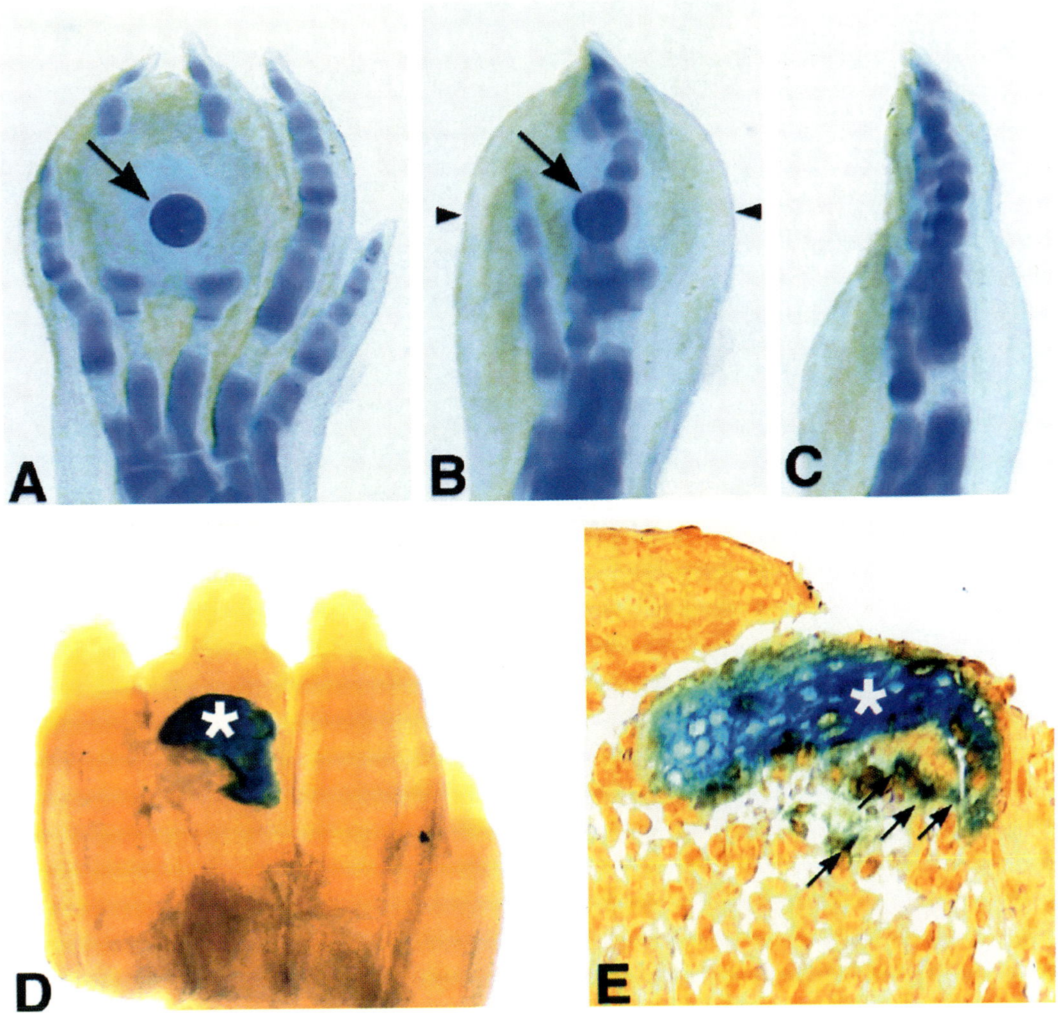

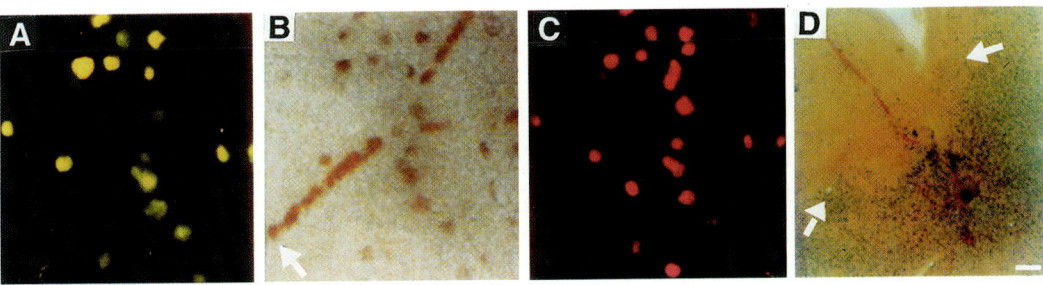

Plate 11, Fig. 2 (*see* discussion in Chapter 46 and full caption on p. 490). **(A)** Example of microcarrier bead-grafting experiment. **(B)** Lateral view of limb shown in **(A)** displaying hypertrophic effect of the TGFβ-1 bead. **(C)** Lateral view of control limb showing normal dorsal-ventral flattening associated with digit formation. **(D)** Example of tissue graft in CD1 mouse embryo. **(E)** Histological section showing B6, 129-TgR and individual cells.

Plate 12, Fig. 1 (*see* discussion in Chapter 45 and full caption on p. 476). Assays for cell ablation: ethidium homodimer assay and X-gal assay.

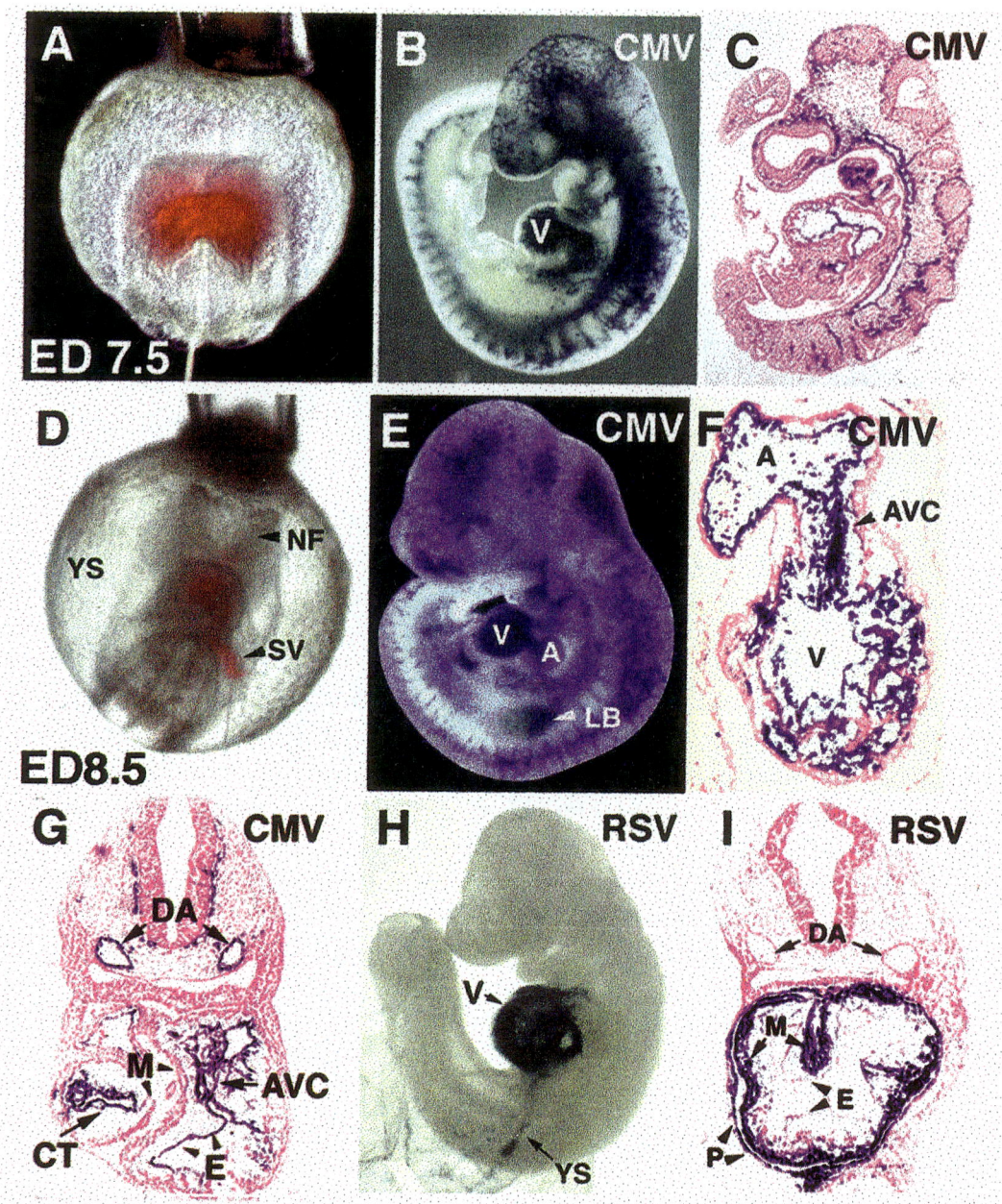

Plate 13, Fig. 1 (*see* discussion in Chapter 51, and full caption on p. 538). (**A**) Viral constructs containing *lac*Z reporter gene. (**B**) Embryos injected with particles of virus. (**C**) Sagittal section through embryo shown in (**B**). (**D**) E8.75 embryos injected with particles of recombinant virus containing *lac*Z reporter gene. (**E**) β-Galactosidase activity detected throughout vascular of developing embryo. (**F**) Sagittal section through heart of embryo shown in (**H**). (**G**) Cross section of heart showing β-galactosidase activity. (**H**) Whole-mount preparations of embryos injected with virus containing RSV-*lac*Z construct. (**I**) Cross section through heart confirming β-galactosidase activity throughout pericardium and myocardium.

4. Notes

1. The zebrafish need to be well cared for to ensure a reliable source of eggs. (Consult **ref. *10*** for details on stock maintenance.) They do best when the water quality is good, they are fed a variety of foods, and the tank light cycle/temperature is maintained. Generally, better quality eggs come from young adult fish (6 mo–1.5 yr).

2. All solutions must be RNase free to avoid degradation of the product. Estimate mRNA concentrations by optical density and gel electrophoresis. Capped mRNA can be synthesized using the mMessage mMachine transcription kit (Ambion Inc., Austin, TX).

3. Test several different parameters on the needle puller to generate a thin tip, yet strong enough to penetrate the chorion. A finer needle can be used on dechorionated embryos but these eggs are more fragile. The embryos can be manually dechorionated by using fine forceps to peel away the translucent membrane surrounding the egg.

4. A simple injection dish can be made by pouring 1.5% agarose into a Petri dish and forming small depressions by overlaying the agarose with 1-mm capillaries. After the agarose has hardened, fill the depressions with enough egg medium to cover the embryos.

5. It will take approx 45 min before the next cell division, depending on the temperature in the injection room. By placing the injected embryos in a heated water bath (approx 28.5°C), the zebrafish will develop faster. The embryos will develop slower at room temperature, but be sure that the room is not too cool. Extended periods of time at 18°C will result in abnormal development.

6. Stimulation of the PI cycle was accomplished by expression of exogenous receptors. The 5-HT$_{1c}$R is a G-protein-coupled receptor that activates the PI cycle in a ligand-dependent manner *(13)*. Mianserin is required to prevent endogenous activity before image analysis. The example used in this chapter is the 5-HT$_{1c}$R receptor; exogenous expression of Xwnt-5A mRNA also stimulates $[Ca^{2+}]_i$ release *(5)*. The procedure for assaying Xwnt-5A is similar except that there is no requirement for 5-HT or mianserin in the egg medium.

7. A small gold wire loop is needed to position the embryo in an inverted orientation. Use fine forceps to make a small rip in the chorion, position the embryo blastodisc down and gently overlay the loop to hold the embryo in place.

8. Fluorescence intensities should be selected such that they are not too faint, to avoid autofluorescence, or too bright, to avoid buffering intracellular calcium. Autofluorescence will be apparent if the yolk has a similar intensity as the blastodisc. Be extremely gentle when sliding the filter cube so as not to rotate the embryo.

9. Introducing the mRNA into a subset of cells is a useful control for activation of $[Ca^{2+}]_i$ release, because it is only seen in the region of TxR distribution. Globin mRNA as well as Xwnt8 serve as useful negative controls. Xwnt8 does not generate $[Ca^{2+}]_i$ release and has the advantage in that mRNA function can be assayed by altered *goosecoid* expression *(5,14)*.

10. Images can be collected at shorter intervals (i.e., 5–10 s) but will require more storage space for the raw data. The average file size for one image is 670k bytes of information.

11. If interested in whole-mount *in situ* hybridization, allow the injected embryos to develop to the appropriate stage and then place in fixative. For details on the generation of riboprobe and embryo processing, consult the reference for the particular marker used. Two useful markers are goosecoid (for the dorsal shield, altered in Xwnt8-injected embryos *[14]*) or Krox-20 (to assess alterations caused by perturbations of cell movement *[15]*), altered in Xwnt-5A- and 5-HT$_{1c}$R-expressing embryos *(5)*.

12. The resulting ratio images are centered in the frame-store such that each image in the time course corresponds pixel by pixel (up to 4095 intensity levels, 658 × 517 pixels) to the

next image. The computer has converted the signal into numerical data. Successive images are subtracted for the previous image to generate the subtractive image.

13. Transients are manually defined to include regions of the blastodisc that are approximately the size of a cell and that are increasing in intensity. Excluded are regions that remain the same intensity or decrease as well as regions that are too small. The digitized data from the subtractive image defined as a transient is converted into an integer.

14. The $5\text{-HT}_{1c}\text{R}$ mRNA was nonhomogeneously distributed in the embryo. In the topographical representation, pseudocolor assigns a distinct color and peak for each value. The height of the peaks and color range show how many transients occurred at that position along the blastodisc. In the plot of the series of transients as a function of time, the dip in the number of transients at 10 and 25 min is more likely to be a function of the stability of the 5-HT ligand in the medium. It is best to continually add fresh ligand during the course of the experiment.

Acknowledgments

Thanks to R. T. Moon and W. B. Busa for advice and help on the image analysis. This work was supported by United States Public Health Service awards GM35463 and GM17484 from the National Institutes of Health.

References

1. Hayashi, H. and Miyata, H. (1994) Fluorescence imaging of intracellular Ca^{2+}. *J. Pharmacol. Toxicol. Methods* **31,** 1–10.

2. McCormack, J. G. and Cobbold, P. H. (eds.) (1991) *Cellular Calcium: A Practical Approach.* IRL, Oxford, UK.

3. Chang, D. C. and Meng, C. (1995) A localized elevation of cytosolic free calcium is associated with cytokinesis in the zebrafish embryo. *J. Cell Biol.* **131,** 1539–1545.

4. Reinhard, E., Yokoe, H., Niebling, K. R., Allbrotton, N. L., Kuhn, M. A., and Meyer, T. (1995) Localized calcium signals in early zebrafish development. *Dev. Biol.* **170,** 50–61.

5. Slusarski, D. C., Yang-Snyder, J., Busa, W. B., and Moon, R. T. (1997) Modulation of embryonic intracellular Ca^{2+} signaling by Wnt-5A. *Dev. Biol.* **185,** 114–120.

6. Ault, K. T., Durmowicz, G., Harger, P. L., Galione, A., and Busa, W. B. (1996) Modulation of *Xenopus* embryo mesoderm-specific gene expression and dorso-anterior patterning by receptors that activate the phosphatidylinositol cycle signal transduction pathway. *Development* **122,** 2033–2041.

7. Berridge, M. J. and Irvine, R. I. (1989) Inositol phosphates and cell signalling. *Nature* **341,** 197–205.

8. Berridge, M. J. (1993) Inositol trisphosphate and calcium signaling. *Nature* **361,** 315–325.

9. Slusarski, D. C., Corces, V. G., and Moon, R. T. (1997) Interaction of Wnt and a Frizzled homologue triggers G-protein-linked phosphatidylinositol signalling. *Nature* **390,** 410–413.

10. Westerfield, M. (1993) *The Zebrafish Book: A Guide for the Laboratory Use of the Zebrafish (Brachydanio rerio)* University of Oregon Press, Eugene.

11. Julius, D., MacDermott, A. B., Axel, R., and Jessell, T. M. (1988) Molecular characterization of a functional cDNA encoding the serotonin 1c receptor. *Science* **241,** 558–564.

12. Grynkiewicz, G., Poenie, M., and Tsien, R. Y. (1985) A new generation of Ca^{2+} indicators with greatly improved fluorescence properties. *J. Biol. Chem.* **260,** 3440–3450.

13. Conn, P. J., Sanders-Buch, E., Hoffman, B. J., and Hartig, P. R. (1986) A unique serotonin receptor in choroid plexus is linked to phosphatidylinositol turnover. *Proc. Natl. Acad. Sci. USA* **83,** 4086–4088.

14. Kelly, G. M., Greenstein, P., Erezyilmaz, D. F., and Moon, R. T. (1995) Zebrafish *wnt-8* and *wnt-8b* share a common activity but are involved in distinct developmental pathways. *Development* **121,** 1787–1799.

15. Ungar, A. R. and Moon, R. T. (1995) *Wnt4* affects morphogenesis when misexpressed in the zebrafish embryo. *Mech. Dev.* **52,** 153–164.

16. Lechleiter, J., Girard, S., Peralta, E., and Clapham, D. (1991) Spiral calcium wave propagation and annihilation in *Xenopus laevis* oocytes. *Science* **252,** 123–126.

26

Acquisition, Display, and Analysis of Digital Three-Dimensional Time-Lapse (Four-Dimensional) Data Sets Using Free Software Applications

Charles F. Thomas and John G. White

1. Introduction

The ability to capture and visualize information throughout the volume of a living three-dimensional (3-D) specimen as it changes over time (four-dimensional [4-D] imaging) has been the goal of many biological researchers. In practice, this is usually accomplished by gathering a series of 3-D data sets or "timepoints" from a given sample where time is allowed to pass between each 3-D acquisition.

Just as a book can be divided into individual pages that are more easily grasped than the whole, it is common practice to gather each 3-D data stack as a series of 2-D "slices" from the top to the bottom of the specimen. This can be accomplished by using some type of noninvasive "image sectioning" microscopy (e.g., differential interference contrast [DIC or Nomarski]) microscopy *(1)*, confocal fluorescence microscopy *(2)*, or multiphoton microscopy *(3)*, which allows a single focal plane of the sample to be imaged at a time moving the imaging plane stepwise through the sample from the top to the bottom with a motorized focus motor. Slices are imaged, digitized, and stored to the disk during the process.

As early as 1960, Gustafson and Kinnander developed a rudimentary setup that allowed them to capture and analyze time-lapse information from multiple focal planes of developing sea urchin embryos *(4)* onto motion-picture film. However, the demands of 4-D imaging are such that it was not really practical without the use of computer technology.

Recently, software applications have become available that enable this type of study to be performed using relatively inexpensive laboratory computers. This chapter will discuss the use of one such system *(5)* that has been developed by the Integrated Microscopy Resource (IMR) at the University of Wisconsin-Madison. The system consists of three separate software applications that can work in concert or independently. Hardware and software considerations will also be addressed as well as protocols for the *acquisition* of 4-D data sets, the *translation* of these potentially enormous data sets

From: *Methods in Molecular Biology, Vol. 135: Developmental Biology Protocols, Vol. I*
Edited by: R. S. Tuan and C. W. Lo © Humana Press Inc., Totowa, NJ

into a format that is more easily handled, and the *viewing and analysis* of these data sets in a straightforward manner.

2. Materials

2.1. Data Acquisition

1. Software: Software for the automated acquisition of 4-D data sets is available without charge from the IMR's 4-D Web page at <http://www.loci.wisc.edu/4d/native/4d.html> (*see* **Subheading 3.1.** and **Notes 1** and **2**).
2. Microscope and optics: We have successfully used both the Nikon Optiphot and the Nikon Diaphot 200 (Nikon/Image Systems, Columbia, MD) in our lab for DIC and multiphoton imaging of a variety of samples. A similar quality microscope with comparable capabilities should be sufficient (*see* **Note 3**).
3. Video camera (detector): Several companies offer video cameras that will work for 4-D imaging. We use an 8-bit, 640- × 480-line Vidicon camera made by Sierra Scientific Inc. (Sunnyvale, CA), but any similar video camera should suffice (*see* **Note 4**).
4. Stage control (Z-focus) motor: There are a variety of different Z-focus motors available that will do the basic job of moving a microscope stage up and down to change the plane of focus. Several manufacturers offer units that can be applied to 4-D imaging (*see* **Notes 5** and **6**).
5. Image digitization ("framegrabber") card: The "4-D Grabber" software expects to control a framegrabber card that will digitize the incoming video signal from the video camera into a file that can be stored to the computer's hard disk. There are several framegrabber cards in a variety of price ranges that can be used (*see* **Notes 7** and **8**).
6. In-line image processors: The use of an in-line image processor such as the Dage DSP-2000 (Dage Inc., Michigan City, IN) can enhance image contrast and reduce noise through the use of frame-averaging techniques. Some framegrabber cards also support contrast enhancement and frame-averaging functions. Such image processing is not required, but has advantages in terms of getting the most information out of the images and in terms of compression efficiency (*see* **Note 9**).
7. Monitors: A useful component of a 4-D acquisition setup is a second video monitor to display the video image coming from the microscope. This is useful for monitoring the sample between 3-D timepoint acquisitions. Any monitor that accepts RGB, BNC, or coaxial video input is suitable.
8. Computer: The 4-D Grabber software is designed to run on a Quicktime-equipped (*see* **Note 10** for information on Quicktime) Macintosh personal computer [Cupertino, CA] (*see* **Notes 11** and **12**). A hard drive large enough to hold the system software plus approx 1.5X the size of the 4-D data set to be captured should be considered a minimum (*see* **Note 13**).
9. Archival media: The 4-D acquisition process can result in enormous amounts of data. Apart from the hard disk space necessary to store the images during acquisition, an archival device will also be necessary to store the data sets after they are removed from the hard drive. We have recently switched from rewritable magneto-optical drives to writing our own CD-ROMs (*see* **Note 14**).

2.2. Data Translation

1. Software: Software for the translation of 4-D data sets is available without charge from the IMR's 4-D Web page (*see also* **Subheading 3.2.**).
2. Computer: The same computer requirements listed for **Subheading 2.1.** will suffice for the data translation process (*see* **Notes 11** and **12**).

2.3. Data Viewing and Analysis

1. Software: Software for viewing and analysis of 4-D data sets is available without charge from the IMR's 4-D Web page (*see also* **Subheading 3.3.**).
2. Computer: The same computer requirements listed for **Subheading 2.1.** will suffice for the data translation process (*see* **Notes 11** and **12**).

3. Methods

3.1. Data Acquisition

The IMR has developed software for 4-D acquisition as part of its suite of 4-D applications. The 4-D Grabber application is available free of charge from the IMR's 4-D Web page and is designed to run on Macintosh personal computers that are Quicktime capable. It will collect data sets using a standard video camera attached to the microscope, a framegrabber board to digitize the incoming video signal, and a focus motor. It will also control an optional illumination shutter to reduce radiation damage from the light source. The 4-D Grabber software was designed for use with DIC microscopy, but it is possible to use it to collect brightfield, darkfield, phase, or standard fluorescence microscopy data sets (although these imaging techniques lack the exclusion of out-of-focus light that the so-called "image sectioning" microscopies have).

1. Place your prepared sample onto the microscope stage (*see* **Note 15**). Make sure your cameras, light sources, stage motors, and so forth are powered up and ready to go and that you have an acceptable image being sent to the video camera.
2. Launch the 4-D Grabber software application by double-clicking on its icon.
3. Choose the appropriate stage motor from the "Select Z Motor" submenu of the Z Motor menu. If you are not using a stage motor and intend only on collecting information from a single focal plane of the sample over time, you can ignore this step.
4. Select "Start Capturing" from the Video menu. You should see the image from the microscope appear on the computer screen (*see* **Note 16**).
5. Using the mouse, draw a box around the region of interest (ROI) you wish to capture. This can be anything from a small box to the entire field of view (*see* **Note 17**).
6. Choose "Start Acquisition" from the Data Acquisition menu. You will be prompted for the following information:
   a. Number of first timepoint? (This is usually set to 1.)
   b. Number of timepoints to acquire? If in doubt, set this to a very high number and plan on being around to abort the acquisition when it has gone far enough.
   c. Use variable time delay? Selecting "No" here will allow you to choose a delay between timepoints that will be used for the duration of the experiment (e.g., 30 s). Choosing "Yes" will bring up a dialog box that will enable you to tailor the acquisition time curve to the specimen's event curve (*see* **Note 18** and **Fig. 1**).
   d. Use illumination shutter? Select "Yes" if you have an illumination shutter and wish to use it. Select "No" if you do not have one or do not want to use it.
   e. Do Z series at each timepoint? Select "No" if you wish to image only a single focal plane as it changes over time or if you have no stage motor. (In this case, proceed to **step 7**.) Selecting "Yes" will bring up the z-motor control dialog (**Fig. 2**).

   Clicking on the "DIC Shutter" box will either open or close the illumination shutter (if present). If the light bulb is on, the shutter is open. The stage can be moved up and down by using the mouse to turn the "Focus Control." The up and down arrows to the side of the focus knob will move you up or down one increment (set with the "Set Increment" command located under the Z Motor menu).

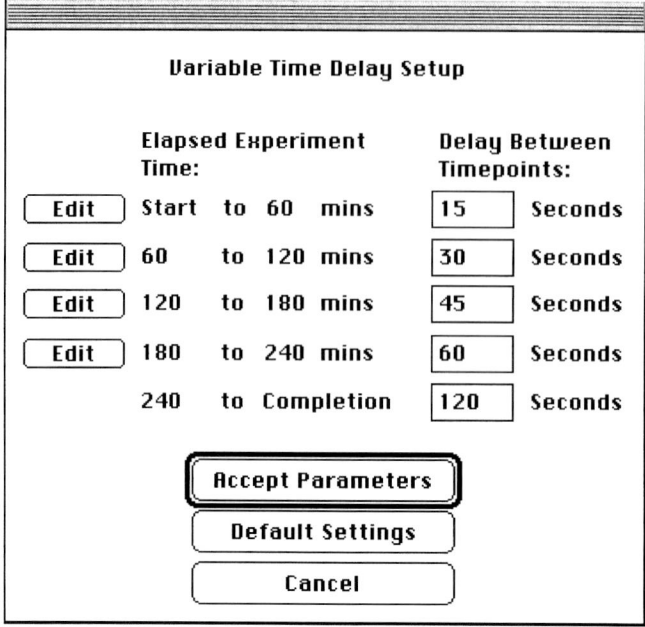

Fig. 1. The "Variable Time Delay" setup window of the "4-D Grabber" software. This window allows the user to change the acquisition interval during the course of the experiment, matching it to the dynamic events that are to be captured. In this way, the optimal use of available hard disk space can be made.

Move the stage until the top of the sample is in focus. Click the "Set Top" box. The position of the top should appear in the box above the "Set Top" box. Move the stage until the bottom of the sample is in focus. Click the "Set Bottom" box. The position of the bottom should appear in the box above the "Set Bottom" box.

If you wish to tell the software how many slices to take, click the "Provide # of Slices" box. Given the number of slices, the software will determine the increment to step the stage motor between slices. If you wish to tell the software the increment to step between slices, click the "Provide Increment" box. Given the increment, the software will determine the number of slices to take.

7. You will be prompted to choose the file format for the data set to be saved. There are currently two choices. Data can be saved as raw PICT/PICS files or can be gathered directly to compressed QuickTime movies (*see* **Note 19**).

8. You will be prompted as to whether or not you wish to annotate your images. Choosing "Yes" will write the elapsed experiment time and focal plane number onto the collected images. This can be helpful for later analysis (*see* **Note 20**).

9. You will be prompted to choose a root name for your timepoints or movies. If you choose Cell, for example, your raw timepoints will be named Cell.001, Cell.002, and so forth, or your movies will be named Cell01, Cell02, and so forth.

10. You will then be prompted as to where to save your movies. Choose a location with a sufficient amount of free hard disk space to store the data set (*see* **Notes 13** and **21**).

11. If you have elected to gather your data set as a group of Quicktime movies, you will be prompted to set up your choice of compressors and compression levels (*see* **Note 22** and **Fig. 3**).

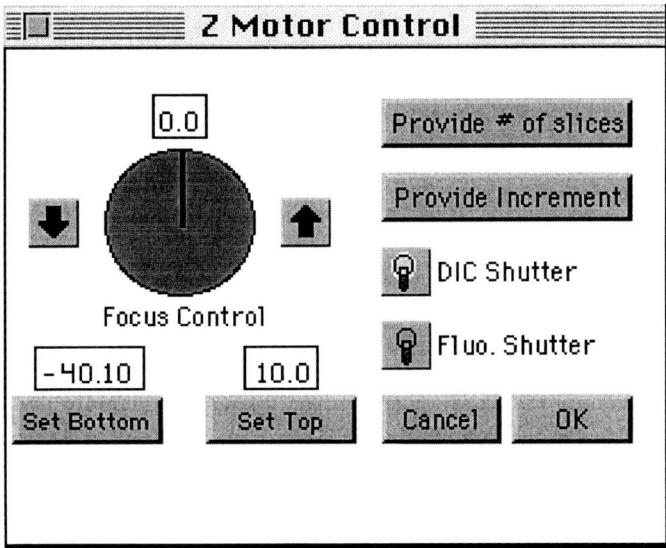

Fig. 2. The "Z-Motor Control" window from the "4-D Grabber" application. From this window, the user can control the illumination shutter and z-focus motor of the microscope. From here the top and the bottom of the sample are selected, as well as the number of slices that the user wishes to capture from the sample. A handy focus-knob control makes operation of the stage control intuitive.

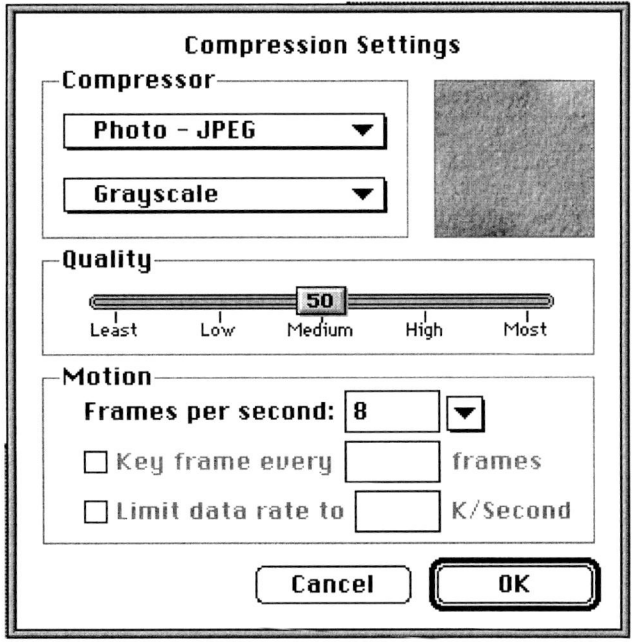

Fig. 3. The "Compression Settings" dialog from the "4-D Turnaround" and "4-D Grabber" applications. From this dialog, the user can select from various compressors and also select the degree of compression so as to make the optimal trade-off between image quality and data set size.

12. Following all of these dialogs, the data acquisition will start. The z-motor will move the stage to the top of the sample, the shutter will open (if applicable), and the focal planes will be digitized one-by-one until the bottom of the sample is reached. The shutter will close (if applicable), and you will be advised of the progress of your acquisition by an information dialog box between timepoints. If all goes well, this process will continue until all timepoints have been gathered or until you abort the acquisition when it has gone far enough.

13. If you have chosen to collect your data set as a Quicktime data set, you will be prompted to save a "4-D Format File" for the experiment. This file will be used later in the data viewing and analysis process.

3.2. Data Translation

Not all 4-D data sets will be collected with the 4-D Grabber software. At the IMR, for example, the 4-D Grabber application collects data sets from our 4-D DIC workstation, whereas control macros collect 4-D data sets from the confocal and multiphoton microscopes.

For data sets that were not collected using 4-D Grabber, the IMR has developed the "4-D Turnaround" application to process raw data sets collected in TIFF, PICT, PICS, or Bio-Rad's PIC format into Quicktime movies. Subsequently, these movies can be loaded into the "4-D Viewer" application for viewing and analysis (*see* **Subheading 3.3.**). The Macintosh application is available free of charge from the IMR's 4-D Web site.

At the same time that the data sets are being processed into movies, the user's choice of compression algorithms can be applied to reduce the size of the data set (*see* **Note 21**). The steps involved in processing a raw data set into a Quicktime data set for use in the 4-D Viewer are as follows:

1. Make sure your data set is prepared for translation (*see* **Note 23**).
2. Double-click on the 4-D Turnaround application's icon to start the program.
3. Under the "Turnaround" menu, select "Set Maximum Focal Planes." This should be set to a number at or slightly above the number of focal planes you intend on processing.
4. Under the Turnaround menu, select "Convert Data Sets."
5. When prompted, open the first timepoint of the 4-D data set. This file can be either a 3-D data stack or a 2-D single image. The software recognizes files in TIFF, PICT, PICS, and Bio-Rad's PIC image formats.
6. Specify the number of timepoints to be converted and which focal planes of these files to convert (the default setting is all of them).
7. Set up your choice of compressors and compression levels (*see* **Note 22** and **Fig. 3**).
8. The application will then process each timepoint in the 4-D data set (*see* **Note 24**).
9. When all timepoints have been compressed and added to their respective movies, the movies are saved to the hard disk and a 4-D Format File is created. This file contains information about the 4-D data set and allows the 4-D Viewer application to open it correctly.

3.3. Data Viewing and Analysis

The IMR has developed the 4-D Viewer application to allow the straightforward viewing, analysis, and annotation of time-lapse data sets. It is a Macintosh application that works with Quicktime data sets of the kind generated by both the "4-D Grabber" and 4-D Turnaround program and is available free of charge from the IMR's 4-D Web site. 4-D Viewer allows the user to roam through data sets both forward and backward

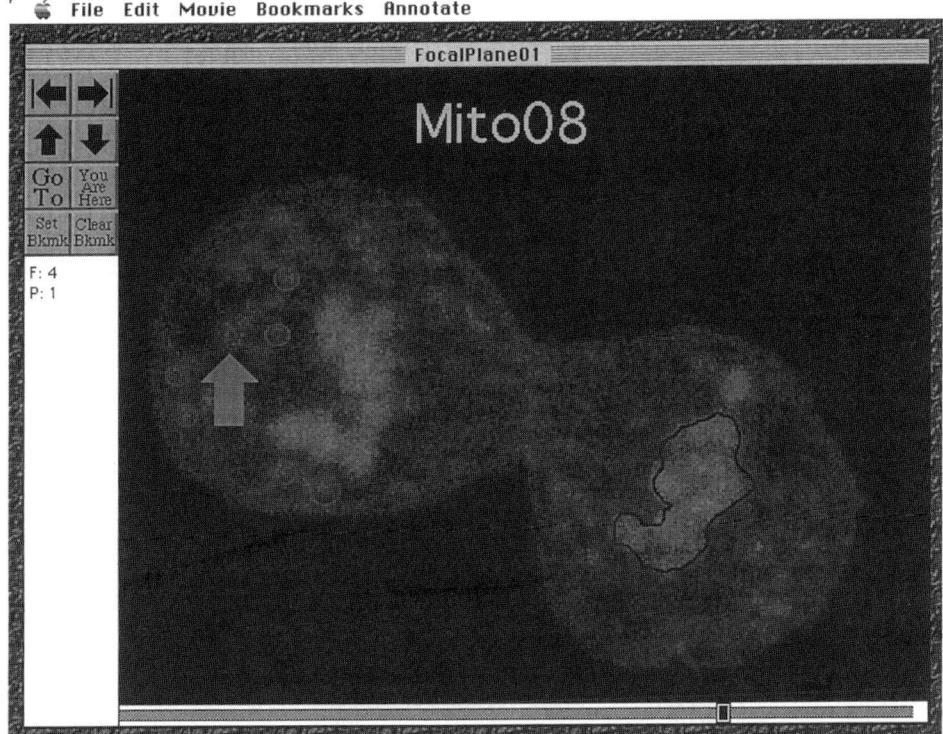

Fig. 4. The "Viewing Window" of the 4-D Viewer application. From this window, the user can navigate up and down through the sample, as well as forward and backward in time to watch changes in the sample. If a given frame of the data set has an associated overlay, it will be displayed.

in time, as well as up and down in focus. Users can set bookmarks, export individual images from the data set for further study, and create dynamic, color overlays that can follow features of interest as they move through 4-D space. The application has a large number of functions and can perform a variety of types of analyses—The basic steps involved in creating a dynamic color overlay to follow a specific feature of interest in a 4-D data set are outlined here.

1. Double-click on the 4-D Viewer's icon to start the application.
2. From the File menu, select Open and open the 4-D Format File for the data set with which you wish to work. The 4-D Format File is created for data sets that were created using either the 4-D Grabber software (in Quicktime mode) or with the 4-D Turnaround application.
3. The data set will be loaded into the viewer (*see* **Fig. 4**).
4. The arrow keys and/or the arrow palette icons can be used to animate the data set backward or forward to view changes in a specific focal plane over time. The same controls allow the user to move up and down in focus through the interior of the sample (*see* **Note 25**).
5. Use of the commands found in the "Bookmarks" menu enable the user to set bookmarks to images that can then be accessed instantly.
6. Select a feature you wish to highlight using the "Annotation Suite" and move backward in time until you find the first image in which it makes an appearance.

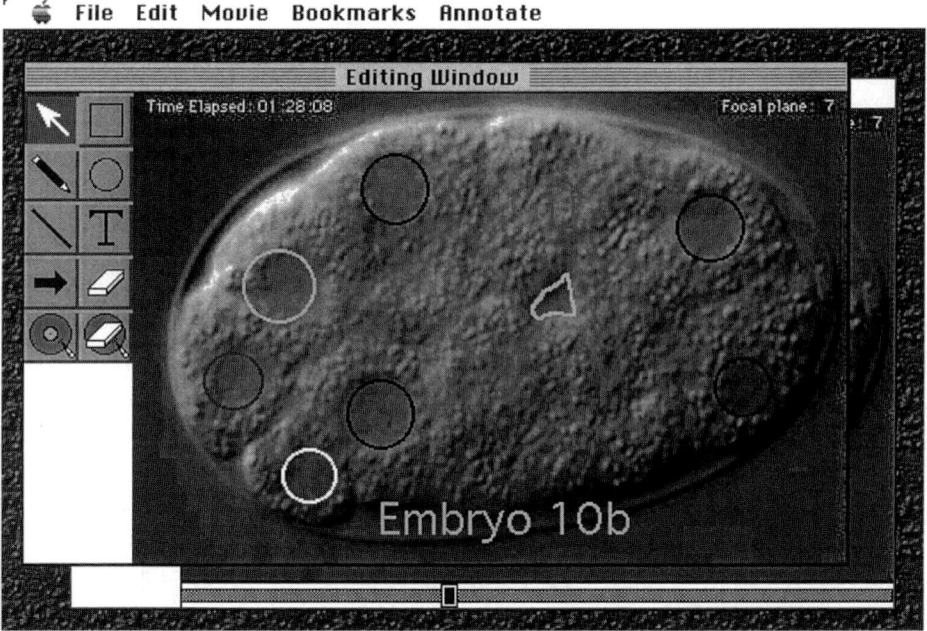

Fig. 5. The "Editing Window" of the 4-D Viewer application. The tool palette can be seen on the left. These tools can be used to create a square, straight line, oval, freehand line, and arrow overlay elements of various sizes and colors that follow specific features of interest in the 4-D data set regardless of where they move in 3-D space. The overlays on this particular data set label nuclei (circles), intracellular space (yellow freehand line), and cytoplasmic granules (arrow). Each overlay object can have a hidden text box associated with it that can elucidate the feature of interest.

7. Select "Make Editing Window" from the "Annotate" menu to bring up the Editing Window (*see* **Fig. 5**).

8. Click on one of the "Object" icons on the left of the Editing Window to select a tool for creating objects. You can choose from square, circle, straight line, freehand line, arrow, or text objects (*see* **Note 26**).

9. Use the mouse to position the object over the feature of interest on the image (*see* **Note 27**).

10. When you are satisfied with the position of the object, select "Save Overlay To Disk" from the Annotate menu. This will save the overlay and bring up the next sequential image in the data set.

11. Repeat **steps 9** and **10** as necessary. If the feature of interest moves up or down in focus, you can follow it using the arrow keys.

12. When you have finished highlighting the feature of interest, repeat **steps 6–11** for other features of interest (*see* **Note 28**).

13. You can close the Editing Window at any time and roam through the data set viewing any overlays you have created (*see* **Note 29**).

4. Notes

1. It must be pointed out that using the 4-D Grabber application is only one way that a 4-D acquisition can be performed. There are other ways to control the basic process of digitizing slices from the top to the bottom of a 3-D sample by using "image-sectioning" micro-

scopy, then repeating the process at subsequent timepoints. For confocal or multiphoton microscopy, a set of macros (batch files of system control instructions) usually automate the process of gathering 4-D data sets. Additionally, some of the more flexible image processing programs (e.g., the freeware application NIH-Image, written by Wayne Rasband and available from http://rsb.info.nih.gov/nih-image/ and the Metamorph imaging system from Universal Imaging Corporation: <http://www.image1.com>) allow automation of their functions to permit 4-D acquisition with a variety of hardware components.

2. The 4-D Grabber, 4-D Turnaround, and 4-D Viewer applications are written for use on Macintosh personal computers. They may be able to be run on IBM PCs through the use of Mac OS emulators (e.g., "Executor" from Ardi Corp. <http://www.ardi.com>). The IMR is currently investigating the possibility of porting these applications to the multiplatform computing language Java (Sun Microsystems), which would enable them to run on Unix, IBM, and Mac platforms. The first versions of these applications were scheduled for release in early 1999. More information can be obtained from the IMR's Web site: http://www.loci.wisc.edu/4d/java/4djava.html.

3. The microscope for 4-D imaging should have sufficient resolution and the quality and type of optics needed to visualize the structures of interest. For example, for fluorescence imaging, specific light sources and filter sets may be required. DIC imaging will require special polarizers, prisms, and analyzers. There also needs to be a way to get an image from the microscope to an appropriate detector (*see* **Note 4**). The stage of the microscope should be stable enough that your z-motor will be able to return to the same spot every time. A 4-D acquisition run may involve thousands of iterations of stepping from the top to the bottom of the sample. If there is any drift in the microscope's focus mechanism, this will soon allow the specimen to drift right out of the microscope's field of view.

   Additionally, you need to consider the needs of your sample and how you intend to mount it. Certain samples will require special environmental chambers to maintain temperature, humidity, and atmospheric gases at specific levels. Other samples may require long working-distance lenses and condensers to work with welled plates, depression slides, or Petri dishes. Make sure the microscope you choose is able to accommodate the needs of your sample.

4. The video camera is the unit that will take the image from your microscope and turn it into either an analog or digital signal. Video cameras have a variety of different resolutions (width and height in pixels), bit depths (number of gray levels that are captured), output signal formats, sensitivities, and frame rates. The 4-D Grabber software requires a camera that provides a video output signal that will subsequently be digitized by the framegrabber card. Other acquisition setups may have different requirements.

5. Some things to think about when choosing a stage motor are:
   a. Stability/accuracy—As previously discussed, it is important to have excellent repeatability when you are going from the top to the bottom of a specimen hundreds or thousands of times. How well your stage motor is designed and engineered has a lot to do with your ability to get repeatable data sets.
   b. Precision—You should think about the minimum step size that your experiment calls for and the minimum step size that the motor can accomplish. Most z-motors on the market will be accurate down to between 0.1 and 0.01 μm.
   c. Speed—The motor must be fast enough to keep up with your framegrabber. You do not want your software waiting around while your stage motor is moving your sample. This can cause delays, or worse, can cause motion-blurring of your sample.
   d. Method of interface with the microscope—Slippage at the connection of the stage motor to the microscope stage control can be a source of great frustration when per-

forming a 4-D data-acquisition run. Some stage motors simple have a friction cup adapter to turn the focus knob. Some actually use sturdy metal shafts that clamp or are bolted into the internal focus mechanism. The design of the microscope/stage motor interface is important when considering a stage motor.

e. Method of control—Make sure your stage motor is controllable by your acquisition software.

6. The 4-D Grabber software is currently configured to control the following z-motors: Ludl Mac-1000 and Ludl Mac-2000 (Ludl Electronic Products, Hawthorne, NY), ASI Model 85 (Applied Scientific Instruments, Eugene, OR), and the Leica DM-RXA series stepper motor (Leica Microsystems Inc., Deerfield, IL).

7. The framegrabber card usually resides in the acquisition computer and takes an analog video input signal, digitizing it for storage and manipulation. You must be sure that your video camera is supplying a signal that is compatible with the framegrabber card you've chosen. Older Macintosh computers use an interface bus called "Nubus," whereas the newer Power PC (and many IBM) computers use the faster "PCI" bus. Be sure to get the correct type of card for your computer's interface bus. The newer PCI bus cards are much faster than the outdated Nubus cards and as such are more attractive for 4-D imaging.

8. The 4-D Grabber software currently supports the following framegrabber cards: Scion LG-3 (Nubus and PCI), Scion VG-5 (Nubus and PCI), Scion AG-5 (Nubus and PCI), and Quicktime-compatible video cards including the built-in Macintosh AV cards.

9. The efficiency of most compression algorithms (including the JPEG compressor favored for our microscopy studies) depends on the complexity of the image. Because noise is interpreted by the compressor as "detail" that needs to be retained, a noisy image will be compressed less efficiently than an image that has had its noise reduced through the use of frame averaging or image filtering. Some framegrabber cards (such as the Scion AG-5) have contrast-enhancement and frame-averaging functions built-in and would obviate the need for a separate in-line processor.

10. Quicktime is a system extension available for Macintosh and Windows that allows the viewing and creation of sequential image animations or "movies" on the computer. More information on Quicktime can be obtained at <http://www.apple.com>. Quicktime also supports a number of built-in image compression/decompression (codec) algorithms, which can be applied to the movie as it is being created to reduce the size of image information.

11. Because the computer will control the hardware during 4-D image acquisition, it is central to any 4-D setup. The computer must be able to run the software and support the hardware intended for use for 4-D data acquisition, manipulation, and visualization. The computer will also store, manipulate, display, and analyze the 4-D data sets after they have been acquired. The maximum acquisition rate is very dependent on the speed of the computer's hard disk access time and the framegrabber bus speed. A faster processor will work with images more efficiently during the viewing/analysis process. Some users may elect to do their data acquisition on a different computer than the one that runs their viewing and analysis software. You'll pay more for A computer that is fast, has a lot of RAM, and has a big hard drive, so it is important to be aware of your hardware and software needs.

12. The IMR's 4-D imaging applications require a Macintosh II or better that is Quicktime equipped. 16 MB of RAM should be considered a minimum. A faster Power PC-type computer with 32 MB or more of RAM, a PCI bus, a CD-ROM, and at least 1 GB of hard disk space would be optimal.

13. To determine the bytes of disk storage space necessary to hold a 4-D data set, the following formula can be used:

$$\{[H_e + (B_p * W * H * S_l)] * T_p\}/C_r$$

Where H_e is the size of the image header in bytes, B_p is the number of bytes per pixel (e.g., an 8-bit image has 1 byte per pixel), W is the width of the image in pixels, H is the height of the image in pixels, S_l is the number of 2-D slices in each 3-D timepoint, T_p is the total number of 3-D timepoints that one wishes to acquire, and C_r is the compression ratio (if applicable) provided by any compression algorithm being used, assuming it is applied prior to saving the images to the hard disk.

14. There are many options for archival media: DAT tape, CD-ROM writers, zip drives, magneto-optical drives, Syquest drives, and others. The main issues are cost (of the drive and of the media), storage capacity, access speed, and volatility (how long the data can be expected to be safely stored). Right now, CD-ROM technology scores very well in most of these categories. Rewritable CD-ROMs hold up to 640 MB of data each, cost only a few dollars each, and are very robust for long-term storage. This makes them a very attractive archival option. However, in the world of computers, there are always bigger, better, and faster ways of storing media evolving. It is worth it to try to keep up with the latest developments.

15. Because there are so many potential samples and so many different methods of performing 4-D imaging, rather than offer specific protocols for sample mounting and preparation, some general guidelines will be presented.

A 4-D-data-acquisition experiment involves the repetitive collection of information from a specific 3-D region of interest using a predefined number of optical sections. It is therefore important to select a sample that is not going to move or grow out of the region being imaged for the duration of the experiment. For example, the embryo of the nematode *Caenorhabditis elegans* develops from a one-celled stage through to a fully formed adult within the confines of a 45×25-μm egg shell, thereby making it an ideal candidate for 4-D study. A mouse embryo, on the other hand, would be difficult to image for its entire developmental cycle, because it grows substantially from fertilization to fetus. However, a portion of that same process (e.g., the changes occurring as the embryo evolves from the one-cell to the eight-cell stage) could be studied for as long as they occurred in a fixed 3-D field of view.

The specimen should be mounted in such a way that it will not drift laterally or axially during the course of the imaging process. Such drift can result in lost information as the sample moves off the screen and can make it nearly impossible to compare information from different timepoints because they do not show the same area.

Another extremely important consideration is the ability to keep the sample alive for the duration of the experiment. This can potentially be the most difficult obstacle to overcome. Some samples can be carelessly mounted on a stage at room temperature and be quite happy for hours (e.g., the *C. elegans* nematode, which is dropped onto an agar pad, then sealed under a cover slip for viewing), but other samples have exacting constraints of temperature, humidity, pH, nutrient, waste removal, and atmospheric gases (e.g., hamster embryos, which are extremely sensitive to all these factors). Some experiments require accommodation of the extreme light sensitivity of the sample (e.g., through the use of an illumination shutter) or the minimization of toxic byproducts produced by the interaction of the illumination source with the fluorescent probes being used (e.g., by use of multiphoton techniques). Keeping the sample alive may involve the use of stage heaters or coolers, environmental chambers, media exchangers, illumination shutters, high-sensitivity detectors, and other specialized pieces of hardware.

Protocols for mounting most specimens of interest are already published for studies using standard methods of imaging. These procedures are generally a good starting point for developing a protocol that permits 4-D imaging, although modification and inventiveness may be required, as this is a field largely in its infancy.

16. Once the sample is visible on the computer monitor, it is helpful to use a Sharpie marker pen to outline the position of the sample on the glass of the monitor's screen. This will allow the user to reposition the sample to its original starting point should the stage be bumped or should the sample drift laterally for any reason. Such marks are easily erased after the acquisition process using ethanol and a Kimwipe. Although it may or may not be required (depending on the stability of the sample, its mounting, the microscope stage, and the z-motor), periodic check on your sample to correct for any drifting in either the lateral or axial directions will help ensure a higher quality data set that is more easily analyzed.

17. Keep in mind that the larger the ROI selected, the bigger the resulting data set will be. Try to minimize the amount of "wasted space" by keeping the ROI close to the border of the sample: 1/4" of space between the widest points of the sample and the ROI boundary is a good rule of thumb (*see also* **Note 20**).

18. The rate at which events unfold in a 4-D experiment is not always constant. For example, in the *C. elegans* nematode embryo, events occur very rapidly immediately following fertilization but slow down as the embryo reaches maturity. If one were to collect timepoints at a constant interval throughout the experiment, there would be the risk that the collection interval would either be too long (such that events were missed in the early stages of development), or that the interval would be too short (such that unnecessary amounts of data were being gathered in the later stages of development). The 4-D Grabber's Variable Time Delay feature allows the user to preprogram up to five different acquisition delays to be used during the course of the experiment. For example, one could collect a 3-D data stack every 15 s during the first 45 min of the experiment, then collect once every 30 s for the next 2 h, then collect once every 75 s for the rest of the experiment. In this way, the timing curve of the acquisitions can be matched to the timing of the events being imaged, and neither too little, nor too much, data are collected.

19. Raw data sets are multiple data stacks where each timepoint is saved as a 3-D data set, so there will be one file for each timepoint gathered. Quicktime data sets are a single Quicktime movie for each focal plane, so there will be as many files as there are focal planes. There are advantages and disadvantages to both types of data sets. Raw data sets take up much more hard disk space and will require a separate processing step, to be viewed using the 4-D Viewer application, but will be unprocessed in any way. If you wish to avoid the artifacts some compressors can introduce, the raw data set may be the option you want. Otherwise, the Quicktime movies are smaller, more efficient, and are available for use with the 4-D Viewer application as soon as the data set has been collected.

20. If you choose to annotate the data set, leave an extra 1/2" of space between the sample and the top of the ROI when you select it in **Subheading 3.1.**, **step 5**.

21. You may wish to make a new folder prior to starting the acquisition process to hold the data set so that all the image files are kept together.

22. The Quicktime architecture supports a variety of different compressors (e.g., video, JPEG, Cinepak, Animation, etc.) and levels of compression within each compressor. Each different compressor is designed with particular jobs in mind. For example, a compressor that works well on photographic images may not be the most efficient compressor to use on a series of line drawings or on a text file. Image compressors offer a trade-off between image quality and image size. The more an image is compressed, the smaller the resulting image file, but there may be a loss of image information in the process. It is up to the user to select a compressor and a level of compression that offers a manageable image size but still retains a sufficient level of image quality to enable subsequent analyses to take place.

A good methodology for choosing an image compressor is to do a short translation run on the data set (two or three timepoints only) with each available compressor. For these tests, use the highest compression setting (least image quality). Look for compressors that offer high levels of compression on your particular data set. Among those with acceptable levels of compression, decide which compressor offers the best looking image. Use this compressor to do a similar series of short runs with various levels of compression until an acceptable image quality is found.

The IMR has found that the JPEG compressor (designed by the Joint Photographic Experts Group) on its maximum compression setting offers up to 15:1 compression ratios with acceptable image fidelity when used on micrograph data sets.

23. Your individual timepoint files should be contained in a separate folder. Make sure that they are named properly (e.g., Filename.001, Filename.002, . . . , etc.) and that no sequential numbers are missing. Double check that there is enough room on the hard disk to hold the movie files being created. For example, if you have a 500-MB data set, and you are getting a 10:1 compression ratio from the compression algorithm, you would need to make sure there is at least 50 MB of free space on the hard disk to hold the movies after they are created.

24. When the compression scheme has been specified, the application creates a Quicktime movie for each focal plane in the first timepoint. It then goes through and compresses the image information from focal plane 1 and places it into movie 1, the information from focal plane 2 into movie 2, and so forth. The application then closes the first timepoint and opens and processes the remaining timepoints in a similar manner, appending the data from each subsequent timepoint to the end of its proper movie. The result is a set of Quicktime movies where each movie describes the changes in one focal plane from the beginning to the end of the acquisition period.

25. Rather than working with rendered 3-D models at each timepoint, the method outlined here uses the rapid animation of sequential 2-D images, thereby bypassing some of the pitfalls involved in creating 3-D reconstructions. It is possible to create 3-D reconstructions from each timepoint of a 4-D data set (using a program such as NIH-Image or Voxel View [Vital Images, Fairfield, IA]), then have the "up-and-down" dimension correspond to rotation angle rather than plane-of-focus. Software and protocols to perform such reconstructions from 4-D data sets are available on the World Wide Web *(6)*. However, with 3-D reconstructions, a user must make empirical decisions on which data are to be rendered and which are not, and these decisions can taint experimental conclusions. Also, interior details of a thick sample may be obscured by features closer to the surface in a 3-D model. By using 2-D images, the user can view all the data collected in an experiment without bias.

26. The attributes of the object (e.g., color, size) can be selected prior to placing the object on the image by holding down the "Alt" key and clicking on the tool icon. The object's attributes can also be edited after the object has been placed onto the image.

27. By holding down the Alt key and clicking on an overlay object, a text box appears that can hold comments about what the object is describing or what is happening at the particular timepoint of the overlay. These text descriptions will be saved to the overlay and subsequently searched.

28. Following and annotating each feature of interest separately is one method of overlay creation. Once the user is familiar with the steps involved in creating an overlay, many objects can be created and followed through subsequent timepoints. Objects may be added or deleted as necessary during this process.

29. Overlays may be switched on and off during playback using the "Display Overlays" command under the Preferences submenu of the Annotate menu.

Since the original authoring of this chapter, the Integrated Microscopy Resource (IMR) has become the Laboratory of Optical & Computational Instrumentation (LOCI). The LOCI's Java Initiative to port its 4-D software into the cross-platform programming language Java has already yielded improved versions of "4-D Turnaround" and "4-D Viewer," which run on both the Windows and the Macintosh platforms. More information about this software is available at <http://www.loci.wisc.edu/4d/java/4djava.html>.

The LOCI's original 4-D imaging Web page can now be found at <http://www.loci.wisc.edu/4d/native/4d.html>.

Acknowledgment

This work was supported by NIH Grant DRR-570 to the Integrated Microscopy Resource.

References

1. Nomarski, G. (1955) *J. Phys. Radium* **16,** S9–S13.
2. White, J. G., Amos, W. B., and Fordham, M. (1987) An evaluation of confocal versus conventional imaging of biological structures by fluorescence light microscopy. *J. Cell Biol.* **105,** 41–48.
3. Denk, W., Strickler, J. H., and Webb, W. W. (1990) Two-photon laser scanning fluorescence microscopy. *Science* **248,** 73–76.
4. Gustafson, T. and Kinnander, H. (1960) Cellular mechanisms in morphogenesis of the sea urchin gastrula. *Exp. Cell Res.* **21,** 361–373.
5. Thomas, C., DeVries, P., Hardin, J., and White, J. (1996) Four-dimensional imaging: Computer visualization of 3D movements in living specimens. *Science* **273,** 603–607.
6. Mohler, W. A. and White, J. G. (1998) Stereo-4-D reconstruction and animation from living fluorescent specimens. *BioTechniques* **24,** 1006–1012.

V

CELL LINEAGE ANALYSIS

27

Cell Lineage Analysis

Applications of Green Fluorescent Protein

Magdalena Zernicka-Goetz and Jonathon Pines

1. Introduction

The ideal cell lineage marker is one that can be visualized in living tissues without perturbing development. Exogenously applied dyes are very useful, but in many instances there would be advantages to an endogenously expressed marker. Green fluorescent protein (GFP) appears to have the potential to be ideal for this purpose, and with its advent, lineage analysis is entering a new phase in which cells can now be followed in real time in living embryos. GFP is a 27-kDa protein found in the jellyfish *Aequorea victoria* that absorbs blue light and emits green light. It has the valuable property that the formation of the chromophore is an autocatalytic event that requires no cofactors. Furthermore, the cDNA that encodes GFP has been cloned, and thus GFP can be expressed and will become fluorescent in any organism or cell type. In this chapter, we describe the properties of GFP and the different mutants available to study cell lineages in different organisms. We will mainly concentrate on their application to the analysis of cell lineages in frogs and mice.

2. Materials

The wild-type GFP cDNA is available for research use from Columbia University and a number of companies (Clontech [Palo Alto, CA], Quantum Biotechnologies, Montréal, Canada). However, wild-type GFP is only useful as a cell marker in a restricted number of organisms, such as nematodes, that live at relatively low temperatures, similar to those experienced by the parental jellyfish. This is because GFP must fold up correctly in order to become fluorescent, and the proportion of proteins that fold correctly decreases at higher temperatures *(1)*. At temperatures above 30°C, the majority of GFP molecules misfold and never become fluorescent. A number of mutations can be introduced into GFP to enhance the proportion of molecules that fold correctly at higher temperatures. Other mutations change the spectral properties of GFP, increase its translation in mammalian or plant cells, and enhance the fluorescence of the chromophore.

From: *Methods in Molecular Biology, Vol. 135: Developmental Biology Protocols, Vol. I*
Edited by: R. S. Tuan and C. W. Lo © Humana Press Inc., Totowa, NJ

Table 1
Variants of GFP and Their Spectral Properties

Variant	Amino acid substitution	Excitation max (nm)	Emission max (nm)	Extinction coeff.[a]	Quantum yield	Ref.
Wild-type	—	395 (475)	508	21.0 (7.15)	0.77	*(11)*
P4	Y66H	383	447	13.5	0.21	*(27)*
P4-3	Y66H	381	445	14.0	0.38	*(10)*
	Y145F					
(BFP)						
S65T	S65T	489	511	39.2	0.66	*(3)*
RSGFP4	F64M	490	505	nd	nd	*(26)*
	S65G					
	Q69L					
T303I	T203I	400	512	nd	nd	*(25)*
E222G	E222G	481	506	nd	nd	*(25)*
C3	F100S	~400	~510	nd	nd	*(28)*
	M254T					
	V164A					
GFP-Bex1	S65T	489	511	nd	nd	*(29)*
	V163A					
GFP-Vex1	V163A	400	512	nd	nd	*(29)*
	S202F					
	T203I					
10C	S65G	513	527	36.5	0.63	*(5)*
	V68L					
(YFP)	S72A					
	T203Y					
W7	F64L	433	475			*(5)*
	S65T					
(CFP)	Y66W	(Minor peak 476)	(Minor peak 505)			
	N146I					
	M153T					
	V163A					
	N212K					
GFPA	V163A	472	509	nd	nd	*(1)*
	S175G					

[a]$10^3 \, M^{-1} \, cm^{-1}$.
Numbers in parentheses indicate minor peaks of excitation or emission.

2.1. Thermostable Mutants

A number of different mutations have been described that enhance the correct folding, and therefore the solubility, of GFP at elevated temperatures. These forms of GFP are preferable as markers in mice or frogs. The combination of V163A and S175G enhances the folding of GFP without altering its spectral characteristics *(1)*. Other mutants, such as S65A/V68L/S72A, S65G/S72A, or F64L/S65T increase the efficiency of folding but also change the properties of the chromophore (*see* **Table 1**) *(2)*.

2.2. Wavelength Mutations

The chromophore of wild-type GFP is formed by an autocyclization of S^{65}-Y^{66}-G^{67}. This has two peaks of absorption, a major peak at 395 nm and a minor one at 475 nm. The chromophore emits light with a peak at 509 nm. The mutation I167T enhances the 475-nm peak of absorption to match that at 395 nm *(3)*. This will increase the apparent brightness of GFP when visualized with blue light, which is less damaging than UV light when living cells and organisms are being observed. Considerable effort has been put into finding mutations that preferentially absorb longer wavelength light, and most of these eliminate the peak of absorption at 395 nm (*see* **Table 1**). Of these mutations, the most widely used is S65T, which changes the peak of absorption to 490 nm and emission to 511 nm. The effectiveness of this mutation can be enhanced further with a second mutation, F64L *(4)*, which greatly increases the proportion of molecules that fold up correctly and thus become fluorescent.

A more recent combination of mutations, S65G/V68L/S72A/T203Y, originally called 10C *(5)* but now more commonly referred to as yellow fluorescent protein (YFP), alters the excitation maximum to 515 nm and the emission maximum to 525 nm (so it still looks green to the naked eye). The emission spectrum of YFP has a very broad shoulder toward the red part of the spectrum and therefore can be distinguished from GFP with filters that collect light at greater than 530 nm. (Indeed, it can be seen with a number of standard rhodamine/tetramethylrhodamine isothionate filter sets.) This form of GFP is almost as bright as GFP and is photostable; therefore, it may prove to be useful as a lineage marker.

There are two other mutations that generate proteins which can be distinguished from GFP and YFP, but are of more limited use because they are less bright. These are Cyan FP (W7) *(5)*, which has an excitation maximum of 434 nm and an emission maximum of 452 nm, and Blue FP (Y66H/Y145F), which has further drawbacks of being photosensitive and very easily bleached.

2.3. Codon Usage and Cryptic Splice Sites

The efficiency with which GFP is made in mammalian cells can be increased by altering the codon usage of the cDNA to match that of human cells *(6)*. This increases the GFP signal in mammalian cells three- to fourfold. Furthermore, there is a cryptic splice site recognized in plant cells that must be removed in order to use GFP as a marker in, for example, Arabidopsis *(7)*.

2.4. Fusion Proteins

GFP can also be used as a fluorescent "tag" to localize other proteins within cells. Because GFP folds up as an autonomous domain, it retains its fluorescent properties as part of a fusion protein. In this respect, it is important to mention that the whole of the GFP molecule must be used; truncating the protein by more than six amino acids at the amino terminus, or by nine amino acids at the carboxyl terminus, destroys its fluorescence. The structural basis for this observation became apparent when the crystal structure of GFP was solved *(5)*. This revealed that the chromophore of the protein lies within a compact "β-barrel" structure, which will be disrupted by deletion of more than a few amino acids at either end of the protein.

GFP seems to have remarkably little effect on the localization of other proteins. It can even be incorporated into microtubules as part of a fusion protein with tubulin. By itself, GFP is distributed fairly evenly between the cytoplasm and the nucleus. GFP has been targeted to the nucleus, to mitochondria, and to the endoplasmic reticulum, and has remained fluorescent in all these locations. Indeed, transgenic plant cells are more healthy when marked with GFP that is localized in the endoplasmic reticulum (ER) compared with wild-type GFP. However, this does not appear to be necessary in transgenic animal cells.

2.5. GFP RNA

Microinjection of synthetic RNA encoding GFP can be very useful in cell lineage studies. In this case, the important parameters are the stability and translational efficency of the RNA. Thus the initiating methionine of the GFP should be placed in a good context for translation, the 3' untranslated region (UTR) should be long enough to protect against degradation, and the RNA should be efficiently capped. We have used the RN3 vector, which has the human β-globin 5' and 3' UTRs, and synthesized the RNA using the Message Machine in vitro transcription system (Ambion, Austin, TX) *(8,9)*.

2.6. GFP cDNAs

The cDNAs for various GFP mutants are commercially available (Clontech, Quantum Biotechnologies). These include both wavelength mutations and codon-optimized versions. Clontech and New England Biolabs (Beverly, MA) also supply anti-GFP antibodies. Antibodies are useful to determine whether any difficulties in visualizing GFP are at the level of protein expression or in the posttranslational cyclization. They can also be used to detect GFP after procedures that destroy its fluorescence, such as *in situ* hybridization.

In our studies on cell lineages in *Xenopus* and in the mouse *(8,9)*, we have used a modified form of GFP (MmGFP) with the following mutations: Val163Ala and Ser175Gly to increase the thermotolerance *(1)* and Ile167Thr *(3)* and Phe64Leu/Ser65Thr *(4)* to enhance absorption in the blue part of the spectrum and also to increase solubility at higher temperatures. The cDNA encoding MmGFP also has a number of codon usage changes *(1)*. This mutant is both thermotolerant and bleaches very slowly. Moreover, when expressed as synthetic mRNA (from the RN3 vector), fluorescence can be detected within 30 min of introducing the RNA into cells. We have also used the blue derivative of MmGFP (changing Phe64L/Ser65Thr to Phe64/Tyr66His and Tyr145His) *(10)*. However, this is of limited value as a cell marker in our hands because this mutant fades quickly, and the UV light required to illuminate it is much more damaging to living cells.

3. Methods

3.1. Introducing GFP into Cells

3.1.1. DNA

GFP has been introduced as a transgene in a number of organisms: nematodes *(1)*, *Dictyostelium (12,13)*, yeast *(14,15)*, *Drosophila (16,17)*, Zebrafish *(18)*, *Arabidopsis*, *Xenopus (8,19)*, and mice *(9)*.

Our experience has been with generating transgenic mice. We have introduced MmGFP into ES cells by electroporation and into the pronuclei of fertilized eggs by microinjection. We have introduced MmGFP under the control of the Cdc2 *(9)* and the EF1α promoters (unpublished results).

CCB embryonic stem (ES) cells were electroporated with a 1:10 ratio of GFP plasmid to a neomycin-resistance selection marker plasmid and were selected for 14 d in 200 μg/mL G418-containing medium. After this time, positive colonies could be identified by epifluorescence microscopy using a fluorescein (FITC) filter set (*see* **Subheading 3.2.1.**). To follow the fate of ES cells expressing GFP we either injected them into the blastocyst of an unlabeled host embryo or aggregated them with unlabeled cells of a morula stage embryo that was subsequently cultured to the blastocyst stage in vitro. Blastocysts in which we could visualize GFP expressing cells by epifluorescence were transferred to foster mothers. The techniques to electroporate, culture, and aggregate ES cells, and transfer to foster mothers, are all standard techniques.

3.1.2. RNA

We have labeled *Xenopus* and mouse cells using synthetic GFP RNA. Whereas injection of RNA into *Xenopus* eggs and embryos is a fairly widely used technique for which good descriptions are available, cytoplasmic injection into mouse blastomeres to mark them for lineage analysis is a much less well described technique. Therefore we will outline our protocol here.

Lineage labeling of single mouse blastomeres can be achieved by injecting synthetic RNA encoding GFP into the cytoplasm. We performed all microinjections using an inverted microscope with bright field illumination. Embryos were placed in a drop of M2 medium supplemented with 4 mg/mL BSA (M2 + BSA) *(20)* in the center of a depression slide and covered in light paraffin oil. Embryos were immobilized in the drop of medium by applying gentle suction to a heat polished holding pipet via a microinjector (Narashige, Tokyo, Japan) filled with heavy paraffin oil. Injection pipets were made from capillaries with an internal filament (Clark Biomedical Supplies, Reading, UK) pulled on an automatic micropipet puller (Sutter Instruments, Novato, CA). RNA was loaded into the back of the capillary. For injections of RNA into blastomeres of cleavage stage embryos, the tip of the injecting needle was slightly broken against the holding pipet just before injection. For small cells requiring use of fine needles for high-injection survival rates, improved membrane penetration can be achieved by using negative capacitance without breaking the tip of the needles *(21)* (Zernicka-Goetz, M., Pedersen, R. A., and Evans, M. J., unpublished observations). With the exception of the RNA solution, the entire system is air filled. Microinjections were performed using a constant flow system (Transjector 5246, Eppendorf [Hamburg, Germany]) in which the RNA solution is under constant positive pressure. This is important to prevent back flow of culture medium or cytoplasmic contents into the injection needle, which might degrade the RNA. Constant pressure also helps to prevent clogging of the injection needle. RNA was dissolved in RNAse-free water (previously treated with diethyl pyrocarbonate and autoclaved) at a concentration of 1 μg/μL. The injection needle was slowly introduced into the blastomere and injection pressure was applied. The pressure of injection was chosen depending on the size of the tip of the needle. As soon as the cytoplasm became visibly more translucent around the tip of the needle, the needle was pulled out of the cell. After injection, embryos were placed in KSOM embryo culture medium *(22)* at 37°C, 5% CO_2. When labeling individual cells of preimplantation mouse embryos, one should be aware that intercellular bridges can remain between sister cells throughout or even beyond the succeeding cell cycle. For example, in order to label the descendants of only one blastomere at the 8-cell stage, the injection should be done either just before or just after division to the 16-cell stage *(23)*. The advantage to using GFP in this case is that the number of injected cells can easily be seen by epifluorescence soon after injection. MmGFP fluorescence can be detected as soon as 40 min after injection of its mRNA, and the signal increases during the following few hours.

3.2. Visualizing GFP (see Notes 4.1–4.3)

3.2.1. Filter Sets

Although some types of GFP can be detected with a "standard FITC" filter set, filter sets vary from company to company, are not necessarily optimized for GFP, or indeed may not be suitable for all types of GFP. Therefore, it is advisable to invest in a custom filter set (Chroma Technology, Brattleboro, VT; Omega Optical, Brattleboro) designed specifically for the GFP and the experimental system to be used. The choice of filter set will depend on the spectral characteristics of the GFP being used and on whether the GFP signal needs to be distinguished from another dye. A filter set is made up of an excitation filter, a dichroic mirror that separates the excitation (incident) light from the emitted light, and an emission filter. Filters are referred to by the wavelength at which they transmit light. In some gene-expression or cell-lineage analyses, GFP will be the only fluorescent signal that needs to be detected. In this case, the filter set should be chosen to maximize the GFP signal and to distinguish it from any autofluorescence. Therefore, an excitation filter designed for the peak of absorbance should be used. The range of wavelengths around the peak of absorption (the bandwidth) transmitted by the filter can be specified. With living tissues, it is best to minimize this to reduce any photo damage while still maximally exciting GFP. We use filters with a bandwidth of 20 nm. The dichroic mirror will normally cut off light approx halfway between the excitation and the emission light wavelengths. The emission filter will determine how much light reaches the detection device (a camera or the eye) and therefore should have as wide a bandwidth as possible. If GFP is the only fluorescent signal, then one can choose a long-pass filter, which lets through all light above a certain wavelength. This can be useful for GFP because it allows the eye (but not a monochrome detector such as a video or single-chip CCD camera) to distinguish between the green color of GFP and, for example, the more yellow autofluorescence of cellular flavin molecules, which are also excited by blue light. However, in multiple labeling experiments, the bandwidth of the excitation filter will be determined by the spectral characteristics of the other dyes.

For example: The peak of absorbance of MmGFP is 490 nm and emission is at 510 nm. Therefore, for a single labeling experiment, we would use the following filter set: excitation filter: 470 nm, 20-nm bandwidth; dichroic mirror, 505 nm; emission filter, 510-nm long pass. (If another, longer wavelength dye is also being used, we would use a 530-nm filter with a 20-nm bandwidth.)

3.2.2. Microscopy

Cell lineage analysis necessarily requires three-dimensional (3-D) analysis of embryos through time (4-D analysis). This is optimally achieved using a computerized microscope that is able to take a series of Z sections at set time intervals and, for mammalian embryos, a temperature-controlled stage. Some confocal microscopes have this facility, and epifluorescence microscopes can be set up to do this by using a computer-controlled Z-axis drive and a fluorescence shutter to minimize photodamage and bleaching. These systems are beyond the scope of this review but guides can be found in *(24)*. Alternatively, samples can be manually imaged at specified intervals. We have imaged GFP expressing cells in the same living *Xenopus* embryo over several days using either epifluorescence or confocal microscopes without perturbing development *(8)*. In the mouse, we have followed blastomere lineages using confocal microscopy throughout preimplantation development in vitro *(9)*.

Imaging living mouse embryos is most easily achieved in the culture dish on an inverted microscope, preferably using a specially designed dish with a cover slip on the bottom. However, this can also be done on an upright microscope. This requires the embryos to be placed under a cover slip on a depression slide.

3.2.3. Fixation

The great advantage to GFP is that it can be visualized in living cells. This is essential to gain a dynamic view of a process over time, but it is also an advantage when only analyzing the endpoint of an experiment, because there is no need to fix and stain cells and therefore no risk of introducing artifacts by the fixation process. However, costaining for other markers may require fixed tissue. Some fixation conditions do not preserve GFP fluorescence very well. These include any regimes that involve reducing conditions because the GFP chromophore is formed by oxidation. In addition, some organic solvents, such as acetone, appear to reduce GFP fluorescence. GFP fluorescence is resistant to standard paraformaldehyde and glutaraldehyde cross-linking protocols and to protease treatment.

4. Notes

4.1. Lack of Signal

1. Check whether the DNA is present by PCR or whether RNA is present by reverse transcriptase PCR. The latter technique will also show whether full-length GFP mRNA is being made or whether the RNA is being spliced. For RNA injections, it is very important to use RNase-free solutions and equipment to prevent degradation, and a constant pressure injection system to prevent medium (and RNase) being aspirated into the injection needle.
2. Check whether the protein is present by using antibodies for immunohistochemistry or immunoblotting. If no protein is detectable but the RNA is present, the codon usage may be incorrect. Alternatively, the protein may be detrimental to the cell. This was an important factor in plant cells that was overcome by using a form of GFP targeted to an organelle (the ER).
3. If the protein is present but is not detectably fluorescent, it could be because the wrong form of GFP is being used (e.g., a thermosensitive form in a mammalian system). Alternatively, the wrong filter set may be being used, or a more sensitive detector may be required.
4. If fluorescence disappears after fixation, another regime should be tried. In addition, certain sealants, such as nail varnish, can reduce GFP fluorescence.

4.2. Photobleaching

GFP is comparatively resistant to photobleaching but will fade with sustained illumination. This state may be reversible by further illumination at 405 nm.

4.3. Autofluorescence

Cells and tissues can contain chemicals that are also excited by blue light. This is especially true of tissues such as the gut and yolk. Autofluorescence can usually be distinguished from GFP fluorescence because it is present across a much wider part of the spectrum. Thus, autofluorescence can usually be seen with a "rhodamine/TRITC filter set." With a long-pass emission filter, GFP will appear as a very characteristic shade of green, whereas autofluorescence will be yellow-green or yellow.

Acknowledgments

We are grateful to Professor Martin Evans for his help and advice, and to the Lister Institute of Preventative Medicine, the Wellcome Trust and the Cancer Research Campaign for Financial Support.

References

1. Siemering, K. R., Golbik, R., Sever, R., and Haseloff, J. (1996) Mutations that suppress the thermosensitivity of green fluorescent protein. *Curr. Biol.* **6,** 1653–1663.
2. Cubitt, A. B., Heim, R., Adams, S. R., Boyd, A. E., Gross, L. A., and Tsien, R. Y. (1995) Understanding, improving and using green fluorescent proteins. *Trends Biochem. Sci.* **20,** 448–455.
3. Heim, R., Cubitt, A. B., and Tsien, R. Y. (1995) Improved green fluorescence. *Nature* **373,** 663–664.
4. Cormack, B. P., Valdivia, R. H., and Falkow, S. (1996) FACS-optimized mutants of the green fluorescent protein (GFP). *Gene* **173,** 33–38.
5. Ormö, M., Cubitt, A. B., Kallio, K., Gross, L. A., Tsien, R. Y., and Remington, S. J. (1996) Crystal structure of the Aequorea victoria green fluorescent protein. *Science* **273,** 1392–1395.
6. Zolotukhin, S., Potter, M., Hauswirth, W. W., Guy, J., and Muzyczka, N. (1996) A "humanized" green fluorescent protein cDNA adapted for high-level expression in mammalian cells. *J. Virol.* **70,** 4646–4654.
7. Haseloff, J., Siemering, K. R., Prasher, D. C., and Hodge, S. (1997) Removal of a cryptic intron and subcellular localisation of green fluorescent protein are required to mark transgenic Arabydopsis plants brightly. *Proc. Natl. Acad. Sci. USA* **94,** 2122–2127.
8. Zernicka-Goetz, M., Pines, J., Ryan, K., Siemering, K. R., Haseloff, J., and Gurdon, J. B. (1996) An indelible lineage marker for Xenopus using a mutated green fluorescent protein. *Development* **122,** 3719–3724.
9. Zernicka-Goetz, M., Pines, J., Dixon, J., Hunter, S., Siemering, K. R., Haseloff, J., and Evans, M. J. (1997) Following cell fate in the living mouse embryo. *Development* **124,** 1133–1137.
10. Heim, R. and Tsien, R. Y. (1996) Engineering green fluorescent protein for improved brightness, longer wavelengths and fluorescence resonance energy transfer. *Curr. Biol.* **6,** 178–182.
11. Chalfie, M., Tu, Y., Euskirchen, G., Ward, W., and Prasher, D. (1994) Green fluorescent protein as a marker for gene expression. *Science* **263,** 802–805.

12. Gerisch, G., Albrecht, R., Heizer, C., Hodgkinson, S., and Maniak, M. (1995) Chemo-attractant controlled accumulation of coronin at the leading edge of Dictostilium cells monitored using green fluorescent protein-coronin fusion protein. *Curr. Biol.* **5,** 1280–1285.

13. Moores, S. L., Sabry, J. H., and Spudich, J. A. (1996) Myosin dynamics in live Dictyostelium cells. *Proc. Natl. Acad. Sci. USA* **93,** 443–446.

14. Nabeshima, K., Kurooka, H., Takeuchi, M., Kinoshita, K., Nakaseko, Y., and Yanagida, M. (1995) p93dis1, which is required for sister chromatid separation, is a novel micro-tubule and spindle pole body-associating protein phosphorylated at the Cdc2 target sites. *Genes Dev.* **9,** 1572–1585.

15. Doyle, T. and Botstein, D. (1996) Movement of yeast cortical actin cytoskeleton visual-ized in vivo. *Proc. Natl. Acad. Sci. USA* **93,** 3886–3891.

16. Davis, I., Girdham, C. H., and O'Farrell, P. H. (1995) A nuclear GFP that marks nuclei in living Drosophila embryos; maternal supply overcomes a delay in the appearance of zygotic fluorescence. *Dev. Biol.* **170,** 726–729.

17. Kerrebrock, A. W., Moore, D. P., Wu, J. S., and Orr Weaver, T. L. (1995) Mei-S332, a Drosophila protein required for sister-chromatid cohesion, can localize to meiotic cen-tromere regions. *Cell* **83,** 247–256.

18. Amsterdam, A., Lin, S., Moss, L. G., and Hopkins, N. (1996) Requirements for green fluorescent protein detection in transgenic zebrafish embryos. *Gene* **173,** 99–103.

19. Tannahill, D., Bray, S., and Harris, W. A. (1995) A Drosophila E(spl) gene is "neuro-genic" in Xenopus: A green fluorescent protein study. *Dev. Biol.* **168,** 694–697.

20. Whittingham, D. G. (1971) Culture of mouse ova. *J. Reprod. Fert.* **14(suppl),** 7–21.

21. Beddington, R. and Lawson, K. A. (1990) Clonal analysis of cell lineages, in Post-Implantation Mammalian Embryos: A Practical Approach, (Copp, A. J. and Cockcroft, D. L., eds.), IRL, Oxford, UK.

22. Lawitts, J. A. and Biggers, J. D. (1993) Culture of preimplantation embryos. *Methods Enzymol.* **225,** 153–164.

23. Pedersen, R. A., Wu, K., and Balakier, H. (1986) Origin of the inner cell mass in mouse embryos: cell lineage analysis by microinjection. *Dev. Biol.* **117,** 581–595.

24. Inoué, S. and Spring, K. R. (1997) *Video Microscopy*, Plenum, New York.

25. Ehrig, T., O'Kane, D. J., and Prendergast, F. G. (1995) Green-fluorescent protein mutants with altered fluorescence excitation spectra. *FEBS Lett.* **365,** 163–166.

26. Delagrave, S., Hawtin, R. E., Silva, C. M., Yang, M. M., and Youvan, D. C. (1995) Red-shifted excitation mutants of the green fluorescent protein. *Biotechnology* **13,** 151–154.

27. Heim, R., Prasher, D. C., and Tsien, R. Y. (1994) Wavelength mutations and posttransla-tional autoxidation of green fluorescent protein. *Proc. Natl. Acad. Sci. USA* **91,** 12,501–12,504.

28. Crameri, A., Whitehorn, E. A., Tate, E., and Stemmer, W. P. C. (1996) Improved green fluorescent protein by molecular evolution using DNA shuffling. *Nat. Biotechnol.* **14,** 315–319.

29. Anderson, M. T., Tjioe, I. M., Lorincz, M. C., Parks, D. R., Herzenberg, L. A., Nolan, G. P., and Herzenberg, L. A. (1996) Simultaneous fluorescence-activated cell sorter analysis of two distinct transcriptional elements within a single cell using engineered green fluorescent proteins. *Proc. Natl. Acad. Sci. USA* **93,** 8508–8511.

28

Cell Lineage Analysis

X-Inactivation Mosaics

Seong-Seng Tan, Leanne Godinho, and Patrick P. L. Tam

1. Introduction

In female mammals, one of the two X chromosomes in each and every embryonic cell is randomly inactivated during embryogenesis, leaving only one active X chromosome per cell and thereby maintaining dosage parity with males *(1)*. This natural phenomenon, known as X inactivation or lyonization, provides an ideal nonsurgical method of producing mosaicism among otherwise identical cells of the embryo. We have created a line of mice (H253) by pronuclear injection of a DNA fragment containing the *lacZ* reporter gene under the control of an ubiquitously active promoter (HMG CoA reductase; *see* **ref. *2***). Breeding and chromosome hybridization experiments (FISH) with line H253 confirmed that the *lacZ* gene is inserted into the X-chromosome *(3,4)*. Male members of this line express the X-linked *lacZ* gene in all cells of the embryo, including the developing nervous system *(5)*. Ubiquitous expression of the *lacZ* gene produces a nuclear-localized β-gal protein, which, after histochemical reaction with X-gal substrate, is visualized as a blue reaction product visible in the cell nuclei in whole-mount embryos, organs, or tissue sections. In hemizygous females, only one of the two Xs carry the *lacZ* transgene and this *lacZ*-bearing chromosome will be randomly turned off in approx 50% of cells *(3)*. The marking process is indelible and heritable (**Fig. 1A**). The *lacZ* is integrated in the genome and every time a cell divides, the *lacZ* gene and either its active or inactive status is passed onto its progeny, forming a marked clone. The initial process of random inactivation occurs early in embryogenesis and is virtually completed by 9.5 d of gestation (E9.5) before organogenesis commences for most tissues.

Transgenic mosaics offer a number of advantages over other conventional methods of marking embryonic cells. First, all cells are genetically identical and are only functionally different with respect to the developmentally-neutral β-gal protein *(6)*. Second, because X inactivation is essentially random, the starting numbers of marked versus unmarked cells in any given tissue are approximately equal, therefore allowing quantitative estimates of cell mixing and proliferation *(4)*. Third, this mouse line provides an infinite supply of experimental material. In contrast, tissues obtained from

From: *Methods in Molecular Biology, Vol. 135: Developmental Biology Protocols, Vol. I*
Edited by: R. S. Tuan and C. W. Lo © Humana Press Inc., Totowa, NJ

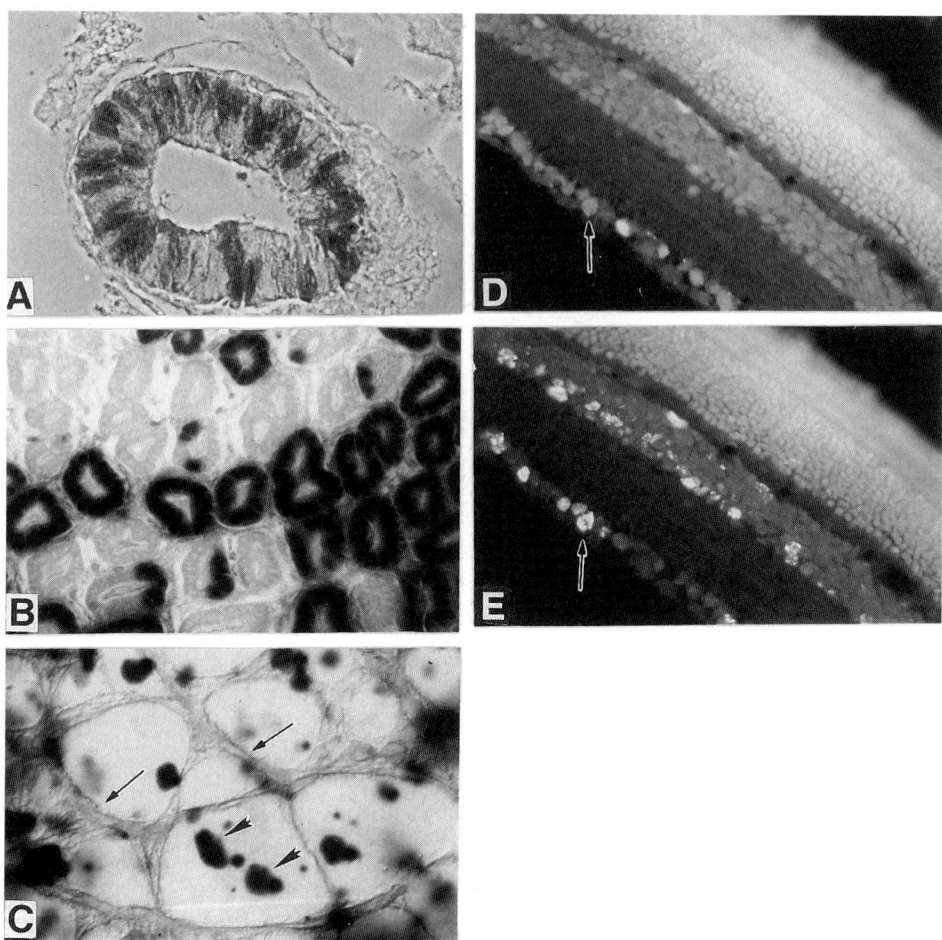

Fig. 1. (**A**) Section of an E6.5 embryo showing that X-inactivation has occurred in the epiblast of a female embryo hemizygous for the *lacZ* transgene. (**B**) Two clonally distinct populations of cells (β-gal-positive and β-gal-negative) are clearly seen in a transverse section of adult stomach glands from hemizygous females. (**C**) The X-gal reaction is compatible with antibody staining for glial fibrillary acidic protein (GFAP). Arrowheads point to ganglion cells in the whole-mounted retina, and glia (small arrows) form a meshwork on the retinal surface. (**D**) Immunocytochemical detection of β-gal using antibodies revealed ubiquitous expression of the transgene in retina of homozygous transgenic mice. The transgene is expressed from either X chromosome in these animals. Arrow indicates the ganglion cell layer. (**E**) Same section as (D) showing double labeling for another antigen (BrdU). Arrow indicates that ganglion cell expressing the transgene has also taken up the thymidine analog.

other cell marking techniques are precious commodities, data collection from each set of experiments is constrained by a variable and limited supply of experimental outcomes.

Most importantly, transgenic mosaics provide a way of distinguishing between clonally related cellular populations and their borders, permitting inferences to be made about cellular dispersion patterns during embryonic development. Thus a blue cell and a white cell lying next to each other clearly cannot be members of the same clone (**Fig. 1B**).

Analysis can therefore be systematically carried out in any tissue of choice across a spectrum of developmental stages (*see* **Note 1**).

This chapter outlines the protocol developed for detection of β-gal in neural tissue of X-inactivation mosaic mice using histochemical and immunocytochemical techniques, and the use of these mice for studying clonal development and cell dispersion patterns in a model system (e.g., retina).

2. Materials

2.1. Buffers

1. Fix buffer: Phosphate buffer (0.1 M, pH 7.3) containing 2 mM MgCl$_2$ and 5 mM EGTA. To make buffer at pH 7.3, make stock solutions A and B. Stock A is 0.2 M solution of NaH$_2$PO$_4$ (27.6 g/L) and stock B is 0.2 M solution of Na$_2$HPO$_4$ (28.4 g/L). Adding 23 mL of stock A to 77 mL of stock B diluted to 200 mL gives 0.1 M buffer. Fix buffer is made by adding 400 μL of 1 M MgCl$_2$ and 2 mL of 0.5 M EGTA to 198 mL of 0.1 M phosphate buffer.
2. Wash buffer: This consists of fix buffer containing 0.01% (w/v) sodium desoxycholate, 0.02% (w/v) Nonidet P-40 (BDH, Poole, UK). To 100 mL fix buffer, add 10 mg (w/v) sodium desoxycholate powder (deoxycholic acid; Sigma D-6750, Sigma, St. Louis, MO) and 20 mg Nonidet NP-40 (1 drop = approx 20 mg).

2.2. Fixative

Paraformaldehyde/glutaraldehyde mixture. Typically, make up a 4% paraformaldehyde/0.2% glutaraldehyde solution for intracardial perfusion or for immersion fixation. To make 4% paraformaldehyde in 0.1 M buffer, add 4 g paraformaldehyde powder into 50 mL water, heat to 60°C while slowly adding drops of 1 M NaOH until clear solution is obtained. Make up to 100 mL with 0.2 M phosphate buffer. Add glutaraldehyde (25% stock) to above to obtain a 0.2% concentration (*see* **Note 2**). Cool in ice bucket for at least 1 h prior to usage. Freshly prepared fixative is preferred although this fixative, stored for up to 48 h at 4°C, has been successfully used for less-critical experiments (*see* **Note 3**).

2.3. X-Gal Reaction Solution

To 10 mL wash buffer, add 250 μL X-gal (4-chloro-5-bromo-3-indolyl-α-D-galactopyranoside (40 mg/mL stock dissolved in dimethylformamide), 100 μL of a 500-mM potassium ferrocyanide solution, and 100 μL of a 500-mM potassium ferricyanide solution (*see* **Note 4**). Prepare just before use.

3. Methods

3.1. Fixing of Embryos

Dissect out embryos of appropriate ages into buffered saline. After removal of extraembryonic membranes, use a heat polished Pasteur pipet to transfer embryos into the cooled fixative (2 mL in a 35-mm Petri dish). For larger embryos, use a flame-polished pipet with a wide mouth. If significant amounts of buffered saline have also been carried over, replace fixative while leaving embryos in the Petri dish (*see* **Note 5**). Leave embryos to fix on a slowly rotating platform at room temperature. Replace fixative with wash buffer and allow to rinse for at least 20 min. Allow for two further changes of wash buffer at 20-min intervals.

3.2. Fixing of Adult Tissues

Transcardial perfusion fixation provides efficient and rapid penetration of fixative into all organs. Animals are anesthetized (avertin, dosage 0.015/g body weight, i.p., *see* **Note 6**) and their thoracic cages opened. Inject 1000 U of heparin (in 0.5 mL of phosphate buffer) into the left ventricle. This is followed by insertion of a blunted 30-gage needle into the left ventricle and rinsing of the circulation with fix buffer, at 100-mm Hg pressure, for 30–60 s. The animal is then perfused with fixative for 10 min, at 100-mm Hg pressure. Selected organs are then dissected out and postfixed for a further 30 min using the same fixative.

3.3. Tissue Sectioning and X-Gal Reaction

Organs are cryoprotected by immersion in 30% sucrose solution (in phosphate buffer) overnight. Sections are cut using a cryostat (10 μm) or a freezing microtome (100–200 μm) and washed 3X in washing buffer, at 20 min intervals. Incubation in X-gal reaction solution is best carried out overnight in a 37°C oven (*see* **Note 7**). To ensure that the sections do not dry out, an adequate volume of reaction solution should be added (e.g., 1 mL for a 35-mm Petri dish). Gentle shaking on a rocking platform is optional; this ensures even penetration of the substrate. An intense blue reaction product should be evident by 12 h. Sections are then rinsed in wash buffer and mounted on slides for dehydration in graded ethanols, cleared, and cover-slipped using Permount. The reaction product is stable in mounted sections. A light bluish background stain is sometimes seen in certain tissues (*see* **Note 8**).

3.4. X-Gal Reaction and Immunocytochemistry

To ascertain and confirm the identity of X-gal (Diagnostic Chemicals Ltd., Charlottetown, Canada) positive cells, it is often necessary to use cell-type specific antibodies. The X-gal reaction product is compatible with a number of commercially raised antibodies, especially those that tolerate a high percentage of glutaraldehyde in the fixative. For example, to reveal astrocytes in the ganglion-cell layer of the retina, rabbit antibodies (DAKO Corp., Carpinteria, CA) to glial fibrillary acidic protein (GFAP) can be used on retina whole mounts following the X-gal reaction (**Fig. 1C**). Briefly, the reacted retina is incubated in 0.25 *M* Tris-HCl buffer (pH 7.4) containing 1% Triton X-100 and 10% goat serum for 1 h. This is followed by primary anti-GFAP antibody (1:100 dilution), diluted with 0.25% Tris-HCl with 1% Triton X-100 for 2 d at 4°C. The tissue is then washed in 0.25 *M* Tris-HCl buffer 3 times, 20 min apart, incubated in biotinylated antirabbit secondary antibody (diluted with 0.25% Tric-HCl with 1% Triton X-100) for 2 d, followed by 3 washes in buffer, at 20-min intervals. In the meantime, 5 μL solution A is mixed with 5 μL solution B (Vectastain) in 500 μL 0.25 *M* Tris-HCl with 5 μL Triton X and allowed stand for 1 h. The tissue is incubated in the A + B mixture for 1 h and washed 3 times, 20 min each. Finally, the tissue is reacted with diaminobenzidine (0.5 mg/mL) in the presence of 0.01% H_2O_2 in 0.25 *M* Tris-HCl.

3.5. Immunocytochemical Detection of β-Galactosidase

The X-gal reaction procedure is often incompatible with other immunocytochemical procedures, resulting in loss of antigenicity. To overcome this, β-gal can be detected using commercial antibodies (Cappel, ICN Australia, Seven Hills, NSW, or Sigma).

To optimize detection of β-gal, glutaraldehyde should be omitted from the fixative, and fixation should be carried out for at least 4 h, preferably overnight. Following fixation and washing in 0.1 *M* phosphate buffer, sections are incubated overnight with rabbit anti-β-gal antibody (Cappel, 1:1000 dilution) in buffer containing 0.2% Triton X-100. Secondary antibodies are biotinylated goat antirabbit IgG (Vector Labs, Burlingame, CA, 1:400 dilution) and avidin-conjugated fluorescein (**Fig. 1D**). Concomitant immunocytochemical detection can be carried out on the same section, using primary antibodies from a second species directed against the antigen of interest (e.g., 5-bromo-2'-deoxyuridine, BrDU) and revealed using avidin-conjugated rhodamine (**Fig. 1E**). The conditions optimal for the second antibody should be established varying antibody dilutions and incubation times.

3.6. Analysis of Clonal Development and Cell Dispersion

Using the retina as a model system, we have used X-inactivation mosaics for studying clonal restriction *(7)*. The retina develops from the optic vesicle, but the transgene is randomly switched off in approx 50% of all proliferating cells of the vesicle by 9.5 d postcoitum (dpc), prior to the onset of retinal neurogenesis. Therefore, all mature retinal neurons will descend from the initial mosaicism of the neuroepithelium, and to the extent that mature retinal cells remain clonally segregated, they should appear as distinct clusters of blue vs white cells. Using this approach, we were able to establish that certain cell classes (rod photoreceptor, Müller, and bipolar cells) are aligned in the radial axis, implying a strictly radial mode of cell dispersion. In contrast, the cone photoreceptor, horizontal, amacrine, and ganglion cells are tangentially displaced with respect to radial columns (*see* **Note 9**).

4. Notes

1. The fact that 50% of all embryonic cells are marked in these mice has both advantages and drawbacks. One significant advantage of X-inactivation mosaics is the ability to draw a global picture of the system under study, the net result being an interplay of coherent growth and cell mixing. However, a major drawback occurs where there is conspicuous mixing that results in blurring of clonal boundaries. Hence, the strength of these mice is that two large groups of cells are unambiguously *not* clonally related, permitting generalizations about clonal segregation or mixing that are not readily drawn from other approaches. Other labeling techniques typically label only a small or tiny proportion of the tissue under study, from which erroneous generalizations may be drawn.
2. Paraformaldehyde is an efficient fixative for rapid tissue penetration but inefficient for X-gal histochemistry. Thus, there is a constant need to trade off a good histochemical reaction for poor tissue preservation. In contrast, the presence of glutaraldehyde increases the intensity of the X-gal reaction. Glutaraldehyde, however, only penetrates slowly into the tissue. Glutaraldehyde can be used at the 0.2–2.0% range, a higher concentration being desirable if, after the X-gal reaction, the tissue is processed for immunocytochemistry using antibodies that have been raised with glutaraldehyde-conjugated haptens. However, because many commercial antibodies do not bind well to tissues fixed with excess glutaraldehyde, it pays to bias the fixation toward conditions that are optimal for the antibody and accept the diminished X-gal reaction intensity as a trade-off.
3. Preparation of fixatives should be carried out inside a well-ventilated fume hood. Vapors from these fixatives are harmful. Gloves and masks should be worn during handling.

4. The X-gal stock and staining solutions should be prepared and stored in glass containers, as the dimethylformamide tends to etch the inside of plastic surfaces, producing a cloudy solution. However, for short transit times, polypropylene containers can be used.

5. Embryos can be fixed for 5 min (preimplantation stages) up to 45 min (12.5 dpc organogenesis stage). Beyond 12.5 dpc, the penetration of the fixative and X-gal reaction solution is inefficient and only superficial structures will be stained. To assist visualization of deeper structures, the embryo can be split into two sagittal halves prior to fixing using a scalpel blade. Excessive fixation will diminish the intensity of the X-gal histochemical reaction. Older embryos tend to produce a higher background staining, especially where the pH of the fixative and X-gal reaction solution exceeds 7.3. Background is evident as a light bluish tinge on the skin and is invariably present in the choroid plexus, a labyrinthic structure developing within the ventricles of the nervous system.

6. Avertin is made by adding 10 g tribromoethanol with 10 mL tertiary amyl alcohol (2-methyl-butan-2-ol) and diluted to 2.5% using boiling phosphate-buffered saline. Stocks should be stored wrapped in foil at 4°C.

7. For deeper penetration of the X-gal reaction substrate, staining should be carried out at 30°C for 48 h. The reaction can be continued further at 4°C for another 48 h for optimal staining. X-gal crystals sometimes appear at the end of the reaction; these can be dissolved using a solution of 0.5% dimethylformamide in phosphate buffer. If an immediate result is required from freshly cut sections, vacuum treatment of sections in warm X-gal solution for 15 min will result in rapid penetration and will provide a blue stain within 1 h.

8. Endogenous activity producing a light blue stain is sometimes seen in choroid plexus, the gut mucosa, sebaceous glands, testis, and lymphoid tissues in bone marrow and spleen. To minimize background, it is important to keep the pH of the fixative and X-gal reaction solution as close to pH 7.3 as possible.

9. The utility of X-inactivation mosaics for these studies depends on the fact that X inactivation of the transgene is essentially completed in most embryonic tissues prior to the onset of organogenesis (from 11.5 dpc onward). Because tissue sections can sometimes provide misleading cell positions from aberrant sectioning angles, it is vital that such analysis be accompanied by whole-mount organ analysis wherever possible. Alternatively, three-dimensional reconstruction from serial sections can be carried out.

Acknowledgment

The authors thank the National Health and Medical Research Council for supporting our work, Frank Weissenborn and Rachael Parkinson for technical assistance, and Peter Rowe for critical comments on the manuscript.

References

1. Lyon, M. (1961) Gene action in the X-chromosome of the mouse (*Mus musculus L.*). *Nat. London* **190,** 372–373.

2. Tam, P. P. L. and Tan, S.-S. (1992) The somitogenetic potential of cells in the primitive streak and the tail bud of the organogenesis-stage mouse embryo. *Development* **115,** 703–715.

3. Tan, S.-S., Williams, E. A., and Tam, P. P. L. (1993) X-chromosome inactivation occurs at different times in different tissues of the post-implantation mouse embryo. *Nat. Genet.* **3,** 170–174.

4. Tan, S.-S., Faulkner-Jones, B. E., Breen, S. J., Walsh, M., Bertram, J. F., and Reese, B. E. (1995) Cell dispersion patterns in different cortical regions studied with an X-inactivated transgenic marker. *Development* **121,** 1029–1039.

5. Tan, S.-S. and Breen, S. J. (1993) Radial mosaicism and tangential dispersion both contribute to mouse neocortex development. *Nature* **362,** 638–640.

6. Beddington, R., Morgernstern, J., Land, H., and Hogan, A. (1989) An *in situ* transgenic enzyme marker for the midgestation mouse embryo and the visualisation of inner cell mass clones during early organogenesis. *Development* **106,** 37–46.

7. Reese, B. E., Harvey, A. R., and Tan, S.-S. (1995) Radial and tangential dispersion patterns in the mouse retina are cell-class specific. *Proc. Natl. Acad. Sci. USA* **92,** 2494–2498.

29

Retroviral Cell Lineage Analysis in the Developing Chick Heart

Robert G. Gourdie, Gang Cheng, Robert P. Thompson, and Takashi Mikawa

1. Introduction

The use of genetically modified retroviruses as agents for gene transfer has provided developmental biology with some of its most elegant and compelling research stories of the last few years. The growing recognition of the utility of retrovirally derived constructs is perhaps unsurprising given that over millennia, evolution has honed the shuttling of genetic material into the native stock-in-trade of this unique class of animal pathogen. One application in which retroviral approaches have particularly shone is in analysis of cell lineage during embryogenesis (reviewed in **refs. *1–5***). In this chapter, we describe a methodology that has been successfully utilized for tracing cell fate in the developing chick heart using replication-defective constructs derived from the avian spleen necrosis virus (SNV) *(6)*. This work has provided a fresh perspective of ventricular muscle revealing it has a compound organization based around the proliferative activity of facets of clonally related myocytes *(7)*. It has also enabled the resolution of a number of longstanding questions in cardiovascular developmental biology, including the origin of cells comprising coronary vascular tissues *(8)* and the cardiac conduction system *(9)*.

Naturally occurring retroviruses have RNA genomes and undertake propagation strategies that involve integration of their own genetic material into the DNA of the infected host *(10)*. This distinctive maneuver is dependent on a virally encoded RNA-to-DNA reverse transcriptase and also requires that the host enters the DNA replication (S) phase of the cell cycle a few hours after viral infection. A practical consequence of this latter requirement is that retroviruses efficiently and stably infect cells actively undergoing division, such as are common in embryos, but are less successful at sustainably targeting quiescent cell populations typical of differentiated tissues. Once integrated, wild-type retroviruses propagate using two main strategies. The first of these tactics involves active production and release of virions to infect new host cells. The second, more passive strategy involves multiplication of the integrated proviral DNA in the normal course of host DNA replication and cell division. This ability to "hitch a ride" with the host genome has been exploited as a tool to study cell lineage by

From: *Methods in Molecular Biology, Vol. 135: Developmental Biology Protocols, Vol. I*
Edited by: R. S. Tuan and C. W. Lo © Humana Press Inc., Totowa, NJ

various workers using a range of different retroviruses. By judicious deletion of viral genes encoding structural proteins, the retrovirus can be inhibited from undertaking active production of new infectious particles but are left able to stably integrate its genetic material into the DNA of the host. Inclusion of a nonfunctional gene within the replication-defective construct, such as *E. coli* β-galactosidase, enables later identification of cells incorporating viral sequence, thus providing a more-or-less permanent marker for tracing the progeny of the originally infected cell.

The principal steps in the generation and use of replication-defective retroviruses are:

1. construction of the viral vector;
2. establishment of an efficient viral packaging cell line; and
3. development of a protocol for viral targeting the cells and/or tissues of interest.

Protocols for the last of these three steps are the primary concern of this techniques chapter. Details of the molecular construction of the SNV-based vectors used here and generation of the packaging cell lines used to propagate these replication-defective viruses can be found in earlier publications *(11)*. For those interested in exploring other method chapters on approaches to retroviral marking of cell lineage, alternative technical guides are detailed in **refs. *1*** and ***4***. The information to be given on equipment is offered as a guide and should not be taken as a literal endorsement of specific brands or models.

2. Materials

1. Equipment for virus packaging, harvesting, and concentration: CO_2 cell culture incubator (Forma Scientific, Marietta, OH); P2 level culture hood (SG-400; The Baker Co., Sanford, ME); ultracentrifuge (RC-70; Kendro Laboratory Products, Newtown, CT) with an AH-629 swinging bucket rotor; refrigerated tabletop centrifuge (TJ-6; Beckman, Coulter Inc., Fullerton, CA); inverted transmission light microscope (CK-2; Olympus, America Inc., Melville, NY); Fisher Scientific water bath (Pittsburgh, PA); liquid nitrogen cryogenic tank and storage canes (Biocane20; Barnstead/Thermolyne, Dubuque, IA).
2. Disposable materials for virus packaging, harvesting, concentration, and titer assay: 100-mm cell culture dishes (cat. no. 3003; Falcon, Los Angeles, CA); 60-mm grid dishes (cat. no. 25011; Corning Glassworks, Corning, NY); Sorvall centrifuge tubes (cat. no. 03141); sterile capped 50-mL conical tubes (cat. no. 2098; Falcon); Nalgene 2-mL cryogenic storage vials (cat. no. 5000-0020, Nalgene, Rochester, NY); screw-cap 2-mL tubes (D-51588; Sarstedt).
3. Solutions for virus packaging cell culture, passaging, and storage: High-glucose Dulbecco's modified Eagle's medium (DMEM; cat. no. 11995-040; Gibco-BRL, Grand Island, NY) containing 7% fetal bovine serum (cat. no. A-111-D; HyClone Laboratories, Logan, UT) and 1% streptomycin and penicillin (cat. no. 15070-063; Gibco-BRL); 0.05% trypsin/EDTA (cat. no. 25300-054; Gibco-BRL); filter-sterilized 10% DMSO (D-2650; Sigma Chemical Co., St. Louis, MO) in culture media; sterile 10X Dulbecco's phosphate-buffered saline (PBS) (cat. no. 14200-075; Gibco-BRL).
4. Solutions for viral titer assay: Culture media and trypsin/EDTA (Gibco-BRL); filter-sterilized (0.2 μm; cat. no. 25925-45; Corning) 100 mg/mL polybrene-hexadimethrine bromide (H9268; Sigma) stock in distilled water (store at 4°C in the dark).
5. X-Gal reaction: PBS (P4417; Sigma), pH 7.4; X-Gal (B71800; Research Products International, Mt. Prospect, IL) stock: 40 mg/mL dissolved in dimethyl formamide (stored at –20°C in dark); X-Gal buffer: in 1X PBS, containing 35 m*M* potassium ferrocyanide [$K_4Fe(CN)_6$-

3H$_2$O] (cat. no. 3114-01; J. T. Baker, Phillipsburg, NJ), 35 mM potassium ferricyanide [K$_3$Fe(CN)$_6$] (cat. no. 3104-01; J. T. Baker), 2 mM MgCl$_2$ (stored at room temperature); X-Gal working solution: X-Gal buffer containing 1 mg/mL X-Gal.

6. Paraformaldehyde fixation: Mix 2 g of paraformaldehyde (T353; Fisher) with 80 mL of distilled water, add 20 μL of 10 M NaOH and heat to 60°C until paraformaldehyde dissolves. Add 10 mL of 10X PBS, adjust pH to 7.4 and bring final vol to 100 mL with distilled water. Store at room temperature and good for 1–2 wk.

7. Equipment for *in ovo* infection (*see* **Fig. 1B**): Humidaire egg incubator (cat. no. 1202; GQF Manufacturing Co.); Leica M10 stereomicroscope (Leica Microsystems Inc., Deerfield, IL); Forstec fiber optic light source; Picospritzer II (General Valve Corp., Fairfield, NJ) pressure regulator and attached N$_2$ tank; Leitz micromanipulator (cat. no. 031626; Leitz); needle holder (cat. no. 50-10-110-1; General Valve); David Kopf micropipet puller (Model 700c; David Kopf Instruments, Tujunga, CA); 1-mm external/ 0.8-mm internal diameter glass micropipet tubes (cat. no. TW100F-4; World Precision Instruments, Sarasota, FL); microloader tips for backfilling glass micropipets with virus (cat. no. 5342 956.003; Eppendorf-5 Prime, Boulder, CO); Parafilm and sticky tape; 1-mL syringe and 30-gage needle; dissection tools, including a tungsten needle, fine forceps, and iridectomy scissors (all instruments purchased from World Precision Instruments); egg cup.

8. Immunohistochemical solutions: Polyfin embedding paraffin (cat. no. 19280-01; Electron Microscopy Sciences, Fort Washington, PA); PBS containing 0.01% sodium azide and 1% bovine serum albumin; TENG-T buffer (10 mM Tris, 5 mM EDTA, 150 mM NaCl, 0.25% gelatin, 0.05% Tween-20, pH 8.0); antibodies against β-galactosidase: Mouse monoclonal antibody from Oncogene (cat. no. OB02; Oncogene Research Products, Cambridge, MA) and rabbit polyclonal antibody from 3 Prime (Boulder, CO) (cat. no. 5307-063100); second antibodies conjugated to biotin, fluorochromes, or peroxidase and streptavidin-conjugates are obtained from various sources, including Dako Corp. (Carpinteria, CA), The Jackson Laboratory (Bar Harbor, ME), and Amersham (Amersham, UK).

3. Methods

1. Virus packaging cell culture conditions: 5% CO$_2$, 95% O$_2$, 37°C.

2. Culture of virus packaging cells: Thaw a frozen aliquot of the D17.2G packaging cell line stably transfected with a replication-defective retroviral vector into a 100-mm culture dish in DMEM culture medium. Change medium after 24 h to wash out remaining DMSO and allow cells to grow to approx 90 % confluence. This typically takes approx 3 days. When the plate is ready to split, aspirate medium, add 3 mL sterile PBS buffer, rinse cells, and remove buffer. Add 2 mL prewarmed (37°C) trypsin/EDTA solution and incubate at 37°C for approx 2 min. A quick check under an inverted light microscope will help determine if cells have detached from the substrate. Remove trypsin/EDTA, add 10 mL fresh DMEM-containing serum to terminate further proteolysis and aliquot cells into new 100-mm dishes in medium.

3. Storage of virus packaging cells: Grow cells to approx 90% confluence on a 100-mm plate and lift from the substrate by trypsinization as described in the previous step. Transfer the cell suspension to a 15-mL conical tube and pellet by centrifugation at 200g for 10 min. Remove supernatant and add DMSO cryoprotectant. For best results, a high cell density (>5 × 10^5 cells/mL) is recommended. Transfer to cryogenic storage vials and allow to freeze slowly in a –20°C freezer (approx 1°C/min). Place the frozen vials in a –70°C freezer for 1–2 h and then transfer to liquid nitrogen for long-term storage.

4. Virus harvesting from packaging cells: Plate approx 10^6 packaging cells into two 100-mm dishes, a cell density that typically results in plates at 100–120% confluence 72 h later.

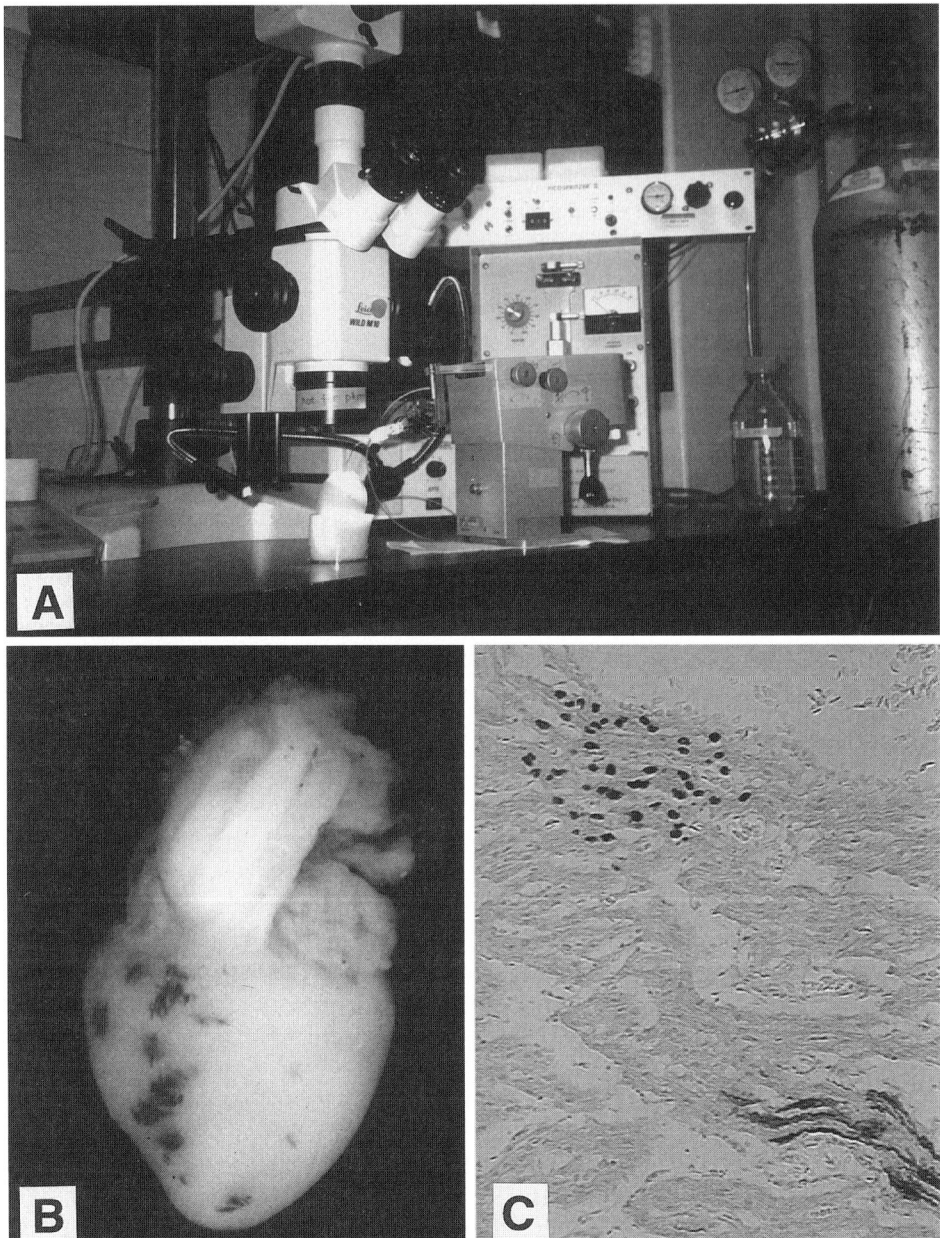

Fig. 1. (*See* color plate 6 appearing after p. 258.) **(A)** Basic virus microinjection equipment includes a stereomicroscope, a micromanipulator, pressure regulator, and light source. **(B)** Whole mount of a heart from a chick embryo, at 14 d of embryonic incubation (ED 14), displaying β-galactosidase-positive sectors (blue) following reaction with the chromogenic substrate X-Gal. These discrete blue zones of cardiac tissue represent clonal groups of myocytes infected with replication-defective retrovirus. **(C)** Two independent groups of myocytes (i.e., clones), each infected with a different replication-defective construct expressing either nuclear-localized (spots) or cytoplasmic (stripes) β-galactosidase activities. Multiple infections with viruses encoding distinctive marker expressions provide a useful test of clonality in cell lineage tracing studies (*see* **Note 12**).

Although not optimal for maintaining the packaging line, over-dense cultures tend to produce virus at higher titers. In parallel, plate at least one separate dish for continuous subconfluent passage of packaging cells (*see* **Note 4**). One day before harvesting retrovirus, aspirate medium and add 15–20 mL of fresh medium to each dish; 24 h later, collect media from the two dishes, transfer into 50-mL conical tubes, and centrifuge at 200g for 10 min at room temperature to remove cellular debris. After low-speed centrifugation, carefully remove the upper phase of the supernatant, leaving the pellet at the bottom of the tube intact. We typically expect viral titers (*see* **step 6**) of approx 10^7 infectious units per 100-mm dish to be obtained from such supernatants.

5. Virus concentration: Ultracentrifugation is used to further concentrate viral particles. Place the harvested supernatant in 36-mL centrifuge tubes and ultracentrifuge at 30,000g for 2–4 h at 25°C. After ultracentrifugation, carefully remove supernatant, leaving a small whitish pellet with no more than 50 µL of accompanying medium. Gently resuspend the pellet in the minimum volume of medium using a 20-µL pipeter. Measure the volume of the suspension using a calibrated 20-µL pipeter during transfer from the centrifuge tube into a sterile 2-mL tube. Add polybrene to a final concentration of either 100 µg/mL (for *in ovo* microinjection) or 10 µg/mL (for titer assay or infection of cells in vitro).

6. Viral titer assay: It is suggested that separate titer assays be carried out for each culture dish from which virus is harvested. The day before assay, plate out uninfected D17 host cells at a concentration of 2×10^5 cells per 60-mm gridded dish. Add 10–100 µL vol of supernatant from virus packaging cells (**step 4**) to host cells in the presence of polybrene at 10 µg/mL. If titering concentrated retrovirus (**step 5**), add 1–10 µL vol of virus-containing suspension instead. After incubating D17 cells with virus for 2 h, aspirate, add fresh medium, and culture for a further 48 h. After 2 d, fix the D17 cells in 2% paraformaldehyde for 10 min and wash with PBS 3X for 5–10 min each. Remove PBS, add X-Gal working solution, and incubate the plate at 37°C for 2 h to overnight in the dark. The virus titer can be determined by counting blue cells per grid from the formula: virions/mL = (number of blue cells $\times$ 500 $\times$ 1000)/added volume in microliters.

7. *In ovo* infection by microinjection (**Fig. 1A**): Chick eggs are incubated at 38°C in a humidified incubator until the required Hamburger and Hamilton (HH) stage is reached. For targeting cardiac tissues, HH stages 12–18 (embryonic days of incubation 2–3.5) are optimal. The glass capillary micropipets used for microinjection are pulled to form needles with inner diameters of approx 10 µm. Backfill the glass needles with 1–3 µL of concentrated virus suspension (containing 100 µg/mL polybrene) using a Gilson 10-µL pipet and a microloader tip. Secure the virus-filled micropipet into the pressure-regulator-coupled needle holder mounted on the micromanipulator. Before microinjection, wipe the external shell of the egg with 70% ethanol and cut a small circular window at the blunt end of the egg. Peel back the shell membrane (chorionic) covering the embryo using a fine pair of tweezers and gently slit open the embryonic membranes and tissues in the path of the glass micropipet with a sharp tungsten needle. After micromanipulating the needle into position under a stereomicroscope, pressure inject 5–20 nL vol of concentrated virus into the embryonic tissues being targeted. Reseal the eggs with tape and return to the incubator to allow further development.

8. X-Gal detection of β-galactosidase (**Fig. 1B**): At the end of the egg incubation period, collect embryos (or dissected parts of embryos) and fix in 2% paraformaldehyde for 3 h. Rinse these samples in three changes of PBS for a 1–2 h and then incubate with X-Gal working solution (Diagnostic Chemicals Ltd., Charlottetown, Prince Edward Island, Canada) at 37°C overnight. Infected tissues should turn blue in the presence of the chromogenic X-Gal substrate, and photographs of whole-mounted specimens can be taken at this time (**Fig. 1B**). Samples will remain stable in 70% ethanol for a few weeks at 4°C.

9. Tissue histology: For further processing, standard paraffin-embedding and sectioning of tissue samples is recommended. Briefly, dehydrate samples at room temperature in two 1-h changes each of 80, 95, 100% ethanol and two subsequent changes of 100% toluene. Following dehydration, place the samples in a 50:50 toluene:paraffin mixture at 60°C and pure paraffin at 60°C for 1 h. Embed samples in fresh paraffin, cut at 5–10 μm and mount on glass slides. Sections can be histologically counterstained with eosin (HT110-3-32; Sigma), cover-slipped and viewed using standard light microscopy protocols.

10. Immuno-detection of β-galactosidase (**Fig. 1C**): Reasons for using immunohistochemical detection of β-galactosidase are briefly discussed in **Note 13**. To achieve optimal immunolabeling, it is best to avoid prior reaction with X-Gal. At the end of the egg incubation period, collect embryonic tissues, fix, paraffin-embed, section, and mount on microscope slides as outlined in **step 9**. Deparaffinize sections by soaking in two changes of 100% xylene for 30 min and partially rehydrate at room temperature in two 5-min changes each of 100, 95, 80, and 70% ethanol. After the graded ethanol sequence, rinse sections in distilled water 2X for 5 min. To block nonspecific staining, immerse sections in TENG-T. If immunoperoxidase staining is being done, a prior treatment with 2% H_2O_2 for 10 min is also required. After completion of blocking steps, rinse sections in PBS and apply β-galactosidase antibodies at 1:150 dilution in PBS/0.1% sodium azide/1% BSA and incubate sections at room temperature overnight. Following treatment with primary antibody, rinse sections in three 5-min changes of PBS and detect localized primary antibody using species-appropriate fluor- or enzyme-tagged second antibodies according to standard immunohistochemical protocols.

4. Notes

1. All procedures must be conducted under appropriate biohazard conditions to avoid exposure, because replication-defective versions of SNV-based viruses have shown to infect a broad range of hosts, including cells from humans and other primates. Normally, provisions for infectious agent biohazards are mandated by law in the country in which the work will be undertaken and should be explored thoroughly with the authorities at your home institution.

2. All solutions and equipment that directly contact the virus should be subjected to autoclaving or equivalent sterilization procedures.

3. Overgrowth of virus packaging cells may decrease viral titer. It is important to split cells before confluence is reached. Periodic checks of viral titer in continuously passaged packaging-cell cultures is probably judicious. Titer tends to decrease over many passages in any case.

4. Packaging cell supernatants should be tested from time-to-time for replication-competent revertants produced by recombination of the vector with endogenous helper viruses. This is an extremely rare event, but its probability rises with continuous passaging of packaging cells. The simplest test is to inoculate D17 host cells as outlined for viral titer assay (**step 6**), and then wash out residual virus by rinsing in three changes of sterile PBS. Subsequently, passage the infected D17 host cells and allow to incubate for a further 5 d. If media harvested from these 5-d cultures is able to infect freshly plated D17 cells, not previously exposed to virus, it is likely that helper virus is present.

5. The virus is extremely fragile and not particularly stable at any temperature, including when frozen. Harvested retrovirus is best used within a few hours. Over 90% of the titer can be lost at room temperature 24 h after harvesting.

6. Tissues should be fixed in paraformaldehyde for as short a time as possible. Long periods in paraformaldehyde reduces X-Gal signal. For hearts from chick embryos incubated for

14–18 d (ED 14–ED 18), we normally fix for 3 h at room temperature. Avoid paraformaldehyde concentrations above 2%, as this may also inhibit X-Gal reaction. Thorough washing with PBS following fixation is also strongly recommended. Gently flushing the ventricular chambers of intact hearts with PBS using a 1-mL syringe and fine hypodermic needle aids removal of residual fixative.

7. 70% ethanol works as an alternative fixative to paraformaldehyde. A number of common fixatives, including Dent's and Bouin's fixative, cannot be used prior to X-Gal reaction.

8. If fixation needs to be avoided completely, X-Gal reaction can be carried out on freshly thawed frozen sections.

9. Delaying color reaction for longer than 24 h after fixation can result in decreased X-Gal signal.

10. Under ideal conditions, retroviral cell lineage analysis depends on analyses of the clonal progeny of an individual cell stably incorporating a single integrated viral genome. This is a difficult ideal to achieve on an egg-to-egg basis during routine microinjection *in ovo*. Nonetheless, some control over numbers of viral particles microinjected into embryos can be achieved by knowledge of two factors: viral titer and volume of viral suspension microinjected. A rule of thumb is that a 1-nL (approx 0.125-mm diameter) droplet from a suspension with a titer of 10^6 should (on average) contain a single SNV particle.

11. Typically, glass needles need replacing after a few rounds of microinjection.

12. For certain types of study, it is necessary to determine whether discrete clusters of virally infected cells represent true clones or are mixtures of cells derived from the infection of multiple mother cells. This is a particular concern when replication-defective retroviruses are used to analyze the differentiation potential of specific cells or populations of cells. Coinfection with viruses mediating distinct spatial patterns of β-galactosidase activity within cells (e.g., nuclear-localized vs cytoplasmic) is a useful control for assessing the degree of mixing between different viral clones at varying levels of tissue infection (*see* **Fig. 1C** and **ref. 9**).

13. Although X-Gal reaction is useful for unequivocal detection of extended domains of β-galactosidase activity in whole-mounted specimens, the extent of diffusion of the chromogenic substrate may vary—particularly in larger pieces of tissue (e.g., ED 18 chick hearts). Similarly, X-Gal reaction may be affected by spatial variations in amount of fixation within larger tissue samples. Although somewhat more time-consuming to perform, immunostaining of histological sections by antibodies against β-galactosidase provides an extremely sensitive assay of the total extent of viral infection within a tissue block (**Fig. 1C**). In our hands, immunohistochemical detection also appears less affected by the vagaries of overfixation than X-Gal-based detection.

14. Dual detection of β-galactosidase and a second antigen of interest is most easily done using double immunohistochemical labeling protocols. Such colocalization of β-galactosidase with a second cellular marker enables powerful phenotypic analyses of differentiation patterns within groups of virally infected cells to be undertaken.

Acknowledgment

This work is supported by grants from the American Heart Association (RGG, TM), March of Dimes Birth Defects Foundation (RGG), National Institutes of Health (HL56728 [RGG], HL50582 [RPT], HL54128 and HL56987 [TM]) and the National Science Foundation (RGG). TM is an Irma T. Hirschl Scholar. RGG is a Basil O'Connor Scholar (March of Dimes) and is the author to whom correspondence should be directed.

References

1. Cepko, C. L., Ryder, E. F., Austin, C. P., Walsh, C., and Fekete, D. M. (1993) Lineage analysis using retrovirus. *Methods Enzymol.* **225,** 933–963.
2. Hyer, J. and Mikawa, T. (1997) Retroviral techniques for studying organogenesis with a focus on heart development. *Mol. Cell. Biochem.* **172,** 23–35.
3. Mikawa, T., Hyer, J., Itoh, N., and Wei, Y. (1996) Retroviral vectors to study cardiovascular development. *Trends Cardiovasc. Med.* **6,** 79–86.
4. Price, J. (1993) Introduction of genes using retroviral vectors, in *Essential Developmental Biology a Practical Approach*, Oxford University Press, Oxford, UK, pp. 179–190.
5. Sanes, J. R. (1989) Analysing cell lineage with a recombinant retrovirus. *Trends Neurosci.* **12,** 21–28.
6. Dougherty, J. P. and Temin, H. M. (1986) High mutation rate of a spleen necrosis virus-based retrovirus vector. *Mol. Cell. Biol.* **168,** 4387–4395.
7. Mikawa, T., Borisov, T. A., Brown, A. M. C., and Fischman, D. A. (1992a) Clonal analysis of cardiac morphogenesis in the chicken embryo using a replication-defective retrovirus: I. Formation of the ventricular myocardium. *Dev. Dyn.* **193,** 11–23.
8. Mikawa, T. and Gourdie, R. G. (1996) Pericardial mesoderm generates a population of coronary smooth muscle cells migrating into the heart along with ingrowth of the epicardial organ. *Dev. Biol.* **174,** 221–232.
9. Gourdie, R. G., Mima, T., Thompson, R. P., and Mikawa, T. (1995) Terminal diversification of the myocyte lineage generates Purkinje fibers of the cardiac conduction system. *Development* **121,** 1423–1431.
10. Varmus, H. E. (1982) Form and function of retroviral proviruses. *Science* **216,** 812–820.
11. Mikawa, T., Fischman, D. A., Dougherty, J. P., and Brown, A. M. C. (1991) *In vivo* analysis of a new LacZ retrovirus vector suitable for lineage marking in avian and other species. *Exp. Cell. Res.* **195,** 516–523.

30

Dynamic Labeling Techniques for Fate Mapping, Testing Cell Commitment, and Following Living Cells in Avian Embryos

Diana K. Darnell, Virginio Garcia-Martinez,
Carmen Lopez-Sanchez, Shipeng Yuan, and Gary C. Schoenwolf

1. Introduction

Specific groups of cells in avian embryos can be marked using a variety of techniques and labels. Which type of technique is used depends on the purpose of the experiment. For example, individual cells can be labeled by iontophoretic injection (*1*) or retroviral infection (*2*), and clones of cells derived from these individuals can be evaluated to gain information on the state of commitment or potency of their labeled ancestor. Protein or RNA can be labeled by immunocytochemistry or *in situ* hybridization, respectively, to assess cell differentiation or to infer the functional role of specific genes from their temporal and spatial patterns of expression. Groups of cells that have been labeled in place with extracellular microinjections of fluorescent dyes, or tissues obtained from whole embryos that have been soaked in fluorescent dyes and subsequently grafted to unlabeled embryos, can be followed over time to yield information on cell fate or commitment. In addition to these considerations that relate to the purpose of the experiment, there are several other considerations when selecting labels, as different labels have different attributes (**Tables 1** and **2**) that may help or hinder the investigator in interpreting experimental results. These include the method of visualization (fluorescence, immunocytochemical, histological), subcellular localization (nuclear, cytoplasmic, membranous), compatibility with other markers or techniques (i.e., for double labeling), specificity (species, regional, tissue), and reliability. This chapter will deal principally with strategy and the labeling techniques used with early avian embryos to ascertain cell fate and commitment and with a number of specific labels that can be selected to facilitate the interpretation of these types of experiments.

Traditionally, studies in avian embryos on ascertaining cell fate and commitment have involved the grafting of cells between embryos, principally the grafting of quail cells to chick embryos, generating quail–chick transplantation chimeras (*see* Chapters 35–37). Following incorporation of the grafted cells and further development of the resulting chimera, the progeny of the grafted cells can be identified and their fates can be determined. Homotopic, isochronic grafts yield information on cell fate, whereas

From: *Methods in Molecular Biology, Vol. 135: Developmental Biology Protocols, Vol. I*
Edited by: R. S. Tuan and C. W. Lo © Humana Press Inc., Totowa, NJ

Table 1
Attributes of Quail-Specific Markers

Marker	Localization	Tissue specificity	Key advantages	Key disadvantages
Feulgen staining	Nuclear	All cells	Excellent histological quality (resulting from use of Vakaet's fixative)	Difficult to distinguish isolated quail cells from surrounding chick cells Can be lost in endothelial cells Cannot be used with living tissue Cannot use to label whole mounts Incompatible with *in situ* hybridization
QCPN	Nuclear	All cells	Easy to distinguish isolated quail cells from surrounding chick cells Compatible with *in situ* hybridization	Cannot be used with living tissue
QH-1	Cytoplasmic	Endothelial/endocardial cells	Specific for quail endothelial cells Compatible with *in situ* hybridization	Cannot be used with living tissue

Table 2
Attributes of Fluorescent Markers

Marker	Localization	Signal strength	Filter set	Tissue specificity of antibody	Key advantages (disadvantages)
CFSE	Cytoplasmic	Fair/strong	Fluorescein	All cells	Permanently labels with antibody/peroxidase and can be seen in sections (Weak fluorescence after culture) (Incompatible with *in situ* hybridization à la Wilkinson)
CRSE	Cytoplasmic	Strong	Rhodamine	All cells	Permanently labels with antibody/peroxidase and can be seen in sections Strong fluorescence after culture Compatible with *in situ* hybridization
DiA	Cell membrane	Strong	Rhodamine	All cells	Strongly colored: allows visual confirmation of the injection (Requires photoconversion to be seen in sections)
DiI	Cell membrane	Strong	Rhodamine	All cells	Strongly colored: allows visual confirmation of the injection (Requires photoconversion to be seen in sections)
DiO	Cell membrane	Strong	Fluorescein or rhodamine	All cells	Strongly colored: allows visual confirmation of the injection (Requires photoconversion to be seen in sections)
Rhodamine 123	Mitochondrial membrane	Strong	Rhodamine	Enhanced in endothelial cells	Permanently labels with antibody/peroxidase and can be seen in sections When used with anti-rhodamine antibody, it labels nonendothelial cells poorly if at all; this can be an advantage or disadvantage depending on the experiment Compatible with *in situ* hybridization

heterotopic or heterochronic grafts yield information on cell commitment. In the classic approach, graft cells are taken from Japanese quail embryos (*Coturnix coturnix japonica*), placed in the chick embryo (*Gallus gallus domesticus*) and identified later using the Feulgen-Rossenbeck histological stain, whereby groups of quail cells are identified by the characteristic appearance of their nucleolus *(3)*. Drawbacks to this technique are: chick cells can also have a similar nucleolar appearance to that of quail cells; not all quail cells in a particular section readily show the characteristic nucleolar marker; and some cell types, such as endothelial cells, may lose the marker, all of which makes the exact identification of individual cells as chick or quail problematic. In addition, the preferred fixative for the Feulgen-Rossenbeck stain, Vakaet's (*see* **Subheading 2.**), is usually incompatible with several other labeling techniques, including immunocytochemistry and *in situ* hybridization. To overcome these drawbacks, the quail cells in quail–chick transplantation chimeras can be identified using a quail-specific antibody *(4)*, QCPN, which labels an epitope in the nucleus, makes individual quail cells easily discernible from their neighboring chick cells and is compatible with other labeling techniques. Both of these methods for labeling quail cells allow most grafted cells to be identified in sections after fixation (and QCPN can be used in fixed, whole mounts as well). A second quail-specific antibody, QH-1, specifically labels the cytoplasm of quail endothelial cells and can be substituted for or used with QCPN when endothelial cells are the subjects of interest *(5)*. The Feulgen-Rossenbeck histological procedure and the immunocytochemical protocol for use with these and other antibodies are described in **Subheadings 3.4.–3.6.** and **3.9.**

Identification of grafted quail cells using either Feulgen staining or antibodies (QCPN, QH-1) does not allow graft cells to be followed in the living embryo. As an alternative or additional procedure, grafts from either chick or quail can be fluorescently labeled and followed by fluorescence microscopy. This can be accomplished by using any of several fluorescent markers: CFSE, CRSE, DiA, DiI, DiO, Rhodamine 123, or some combination of these, either with or without detection of a quail-specific marker. In addition, some of these dyes (CFSE, CRSE, and Rhodamine 123) can be converted into permanent markers after fixation of embryos by using antibodies to fluorescein *(6)* or rhodamine and peroxidase immunocytochemistry (**Table 2**), which allows the grafted cells to be visualized by conventional light microscopy in either whole-mount or sectioned embryos. Choosing which dyes to use, like choosing which technique of labeling to use, depends on the purpose of the experiment. For example, if the purpose of the experiment is to follow grafted cells in living chick or quail embryos, one could use chick or quail grafts labeled with any fluorescent marker (i.e., CFSE, CRSE, DiA, DiI, DiO, Rhodamine 123) or any combination of fluorescent markers (**Fig. 1A,B**). Time-lapse video microscopy of embryos with fluorescent grafts can be very informative (**Fig. 2**). The technique of time-lapse video microscopy is described in **Subheading 3.3.** If the purpose is to identify all endothelial cells derived from a graft, one could either transplant quail grafts to chick hosts and subsequently label graft cells with QH-1, or one could use a graft from a donor embryo (either chick or quail) labeled with Rhodamine 123, and use the anti-rhodamine antibody (this antibody binds to rhodamine, but when used in combination with Rhodamine 123, it marks [for an unknown reason] principally rhodamine-labeled endothelial cells—not rhodamine-labeled, nonendothelial cells). Both nuclei and cytoplasm of endothelial cells can be

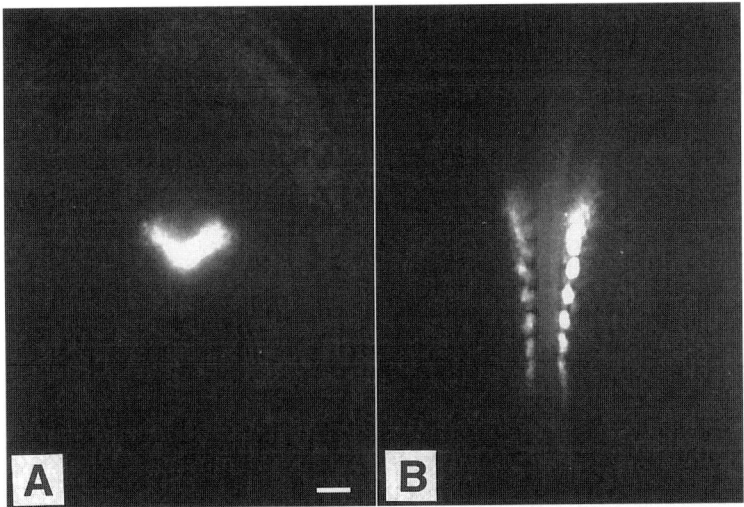

Fig. 1. Use of grafts labeled with fluorescent dyes to follow labeled cells over time. Whole mounts viewed with epifluorescence illumination. (**A**) Two hours after grafting to the primitive streak of an embryo at the late gastrula/early neurula stage. The graft was labeled with CRSE and cells are beginning to migrate bilaterally. (**B**) Same embryo as in (A) 24 h after grafting. Cells have contributed bilaterally to the somites flanking the neural tube. Bar = 100 μm.

highlighted. Cells derived from quail grafts can be detected in chick hosts using QCPN antibody, which marks the nuclei of all quail cells, including endothelial cells (**Fig. 3A**). Sections of such embryos can be cut and examined in detail (**Fig. 3B**). Then, the same sections can be labeled with QH-1 antibody to reveal the cytoplasm of endothelial cells derived from the quail graft (**Fig. 3C**). If the purpose of the experiment is to identify all cells (i.e., both nonendothelial and endothelial cells) derived from a graft, then one could (a) use a quail graft in a chick host and QCPN antibody to label all cells derived from the graft (**Fig. 4A,B**); (b) use a graft (either chick or quail) labeled with CFSE and anti-fluorescein (not illustrated); or (c) use a graft (either chick or quail) labeled with CRSE and anti-rhodamine (not illustrated; note: the anti-rhodamine antibody when used in combination with CRSE labels all cells labeled with CRSE, not just endothelial cells). With labeling experiments, one is basically limited only by one's creativity (**Fig. 5A–F**).

If fluorescent markers are chosen for a particular experiment, one must also choose between grafting cells labeled with these markers and microinjecting these markers into localized regions of embryos, or the two approaches can be combined. Grafting allows one to include quail markers in the experiment; the labeled region is typically larger with grafting than with microinjection and specific *layers* of the embryo can be more precisely targeted with grafting without concern for diffusion of dyes microinjected extracellularly. Injected markers can be used to label only a few or many cells by varying the size of the microinjection, and microinjection avoids the possibility of experimental artifact caused by wounding of the host, delayed healing of the graft, slight differences in the stages of the donor and host or variation in the sites from which the graft is obtained and subsequently placed. The dyes CFSE, CRSE, DiA, DiI, DiO, and Rhodamine 123 all can be used as injected fluorescent markers. Each of these dyes

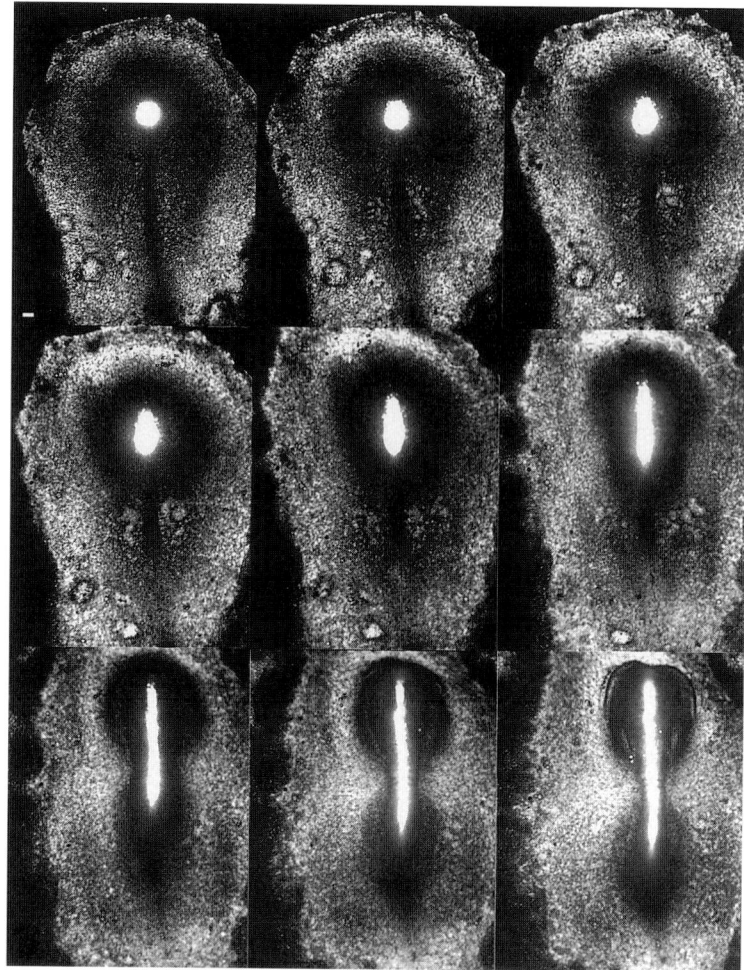

Fig. 2. Use of grafts labeled with fluorescent dyes to follow labeled cells over time. Time-lapse videomicroscopy of whole mounts viewed with epifluorescence illumination. The nine panels show a single embryo after grafting Hensen's node from a donor embryo at the late gastrula/early neurula stage. The donor embryo was labeled with Rhodamine 123 and grafted homotopically and isochronically to an unlabeled host. Images are taken at 1–2-h intervals. The graft contributed labeled cells to three midline populations: the ectodermal floor plate of the neural tube, the mesodermal notochord, and the endodermal roof of the gut. Reprinted (with permission) from **ref. 7**. Bar = 100 μm.

can be injected extracellularly as a small bolus (**Fig. 6A**), after which a localized population of cells in the immediate vicinity of the injection site becomes labeled. The fates and morphological movements of these cells can be traced by following the fluorescent marker over time (**Fig. 6B**). In our hands, membrane-intercalating dyes (DiA, DiI, DiO) give the same labeling pattern after culture as do labels taken into the cells' cytoplasm (CFSE, CRSE, and Rhodamine 123). A mixture of these labels can be injected to take advantage of their different properties. For example, a colored label (e.g., DiI, which has a red color, visible with the naked eye) can be included so that the actual injection

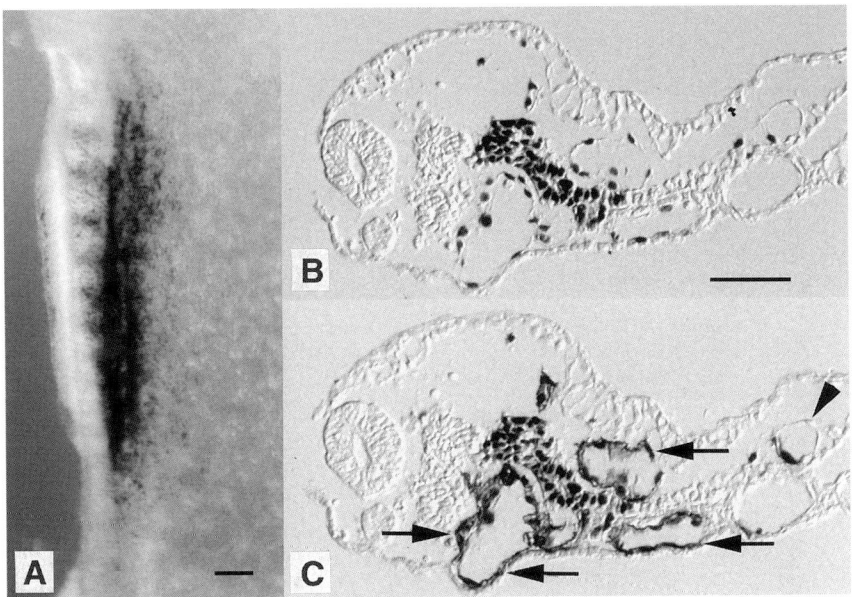

Fig. 3. Labeling nuclei and cytoplasm of quail cells contributing to the cardiovascular system in a chimera. **(A)** The right side of a whole mount 24 h after grafting to the primitive streak at the late gastrula/early neurula stage (the left side has been removed for processing in another manner). The graft was obtained from a quail and placed in a chick, and graft cells were marked after the experiment was terminated with the QCPN antibody to reveal their contributions. **(B)** Transverse section through the trunk of the embryo shown in **Fig. 3A**. Note the antibody labels quail-cell nuclei. **(C)** Same section after relabeling of the section with QH-1 antibody. This antibody labels the cytoplasm of quail-derived endothelial cells. *Arrows* indicate labeled cytoplasm of quail-derived endothelial cells. *Arrowhead* indicates an unlabeled nucleus of an endothelial cell (chick derived); note that its cytoplasm is also unlabeled. (A): Bar = 100 µm; (B,C): Bar = 50 µm.

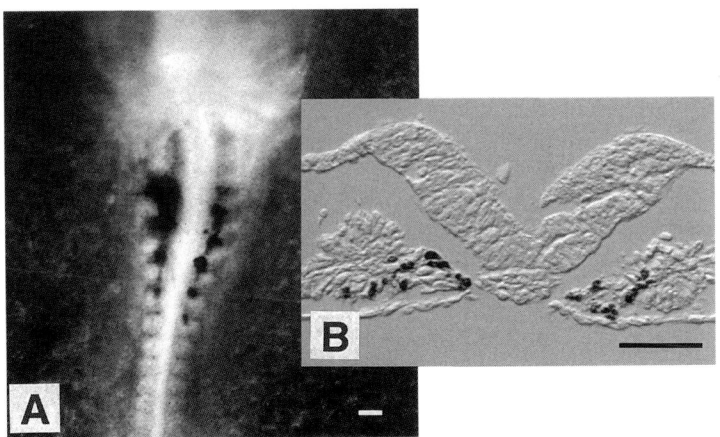

Fig. 4. Grafting cells to map all of the derivatives of the primitive streak. **(A)** Whole mount 24 h after grafting to the primitive streak at the late gastrula/early neurula stage. The graft was obtained from a quail and placed in a chick, and graft cells were marked after the experiment was terminated with the QCPN antibody to reveal all of their contributions (mainly bilaterally to the somites). **(B)** Transverse section through the trunk of the embryo shown in **(A)**. Labeled cells occupy the paired somites. (A): Bar = 100 µm; (B): Bar = 50 µm.

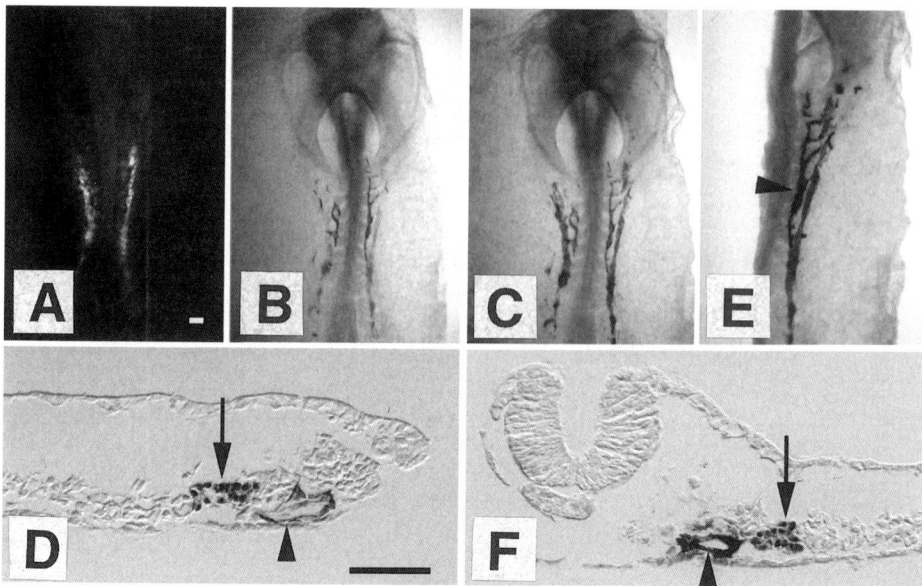

Fig. 5. (*See* color plate 8 appearing after p. 258.) Creativity in the use of dynamic labeling techniques. (**A**) Whole mount 24 h after grafting to the primitive streak at the late gastrula/early neurula stage. The graft was obtained from a quail donor labeled with Rhodamine 123 and CRSE and was placed in an unlabeled chick host. Graft cells are viewed in the living embryo with epifluorescence. (**B**) Same embryo after labeling grafted cells with anti-rhodamine antibody/peroxidase without cobalt-nickel intensification. The *cytoplasm* of grafted cells labels with a brown color. (**C**) Same embryo after labeling with QCPN antibody/peroxidase with cobalt-nickel intensification. Nuclei of grafted cells now label with a black color. The embryo was cut midsagittally with a razor blade. One half was embedded in paraffin and sectioned transversely. (**D**) Transverse section of one half of the embryo shown in **Fig. 5C**. Arrow indicates grafted cells within the intermediate/lateral plate mesoderm; note their black nuclei (QCPN-positive) surrounded by brown cytoplasm (anti-rhodamine-positive, specifically marking cells labeled with CRSE). *Arrowhead* indicates grafted cells forming the dorsal aorta; note their brown cytoplasm (anti-rhodamine-positive, specifically marking cells labeled with Rhodamine 123 and CRSE; nuclei of endothelial cells are highly flattened and continue to be colored black, but they are difficult to identify at this low magnification). (**E**) The other one half of the embryo shown in **Fig. 5C**, which after cutting was labeled with QH-1 antibody/ peroxidase and cobalt-nickel intensification. The more medial grafted cells (*arrowhead*) have turned black. (**F**) Transverse section through the one half shown in (E); note the cytoplasm of the cells contributing to the dorsal aorta is now colored black (*arrowhead*). Cells within the intermediate/lateral plate mesoderm retain their black nuclei and brown cytoplasm (*arrow*). (A–C,E): Bar = 100 μm; (D,F): Bar = 50 μm.

can be monitored visually using conventional light microscopy. Membrane (e.g., DiI) and cytoplasmic (e.g., CFSE) labels, visible with different filter sets, can be mixed for internal verification of the experimental results, ensuring that, for example, membrane dyes are not transferred from cell to cell. A fluorescein-containing label (e.g., CFSE) or rhodamine-containing label (e.g., CRSE or Rhodamine 123) can be included in the mixture so that the embryos can be labeled after culture and fixation using an anti-fluorescein or anti-rhodamine antibody, respectively, and peroxidase immuno-

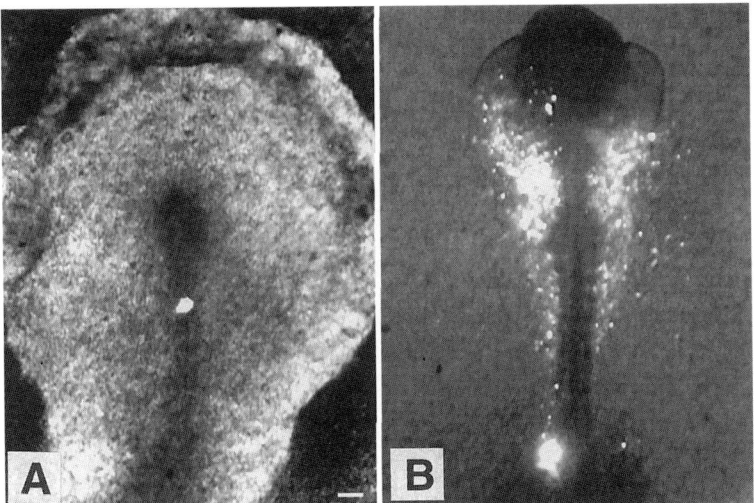

Fig. 6. Fate mapping by microinjection of fluorescent dyes. (**A**) Whole mount immediately after injection of a small bolus of Rhodamine 123 into the primitive streak at the mid-to-late gastrula stage. (**B**) Same embryo after 24 h of culture; fluorescently labeled cells have migrated into the prospective heart mesoderm and lateral plate mesoderm. Bar = 100 μm. Reprinted (with permission) from **ref. 7**.

cytochemistry (*see* **Note 1**). This immunocytochemical reaction provides a permanent label that can be detected in whole mounts or in sections after processing for paraffin histology and is superior to photoconversion, which had previously been used with DiI to deposit an insoluble marker after fluorescent labeling. Photoconversion often yields either no signal or high background and is labor intensive.

The intracellular localization of markers can be used to help differentiate multiple labels from one another when they are used concurrently (or sequentially). For example, to detect colocalized signals from both the progeny of the grafted cells and a specific regional marker, two possibilities exist. One could use grafts labeled with a cytoplasmically localized marker (e.g., CRSE with anti-rhodamine/peroxidase) in conjunction with a regional marker for a nuclear-localized protein (i.e., an antibody to a regionalized transcription factor) so that both markers could be seen in individual cells (not illustrated). Or one could use a quail graft and label quail-cell nuclei (with QCPN, the anti-quail antibody) in conjunction with a cytoplasmically localized marker (i.e., a riboprobe for a regionally expressed mRNA detected by *in situ* hybridization; *see* **refs. 7** and *8*) to demarcate both markers in individual cells (**Fig. 7A**). If two nuclear labels are used (not illustrated) or if two cytoplasmic labels are used (**Fig. 7B**) both labels cannot be visualized in the same cells because labeling overlaps and one label would extinguish the other. Nevertheless, this method could be used, for example, to show that grafted (i.e., peroxidase-labeled) cells contribute to an embryonic rudiment (in the case of **Fig. 7B**, to the notochord), with the identification of the rudiment being based on its characteristic morphology (i.e., the rodlike notochord, which is circular in transverse section), position (i.e., beneath the neural tube), and expression of a specific marker (i.e., *brachyury*).

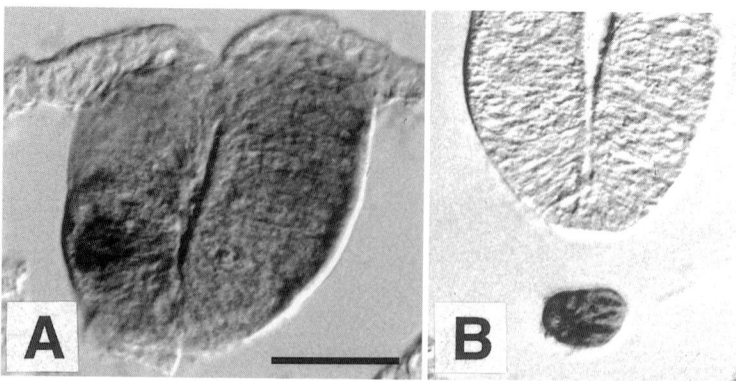

Fig. 7. (*See* color plate 9 appearing after p. 258.) Combination of *in situ* hybridization with immunocytochemical detection of graft cells. (**A**) Transverse section through the neural tube of an embryo showing quail graft cells (brown nuclei) within the neural tube (purple) of an embryo that has been double labeled with the *in situ* probe for the transcription factor *Hoxb-1* and with the anti-quail antibody, QCPN. Reprinted (with permission) from **ref. 8**. *Hoxb-1* was kindly provided by R. Krumlauf. (**B**) Transverse section through the neural tube and notochord of an embryo showing graft cells (brown cytoplasm) within the notochord (purple) of an embryo that has been double labeled with the *in situ* probe for the transcription factor *brachyury* and with the anti-rhodamine antibody (graft was obtained from a donor embryo labeled with CRSE). Because both labels are cytoplasmic, double labeling cannot be detected in individual cells. *Brachyury* was kindly provided by R. Runyan. Bar = 50 μm.

In short, one should select the best labeling techniques based on several criteria and design experiments to maximize the types of information that can be gained with the particular tools at hand.

2. Materials

1. Saline: 123 m*M* NaCl plus 5U/mL penicillin/streptomycin = 7.19 g NaCl/L distilled water, autoclave, cool to room temperature. Immediately prior to use, add 1 mL penicillin/streptomycin to each liter of saline. (Antibiotic can be aliquoted into 1-mL tubes and frozen until needed).
2. Fluorescent dyes (Molecular Probes, Inc., Eugene, OR):
    CFSE (5-[and -6] carboxyfluorescein diacetate, succinimidyl ester; C-1157).
    CRSE (5-carboxytetramethylrhodamine, succinimidyl ester; C-2211 or C-1171).
    DiA (4-[4-dihexadecylaminostyryl]-N-methylpyridinium iodide; D-3883).
    DiI (1,1'-dioctadecyl-3, 3, 3', 3'-tetramethylindocarbocyanine perchlorate; D-282).
    DiO (3,3'-dioctadecyloxacarbocyanine perchlorate; D-275).
    Rhodamine 123 (R-302).
3. Dimethyl sulfoxide (DMSO); e.g., Baker Analyzed reagent, VWR Scientific Products, Salt Lake City, UT.
4. Stock solutions of fluorescent dyes for graft labeling: CRSE, CRSE, and Rhodamine 123. Dissolve 10 mg CFSE, CRSE, or Rhodamine 123 in 1 mL DMSO. Store at –20°C in the dark.
5. Working solutions of fluorescent dyes for labeling donor embryos to obtain fluorescently labeled grafts: Add 50 μL of stock to 10 mL saline. For combined working solutions, add 50 μL of each stock to the same 10 mL saline.

6. Stock solutions of fluorescent dyes for injections: CFSE, CRSE, DiA, DiI, DiO, and Rhodamine 123. For CFSE, CRSE, or Rhodamine 123, dissolve 2.5 mg of each dye in 1 mL DMSO and mix (e.g., DiI and Rhodamine 123) in equal volumes at the time of injection. For DiA, DiI, or DiO dissolve 2.5 mg each dye in 50 μL DMSO then mix with 100% ethanol; mix dyes in equal volumes at time of injection if desired.

7. Phosphate-buffered saline (PBS): 8 g NaCl, 0.2 g KCl, 1.44 g Na_2HPO_4, 0.24 g KH_2PO_4 to 1 L and adjust pH to 7.4, store at 4°C.

8. 10% Triton X-100 in PBS (10 mL Triton X-100 plus 90 mL PBS).

9. Phosphate-buffered Triton (PBT): PBS, 0.2% bovine serum albumin (BSA), 0.1% Triton X-100 (mix 100 mL PBS, 200 mg BSA, 1 mL 10% Triton X-100).

10. Normal goat serum: Heat inactivated at 56°C for 30 min; aliquot and store at –20°C.

11. PBT + N: PBT plus 5% normal goat serum (4 mL PBT plus 200 μL normal goat serum); can be stored for a few days at 4°C.

12. DAB: 1 mg/mL diaminobenzidine (DAB) (cat. no. D-5905; Sigma Chemical Co., St. Louis, MO) in PBT (dissolve 10-mg tablet in 10 mL, filter and aliquot; 1-mL aliquots can be stored at –20°C for months; DAB is a carcinogen and should be handled with gloves and disposed of as a biohazard after inactivation with bleach; decontaminate instruments and work surface with bleach and dispose of used bleach as a biohazard).

13. DAB + PBT (1 mL DAB plus 2 mL PBT).

14. 0.6% Hydrogen peroxide: 30% H_2O_2 (cat. no. H-1009; Sigma); add 25 μL of 30% stock to 1.25 mL distilled water just before use).

15. 1% $CoCl_2 \cdot 6H_2O$ = 10 mg/mL PBT or distilled water.

16. 1% $Ni(NH_4)_2(SO_4)_2$ = 10 mg/mL PBT or distilled water.

17. Antibodies (used diluted in PBT + N):
    a. Anti-fluorescein (primary antibody) = rabbit IgG polyclonal used at 1:200 (cat. no. A-889; Molecular Probes); horseradish peroxidase-conjugated goat anti-rabbit IgG (secondary antibody) used at 1:200 (cat. no. 605-220; Boehringer Mannheim, Indianapolis, IN).
    b. Anti-rhodamine (primary antibody) = rabbit IgG polyclonal used at 1:100 (cat. no. A-6397; Molecular Probes); horseradish peroxidase-conjugated goat anti-rabbit IgG (secondary antibody) used at 1:200 (cat. no. 605-220; Boehringer Mannheim).
    c. QCPN (anti-quail; primary antibody) = mouse IgG1 monoclonal used 1:100 (Developmental Studies Hybridoma Bank, Iowa City, IA); horseradish peroxidase-conjugated goat anti-mouse IgG (secondary antibody) used at 1:200 (115-035-033; Jackson Immunoresearch Laboratories, West Grove, PA).
    d. QH-1 (anti-quail endothelium/endocardium; primary antibody) = mouse IgG monoclonal 1:1000 (Developmental Studies Hybridoma Bank); horseradish peroxidase-conjugated goat anti-mouse IgG (secondary antibody) 1:200 (115-035-033; Jackson Immunoresearch).

18. PBS-PFA = 4% paraformaldehyde in PBS: Mix 1 g paraformaldehyde with 25 mL PBS, heat to 60°C (microwave approx 30 s on high setting), add 1 drop of 2 *N* NaOH, mix to dissolve, refrigerate, and use within 24 h.

19. Vakaet's fixative: 70% Absolute ethanol, 20% formaldehyde [37%] and 10% glacial acetic acid. Store at room temperature in a sealed glass container.

20. Microinjection pipets. From capillary tubing with an internal fiber for rapid loading (e.g., capillaries with Omega dot fiber from Fredrick Haer & Co., Brunswick, ME), pull a microinjection pipet with a 10–12-μm tip (e.g., a Narishige micropipet puller, Narishige USA, Inc., Greenvale, NY). Optimal settings need to be determined empirically for each puller.

21. Time-lapse videomicroscopy: To do conventional light microscopy time-lapse, the following apparatus is required: a video camera attached to a dissecting microscope enclosed in an incubator and a time-lapse video recorder (e.g., Panasonic; Panasonic Industrial Co.,

Seacaucus, NJ) regulated by an external timer (e.g., a Dayton Timewatch garden timer; Dayton Electric Manufacturing Co., Chicago, IL). For fluorescence time-lapse, the following apparatus is required in addition to a dissecting microscope enclosed in an incubator: an intensified video camera (e.g., Optronix camera, S & M Microscopes, Inc., Boise, ID), a shutter driver to coordinate the action of the fluorescence shutter and recording device (e.g., Uniblitz shutter driver, Vincent Associates, Rochester, NY), and an image processor to collect the images (e.g., a Macintosh Quadra 840; Apple Computer, Inc., Cupertino, CA; running Axovideo 2.0 software, Axon Instruments, Inc., Foster City, CA); the video recorder and external timer are not required because images are collected to the computer's hard disk and Axovideo serves as the timer. Embryos are transilluminated by a light source (preferably fiber optics) in the stage of the microscope and warmed by a 38°C warm-air source that blows into the Plexiglas enclosure containing the microscope. To preserve the fluorescent label, the room should be dark.

22. Acid-etched slides: Soak a tray of glass slides in an acid bath solution (10% potassium dichromate and 10% concentrated sulfuric acid) for at least 1 h (many trays can be made up ahead of time and stored in acid bath until needed). On the day of use, rinse slides with distilled water until acid is no longer detectable. Allow slides to air dry and use immediately or for up to 1 wk if stored covered to kept dust free.

23. For labeling microscope slides, we use a Manomark marker (Monostat, New York, NY) to write on the frosted end of the slide, and then we flame the frosted end containing the writing by passing it through the flame of an alcohol burner a few times. (This minimizes the loss of the ink during later processing through Histosol (National Diagnostics, Manville, NJ) and ethanol, and eliminates the problem of carbon particles on the sections, often found when pencil is used for labeling the slides.)

24. Schiff's reagent: Mix 6 g pararosanilin hydrochloride, 9 g sodium bisulfite ($NaHSO_3$), 567 mL distilled water and 6 mL concentrated HCl in a 1-L beaker covered with Parafilm (VWR Scientific Products) or aluminum foil. Stir for at least 1 h on a magnetic stirrer. After stirring, place beaker in a dark cabinet and allow it to stand overnight at room temperature. The next day, decolorize by adding 9 g decolorizing carbon (Fisher*brand*; Fisher Scientific, Pittsburgh, PA) and stirring on a magnetic stirrer for 30 min; then filter through Whatman #1 filter paper (Whatman Corp., Clifton, NJ). Stain should be clear or straw color. If pink, add more decolorizing carbon, stir thoroughly, and filter as above. At this point, stain is ready for use. Schiff's reagent must be at 23–24°C for staining. To store Schiff's reagent between uses, keep it in a tightly closed, brown bottle in the refrigerator. To determine if Schiff's reagent has lost its potency, add a few drops of Schiff's reagent with a pipet to 10 mL of formaldehyde solution (37–40%). If the stain is still good, it changes color rapidly to reddish purple; however, if the color changes slowly and becomes blue-purple, the reagent is breaking down. (A white precipitate sometimes forms while the Schiff's reagent is stored, but this does not seem to affect staining.) If stored properly, Schiff's reagent should be active for at least 6 wk.

25. Bisulfite solution = 4.0 g sodium bisulfite ($NaHSO_3$), 40 mL of 1 *N* HCl and 760 mL distilled water; always use fresh bisulfite solution.

26. Fast Green FCF (Sigma, St. Louis, MO) in 95% ethanol for 30 s. (The older the Fast Green, the quicker the sections will stain.)

3. Methods

1. Labeling of donor embryos to obtain fluorescent grafts. Set up donor (chick or quail) embryos for New culture (*see* Chapter 5). Quail embryos require smaller (15-mm) rings. Add fluorescent label to the donor embryos by gently pipetting a few drops of working dye solution (single dye or dye combination) onto the embryo and culture in the dark for

1 h at 38°C. After 1 h, remove the dye and rinse the embryo with saline to remove excess dye. Isolate the graft and transfer it to a drop of saline for additional washing. Transfer the graft to the host as described in Chapter 35. Examine (and photograph or videotape) embryos at time 0 (and after culture) using a fluorescence microscope and appropriate filter sets. Culture grafted embryos for up to 24 h at 38°C in a humidified chamber in the dark (i.e., cover the humidified chamber with aluminum foil).

2. Microinjection of fluorescent dyes. (This can be done instead of or in addition to the previous step.) Back fill the microinjection pipet with the selected dye or dye combination using a Microfil 28 AWG needle (World Precision Instruments, Inc., Sarasota, FL). For the actual injection, use a micromanipulator (e.g., a Narishige hydraulic micromanipulator; Narishige USA, Inc., Greenvale, NY) attached to a Picospritzer II (General Valve Corp., Fairfield, NJ) or similar delivery device for pressure-driven injection. Injection pressure and duration need to be determined empirically for each needle, but they will be approx 60 psi and 5–15 ms, respectively. Test the needle by making a practice injection into a dish of sterile saline. One activation should deliver a small but visible bolus of dye if a colored dye is present (e.g., DiI). For the actual injection, manually lower the microinjection pipet until it is near the surface of the embryo (or use a course micromanipulator). Lower it the remaining distance using a fine (hydraulic) micromanipulator. Pierce the tissue and use the Picospritzer to deliver some dye. Sometimes the Picospritzer needs to be activated several times to deliver the desired amount of dye into the tissue. It is better to deliver too little dye per injection than too much; in the latter case, embryos should be discarded. It should be possible to use one pipet to inject several embryos (*see* **Note 2**). Withdraw the pipet, record the injection location and size using a fluorescence microscope fitted with an intensified video camera (e.g., a Dage-MTI camera; Dage-MTI, Inc., Michigan City, IN), image processor for image averaging (e.g., a Dage-MTI image processor) and video recorder. Place the embryo back into the incubator and continue with the next embryo. Culture embryos for up to 24 h at 38°C in a humidified chamber in the dark.

3. Time-lapse video microscopy of (fluorescently labeled) embryos in culture. To begin recording, turn on the (a) fluorescent light source, (b) chamber heat source, (c) microscope illuminator (located within the stage of the microscope), (d) shutter driver, (e) computer, and (f) video camera. When using a mercury bulb for fluorescence, make sure that you turn on its power source before turning on any other electronic component to avoid damage to the electronics of the latter; likewise, when turning off the system, turn off all other components before turning off the power source for the mercury bulb. Using Axovideo, set the program to collect images approximately every 4 min. Seal the culture dish containing the grafted or injected embryo by stretching Parafilm around its perimeter, overlapping the interface between the lid and base of the dish. Place on the culture dish on the microscope's stage. Orient the embryo/camera, focus, and adjust contrast and magnification based on the image on appearing on the monitor (embryo may "drift" during growth, so do not magnify to fill the frame of the camera). Videotape for up to 24 h in a dark room to yield a time-lapse tape of a short duration.

4. Fixation of embryos. Remove the embryo and ring from the culture dish and place them in a disposable plastic dish for fixation. Unwrap the vitelline membrane from the ring and remove the ring (wash and sterilize the ring for subsequent reuse). Detach the embryo from the vitelline membrane, split the vitelline, and gently slide it from beneath the embryo. Add fixative gently, dropwise onto the embryo, until the embryo is immersed. Cover, label, and place at 4°C for 2–24 h. For *in situ* hybridization and most immunocytochemistry (except anti-fluorescein; *see* **Note 1**), use 4% PBS-PFA; for Feulgen-Rossenbeck, use Vakaet's fixative.

5. Storage of fixed embryos prior to further processing. Cut embryo away from the extraembryonic tissues (if desired) using a scalpel. For Feulgen-Rossenbeck staining, rinse and store tissue in 70% ethanol and skip to **step 7**. For immunocytochemistry, we typically store embryos for up to several days in PBS at 4°C and do our immunocytochemistry in 96-well plates. For embryos that will be subjected to *in situ* hybridization prior to immunocytochemistry, dehydrate immediately into 100% methanol and store at –20°C (*see* **Note 3**).

6. Immunocytochemistry. Rinse repeatedly with PBS and then wash embryos 2X for 5 min and 1X for 30 min in PBT. (All washes are done on an orbital shaker to reduce background.) Incubate 1X for 30 min in PBT + N. Remove and replace with appropriately diluted primary antibody. Incubate overnight at 4°C (or for 4 h at room temperature). Wash 3X for 5 min and 4X for 30 min with PBT. Incubate 30 min with PBT + N. Remove and replace with appropriately diluted peroxidase-conjugated secondary antibody. Incubate overnight at 4°C (or for 4 h at room temperature). Wash 3X for 5 min and 4X for 30 min with PBT. Mix 2 mL PBT with 1 mL DAB solution (add 75 μL 1% $CoCl_2$ and 60 μL 1% $Ni(NH_4)_2(SO_4)_2$ if intensified, black reaction product is desired or if double-immunocytochemistry will be done; *see* **Note 4**). Incubate embryos 1X for 10 min in DAB-PBT (use 2 mL for incubation and save 1 mL for the next step). Add 5 μL of a 0.6% solution of fresh H_2O_2 to 1 mL DAB-PBT, mix, and pour into a watch glass or small dish. Place embryos into this solution a few at a time and observe the reaction as it proceeds under the microscope (on a white background with epi-illumination) until labeling reaches the desired intensity. Stop the reaction by returning the embryos to PBT or PBS and rinsing them several times.

7. Paraffin embedding (*see* **Note 5**). Transfer each embryo to a glass scintillation vial containing PBS. Remove the liquid from each vial with a pipet and wash embryos 2X with PBS for 5 min each. Dehydrate through one 5-min change each of 35, 50, and 70% ethanol in PBS. (If embryos came directly from Vakaet's fix, then start the procedure here.) Continue dehydrating tissue through two changes of 95% ethanol and two changes of 100% ethanol for 5 min each. Clear tissue in two changes of Histosol, 5 min each (*see* **Note 6**). Infiltrate tissue with Paraplast X-tra (VWR Scientific) through three changes, 30 min each, in a 60°C oven. Embed each embryo in fresh Paraplast X-tra in an embedding mold. To do this, remove the tissue from the vial using a prewarmed spatula, fill a mold with Paraplast X-tra and place the tissue into the mold, orienting it according to the desired plane of sectioning. Place the mold on a cold plate and allow the block to stand overnight at room temperature.

8. Paraffin sectioning and mounting. Set the slide warmer at 40°C. Set the microtome for the desired section thickness. Select a block of paraffin containing the embryo to be sectioned, peel away the embedding mold from the block, and trim the block face to a small trapezoid shape using a sharp single-edged razor blade. Make sure that the top and bottom of the block face are parallel to each other, the bottom edge of the block face is parallel to the microtome blade, and the block face is aligned with the microtome blade. Cut a long ribbon of sections (*see* **Note 7**), and use fine paintbrushes to transfer the ribbon onto black paper placed adjacent to the microtome. Continue sectioning until the embryo has been completely sectioned, keeping ribbons in order. Use a razor blade to subdivide the longer ribbons into smaller ribbons roughly 4 cm long, making sure that the smaller ribbons remain in the order in which they were cut. Label the frosted end of each slide with the specimen number, section thickness, and its rank in the series so that slides may be identified with respect to the sections they contain. Place the labeled slide on a 40°C slide warmer. Using a 10-mL syringe, cover the slide with boiled, distilled water. Using forceps and a fine paint brush, transfer the first short ribbon of paraffin sections to the top edge of the slide, with the first section that was cut with the microtome placed near the frosted end

of the slide. Ribbons will expand in the warm water, so do not cut them too long. Place as many ribbons as will fit onto the slide. Ribbons can be repositioned using forceps or a paintbrush as long as the ribbons are still floating on the surface of the water. Remove the slide from the warmer and drain away excess water by placing a tissue against the edge of the slide and letting capillary action remove the water. Be careful not to touch a ribbon with the tissue or it will also come off. Return the slide to the warmer and repeat until all ribbons are mounted and all blocks are sectioned. Slides should remain on the slide warmer for 48 h prior to mounting a cover slip or Feulgen-Rossenbeck staining (*see* **step 9**). To mount a cover slip, remove paraffin with two changes of Histosol, 1X for 10 min and 1X for 5 min, then pass slides through this series for 5 min each: 100% ethanol, 100% ethanol, 95% ethanol, 95% ethanol, 100% ethanol, 100% ethanol, Histosol, and Histosol. Add Pro-Texx mounting media (VWR Scientific) and a cover slip. If embryos were subjected to *in situ* hybridization, it is best to photograph sections immediately, because the signal may diminish over time. Allow slides to remain flat for at least 48 h so that the mounting media has time to dry before slides are filed away.

9. Feulgen-Rossenbeck staining (*see* **Note 8**). Deparaffinize and hydrate sections to water with these changes: 10 min Histosol, 5 min Histosol, 2 min each of ethanol solutions (100, 100, 95, 95, 70, 50, 35%), and 2 min each of 2 changes of distilled water. Hydrolyze sections in three changes of *freshly prepared* 1 N HCl: 2 min at room temperature, 6 min at 60°C, and 2 additional minutes at room temperature; follow this with 1 change (2 min) of clean distilled water. Stain in Schiff's reagent for 1 h at 24°C. Bleach sections in three changes of fresh bisulfite solution, 2 min each. Replace with *fresh* bisulfite solution after *every rack* of slides. Rinse racks of slides in running tap water for 5 min. At this point, check the sections with a microscope to see if nuclei are stained bright pink. If not, place slides back into Schiff's reagent, bisulfite solutions, and running tap water as above and then recheck the staining of the nuclei. When the nuclei are sufficiently stained, dehydrate the sections through a graded series of ethanols up to 90% ethanol (i.e., 35, 50, 70, and 90%), 2 min each. Counterstain sections with 0.1% Fast Green FCF in 95% ethanol for 30 s. (The older the fast green, the faster the sections will stain.) Complete the dehydration procedure (i.e., ethanol, 95, 95, 100, 100%; 2 min each). Clear with two changes of Histosol (first change, 2 min; second change, 3 min). Add cover slips using Pro-Texx or similar mounting medium and allow slides to dry.

4. Notes

1. Anti-fluorescein antibody used in conjunction with CFSE labeling yields a similar pattern to QCPN labeling of quail grafts or CRSE labeling with anti-rhodamine antibody; however, a different fixative is required. Instead of PBS-PFA, fix in PEM-FA. PEM is 0.1 M Pipes (1.73 g/50 mL), 2 mM EGTA (38.5 mg/50 mL) and 1 mM $MgSO_4$ (12.3 mg/50 mL); adjust the pH to 6.95 with concentrated HCl; PEM can be stored at 4°C for several weeks. Just prior to fixation, mix 9: 1 PEM:37% formaldehyde (FA); use within 24 h and store refrigerated.

2. If an injection pipet clogs with debris, it can sometimes be cleared by wiping the tip with another injection pipet or by temporarily increasing the injection pressure or the duration of the air pulse and activating the Picospritzer a few times.

3. If *in situ* hybridization will be used in conjunction with other labels, embryos should be rinsed with PBS with 0.1% Tween-20 and dehydrated into methanol for storage at **step 5**. For double labeling (*in situ* hybridization/immunocytochemistry), the *in situ* protocol of Wilkinson (*9*) is compatible with and should precede the immunocytochemical protocol presented (one exception: anti-fluorescein labeling is not compatible with Wilkinson's

protocol but may be with the protocol of Nieto et al.; *see* **ref. *10***). For this double labeling, do not intensify the immunocytochemistry reaction; color discrimination between the peroxidase reaction product and the *in situ* hybridization reaction product is better without intensification (brown and purple, respectively). Embryos should be postfixed with 4% PFA overnight at 4°C prior to embedding to prevent loss of the *in situ* label and to improve section quality.

4. For double labeling with two peroxidase-conjugated antibodies (immunocytochemistry/immunocytochemistry), reactions should be done sequentially, using first the antibody yielding the weakest signal with intensification. After rinsing, the second immunocytochemistry reaction can be done using the antibody yielding the stronger signal, without intensification. The two labels will be black and brown, respectively.

5. Paraffin histology allows every section to be collected in a quick, organized fashion. Different fixatives differ in the ability to provide high-quality histology. PBS-PFA gives acceptable results with paraffin sectioning. Vakaet's fix, used with the Feulgen-Rossenbeck staining procedure, gives superior histology, but it is often incompatible with immunocytochemistry and *in situ* hybridization. Never heat paraffin above 60°C, or you may reach its flash point. Also, temperatures greater than 60°C will adversely effect histological quality, especially with early embryos.

6. For embryos processed for *in situ* hybridization, the Histosol step prior to embedding should be substituted with two 30 min changes of isopropanol at room temperature, followed by a 1:1 isopropanol:Paraplast X-tra (VWR Scientific) infiltration at 60°C for 30 min.

7. Sections are usually cut at 5–15 µm (we routinely use 7 µm for immunocytochemistry and 15 µm for *in situ* hybridization), and sections from whole embryos at early stages fit onto a few slides. If sections are not forming ribbons, the block face may need to be retrimmed or the microtome blade moved to a fresh area. Chilling the block may also help. Having an Bunsen burner or other small flame going nearby may diminish the effects of static electricity.

8. Prepare Schiff's reagent in advance. We get the best results when we make the Schiff's reagent ourselves rather than buying a commercially available stock solution. For this procedure, embryos should be fixed in Vakaet's fixative (70% absolute ethanol, 20% formaldehyde [37%] and 10% glacial acetic acid). Strictly aldehyde-based fixatives (e.g., PBS-PFA) may be used, but the timing and concentration of the Schiff's reaction must be reduced and the nucleolar labeling is less reliable.

Acknowledgments

Original work described herein from the Schoenwolf laboratory was supported by NIH Grants NS 18112 and HD 28845, Grant no. 94-0431 from the FIS, and Grant no. PR95-243 from the DGICYT to V. G. M; C. L. S was supported in part by a grant from the Junta de Extremadura. D. K. D. was supported in part by NIH Developmental Biology Training Grant no. HD 07491.

References

1. Fraser, S. E. (1996) Iontophoretic dye labeling of embryonic cells, in *Methods in Cell Biology, Vol. 51, Methods in Avian Embryology* (Bronner-Fraser, M., ed.), Academic, New York, pp. 147–161.

2. Cepko, C. (1988) Retrovirus vectors and their applications in neurobiology. *Neuron* **1,** 345–353.

3. Le Douarin, N. (1973) A Feulgen-positive nucleolus. *Exp. Cell Res.* **77,** 459–468.

4. Inagaki, T., Garcia-Martinez, V., and Schoenwolf, G. C. (1993) Regulative ability of the prospective cardiogenic and vasculogenic areas of the primitive streak during avian gastrulation. *Dev. Dyn.* **197,** 57–68.

5. Garcia-Martinez, V. and Schoenwolf, G. C. (1993) Primitive-streak origin of the cardiovascular system in avian embryos. *Dev. Biol.* **159,** 706–719.

6. Garton, H. J. L. and Schoenwolf, G. C. (1996) Improving the efficacy of fluorescent labeling for histological tracking of cells in early mammalian and avian embryos. *Anat. Rec.* **244,** 112–117.

7. Daston, G. P. (ed.) (1996) *Molecular and Cellular Methods in Developmental Toxicology.* CRC, New York, pp. 240, 255.

8. Garcia-Martinez, V., Darnell, D. K., Lopez-Sanchez, C., Sosic, D., Olson, E. N., and Schoenwolf, G. C. (1997) State of commitment of prospective neural plate and prospective mesoderm in late gastrula/early neurula stages of avian embryos. *Dev. Biol.* **181,** 102–115.

9. Wilkinson, D. G. (1993) In Situ *Hybridization. A Practical Approach.* IRL at Oxford University Press, Oxford, UK, pp. 1–163.

10. Nieto, M. A., Patel, K., and Wilkinson, D. G. (1996) *In situ* hybridization analysis of chick embryos in whole mount and tissue sections, in *Methods in Cell Biology, Vol. 51, Methods in Avian Embryology* (Bronner-Fraser, M., ed.), Academic, New York, pp. 220–236.

31

Cell Lineage Analysis

Videomicroscopy Techniques

Paul J. Heid and Jeff Hardin

1. Introduction

Complete or partial embryonic cell lineages are available for several animal model systems. In the case of the nematode *Caenorhabditis elegans*, the entire embryonic cell lineage has been determined and is largely invariant *(1)*. This makes lineage analysis a potentially useful tool for assessing mutant phenotypes in *C. elegans*. Indeed, lineage analysis of some mutants has shown that one cell can be transformed into a different cell resulting in duplication or absence of certain tissues *(2,3)* (**Fig. 1**).

The wild-type lineage was originally determined by direct observation under a light microscope. This was a very slow process since only one or two of the >500 total embryonic cells could be followed per embryo. Determining the lineage of mutants by direct observation would be even slower and more difficult than in wild-type for at least two reasons:

1. Many of the mutants that display lineage defects will be recessive zygotic lethal mutations. These mutations must be maintained as heterozygotes, which results in only one of four embryos displaying a lineage defect.
2. Mutant phenotypes are not always 100% penetrant, which means every lineage would have to be followed multiple times.

The advent of four-dimensional (4-D) microscopy *(4)* makes cell lineage analysis of mutants much more practical. This technique allows development of embryos to be recorded in three dimensions (3-D) over time. Multiple cells can then be followed in a single embryo, making lineage analysis much simpler and faster. This technique can be used for a variety of other experiments, including analysis of morphogenetic movements *(5)*, analysis of cytoplasmic flow in early embryos *(4)*, cell ablation studies *(5)*, and analysis of migration of individual cells *(6,7)*. Here we will describe how we prepare mounts, record embryos, and analyze recordings to determine the cell lineage of mutant *C. elegans* embryos. Although some of the techniques described here are specific to *C. elegans*, the apparatus and software can be adapted easily to a variety of experimental uses.

From: *Methods in Molecular Biology, Vol. 135: Developmental Biology Protocols, Vol. I*
Edited by: R. S. Tuan and C. W. Lo © Humana Press Inc., Totowa, NJ

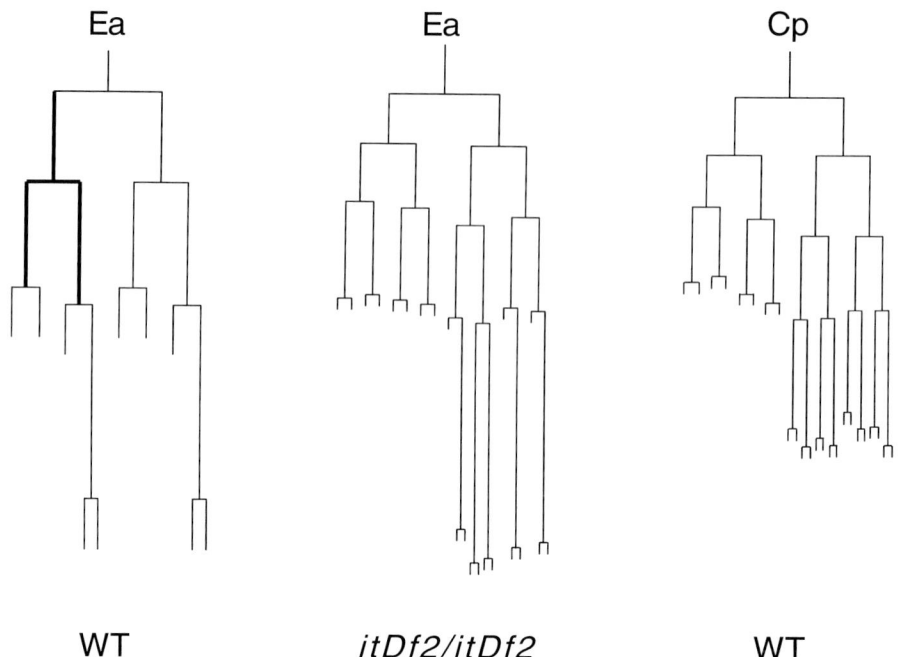

Fig. 1. Examples of lineage trees from wild-type and *itDf2* embryos. This figure is modified from **ref. 2**. The E lineage in *itDf2* homozygous embryos is different than the E lineage in wild-type embryos, but similar to the wild-type C lineage *(2)*. Bold lines indicate the part of the wild-type lineage that was followed in **Fig. 3**. The E lineage produces intestine in wild-type animals and the C lineage normally produces epidermal and muscle cells *(1)*. *itDf2* homozygotes produce muscle and epidermis from the E lineage instead of intestine *(2)*. This is consistent with the observed lineage transformation shown here.

2. Materials

2.1. Embryo Preparation

1. 3-mm single-depression microslide. (VWR Scientific, West Chester, PA).
2. M9 solution: 3 g KH_2PO_4; 6 g Na_2HPO_4; 5 g NaCl; 1 mL of 1 M $MgSO_4$; 1 L H_2O
3. Scalpel (Feather Disposable, #15, curved blade).
4. Mouth pipet. A 20-μL micropipet (Fisher Scientific, Pittsburgh, PA) is drawn out over a flame and broken in half; only one half is used. A small piece is broken off of the tip to provide an opening that is >40 μm in diameter. The capillary tube is inserted into one of the plastic adapters supplied with the capillary tubes. This assembly is inserted into a piece of rubber tubing. A second plastic adapter is inserted into the other end of the tubing and is used as a mouthpiece to apply suction.
5. 5% agar (Difco Bacto-Agar, Difco, Detroit, MI).
6. Microscope slides (Fisher*brand* Colorfrost [Fisher Scientific], precleaned, $25 \times 75 \times 1$ mm).
7. Eyelash brush. This consists of an eyelash glued to the end of a toothpick. (glue: Devcon Duco® Cement, Danvers, MA).
8. 18×18-mm cover slips (Corning No. $1\frac{1}{2}$, Corning Glassworks, Corning, NY).
9. Platinum wire pick. A one inch long piece of platinum wire (Fisher, 32 g) is inserted into the end of a short Pasteur pipet and held over a flame until the glass melts around the wire holding it in place. A pliers with a flat face is used to press the tip of the wire flat.

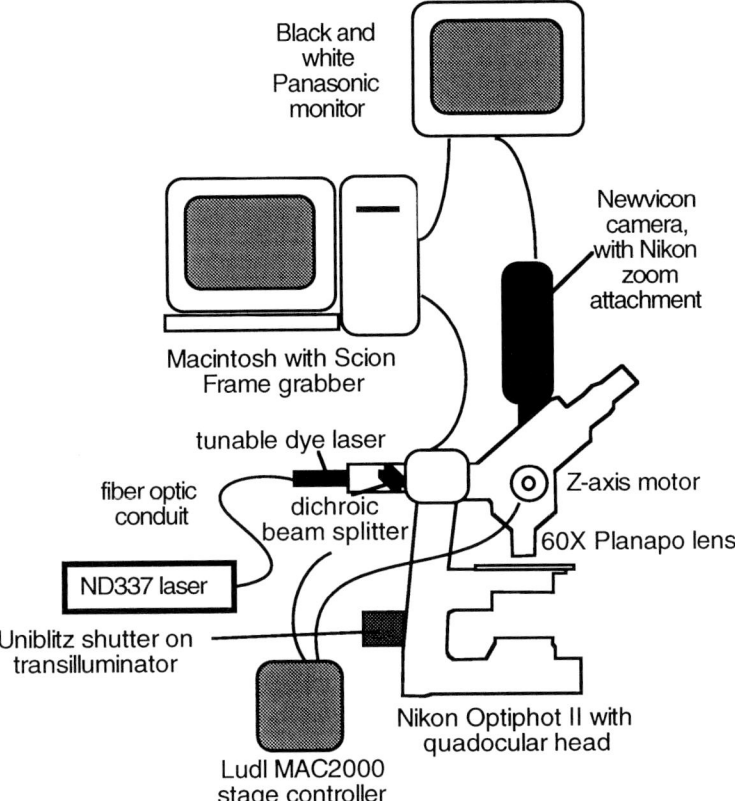

Fig. 2. Schematic diagram of integrated 4-D microscopy setup. The video camera with zoom attachment sends signals to a TV monitor, which is used to set up optics. The video signal is then sent to the computer and optics on the computer screen are adjusted to match the monitor. Parameters are then specified for the modified version of NIH Image, which is used to control the Ludl box and the shutter. Also shown is a fiber-optic delivered laser (Photonic Instruments, Chicago, IL), which allows ablations to be performed followed by immediate imaging.

2.2. Recording and Lineage Analysis

1. Nikon Optiphot-II microscope equipped with differential interference contrast optics, a 60X oil immersion 1.4 NA planapochromat lens (Nikon Corp., Tokyo, Japan), a LUDl Z-axis stage controller operated by a LUDl Mac2000 control box (LUDl Electronic Products Ltd., Hawthorne, NY), and a Uniblitz electronic shutter on the transilluminator port (Vincent Associates, Rochester, NY) (**Fig. 2**).
2. Shutter and Z-axis motor are controlled by serial cable connections to a Macintosh computer and a Scion LG3 8-bit frame grabber (Scion Corp., Frederick, MD) (**Fig. 2**).
3. Device control and image acquisition are accomplished using NIH Image. NIH Image is a public domain image analysis program written by Wayne Rasband available via anonymous ftp from www.zippy.nimh.nih.gov. A modified version of NIH Image is used for QuickTime movie construction and playback, and is available on request.
4. Newvicon video camera (Dage-MTI [Michigan City, IA], 13.5 W, 120 V) with a Nikon zoom attachment (**Fig. 2**).
5. Panasonic black-and-white TV monitor (**Fig. 2**).

3. Methods

3.1. Embryo Preparation

1. Five to 10 young adult *C. elegans* hermaphrodites are removed from plates with a platinum wire pick and deposited into a depression slide containing approx 0.5 mL of M9 buffer.

2. Animals are cut in half transversely with a scalpel midway along the A-P axis of the animal to release embryos into the solution. This is done under a dissecting microscope (*see* **Note 1**).

3. A strip of colored laboratory label tape is placed on the bottom of two slides. These slides are placed on a hard flat surface with an untaped third slide between and parallel to them. Two or three drops of melted 5% agar are put on the center of the third (middle) slide, and a fourth slide is placed over the agar perpendicular to the third slide. The top slide is pressed down firmly to produce a thin agar pad between slides 3 and 4. The slides are carefully pulled apart such that the agar pad is left on the center of one slide.

4. Under the dissecting microscope, 8–10 2-cell embryos are collected from the M9 solution with a mouth pipetter into a small volume (<20 µL) and deposited onto a corner of the agar pad.

5. Embryos are brushed away from the main puddle of liquid toward the center of the pad using an eyelash. Typically two embryos are placed side by side; this allows recording of two embryos at a time. The liquid surrounding the embryos is allowed to dry, which embeds the embryos in the agar.

6. Once the liquid has dried a drop of M9 solution is put on top of the embryos and a cover slip is gently placed over the embryos. It is important to avoid getting air bubbles near the embryos; air bubbles make it impossible to obtain good optics later in the procedure.

7. Excess agar is trimmed away from the edges of the cover slip with a razor blade; care must be taken not to move the cover slip. Melted Vaseline is applied to the edges of the cover slip with a paint brush. This provides a seal that keeps the agar and embryos from drying out (*see* **Note 2**).

3.2. Recording and Lineage Analysis

1. The slide is placed on the stage of the microscope with a drop of immersion oil between the slide and the condenser lens. A 10X objective is used to locate embryos. Embryos are centered in the field of view, a drop of immersion oil is put on the cover slip and the 60X objective is swung into place (*see* **Note 3**).

2. Optics are optimized by focusing the condenser lens and rotating the Nomarski prism until the desired level of contrast is obtained. The microscope lamp is never set above 6 (50% of maximum) to avoid heat damage to the embryos.

3. Light is then sent to the video camera and embryos are observed on a TV monitor. The camera is rotated so that the anterior/posterior (A/P) axis of the embryos is vertical. If only a single embryo is used the A/P axis should be horizontal on the monitor. The zoom is then adjusted until the embryos fill the screen without any edges being cut off.

4. Contrast and gain knobs on both the camera and the monitor are adjusted to obtain the best image. We prefer a high-contrast image to make the nuclei stand out but not to the extent that they look distorted.

5. To set up a recording, NIH Image is opened, the image on the monitor is captured and displayed on the computer screen, and contrast and gain are adjusted on the computer until the image resembles what is observed on the TV monitor.

6. The 4-D recording macro is then invoked, using the appropriate menu option in NIH Image. Parameters for the recording are specified including number of focal planes, distance between focal planes (micrometer), time interval between z-series (seconds), number of timepoints and root file name. The software uses all of this information and issues the appropriate serial commands to control the focus motor (telling it how often and what distance to move), and the shutter (so that light is not hitting the embryos between z-series acquisitions). The focus is adjusted so that the top surface of the embryo is in focus, the shutter is closed, and the recording is started. The motor automatically resets to the top focal plane after each scan. For a lineage quality, recording scans are typically taken starting with a four-cell embryo every 45 s, 23–25 focal planes and 1.0 μm apart, for approx 6 h.

7. Images are stored directly on the hard drive, which requires approx 1.5 GB of hard drive space. There are options for reducing the amount of space required for lineage recordings (*see* **Note 4**). Files are later compressed and transferred either to recordable CDs or magneto-optical disks.

8. Images can be played back as movies within a specially modified version of NIH Image. A single focal plane can be followed forward or backward in time. At each timepoint, all of the different focal planes of a recording can be scanned. This allows us to follow a specific cell throughout the recording, even though it may move through different focal planes.

9. To analyze the lineage, a cell of interest is selected at the beginning of the recording and followed through time until it divides (**Fig. 3**). The division is carefully followed to make sure that both daughters can be clearly identified. The time, focal plane, and location at which the cell divided are documented so that it can be returned to later without having to retrace the entire lineage. This is accomplished by placing transparency film over the computer monitor and tracing the positions of the nuclei immediately after the division.

   One of the daughter cells is then chosen and monitored until it divides. One daughter cell from each successive division is followed until the end of the recording is reached or the cell is lost. Because the positions of cells are charted after each division, it is possible to go back to earlier timepoints in the recording to pick up a sister cell and follow its lineage. This can be done until all of the cells of interest have been lineaged (*see* **Note 5**).

10. Lineage trees are then drawn either manually or using available commercial software that can be used to trace and plot lineage trees semiautomatically (*7*). Lineages are then compared to the wild-type lineage to assess any differences (**Fig. 1**) (*see* **Notes 6** and **7**).

4. Notes

1. Cutting worms with a scalpel in liquid can be difficult because the worms are constantly moving. Pinning them against the glass with the blade helps. Also, lifting the scalpel blade in and out of the liquid causes the animals to move. Once the blade is submerged it should be kept under water until cutting is finished. Finally, serotonin may be added to the M9 solution; this drug causes the worms to lay eggs quickly so young embryos can be obtained without cutting worms (*9*).

2. The steps from making the agar pad to sealing the slide with Vaseline must be done as quickly as possible to prevent the pad from drying out. Ideally, the entire process from collecting two cell embryos to starting a recording should be done in 30 min or less. This ensures that embryos will be four or eight cells when the recording is started, which makes identification of cells easier when lineaging.

3. Any source of vibration will cause blurring of images. If vibration is a problem an isolation table can be purchased to stabilize the microscope, or the microscope can be placed on an apparatus as simple as a large, heavy board on top of inflatable rubber donuts to eliminate vibration.

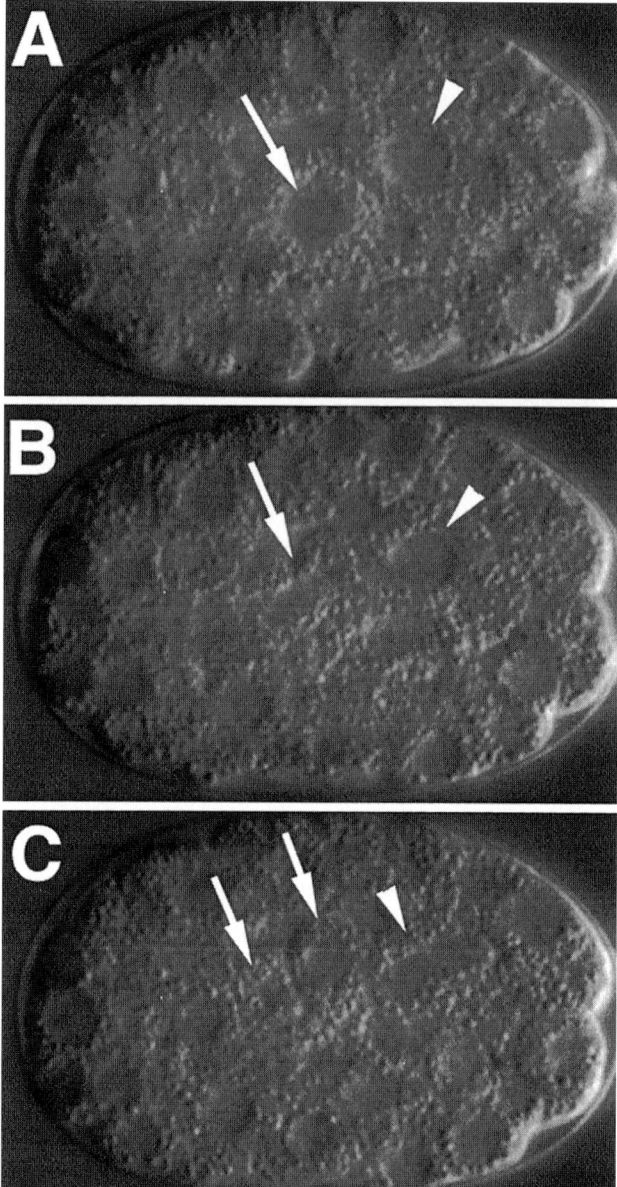

Fig. 3. Images of a wild-type embryo recorded for lineage analysis with the integrated 4-D microscopy system. Anterior is to the left and dorsal is to the top. Images display how the lineage of the cell Eal can be followed. Standard nomenclature was used to name cells (1). (A) Arrow indicates nucleus of Eal and arrowhead indicates nucleus of Epl as determined by lineage analysis, starting with a 28-cell embryo in which the precursor cells Ea and Ep can easily be identified. (B) The same embryo approx 5 min later; Eal is dividing (arrow indicates furrow between forming cells) and Epl has not yet started to divide (arrowhead indicates nucleus). (C) Same embryo approx 2 min later; arrows indicate the progeny of Eal, the anterior daughter Eala (left) and the posterior daughter Ealp (right). Only the nucleus of Ealp can be distinguished in the focal plane shown. Epl is in the process of dividing (arrowhead). At this point, to continue lineaging, either Eala or Ealp would be followed through the recording until it divides. The other cell would be ignored until later but could easily be identified if time, location on the screen, and focal plane of the division are all documented.

4. Several things can be done to reduce the amount of hard drive space that a lineage recording occupies. Early in development, cells are bigger and divide more slowly, so it is possible to record using only 15 focal planes instead of 25 and scans that are 1–1.5 min apart instead of 45 s. With the described system this can only be accomplished by setting up two separate recordings. The second recording has to be set up quickly so that cells do not divide or move during setup. There are other NIH Image macros and standalone software for this recording system (*10*) that allow change of parameters during a single recording. Available standalone software compresses the images before storing them to the hard drive, which significantly reduces the amount of space occupied (*11*) (*see* Chapter 26). Finally, because early embryonic development is well described, it is possible to wait until the embryos have 28 cells to start recording. Cell position is essentially invariant until this time, and diagrams exist showing the position of each cell up to this stage (J. Priess, personal communication). This saves about 1 h of recording time. The diagrams are available on request.

5. During a recording, embryos turn either ventral side up or dorsal side up. Embryos that are left side up at the 4- or 28-cell stage usually become dorsal views; embryos that are right side up tend to become ventral views. It is often difficult to lineage deep into the embryo, especially in older embryos. If comprehensive lineaging of several cells is necessary, it is helpful to do separate recordings of both dorsal- and ventral-view embryos. This ensures that every cell is in the top half of the embryo in one recording or the other.

6. For many other types of experiments in *C. elegans*, including following cytoplasmic streaming in early embryos, cell migrations, and following development of embryos in which cells have been killed with a laser or embryos that have been laser-permeabilized and treated with drugs; 4-D microscopy can be used. Many of the procedures described in this chapter can be modified for different experiments. For example, laser-permeabilized embryos would explode on the agar mounts described here and have to be mounted on poly-lysine coated cover slips and placed over a slide with grease feet (*5*) or a depression slide (*12*) to prevent compression. This type of experiment also requires that a laser be attached to the microscope. For an example of the use of our integrated 4-D workstation for this sort of experiment, *see* **ref. 5**. A schematic diagram of this system is shown in **Fig. 1**.

   To look at early events in single-cell *C. elegans* embryos, M9 solution cannot be used, because embryos do not have completely developed egg shells. More complex solutions that better mimic the internal conditions of the embryo must be used for these experiments (*13*).

   Our laboratory has also used 4-D microscopy to analyze gastrulation events in normal and laser-ablated sea urchin embryos. Conditions for mounting embryos are significantly different than those described here (*14*) (*see* Chapters 2 and 3).

7. A variety of systems exist for doing 4-D recordings (*4,5,7,9,15–17*). The original analog system was developed by John White (*4*). Advantage of the analog system are that recorded images can be played back at high speed and image quality is good. The main disadvantage of this system is that recordings cannot be deleted and storage discs become a major expense. Software has been developed for use with the original analog system that allows computer-generated lineage trees and 3-D reconstructions of nuclear positions to be produced (*7*).

   More recently, several digital systems have been developed, including the one described here. These have the advantage of using erasable media. Playback of recordings is slower (though improving) and image quality may be slightly lower than analog systems using existing technology. However, new software and better computers are starting to overcome these problems (*see* **ref. 10** for discussion).

Acknowledgments

We would like to thank Wayne Rasband for sharing the original version of NIH Image.

References

1. Sulston, J. E., Schierenberg, E., White, J. G., and Thomson, J. N. (1983) The embryonic cell lineage of the nematode *C. elegans. Dev. Biol.* **100,** 64–119.

2. Zhu, J., Hill, R. J., Heid, P. J., Fukuyama, M., Sugimoto, A., Priess, J. R., and Rothman, J. H. (1997) *end-1* encodes an apparent GATA factor that specifies the endoderm precursor in *Caenorhabditis elegans* embryos. *Genes Dev.* **11,** 2883–2896.

3. Draper, B. W., Mello, C. C., Bowerman, B., Hardin, J., and Priess, J. R. (1996) MEX-3 is a KH domain protein that regulates blastomere identity in early *C. elegans* embryos. *Cell* **87,** 205–216.

4. Hird, S. N. and White, J. G. (1993) Cortical and cytoplasmic flow polarity in early embryonic cells of *Caenorhabiditis elegans. J. Cell Biol.* **121,** 1343–1355.

5. Williams-Masson, E. M., Malik, A. N., and Hardin, J. (1997) An actin-mediated two-step mechanism is required for ventral enclosure of the *C. elegans* hypodermis. *Development* **124,** 2889–2901.

6. Ferguson, K. C., Heid, P. J., and Rothman, J. H. (1996) The SL1 *trans*-spliced leader RNA performs an essential embryonic function in *Caenorhabditis elegans* that can also be supplied by SL2 RNA. *Genes Dev.* **10,** 1543–1556.

7. Schnabel, R., Hutter, H., Moerman, D., and Schnabel, H. (1997) Assessing normal embryogenesis in *Caenorhabditis elegans* using a 4D microscope: variability of development and regional specification. *Dev. Biol.* **184,** 234–265.

8. Lowe, C. J. and Wray, G. A. (1999) Rearing larvae of sea urchins and sea stars for developmental studies. *Developmental Biology Protocols*, Methods in Molecular Biology, vol. 135 (Tuan, R. S. and Lo, C. W., eds.), Humana Press, Totowa, NJ.

9. Croll, N. A. (1975) Indolealkylamines in the coordination of nematode behavioral activities. *Can. J. Zool.* **53,** 894–903.

10. Thomas, C., DeVries, P., Hardin, J., and White, J. (1996) Four-dimensional imaging: computer visualization of 3D movements in living specimens. *Science* **273,** 603–607.

11. Ariizumi, T., Takano, K., Asashima, M., and Malacinski, G. M. (1999) Bioassays of inductive interactions of amphibian development. *Developmental Biology Protocols*, Methods in Molecular Biology, vol. 135 (Tuan, R. S. and Lo, C. W., eds.), Humana Press, Totowa, NJ.

12. Priess, J. R. and Hirsh, D. I. (1986) *C. elegans* morphogenesis: The role of the cytoskeleton in elongation of the embryo. *Dev. Biol.* **117,** 156–173.

13. Edgar, L. G. (1995) Blastomere culture and analysis, in Caenorhabditis elegans: *Modern Biological Analysis of an Organism* (Epstein, H. G. and Shakes, D. C., eds.), Academic, San Diego, pp. 303–320.

14. Rogers, J. M. and Narotsky, M. G. (1999) Examination of the axial skeleton of fetal rodents. *Developmental Biology Protocols*, Methods in Molecular Biology, vol. 135 (Tuan, R. S. and Lo, C. W., eds.), Humana Press, Totowa, NJ.

15. Fire, A. (1994) A four-dimensional digital image archiving system for cell lineage tracing and retrospective embryology. *Comput. Appl. Biosci.* **10,** 443–447.

16. Soll, D. R., Voss, E., Varnum-Finney, B., and Wessels, D. (1988) "Dynamic Morphology System": A method for quantitating changes in shape, pseudopod formation, and motion in normal and mutant amoebae of *Dictyostelium discoideum. J. Cell. Biochem.* **37,** 177–192.

17. Minden, J. S., Agard, D. A., Sedat, J. W., and Alberts, B. M. (1989) Direct cell lineage analysis in *Drosophila melanogaster* by time-lapse, three-dimensional optical microscopy of living embryos. *J. Cell. Biol.* **109,** 505–516.

32

Cell Lineage Analysis in *Xenopus* Embryos

Sally A. Moody

1. Introduction

Cell lineage studies reveal what kinds of tissues descend from a single cell or specific region of an embryo. By defining precisely from which cells the various tissues and organs arise one can elucidate the mechanisms that control body organization, understand morphogenetic movements, and test the influence of exogenously applied gene products on these events. Because of easy accessibility, complete fate maps of the early cleavage stages of *Xenopus* have been published *(1–4)*. However, fate maps only describe the developmental path taken by a cell under normal, intact embryo conditions. Such studies cannot describe the full developmental potential of a cell or the times or mechanisms by which its fate is determined. The fate expressed by a cell, that is, the different tissue types that descend from it, is usually influenced by a number of factors, which may include maternal determinant molecules, cell–cell interactions, growth factor signals, and position within a morphogen gradient. Cell lineage tracing in *Xenopus*, therefore, is an essential technique to test the fate of a cell as it develops under novel experimental conditions *(5)*. Lineage tracing also is an important tool for labeling host tissues for use in tissue recombinant experiments. The ability to recognize the origin of embryonic tissues was critical for interpreting the pioneering experiments of embryonic inductions *(6,7)*. These early studies used pigmentation differences between donor and host species, but modern lineage labeling is more long lasting and reliable and allows tissues to be recombined within the same species.

A lineage label is a tracer that marks not only the cell of interest, but also all of its descendants. A molecule must meet several requirements to be a successful lineage tracer:

1. It must be nontoxic and nonreactive so that it does not change the developmental fate of the labeled cell.
2. It must be small enough to diffuse quickly through the injected cell before it divides, so that all of the descendants of that cell will be evenly and completely labeled.
3. It must be large enough not to pass through the gap junctions found between adjacent blastomeres *(8)*.
4. It must be detectable throughout development and not be diluted by cell division or intracellular degradation.

From: *Methods in Molecular Biology, Vol. 135: Developmental Biology Protocols, Vol. I*
Edited by: R. S. Tuan and C. W. Lo © Humana Press Inc., Totowa, NJ

5. It should be easily detectable by simple histological procedures. These requirements have been fulfilled by two classes of molecules: horseradish peroxidase (HRP) and fluorescent-dextrans *(9–12)*. HRP is a plant enzyme that has no natural substrate in animal cells and in frog is not recognized by the embryonic lysosomal compartment until late tadpole stages. Dextrans are hydrophilic polysaccharides that are biologically inert and resistant to a cell's endogenous glycosidases.

Another type of marker is mRNA encoding a tracer molecule. Because mRNAs do not diffuse as well as HRP or dextrans and therefore do not always mark the entire lineage of the injected cell, mRNAs are less accurate as lineage tracers. But, when mixed with a test mRNA, they are ideal for marking those cells that express the exogenously provided transcript. Two commonly used mRNA tracers are β-galactosidase (β-Gal; *see* **ref. *13***) and green fluorescent protein (GFP; *see* **ref. *14***). Both proteins are derived from nonvertebrates (β-Gal from bacteria and GFP from jellyfish), can be distinguished from endogenous vertebrate proteins, are too large to diffuse through gap junctions and have no known deleterious effects on developing vertebrate cells.

2. Materials

2.1. Equipment

1. PLI-100 Pico-Injector (Medical Systems Corp., Greenvale, NY), Narashige IM300 Microinjector (Narashige, Greenvale, NY) or Drummond Nanoject (Drummond Scientific, Broomall, PA) are the most common.
2. Programmable micropipet puller. A vertical puller, which makes very long, flexible tips, may suffice for injections of oocytes. A horizontal puller is required to manufacture the short, fine tips necessary for injections of fertilized eggs and embryos. Borosilicate capillary glass with an outer diameter of approx 0.8–1.0 mm. Glass should *not* contain a filament and may need to be siliconized (for backfilling only) or autoclaved (RNA injections only) before being pulled.
3. Dissection stereomicroscope placed on a steel plate in an area that is level and free of vibrations. Illumination should come from a fiber optic lamp, *not* by transillumination through the base of the microscope.
4. A micromanipulator with x-, y- and z-axes, mounted on a magnetic base secured to the steel plate.
5. Injection dish: This can be a 35-mm Petri dish filled with a base of blue, nontoxic, nondrying modeling clay (such as plasticine) in which 1.5-mm depressions have been made with a glass ball (crafted by melting the tip of a Pasteur pipet). Alternatively, affix a piece of Nitex mesh (cat. no. #8-670-176, Fisher Scientific, Pittsburgh, PA) to the bottom of a Petri dish with a few drops of chloroform or Superglue. The grids of the mesh are about the same diameter as the embryo (*see* **Note 1**).
6. Cryostat with microtome blade.
7. Tabletop centrifuge for 1.5-mL tubes.
8. Glass microscope slides coated with 2% gelatin (Type A, 300 Bloom), 0.08% chromium potassium sulfate.
9. Stage micrometer.
10. Fine sharpened forceps (e.g., Dumont #5 Biologie, Moria Instruments, Paris, France).
11. Hair loop: This can be crafted by placing both ends of a fine hair (approx 10 cm long) into the narrow tip of a 6-in Pasteur pipet to form a 2–3-mm loop. Seal the hair in place with melted wax.

2.2. Lineage Tracers

1. HRP: 5% horseradish peroxidase (Boehringer #814 393, EIA grade, 90% isoenzyme C, Roche Molecular Biochemicals, Indianapolis, IN) in sterile, distilled water (*see* **Note 2**). Dissolve in a small volume (5 mg in 100 μL). Transfer to a microfiltration tube (e.g., Costar Spin-X cellulose acetate device [Costar, Cambridge, MA] for aqueous solutions [Costar #8160; Fisher #07-200-385, Fisher Scientific, Pittsburgh, PA]). Spin at top speed in tabletop microcentrifuge for 15–20 min at room temperature (*see* **Note 3**). Aliquot the flow-through in small volumes (5–10 μL) and store at –20°C for up to 6 mo (*see* **Note 4**).

2. KCl: 0.2 *N* KCl, pH 6.8. Make 100 mL and adjust pH to 6.8 with 0.05 *M* KOH. The pH will overshoot and then gradually fall, so it can take about 3 h to stabilize at 6.8 (*see* **Note 5**). Filter sterilize and store at room temperature for several months.

3. Fluorescent dextrans: Use dextran-amines conjugated to fluorochromes (D-1863, Texas Red [Molecular Probes, Eugene, OR], 10,000 MW; D1817, tetramethyl rhodamine, 10,000 MW; D1845, fluorescein, 40,000 MW; Molecular Probes, Eugene, OR) (*see* **Notes 6** and **7**). Make approx 100 μL of a 0.5–1.0% solution of fluorescent dextran in the 0.2 *N* KCl solution (*see* **Note 8**). Vortex vigorously to dissolve. Transfer the dye solution to a microfiltration tube, filter, and store as described for HRP, but keep in the dark (*see* **Notes 3** and **4**).

4. β-Gal and GFP vectors can be purchased from a number of companies (e.g., #6047-1 and 6089-1, Clontech Labs, Palo Alto, CA). The coding region will need to be subcloned into a *Xenopus*-appropriate expression vector (e.g., pSP64RI or pCS2⁺), so that mRNA can be transcribed with upstream and downstream sequences allowing efficient translation in blastomeres after intracellular injection. Alternatively, *Xenopus* vectors already containing β-Gal or GFP can be obtained from a number of labs (search the *Xenopus* Molecular Marker Resource page on the World Wide Web, http://vize222.zo.utexas.edu). Injection RNA should be synthesized by in vitro transcription methods, as described in **ref. *13***, dissolved in sterile, RNase-free distilled water at a concentration of approx 1.0–1.5 μg/mL. Stocks should be stored at –80°C for approx 6 mo.

2.3. Embryo Production and Culture Solutions

1. HCG: human chorionic gonadotropin made with sterile water at a concentration of 100 IU/mL. Should be refrigerated and used within 1 mo.

2. Dejellying solutions: **Either** 2% cysteine hydrochloride (aqueous) pH to 8.1 by adding 10 *M* NaOH. Should be made fresh each day **OR** 0.05 *M* HEPES sodium salt stock, pH 8.9 and 6.5% dithiothreitol (DTT) stock. Refrigerate. Can be stored for months. On day of use, mix 10 mL HEPES stock, 2 mL DTT stock and 98 mL distilled water (*see* **Note 9**).

3. Ficoll (Amersham Pharmacia Biotech, Piscataway, NJ): 3–5% solution in either Steinberg's, 0.5X MMR, or 0.5X MBS. Make fresh each week, filter sterilize, and store in refrigerator. Warm to room temperature before use (*see* **Note 10**).

4. Steinberg's solution: 60 m*M* NaCl, 0.67 m*M* KCl, 0.83 m*M* $MgSO_4$, 0.34 m*M* $Ca(NO_3)_2$, 4 m*M* Tris-HCl and 0.66 m*M* Tris base, pH 7.4. Filter sterilize, store refrigerated, and use within 2 mo (*see* **Note 11**).

5. Marc's Modified Ringers (MMR): 100 m*M* NaCl, 2 m*M* KCl; 1 m*M* $MgSO_4$; 2 m*M* $CaCl_2$; 5 m*M* HEPES, pH 7.8; 0.1 m*M* EDTA, in distilled water. Filter sterilize and store at room temperature. Can be stored for months (*see* **Note 11**).

6. Modified Barth's Solution (MBS): 88 m*M* NaCl; 1 m*M* KCl, 1 m*M* $MgSO_4$, 0.7 m*M* $CaCl_2$, 5 m*M* HEPES, pH 7.8, 2.5 m*M* $NaHCO_3$ in distilled water. Filter sterilize and store at room temperature for approx 1 wk (*see* **Note 11**).

2.4. General Solutions

1. Phosphate buffer (PB): 0.1 *M*, pH 7.4. Make 0.2 *M* stocks of monobasic (Na_2HPO_4) PB and of dibasic (NaH_2PO_4) PB. Mix 4 parts monobasic PB stock, 1 part dibasic PB stock and 5 parts distilled water to reach final concentration and pH.
2. Phosphate-buffered saline (PBS): 0.1 *M* PB, 0.9% NaCl. Mix 40 mL monobasic PB stock and 10 mL dibasic PB stock. Dissolve 0.9 g NaCl, then bring up to 100-mL vol with distilled water (*see* **Note 12**).
3. Tris/glycerol: 1 part 0.1 *M* Trizma, pH 7.4–7.6, 9 parts glycerol.
4. Heavy mineral oil, sterile, RNase-free.
5. 30% hydrogen peroxide stock.
6. Graded series of ethanol or methanol (35, 50, 70, 80, 95, 100%).

2.5. Solutions for Fixation

1. For HRP-labeled specimens: 4% paraformaldehyde, 1% glutaraldehyde, 0.5% dimethyl-sulfoxide (DMSO) in 0.1 *M* PB. Add 4 g paraformaldehyde to 40 mL dH_2O. Stir constantly and heat to 60°C in a fumehood with the beaker covered with foil. Do not let the temperature rise above 65°C. Add 1 *N* NaOH dropwise until the solution clears. Cool solution on ice to room temperature. Add 2 mL of 50% EM-grade glutaraldehyde and 0.5 mL DMSO and mix thoroughly. Add 40 mL monobasic PB stock and 10 mL dibasic PB stock. Bring up to 100-mL vol with distilled water. Mix thoroughly and store in the refrigerator. Can be stored for months (*see* **Note 13**).
2. For β-Gal-labeled specimens: 2% paraformaldehyde, 0.2% glutaraldehyde, 0.02% NP-40, 0.01% sodium deoxycholate, in 0.1 *M* PB. Dissolve, clear, and cool paraformaldehyde, as previously described. Add 0.4 mL of 50% glutaraldehyde, 20 μL NP-40 and 10 μL sodium deoxycholate, mixing thoroughly after each addition. Add 40 mL of 0.2 *M* monobasic PB stock, 10 mL of 0.2 *M* dibasic PB stock and bring to 100 mL with distilled water. Mix thoroughly and store in the refrigerator. Can be stored for months (*see* **Note 13**).
3. For fluorescent-dextran and GFP-labeled specimens: 4% paraformaldehyde in PBS. Dissolve, clear, and cool paraformaldehyde, as previously described. Add 0.9 g NaCl. Add 40 mL of 0.2 *M* monobasic PB stock, 10 mL of 0.2 *M* dibasic PB stock and bring to 100 mL with distilled water. Can be stored at –20°C in small, single-use aliquots for weeks (*see* **Notes 12**, **14**, and **15**).
4. MS222: 0.3% methanesulfonate salt (3-aminobenzoic acid ethyl ester or tricaine).

2.6. Histochemical Solutions

1. β-Gal reaction buffer: 20 m*M* $K_3Fe(CN)_6$, 20 m*M* $K_4Fe(CN)_6$, 2 m*M* $MgCl_2$, 1 mg/mL X-gal (5-bromo-4-chloro-3-indolyl-β-D-galactopyranoside) in 0.1 *M* PB. Must be made fresh.
2. HRP reaction buffer: 1.0% diaminobenzidine (DAB; e.g., Sigma #D9015, 3,3'-diamino-benzidine tetrahydrochloride, Sigma Chemical Co., St. Louis, MO) stock in PB. Can be stored frozen in single use aliquots for months (*see* **Notes 16** and **17**).
3. Clearing solution: 1 part benzyl benzoate, 2 parts benzyl alcohol (BB/BA).

3. Methods

3.1. Embryo Production and Collection

1. Two major methods are used to obtain fertilized eggs: natural matings and in vitro fertilization. For both methods, adult frogs are primed by hormone injections. Typically, males receive an injection of 300–500 IU of HCG 2 d before the experiment and again 12–14 h

Fig. 1. Side view of plan for a tank designed to facilitate the collection of naturally fertilized eggs. The tank is constructed of opaque Plexiglass. It is divided into two chambers. In one chamber (left side) the mating frogs sit on a stiff plastic grid floor (crosshatched) through which fertilized eggs (small dots) can fall. They fall into square Petri dishes that fit tightly into a drawer that occupies the space below the grid. A tightly fitting lid, with air holes drilled through, covers the top of this chamber to prevent the frogs from escaping. To collect eggs throughout the day without disturbing the frogs, one simply reaches into the chamber on the right side, pulls open the drawer and removes the Petri dishes. By replacing the dishes each time, one can collect eggs multiple times.

    before the experiment. Females receive an injection of 600–800 IU 12–14 h before the experiment. Details for how to inject frogs can be found in **ref.** *15* (*see* **Notes 18** and **19**).

2. For natural mating, place the male and female frogs in a 15-gal tank filled with 8 gal of 10% culture medium (Steinberg's, MMR, MBS; *see* **Note 11**) 12 h prior to the time when fertilized eggs are desired. The bottom of the tank should contain square Petri dishes covered with a stiff plastic screen. The frogs should be left in the dark (we drape the chamber with black cloth) for the next 24 h. As eggs are laid, they drop through the plastic screen into the square Petri dishes and can be collected throughout the day. We use a specially constructed Plexiglass chamber for this purpose (**Fig. 1**).

3. For in vitro fertilization, the mature eggs are gently squeezed from the hormone-treated female into Petri dishes. Males are anesthetized by submersion in an ice bath with MS222, sacrificed, and their testes removed. The testes are minced, and the released sperm are added to the eggs. Details for this method can be found in *(16)*.

4. Remove the jelly coats from fertilized eggs that have just begun to cleave by gently swirling the eggs in 4X vol of dejellying solution either for 5–15 min (if using cysteine) or no longer than 4 min (if using HEPES/DTT). After the jelly coats are free, immediately wash embryos in four changes of embryo culture medium (either Steinberg's, MMR, or MBS) (*see* **Note 20**).

5. Transfer embryos to fresh embryo culture medium in a clean Petri dish. Store at 18–20°C until they begin to cleave.

3.2. Making Micropipets

1. Glass capillary tubes are pulled into fine tips strong enough to puncture the vitelline membrane, yet fine enough to cause minimal damage to the injected cell. Using a specialized

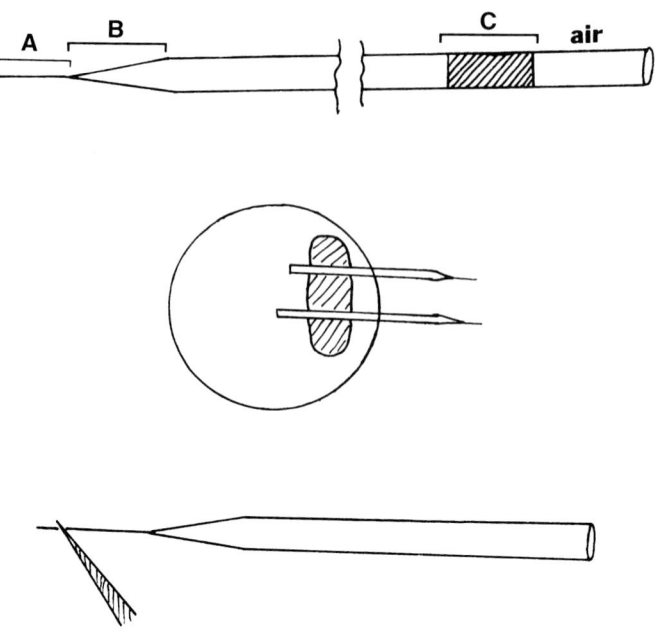

Fig. 2. Top: A glass micropipet should have a very fine secondary taper (**A**) and a short primary taper (**B**) each of which should be about 3–4 mm in length. (**C**) denotes where to place tracer when backfilling the micropipet. The dead airspace behind the tracer must be filled with heavy mineral oil if using a hydraulic microinjection system. Middle: To manually break the tip of the micropipet, hold it in place by pressing the distal end into clay (crosshatched) mounted on the lid of a 35-mm Petri dish (circle). Bottom: While viewing through the microscope, snip off the very tip of the micropipet at a 45° angle with fine forceps.

    micropipet puller, adjust parameters (e.g., heat, pull time, pull strength, etc.) according to manufacturer's instructions to fashion a micropipet with dimensions approximating those in **Fig. 2**.

2. Bevel the fine tip of the micropipet (which is fused by the heat of the pull) either with a commercial beveling device (e.g., Narashige EG-3 grinder) or by manually breaking the tip with fine forceps. To perform the latter, hold the glass micropipet in a strip of clay attached to the top of a 35-mm Petri dish, place it under a stereomicroscope and focus on the tip at high magnification. Snip off the very tip at about a 45° angle (**Fig. 2**).

3. Measure the outer diameter of the tip under a compound microscope using either a stage or eyepiece micrometer. Ten to 15 µm is ideal.

3.3. Filling the Micropipet

The method used depends on the type of microinjection equipment.

1. Backfilling is used if the injection equipment is not capable of drawing fluid into the tip of the micropipet (e.g., hydraulic systems and pressure systems that are not designed to back draw). Capillary glass should be siliconized before the micropipets are pulled. Place a small aliquot of tracer (5–10 µL) on a clean, nonabsorbent surface (e.g., Parafilm (Ted Pella Inc., Redding, CA) stretched over a 35-mm Petri dish). Using either a Hamilton syringe (#701-RN [Fisher Hamilton, Reno, NY]) or one provided by the microinjection apparatus company, draw up the aliquot into the syringe. This should be done under a

stereomicroscope to visually monitor that neither dust particles nor air bubbles are sucked into the needle. Insert the syringe needle into the open end of a micropipet whose tip has already been broken. Place the tip of the syringe needle about 10 mm from the open end and slowly deliver the aliquot. Make sure not to introduce air bubbles in the dye. For some equipment, these micropipets can be mounted directly onto the injection apparatus. For hydraulic systems, heavy mineral oil must be delivered in the same manner, replacing the dead airspace behind the dye aliquot (**Fig. 2**). Oil, completely free of air bubbles, should be in a 1-cc tuberculin syringe fitted with a 26-g needle.

2. Front filling is used by many air-pressure microinjection systems that have sufficient internal pressure to simply suck the solution into the micropipet through the tip. Place a small droplet of solution on Parafilm, as described previously. Mount the micropipet onto the injection apparatus and secure onto the micromanipulator. Under the stereomicroscope, submerge the tip in the center of the droplet, and activate the "fill" mode according to manufacturer's instructions. When filled, submerge the tip in a dish of culture medium so the tracer does not dry within the tip and clog it.

3.4. Calibration

Each micropipet must be calibrated, because the volume delivered per unit pressure depends on the inner diameter of the tip, which varies with every pipet. There are two common methods.

1. Place the micropipet on a fine ruler, and starting at the shoulder of the primary taper (**Fig. 2**), mark several 1-mm lengths along the shank with a permanent marker. Fill the micropipet with about 5 µL sterile water, as described in **Subheading 3.3.** Watching the meniscus of the water through the microscope, measure how many deliveries it takes to move the column of water 1 mm. Repeat a few times to assure consistency. The volume of the column of water is 1 mm X Πr^2, where $r = 0.5$ the inner diameter (ID) of the capillary glass. To determine the volume of each delivery, divide the volume of the column by the number of deliveries it took to move the meniscus 1 mm. Adjust the delivery time according to manufacturer's instructions to deliver desired amount (typically 0.5–5.0 nL) (*see* **Notes 21** and **22**). Expel the water and fill micropipet with tracer solution.

2. Fill micropipet with about 5 µL of sterile water. Place a droplet of sterile heavy mineral oil on the calibration lines of a stage micrometer. Under the stereomicroscope, lower the micropipet tip into the oil, just above the calibration lines. Expel a test droplet. The oil causes the aqueous solution to form a ball at the end of the micropipet tip. Measure the diameter of this ball with the stage micrometer and calculate volume by following equation: $4/3\Pi r^3$. Repeat several times to make sure the measurements are consistent. Adjust delivery time on the microinjection apparatus accordingly (*see* **Notes 21** and **22**).

3. After calibrating with water, expel the water and fill with desired lineage tracer. Just before filling the micropipet, briefly (20–30 s) spin the tracer to pellet any particulate material that might clog the tip.

3.5. Identifying Blastomeres

1. If the experimental goal is to globally express the RNA or to completely label the embryo with lineage tracer for tissue recombination experiments, then inject both blastomeres of the 2-cell embryo. Cleavage furrow patterns will not matter for these experiments. Injections should be placed just animal to the equator to facilitate diffusion of the marker (**Fig. 3**).

2. If your study requires localization of the marker to specific tissues or regions, it is essential to know where the dorsal side of the embryo will be (*see* **Note 23**). For in vitro fertilized eggs, the sperm entry point (SEP) should be marked; the dorsal midline develops

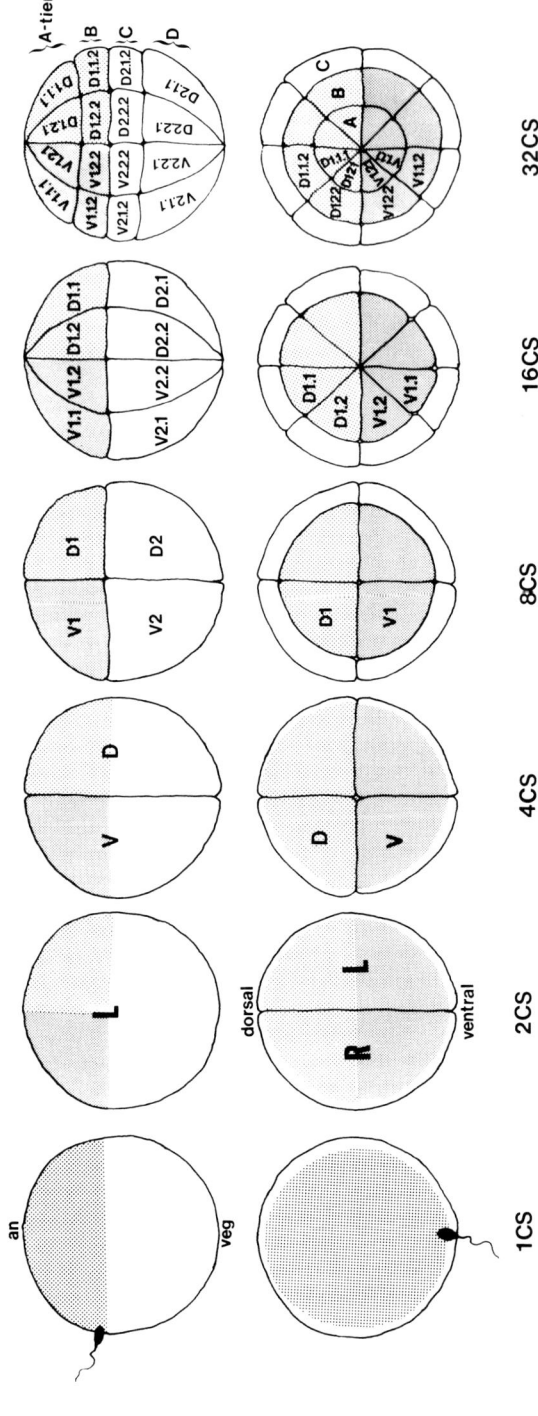

Fig. 3. Stereotypic cleavage patterns of embryos at each cell division from the 1-cell stage (1CS) to the 32-cell stage (32CS). Top row is a left side view, with animal pole (an) to the top and vegetal pole (veg) to the bottom. Bottom row is the corresponding animal pole views of each stage. Right (R) and left (L) sides are indicated. Shading indicates the pigmented animal hemisphere. Sperm indicate their point of entry (SEP) at the 1CS. In response, at the 2CS the pigmentation becomes asymmetric because of cortical rotation (17). The dorsal side is indicated either as the region 180° opposite the SEP (17) or by the lightly pigmented region of the animal hemisphere if bisected by the first cleavage furrow (18,19). Blastomere nomenclature is according to Jacobson and Hirose (23,24). The nomenclature for 32CS blastomere tiers, as designated by Nakamura and Kishiyama (22), are indicated to the right of 32CS embryos (see Note 27).

338

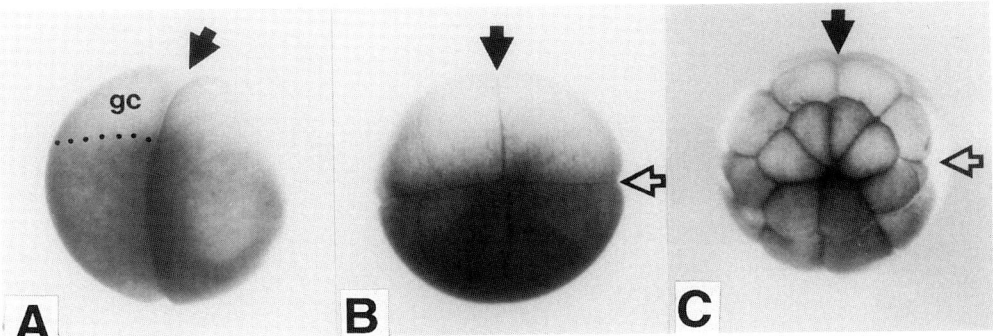

Fig. 4. Early cleavage stage embryos selected for pigmentation and cleavage patterns consistent with the fate maps *(1–4)*. **(A)** Embryos should be selected if the first cleavage furrow (arrow) bisects the lightly pigmented region of the animal hemisphere (gc, gray crescent). **(B)** In embryos like that depicted in (A), the second cleavage furrow (open arrow) will separate dorsal (light) from ventral (dark) blastomeres. **(C)** Stereotypic pattern of cleavages at the 32-cell stage result in identifiable blastomeres (compare to nomenclature in **Fig. 3**). Black arrow depicts the first cleavage furrow and open arrow depicts the second cleavage furrow.

about 180° opposite the SEP *(17)*. For naturally fertilized eggs, the dorsal side of the embryo can be predicted very accurately (>90%) by noting the orientation of the first cleavage furrow *(18,19)*. At fertilization, the animal hemisphere pigmentation begins to contract towards the SEP on the ventral side, causing the dorsal equatorial region to become less pigmented (**Fig. 3**). If the first cleavage furrow bisects this lighter area equally between the two daughter cells, then that lighter area can be used as the indicator of the dorsal side, and the first cleavage furrow will indicate the midsagittal plane (*see* **refs. *18*** and *19*; **Fig. 4**; *see* **Note 24**). Next, the embryos must be selected for regular cleavage furrows (*see* **Note 25**). These are found in a smaller and smaller percentage of embryos as cell divisions proceed. Embryos should be selected at each cleavage stage for their adherence to the patterns used for the published fate maps (**Figs. 3** and **4**; *see* **Notes 26** and **27**).

3.6. Microinjection

1. Place several embryos in an injection dish filled with either culture medium or Ficoll solution (*see* **Note 10**). With a hair loop or blunt forceps gently angle the embryos so that the desired cell is facing the micropipet (**Fig. 5**).
2. Using a micromanipulator, advance the tip of the micropipet toward the blastomere to be injected. To prevent ripping the cell membrane, angle the embryo and adjust the micromanipulator so the micropipet tip will be nearly perpendicular to the center of the cell (**Fig. 5**).
3. When the tip touches the vitelline membrane, there will be a little resistance. Advance the micropipet with the z-axis control knob of the manipulator into the target blastomere (*see* **Note 28**). Do not advance deeply into the cell in order to avoid damaging the nucleus (**Fig. 5**).
4. Deliver the tracer solution (approx 1 nL) according to the equipment used (*see* **Notes 21, 22,** and **29**). Wait about 10 s before removing the micropipet from the cell to prevent tracer from leaking out or blebs from forming. Do not move the embryo for several minutes, but go on to the next embryo in the dish.
5. After all embryos in the dish are injected, transfer the group to a 35-mm or 60-mm Petri dish containing either Ficoll or culture medium.

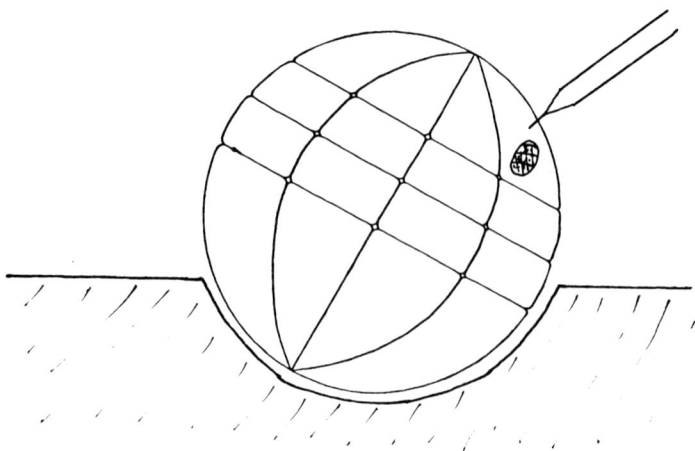

Fig. 5. Position the embryo in a well (hatching) in the injection dish so that the blastomere to be injected is facing the micropipet. The least damage will occur if the tip punctures the cell at a perpendicular angle and enters to only a short depth, avoiding the nucleus (darkened oval).

3.7. Embryo Culture

1. If embryos are cultured in Ficoll solution for injection, they can remain there no longer than 2 h. Transfer embryos to fresh culture medium in clean Petri dishes.
2. About 4–5 h after injection, transfer embryos to diluted culture medium (50% Steinberg's, 10% MMR, or 10% MBS).
3. Place no more than 10 embryos in a 35-mm Petri dish (20 per 60-mm dish), and fill the dish at least 70% full of solution.
4. Embryos injected with enzyme tracers can be raised on the laboratory bench at room temperature (20–22°C) or in an 18–20°C incubator. Embryos injected with fluorescent tracers should be raised in the dark either on the bench beneath an opaque plastic cover, such as the top of a black slide box, or in an incubator.
5. Change medium daily (50% Steinberg's, 10% MMR, or 10% MBS) and remove any dead embryos or debris to prevent bacterial infections (*see* **Notes 30** and **31**). By 4 d after fertilization, the culture medium should be diluted further (10% Steinberg's, 2% MMR, 2% MBS).
6. Culture embryos until the tissues of interest differentiate (*see* **Note 32**).
7. Live embryos labeled with fluorescent tracers can be viewed using epifluorescence or laser-confocal microscopy (*see* **Note 33**).

3.8. Fixation

1. To detect enzyme tracers or to cut tissue sections, embryos must be fixed in the appropriate solution.
2. If embryos are free-swimming tadpoles, anesthetize them before fixation by cooling on ice or adding several drops of MS222 to the culture medium.
3. Pick up embryos in a small volume of culture medium with a plastic transfer pipet and drop into a large volume of fixative in a vial with a tight-sealing cap. Use approx 40X vol of fixative per vol of embryo.
4. Place vials on a rotator for about 1 h at room temperature and then store overnight in the refrigerator. Embryos older than stage 38 should have a hole punctured in the skin to

promote infusion of the fixative; we snip off the tip of the tail. Total fixation time should be 6–12 h, depending on the size of the embryo.

5. Transfer to a fresh vial containing PB (for enzyme tracers) or PBS (for fluorescent tracers) and store in the refrigerator no longer than 1 wk.

3.9. Whole-Mount Preparations

1. Embryos labeled with fluorescent dextrans or GFP can be mounted in PBS in depression slides or in wells cut into an agar or Sylgard bed on a microscope slide. They are simply viewed under epifluorescence illumination or with laser confocal microscopy. Because the embryos are opaque, only surface labeling will be clearly visible (*see* **Notes 34** and **35**).

2. β-Gal-labeled embryos should be washed 3X in PB. Incubate at 37°C in reaction buffer for 1–24 h, depending on the strength of the enzyme activity. Rinse embryos 3X in PB, and refix for 1 h on rotator to stabilize the reaction product. Rinse 1X in PB. These can be viewed under epi-illumination on a dissecting or compound microscope or cleared (*see* **step 4**) for transillumination. Embryos can be dehydrated through a graded series to 100% methanol (or ethanol) for long-term storage (*see* **Notes 36–38**).

3. HRP-labeled embryos should be transferred to tightly sealing screw-cap vials and washed 3X in PB. Dilute DAB stock to 0.1% and add 1.0 mL to each vial. Agitate for 30 min at room temperature. Dilute stock hydrogen peroxide to 0.3%, and add 3.3 µL of the dilution to each vial. Agitate for another 15–30 min, frequently checking the reaction under a stereomicroscope. Stop the reaction by washing 5X 5 min with PB. These preparations can be viewed under epi-illumination on a dissecting or compound microscope or cleared (*see* **step 4**) for trans-illumination. Embryos can be dehydrated through a graded series to 100% methanol (or ethanol) for long-term storage (*see* **Notes 36–38**).

4. Clearing embryos: Embryos in which a stable histochemical reaction product has been fixed in place can be cleared to reveal the three-dimensional patterns of the labeled cells, including those in the internal organs. After histochemical reaction and postfixation, wash 2X in PB and dehydrate in a graded series of ethanol (30 min each in 25, 50, 75, 95, and 2X 15 min of 100%). Transfer to BB/BA solution in a depression slide or spot plate and view under microscope. HRP specimens can be stored in BB/BA. To store β-Gal specimens after viewing, they should be washed in 100% methanol (or ethanol) several times (they will again become opaque) and stored in 100% methanol (or ethanol) in the freezer (*see* **Note 39**).

3.10. Tissue Preparations

1. For tissue sectioning: Wash embryos in PB (for enzyme tracers) or PBS (for fluorescent tracers) containing 5% sucrose overnight. One hour before embedding, wash in PB/PBS containing 15% sucrose on a rotator. Place each embryo to be sectioned in a small volume of embedding material (e.g., O.C.T. Compound, VWR Scientific, West Chester, PA, TissueTek #4583, Sakura Finetek USA Inc., Torrance, CA) to remove excess PB/PBS. Mount embryo in a mold filled with embedding medium and freeze in cryostat. When frozen, section at 10–20 µm. Pick up sections on gelatin-coated slides and store in the freezer until ready for processing. If the tissue is fluorescent, it should be sectioned no thicker than 14 µm and the slides stored in the dark (*see* **Notes 40** and **41**).

2. Preparation of slides containing fluorescent tracers: Slides should be dried on a warming plate (37°C) for 20–30 min. They can be refixed (4% paraformaldehyde) onto the slides for 5–10 min to prevent sections from falling off during further processing. Wash slides 3X for 5 min in PBS. If double-labeling with immunofluorescence techniques is desired, it can be started at this point (*see* **Note 40**). Mount glass cover slips with an aqueous,

nonfluorescing mounting medium, such as Tris/Glycerol or one of several that are commercially available (*see* **Note 42**). To preserve the fluorescent signal, slides should be stored in the refrigerator or freezer in the dark (*see* **Note 41**).

3. Preparation of slides containing β-Gal-labeled tissue: Wash embryos 3X in PB and incubate at 37°C in reaction buffer for 10 min to 2 h, depending on the strength of the enzyme activity. Rinse slides 3X in PB, and refix for 10 min to stabilize the reaction product. Rinse slides 1X in PB, dehydrate in a series of ethanol (1 min each in 35, 50, 70, 80, 95%, 2X 1 min in 100%), clear in toluene (2X 1 min) and mount glass cover slips with a permanent medium. Allow slides to dry for at least 24 h, then box to keep clean. These can be stored and viewed for years.

4. Preparation of slides containing HRP-labeled tissue: Wash slides 2X in PB. Dilute DAB stock to 0.0125% (2.5-mL stock in 200 mL PB), add 6 μL stock hydrogen peroxide (to 200-mL reaction buffer) and submerge slides. Allow reaction to proceed at room temperature (usually 7–10 min). Check reaction under a microscope after approx 5 min. Stop reaction by rinsing 2X for 2 min in PB. Dehydrate through a graded series of ethanol, clear in two washes of toluene and mount glass cover slips with a permanent medium. Allow slides to dry for at least 24 h, then box to keep clean. These can be stored and viewed for years (*see* **Notes 16** and **17**).

4. Notes

1. Nitex-fitted injection dishes (Fisher Scientific) should be soaked in several changes of distilled water for several days before use to remove any toxic contaminants. Test the dish for toxicity by growing control embryos overnight in its wells.

2. Many isoforms of HRP are commercially available. The preparations may contain contaminants that are toxic (*see* **Note 29**), so each lot should be tested.

3. Microfiltration of HRP and dextrans is important to make the solution sterile and free of particles that would clog the micropipet tip.

4. Small aliquots (5 μL) of HRP and dextrans can be stored in sterile capillary tubes that are sealed at the ends with hematocrit tube-sealing compound to prevent evaporation. Longer storage than recommended is possible, but the dye can become toxic (*see* **Note 29**).

5. When adjusting the pH of the KCl solution, go slowly or you will overshoot. The pH is important for the intracellular health of the cell to be injected, so it must be adjusted carefully. Fluorescent tracers can be made in water, but they are far less stable and more toxic.

6. Purchase dextrans that are conjugated to lysine. The amino acid side chains allow them to be crosslinked to intracellular proteins by formaldehyde.

7. Purchase dextrans that are 10,000 or 40,000 MW for maximum diffusion within the cytoplasm of the injected cell. Larger molecules might not label the entire lineage.

8. The concentration of the tracer recommended for injection is based on a 1-nL injection vol: 1% fluorescein-dextran, 0.5% Texas Red-dextran, 1% rhodamine-dextran, 5% HRP, and 300–500 μg/mL of β-Gal or GFP mRNA.

9. The high pH of the dejellying solution is critical for it to work properly. A lowered pH will cause the dejellying process to take too long, causing damage to the embryos.

10. Incubating embryos in Ficoll (Amersham Pharmacia) causes the vitelline membrane to collapse onto the cell surface. This is an advantage during microinjection, because it exerts pressure on the puncture hole, preventing leakage of cytoplasm and the label. Do not use Ficoll if you plan to subsequently remove the vitelline membrane to perform a transplantation or dissection, because the vitelline membrane will collapse so tightly against the blastomeres that its manual removal without damaging the embryo is virtually impossible. A range of 3–5% Ficoll is used by different laboratories.

11. Steinberg's, MMR, and MBS are virtually interchangeable. Different labs prefer one to the other, mostly because of laboratory history. MBS has a shorter shelf life because it is buffered with bicarbonate.

12. Salt is added to fixatives and buffers for fluorochrome-labeled embryos to stabilize the tracer, lower background autofluorescence and improve antibody specificity if these tissues are to be labeled subsequently for proteins.

13. HRP and β-Gal are very hardy enzymes and their activity can withstand detergents and strong cross-linking agents in the fixatives. These additives promote fixative diffusion into the embryo and therefore prevent any breakdown of the tracer. Because detergents will come out of solution if frozen, these fixatives are stored in the refrigerator.

14. Detergents either autofluoresce or quench the fluorescence of fluorochromes. Glutaraldehyde also causes tissue to strongly autofluoresce. Therefore, these reagents are not included in fixatives for fluorochrome-labeled embryos.

15. Commercially purchased formaldehyde solution contains many breakdown products that either fluoresce themselves or cause tissues to autofluoresce. Therefore, fixative should be freshly made with granular paraformaldehyde. Single-use aliquots of fresh paraformaldehyde fixative for fluorochrome-labeled embryos are frozen to deter chemical degradation.

16. DAB is a carcinogen. It is safest to purchase it in a premeasured "isopac" and make up the stock solution in this bottle. For safe handling, the stock can be frozen in aliquots and stored for several months.

17. When working with DAB, always wear gloves and work on bench paper in case of spills. Keep a beaker of bleach nearby to deactivate any spilled DAB, and rinse all containers and surfaces with bleach after the DAB reaction is completed. Any liquid DAB waste should be deactivated with fresh bleach and stored in a designated waste container. The DAB solutions should turn dark brown when oxidized by the bleach and will eventually clear when the DAB has been completely broken down.

18. In vitro fertilization is ideal for obtaining large numbers of embryos on demand, synchronized to the same stage of development. However, it requires sacrificing the male frog, and the embryos do not always cleave in regular patterns that match the fate maps. Natural fertilization frequently produces regular cleavage patterns and was used for all the fate maps from our lab *(2–4,18)*. Natural fertilization provides developmental stages of embryos spread out over a long time period, which is advantageous when complex manipulations are planned. However, frogs do not always mate successfully on a time frame convenient to the experimenter's schedule.

19. Allow female frogs to rest at least 6–8 wk between hormone treatment to prevent stress and to allow them to replenish egg supplies.

20. Dejellying must be performed carefully. Do not dejelly prior to the appearance of the first cleavage or the dejellying solution will disable sperm. Do not agitate the eggs, as this can cause polyspermy and very irregular cleavages; eggs should be gently swirled at intervals. Watch for signs that the jelly is falling off the eggs; the eggs will touch one another, rather than being separated by their coats. As soon as this happens, wash away the dejellying solution, as it is quite toxic to the embryos.

21. A very small volume of the tracer molecule needs to be pressure-injected inside the blastomere without damaging that cell. There are several kinds of microinjection apparatus commercially available that range in complexity, but the most important feature to be considered is the typical injection volumes that you will use. Injection volumes into oocytes, fertilized eggs and 2-cell embryos can be as large as 10 nL (or sometimes larger), and thus simple equipment will suffice (e.g., Drummond Scientific "Nanoject" (Broomall, PA), which delivers a fixed 4 nL as its smallest volume). For older stages it is best to keep injection volumes to about 1 nL per blastomere.

22. Although some laboratories report injecting 4–10 nL of mRNA into each blastomere of a 2-cell embryo, these volumes are cytotoxic if using HRP or dextran tracers at the concentrations suggested above. Also, at stages later than two cells, large-volume injections of either mRNAs or tracers are not well tolerated by the cells and can result in artifactual fate changes. We found that injecting >10 nL of tracer into some 16-cell blastomeres can have significant effects on cell fate (e.g., drive epidermal lineages into brain) *(20)* and thus should be avoided.

23. The dorsal side of the embryo can be identified by marking the sperm entry point with a vital dye *(21)*, by tipping in vitro fertilized eggs and marking one side *(21)*, or by selecting embryos in which the first cleavage furrow bisects the gray crescent *(18,19)*.

24. One can also select embryos at the early part of the 4-cell cleavage, when the first and the second furrows at the vegetal pole can be distinguished; the first furrow should be complete and the second furrow not yet complete. If, however, you wait until the end of the 4-cell stage to select embryos, when you no longer can discriminate between the first and second cleavage furrows, the lightly pigmented cells may be dorsal ones in only approx 70% of embryos *(18,19)*.

25. Observe embryos frequently until they reach the required cleavage stage to ensure that cleavage furrows are dividing the cytoplasm in a regular pattern. The stereotyped pattern used for the fate maps is illustrated in **Fig. 3**. If spatial localization is critical for the interpretation of the experiment, select embryos that adhere to this ideal pattern, at least on the side of the embryo to be injected. Each cleavage cycle takes 20–30 min, depending on the temperature at which they are stored.

26. Not every blastomere in the embryo has to be "perfect." If you are targeting one specific cell, only that cell needs to cleave according to the ideal pattern.

27. Regarding blastomere nomenclature: Nakamura and Kishiyama *(22)* presented a very simple nomenclature for the 32-cell embryo that is simple to remember, because the tiers are labeled A–D (animal to vegetal) and the rows are labeled 1–4 (dorsal midline to ventral midline) (**Fig. 3**). When Jacobson and Hirose *(23,24)* began to map the nervous system lineages at all of the different cleavage stages, they devised a plan similar to those used in sea urchins and ascidians, which would relate the cells to their mothers, grandmothers, and descendants. Although these numbers and letters are harder to remember, there is a logic to the system that communicates lineal relationships. All cells starting with "D" are on the dorsal side of the embryo, and all cells starting with "V" are on the ventral side. A number is then added at each cleavage stage, which denotes the position of the blastomere with regard to poles and the midlines (**Fig. 3**; *see* **ref. 25**).

28. If the micropipet encounters a lot of resistance at the cell surface, the tip is blunt and will likely damage the cell. The puncture hole may be so large that cytoplasm (and the tracer) will leak out, or blebs will form at the puncture site. Discard these embryos (*see* **Note 21**). Rebreak the tip and calibrate again or use a new micropipet. If tip bends when it touches the vitelline membrane, the taper is too long. Adjust the programming of the pipet puller to forge a stronger tip.

29. If the volume or concentration of tracer is too large, the injected cell or some of its descendants will stop dividing. At the extreme, shortly after injection there will be one or two large cells in a field of smaller ones. However, damaged cells may not be noticed until a later point in development. There may be larger than normal, labeled cells incorporated into the organs of the embryo, or labeled cells that are the correct size, but spherical, rather than differentiated, in shape. This can happen to the entire clone or to only a subset of the clone. Often these cells will move to the correct regions of the embryo but will never differentiate. Another sign of damage is the accumulation of labeled cells in the spaces within the embryo (i.e., the central canal and ventricles of the nervous system and the

lumen of the gut, liver, and heart). These damaged cells probably dissociated from the rest of the embryo during gastrulation movements and accumulated wherever space appeared. If any of these signs of damage occur, discard the embryos and inject a smaller volume or concentration of tracer in the next experiment.

30. If bacterial infections are a problem, culture media can be supplemented with 0.1% gentamicin. This addition will reduce the storage life of the medium to approx 1 wk and may lower the pH.

31. Xenopus embryos prior to gastrulation prefer the pH to be approx 7.8. If the pH of the medium dips below 7.4, make up fresh solutions.

32. HRP and fluorescent-dextran tracers are detectable immediately. Tracer mRNAs require 2–3 h for adequate amounts of protein to be synthesized at detectable levels. All four tracer molecules can be detected in the descendants at least through tadpole stages 45–48. Because Xenopus cells decrease in size by cell division up through blastula stages (during which there are basically no G phases in the cell cycle), the originally injected concentration of tracer remains stable. The growth of the embryo after neurulation does not appear to significantly dilute the tracers. However, once tadpoles begin to feed, HRP and dextran labeling can become granular and uneven, suggesting that they are being packaged into lysosomes. Injected tracer mRNAs are probably degraded by the end of gastrulation, but both β-Gal and GFP proteins are very stable and can be detected at least through stage 45.

33. Excitation of the fluorochrome releases free radicals that can damage living cells, so view living embryos under low-light conditions or very, very briefly.

34. With fluorescent tracers, only cells near the surface can be analyzed in whole-mount preparations. Intracellular yolk platelets quench the fluorescence from the deeper tissues. Therefore, sectioning the tissue is necessary for analyzing the deep cells of a clone.

35. Fluorescence will be destroyed by the organic solvents used to store or clear embryos prepared with a histochemical reaction.

36. Enzymes (HRP and β-Gal) are detected by fixing them in place with aldehydes and providing them with a substrate that, when being altered by enzymatic action, becomes an insoluble, colored precipitate. This precipitate therefore indicates the cellular location of the enzyme, which was either directly injected or encoded for by the injected mRNA. Good fixation is essential for correct localization of the histochemical reaction product.

37. Anti-HRP and anti–β-Gal antibodies are commercially available if double-labeling with another protein marker is desired *(13)*.

38. Enzyme histochemical reactions provide essentially permanent specimens that can be referenced for years.

39. Clearing greatly improves the three-dimensional visualization of the staining pattern and reveals deep members of the clones. However, β-Gal specimens should not be stored in the clearing solution because the reaction product will fade after a few days. BB/BA will corrode any surface other than glass, especially microscope parts. Clean up spills with 100% ethanol.

40. Fluorescent tracers can be combined with other fluorescent methods, such as UV-excitable nuclear markers, fluorescent streptavidin to detect biotin-labeled compounds, and the immunological detection of cell type-specific proteins.

41. Fluorescent lineage tracers are not permanent, but they can be extremely hardy if the specimens are stored in the dark and refrigerated or frozen. For example, we have viewed 2-yr-old dextran and GFP-labeled tissue sections stored at –80°C with no detectable diminution of the signal.

42. To improve the life-span of fluorescent preparation, they must be mounted in a buffered, aqueous medium. Fluorescein absorption is especially sensitive to acidic pH, so the pH of the mounting medium must be between 7.4 and 7.6.

Acknowledgments

I wish to thank Kristy Kenyon and Petra Pandur for their advice on this protocol. This work is supported in part by NIH Grants EY10096 and NS23158.

References

1. Dale, L. and Slack, J. M. W. (1987) Fate map of the 32-cell stage of Xenopus laevis. *Development* **100**, 279–295.
2. Moody, S. A. (1987) Fates of the blastomeres of the 16-cell stage Xenopus embryo. *Dev. Biol.* **119**, 560–578.
3. Moody, S. A. (1987) Fates of the blastomeres of the 32-cell stage Xenopus embryo. *Dev. Biol.* **122**, 300–319.
4. Moody, S. A. and Kline, M. J. (1990) Segregation of fate during cleavage of frog (Xenopus laevis) blastomeres. *Anat. Embryol.* **182**, 347–362.
5. Moody, S. A. (1999) Testing the cell fate commitment of single blastomeres in *Xenopus laevis,* in *Advances in Molecular Biology: A Comparative Methods Approach to the Study of Oocytes and Embryos* (Richter, J., ed.), Oxford University Press, Oxford, UK, pp. 355–381.
6. Spemann, H. and Mangold, H. (1924) Induction of embryonic primordia by implantation of organizers from a different species, in *Foundations of Experimental Embryology* (Willier, B. H. and Oppenheimer, J. M., eds.), Hafner, New York, pp. 144–184.
7. Nieuwkoop, P. D. (1973) The "organization center" of the amphibian embryo: Its origin, spatial organization and morphogenetic action. *Adv. Morphogen.* **10**, 1–39.
8. Guthrie, S., Turin, L., and Warner, A. E. (1988) Patterns of junctional communication during development of the early amphibian embryo. *Development* **103**, 769–783.
9. Weisblat, D. A., Sawyer, R. T., and Stent, G. S. (1978) Cell lineage analysis by intracellular injection of a tracer enzyme. *Science* **202**, 1295–1298.
10. Jacobson, M. (1985) Clonal analysis and cell lineages of the vertebrate nervous system. *Annu. Rev. Neurosci.* **8**, 71–102.
11. Stent, G. S. and Weisblat, D. A. (1985) Cell lineage in the development of invertebrate nervous systems. *Annu. Rev. Neurosci.* **8**, 45–70.
12. Gimlich, R. L. and Braun, J. (1985) Improved fluorescent compounds for tracing cell lineage. *Dev. Biol.* **109**, 509–514.
13. Vize, P. D., Melton, D. A., Hemmati-Brivanlou, A., and Harland, R. M. (1991) Assays for gene function in developing Xenopus embryos. *Methods Cell Biol.* **36**, 367–387.
14. Chalfie, M., Tu, Y., Euskirchen, G., Ward, W. W., and Prasher, D. C. (1994) Green fluorescent protein as a marker for gene expression. *Science* **263**, 802–805.
15. Etheridge, A. L. and Richter, S. M. A. (1978) *Xenopus laevis: Rearing and Breeding the African Clawed Frog.* Nasco, Ft. Atkinson, WI.
16. Heasman, J., Holwill, S., and Wylie, C. C. (1991) Fertilization of cultured Xenopus oocytes and use in studies of maternally inherited molecules. *Methods Cell Biol.* **36**, 213–230.
17. Vincent, J.-P. and Gerhart, J. C. (1987) Subcortical rotation in Xenopus eggs: An early step in embryonic axis specification. *Dev. Biol.* **123**, 526–539.
18. Klein, S. L. (1987) The first cleavage furrow demarcates the dorsal-ventral axis in Xenopus embryos. *Dev. Biol.* **120**, 299–304.
19. Masho, R. (1990) Close correlation between the first cleavage plane and the body axis in early Xenopus embryos. *Dev. Growth Differ.* **32**, 57–64.
20. Hainski, A. M. and Moody, S. A. (1992) Xenopus maternal RNAs from a dorsal animal blastomere induce a secondary axis in host embryos. *Development* **116**, 347–355.
21. Peng, H. B. (1991) Appendix A: Solutions and protocols. *Methods Cell Biol.* **36**, 657–662.

22. Nakamura, O. and Kishiyama, K. (1971) Prospective fates of blastomeres at the 32-cell stage of Xenopus laevis embryos. *Proc. Japan Acad.* **47,** 407–412.

23. Hirose, G. and Jacobson, M. (1979) Clonal organization of the central nervous system of the frog. I. Clones stemming from individual blastomeres of the 16-cell and earlier stages. *Dev. Biol.* **71,** 191–202.

24. Jacobson, M. and Hirose, G. (1981) Clonal organization of the central nervous system of the frog. II. Clones stemming from individual blastomeres of the 32- and 64-cell stages. *J. Neurosci.* **1,** 271–284.

25. Sullivan, S. A., Moore, K. B., and Moody, S. A. (1998) Early events in frog blastomere fate determination, in *Cell Lineage and Fate Determination* (Moody, S. A., ed.), Academic, New York, pp. 297–321.

33

Photoactivatable (Caged) Fluorescein as a Cell Tracer for Fate Mapping in the Zebrafish Embryo

David J. Kozlowski and Eric S. Weinberg

1. Introduction

Classical fate mapping approaches in any embryo typically involve labeling of single cells with nondiffusible dyes by pressure or iontophoretic microinjection followed by tracing the fate(s) of those cells and their descendants throughout development. The limitation of any approach that involves microinjection as a means of cell labeling is that, as development progresses, cells typically decrease in size and become less accessible to mechanical manipulation. The approach presented here utilizes local activation of a photoactivatable (caged) dye that becomes fluorescent when illuminated with specific wavelengths of ultraviolet light (*1*).

At the one-cell stage, embryos are injected with a nonfluorescent caged dye, which will then be present in all cells of the developing embryo (*2*). By constraining the diameter of an ultraviolet light beam of the uncaging wavelength, specific embryonic regions, ranging in size from 1–2 cells to >20 cells, can be irradiated and fluorescently labeled at any site and at any time during development. The original position, morphogenetic movements, and ultimate fate(s) of these labeled cells can be followed and documented in live embryos by conventional fluorescence microscopy. In addition, immunohistochemical detection of cells containing uncaged fluorescein can provide a stable and permanent record of the developmental fates of labeled cells. This approach allows rapid generation of high-resolution fate maps of both early- and late-developing embryonic structures (*3*).

2. Materials

1. Buffers
   a. Injection buffer: 2X stock is 0.5% phenol red, 240 mM KCl, 40 mM HEPES-NaOH, pH 7.5.
   b. Embryo water: 60 mg Instant Ocean® (Aquarium Systems, Inc. Mentor, OH) per liter of deionized water. Buffer to pH 7.0 with $NaHCO_3$ (sodium bicarbonate, Fisher Scientific Co., Pittsburgh, PA).
   c. Phosphate-buffered saline (PBS): pH 7.4: composition per liter, 8 g NaCl, 0.2 g KCl, 0.61 g Na_2HPO_4, 0.2 g KH_2PO_4.

From: *Methods in Molecular Biology, Vol. 135: Developmental Biology Protocols, Vol. I*
Edited by: R. S. Tuan and C. W. Lo © Humana Press Inc., Totowa, NJ

    d. PBST: PBS containing 0.2% polyoxyethylene-sorbitan monolaurate (Tween-20, Sigma Chemical Co., St. Louis, MO).

    e. PBST-NCS: PBST containing 10% heat-inactivated newborn calf serum (Gibco-BRL, Gaithersburg, MD) and 1% dimethyl sulfoxide (DMSO, Sigma).

    f. Alkaline phosphatase reaction buffer: 100 mM Tris-HCl, pH 9.5, 100 mM NaCl, 50 mM MgCl$_2$.

2. Caged fluorescein: Stock solutions are 10% anionic DMNB-caged fluorescein (D-3310, 10,000 MW; Molecular Probes, Eugene, OR) or lysine fixable DMNB-caged fluorescein (D-7146, 10,000 MW; Molecular Probes) in DEPC-treated water. Stocks are spin filtered (0.22 μm spin filter, UFC 30GU25, Millipore Corp., Bedford, MA) and stored in aliquots at –70°C. Working solutions are 2.5% (anionic) or 5% (lysine-fixable) caged fluorescein (*see* **Note 1**) in DEPC-treated water and injection buffer, aliquoted and stored at –20°C. Working stocks are centrifuged at 4°C prior to each use. Each aliquot can be frozen and thawed several times without apparent loss of activity.

3. Microinjection setup: Agarose molds used to hold embryos during microinjection *(4)* are prepared the night before and equilibrated to 28°C before use. Embryos are pressure injected with an N$_2$-driven injector (Model PLI-90; Medical Systems Corp., Greenvale, NY) under a standard dissecting microscope (Leica MZ12; Leica Microsystems, Deerfield, IL).

4. Embryo viewing chamber: To view and manipulate embryos during the uncaging process, a microscope slide is covered with 7–10 layers of electrical tape from which a central hole (2 × 2 cm) is cut. The exact number of layers of tape (and thus the thickness of the chamber) needs to be determined empirically with the following goal: When an embryo (still in the chorion) is placed in the chamber and a cover slip is placed on top, the chorion becomes slightly compressed. By sliding the cover slip, the embryo (via the compressed chorion) can be rolled and precisely oriented while being viewed on the microscope stage.

5. Compound microscope equipped with epifluorescent illumination, including FITC or I3 (Leica; *see* **Notes 6** and **7**) and standard DAPI filter sets. Photolysis of the fluorescein caging group occurs in milliseconds at UV wavelengths less than 360 nm *(1)*, which are achieved with a standard 100-W Hg-arc lamp light source and DAPI filter set.

6. Pinhole device: To limit the diameter of the uncaging epifluorescent light beam, a stainless steel pinhole (100 μm precision pinhole, F36392, Edmund Scientific, Barrington, NJ) is placed in the epifluorescent illumination pathway (*see* **Note 4**).

7. Anesthetic: 25X stock is 4 mg tricaine (3-aminobenzoic acid ethyl ester (A-5040; Sigma) per milliliter of 20 mM Tris, pH 7.5. Dilute in embryo water for use with embryos older than 22 h.

8. Fixative: 4% paraformaldehyde solution is prepared by dissolving 2 g paraformaldehyde (Baker, Phillipsburg, NJ) in 50 mL of PBS (heating at 65°C will facilitate its going into solution). Store unused portions at –20°C in the dark.

9. Rehydration: Embryos are rehydrated in a graded MeOH series (MeOH:PBS 75:25, 50:50, 25:75, respectively).

10. Acetone (Aldrich, Milwaukee, WI) maintained at –20°C.

11. Alkaline phosphatase conjugated antifluorescein antibody (Fab fragment, Boehringer-Mannheim, Mannheim, Germany).

12. Endogenous alkaline phosphatase inhibitor: levamisole (Sigma).

13. Alkaline phosphatase substrate: 5-bromo-4-chloro-3-indolyl-phosphate (BCIP; Boehringer-Mannheim) and 4-nitroblue tetrazolium chloride (NBT; Boehringer-Mannheim). Stock solutions are 50 mg BCIP per milliliter of N,N,-dimethyl formamide (DMF; Fisher Scientific) and 75 mg NBT per milliliter of 70% DMF.

3. Methods

1. Microinjection of newly fertilized eggs: Embryos of the one-to-four cell stage are collected from breeding chambers (*5*) and rinsed several times with embryo water to remove food and fecal debris. With a wide-bore Pasteur pipet (#13-678-30; Fisher Scientific), embryos are transferred to the grooves of a previously prepared agarose injection mold. 1–3 nL of a buffer-caged fluorescein mixture is pressure microinjected into the yolk cell or a single blastomere of each embryo (*see* **Note 2**). It is important for subsequent manipulations in the viewing chamber that the chorion remains intact (i.e., not wrinkled or collapsed). Following microinjection, embryos are collected into a 35 × 10 mm Petri dish with 28°C embryo water and incubated in the dark at 28°C until they reach 50% epiboly (shield stage) or another developmental stage of interest (*see* **Note 3**).

2. Local uncaging of the dye: A shield-stage embryo is placed on the viewing chamber, in embryo water, using a wide-bore Pasteur pipet. Gently press a cover slip over the chamber and blot away excess water with a Kimwipe (Kimberly-Clark, Roswell, GA). Using a compound microscope, tungsten illumination, and a low-power objective, orient the embryo (by gently sliding the cover slip) so that the cells to be irradiated are located at the topmost aspect of the embryo. Switching to a higher power objective, place the cells to be irradiated in sharp focus.

   To locate and aim the illuminating UV light beam, place the pinhole device in the epifluorescent light pathway (*see* **Notes 4** and **5**), and by looking through the FITC filter set, locate the position of the UV light beam (you will see a dull-green fluorescent spot) on the embryo. Move the stage so that the cells (still in focus) to be irradiated are incident to the light beam. To uncage the dye, switch to the DAPI filter set and irradiate the cells for 2–4 s. To view cells containing the uncaged fluorescein, remove the pinhole and observe with FITC optics (*see* **Note 6**). Document the position of the uncaged spot (*see* **step 3**) and allow the embryo(s) to develop in the dark at 28°C until the fate(s) of the uncaged cells are to be determined.

3. Documentation: Critical to the resolution of any fate map is the ability to precisely locate the original position of uncaged cells. Using a 35-mm or CCD camera, carefully document the position of the uncaged spot while simultaneously viewing under brightfield and I3 fluorescence optics (alternatively, sequential brightfield and FITC fluorescence optics can be used; *see* **Note 7**). Accurate documentation of the position of the uncaged cells in the embryo will often necessitate several views. For embryos uncaged at the shield stage we routinely document both an animal view (to determine the dorsoventral position of the spot) and a lateral view (to determine the animal-marginal position of the spot) of the embryo (**Fig. 1A,B**; *see* **Note 8**).

   To compile data from several experiments with the least amount of error, photos (also negatives or slides) of documented embryos are scanned into a graphics program (e.g., Adobe Photoshop®) so that the position of cells containing the uncaged fluorescein can be transferred to a superimposable polar coordinate grid. Through the graphic program, the size of the grid can be adjusted for minor variations in embryo size or shape so that grids from many embryos can be normalized to a single reference image that will become the basis of the 6-h fate map.

   The ultimate fate(s) of the uncaged cells must also be determined and documented. To view live embryos older than 22 h (which spontaneously twitch), it is necessary to use an anesthetic to limit movement during photographic exposures. These embryos are mounted (in the chorion if unhatched) in tricaine on the viewing chamber. Again, by sliding the cover slip, the embryo can be manipulated so that any part of the embryo can be brought into view.

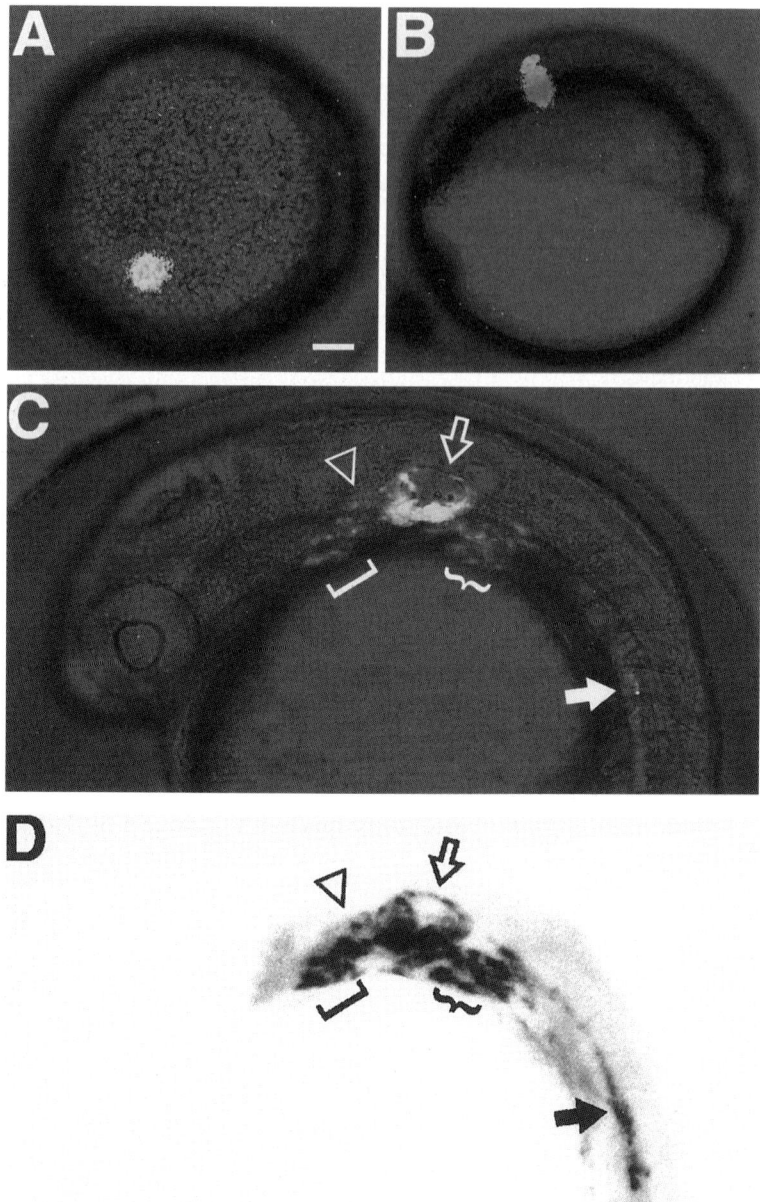

Fig. 1. (*See* color plate 10 appearing after p. 258.) Detection of uncaged fluorescein (lysine-fixable) in zebrafish embryos: **(A)**–**(C)** Live embryos viewed with combined I3 epifluorescent and brightfield illumination (*see* **Note 7**). **(A)** Animal and **(B)** lateral (respectively, dorsal/shield to right) views of a 6-h embryo immediately after uncaging. Cells containing active (uncaged) fluorescein are yellow-green. (C) Lateral view (anterior to left, dorsal to top) of 24-h embryo, demonstrating that cells originally labeled at 6 h (yellow-green) now populate the otic vesicle (*hollow arrow*), anterior (*triangle*), and posterior (*solid arrow*) lateral line primordia, anterior arches (*bracket*) and branchial arches (*curved bracket*). **(D)** Embryo from (C) fixed and stained with an alkaline phosphatase conjugated antifluorescein antibody to detect cells containing uncaged fluorescein. Cells containing uncaged fluorescein are purple. Symbols and orientation as in (C). Bar = 100 µm.

The signal from the uncaged fluorescein remains strong through at least 48 h of development (shown for 24 h in **Fig. 1C**), although the background (due to autofluorescence and prolonged viewing with high-power objectives) does increase through development. We recommend limited viewing of the uncaged cells, as there appears to be low-level photoactivation of the caged dye at wavelengths >360 nm. For a permanent record of the fate(s) of embryonic cells containing uncaged dye, embryos injected with the lysine-fixable caged fluorescein can be fixed and uncaged fluorescein detected immunohistochemically.

4. Embryo fixation, dehydration and storage: Individual embryos (one per well of a 24-well microtiter plate with an additional well containing uninjected "junk" embryos) are fixed in 4% paraformaldehyde for 2 h at room temperature (or 4°C overnight). Rinse embryos 3X in PBS and manually dechorionate (if necessary) in PBS. Dehydrate and store embryos in MeOH at –20°C.

5. Rehydration and permeabilization of embryos: Embryos are rehydrated as follows: 5 min each in a decreasing MeOH series (MeOH:PBS 75:25, 50:50, 25:75, respectively) and washed 2X (5 min each) with PBS. Incubate embryos in –20°C acetone for 7 min at room temperature, then wash 3X (5 min each) in PBS.

6. Detection and development of uncaged fluorescein signal: Preabsorb 1° antibody by incubating uninjected "junk" embryos with antifluorescein antibody (1:400) in PBST-NCS for 4 h at room temperature (or overnight at 4°C). Block the injected embryos in PBST-NCS for 1 h at room temperature. Incubate injected embryos with preabsorbed antifluorescein antibody (1:4000) in PBST-NCS at 4°C overnight. Wash embryos 4X (15 min each) with PBST. Equilibrate embryos 2X (15 min each) with alkaline phosphatase reaction buffer containing 2 mM levamisole. Stain the embryos by incubating in new alkaline phosphatase reaction buffer containing 3.5 μL BCIP and 4.5 μL NBT stocks per milliliter of AP reaction buffer containing levamisole. Allow the reaction to continue for 15–60 min (assessing reaction progress every 5–10 min) in the dark at room temperature. Stop the reaction by washing 3X with PBS. Refix embryos in 4% paraformaldehyde for 2 h at room temperature (or 4°C overnight). Wash 3X with PBS. Mount embryos in >80% glycerol in PBS and view under a dissecting microscope with reflected light. The signal appears as distinct purple deposits (**Fig. 1D**).

4. Notes

1. The lysine-fixable caged fluorescein appears to be less bright than the anionic-caged fluorescein. Of these two caged-fluorescein products, the lot number for the fixable-caged fluorescein indicates that there are fewer (approx 30%) conjugated fluorescein molecules per mole of dextran than the anionic version.

2. The early blastomeres of a zebrafish embryo are syncytial until the 5th cleavage (*2*) so that injection into the yolk or a single cell results in the labeling of all the early blastomeres and their descendants.

3. The shield stage is the first time in zebrafish development when the dorsoventral axis is morphologically apparent. This stage is just after the time when cell fates are generally restricted (*6*) and a fate map is possible to construct. The use of uncaged fluorescein allows fate mapping without the time-consuming and sometimes difficult single-cell injections utilized in previous studies (*6,7*).

4. The stainless steel circle containing a pinhole is cut to fit within the recess located over the field aperture of the reflected light module of a Leica DMRXA compound microscope (on some microscopes, a diaphragm in the epifluorescent light path may close down to a sufficiently small aperture for this purpose). Because the pinhole will not be perfectly

centered nor easily centrable, it is necessary to empirically determine the size (diameter) and position of the illuminating beam with respect to the field reticule. Often the pinhole will not be in the center of the light path, so that the relative location of UV light beam as viewed through FITC optics will not be in exact register as when viewed through DAPI optics. If this is the case, a test embryo is used to determine the location of the UV light beam (with respect to the reticule) when viewed through DAPI optics.

Alternatively, more consistent and faster results (i.e., size and location of pinholes) are achieved with a fabricated device containing interchangeable pinholes. We have had excellent results with one that employs a sliding bar mechanism containing variously sized pinholes located along the length of a moveable bar. Set screws allow each pinhole to be aligned within the field reticule and greatly increases the speed at which the uncaging beam can be aimed and embryos accurately uncaged.

5. When light enters the objective via the incident light path, the objective acts like a condenser decreasing the diameter of the light beam. Thus, the diameter of the uncaging light beam can be manipulated by changing the power of the objective used.

6. We routinely obtain stronger fluorescence signal (at the expense of higher background) using an I3 long-pass filter set (Leica). Either the FITC or I3 filter set will give satisfactory results (but also *see* **Note 7**).

7. To obtain both fluorescent and brightfield images on a single photographic exposure, place a red filter over the transmitted white light (tungsten) source and then adjust its intensity until it is about equal to the intensity of the observed fluorescence with I3 epifluorescent illumination. This will give the brightfield image a red cast with a green fluorescent signal (note that this only works with a long-pass emission filter). We routinely use 35-mm film (800 ASA) to document embryos. Photographic exposure time is determined automatically through an automatic exposure meter (Leica MPS 48).

8. The greatest source of error in documenting the position of labeled cells appears to be from embryos that are not perfectly aligned in the documented animal or lateral views. Misalignment of the former view induces errors in estimation of latitude, whereas misalignment in the latter view affects estimation of lines of longitude. It is often necessary to use both views of an embryo to correct for potential errors in estimation of exact cell location.

Acknowledgment

The authors thank Tohru Murakami and John Kanki for their help in the early development of this protocol, Alvin Chin for critical reading of this manuscript, and Robert Ho for originally suggesting to us the use of photoactivatable dyes.

References

1. Haugland, R. P. (1996) *Handbook of Fluorescent Probes and Research Chemicals*, 6th ed. Molecular Probes, Eugene, OR.
2. Kimmel, C. B. and Law, R. D. (1985) Cell lineage of zebrafish blastomeres. I. Cleavage pattern and cytoplasmic bridges between cells. *Dev. Biol.* **108,** 78.
3. Kozlowski, D. J., Murakami, T., Ho, R. K., and Weinberg, E. S. (1997) Regional cell movement and tissue patterning in the zebrafish embryo revealed by fate mapping with caged fluorescein. *Biochem. Cell Biol.* **75,** 551–562.
4. Weinberg, E. S. (1993) A device to hold zebrafish embryos during microinjection, in *The Zebrafish Book, A Guide for the Laboratory Use of Zebrafish* (Brachydanio rerio), (Westerfield, M., ed.), University of Oregon Press, Eugene, OR, pp. 5.1–5.4.

5. Westerfield, M., ed. (1993) *The Zebrafish Book, A Guide for the Laboratory Use of Zebrafish* (Brachydanio rerio), University of Oregon Press, Eugene, OR, pp. 2.8, 2.9.
6. Kimmel, C. B., Warga, R. M., and Schilling, T. F. (1990) Origin and organization of the zebrafish fate map. *Development* **108,** 581.
7. Woo, K. and Fraser, S. E. (1995) Order and coherence in the fate map of the zebrafish nervous system. *Development* **121,** 259.

34

Carboxyfluorescein as a Marker at Both Light and Electron Microscope Levels to Follow Cell Lineage in the Embryo

Dazhong Sun, C. May Griffith, and Elizabeth D. Hay

1. Introduction

Lipid-soluble dyes have been found to be useful markers for studying cell lineages in embryos (1,2). They have the advantage of diffusing into or across the plasmalemma, eliminating the need for microinjection. DiI (1,1 dioctadecyl 3,3,3',3' tetramethylindocarbocyanine perchlorate) is irreversibly incorporated into the plasmalemma and has been used extensively to follow cell fates, but it has the disadvantage of not surviving fixation or wax embedding (2–4). The frozen sections of unfixed tissue that must be examined lack cytological detail. On the other hand, 5, 6 carboxy 2',7' dichlorofluorescein diacetate succinimidyl ester (CCSFE), a carboxyfluorescein derivative that resists bleaching (5) better than the monochloro-form (1), is a stable cytoplasmic marker that stands up to formaldehyde fixation and wax embedding. In this report, we show that CCSFE also tolerates glutaraldehyde and osmium tetroxide fixation, embedding in epoxy resin, sectioning, and transmission electron microscopy (TEM).

Our studies using CCSFE have been carried out on the developing mouse and chicken palates with the objective of proving that the adherent medial edge epithelia (MEE) of the palatal shelves form a midline seam that transforms to mesenchyme, bringing about palatal fusion (5,6). In order to specifically label the MEE, the lower jaw of the embryo is removed and the upper head exposed to CCSFE, which enters but does not cross the MEE of the prefusion palatal shelves (**Fig. 1A**). The incorporated CCSFE occurs in fluorescent bodies (arrowheads, **Fig. 1A**) that appear as electron-dense bodies by TEM (arrowheads, **Fig. 1B**). The palatal shelves placed in close apposition in vitro adhere to form a midline seam containing fluorescent bodies we first referred to as phagosomes (5), but they are not. They do not contain acid phosphatase (compare arrowheads in **Fig. 2A,B**).

We now refer to the carboxyfluorescein-containing bodies as "isolation bodies," because they clearly represent a mechanism by which the cell isolates carboxyfluorescein from the rest of the cytoplasm. When CCSFE diffuses across the plasmalemma into the cytoplasm, it is cleaved by intracellular esterases into a water-soluble compound that cannot recross the plasmalemma. These potentially toxic, small

From: *Methods in Molecular Biology, Vol. 135: Developmental Biology Protocols, Vol. I*
Edited by: R. S. Tuan and C. W. Lo © Humana Press Inc., Totowa, NJ

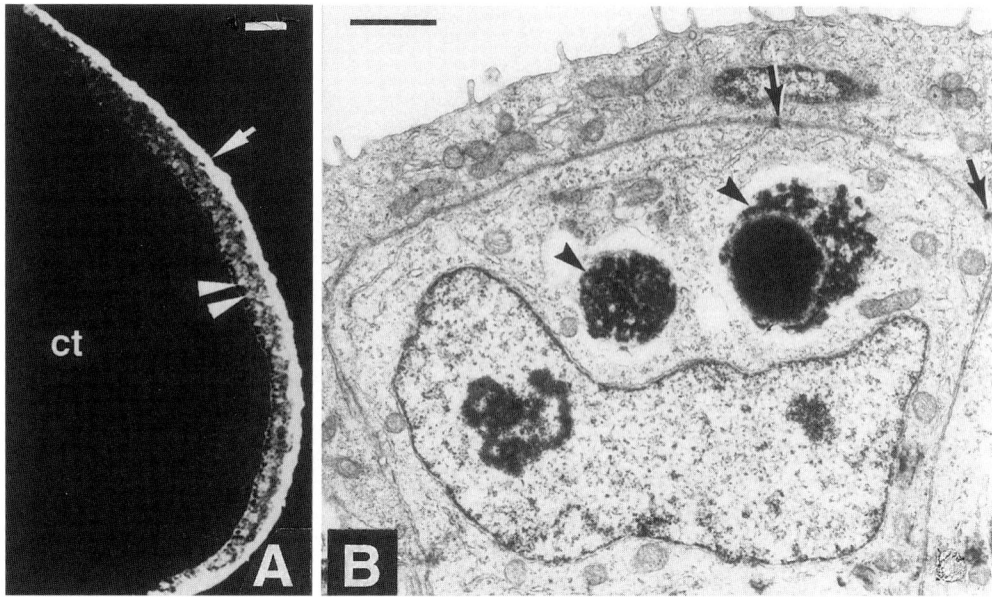

Fig. 1. Single avian palatal shelves labeled with CCSFE for 1 h. **(A)** Periderm is intensely fluorescent by confocal microscopy (arrow) and basal layer less so, as the molecule reaches it through gap junctions. Fluorescent isolation bodies are numerous (arrowheads). Connective tissue (ct) is not labeled. **(B)** By TEM, isolation bodies (arrowheads) can be seen to be enclosed by membrane (open arrow). Arrows = desmosomes. Bars, A = 100 μm; B = 1 μm (from **ref. 6**).

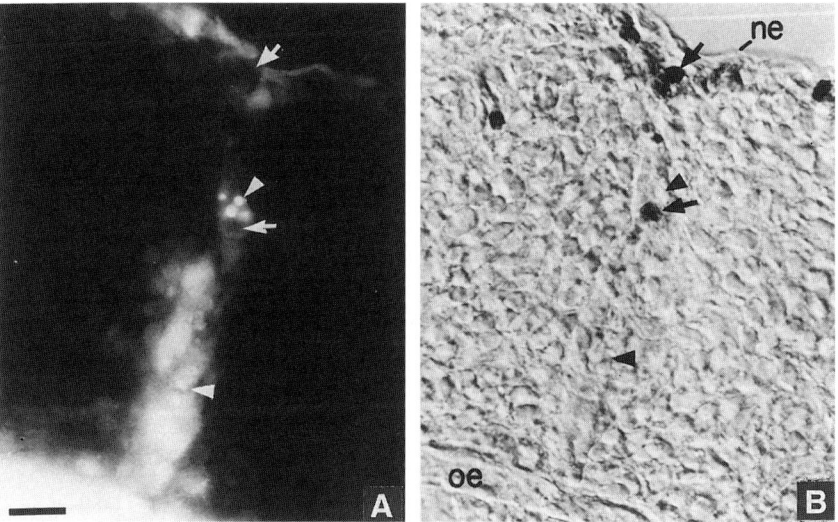

Fig. 2. Adherent mouse palate has formed a midline seam at 24 h after CCSFE labeling of the single palatal shelves. **(A)** Isolation bodies (arrowheads) can be seen against a background of fluorescence in the seam. Lysosomes (arrows) are not fluorescent. **(B)** The same section has been stained by the Gomori method for acid phosphatase. The isolation bodies (arrowheads) are not stained. Lysosomes (arrows) are stained. ne = nasal epithelium. oe = oral epithelium. Bar = 25 μm (from **ref. 5**).

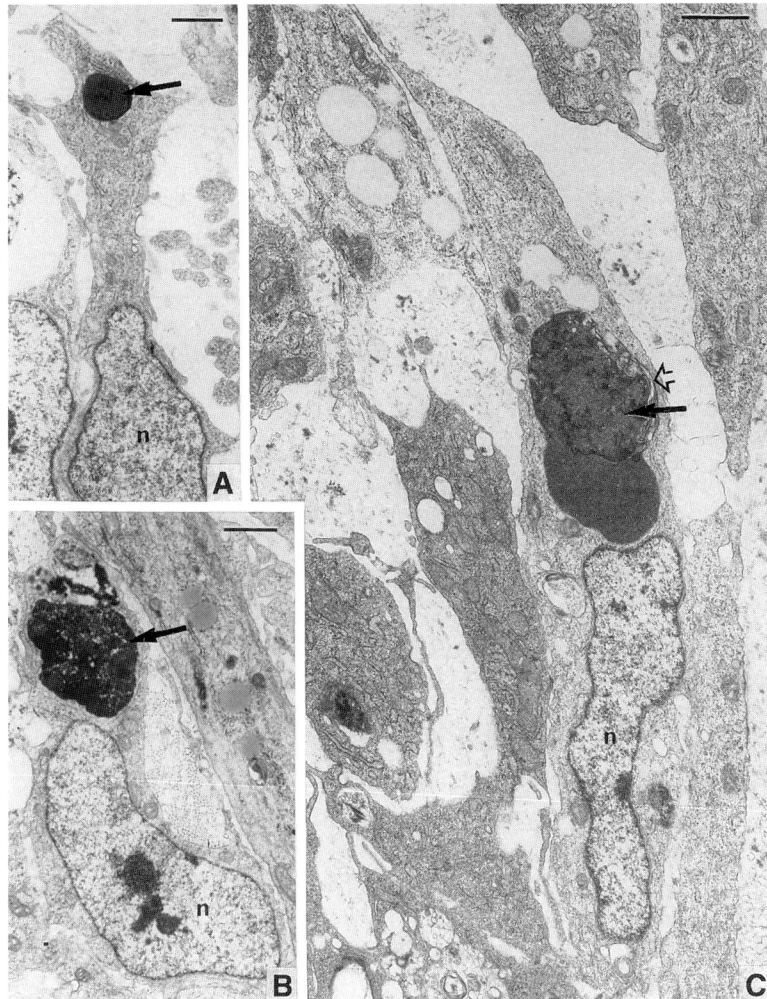

Fig. 3. Mesenchymal cells derived from the carboxyfluorescein labeled MEE are identified as fibroblasts by their fine structure. By 72 h in vitro, the midline seam has transformed completely to fibroblasts in avian palates treated with TGFβ3. Isolation bodies (arrows) are enclosed in membrane (open arrow). n = nucleus. Bars = 1 μm (from **ref. 6**).

fluorescein molecules are not proteins and do not enter the lysosome degradation pathway. Instead, they are incorporated into electron-dense bodies surrounded by cell membrane (open arrows, **Figs. 1B** and **3C**), where they persist indefinitely *(5,6)*. During cell division, they do not divide, but pass to one or the other of the two resulting cells (arrows, **Fig. 3A–C**), thus retaining their initial size and permitting the mesenchymal progeny of the MEE containing them to be identified easily at subsequent time points in vitro. Moreover, after injection of CCSFE into the amniotic cavity of embryos in utero, labeled fibroblasts derived from the MEE can be identified 4 d later at the light microscopic level (**Fig. 4**) and by TEM *(5)* in vivo.

The great advantage of using TEM with CCSFE is the clarity of identification of the labeled cell type. Epithelia can be recognized not only by their location (**Fig. 1A**), but

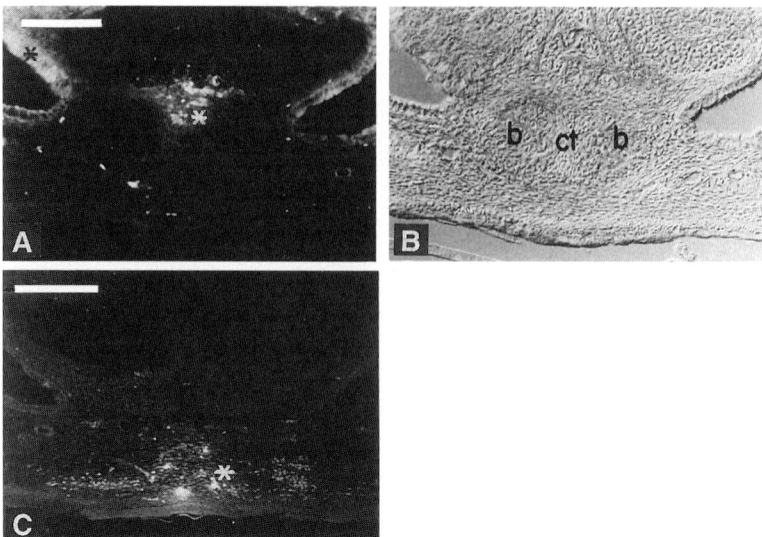

Fig. 4. CCSFE was injected into the amniotic cavity to label the palate MEE in vivo. After 4 d, palatal shelves have fused with nasal septum (A, B) and each other (C) and formed labeled fibroblasts (*) by epithelial-mesenchymal transformation. b = bone; ct = connective tissue. Bars = 100 µm (from **ref. 5**).

also by their contiguity, microvilli, desmosomes (arrows, **Fig. 1B**), and the other characteristics of this tissue phenotype. The mesenchymal cells, which in this case derived from the MEE, can be recognized as fibroblasts by the elongate cell shape, pseudopodia and filopodia, and abundant RER revealed by TEM (**Fig. 3**). Both CCSFE and DiI are very useful markers to follow epithelial-mesenchymal transformation because it is so easy to administer them to the epithelium. However, only CCSFE provides the cytological detail that can be obtained by fixing and embedding. We believe the TEM technique enormously extends the usefulness of this cell marker, and here we present the methods for its use with CCSFE in detail.

2. Materials

1. Buffers:
   a. Hanks' balanced salt solution (HBSS), pH 7.4, containing 3.5% $NaHCO_3$. Composition per 500 mL: 50 mL 10X HBSS (Gibco-BRL, Grand Island, NY), 0.175 g $NaHCO_3$. Warm up to 37°C prior to use.
   b. Phosphate-buffered saline (PBS), pH 7.4. Composition per liter: 7.2 g NaCl, 1.28 g Na_2HPO_4, 0.43 g KH_2PO_4.
   c. 0.2 M cacodylate buffer, pH 7.4. Composition per 250 mL: 10.05 g sodium cacodylic acid (Sigma Chemical Co., St. Louis, MO).
   d. 0.1 M sodium acetate buffer, pH 5.2, containing 7.5% sucrose.
2. Culture medium and gel:
   a. Dulbecco's modified Eagle's medium/Ham's nutrient mixture F12 culture medium (DMEM/F12, Gibco-BRL), pH 7.4. Add 5% fetal bovine serum, 3.15 mg/mL glucose, 2.5 mM L-glutamine, 0.055 mg/mL sodium pyruvate, 0.25 mg/mL ascorbic acid, and 1% penicillin-streptomycin. Warm up to 37°C prior to use.

  b. Agar gel for culturing: 0.5% agar gel in DMEM/F12 culture medium with additives as described above. Make fresh every time.
 3. Carboxyfluorescein solutions: Prepare CCFSE (Molecular Probes, Inc., Eugene, OR) as a 10-mM (6.26 mg/mL) stock solution in dimethyl sulphoxide (DMSO). Aliquot and store the solution at –20°C. Warm up at room temperature, dilute the stock 1:500 in prewarmed HBSS prior to use. Make fresh every time.
 4. Fixatives:
  a. 4% paraformaldehyde in PBS. For the stock, add 8 g of paraformaldehyde (purified grade, Fisher Scientific, Pittsburgh, PA) to 100 mL of distilled water, stir, and heat to 60°C, adding a few drops of 10 M NaOH to dissolve the paraformaldehyde. For the working solution, dilute this stock with an equal volume of 2X PBS, bringing the final concentration of paraformaldehyde to 4% prior to use. Aliquot and store the stock for up to a month at 4°C or longer at –20°C. If the pH remains neutral, the molecule has not turned to formic acid, so the fixative is good. (Formic acid is not a fixative.)
  b. 4% paraformaldehyde plus 4% glutaraldehyde in 0.1 M cacodylate buffer. Composition of working solution per 50 mL: 8 mL of 25% glutaraldehyde stock (Grade 1, Sigma), 25 mL of 8% formaldehyde stock, bringing to the final volume with 0.2 M cacodylate buffer. Make the amount needed to fix fresh from stock each time.
  c. 1% osmium tetroxide in 0.1 M cacodylate buffer. Add 1 g of osmium tetroxide (Merck & Co., Inc., Rahway, NJ) to 50 mL distilled water and stir with a glass rod until it dissolves. Osmium tetroxide is very toxic, so protect yourself while preparing the solution. The best way to prepare it is to put the vial containing the dry OsO_4 in a glass bottle and break the vial with a glass rod, then add distilled water to make the stock. Store this stock at 4°C (to avoid light, cover with aluminum foil). Dilute the solution needed with equal volume of 0.2 M cacodylate buffer prior to each use.
 5. Avertin is an aqueous solution containing 0.5 g of 2,2,2 tribromoethanol in 0.31 mL of 2-methyl-2 butanol (Aldrich Chemical Co., Milwaukee, WI). Store at 4°C. To 0.31 mL, add 40 mL distilled water and warm up to 37°C prior to use. Store at 4°C.
 6. Miscellaneous reagents for embedding: 100%, 95%, 70% ethanol; xylene (Sigma); propylene oxide (Fisher Scientific); epon/araldite. For the latter, mix 10 parts epon #812 (Electron Microscopy Sciences, Ft. Washington, PA), 10 parts araldite (Electron Microscopy Sciences), 24 parts DDSA (Electron Microscopy Sciences). Mix well by warming stock up in a 60°C oven and shaking vigorously. After it is thoroughly mixed, the stock can be stored in the refrigerator. Warm up at 60°C prior to use. For embedding, use stock plastic containing 2% DMP-30 (Ladd Research Industries, Inc., Burlington, VT). Dilute this solution with propylene oxide for "half-and-half." Make both solutions fresh and mix well.
 7. Staining solutions:
  a. 2% uranyl acetate (Electron Microscopy Sciences) in water. Make fresh every time.
  b. Lead citrate (Electron Microscopy Sciences): dissolve 20 mg lead citrate in 10 mL distilled water, add 4–5 drops 5 M NaOH, and store the solution into a 10-mL syringe attached to a filter and a needle.
  c. Gomori stain contains 0.12% lead nitrate and 0.3% β-glycerophosphate in 0.1 M sodium acetate.

3. Methods

 1. In vitro labeling and culture of the tissue
  a. Place embryos in warm HBSS and sacrifice by decapitation. Dissect the tissue to be labeled from embryo (*see* **Note 1**) and place in warm HBSS containing CCFSE (1:500 dilution) for 1 h at 37°C, then rinse several times in fresh HBSS (*see* **Note 2**).

    b. Place the tissue on agar gel in a Falcon 3037 organ dish, barely covering with culture medium. Maintain rodent tissues in a humidified incubator at 37°C, under 5% CO_2 and 20–40% O_2. Change the medium every 24 h.

2. In vivo labeling

    a. Anesthetize pregnant mice by intraperitoneal injection of prewarmed Avertin, 0.1 mL per 5 g body weight. Fix mouse on a board and make a midline incision through the abdominal wall under sterile conditions. Pull out the uterine horns and inject CCFSE into amniotic cavities of the embryos through the uterine wall (approx 0.5 μL/embryo). Replace uterine horns and flush abdominal cavity with sterile saline before closing (*see* **Note 3**).

    b. Allow the pregnancy to continue for the desired period, sacrifice the dam, and collect embryos for fixing tissue.

3. Light microscopy following CCFSE administration

    a. Fix the tissue with 4% paraformaldehyde in PBS at room temperature (RT, time varies with size of tissue from 1/2 to 2 h), then rinse with PBS 2X. Dehydrate through 70%, 95%, and 100% ethanol. To ensure complete dehydration, change the 100% solution 3X and open a fresh bottle for the last change. Clear the tissue with three changes of xylene. After the last change, add an equal volume of paraffin chips to the xylene and infiltrate the tissue in this "half-and-half" solution overnight at room temperature, then place in 60°C oven. After the paraffin has melted, change 2X with melted 100% paraffin at 60°C (1 h each change). Embed the tissue in fresh paraffin with proper orientation in a plastic or paper mold.

    b. Cut the paraffin block on a microtome and collect serial sections of 7–10 μm thickness. Float the sections on water on superfrost slides (Fisher Scientific) at 37°C until flat. Then dry them onto the slides, deparaffinize in xylene, and rehydrate through ethanol into H_2O. Add water-soluble mounting medium and cover slips. Examine the sections for fluorescence with a light microscope (confocal if possible). Store rest of the slides at 4°C in the dark.

4. Electron microscopy following CCFSE administration

    a. Fix the tissue at RT with 4% glutaraldehyde and 4% paraformaldehyde in 0.1 *M* cacodylate buffer, pH 7.4, for 1/2–2 h (depending on size), then rinse in 0.1 *M* cacodylate buffer 2–3X. Post-fix the tissue for 1 h on ice with 1% osmium tetroxide in 0.1 *M* cacodylate buffer, then rinse 2X in deionized water, and stain en bloc in 2% (or saturated) uranyl acetate at RT for 1 h. Rinse 1X in deionized water for 10 min and dehydrate immediately through 70%, 95%, and 100% ethanol (3 changes of 100% ethanol with the last change from a freshly opened bottle). Put the tissue in propylene oxide for 30 min while making up propylene oxide-epon/araldite "half-and-half." Then let the tissue infiltrate in the "half-and-half" for 2 h to overnight. Gentle agitation improves infiltration. Embed the tissue in freshly mixed epon/araldite and solidify in a 60°C oven for 8–24 h.

    b. Trim the plastic block to a desirable size and cut thin sections on a microtome. Stain the sections on metal grids (Electron Microscopy Sciences) with 0.2% lead citrate (1 min) and examine with a transmission electron microscope (*see* **Note 4**).

5. Gomori stain for lysosomal acid phosphatase: Fix the tissue in cold 4% formaldehyde overnight, and rinse in 0.1 *M* sodium acetate buffer. React with preheated (60°C) Gomori stain at 37°C for 1 h. Rinse the tissue with buffer and visualize acid-phosphatase-positive cells by reaction with a 1% ammonium sulfide solution. Briefly fix the tissue in formaldehyde again and process for paraffin sectioning. Control is reacted in a mixture that lacks the β-glycerophosphate.

4. Notes

1. CCFSE is mainly used to label epithelium, which it penetrates easily from the outside of the block, but does not cross *(5)*. Be careful during dissecting not to leave a cut face near the epithelium to be labeled, so as not to expose the underlying connective tissue to CCFSE. For example, to label palatal MEE, the whole upper head is placed in CCFSE rather than just the palatal shelves, and the shelves are not dissected until all free CCFSE is rinsed away.

2. Always prewarm HBSS containing CCFSE to 37°C. Gently shaking the dish will also help penetration of CCFSE. Always rinse well so that no CCFSE residue remains to label the tissue during further processing.

3. The surgical procedure described here is for rodent embryos. Anesthesia should be sufficient and, ideally, the operation should be performed in a heated hood. Do not leave the uterus exposed to air (cover with warm HBSS) and be as gentle as possible. Coldness, dryness, and extensive manipulation will cause premature death of the embryos.

4. One of the major problems is how to distinguish CCFSE-isolation bodies from lysosomes when using this technique at the TEM level. When the tissue is fixed in vivo, dying cells are not usually seen, but lysosomes may be present in cultured cells. At the light microscopic level, fluorescence identifies isolation bodies, but it can also be shown, by staining the same sections with the Gomori stain for acid phosphatase, that the fluorescent CCFSE-isolation bodies in the cytoplasm are not lysosomes (**Fig. 2**). At TEM level, several criteria can be used to distinguish these membrane-bound bodies. First of all, the density of CCFSE-isolation bodies (**Figs. 1** and **3**) is more homogenous than that of lysosomes. Second, the cells containing CCFSE-isolation bodies are seen to be healthy looking with euchromatic nuclei and normal-appearing organelles (**Fig. 3**). Third, a control sample that is not labeled with CCFSE should be fixed and examined at the same time as the experimental at the TEM level to show that lysosomes are not a feature of the particular tissue being examined *(5)*.

Acknowledgment

This contribution was supported by the National Institute for Dental Research (DE 11142).

References

1. Bronner-Fraser, M. (1985) Alterations in neural crest migration by a monoclonal antibody that affects cell adhesion. *J. Cell Biol.* **101,** 610–617.
2. Serbedzija, G. N., Fraser, S. E., and Bronner-Fraser, M. (1990) Pathways of trunk neural crest cell migration in the mouse embryo as revealed by vital dye labeling. *Development* **108,** 605–612.
3. Carette, M. J. M. and Ferguson, M. J. W. (1992) The fate of medial edge epithelial cells during palatal fusion in vitro: Analysis by DiI labeling and confocal microscopy. *Development* **114,** 379–388.
4. Shuler, C. F., Halpern, D. E., Guo, Y., and Sank, A. C. (1992) Medial edge epithelium fate traced by cell lineage analysis during epithelial-mesenchymal transformation. *Dev. Biol.* **154,** 318–330.
5. Griffith, C. M. and Hay, E. D. (1992) Epithelial-mesenchymal transformation during palatal fusion: Carboxyfluorescein traces cells at light and electron microscopic levels. *Development* **116,** 1087–1099.
6. Sun, D., Vanderburg, C. R., Odierna, G. S., and Hay, E. D. (1998) TGFβ3 promotes transformation of chicken palate medial edge epithelium to mesenchyme in vitro. *Development* **125,** 95–105.

VI

CHIMERAS

35

Transplantation Chimeras

Use in Analyzing Mechanisms of Avian Development

Diana K. Darnell and Gary C. Schoenwolf

1. Introduction

The word chimera originally referred to a mythological beast with the head of a lion, the body of a goat, and the tail of a serpent, but it has come to mean any individual made up of the parts of more than one individual. Transplantation chimeras, which can be made in many species and sometimes between species, are formed when tissue is grafted from one embryo (the donor) to another (the host) and permitted to incorporate. This technique allows one to follow a specific group of cells (the graft) through a period of development and to determine the fates and locations of their progeny. To allow graft cells to be distinguished from the cells of the host after culture, the graft is usually labeled with vital fluorescent dyes (*see* **Note 1**) or after fixation by using immunocytochemistry or a histological stain. Because chimeras can be made between a host and donor of the same stage (isochronic) or different stages (heterochronic), between similar tissues or regions (homotopic) or different tissues or regions (heterotopic), and between members of the same species (intraspecific) or different species (interspecific), minor variations in the technique allow one to investigate several parameters of development including cell fate, potency, commitment, evolutionary conservation of signals, and cell movement. For example, isochronic, homotopic grafting is used for fate mapping (and cell movement) studies *(1–3)*, heterotopic and heterochronic grafting yield information about competency and commitment *(3–11)*, and interspecific grafts can demonstrate evolutionary conservation of signaling pathways *(12–14)* and may eventually lead to the analysis of the behavior of mutant mouse-embryo cells in chick-embryo hosts. The classic technique using avian chimeras involves grafting quail tissue to chick embryos (interpretation of these experiments is based on the assumption that grafts between these two similar species are equivalent to grafts from within a single species) and identifying the grafted cells after sectioning by using the Feulgen-Rossenbeck histological stain *(15)*. With this technique, groups of quail cells can be identified by the characteristic appearance of their nucleolus. Drawbacks to this technique are that chick cells also have a similar nucleolar appearance to that of quail cells; not all quail cells in a particular section readily show the characteristic nucleolar

From: *Methods in Molecular Biology, Vol. 135: Developmental Biology Protocols, Vol. I*
Edited by: R. S. Tuan and C. W. Lo © Humana Press Inc., Totowa, NJ

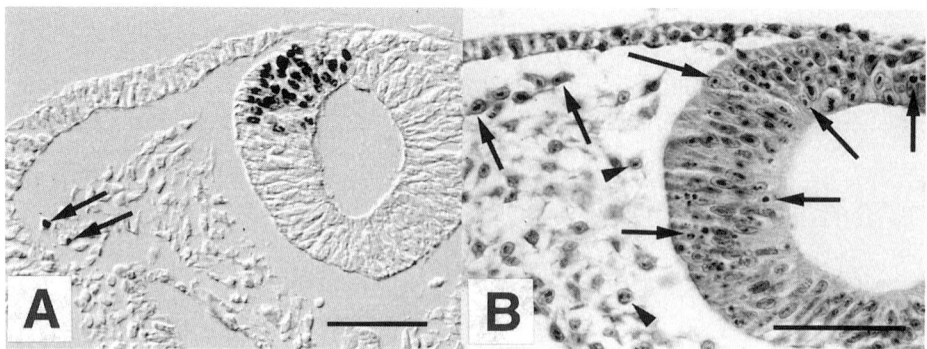

Fig. 1. Transverse sections of chimeras collected 24 h after grafting quail cells into chick host embryos. (**A**) The chimera was labeled immunocytochemically with QCPN and subsequently sectioned. Quail cells mainly form a patch within the dorsolateral wall of the neural tube. Other quail cells within the mesenchyme are indicated by *arrows*. (**B**) Section of a chimera stained with the Feulgen-Rossenbeck procedure. Quail cells are indicated by *arrows*. Bar = 100 μm. Reprinted with permission from **ref. *17***.

marker; and some cell types, such as endothelial cells, may lose the marker, all of which makes the exact identification of individual cells as chick or quail problematic. We find that immunocytochemistry with the anti-quail antibody (QCPN; *see* **ref. *16***) makes *every* quail cell easily discernible from the neighboring chick cells and thus is preferable to the histological stain (**Fig. 1A,B**). However, the Feulgen-Rossenbeck procedure may have advantages for some experiments using multiple labels *(6)*.

Transplantation chimeras can be made using host embryos developing *in ovo*, but for gastrula and neurula stages, the majority of such grafting experiments are done in New culture (*see* Chapter 5). New culture allows grafts to be followed for approx 1 d of development, after which the fates of their cells, the effects of grafted cells on development of the host, and/or the level of commitment of grafted cells to a particular prospective fate can be determined. Transplantation chimeras using New-cultured embryos as hosts will be described.

2. Materials

1. Saline: 123 m*M* NaCl, autoclaved. Add 1 mL penicillin/streptomycin (Gibco-BRL, Grand Island, NY) at the time of use.
2. Culture plates (*see* Chapter 5 for detailed instructions): For 40 plates, collect 60 mL of thin albumen from unincubated eggs and warm to 49°C. Dissolve a 0.6% (0.36 g) solution of Bacto-Agar (0140-01; Difco Laboratories, Detroit, MI) in 60 mL saline and cool to 49°C. Mix albumen and agar thoroughly, and pipet approx 2.5 mL agar-albumen into each 35 × 10-mm culture dish.
3. Humidified chamber (e.g., a 150-mm Petri dish with a wet filter paper in the bottom).
4. Fluorescent dyes (*see* Chapter 30): RRF (<u>R</u>hodamine 123 [cat. no. R-302], C<u>R</u>SE [cat. no. C-2211; a rhodamine-containing dye], C<u>F</u>SE [cat. no. C-1157; a fluorescein-containing dye]; Molecular Probes, Eugene, OR). Mix a 1% solution (5 mg/500 μL) of each dye separately in DMSO. For the working solution, add 50 μL of each stock solution to 10 mL saline, mix and aliquot into 10 microfuge tubes. Freeze extra stock and working solution. Keep dyes, stocks, and working solution in the dark.

5. Enzymes:
   a. Collagenase (1000–2000 IU/mL in Tris buffer, pH 7.1), Dispase (0.24 U/mL), or trypsin/EDTA/glycine (0.4%).
   b. To make trypsin/EDTA/glycine, mix 0.4 g trypsin (152-15-9; Difco Laboratories), 0.4 g ethylenediaminetetraacetic acid (EDTA) and 7.5 g glycine with 60 mL distilled water; stir for 1 h. Add 10 *N* NaOH dropwise until the EDTA goes into solution, then 2 *N* HCl dropwise until the pH is 7.2. Bring the volume to 100 mL with distilled water.

6. Microtools:
   a. Eyebrow hair tool: an eyebrow hair with its distal end attached to a dowel with nail polish; serves as a micromover.
   b. Needles: pulled glass microelectrodes, electrochemically sharpened tungsten needles, or small cactus needles (spines) attached to a dowel with nail polish; serve as scalpels.
   c. "Hole punch": Flame the center of a 100-µL glass capillary tube until soft, remove the glass from the flame and pull the two ends away from one another; score the tube with a diamond pencil at the desired diameter (e.g., 50–200 µm), and break off the tip by pulling the two pieces straight apart (like taking the cap off a pen); select pipets with straight tips, and polish these on abrasive film (cat. no. 6775E38; Thomas Scientific, Swedesboro, NJ) to obtain an even edge; the hole punch is used on the blastoderm in cookie-cutter fashion.
   d. Transfer pipets: "hole punch" capillary tubes with wider openings.
   e. Flexible rubber tubing and a mouthpiece for mouth-pipetting with "hole punch" and transfer pipets.

7. Antibodies:
   a. Anti-rhodamine (A-6397, rabbit IgG polyclonal; Molecular Probes): use 1:100 and use goat-anti-rabbit IgG secondary at 1:200.
   b. Antifluorescein antibody (A-889; rabbit IgG, polyclonal; Molecular Probes): use 1:200 and use goat-antirabbit IgG secondary at 1:200.
   c. Antiquail (QCPN: IgG1, monoclonal; Developmental Studies Hybridoma Bank, Iowa City, IA): use 1:100 and use goat-antimouse IgG secondary at 1:200.

3. Methods

1. Incubate eggs to the desired stage(s) (*see* Chapter 4).
2. Place culture plates into a humidified chamber and place this in a plate incubator at 38°C.
3. Put donor embryos into New culture (*see* Chapter 5). Drop enough RRF fluorescent dye mix onto donor embryos to cover them (approx 5 drops), and incubate embryos for 1 h at 38°C in the dark (*see* **Note 1**).
4. Set up host embryos for New culture (*see* **Note 2** and Chapter 5).
5. Extirpate the region from the donor embryo that is to be grafted by using a needle or hole punch (*see* **Notes 3–5** for variations in grafting). To obtain a single-layered graft, cut through the entire blastoderm to isolate the graft and then separate the layers of the graft with a brief treatment in enzyme. Rinse the graft in several changes of saline to stop the enzymatic reaction and to wash away the enzyme or excess dye or both (*see* **Note 6**).
6. Remove a region of similar size and shape from the host by using a needle or the same hole punch used for the donor. Single layers can be removed from the host by careful dissection. For single-layer grafts into the gastrula or early neurula epiblast (the layer next to the vitelline membrane), cover the cultured host embryo with saline, cut through the hypoblast/endoderm just peripheral to the boundary between the area pellucida and area opaca and carefully reflect the hypoblast/endoderm, using an eyebrow hair tool as a blunt dissecting instrument. Remove the region from the host epiblast that will be replaced by the graft.

7. Transfer the graft for the donor to the host using a transfer pipet, and manipulate the graft into the hole in the host, using an eyebrow hair as a micromover.

8. In the case of an epiblast graft, replace the hypoblast/endoderm by rolling it back across the graft with the eyebrow hair, then remove the excess saline.

9. If possible, take a time-0 image (photograph or video image) of the embryo under epifluorescence illumination for documentation.

10. Culture the embryos in a humidified chamber at 38°C until the desired stage is reached.

11. After culture, again take an image for documentation.

12. Chimeras can be fixed and treated for whole-mount immunocytochemistry using an anti-rhodamine or anti-fluorescein antibody or with QCPN, if quail donors were used; subsequently, chimeras can be embedded in paraffin and sectioned to show in detail the tissue location of graft cells.

4. Notes

1. We use a mixture of fluorescent dyes (RRF) to follow the cells by fluorescent microscopy and to allow the use of either anti-rhodamine or anti-fluorescein antibodies to identify the grafted cells in whole-mount and sectioned tissue. Two rhodamine-containing dyes are included, because neither alone labels all types of graft cells equally, but together all cells are labeled and the fluorescent signal is strong. CFSE is not used alone because its fluorescent tag is often too weak to pick up after culture, when cells have undergo extensive dispersion (*see* Chapter 30).

2. Extirpation can be done using a needle (a pulled glass microelectrode, a tungsten needle, or a cactus needle [spine] attached to a small dowel) or a "hole punch." A hole punch generates circular plugs and these facilitate healing of grafts into the epiblast and make reproducible sizes of graft holes in both donor and host and between host embryos (sometimes desirable).

3. Grafts can also be inserted into the mesodermal layer either by retracting the endoderm, placing the graft on the mesoderm and covering the graft with endoderm, or by creating a tunnel between the endoderm and epiblast with the micromover tool and pushing the graft into the tunnel to the desired location. The latter technique allows for more control over graft location, as the graft is less likely to float around than when the endoderm has been reflected back.

4. When grafts are made in all three germ layers at the same time, the possibility of healing problems are greatly increased. Two strategies to minimize this are to cover the graft with an unlabeled piece of endoderm from another embryo (a bandage), and to leave the grafted host at room temperature for several minutes to give the gap between donor and host a chance to close before placing the embryo back into the incubator. Neither step is necessary if the grafts are placed into the primitive streak.

5. Use enzymes conservatively. Practice with each batch of enzyme to establish the correct timing and concentration. Stop the reaction by moving the tissue through several changes of saline before complete separation of layers is observed. It is better to wait an extra minute using lower concentrations than have a very fast reaction at higher concentrations, which could disintegrate your tissue before you can stop the reaction.

Acknowledgments

Original work described herein from the Schoenwolf laboratory was supported by the National Institutes of Health grant nos. NS 18112 and HD 28845. DKD was supported in part by NIH Developmental Biology Training grant no. HD 07491.

References

1. Schoenwolf, G. C., Garcia-Martinez, V., and Dias, M. S. (1992) Mesoderm movement and fate during avian gastrulation and neurulation. *Dev. Dynamics* **193,** 235–248.
2. Garcia-Martinez, V., Alvarez, I. S., and Schoenwolf, G. C. (1993) Locations of the ectodermal and nonectodermal subdivisions of the epiblast at stages 3 and 4 of avian gastrulation and neurulation. *J. Exp. Zool.* **267,** 431–446.
3. Schoenwolf, G. and Alvarez, I. S. (1991) Specification of neurepithelium and surface epithelium in avian transplantation chimeras. *Development* **112,** 713–722.
4. Alvarez, I. S. and Schoenwolf, G. C. (1991) Patterns of neurepithelial cell rearrangement during avian neurulation are determined prior to notochordal inductive interactions. *Dev. Biol.* **143,** 78–92.
5. Bally-Cuif, L. and Wassef, M. (1994) Ectopic induction and reorganization of Wnt-1 expression in quail/chick chimeras. *Development* **120,** 3379–3394.
6. Darnell, D. K. and Schoenwolf, G. C. (1995) Dorsoventral patterning of the avian mesencephalon/metencephalon: Role of the notochord and floor plate in suppressing *Engrailed-2*. *J. Neurobiol.* **26,** 62–74.
7. Garcia-Martinez, V. and Schoenwolf, G. C. (1992) Positional control of mesoderm movement and fate during avian gastrulation and neurulation. *Dev. Dynamics* **193,** 249–256.
8. Martinez, S. and Alvarado-Mallart, R.-M. (1990) Expression of the homeobox *Chick-en* gene in chick-quail chimeras with inverted mes-metencephalic grafts. *Dev. Biol.* **139,** 432–436.
9. Nakamura, H. (1990) Do CNS anlagen have plasticity in differentiation? Analysis in quail-chick chimera. *Brain Res.* **511,** 122–128.
10. Nakamura, H., Nakano, K. E., Igawa, H. H., Takagi, S., and Fujisawa, H. (1986) Plasticity and rigidity of differentiation of brain vesicles studied in quail-chick chimeras. *Differentiation* **19,** 187–193.
11. Selleck, M. and, Stern, C. D. (1992) Commitment of mesoderm cells in Hensen's node of the chick embryo to notochord and somite. *Development* **114,** 403–415.
12. Davidson, D. R., Crawley, A., Hill, R. E., and Tickle, C. (1991) Position-dependent expression of two related homeobox genes in developing vertebrate limbs. *Nature* **352,** 429–431.
13. Fontaine-Perus, J., Jarno, V., Fournier le Ray, C., Li, Z., and Paulin, D. (1995) Mouse chick chimera: A new model to study the in ovo developmental potentialities of mammalian somites. *Development* **121,** 1705–1718.
14. Itasaki, N., Sharpe, J., Morrison, A., and Krumlauf, R. (1996) Reprogramming *Hox* expression in the vertebrate hindbrain: influence of paraxial mesoderm and rhombomere transposition. *Neuron* **16,** 487–500.
15. Le Douarin, N. (1973) A Feulgen-positive nucleolus. *Exp. Cell Res.* **77,** 459–468.
16. Inagaki, T., Garcia-Martinez, V., and Schoenwolf, G. C. (1993) Regulative ability of the prospective cardiogenic and vasculogenic areas of the primitive streak during avian gastrulation. *Dev. Dyn.* **197,** 57–68.
17. Daston, G. P., ed. (1996) *Molecular and Cellular Methods in Developmental Toxicology*, CRC, New York, p. 243.

36

Interspecific Chimeras in Avian Embryos

Nicole M. Le Douarin, Françoise Dieterlen-Lièvre, Marie-Aimée Teillet, and Catherine Ziller

1. Introduction

Avian embryos provide all sorts of opportunities for the study of higher vertebrate development, especially because an experimental model, in which cells from two species are combined, has been proposed as the basis of a marking technique *(1,2)*. Cell-marking techniques are essential in developmental biology to trace cell precursors and their progeny and thus establish their behavior and fate. These markers must be nonleaky and stable as cells divide and must not interfere with normal development. The quail–chick labeling method meets all these requirements.

The principle of the method *(1,2)* is based on the presence, in all embryonic and adult cells of the quail (*Coturnix coturnix japonica*), of condensed heterochromatin in the form of one (sometimes two or more, depending on the cell types) large mass(es) associated to the nucleolus. This organelle is thus strongly positive after DNA staining, usually carried out according to the Feulgen and Rossenbeck technique *(3)*. In chick (*Gallus gallus domesticus*) cells, like in most of animal species, small chromocenters are dispersed in the nucleoplasm. In combinations, quail cells are readily recognizable by the structure of their nucleus, which thus provides a permanent natural marker (**Fig. 1**).

Surgically constructed quail–chick chimeras have been used to follow the fate of definite embryonic territories during development and to discover the origin of the different groups of cells concurring in organ development. Investigations were carried out on the neural crest *(4)* and the neural primordium *(5,6)*, the blood and immune system, the vascular system, limb buds, skin, visceral organs, and so forth. This type of study implies that the developmental processes unfold in the chimeras as close as possible to the normal course. To achieve this, quail tissues are transplanted into chick embryos (or vice versa) *in ovo* and, rather than adding cells to normal embryos, a precise territory is removed from the host and replaced as exactly as possible by an equivalent region dissected form the donor at the same developmental stage. Quail and chick are closely related in taxonomy, although their incubation time (17 d for the quail, 21 d for the chick) as well as their size at birth (approx 10 g for the quail and 30 g for the chick) differ. However, during the first days of incubation, when most of the important events in embryogenesis take place, the size of quail and chick embryos and the

From: *Methods in Molecular Biology, Vol. 135: Developmental Biology Protocols, Vol. I*
Edited by: R. S. Tuan and C. W. Lo © Humana Press Inc., Totowa, NJ

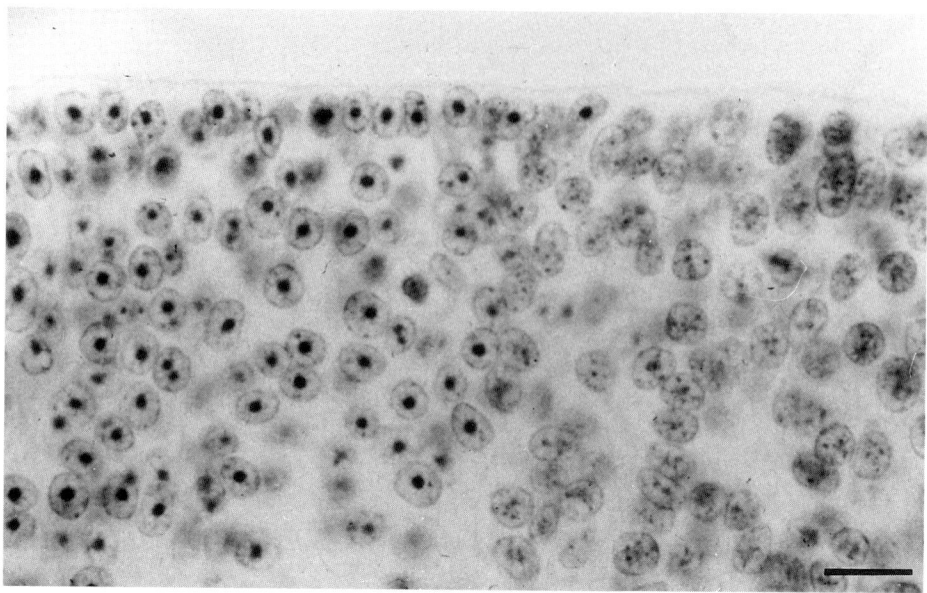

Fig. 1. Feulgen-Rossenbeck staining of a 5-µm thick section of a chimeric neuroepithelium at 4 d of incubation. Quail nuclei, to the left, are distinguishable from chick nuclei by their nucleolus, which displays a conspicuous heterochromatin clump. Scale bar = 10 µm.

chronology of their development are nearly alike. Bothway grafts and chronological studies of development in controls are useful to avoid any bias resulting from differential processes between chick and quail embryos.

Quail–chick chimeras can hatch and survive in good health during a certain time, limited however by the appearance of an immunological reaction mounted by the host. This occurs only after birth, when the immune system has built up, the exact timing depending on the nature of the graft. For neural grafts, a long delay (1–2 mo) is observed between the onset of immune maturity and the rejection *(7)*. For neural grafts associated with other tissues and for grafts of nonneural tissue, rejection occurs as soon as maturation of the immune system is achieved *(8,9)*.

Grafts need not be isochronic and isotopic. Some developmental processes can be investigated by means of heterotopic, or heterochronic grafts with or without previous ablation of tissues. For instance, such a strategy was used to test the degree of determination of neural crest cells and their derivatives *(4,10)* and to demonstrate the period of colonization of primary lymphoid organ rudiments (thymus and bursa of Fabricius) by hemopoietic cells in birds *(11)*.

For many years, the analysis of the quail-chick chimeras was based on the differential staining of the nucleus by either the Feulgen and Rossenbeck nuclear reaction (**Fig. 1**) or any other method revealing specifically the DNA profiles in light or electron microscopy. Significant advantages have been gained with the advent of species-specific antibodies recognizing either quail or chick cells of one or several types (*see* **Subheading 3.5.3.** and **Fig. 2A**). Several quail or chick specific nucleic probes are also available, allowing gene activities to be distinguished between the two species at the single-cell level (*see* **Subheading 3.5.4.** and **Fig. 2B**).

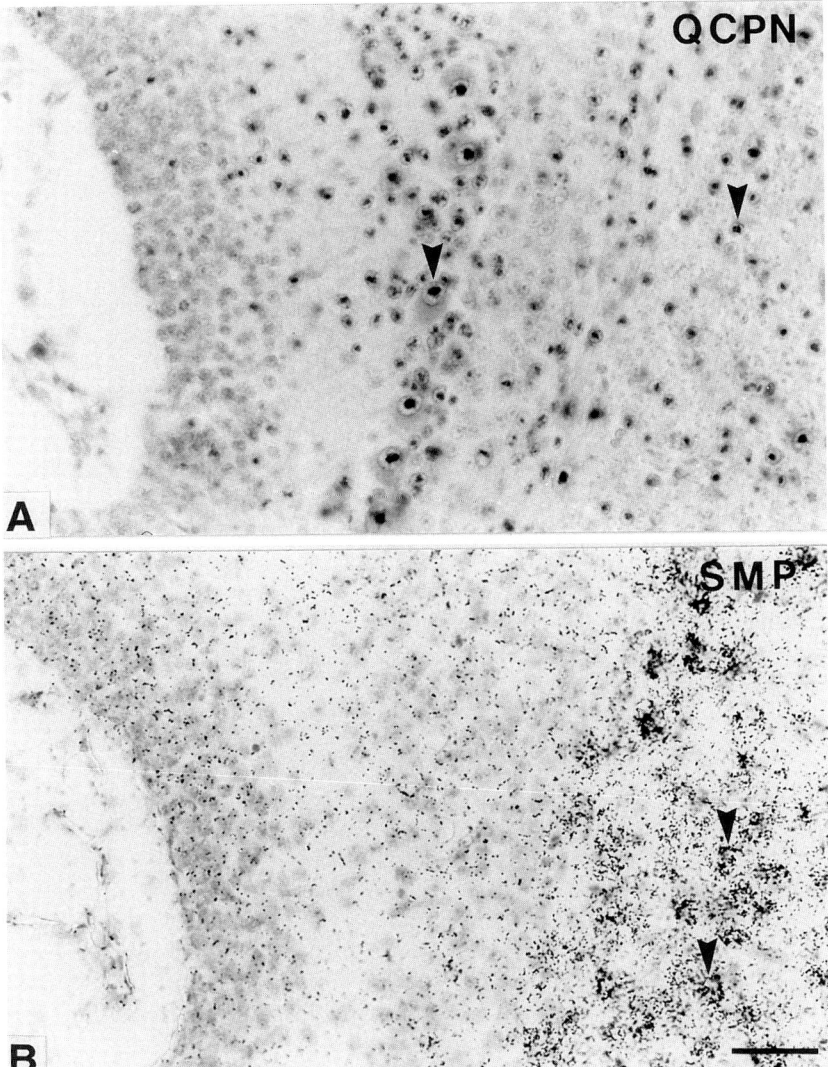

Fig. 2. Quail cells are differentially identified in 5 μm sections of a chimeric cerebellum (arrowheads), with the quail specific QCPN Mab (**A**) and with the SMP nucleic probe (**B**), which hybridizes specifically with the quail oligodendrocytes. Scale bar = 70 μm.

In this chapter, we describe the protocols of several representative examples of quail–chick chimeras. These particular protocols can be adapted to the graft of any other type of tissue.

2. Materials

2.1. Egg Incubation

1. Fertilized chick and quail eggs (*see* **Note 1**).
2. Humidified ventilated incubators equipped with time programmers. Time programmers are useful to obtain very precise stages of development (*see* **Note 2**).

3. Egg holders (*see* **Note 3**).
4. Developmental tables of Hamburger and Hamilton for chick embryo development *(12)* and of Zacchei for quail embryo development *(13)* (*see* **Note 4**).

2.2. Preparation of Host and Donor Embryos and Grafts

1. 70% Ethanol for sterilization.
2. Disposable syringes (1 or 2 mL) and needles (0.6–0.8 mm in diameter).
3. Transparent Scotch tape (5 cm wide) (*see* **Note 5**).
4. Paraffin 60°C and a thin paintbrush.
5. Physiological liquid: Phosphate-buffered saline (PBS) or Tyrode solution (pH 7.4) (*see* **Note 6**).
6. Antibiotics: penicillin and streptomycin (*see* **Note 7**).
7. Proteolytic enzymes (pancreatin) (*see* **Note 7**).
8. Bovine serum (*see* **Note 7**).
9. Pasteur pipets.
10. Glass micropipets hand drawn from Pasteur pipets and equipped with plastic tubes for mouth use (Tygon, Poly-Labo, Strasbourg, France) (*see* **Note 8**).
11. Small glass dishes (*salières*), normal or containing a plastic base for dissections (rhodorsyl, Rhône-Poulenc, Paris, France) (*see* **Note 9**) and insect pins.
12. Indian ink (Pelikan for drawing) (*see* **Note 10**).
13. Microscalpels and their holders (*see* **Note 11**).
14. Small curved scissors, iridectomy (Pascheff-Wolff) scissors and thin forceps (Swiss Dumont n°5 forceps) (Moria-instruments, Paris).
15. Optical equipment : stereomicroscope equipped with a zoom from X6 to X50.
16. Optic fibers for object illumination (*see* **Note 12**).

2.3. Histological Analysis of the Chimeric Embryo Tissues and Organs

1. Carnoy fluid, 4% paraformaldehyde in PBS, Bouin fluid, etc. for fixing of the tissues (*see* **Note 13**).
2. Materials for paraffin embedding.
3. Reagents for Feulgen and Rossenbeck staining (*3,14*; *see* **Note 14**).
4. Immunohistochemistry reagents, species-specific and/or cell type-specific antibodies.
5. Reagents for *in situ* hybridization with species-specific nucleic probes.

3. Methods
3.1. Preparation of the Host and Donor Embryos

1. *Incubations*: Quail and chick eggs are incubated with their long axis horizontal during the time necessary to obtain the stages adequate for the experiments: 36–48 h for the experiments described here (*see* **Note 4**).
2. *Environmental conditions of the experiments*: The experiments have to be performed under relatively sterile conditions. Egg shells are cleaned rapidly with 70% ethanol and dried. Instruments are sterilized in a dry oven (1 ½ h at 120°C). Sterile physiological liquids are supplemented with antibiotics (10–20 IU/mL). Experiments are performed in a clean separate room but never under forced air apparatus (*see* **Notes 6** and **12**).
3. *Opening the eggs*: The blastoderm develops on the top of the yolk against the shell membrane. In order to separate the blastoderm from the shell membrane before opening a window in the shell, a small quantity of albumen (approx 0.3 mL) is removed at the small end of the egg, using a syringe. Another method consists in perforating the egg shell at the level of the air chamber, then rolling the egg horizontally several times. These manipulations are sufficient to loosen the blastoderm from the shell membrane and prevent the

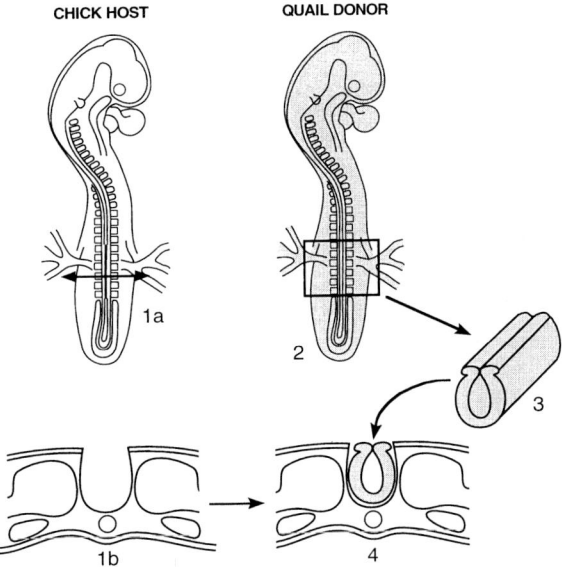

Fig. 3. Orthotopic quail–chick transplantation of the neural tube. A 2-d-old chick embryo *in ovo* (**1a**) is deprived of a segment of its neural tube by microsurgery (**1b**) at the level of the 5–6 last-formed somites. The corresponding region of the blastoderm of a quail embryo at the same stage (2) is enzymatically dissociated. The isolated quail neural tube segment (3) is orthotopically grafted into the recipient chick embryo (4).

    window through the shell from damaging the embryo. The small holes through the egg shell are obturated with a piece of tape or a drop of paraffin (*see* **Note 5**).

4. *Contrasting the embryos in ovo*: India ink, diluted 1:1 in PBS or Tyrode supplemented with antibiotics is injected under the blastoderm (donor or recipient) using a glass micropipet mounted with a plastic tube for mouth use (*see* **Fig. 4** and **Note 10**).

5. *Gaining access to the embryos in ovo*: The vitelline membrane that covers the embryo is slitted out with a microscalpel on the spot where microsurgery is to be performed.

6. *Explanting the donor blastoderms*: In certain cases the donor blastoderm is cut out from the egg with Pascheff-Wolff scissors, washed free of vitellus in PBS or Tyrode supplemented with antibiotics, transferred onto a dish with a black plastic base, and pinned out for dissection.

3.2. The Grafts

3.2.1. Neural Tube Transplantations

    Orthotopic transplantations of fragments of neural tube (**Fig. 3**) have allowed the construction of a neural crest fate map *(4)* and the detection of crest cell migration pathways *(15)*. The rules of this operation are based on the fact that migration of neural crest cells starts in the cephalic region, then moves progressively caudalward when the neural tube forms. The interspecific graft is performed at a level where crest cells are still present at the apex of the neural tube i.e., in the neural folds in the cephalic area, at the level of the last formed somites in the cervical and thoracic regions, and at the level of the segmental plate in the lumbosacral region. The replacement should concern a length of no more than 5–6 somites to take into account the rostrocaudal differential state of evolution of the neural crest. Donor and host embryos are strictly stage matched (*see* **Note 4**).

1. *Excision of the host neural tube*: The selected neural tube fragment is excised from the host embryo by *in ovo* microsurgery. A bilateral longitudinal slit through the ectoderm and between the neural tube and the adjacent paraxial mesoderm, at the level selected, is performed with a microscalpel. The neural tube is then gently separated from the neighboring mesoderm and cut out transversally, rostrally, and caudally without damaging the underlying notochord and endoderm. The fragment of neural tube is then progressively severed from the notochord and finally sucked out using a calibrated glass micropipet (*see* **Note 8**).

2. *Preparation of the graft*: The transverse region of the stage-matched donor embryo comprising the equivalent fragment of neural tube plus surrounding tissues (ectoderm, endoderm, and mesoderm) is retrieved with iridectomy scissors and subjected in vitro to enzymatic digestion (pancreatin [Gibco-BRL, Grand Island, NY] one-third in PBS or Tyrode) during 5–10 min on ice or at room temperature according to the stage of the embryo (*see* **Note 7**). Then tissues are dissociated using two smooth microscalpels (*see* **Note 11**) and finally, the isolated neural tube fragment is rinsed with PBS or Tyrode supplemented with bovine serum to inhibit the action of the proteolytic enzymes. It is then ready to be grafted.

3. *Grafting procedure*: The donor neural tube is transferred to the host embryo using a calibrated glass micropipet and placed in the groove resulting from the excision, in the normal rostrocaudal and dorsoventral orientation (*see* **Note 15**).

Heterotopic, heterochronic grafting was implemented to study whether the fate of neural crest cells is specified before their emigration *(10)*. The graft was taken at a more rostral or more caudal level than the grafting level in the host. Depending on the latter, the donor embryo was older or younger than the recipient (*see* **Note 16**).

Partial dorsoventral orthotopic graftings have also been performed in order to pinpoint possible early segregation of precursors in the neural tube *(16)*.

3.2.2. Orthotopic Transplantations of Brain Vesicles

This operation has been devised to label defined regions of the neuroepithelium and thus to study cell migrations and morphogenetic movements during brain development *(17,18)* and *(5)* (**Fig. 4**). It also allowed the transfer of genetic behavioral or functional traits from donor to recipient in either xenogeneic or isogeneic combinations *(9,19,20)*.

1. *Excision of brain vesicles from donor and host embryos*: Homologous brain vesicles are excised microsurgically in both stage-matched donors and recipients (*see* **Note 17**). The dorsal ectoderm is slit precisely at the limit between the neural tissue and the cephalic mesenchyme on each side of the selected part of the brain. The neural epithelium is loosened from the cephalic mesenchyme, cut out transversally at the chosen rostral and caudal levels, and finally severed from the underlying notochord.

2. *Exchange of brain vesicles*: Brain vesicles from the quail to the chick or vice versa (or from a mutant to a normal chick embryo) are transferred in a calibrated glass micropipet. The piece of neural tissue is inserted into the groove left after the excision, preserving the original rostrocaudal and dorsoventral orientation, then adjusted (*see* **Note 18**).

3. *Modifications of the technique* consist of orthotopic partial dorsal or dorsolateral grafts of brain vesicles *(18,21–23)*. Heterotopic grafts have also been performed to study specific problems *(21,24–26)*.

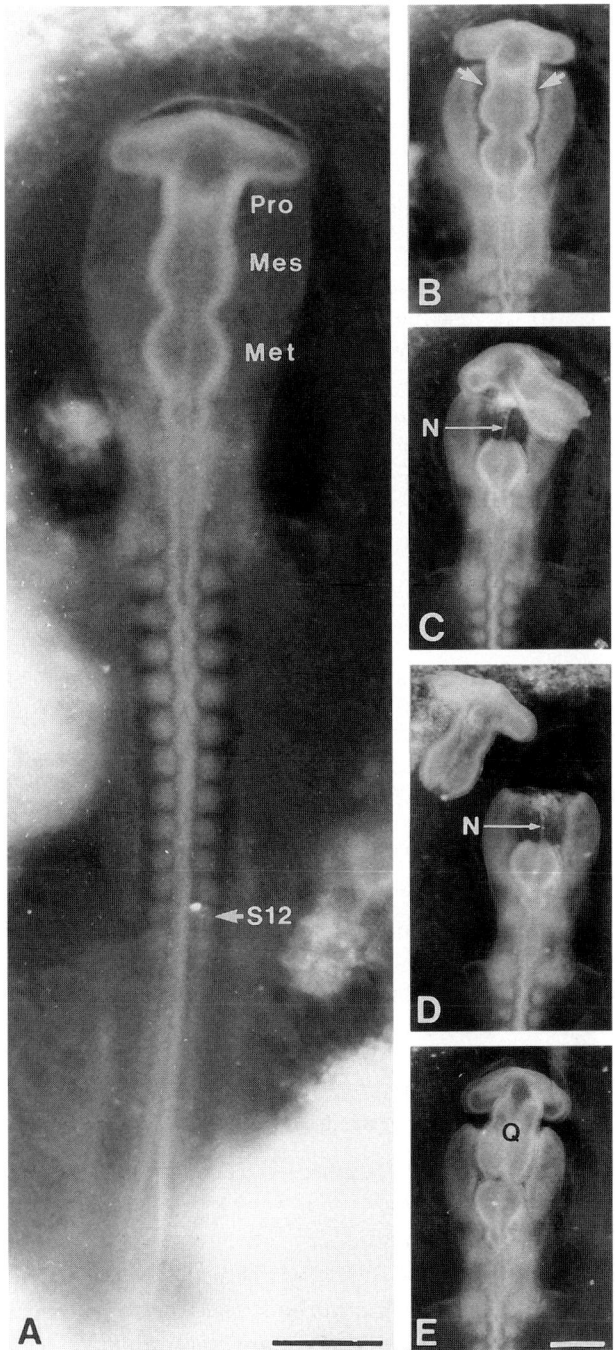

Fig. 4. Orthotopic transplantation of brain vesicles. **(A)** A 12-somite-stage chick embryo *in ovo* after injection of India ink underneath the blastoderm. **(B)** Longitudinal incisions are made between the cephalic neural tube and the paraxial mesenchyme to delimit the brain excision (arrows). **(C)** A transverse incision at the level of the mesencephalo-metencephalic constriction separates the prosencephalon and the mesencephalon from the mesoderm and endoderm. The notochord is then visible. **(D)** The equivalent quail brain vesicles are grafted into the chick recipient. Pro = prosencephalon; Mes = mesencephalon; Met = metencephalon; S12 = 12th somite; N = notochord; Q = quail brain vesicles. Scale bars = 0.05 mm.

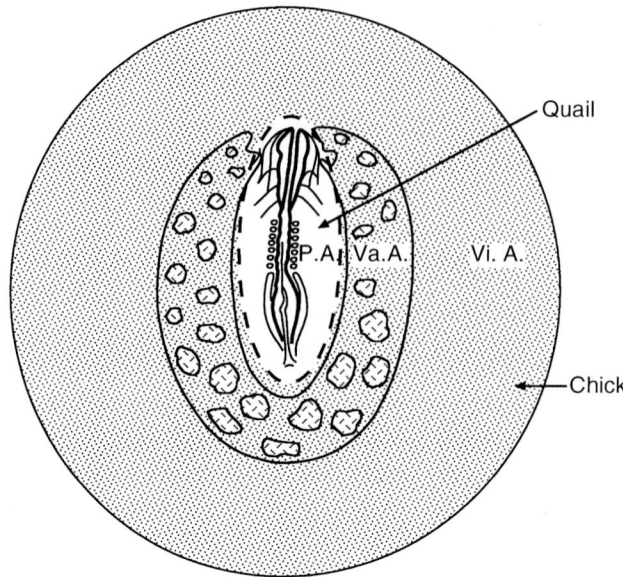

Fig. 5. Construction of a quail–chick yolk sac chimera. The stippled line indicates the suture between the two components. P. A. = pellucid area; Va. A. = vascular area; Vi. A. = vitelline area.

3.3.3. Neural Fold and Neural Plate Transplantations

Orthotopic and isochronic grafts served to map presumptive territories in the early rostral or caudal neural primordium *(6,27–30)*.

1. *Excision of accurate levels from the neural fold or neural plate in the chick host*: Very thin microscalpels (insect pins or steel needles sharpened on an oil stone) are used to excise precise segments of the folds and neural plate in 0–5-somite-stage chick embryos *in ovo*. An ocular micrometer is used to measure the pieces of tissue to be removed.
2. *Excision of equivalent pieces of tissue from the quail donor*: The grafts are excised from stage-matched quail in vitro using the same method. They are not treated with enzymes.
3. *Grafting*: The pieces of quail tissue are grafted orthotopically onto the chick.

Heterotopic grafts have also been performed to establish the degree of autonomy of rhombencephalon territories *(31,32)*.

Simple excisions combined with adjacent orthotopic grafts have been performed in order to identify neural crest cells differentiating at the level of the excision *(33)*.

The quail–chick transplantation method can be applied to any kind of embryonic rudiment, tissue, or organ. Orthotopic and heterotopic transplantations of somites, lateral meso-derm, and limb territories have led to a precise knowledge of the origin of skeletal muscles and surrounding tissues *(34)*. The fate maps of the early cephalic neurectoderm and mesoderm *(30)*, of the tail bud *(35)* and of Hensen's node *(6)* have been established by orthotopic grafting of small and precisely defined territories from one species to the other.

An interesting modality of the method consists in constructing yolk sac chimeras *(36)* by microsurgically suturing a 2-d-old quail embryo (restricted to the embryonic body itself and the surrounding area pellucida) onto a stage-matched recipient chick extra-embryonic area, from which the embryo has been excised (**Fig. 5**). These chimeras allowed to demonstrate that hemopoietic organs are colonized by hemopoietic stem cells from the embryo proper, not from the yolk sac *(37,38)*.

3.4. Sealing the Eggs and Postincubation

When the grafting operations are completed, the window in the egg shell is sealed with a piece of tape and the eggs reincubated in a horizontal stable position (*see* **Notes 5** and **19**).

3.5. Analysis of the Grafts

3.5.1. Fixation of the Experimental Embryos

Host embryos can be fixed from several hours after the operation to several days after hatching according to the experimental design (*see* **Note 20**). Carnoy fluid is one of the most useful fixing fluids, because it allows Feulgen and Rossenbeck staining *(14)*, immunohistochemistry, and *in situ* hybridization to be performed on consecutive paraffin sections. Fixation with 4% paraformaldehyde is appropriate for whole-mount immunohistochemistry, *in situ* hybridization, and all other methods requiring cryostat sectioning.

3.5.2. Feulgen and Rossenbeck Staining

The Feulgen and Rossenbeck nucleal reaction *(3,14)* (*see* **Note 21**) is applied on 5 μm serial sections.

3.5.3. Species-Specific Antibodies

1. Two antibodies recognize virtually all cell types in the quail and none in the chick: the polyclonal antibody raised by Lance-Jones and Lagenaur *(39)* and the monoclonal antibody (mAb) QCPN prepared by B. M. Carlson and J. A. Carlson, which is available at the Developmental studies Hybridoma Bank (Department of Biology, University of Iowa, Iowa City, IA). The use of QCPN is easy and can be combined with other antibodies like HNK1 *(40)*, which recognizes neural crest cells, or 13F4 *(41)*, which marks muscle cells and their precursors.
2. Other mAbs are species and cell type specific: MB1 and QH1 *(42,43)*, which recognize a glycosylated epitope carried by surface proteins expressed in quail leucocytes and endothelial cells at the exclusion of any cell type of the chick.
3. Neural chimeras can be analyzed with mAbs, which recognize either neuronal cell bodies or neurites of quail or chick exclusively *(44,45)*.

3.5.4. Species-Specific Nucleic Probes

A growing number of these probes is applied for *in situ* hybridization, on sections, or whole-mount preparations. For example, a chick probe was used to demonstrate the activation of the homeobox gene *goosecoid* in a chick engrafted with quail goosecoid-producing tissues *(46)*. The quail-specific SMP (Schwann cell myelin protein) probe *(47)* allows to distinguish quail oligodendrocytes in chimeric spinal cord (*[16]*, *see* **Fig. 2B**). Chick *Wnt*1 and quail *Wnt*1 probes have been combined to demonstrate the induction of *Wnt*1 in quail–chick chimeras *(48)*.

4. Notes

1. If available, select a rapidly growing strain of chickens in which early stages of development will be phased with the quail embryos. The use of a nonpigmented strain of chickens will allow pigmented quail melanocytes to serve as a second marker. Freshly laid eggs should be stored no more than one week at 15°C.

2. Ideal conditions for chick and quail incubation are 38°C, 45% humidity (first two-thirds of incubation time), 75% (last one-third of incubation time and hatching). Small incubators equipped with time programmers, placed in a 15°C room, will start incubation at a defined time, yielding accurate stages of development.

3. Multiple or individual egg holders and hollowed out wooden slats are useful respectively for preincubation, microsurgical experiment, and postincubation of the chicken eggs.

4. The operations that we describe here are performed at embryonic day 2 (E2) when somites, which can be easily counted, can serve to stage the embryos.

5. In order to disturb gas exchanges through the shell as little as possible, avoid large pieces of tape and folds. The latter would cause air entry and consequently progressive drying of the egg content. Folds in the tape should be flattened carefully against the shell.

6. PBS and Tyrode solutions supplemented with antibiotics (10–20 IU/mL) are directly deposited on the blastoderm, to moisten it when needed.

7. Pancreatin (Gibco-BRL) is diluted one-third to one-sixth with PBS or Tyrode solution. In this way, tissue dissociation can be easily controlled. Titer and temperature are adapted to the developmental stage of the tissues. The younger the tissues, the lower the titer and temperature; for example, tissues from 10-somite stage embryos will be treated with 20% pancreatin in Tyrode on ice, whereas tissues from 20-somite stage embryos will be treated with 30% pancreatin at room temperature. Tyrode solution supplemented with bovine serum will serve to arrest enzyme action.

8. Glass micropipets hand drawn from Pasteur pipets are curved and calibrated according to use—injection of liquids or transfer of pieces of tissues. Calibration of the micropipet according to the size of the rudiment to be transplanted (for instance, neural tube versus brain vesicle) is an important requirement.

9. The Rhodorsyl base is now preferred to the paraffin base because it can be either black or perfectly transparent, whether animal carbon is added or not to the commercial preparation. Moreover it can be sterilized as often as necessary in dry oven and is not damaged by insect pins.

10. India ink has to be tested for toxicity before use and must be used without excess. The injecting pipet is introduced obliquely through the blastodisc in the extraembryonic area.

11. Microscalpels have to be perfectly adapted to each use. Microscalpels, manufactured by stropping and honing steel needles (sewing needles) on an Arkansas oil stone, are the most convenient for excising fragments of neural tube or brain vesicles because they can be both extremely thin and resistant. For dissociating tissues after enzymatic treatment, they must have a smooth tip. Tungsten microscalpels *(49)* or microscalpels made from insect pins are also useful for dissecting very small pieces of tissues. Their preparation is more rapid, but they are more fragile.

12. Formerly used conventional light bulbs with a condenser tend to radiate heat and cause traumatic drying to the embryos during surgery, so that defects in amnion formation and subsequent death are often observed.

13. Zenker fluidlike Carnoy fluid are appropriate fixatives for Feulgen and Rossenbeck staining *(14)*, but they do not allow immunohistochemistry or *in situ* hybridization.

14. A modification of the Feulgen and Rossenbeck classical protocol consists of performing the hydrolysis with 5 N HCl during 20–30 min at room temperature after Carnoy fixation instead of 1 N HCl during 4 to 8 min at 60°C as previously recommended in Gabe *(14)*.

15. Dorsoventral and rostrocaudal orientations of the graft are recognized either by morphological characters or by various labelings, for instance, a precisely localized minute slit.

16. Differences in stage and caudorostral level implicate difference of size. If the fragment of neural tube to be grafted is much bigger than the one that has been removed, it should be resected before grafting.

17. Brain vesicles are transplanted at the 12- to 14-somite stages for the following reasons: brain vesicles, not yet covered by the amnion, are clearly demarcated by constrictions, yet the brain is uncurved; the notochord no longer strongly adheres to the floor of the neural epithelium at this level; the neuroepithelium is not vascularized. Some neural crest cells and cephalic mesoderm are transferred along with the brain vesicles. Their presence does not interfere with the development of the brain and presence of melanocytes in the head feathers of the chimera indicates the level of the graft.

18. Adhesion of the graft to host tissues is promoted by sucking out the excess of physiological liquid added during the operation with a micropipet.

19. Daily gentle manual rocking of the operated eggs can enhance embryo survival. Incubator humidity must grow from 45–75% on the 18th day of incubation if hatching of the operated embryos is hoped.

20. E3–E4 chimeric embryos are fixed as a whole, during 1–3 h at room temperature. Older embryos have to be fixed as fragments and maintained *in vacuo*, during a time that increases with the stage and the size of the tissue pieces or organs. The same conditions will be applied for dehydration and paraffin embedding of the samples.

Acknowledgments

The authors thank Marie-Françoise Meunier, Sophie Gournet, and Francis Beaujean for their help with manuscript preparation.

References

1. Le Douarin, N. (1969) Particularités du noyau interphasique chez la Caille japonaise (*Coturnix coturnix japonica*). Utilisation de ces particularités comme "marquage biologique" dans les recherches sur les interactions tissulaires et les migrations cellulaires au cours de l'ontogenèse. *Bull. Biol. Fr. Belg.* **103**, 435–452.

2. Le Douarin, N. M. (1973) A biological cell labelling technique and its use in experimental Embryology. *Dev. Biol.* **30**, 217–222.

3. Feulgen, R. and Rossenbeck, H. (1924) Mikroskopisch-chemischer Nachweis einer Nukleinsäure von Typus der Thymonukleinsäure und die darauf beruhende elektive Färbung von Zellkernen in mikroskopischen Präparaten. *Hoppe-Seyler's Z. Physiol. Chem.* **135**, 203–248.

4. Le Douarin, N. M. (1982) *The Neural Crest.* Cambridge University Press, Cambridge, UK.

5. Le Douarin, N. M. (1993) Embryonic neural chimaeras in the study of brain development. *Trends Neurosci.* **16**, 64–72.

6. Catala, M., Teillet, M.-A., De Robertis, E. M., and Le Douarin, N. M. (1996) A spinal cord fate map in the avian embryo: while regressing, Hensen's node lays down the notochord and floor plate thus joining the spinal cord lateral walls. *Development* **122**, 2599–2610.

7. Kinutani, M., Coltey, M., and Le Douarin, N. M. (1986) Postnatal development of a demyelinating disease in avian spinal cord chimeras. *Cell* **45**, 307–314.

8. Ohki, H., Martin, C., Corbel, C., Coltey, M., and Le Douarin, N. M. (1987) Tolerance induced by thymic epithelial grafts in birds. *Science* **237**, 1032–1035.

9. Balaban, E., Teillet, M.-A., and Le Douarin, N. M. (1988) Application of the quail-chick chimera system to the study of brain development and behavior. *Science* **241,** 1339–1342.

10. Le Douarin, N. M. and Teillet, M.-A. (1974). Experimental analysis of the migration and differentiation of neuroblasts of the autonomic nervous system and neurectodermal mesenchymal derivatives, using a biological cell marking technique. *Dev. Biol.* **41,** 162–184.

11. Le Douarin, N. M., Dieterlen-Lièvre, F., and Oliver, P. D. (1984) Ontogeny of primary lymphoid organs and lymphoid stem cells. *Am. J. Anat.* **170,** 261–299.

12. Hamburger, V. and Hamilton, H. L. (1951) A series of normal stages in the development of the chick embryo. *J. Morphol.* **88,** 49–92.

13. Zacchei, A. M. (1961) Lo sviluppo embrionale della quaglia giapponese (*Coturnix coturnix japonica*, T. e S.). *Arch. Ital. Anat. Embriol.* **66,** 36–62.

14. Gabe, M. (1968) *Techniques Histologiques.* Masson, Paris.

15. Teillet, M.-A., Kalcheim, C., and Le Douarin, N. M. (1987) Formation of the dorsal root ganglia in the avian embryo: segmental origin and migratory behavior of neural crest progenitor cells. *Dev. Biol.* **120,** 329–347.

16. Cameron-Curry, P. and Le Douarin, N. M. (1995) Oligodendrocyte precursors originate from both the dorsal and the ventral parts of the spinal cord. *Neuron* **15,** 1299–1310.

17. Hallonet, M. E. R., Teillet, M.-A., and Le Douarin, N. M. (1990) A new approach to the development of the cerebellum provided by the quail-chick marker system. *Development* **108,** 19–31.

18. Tan, K. and Le Douarin, N. M. (1991) Development of the nuclei and cell migration in the medulla oblongata. Application of the quail-chick chimera system. *Anat. Embryol.* **183,** 321–343.

19. Teillet, M.-A., Naquet, R., Le Gal La Salle, G., Merat, P., Schuler, B., and Le Douarin, N. M. (1991) Transfer of genetic epilepsy by embryonic brain grafts in the chicken. *Proc. Natl. Acad. Sci. USA* **88,** 6966–6970.

20. Batini, C., Teillet, M.-A., Naquet, R., and Le Douarin, N. M. (1996) Brain chimeras in birds: application to the study of a genetic form of reflex epilepsy. *Trends Neurosci.* **19,** 246–252.

21. Alvarado-Mallart, R. M. and Sotelo, C. (1984) Homotopic and heterotopic transplantations of quail tectal primordia in chick embryos: organization of the retinotectal projections in the chimeric embryos. *Dev. Biol.* **103,** 378–398.

22. Martinez, S. and Alvarado-Mallart, R. M. (1989) Rostral cerebellum originates from the caudal portion of the so-called "mesencephalic" vesicle: a study using chick/quail chimeras. *Eur. J. Neurosci.* **1,** 549–560.

23. Hallonet, M. E. and Le Douarin, N. M. (1993) Tracing neuroepithelial cells of the mesencephalic and metencephalic alar plates during cerebellar ontogeny in quail-chick chimaeras. *Eur. J. Neurosci.* **5,** 1145–1155.

24. Nakamura, H. (1990) Do CNS anlagen have plasticity in differentiation? Analysis in quail-chick chimera. *Brain Res.* **511,** 122–128.

25. Martinez, S. and Alvarado-Mallart, R. M. (1990) Expression of the homeobox *Chick-en* gene in chick-quail chimeras with inverted mes-metencephalic grafts. *Dev. Biol.* **139,** 432–436.

26. Martinez, S., Wassef, M., and Alvarado-Mallart, R. M. (1991) Induction of a mesencephalic phenotype in the 2-day-old chick prosencephalon is preceded by the early expression of the homeobox gene *en. Neuron* **6,** 971–981.

27. Couly, G. F. and Le Douarin, N. M. (1985) Mapping of the early neural primordium in quail-chick chimeras. I. Developmental relationships between placodes, facial ectoderm, and prosencephalon. *Dev. Biol.* **110,** 422–439.

28. Couly, G. F. and Le Douarin, N. M. (1987) Mapping of the early neural primordium in quail-chick chimeras. II. The prosencephalic neural plate and neural folds: implications for the genesis of cephalic human congenital abnormalities. *Dev. Biol.* **120,** 198–214.

29. Couly, G. and Le Douarin, N. M. (1988) The fate map of the cephalic neural primordium at the presomitic to the 3-somite stage in the avian embryo. *Development* **103(Suppl.),** 101–113.

30. Couly, G. F., Coltey, P. M., and Le Douarin, N. M. (1993) The triple origin of skull in higher vertebrates: a study in quail-chick chimeras. *Development* **117,** 409–429.

31. Grapin-Botton, A., Bonnin, M.-A., McNaughton, L. A., Krumlauf, R., and Le Douarin, N. M. (1995) Plasticity of transposed rhombomeres: Hox gene induction is correlated with phenotypic modifications. *Development* **121,** 2707–2721.

32. Grapin-Botton, A., Bonnin, M.-A., and Le Douarin, N. M. (1997) Hox gene induction in the neural tube depends on three parameters: competence, signal supply and paralogue group. *Development* **124,** 849–859.

33. Couly, G., Grapin-Botton, A., Coltey, P., and Le Douarin, N. M. (1996) The regeneration of the cephalic neural crest, a problem revisited: the regenerating cells originate from the contralateral or from the anterior and posterior neural folds. *Development* **122,** 3393–3407.

34. Christ, B. and Ordahl, C. P. (1995) Early stages of chick somite development. *Anat. Embryol.* **191,** 381–396.

35. Catala, M., Teillet, M.-A., and Le Douarin, N. M. (1995) Organization and development of the tail bud analyzed with the quail-chick chimaera system. *Mech. Dev.* **51,** 51–65.

36. Martin, C. (1972) Technique d'explantation *in ovo* de blastodermes d'embryons d'oiseaux. *C. R. Soc. Biol.* **116,** 283–285.

37. Dieterlen-Lièvre, F. (1975) On the origin of haemopoietic stem cells in the avian embryo: an experimental approach. *J. Embryol. Exp. Morphol.* **33,** 607–619.

38. Martin, C., Beaupain, D., and Dieterlen-Lièvre, F. (1978) Developmental relationships between vitelline and intra-embryonic haemopoiesis studied in avian "yolk sac chimaeras." *Cell Diff.* **7,** 115–130.

39. Lance-Jones, C. C. and Lagenaur, C. F. (1987) A new marker for identifying quail cells in embryonic avian chimeras: a quail-specific antiserum. *J. Histochem. Cytochem.* **35,** 771–780.

40. Abo, T. and Balch, C. M. (1981) A differentiation antigen of human NK and K cells identified by a monoclonal antibody (HNK-1). *J. Immunol.* **127,** 1024–1029.

41. Rong, P. M., Ziller, C., Pena-Melian, A., and Le Douarin, N. M. (1987) A monoclonal antibody specific for avian early myogenic cells and differentiated muscle. *Dev. Biol.* **122,** 338–353.

42. Péault, B. M., Thiery, J.-P., and Le Douarin, N. M. (1983) Surface marker for hemopoietic and endothelial cell lineages in quail that is defined by a monoclonal antibody. *Proc. Natl. Acad. Sci. USA* **80,** 2976–2980.

43. Pardanaud, L., Altmann, C., Kitos, P., Dieterlen-Lièvre, F., and Buck, C. A. (1987) Vasculogenesis in the early quail blastodisc as studied with a monoclonal antibody recognizing endothelial cells. *Development* **100,** 339–349.

44. Takagi, S., Toshiaki, T., Kinutani, M., and Fugisawa, H. (1989) Monoclonal antibodies against specific antigens in the chick central nervous system: putative application as a transplantation marker in the quail-chick chimaera. *J. Histochem. Cytochem.* **37,** 177–184.

45. Tanaka, H., Kinutani, M., Agata, A., Takashima, Y., and Obata, K. (1990) Pathfinding during spinal tract formation in quail-chick chimera analysed by species specific monoclonal antibodies. *Development* **110,** 565–571.

46. Izpisùa-Belmonte, J. C., De Robertis, E. M., Storey, K. G., and Stern, C. D. (1993) The homeobox gene *goosecoid* and the origin of organizer cells in the early chick blastoderm. *Cell* **74,** 645–659.

47. Dulac, C., Tropak, M. B., Cameron-Curry, P., Rossier, J., Marshak, D. R., Roder, J., and Le Douarin, N. M. (1992) Molecular characterization of the Schwann cell myelin protein, SMP: structural similarities within the immunoglobulin superfamily. *Neuron* **8,** 323–334.

48. Bally-Cuif, L. and Wassef, M. (1994) Ectopic induction and reorganization of *Wnt-*1 expression in quail-chick chimeras. *Development* **120,** 3379–3394.

49. Conrad, G. W., Bee, J. A., Roche, S. M., and Teillet, M.-A. (1993) Fabrication of microscalpels by electrolysis of tungsten wire in a meniscus. *J. Neurosci. Methods* **50,** 123–127.

37

Quail–Chick Transplantation in the Embryonic Limb Bud

Elizabeth E. LeClair and Rocky S. Tuan

1. Introduction

Chick and quail have been powerful partners in the investigation of avian development. Cells from these two species can develop harmoniously in heterospecific combinations, yet each remains histologically distinct, facilitating the fate mapping of transplanted tissues (1). This property of the quail–chick chimera has been used to track cell position and fate in many embryonic processes including gastrulation (2–4), neural tube formation (5), hematopoiesis (6), and craniofacial development (7–9) among others. One of the most accessible areas for such grafts is the embryonic chick limb bud, site of many pioneering manipulations (10; for review, see **ref. 11**). Quail-cell grafts have been used to investigate the origin of the limb bud from somatopleural mesoderm (12), the contribution of the somites to limb musculature (13–16), and the effect of limb mesoderm on species-specific limb development (17).

This chapter describes a quail tissue-grafting technique to aid the investigation of normal cartilage morphogenesis in later stages of chick hindlimb development. Among vertebrates, the avian limb is particularly rich in embryonic skeletal fusions, including fusions of separate prechondrogenic condensations, fusions of well-differentiated cartilages, and the incorporation of multiple cartilages into single ossifications (18–20). Thus the chick is an ideal system for exploring how different regions of the prechondrogenic limb bud may contribute to embryonic cartilages, and how embryonic cartilages from different regions may combine to form bones. Using previous limb bud fate maps as a guide (21,22), heterospecific grafts can be a powerful technique to track how embryonic cartilage pattern becomes adult limb structure.

In this protocol, a region of anterior or posterior mesenchyme within the apical ectodermal ridge (AER) of the chick leg bud is removed and replaced with an isotopic, isochronic piece of quail tissue. Because the AER remains intact, limb development is rarely perturbed and a normal cartilage pattern results. Thus the fate of the grafts can be used to assess the relative contribution of anterior and posterior limb mesenchyme to embryonic cartilages and adult limb structure. This technique was originally developed to follow the migration and fusion of embryonic cartilages forming the avian tibiotarsus, and so the protocol gives special attention to operations on the chick leg bud, which has been less often used than the wing because of its relative inaccessibility

From: *Methods in Molecular Biology, Vol. 135: Developmental Biology Protocols, Vol. I*
Edited by: R. S. Tuan and C. W. Lo © Humana Press Inc., Totowa, NJ

in the embryo. Most of the experimental details, however, apply to any grafting operation on limb buds and can be adapted to the investigator's particular goals.

2. Materials

1. Surgical Instruments:
   a. Spearpoint microscalpel (#6125; Roboz Surgical Instruments, Rockville, MD)
   b. Two pairs of fine angled forceps (Dumont #5-45; Fine Science Tools, Foster City, CA)
   c. Miniature spatula: Using a hammer and a sturdy block, flatten one end of a short length of 500-µm diameter nickel-chromium wire (Omega Engineering, Stamford, CT). Insert the round end of the wire into a pin holder (#26016-12; Fine Science Tools) and bend the flattened tip with forceps to a comfortable working angle (approx 45°).
   d. Sharpened tungsten needles: Take a fine hypodermic needle tip (#22 gage) and gently blunt the pointed end by dragging it against coarse sandpaper or a sharpening stone. Insert a few inches of thin (0.005 in. diameter) tungsten wire (Ted Pella Inc., Redding, CA) through the barrel of the needle, leaving a short length extending from each end. Generously fill the wide end of the needle base with commercial epoxy and secure it to a wooden applicator stick or a fine diameter plastic pipet, which will serve as a handle. Cut off any excess wire protruding from the base and cover the cut end with additional epoxy to avoid a sharp point.

   When the epoxy is completely dry, trim the tungsten wire to about 1 cm above the end of the blunted hypodermic needle. Sharpen the tungsten tip as follows, wearing rubber gloves throughout the procedure: Prepare a small, well-secured glass vessel, nearly filled with sodium hydroxide (NaOH) saturated in water. Hang a bent paper clip over the side of the vessel, partially immersed, to serve as one electrode. Attach two wires with alligator clips to the negative and positive terminals of a suitable power source (e.g., a 9-V battery or a microscope light). Complete the circuit by attaching one wire to the free end of the immersed paper clip and the other wire to the metal portion of the hypodermic needle (**Fig. 1**). Sharpen by quickly dipping the tungsten tip up and down in the electrified solution. Tiny bubbles forming around the immersed needle means the circuit is complete; if no bubbles appear, check the power supply and all connections. DO NOT bring the electrified needle in contact with the immersed electrode; this will produce a short circuit and an impressively big spark. Periodically disconnect the clips and examine the needle tip under a dissecting microscope to judge your sharpening progress. It is easiest to sharpen a straight wire and then bend it to a suitable shape with forceps. A 90° bend forming a short L-shaped tip is very convenient for most operations, but other types of hooks and points are equally possible.

2. Blue-stained agar plates *(23)*: Prepare a 2% solution of agarose in distilled water and microwave until thoroughly dissolved. Allow the solution to cool slightly and add Nile Blue sulfate powder (2% w:v, #N 5632; Sigma Chemical Co., St. Louis, MO). Swirl until mixed, heating again if necessary. Pour into thin (2–3 mm) layers in 35-mm plastic plates and cover until the agar has solidified. At this point, small stacks of plates can be wrapped tightly in plastic and stored at 4°C for future use. The day before the agar is needed, uncover one plate and leave it overnight in a flowthrough hood. The agar will shrink and crack, forming a dark blue film. Immediately before use, sterilize the top surface of the dried agar with a bit of tissue barely wetted in 70% ethanol, or irradiate the entire plate under UV light.

3. Siliconized, sterile transfer pipets: Trim regular 200-µL plastic micropipet tips with a large pair of scissors so that each tip end is slightly larger than the expected dimensions of the graft. Load the tips into a plastic "refill" plate and in a fumehood dip the cut ends several

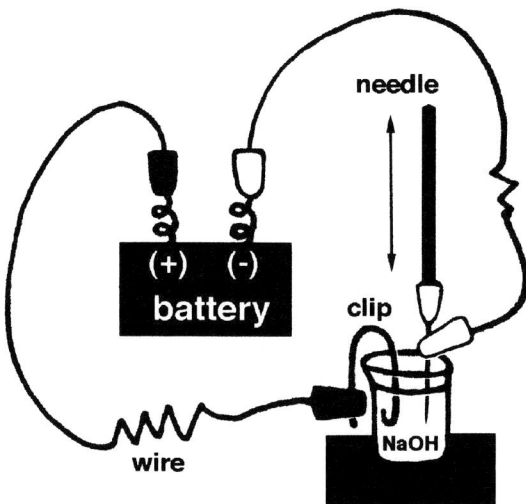

Fig. 1. Circuit for sharpening tungsten needles. Connect one terminal of the power source (e.g., a 9-V battery) to a bent paper clip hanging over the side of a glass vessel filled with saturated sodium hydroxide (NaOH). Connect the other terminal to the metal base of the blunt needle. Sharpen by briefly and repeatedly dipping the tungsten tip into the electrified solution. For further details, *see* **Subheading 2.**

times in Sigmacote (#SL-2; Sigma). This creates a permanent silicone film that prevents graft tissue adhering to the inside of the tip. Drain the tips by touching to a clean piece of filter paper. Allow the tips to dry suspended in the hood for several hours. Autoclave before use.

4. 20X Ringer's saline (sterile): Mix 180 g NaCl, 8.4 g KCl, and 5 g $CaCl_2$ in 1 L of distilled, deionized water (ddH_2O). Autoclave and store stock at room temperature. Make a 1X solution by diluting stock 1:20 in sterile ddH_2O. Autoclave again if desired.

5. Antibiotic solution: 1X Ringer's saline with penicillin G (100 U/mL) and streptomycin sulfate (0.1 mg/mL). Filter before use. Make small quantities and store at 4°C.

6. 20X Phosphate-buffered saline (PBS): Add 40 g NaCl, 4 g KCl, 28.8 g Na_2HPO_4, and 4.8 g KH_2PO_4 to 900 mL ddH_2O until dissolved. Adjust pH to 7.4. Add ddH_2O up to a total volume of 1000 mL, mix well, and store at room temperature. Make a 1X working solution by diluting stock 1:20 in ddH_2O.

7. 4% paraformaldehyde/phosphate buffered saline (PF/PBS): In a fume hood, mix 10 g reagent-grade paraformaldehyde powder (#P-6148; Sigma) with 250 mL 1X PBS (pH 7.4) in a heat-proof bottle or flask. Place the bottle on low heat, stirring until the mixture is completely dissolved. Allow to cool to room temperature before using. PF/PBS should be used within 24 h or stored in small aliquots at –20°C for future use. Always use with adequate ventilation.

8. 1% Acid alcohol: Mix 1:99 v = v concentrated HCl:70% ethanol. Store indefinitely at room temperature.

9. Victoria Blue B solution: Dissolve Victoria Blue B powder (#V 0753; Sigma) in 1% acid alcohol at a final stain concentration of 0.5% (w:v). Stir well for 10 min and filter before use. This solution keeps indefinitely at room temperature and can be reused multiple times without significant loss of stain intensity. When tissue staining becomes noticeably weaker, prepare a new batch.

10. Carnoy's fluid: 6:3:1 100% ethanol:chloroform:glacial acetic acid. Make fresh in small quantities just before use. Do not store.
11. Methyl salicylate (#M 2047; Sigma).
12. Alcian Blue staining solution: Add 4.25 g Alcian Blue 8GX powder (#A 3157; Sigma) to 500 mL of 0.1 *N* HCl. Stir for 1 h and filter before use. Keeps several months at room temperature.
13. Mouse anti-quail primary antibody (QCPN, Developmental Studies Hybridoma Bank, University of Iowa, Iowa City, 319-335-3826).
14. Broad-spectrum immunohistochemical detection kit (Histostain SP, #95-9943-B; Zymed Labs South, San Francisco, CA).
15. Colorimetric detection kit (AEC substrate, #00-2007; Zymed Labs).
16. Aqueous mounting medium (CrystalMount, #M02; Biomeda, Foster City, CA).
17. Permanent mounting medium (Permount, #SP15-100; Fisher Scientific Co., Pittsburgh, PA).
18. Hydrophobic marking pen (PAP pen, #00-8877; Zymed Labs).

3. Method

3.1. Egg Incubation and Shell-Less Culture

Incubate chick eggs for approx 3 d before putting the host embryos in *ex ovo* shell-less (SL) culture (*24,25*; *see* Chapter 6). Under our incubation conditions, 30 chick eggs were placed in the incubator at 5 PM on Friday. On Monday afternoon the eggs were placed in SL culture, and by Tuesday morning the hosts were ready for grafting at Hamburger-Hamilton stage 20-21. For each batch of 30 chick eggs, 24 quail eggs were incubated on the same schedule, except that the quail embryos had their development slowed by removing them from the incubator and leaving them at room temperature for approx 8–10 h on the third day (i.e., 8 AM–6 PM). After this "coasting" period, the quail eggs were returned to the incubator until the time of the operations. Although coasting can increase the stage variability of the quail donors, there is usually enough natural developmental variation in the chick hosts to allow many good matches between host and donor embryos.

3.2. Preparation of Hosts

Take the chick host from the incubator and open the SL culture vessel on the stage of a dissecting microscope set up in a sterile flowthrough hood. Operate only on healthy-looking cultures with intact yolks that are free from blebbing. Using a pair of fine-angled forceps in each hand, gently pinch, lift, and tear the extraembryonic membranes immediately over the leg bud (*see* **Note 1**). Tease these membranes open and spread them apart to clear the operating area (*see* **Note 2**). The first few times this is attempted it may be useful to dye the membranes with a drop of Nile Blue sulfate (2% in sterile water, autoclaved) to better see their position over the embryo.

The remainder of the protocol describes the removal and replacement of posterior limb bud mesenchyme; anterior replacements involve similar maneuvers on the opposite side of the leg bud (**Fig. 2**). For a right-handed operator, either operation is most conveniently performed with the embryo oriented head toward you on the microscope stage, so that the uppermost (right) leg bud extends to the left (**Fig. 2A**). Lefties may find the reverse more comfortable and should alter all following directions accordingly.

Using the spatula in your left hand, gently lift the leg bud and support the tip, making sure it is free of all extraembryonic membranes. At this point, the AER should

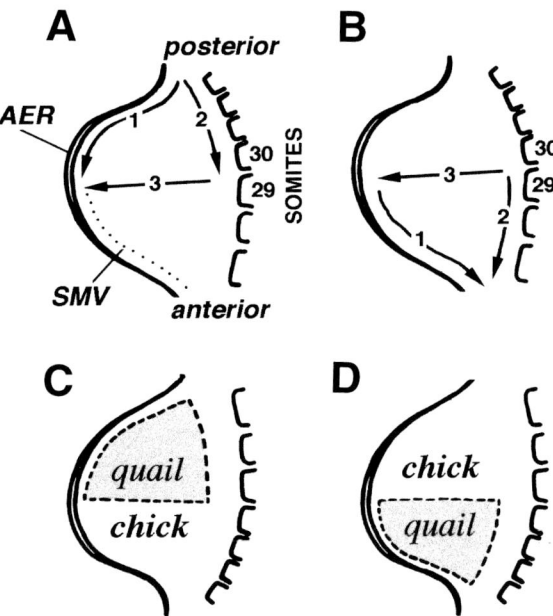

Fig. 2. Diagrams of chick leg bud surgeries, stage 21. Note that the limb buds are shown with the posterior margin uppermost and the anterior margin lowermost, as seen by the operator. (**A**) Suggested sequence of surgical incisions for removing posterior limb bud mesenchyme within the AER. (**B**) A similar procedure is used to remove anterior limb bud mesenchyme. Diagrams of completed posterior (**C**) and anterior (**D**) operations with quail donor tissue in place. For further details, *see* **Subheading 3.2.** AER = apical ectodermal ridge. SMV = submarginal vessel.

become conspicuous and also the submarginal vessel (SMV) immediately proximal to it. Adjust the lighting if necessary so these features are visible. With a sharp tungsten needle in your right hand, begin an incision at the posterior margin where the limb meets the flank, just proximal to the AER (**Fig. 2A, step 1**). Using short pulling strokes, extend the incision along the limb margin, cutting just under the ridge along the SMV. As you cut, the incision usually fills with a small amount of blood, making the path of the needle more visible. Continue cutting to the apex of the limb bud, rotating the culture vessel if necessary to keep your hands in a comfortable position. Continue using the needle or switch to the microscalpel to make a straight second cut along the leg base parallel to the body wall (**Fig. 2A, step 2**). Avoid cutting too close to the body for risk of breaching the major subcardinal vessels; should this happen, the hemorrhagic results will be obvious. Finally, turn the host culture so the tip of the limb bud is facing you, and make a final straight cut along the midline of the bud, approximately at the junction of somites 29 and 30 (**Fig. 2A, step 3**). When the cuts are complete, make sure all of the edges and corners of the tissue block are free by using the tungsten needle as a probe.

Before removing the block, measure the dimensions of the perpendicular cuts (nos. 2 and 3) with an ocular micrometer. There is usually some bleeding around the edges of the block that makes these borders quite visible. Write down these measurements on the plastic lid of the host culture, as you will need them when cutting the same-sized graft. Finally, use forceps or blunt tungsten needles to manipulate the posterior block out of the limb bud, leaving it on the surface of the culture or discarding it. Add 50 µL of antibiotic solution directly over the operated leg, cover the culture, and return the host to the incubator while preparing graft tissue.

3.3. Preparation of Grafts

Dissect several quail donors (stage 20-21) into sterile Ringer's solution at room temperature in a plastic plate. Examine the embryos to select one at the desired stage. Remove the extraembryonic membranes and excise the caudal part of the trunk containing the leg buds. Gripping the trunk with forceps, stabilize the right leg bud against the plate bottom. Using the microscalpel or tungsten needle, strip off the AER and the immediately underlying mesenchyme. Cut the limb bud in half along the proximal–distal axis, using the somite boundaries (29/30) as a reference point to find the midline. At this point, moisten a piece of blue-stained agar and make an orientation mark on some part of the graft ectoderm (e.g., the posterior dorsal corner) to guide proper positioning in the host (*see* **Note 3**). Finally, remove the posterior piece of tissue from the donor embryo by cutting all the way through the limb tissue parallel to the limb base. Use the ocular micrometer as you work to make the block the right size before you cut it free from the donor, as this is far easier than trying to trim a free-floating graft.

3.4. Transferring Graft to Host

Pick up the stained graft with a siliconized sterile pipet tip carrying a minimal amount of fluid (approx 10 µL). Gently eject the tissue onto the surface of the host. Manipulate the graft into place with fine forceps or blunt needles, orienting it according to the reference mark. Replace the lid on the culture and return the host to the incubator for further development.

3.5. Embryo Harvesting

Allow the embryo to develop up to the desired stage (*see* **Note 4**). Before removing the embryo from the culture, gently tear the extraembryonic membranes free from the hindlimb region, as these sometimes adhere to the operated leg and cause damage to the specimen.

3.6. Whole-Mount Stain for Cartilage

Visualizing all of the cartilages in the operated limb is useful for confirming that the surgery does not radically perturb the expected skeletal pattern. In our experience, >95% of chimeric limbs produced by this technique have a perfectly normal cartilage pattern, although the operated limb is usually smaller than the contralateral control (**Fig. 3A**). Fix freshly dissected tissue in 4% PF/PBS overnight. Rinse in water (2 × 5 min) and dehydrate in 70% ethanol (1 h). Equilibrate in 1% acid alcohol (1 h) and stain in Victoria Blue B solution (2–3 h). Destain in 70% ethanol (4 × 15 min with agitation, then overnight on a shaker), with additional changes of ethanol until the tissue is whitish and the alcohol is completely clear. Dehydrate to 95% ethanol (1 h) and then 100% ethanol (overnight) and clear in methyl salicylate (*see* **Note 5**). Observe the overall cartilage pattern under a dissecting scope using transmitted light (**Fig. 3B**). Cleared limbs

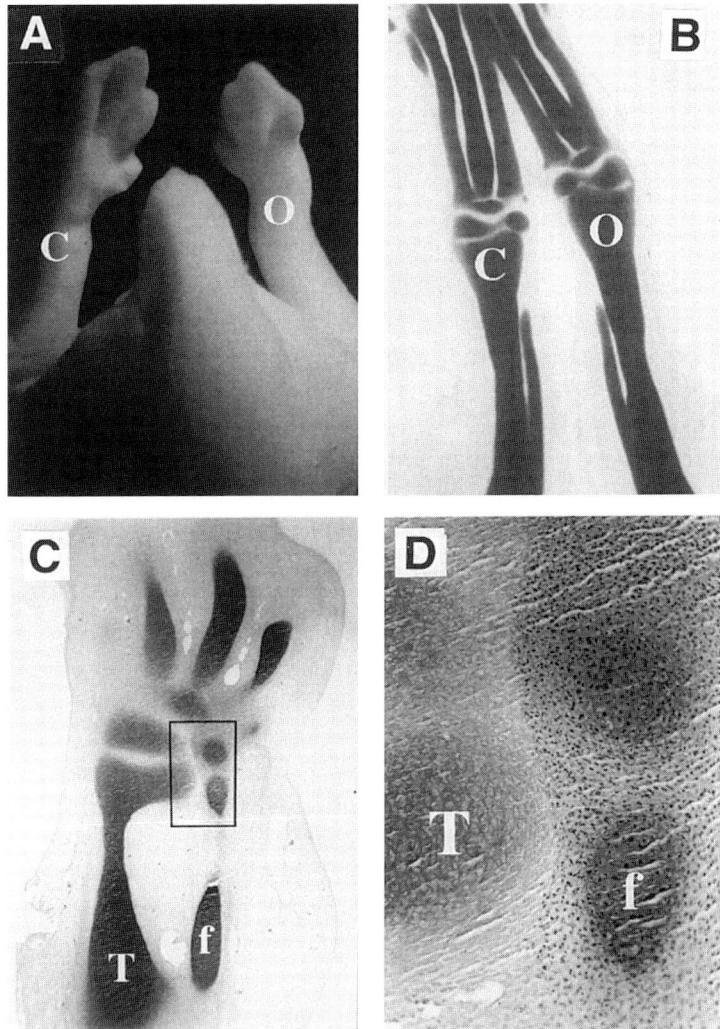

Fig. 3. Experimental results of quail–chick grafting procedures. (**A**) Anterior replacement, harvested at day 8. The operated limb (o) forms a normally flexed foot with the proper number of digits, though it is slightly smaller than the unoperated control (c). (**B**) Anterior replacement (day 10) stained in whole-mount for cartilage with Victoria Blue B. Both control (c) and operated (o) limbs show normal skeletal patterning. (**C**) Posterior replacement (day 7), sectioned and sequentially stained with Alcian Blue and QCPN, a quail-specific antibody. (**D**) Detail of boxed area in C. The embryonic tibia (T) is derived from anterior limb mesenchyme of the chick host. The embryonic fibula (f) and other cartilages derived from the posterior quail graft stain positively with QCPN, showing dark-stained nuclei.

can be stored indefinitely at room temperature or transferred directly to molten paraffin for standard histological processing.

3.7. Sectioning and Staining Cartilage Matrix and Quail Cells

The most unambiguous way to identify grafted quail cells is to use a quail-specific primary antibody (QCPN = mouse anti-quail; Developmental Studies Hybridoma Bank) followed by immunohistochemical detection with a biotin-conjugated secondary antibody (e.g., biotinylated goat anti-mouse) and an avidin conjugated enzyme (e.g., avidin-peroxidase). To localize the graft cells in embryonic cartilages, one can sequentially stain for cartilage proteoglycan and for quail cells in the same tissue sections. Fix harvested limbs 1–2 h at 4°C in Carnoy's fluid or another alcohol-based fixative. (The QCPN antibody is not compatible with aldehydes.) Transfer tissue directly to 100% ethanol at 4°C overnight. At this point, the tissue can be stored in cold ethanol for several days before embedding. Clear in amyl acetate (1 h, 56°C) and equilibrate in several changes of paraffin (Paraplast X-tra [VWR Scientific, Denver, CO] 56°C, 3 × 1 h, then overnight). The next day, embed in plastic molds for wax sectioning at 8 μm.

To dewax paraffin sections, immerse in xylene (2 × 1 min), 100% ethanol (1 min) and rehydrate in a graded ethanol series to distilled, deionized water (95, 75, 50% ethanol, ddH$_2$O; 1 min each). Immerse in 3% glacial acetic acid (3 min), then stain in Alcian Blue solution for 10–15 min, depending on the desired intensity. Rinse in 1% acid alcohol (2 × 3 min), then distilled water (1 × 2 min). At this point, the slides can be briefly examined under a microscope to see the embryonic cartilages (**Fig. 3C**) and to select specific sections for immunohistochemical staining.

Circle sections to be stained with a hydrophobic marking pen and block with 10% goat serum in PBS (10 min). Drain slides (do not rinse) and apply QCPN diluted 1:500–1:1000 in PBS. Incubate 1 h in a humid chamber at room temperature. Rinse in PBS (3 × 2 min) and detect the primary antibody by applying the biotinylated secondary antibody, avidin-peroxidase conjugate, and peroxidase substrate according to manufacturer's instructions (Histostain SP kit and AEC substrate kit; Zymed). After 5–10 min of color development, quail nuclei should stain deep red; chick nuclei will be colorless (**Fig. 3D**; *see* **Note 6**). Stop the reaction by rinsing in distilled water. Cover stained sections with a few drops of CrystalMount and bake the slides flat at 80°C for 1 h. When the slides are cool, they can be conventionally cover-slipped with a few drops of Permount thinned in xylene or Histoclear (Amresco, Solon, OH). The resulting sections (**Fig. 3D**) show the contribution of grafted quail cells to the embryonic cartilages.

4. Notes

1. Surgical instruments: Although the chick embryo is quite resistant to infection, precautions should be taken to sterilize instruments that contact it. Solid metal instruments can be autoclaved in an instrument bag or foil. Homemade instruments and tungsten needles can be dipped in 70% ethanol or UV-irradiated before use. Needles that have just been electrolytically sharpened are essentially sterile. During operations, it is worthwhile to periodically dip instruments in 70% ethanol, then dip them again in a shallow dish of sterile saline to remove any excess ethanol before touching the embryo.

2. Leg bud operations: The less-accessible leg bud requires special care in host operations. When opening the membranes over the leg bud, do not pull too vigorously or major vessels may rupture in the region of the allantois. Avoid puncturing the allantois itself. The curved, flattened spatula (*see* **Note 1**) is essential for lifting the leg bud and protecting the underlying allantois from sharp instruments.

3. Agar plates: One plate provides sufficient dye for dozens (if not hundreds) of operations. The same plate can be used over months if kept sealed and dry after each use. When staining, if the agar chip is too dry, it can adhere to the ectoderm and damage the graft. Be sure each chip is sufficiently hydrated by dipping in sterile water or saline for several

seconds before use. This step also allows an initial cloud of blue dye to escape, preventing obscuration of the operating area. Use only deeply blue stained portions of agar. Avoid pieces with a greenish or brownish iridescent appearance, caused by separation of the dye from the agar substrate. In such cases, prepare a new plate. Finally, Nile Blue dye is highly soluble in water and can fade quickly after application. To prevent loss of informative stains, color the initial mark darkly and orient the graft in the host as soon as possible.

4. Postsurgical survival: Very few published studies will report the number of embryos that simply die after surgical intervention. In our hands, approx 50% of operated embryos (day 4, stage 20-21) survived to harvesting (days 7–10). Given the complexity and time-consuming nature of these operations, this estimate should be kept in mind when planning studies requiring large numbers of experimental surgeries.

5. Methyl salicylate: This liquid will dissolve certain types of plastic. For clearing and long-term storage, transfer the specimen to a glass vessel or a resistant type of plastic vial (e.g., #03-337-1; Fisher).

6. Immunostaining: This procedure omits a hydrogen peroxide preincubation commonly used to quench endogenous peroxidases that might interfere with enzymatic detection. Typically, the protocol produces no appreciable background that would interfere with proper identification of grafted tissues. Some nonspecific color development can appear in blood cells (which have endogenous peroxidases), but by their histology and location these cells are hardly to be confused with quail cell nuclei. If background staining is problematic, a 5-min incubation with 1:9 hydrogen peroxide:absolute methanol, followed by a PBS rinse (3 × 2 min) can be added before blocking with goat serum.

Acknowledgments

Peter Alexander and James Sanzo provided enthusiastic discussions and demonstrations of chick embryo surgery, Karen Jensen (Developmental Studies Hybridoma Bank) advised on several batches of the QCPN antibody, and Dr. John Saunders, Jr. generously corresponded on the topic of limb grafting. This work was supported in part by the NIH Grant nos. HD 15822, ES 07005, and DE 11327. E. E. L. is a recipient of an NRSA Postdoctoral Traineeship no. AR 07583 and an NRSA Postdoctoral Fellowship no. HD 08368.

References

1. LeDouarin, N., Dieterlen-Lievre, F., and Teillet, M.-A. (1996) Quail-chick transplantations, in *Methods in Avian Embryology*, (Bronner-Fraser, M., ed.), Academic, New York, pp. 24–59.

2. Ooi, V. E. C., Sanders, E. J., and Bellairs, R. (1986) The contribution of the primitive streak to the somites in the avian embryo. *J. Embryol. Exp. Morphol.* **92**, 193–206.

3. Veini, M. and Bellairs, R. (1991) Early mesoderm differentiation in the chick embryo. *Anat. Embryol.* **183**, 143–149.

4. Schoenwolf, G. C., Garcia-Martinez, V., and Dias, M. S. (1992) Mesoderm movement and fate during avian gastrulation and neurulation. *Dev. Dyn.* **193**, 235–248.

5. LeDouarin, N. M., Catala, M., and Batini, C. (1997) Embryonic neural chimeras in the study of vertebrate brain and head development. *Int. Rev. Cytol.* **175**, 241–309.

6. Dieterlein-Lievre, F., Godin, I., and Parnadaud, L. (1997) Where do hematopoietic cells come from? *Int. Arch. Allergy Immunol.* **112**, 3–8.

7. Noden, D. M. (1991) Vertebrate craniofacial development: the relation between ontogenetic process and morphological outcome. *Brain Behav. Evol.* **38**, 190–225.

8. Couly, G. F., Coltey, P. M., and LeDouarin, N. M. (1992) The developmental fate of the cephalic mesoderm in quail-chick chimeras. *Development* **114,** 1–15.
 9. Couly, G. F., Coltey, P. M., and LeDouarin, N. M. (1993) The triple origin of the skull in higher vertebrates: A study in quail-chick chimeras. *Development* **117,** 409–429.
10. Saunders, J. W., Jr. and Gaessling, M. (1968) Ectoderm-mesenchymal interactions in the origin of limb symmetry, in *Epithelial-Mesenchyme Interactions* (Fleischmajer, R. and Billingham, R. E., eds.), Williams & Wilkins, Baltimore, MD, pp. 78–97.
11. Saunders, J. W., Jr. (1996) Operations on limb buds of avian embryos, in *Methods in Avian Embryology* (Bronner-Fraser, M., ed.), Academic, New York, pp. 125–145.
12. Geduspan, J. S. and Solursh, M. (1992) Cellular contribution of the different regions of the somatopleure to the developing limb. *Dev. Dyn.* **195,** 177–87.
13. Beresford, B., LeLievre, C., and Rathbone, M. P. (1978) Chimaera studies on the origin and formation of the pectoral musculature of the avian embryo. *J. Exp. Zool.* **205,** 321–326.
14. Lance-Jones, C. (1988) The somitic level of origin of embryonic chick hindlimb muscles. *Dev. Biol.* **126,** 394–407.
15. Schramm, C. and Solursh, M. (1990) The formation of premuscle masses during chick wing bud development. *Anat. Embryol.* **182,** 235–247.
16. Zhi, Q., Huang, R., Christ, B., and Brand-Saberi, B. (1996) Participation of individual brachial somites in skeletal muscles of the avian distal wing. *Anat. Embryol.* **194,** 327–339.
17. Ohki-Hamazaki, H., Katsumata, T., Tsuakamoto, Y., Wada, N., and Kimura, I. (1997) Control of the limb bud outgrowth in the quail-chick chimera. *Dev. Dyn.* **208,** 85–91.
18. Montagna, W. (1945) A re-investigation of the development of the wing of the fowl. *J. Morphol.* **76,** 87–113.
19. Shubin, N. H. and Alberch, P. (1986) A morphogenetic approach to the origin and basic organization of the tetrapod limb. *Evol. Biol.* **20,** 319–387.
20. Müller, G. B. a. S., Jr. (1989) Ontogeny of the syndesmosis tibiofibularis and the evolution of the bird hindlimb: A caenogenetic feature triggers phenotypic novelty. *Anat. Embryol.* **179,** 327–339.
21. Hampé, A. (1960) La compétition entre les éléments osseux du zeugopode de Poulet. *J. Embryol. Exp. Morphol.* **8,** 241–245.
22. Stark, R. J. and Searls, R. L. (1973) A description of chick wing bud development and a model of limb morphogenesis. *Dev. Biol.* **33,** 138–153.
23. Hamburger, V. (1973) *A Manual of Experimental Embryology*. University of Chicago Press, Chicago, IL.
24. Tuan, R. S. (1983) Supplemented eggshell restores calcium transport in chorioallantoic membrane of cultured shell-less chick embryos. *J. Embryol. Exp. Morphol.* **74,** 119–131.
25. Tuan, R. S. and Ono, T. (1986) Regulation of extraembryonic calcium mobilization by the developing chick embryo. *J. Embryol. Exp. Morphol.* **97,** 63–74.

38

Mouse Chimeras and the Analysis of Development

Richard L. Gardner and Timothy J. Davies

1. Introduction

Because transgenesis via embryonic stem (ES) cells has become so topical in recent years, there have been a plethora of technical accounts of how to produce chimeras by introducing these cells into the preimplantation mouse conceptus, both using and avoiding micromanipulation (1–5). In the overwhelming majority of cases this type of chimerism is regarded simply as a means to an end, namely establishing and perpetuating in vivo modifications to the genome that have been produced in vitro. However, it is also being exploited to a limited extent to assess the developmental potential of homozygous mutant ES cells in wild-type embryos and vice versa (6,7).

Less attention, however, has been paid to updating approaches to the historically older strategy of making chimeras that are composed entirely of cells obtained directly from different preimplantation conceptuses. This type of chimerism has been used extensively for addressing a variety of issues relating to both normal and abnormal development as well as numerous aspects of adult physiology and pathology. The two stages of development used most commonly for making conventional as opposed to ES cell chimeras are the early morula (aggregation chimeras) and the expanding blastocyst when it is still encased in an intact zona pellucida (injection chimeras).

The technical requirements for making aggregation chimeras are very simple, the essential skill being that of good mouth-controlled pipetting of conceptuses between dishes and, if possible, availability of a warm stage to maintain conceptuses at 37°C during manipulation and inspection. This contrasts with injection chimeras whose production not only depends on the use of micromanipulators but also on the fashioning of elaborate microinstruments from glass capillary. Given the very marked disparity in investment of time and resources needed to gain proficiency with the two approaches, it is worthwhile briefly to consider their relative merits.

1.1. Choosing Between Aggregation and Injection Chimeras

Aggregation is normally undertaken at a stage when the fate of most if not all blastomeres has yet to be specified. Therefore, providing that the cells of one of the participating conceptuses are not compromised genetically in a way that precludes their differentiating into particular tissues, those of both should commonly contribute to all

From: *Methods in Molecular Biology, Vol. 135: Developmental Biology Protocols, Vol. I*
Edited by: R. S. Tuan and C. W. Lo © Humana Press Inc., Totowa, NJ

embryonic and extra-embryonic components of the resulting composite. In many studies, there is simply a need to obtain fetuses or offspring, all or most of whose tissues are composed of cells of two different genotypes. This includes cases in which the aim is to determine whether a particular gene behaves in a cell autonomous manner *(8)* or to obtain offspring from a lethal mutant in which the germline is unaffected *(9)*. In such circumstances, the aggregation technique is entirely appropriate, and there would be no case for attempting to master the intricacies of blastocyst injection. A possible complication with aggregation is failure to achieve a balanced range of contributions of the two genotypes among a series of chimeras, because one genotype consistently outcompetes the other. This issue has been analyzed for the greatest number of common strain combinations by Mullen and Whitten *(10)*, whose study should be consulted if either balanced or grossly unbalanced chimerism is desired.

However, where one wishes to establish the cell type of origin of a particular tissue or ensure that chimerism is restricted to the epiblast, primitive endodermal, or the trophectodermal lineage, then, at present, blastocyst injection offers the only incisive way of achieving such ends. Various attempts have been made to harness the aggregation technique for obtaining chimerism that is restricted to only one or two of the three cell lineages that are established by the late blastocyst stage. The simplest approach, based on the findings of Graham and Deussen *(11)* that faster developing blastomeres make a greater contribution to the inner cell mass (ICM) than do slower developing ones, is to aggregate asynchronous cleavage stages. However, aggregation of an 8-cell stage with a 4-cell stage ensures only a bias toward greater contribution of the 8-cell to ICM derivatives rather than their exclusive origin from the latter *(12)*. Perturbing the development of one of the participating conceptuses more radically by making it tetraploid through exposure to cytochalasin at the 2-cell stage and then aggregating it with a normal 8-cell has also been investigated as a way of securing chimerism that is restricted to extra-embryonic tissues. However, even though the resulting fetuses can be composed entirely of cells of the diploid component, it is still uncertain whether tetraploid cells typically fail to contribute to the epiblast from the outset or are outcompeted later *(13)*. For certain types of chimera studies, it could be important to know which is the case.

2. Media, Reagents, Materials, and Equipment

2.1. Media

The following six media are used in the various protocols to be discussed:

1. Ca^{2+}- and Mg^{2+}-free Tyrode's saline (CMFT) is used for blastocyst dissection and acidified Tyrode's saline (AT) for removing the zona pellucida (ZP) *(14)* (**Table 1**).
2. MTF *(15)* is used for culture of morulae at 37°C in a gas phase of 5% CO_2 in air (**Table 2**). Microdrops of this medium are prepared under light liquid paraffin oil in disposable tissue culture dishes. (The KSOM medium developed by Lawitts and Biggers et al. *(16)* is an alternative to MTF, which is claimed to give even better preimplantation development in vitro.)
3. MTF-HEPES (**Table 2**) is a variant of MTF that is used for the recovery, storage at room temperature, and manipulation of morulae and blastocysts, as well as isolated tissues and cells derived therefrom.

Table 1
Acidified Tyrode's Saline (AT) and Ca²⁺/Mg²⁺-Free Tyrode's (CMFT)

	Tyrode's saline[a] (g/100 mL)	Ca^{2+}/mg^{2+}-free Tyrode's[a] (g/100 mL)
NaCl	0.800	0.8000
KCl	0.020	0.0300
Glucose	0.100	0.2000
$CaCl_2 \cdot 2H_2O$	0.024	—
$MgCl_2 \cdot 6H_2O$	0.010	—
Polyvinyl-pyrrolidone (PVP-10)	0.400	—
$NaH_2PO_4 \cdot 2H_2O$	—	0.0050
KH_2PO_4	—	0.0025
$NaHCO_3$	—	0.1000

[a]Ingredients dissolved in analytical-grade water, sterile-filtered, and stored in TC-grade plastic container at 4°C.

[b]To acidify the AT, titrate 10 mL of saline against 1 N HCl solution until the pH falls below 3.0 (often in 25 μL).

Table 2
MTF and MTF-HEPES

	MTF (100 mL)	MTF-HEPES (100 mL)
H_2O	58.0 mL	58.0 mL
Salts ($10X_s$ stock)	10.0 mL	10.0 mL
Calcium chloride stock	1.0 mL	1.0 mL
Lactate stock	10.0 mL	10.0 mL
Bicarbonate stock	10.0 mL	2.0 mL
Pyruvate stock	1.0 mL	1.0 mL
HEPES stock	—	8.0 mL
Antibiotic stock	10.0 mL	10.0 mL
BSA	400.0 mg	300.0 mg

4. Ca²⁺-free OC-medium *(17)* plus ethylene glycol tetra-acetic acid (EGTA) (CF-OC) is used to aid dissociation of tissues following their exposure to pronase.
5. PB1 medium *(18)* diluted 10X with Dulbecco A phosphate-buffered saline (PBS) is used for diluting both the antisera and complement used for immunosurgery, and also for the rinses.
6. Alpha medium plus 10% fetal calf serum (α + fcs) *(19)*.

2.2. Stock Solutions

1. Salts: $10X_s$ Conc. (per 100 mL H_2O): 6.67 g $NaCl_2$, 0.356 g KCl, 0.162 g KH_2PO_4, 0.293 g $MgSO_4 \cdot 7H_2O$, 0.612 g glucose. Sterile filter and store at 4°C.
2. Calcium chloride stock: $100X_s$ Conc. (per 10 mL H_2O): 0.251 g $CaCl_2 \cdot 2H_2O$. Sterile filter and store at 4°C.
3. Lactate stock: $10X_s$ Conc.: 0.172 mL dL-lactic acid, 25 mL H_2O. Make up fresh each week, store at 4°C.

4. Bicarbonate stock: $10X_s$ Conc. (per 25 mL H_2O): 0.525 g $NaHCO_3$, 0.0025 g Phenol Red. Make up fresh each week, store at 4°C.

5. Pyruvate stock: $100X_s$ (per 10 mL H_2O): 0.0407 g pyruvate. Make up fresh each week, store at 4°C.

6. HEPES stock: $10X_s$ Conc. (per 100 mL H_2O): 6.5 g HEPES, 0.01 g Phenol Red. pH to 7.4, sterile filter, and store at 4°C.

7. Antibiotics stock: $10X_s$ Conc. (per 100 mL H_2O): 0.06 g Penicillin, 0.05 g streptomycin sulphate. Sterile filter and store as frozen aliquots.

8. Ca^{2+}-free OC medium (CF OC medium, per 10.0 mL): 9.9 mL OC-Ca^{2+} Stock 1, 0.1 mL OC-Ca^{2+} Stock 2, 0.021 g $NaHCO_3$, 0.002 g EGTA. Sterile filter. Preequilibrate for 1 h before use.

9. OC-Ca^{2+} Stock 1 (per 100 mL): 0.554 g NaCl, 0.0356 g KCl, 0.0162 g KH_2PO_4, 0.0294 g $MgSO_4 \cdot 7H_2O$, 0.1 g glucose, 1.068 mL Na-lactic acid, 0.005 g streptomycin, 0.001 g Phenol Red. 0.1 g BSA. Dissolve in 100 mL BDH Analar water and sterile filter. Can be stored for several months in a tightly stoppered bottle at 4°C.

10. OC-Ca^{2+} Stock 2 (per 10 mL): 0.028 g Na-pyruvic acid, 0.088 g NaCl, 0.060 g penicillin. Dissolve in 10 mL BDH Analar water and sterile filter. Can be stored frozen for several months and may be repeatedly thawed and frozen without detrimental affect.

11. PB1 medium: 8.0 g/L NaCl, 0.2 g/L KCl, 0.1 g/L $MgCl_2 \cdot 6H_2O$, 0.51 g/L Na_2HPO_4, 0.2 g/L KH_2PO_4, 0.1 g/L $CaCL_2 \cdot 2H_2O$, 1.0 g/L glucose, 0.043 g/L Na-pyruvate, 0.06 g/L penicillin, 0.01 g/L Phenol Red. Dissolve in Analar water and sterile filter, store in tightly stoppered TC plastic flasks. Can be stored for several months at 4°C. Before use, add 10% heat-inactivated FCS and sterile filter.

2.3. Reagents

Analar water (BDH, Poole, UK).
Bovine serum albumen (BSA) (A-3311; Sigma, St. Louis, MO).
5-bromo-4-chloro-3-indolyl-β-D-galactopyranoside (X-gal) (B-4252; Sigma).
Deoxycholic acid (D-6750; Sigma).
Dimethyldichlorosilane (cat. no. 331646A; BDH).
DL-lactic acid (L-4263; Sigma).
Dulbecco A phosphate-buffered saline (PBS) (Oxoid Ltd., Basingstoke, Hampshire, UK).
Ethylene glycol *bis*-(beta-amino ethyl ether)*N,N,N',N'*-tetra-acetic acid. EGTA (E-4378; Sigma).
Fructose-6-phosphate (F6P) (F-3627; Sigma).
Glucose-6-phosphate dehydrogenase (G6PD) (G-8878; Sigma).
Glutaraldehyde (cat. no. 36080; BDH).
Heavy liquid paraffin (HLP) (cat. no. 294375J; BDH). This needs to be tested for toxicity.
HEPES (H-0763; Sigma).
High vacuum grease (Cole Parmer, London, UK).
Light liquid paraffin oil (LLP) (cat. no. 294365H; BDH).
MEM Alpha medium (cat. no. 072-01900A; Life Technologies, Scotland, UK) plus 55 mg/L of each of penicillin and streptomycin.
Nicotinamide adenine dinucleotide phosphate (NADP) (N-0505; Sigma).
Nitro blue tetrazolium (NBT) (N-6876; Sigma).
Nonidet P-40 (N-6507; Sigma).
Normal non-heat-inactivated rat serum (NRS).
Pancreatin (cat. no. 0296-17; Difco, West Molesey, UK).
Penicillin (P-3032; Sigma).
Phenazine methosulfate (PMS) (P-9625; Sigma).
Polyvinyl-pyrrolidone, average mol wt 10,000 (PVP 10) (P-2307; Sigma).

Potassium ferricyanide hexahydrate (P-8131; Sigma).
Potassium ferrocyanide (P-9387; Sigma).
Pronase (cat. no. 537088; Calbiochem-Novabiochem, San Diego, CA).
Pyruvic acid (P-4562; Sigma).
Rabbit antimouse serum (RAM).
Spermidine trihydrochloride (S-2501; Sigma).
Streptomycin sulphate (S-9137; Sigma).
Trypsin (cat. no. 0152-15-9; Difco).

2.4. Materials and Equipment

Capillary tubing, thick-walled, with 1-mm outer diameter (od) and 0.56-mm inner diameter (id) (Leitz Microsystems, Deerfield, IL).

Capillary tubing, thin-walled, with 1-mm od and 0.85-mm id (Drummond Scientific, Broomall, PA).

Cold stage for microscope (can be made in the laboratory or associated workshop).

Cover slips, 22 × 32-mm No. 1 glass, for Puliv chambers (E. Leitz [Instruments] Co. Ltd., Luton, UK).

De Fonbrune microforge (Beaudouin, Paris, France).

De Fonbrune suction-and force pump microinjector (Beaudouin).

Dishes, disposable plastic 35-mm tissue-culture (Falcon, UK).

Dishes, disposable plastic 30-mm bacteriological (Sterilin) (Bibby Sterilin Ltd., Staffordshire, UK).

Electrode puller (e.g., Sutter Instruments, Novato, CA).

Electrophoresis apparatus and sample dispensers (Helena Laboratories, Beaumont, TX).

Extension arms for screw controls on double-instrument holders, unlockable (not available commercially).

Extension arm for attaching Leitz single-instrument holder to stereotaxic manipulator (not available commercially).

Filters, disposable syringe-fitting, 0.22 μm (Millex) (Millipore Corp., Bedford, MA).

Fine scissors plus two pairs of No. 5 watchmaker's forceps.

Gas microburner (e.g., prepared from 19-gage syringe needle) (not available commercially).

Instrument holders for micromanipulators, two double and one single (Leitz).

Instrument tubes, several, plus caps, and brass and silicone rubber washers (Leitz).

Micromanipulator baseplate (Leitz or Micro-Instruments, Oxford, UK).

Micromanipulators, left and right pair (Leitz).

Micromanipulator, stereotaxic (e.g., Prior, UK).

Microscope, stereo-binocular dissecting type, with incident and transmitted illumination (E. Leitz).

Microscope, upright compound type with fixed stage, barrel focusing, and erected optics (Leitz Laborlux—no longer made, only available second-hand or through Micro-Instruments Microtec II).

Micrometer syringe microinjector (Narishige, Tokyo, Japan).

Pasteur pipets: plugged, siliconized, and nonsiliconized.

Plastic tubing, thick-walled translucent, for connecting microinjectors to instrument tubes.

Puliv oil-filled manipulation chambers (can be made in the laboratory or associated workshop).

Thermocirculator/chiller unit (Churchill, UK).

Titan III cellulose acetate plates (Helena Laboratories).

Warm stage for dissecting microscope (can be made in the laboratory or associated workshop).

3. Methods

3.1. Making Chimeras by Morula Aggregation

The early morula is favored for making aggregation chimeras because blastomeres are both relatively adhesive and motile before the formation of junctional complexes between the outer, future trophectodermal, cells has progressed very far. The standard approach for producing such chimeras is to place developmentally synchronous, intact early morulae of different genotypes together in pairs, having first denuded them of the ZP. Successful aggregation leads to the formation of single integrated blastocysts that are composed of approx two times the normal number of cells. An ill-understood regulative process occurs following transfer of these giant blastocysts to the uterus whereby their cell number is reduced to that of standard conceptuses by the onset of gastrulation. As discussed elsewhere, this could be an important factor affecting both the balance and frequency of chimerism among fetuses and offspring *(20)*. A simple way to avoid the complications of size regulation is to aggregate half morulae in pairs, their bisection being relatively simple once they have been partially dissociated.

As previously noted, the early morula with a total cell number of between 8 and 16 is most convenient for this purpose. This stage of development is reached by the afternoon of the third day of pregnancy in mice kept under standard lighting conditions, the day of the vaginal plug being counted as the first day postcoitum (dpc). At this stage, conceptuses typically lie either in the lower isthmic segment of each oviduct or within the uterotubal junction. Both these regions of the female tract have a narrow lumen and thick muscular walls that tend to make the recovery of conceptuses without damage more difficult than earlier or later in preimplantation development.

3.1.1. Recovery, Aggregation, and Culture (see **Note 1**)

1. Isolate the oviduct together with the adjacent 3–4 mm of the uterus from both the ovary and the remainder of the tract. Using a Pasteur pipet with a fine, flame-polished tip that can be inserted readily into the fimbrial (ovarian) end of the oviduct, flush MTF-HEPES (**Table 2**) toward the uterus. If the yield is poor, the remainder of the uterus should also be flushed. If flushing fails altogether to yield any conceptuses, cut the isthmus and uterotubal junction into fragments and then tease them apart.
2. Rinse the morulae in fresh medium before transferring them to a drop of freshly acidified and filtered Tyrode's (AT) saline (**Table 1**) for the minimal time necessary to denude them of the ZP. As soon as the zona has disappeared, transfer the morulae immediately to MTF-HEPES to neutralize the acidity.
3. Place a single denuded morula in each of a series of drops of MTF medium under light paraffin oil in a tissue culture dish that has been pre-equilibrated for several hours at 37°C under a gas phase of 5% CO_2 in air.
4. Repeat steps 1 and 2 for recovering morulae of the second genotype to be used for making chimeras, adding one to each culture drop that already contains a morula of the first genotype. Following reincubation of the culture dish for up to 1 h, gently push each pair of morula together with the tip of blunt glass probe until they adhere.
5. Briefly inspect the culture after incubation for a further hour to check that aggregation is underway and, if necessary, to replace any still nonadherent pairs in contact. Once all pairs have begun to aggregate, reincubate them until the early afternoon of the following day, when preparations for transferring them to the uteri of third dpc pseudopregnant

recipient females have been completed. At this stage, discard any aggregates that have failed to form single well-integrated late morulae or blastocysts.

6. Transfer the remainder to the uteri of third dpc pseudopregnant recipient females at a number per horn that does not exceed one half the mean litter size of the strain used as recipients by more than 1 or 2.

3.1.2. Transfer to the Uteri
of Pseudopregnant Recipient Females (see **Note 2**)

Two considerations are critical for the success of this operation. The first is the stage of the recipient, which should not be earlier than the afternoon of the third dpc or later than the morning of the fourth dpc *(22)*. Third dpc recipients tend to give better results than fourth dpc recipients with conceptuses that have been cultured from the morula stage or held in vitro for some hours as blastocysts. The second consideration is limiting the volume of fluid injected into each uterine horn when transplanting the conceptuses and, particularly, taking care to exclude air bubbles. Details of the preparation of transfer pipets that can be mouth operated to give very good volume control are provided in **Fig. 1**. Apart from a sharp, fine pair of scissors and two pairs of watchmaker's forceps, the other item required is a thin, sharp-tipped, sewing needle for puncturing the uterine horn. For convenient handling, the latter should be mounted in a glass or wood holder.

There are an increasing number of agents that can be used either singly or in combination to produce short-term anesthesia of mice following intraperitoneal injection. One that is still widely used, despite ceasing to be available commercially many years ago, is Avertin *(23)*. There is, however, no general consensus regarding the optimal short-term general anesthetic for mice. If in doubt, it is wise to seek veterinary advice.

1. Weigh each pseudopregnant female to be able to administer the appropriate dose of anesthetic for short-term general anesthesia.
2. Once anesthesia has been produced, orient the mouse under a suitable dissecting microscope with its tail pointing directly away, and wet the dorsal fur with a tissue soaked in alcohol.
3. Part the coat mid-dorsally in a line extending forward by approx 1.5 cm from the sacrum. Incise the skin along this line and then pull the right cut surface toward the right to obtain a clear view of the translucent dorso-lateral body wall.
4. Once the ovary has been identified, make an incision in the body wall large enough to allow it to be exteriorized readily by gripping its fat-pad with forceps.
5. Check for the presence of multiple active (red) corpora lutea before charging the transfer pipet with conceptuses as described in **Fig. 1**.
6. Hold a pair of forceps in the left hand and both the transfer pipet plus needle in the right hand. While using the forceps to grip the oviductal end of the horn by its mesometrial region, push the needle into the horn in this region with its axis parallel to the horn until it is clearly in the lumen. Withdraw the needle without losing sight of the wound and insert the tip of the transfer pipet so that it extends well down the lumen.
7. Gently eject the blastocysts while slowly withdrawing the pipet. Stop ejecting fluid from the pipet once the marker air bubbles are clearly below the bend (*see* **Fig. 1** legend).
8. Immediately check the pipet for the presence of any retained conceptuses by ejecting further fluid into a dish.
9. Either repeat **steps 4–8** for the left tract or close the skin wound with sutures or wound clips.

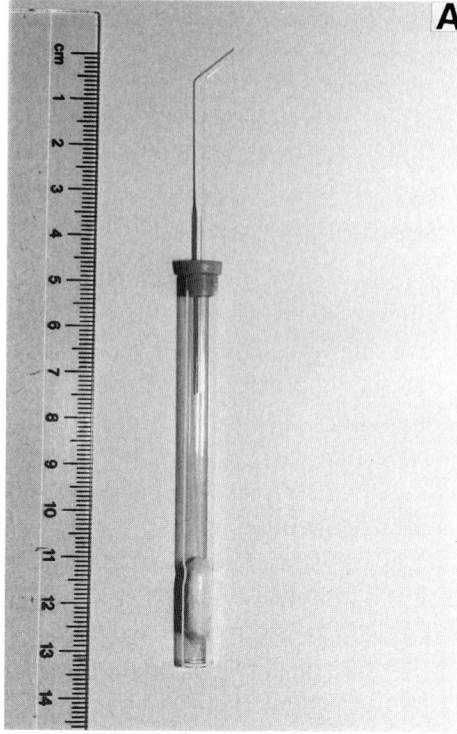

Fig. 1. **(A)** Photograph of a blastocyst-transfer pipet plus holder. The holder is the stem of a Pasteur pipet plugged distally with the silicone rubber bung from a Microcap dispenser that has a small hole through its center for tight insertion of the transfer pipet. The proximal end of the holder plugged with cotton wool is connected to flexible plastic tubing for mouth operation. **(B)** Diagram showing details of transfer pipet, which is "braked" by dipping the proximal end in heavy paraffin oil until the oil reaches the bend. A column of MTF-HEPES interrupted by two air bubbles is then sucked into the distal end until the bubbles are just proximal to the bend. Finally, the blastocysts are drawn into the tip in a minimal volume of medium. The tip of the pipet is then inserted into the uterine lumen and fluid ejected until the oil drops are just distal to the bend.

3.2. Making Chimeras by Blastocyst Injection or Reconstitution

This requires the availability of good-quality micromanipulators. Transplantation of dissociated cells can be accomplished with just two pipets, as is done almost universally with ES cells. One of the pipets is used to immobilize the blastocyst by suction, and the other is used to introduce the cells into the blastocoelic cavity. A blunt-tipped pipet can only be made to penetrate the trophectoderm if it is advanced very rapidly in a way that can damage a delicate micromanipulator *(24)*. Hence, the almost universal practice is to bevel the tip of the cell injection pipet, which can be done in a number of different ways. Since several very detailed protocols for this two-microinstrument approach for making chimeras at the blastocyst stage have been published *(1,4)*, the technique will not be discussed further here. Instead, we will confine ourselves to a more versatile approach, which, as well as enabling injection of dissociated cells into a wider range of blastocyst stages than is possible with a beveled pipet, can also be used to transplant whole tissues and even reconstitute blastocysts from their component tissues.

Transplanting intact tissue such as a whole ICM into the blastocoele depends on making a larger hole in the trophectoderm than can be achieved with a beveled pipet. The basic approach is to immobilize the blastocyst by applying gentle suction to its embryonic (ICM) pole while using the apposed tips of two sharp needles to penetrate the trophectoderm at the abembryonic pole. These needle tips are then moved apart to create a slit in the trophectoderm of up to two thirds the blastocyst's diameter in width. A third blunt-tipped needle is maneuvered into the slit, which is converted into a triangular hole by raising this needle while correspondingly lowering the pair of sharp-tipped ones. The donor tissue, which is held by suction on the tip of a second pipet, is then inserted through the hole and released in the blastocoele *(25)*. Hence, for tissue as opposed to cell transplantation (*see* **Subheading 3.2.5.1.**), a total of five microinstruments are required, two pipets and three needles, and these have to be arranged so that the immobilizing pipet approaches the blastocyst from one side and all three needles plus the tissue-holding pipet from the other. It is, furthermore, essential to be able to move the paired needles either together or independently, as also the third needle and tissue-holding pipet. These diverse requirements can be met using the setup described in the next section.

3.2.1. Micromanipulator Assembly

This consists of a pair of Leitz micromanipulators mounted on their standard baseplate with a Leitz Laborlux microscope with erected optics and fixed stage that can be cooled to approx 5°C set between them. A third simple stereotaxic-type Prior micromanipulator is attached to the baseplate in front of the left Leitz micromanipulator. This third micromanipulator has an extension arm bearing a single Leitz instrument holder that enables the blastocyst-holding pipet to be aligned between the pair of microinstruments carried on the left Leitz micromanipulator (**Fig. 2**). Each Leitz micromanipulator carries a double-instrument holder one or more of whose screw controls for the relative positioning of the two microinstruments are fitted with releasable extension arms (*see* **Fig. 3**). Such arms allow one microinstrument to be moved independently of the other with precision actually during micromanipulation rather than, as originally intended, simply during their initial alignment. In the case of the left Leitz unit, which carries the pair of sharp-tipped needles, it is the separating or scissor movement that has to be operated during manipulation to make a slit in the trophectoderm, while for the right unit the back-and-forth movement is required to move the injection pipet forward relative to the third needle. The double-instrument holder on the left

Fig. 2. Three views of the micromanipulator assembly. **(A)** General view from above. **(B)** Closer view from front right showing arrangement of microinstruments in their holders. **(C)** Closer view from the front left showing how terminal segment of the blastocyst-holding pipet is aligned between the pair of recurved needles by means of a Leitz single-instrument holder attached to an arm extending from a simple stereotaxis (Prior) micromanipulator. In this assembly, the arms for controlling the relative movement of microinstruments held in the Leitz double-instrument holders are secured by screwing them into the heads of the control screws. A more elegant way of securing them reversibly is shown in **Fig. 3**. Abbreviations: ABFC = extension arm for back-and-forth control of injection pipet; ASC = extension arm for scissor movement of the front relative to the back recurved needle; AVC = extension arm for vertical

(Figure 2 caption continued) adjustment of third needle relative to injection pipet; BRN = back recurved needle; CT = connector tubing between microstage and Churchill thermocirculator/chiller unit; DF = De Fonbrune suction-and-force pump microinjector connected to injection pipet; DIH= double-instrument holder; FRN = front recurved needle; HP = holding pipet; IP = injection pipet; LLM = left Leitz micromanipulator; PM = Prior stereotaxic micromanipulator; PME = extension arm from Prior micromanipulator; RLM = right Leitz micromanipulator; SIH = Leitz single-instrument holder; SM = syringe-type microinjector connected to holding pipet; TN = third needle.

manipulator is oriented with its scissor-action toward the front and the one on the right manipulator with the back-and-forth control toward the front where the injection pipet is inserted. A simple Agla or Narishige micrometer-syringe device is used to operate the holding pipet and a more robust De Fonbrune suction-and-force pump the injection pipet (**Fig. 2**). Both microinjectors systems are filled throughout with heavy-grade paraffin oil.

All manipulations described in this chapter are done at approx 5–10°C in small drops of medium attached to the undersurface of No. 1 cover slips that have been acid-cleaned, siliconized by dipping in dimethyldichlorosilane and then oven-sterilized. The cover slips are secured to the top of open-sided Puliv-type (Leitz) chambers with high-

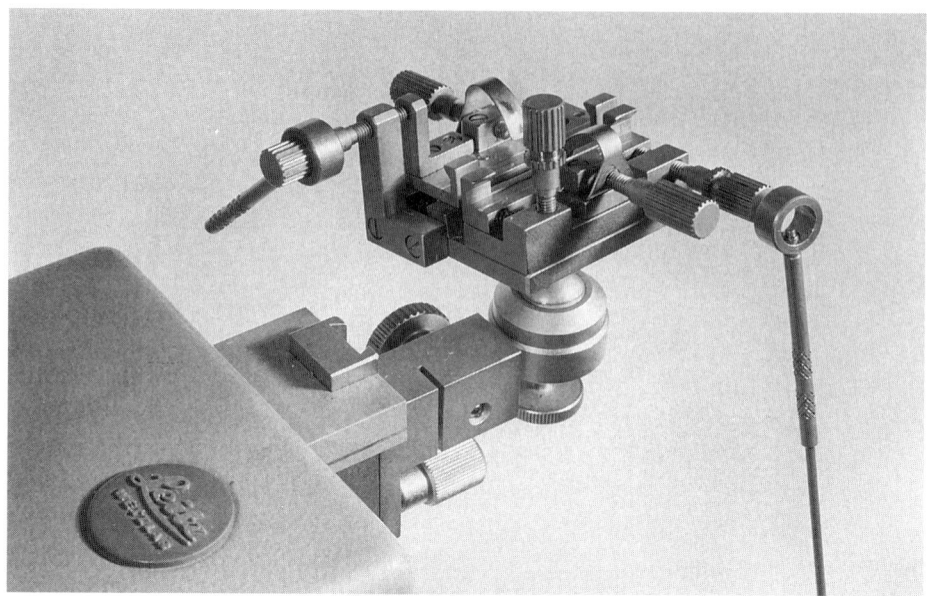

Fig. 3. Double-instrument holder on Left Leitz micromanipulator showing arm secured to the movement for back-and-forth adjustment of the back recurved needle relative to the front one. A second arm placed over the scissor movement control is shown loosened. Once the front recurved needle is positioned correctly, the arm is locked by screwing it into the ring so that it grips the control screw.

vacuum grease, drops of medium are deposited in a line on their undersurface with pulled-out Pasteur pipet, and the chambers then filled with heavy-grade paraffin oil (*see* **Fig. 4**). Not only does this arrangement ensure better optics than can be obtained with a dish placed on an inverted microscope, but it also obviates the need to introduce multiple bends into each microinstrument *(1,4)*.

3.2.2. Preparation of Microinstruments

As noted earlier, it is necessary to have all microinstruments except the holding pipet on the same side of the blastocyst (**Fig. 2**), and it is this requirement that constitutes the principal complication in preparing the four or five microinstruments needed for this approach to blastocyst injection. Both an electrode puller and a microforge are necessary for preparing the microinstruments.

The holding pipet is carried on the simple stereotaxic manipulator because it requires the least sophisticated control. It is pulled by hand over a gas microburner to obtain a long gradually tapering tip, which is then broken perpendicular to the long axis on a microforge at an od of approx 100 μm. The tip is then heat-polished to an apparent id of approx 25–30 μm. (Because of its circular profile, the id of a glass capillary viewed from the side view differs from the value obtained when viewing it end on.) Finally, the unpulled stem of the pipet is bent through an angle of approx 30° some 4 cm from the tip using the microburner, so that its proximal end can be held well clear of the left Leitz manipulator unit when its tip lies between and parallel to the latter's two instrument holders (*see* **Fig. 2**).

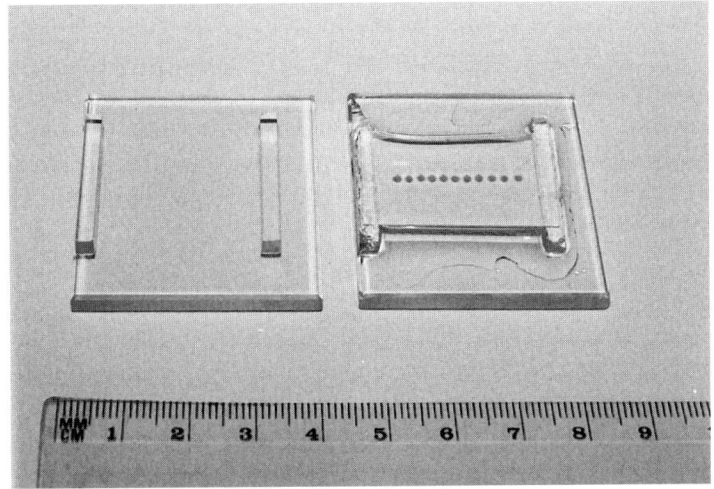

Fig. 4. An empty Puliv chamber is shown on the left and a filled one on the right. To fill a chamber, the top of each wall is smeared with Vaseline or high-vacuum grease to hold the sterile siliconized 32 × 22-mm cover slip in place. A row of drops of MTF-HEPES is then dispensed on the undersurface of the cover slip as shown, using a finely pulled Pasteur pipet. Accurate placement of drops is aided by resting the stem of the pipet on the stage of the dissecting microscope. Finally, the chamber is filled completely with heavy paraffin oil from a second Pasteur pipet, which is brought up against the undersurface of the cover slip well away from the drops. This pipet should be filled with oil in advance so that the chamber can be filled immediately after the drops have been dispensed, thereby avoiding concentration of the medium through evaporation.

The characteristics of the injection pipet depend on whether intact tissue or dissociated cells are to be transplanted into the blastocoele. Even when prepared from thin-walled capillary, pipets of a size that can readily accommodate an isolated ICM, epiblast, or primitive endoderm inside them are not easy to insert into the blastocoele, notwithstanding the use of three needles to open the trophectoderm. Hence, the strongly preferred approach is to hold the tissue by suction against the end of a pipet that is fashioned like a blastocyst-holding pipet but with a tip od of not more than 35 μm and tip id of approx 15–20 μm. Not only is this less likely to damage the donor tissue, but it is also simpler to maneuver a deformable mass of cells than an equivalent-sized pipet into the blastocoele. For transplanting dissociated cells, use of only the two sharp-tipped needles is practicable providing the tip diameter of injection pipet is kept small. This can be achieved by using capillary whose wall thickness is approx 15–20% of the od. This is pulled in an electrode puller to give a long, gradually tapering tip that is then broken and polished on a microforge to give a tip id slightly greater than that of the cells to be injected.

All three needles are made by solidifying the central region of a piece of thick-walled glass tubing (e.g., Leitz capillary) by rotating it in a microburner flame. The solidified central region is then positioned within the heating filament of an electrode puller to produce a gradual taper to a sharp tip. Making the tips of the needles solid both enhances their strength and avoids any risk of retention of fluid when they are dipped in the siliconizing agent or in alcohol for sterilization immediately before use. The tip of the third needle is heat-polished to a smooth rounded form by holding it vertically downward in a microforge and raising the tip of a fine platinum filament toward it from below. A slight bend is then made about 0.5 mm proximal to the tip so that the most distal segment of this needle can run approximately parallel to the injection pipet after insertion into the double instrument head on the right-hand Leitz micromanipulator (**Fig. 2**).

Because the paired needles carried on the left Leitz manipulator must have their sharp tips approach the blastocyst from the right (**Fig. 2**), they have to be recurved. The member of the pair that occupies the front position on the double-instrument holder will be referred to hereafter as the *front* recurved needle and the one that occupies the back position as the *back* recurved needle.

Successive steps in the preparation of a back recurved needle are illustrated in **Fig. 5** and detailed in its legend; the precise ways in which these steps differ for a front recurved needle are illustrated as well.

3.2.3. Setting Up the Microinstruments (see **Note 3**)

1. The back recurved needle is fitted into its housing in the double-instrument holder on the left micromanipulator and then rotated until its penultimate segment between the first and second right-angle bends is vertical with the tip segment uppermost. Once it has been secured tightly, the scissor movement is opened widely before the front needle is inserted, rotated until correctly oriented, and secured. All three independent movement controls are then used to bring the tips of the two needles together. The tip of the front needle should lie slightly to the left of that of the back needle when the two are together. This is because the scissor movement causes the tip of the front needle to move toward the right as it is separated from the back one and so tends to withdraw from the blastocoele. Once the needle tips are correctly positioned, the extension arm should be secured on the scissor-movement screw.
2. The blastocyst-holding pipet is then inserted into a Leitz single instrument holder attached to the third simple micromanipulator in such a way as to descend from above the left Leitz manipulator unit and is then aligned between the pair of recurved needles.
3. The injection pipet is inserted in the front housing on the right manipulator unit; if its tip is not coaxial with its base, rotate until the tip points upward before it is secured.
4. The third needle is then fitted in the rear instrument housing where a small piece of folded foil is first inserted to ensure that its distal end can lie above that of the injection pipet. It is then rotated until its distal segment is parallel with the injection pipet, secured tightly, and the independent movement controls used to position its tip immediately above and projecting slightly beyond that of the injection pipet.
5. Finally, an extension arm is secured on the back-and-forth control to enable the tip of the injection pipet to be moved forward and backward relative to the third needle. As noted earlier, the third needle can be dispensed with when dissociated cells rather than tissue fragments are to be transplanted.

3.2.4. Preparation of Cells or Tissue for Injection

Blastocysts are harder to flush from the uterus without damage once they have lost the zona. Recovery is aided by ensuring that the tip of the flushing device, whether a Pasteur pipet or a syringe needle, fits tightly into the cervical end of uterine horns. By initially holding its oviductal end closed with a pair of watchmaker's forceps, each

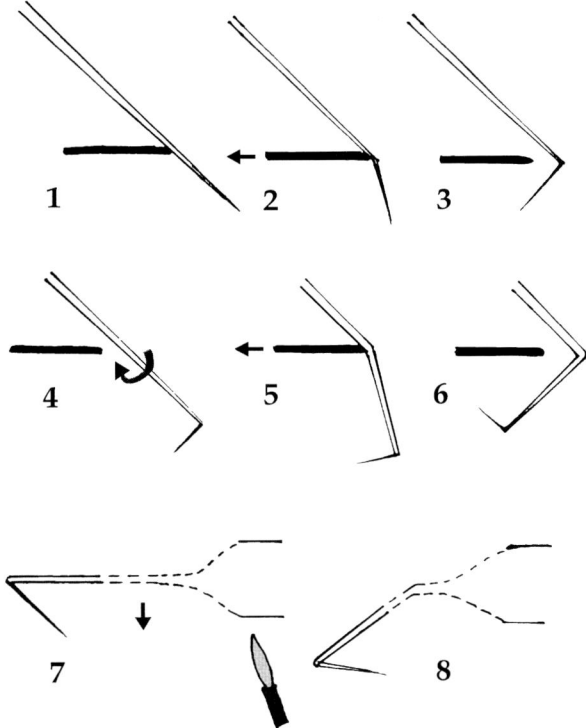

Fig. 5. Diagrams of stages in the preparation of a back recurved needle. In **Stage 1**, a pulled sharp, solid-tipped needle is arranged on the microforge as shown and the flattened apex of a fine platinum filament (0.125-mm-diameter wire beaten out to a tip thickness of approx 20 µm) is brought into contact with one side of it approx 0.3 mm from the tip. The rheostat should be adjusted very carefully so that the filament can be attached firmly enough to the glass to allow the distal segment of the needle to be pulled through a right angle (**Stage 2**) but can be disengaged on lowering the temperature thereafter (**Stage 3**). The axis of the needle is then rotated clockwise through 30–40° (**Stage 4**) before being pulled through a second right-angle bend approx 0.6 mm proximal to the first (**Stages 5** and **6**). The needle is then reoriented horizontally in the microforge with its penultimate segment also horizontal but pointing directly toward the rear of the microforge (**Stage 7**). The small flame of a gas microburner is then brought toward the underside of the needle just proximal to its pulled region so that the latter gradually bends downward. This is continued until the most distal segment of the needle is, as shown, not quite horizontal (**Stage 8**). Front recurved needles are prepared similarly, except they are rotated counterclockwise before making the second right-angle bend, and the latter is made 0.8 mm rather than 0.6 mm proximal to the first.

When making the first two bends it is better to err on the side of having the filament detach repeatedly as it is used to pull the needle through a right angle than to risk its becoming attached so firmly that the end of the needle breaks when the heating is switched off. If the third bend is excessive, it can be reduced by rotating the needle through precisely 180° and applying the flame to what was originally its upper surface until the overshoot is corrected. The distance between the first and second bend is greater for the front than the back needle because the independent vertical control can only move the distal end of the front needle *down* from the horizontal. This differential therefore ensures that the tips of the two recurved needles can be adjusted to the same vertical level even if one instrument holder is slightly bent or the distal part of a needle is not precisely coaxial with its proximal region.

horn can thus be inflated gradually with flushing medium, thereby transiently opening up the nascent crypts investing the blastocysts before allowing the medium to exit by releasing the forceps. Often, additional blastocysts can be recovered by reflushing the horn one or more times from opposite ends, though the yield is seldom as good as for earlier blastocysts. Furthermore, blastocysts recovered in subsequent flushes are more likely to be collapsed or have obviously torn trophectoderm than those recovered in the first, and are therefore generally harder to dissect.

The following details relate to the use of preimplantation and implanting blastocysts and early postimplantation conceptuses as donors. ICM tissue can be isolated readily from preimplantation blastocysts by immunosurgery. However, this technique is less reliable for advanced blastocysts for which it may be necessary to resort to microdissection. Serial microdissection is the only viable approach if epiblast or primitive endoderm is to be used for transplantation, because the boundary between these tissues is very difficult to resolve once the ICM has been divested of trophectoderm. Furthermore, the primitive endoderm becomes multilayered early in its differentiation, so exposure of isolated advanced ICMs to immunosurgery cannot be used to obtain pure epiblast *(26)*. Once isolated, the donor tissue can either be transplanted whole using the three-needle approach or dissociated into a single-cell suspension. Thus far, colonization of host blastocysts with cells from early postimplantation conceptuses has only proved successful for primitive/visceral endoderm and for extra-embryonic ectoderm *(27,28)*. Despite repeated attempts in several different laboratories, postimplantation epiblast has consistently yielded negative results. Even when cells from epiblasts of implanting blastocysts are transplanted, the rate of chimerism is markedly lower for 4.75 dpc than for 4.5 dpc donors *(29)*.

3.2.4.1. Immunosurgical Isolation of ICMs from Preimplantation Blastocysts *(30)* (see Note 4)

Reagents required (*see* **Subheading 2.** for abbreviations): AT, RAM, PBS containing 4 mg/mL BSA, NRS, α + fcs. Both the RAM and the NRS should be diluted 10X with the PB1 + PBS before use. The NRS should be kept frozen until immediately before use.

1. Pre-equilibrate drops of α + fcs under light paraffin oil in a tissue culture dish for 1–2 h in an incubator with a gas phase of 5% CO_2 in air.
2. Expose blastocysts to AT, as detailed earlier for morulae (*see* **Subheading 3.1.1.**), and incubate them in preequilibrated α + fcs for at least 1 h thereafter.
3. Transfer blastocysts to the freshly diluted RAM at 4°C for 45 min.
4. Transfer blastocysts through three large-volume rinses of PB1 + PBS at approx 1 min intervals to wash away all unbound antibody. Leave them in the third rinse while thawing and diluting the NRS.
5. Transfer the blastocyst to the diluted NRS and incubate at 37°C for 6–8 min. Thereafter, check that each of the blastocysts shows a halo of lysing trophectoderm cells. (If not, repeat this step using a fresh aliquot of NRS.)
6. Transfer blastocysts from the diluted NRS to α + fcs via a rinse in MTF-HEPES and incubate for at least 45 min.
7. Pull a siliconized Pasteur pipet to a tip whose diameter is one third to one half that of an intact blastocyst and is carefully heat-polished to eliminate any sharp projections.
8. Transfer blastocysts to MTF-HEPES and then pipet each repeatedly in and out of the Pasteur pipet until the ICM is released from its investing trophectodermal debris. There-

after, the ICMs can either be stored intact in MTF-HEPES at room temperature until preparations for their transplantation are completed or dissociated as described next.

3.2.4.2. MICROSURGICAL ISOLATION OF ENTIRE ICMs OR EPIBLAST VERSUS PRIMITIVE ENDODERM FROM IMPLANTING OR DELAYED-IMPLANTING BLASTOCYSTS *(26)*

Before recovering the blastocysts, add 25 mg trypsin and 125 mg pancreatin to 5 mL of CMFT (*see* **Table 1**), mix thoroughly, and then spin at 2500 rpm (1000*g*) for 25 min in a standard bench centrifuge. Filter the supernatant through a 0.22-μm filter and deposit one large drop per 30-mm bacteriological dish, cover with light paraffin oil (LLP), and refrigerate at approx 4°C.

Microdissection is done in a Puliv chamber with one blastocyst per hanging drop of MTF-HEPES. A solid, sharp-tipped siliconized glass needle is mounted in each Leitz micromanipulator with the shank tilted slightly upward so that its tip can make firm contact with the underside of the cover slip.

The procedure is as follows:

1. Stab each blastocyst with the needles toward its abembryonic pole, so that it can be raised on the tip of one of them until it is in contact with the undersurface of the cover slip.
2. Use the needles working against the cover slip to open the mural trophectoderm widely, well away from the ICM.
3. When a series of blastocysts have been opened thus, transfer them to the cooled trypsin/pancreatin in CMFT and refrigerate for 22–24 min.
4. Return blastocysts individually to chamber drops via a brief rinse in MTF-HEPES.
5. With one needle pushed into the trophectoderm well clear of the ICM, raise the blastocyst from the floor of the drop. Then, using both needles, spread the trophectoderm out against the cover slip with its outer surface uppermost. The aim is to flatten out the trophectoderm with the ICM projecting downward centrally so that the two tissues resemble, respectively, the white and yolk of a fried egg.
6. Next, for isolating the entire ICM, pin the trophectoderm against the cover slip close to one edge of the ICM with one needle and use the side of the other needle, which is adjusted to a fractionally lower level than the first, to gently scrape the ICM away from the overlying polar trophectoderm (**Fig. 6**). The same lateral scraping movement is also used for dissecting the ICM into primitive endoderm versus epiblast. However, in this case, the scraping needle has to be lowered slightly during the scraping movement in order to peel the investing endoderm away from the underlying epiblast. Once the endoderm begins to peel away, the bipartite structure of the ICM becomes obvious, and the boundary between the two tissue can usually be readily discerned. After the endoderm has been removed, the epiblast is scraped from the trophectoderm like an intact ICM. It is important to do the scraping stage of the operation at a high-enough magnification (approx ×500) to ensure a limited depth of focus and to monitor its progress very carefully to ensure that as the ICM or epiblast is removed, the polar trophectoderm monolayer remains intact. Development of facility in switching back and forth from the manipulator's joystick to its fine vertical control is vital for the success of these manipulations.

Once isolated, ICMs, epiblasts, or primitive endoderms are stored at room temperature in MTF-HEPES until transplantation. For triple tissue reconstitution of blastocysts, epiblasts isolated from blastocysts of one genotype are paired with primitive endoderms from blastocysts of a different genotype in pre-equilibrated drops of α + fcs

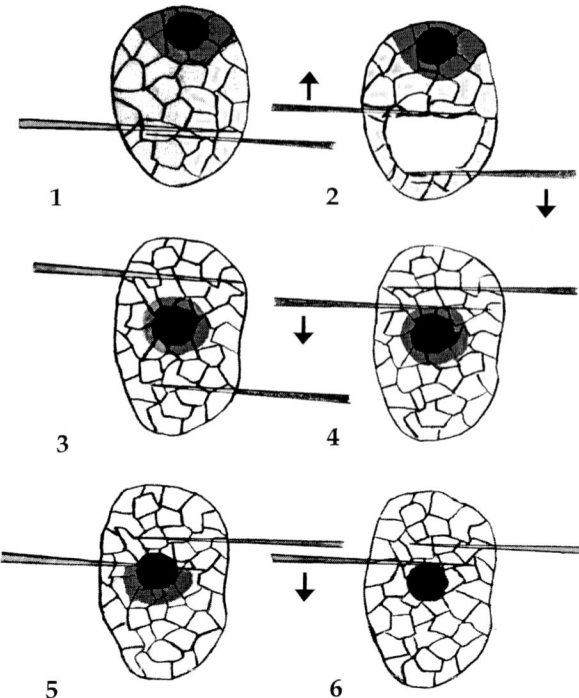

Fig. 6. Diagrams of stages in blastocyst dissection. 1. The two needles are inserted through the distal mural trophectoderm and the blastocyst then raised against the undersurface of the cover slip. 2. Working against the cover slip, the needles are parted to tear the mural trophecto-derm open wide, well away from the ICM. Once opened thus, blastocysts are incubated in trypsin-pancreatin for 22–24 min at 4°C, rinsed in MTF-HEPES, and then returned individu-ally to drops in the Puliv chamber. 3. Each blastocyst is again raised against the undersurface of the cover slip with the outer surface of its trophectoderm uppermost and stretched out. 4. Keep-ing one needle in place securely pinning the trophectoderm against the cover slip, the other is moved to the same side, lowered slightly, and then used to scrape the endoderm from the ICM. The level of this needle should be adjusted very carefully to ensure that it does not detach the entire ICM. 5. Once it is evident that only the endoderm is peeling away, this process should be completed. 6. Finally, the needle should be returned to the position shown in **Stage 4**, adjusted upward slightly, and then used to scrape away the epiblast while making sure that the overlying polar trophectoderm remains intact.

and incubated with inspection at intervals, re-pairing if necessary until they are firmly attached. Such recombinant ICMs are finally transplanted into host blastocysts whose own ICM has been exteriorized for excision by herniation through a slit in the zona (*see* **Subheading 3.2.6.**).

3.2.4.3. DISSOCIATION OF ICMS, EPIBLAST, AND PRIMITIVE ENDODERMS

Small bacteriological dishes containing drops of CF-OC medium supplemented with 0.02% (w/v) EGTA are pre-equilibrated in culture, while a 0.25% (w/v) solution of pronase is made up in PBS, passed through a 0.22-μm filter, and dispensed as drops under oil in a small bacteriological dish, which is then refrigerated.

1. Transfer tissue fragments into cooled pronase, which is then refrigerated for 8–15 min according to type of tissue; primitive endoderm requires a longer time than epiblast or early ICM.
2. Rinse tissue fragments in a drop of the pre-equilibrated CF-OC medium plus EGTA, transfer individually to fresh drops, and return dish to the incubator.
3. Inspect at intervals of 15 min. Tissues are ready for dissociation when they have adopted a looser mulberry-like appearance through the partial separation and rounding up of their constituent cells.
4. Transfer loosened tissue to a manipulation chamber drop of MTF-HEPES and dissociate by repeated aspiration into a heat-sterilized, siliconized pipet whose aperture has been heat-polished to approximately one half the diameter of the tissue fragments on a microforge. To obtain a satisfactory proportion of single cells, it may be necessary to resort to an even finer tipped pipet. This is particularly the case with the primitive endoderm, which tends to be refractory to complete dissociation. It is therefore worthwhile preparing a good stock of pipets with a range of aperture sizes. Such pipets can be controlled by mouth by inserting them into the same type of holder described for blastocyst transfer pipets (*see* **Fig. 1**). Resistant pairs of cells are often likely to be recently formed sister cells whose separation is often difficult to achieve without lysing one or both. Residual fragments, debris, and, if necessary, resistant pairs should be removed from the hanging drops before the microinstruments are introduced into them.

3.2.5. Blastocyst Injection

A manipulation chamber with hanging drops is prepared and charged with both host blastocysts and donor cells or tissue fragments. The dissociated cells are normally placed in separate drops from the blastocysts, and tissue fragments in the same drops.

All the microinstruments are withdrawn well clear of the center of the microscope stage so that the micromanipulation chamber can be introduced and the microscope focused on a blastocyst. The chamber is then removed, and all the microinstruments are returned to the microscope, so that the height of their tips can be adjusted to the focal plane of the blastocyst. They are withdrawn once again, the chamber is reintroduced and the microinstruments then brought into the hanging drop. At this stage, the mutual alignment of the tips of the two recurved needles should be checked at a magnification of at least ×300. In addition, where a third needle is in use, it is important to ensure that the height of its tip above that of the injection pipet is just sufficient to clear a tissue fragment immobilized on the latter.

3.2.5.1. Tissue Transplantation and Blastocyst Reconstitution (*see* **Note** 5)

For transplanting tissue into the blastocoele of standard blastocysts or, in the case of reconstitution, a blastocyst whose ICM has been induced to herniate through a slit in the zona pellucida, the operation is as follows:

1. Use the third needle to orient the host blastocyst and push it against the tip of the holding pipet so that it can be immobilized over the ICM region, viewing at a magnification of approx ×100.
2. Lift the holding pipet to clear the blastocyst from floor of the drop, raise the recurved needles so that their apposed tips are level with the blastocyst and close to its right side. After gently immobilizing a tissue fragment by suction on the tip of the injection pipet

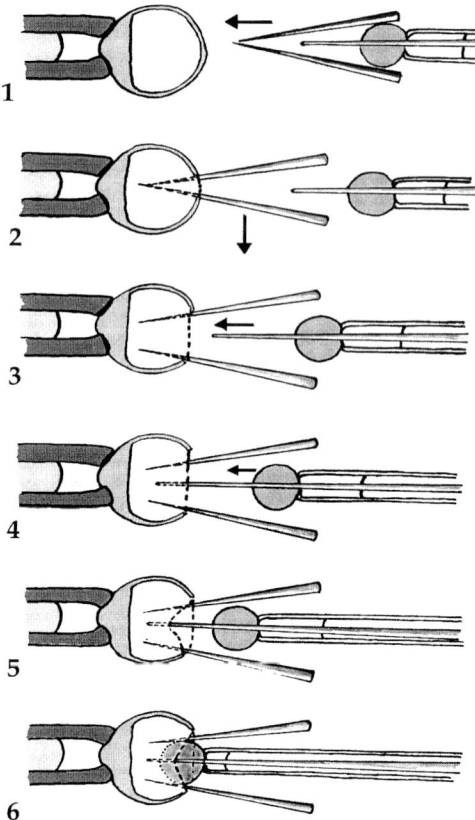

Fig. 7. Stages in the injection of tissue into the blastocyst. 1. The blastocyst is held firmly on the holding pipet via its embryonic pole and then lifted clear of the floor of the drop. All the microinstruments are then raised to the midplane of the blastocyst, as shown, before the apposed tips of the pair of recurved needles are pushed well into the blastocoele to the rear of the abembryonic pole, using the joystick of the left Leitz micromanipulator. 2. The extension arm on the scissor control is then moved to the right to part the needles and thus create a horizontal tear in the trophectoderm that is obviously wider than the tissue to be injected. 3. The third needle plus injection pipet are moved to the left with the joystick on the right Leitz micromanipulator. 4. The tip of the third needle is then maneuvered into the slit in the abembryonic polar trophectoderm. 5. Using the fine vertical controls, the third needle plus injection pipet are raised and the paired needles correspondingly lowered to convert the slit in the trophectoderm into a triangular hole, which should be centered on the tissue held on the tip of the injection pipet. 6. Finally, the back-and-forth movement is used to advance the injection pipet relative to the third needle and thus to push the tissue into the blastocoele. The tissue is then released from the injection pipet and the needles withdrawn before the blastocyst is freed from the holding pipet.

similarly raise this pipet together with the third needle and position them in the angle between the recurved needle and as close as possible to their tips.

3. Switch to a magnification of at least ×300, then adjust the position of the pair of recurved needles so that their apposed tips are at the midplane of the blastocyst and somewhat behind its center (**Fig. 7**). Bring the third needle to the same height as the recurved needle tips with its tip close to theirs.

4. Use the left manipulator joystick to push the apposed tips of the recurved needles through the abembryonic polar trophectoderm until they are close to the blastocoelic surface of the ICM.

5. Move the extension arm on the scissor action to the right so that the tip of the front recurved needle moves forward away from that of the back one to make a slit in the trophectoderm that approaches the diameter of the blastocyst in width.

6. While slightly reversing the scissor movement of the front recurved needle, tentatively move the third needle forward altering its height slightly if necessary. When its smooth tip is clearly located in the slit in the trophectoderm, move it forward until it is also well inside the blastocoele.

7. Use the fine rather than the coarse vertical controls on the two micromanipulators simultaneously to raise the third needle and to lower the recurved pair, thus converting the slit in the abembryonic mural trophectoderm into a triangular hole.

8. Carefully move the tip of the injection pipet forward relative to the third needle so as to push the attached tissue fragment into the blastocoele. Release the tissue once it is securely in the blastocoele, and withdraw the injection pipet and third needle together using the right manipulator joystick. Bring the tips of the recurved needles together, withdraw them from the blastocoele and, finally, release the blastocyst from the holding pipet.

Following injection the blastocysts are normally cultured for 1–2 h before transfer to facilitate integration of the donor tissue and allow healing of the wound in the trophectoderm. In the case of blastocyst reconstitution, where an ICM has been transplanted into a blastocyst whose own ICM has previously been induced to herniate through a slit in the ZP (*see* **Subheading 3.2.6.**), the host ICM must then be removed. This is done using a pair of solid, sharp-tipped needles, one of which is held in each manipulator. The needles are secured with their tips pointing slightly upward so that they can operate against the cover slip of the manipulation chamber. The needles are pushed into the exteriorized ICM of the blastocyst, so that it can be raised against the undersurface of the chamber cover slip. There they are used crosswise to squeeze, saw, or tear the herniated tissue from the remainder of the blastocyst, at the same time carefully avoiding exteriorization or damage to the mural trophectoderm or donor ICM tissue contained within the zona. Thereafter, such reconstituted blastocysts are also cultured for a limited period prior to transfer to the uterus.

3.2.5.2. SINGLE OR MULTIPLE DISSOCIATED CELLS (*SEE* NOTES 6–8)

The technique is essentially the same as previously described in detail for the transplantation of tissues. The only differences are that the third needle is omitted and the pipet is inserted into the blastocoele, as the donor cells are drawn into this injection pipet rather than held on its tip. Once a wide slit has been made in the trophectoderm with the recurved needles, the separation of their tips is reduced somewhat more than for tissue transplantation so that the slit can relax to a wider aperture. This facilitates insertion of the injection pipet, which inevitably has a greater tip diameter than the third needle employed in tissue transplantation. Again, the slit should be carefully probed with the tip of the pipet, using slight vertical adjustment if necessary, until one is confident of achieving unimpeded entry. If this precaution is neglected, strands of tissue debris may block the aperture of the pipet once it has been pushed into the blastocoele and thus prevent ready ejection of the donor cell or cells. Providing the recurved needle tips are kept well separated, the pipet tip can be moved quite freely within the blastocoele, thus allowing placement of individual cells at chosen sites on the surface of the ICM. Alternatively, the pipet can be pushed well into the ICM before the cells are ejected. Particularly in the MF1 strain, even multiple donor cells tend to remain in the future epiblast region of the host ICM following deep injection.

3.2.6. Preparation of Trophectoderm Donor Blastocysts for Blastocyst Reconstitution (31)

The blastocysts need to be in the process of expanding but prior to the stage when thinning of the ZP becomes obvious. Earlier blastocysts in which the volume of the blastocoele is no bigger than the ICM should be cultured in α + fcs for 1–3 h before operation.

1. Mount a blastocyst-holding pipet on the left Leitz manipulator and a fine but smooth-tipped solid needle on the right manipulator. Then place one blastocyst in each drop of a Puliv chamber (**Fig. 4**).
2. At a magnification of ×100–200, orient the blastocyst with the needle so that it can be immobilized on its side on the holding pipet with the ICM toward the rear.
3. Lift the blastocyst from the floor of the drop until it just makes contact with the cover slip and, with the magnification at approx ×500, focus on its midplane. If the blastocoelic surface of the ICM is obviously not vertical, the blastocyst can be rotated appropriately with the needle while slightly reducing the suction applied to the holding pipet.
4. Push the tip of the needle into the zona tangentially well to the side of the polar region away from the holding pipet, and then move it gently forward round the center of the region so that its tip emerges well to the other side. Providing the needle is smooth-tipped and not too fine, it should be possible to accomplish this maneuver without damaging the trophectoderm.
5. Release the blastocyst from the holding pipet so that the needle can be pushed against the side of the pipet until it has cut through the zona, and the blastocyst falls to the floor of the drop.
6. Incubate the blastocysts in α + fcs until the entire ICM region has herniated through the slit in the zona, which should occur within 1–3 h.

A hairline slit in the zona that extends across at least two thirds of the polar region not only allows fairly rapid herniation of the ICM region but also ensures that the latter is connected to the remainder of the blastocyst by a narrow waist. The donor ICM can then be transplanted into the mural trophectoderm vesicle remaining inside the zona with little risk of the host ICM sinking back inside the zona, and the host ICM excised with needles, as described earlier.

3.3. Analysis of Chimeras

Obviously, the type of analysis undertaken will depend on the purpose for which the chimeras are made. Investigation of the effects on chimeric combination with wild-type conceptuses of those carrying mutations that affect adult physiology, behavior, or susceptibility to neoplasia or other diseases will require extended observation before examination of the distribution of chimerism is undertaken. The use of chimerism to obtain homozygous or hemizygous gametes will necessitate breeding alone. In other cases where the aim is to ascertain the normal lineage of cells or to characterize a developmental mutation, prenatal analysis is likely to be indicated, the overall distribution of chimerism being of primary interest.

A growing list of genes, both variants of native ones and artificial transgenic constructs, are now available as cell markers for examining the distribution of cells derived from the component conceptuses in chimeras *(32)*. Most of these suffer from various shortcomings. Some can only provide information on the relative proportions of cells of different genotype in a sample rather than their distribution. This is the case for the electrophoretically resolvable allozymes of the ubiquitously expressed enzyme glucose phosphate isomerase (GPI) *(33)*. Many others that can be used to visualize individual cells *in situ* are either expressed in only some tissues *(34)*, or their method of detection is such that they cannot be identified reliably in all cells that carry them *(35)*. The most favored *in situ* genetic cell marker available at present is the ROSA-26 transgene *(36)*, which seems to give expression of *E. coli* β-galactosidase in all cells of heterozygous and homozygous conceptuses from shortly after implantation. Staining for this enzyme requires prior fixation and permeabilization of cells, thereby placing limits on the size of specimens in which the distribution of chimerism can be scored with this marker. In fact, because it does not suffer from this limitation, GPI is still the marker of choice for assessing the gross level and distribution of chimerism both pre- and postnatally. Prenatally, many membranes and organs can be dissected into their constituent tissues for allozymal analysis, either directly (as with separation of parietal endoderm from mural trophoblast) or (in the case of visceral endoderm vs adjacent epiblast or later, visceral mesoderm) following incubation in the same mixture of trypsin/pancreatin in cold CMFT used to loosen ICM tissues in later blastocysts. Postnatally, the scope for resolving the distribution of chimerism with respect to constituent tissues within organs by GPI is much more limited. Here there is a strong case for simultaneous use of both different allozymes of GPI and presence versus absence of the ROSA-26 transgene as markers. This enables samples of organs to be examined for the tissue distribution as well as overall levels of chimerism. It is obviously vital to undertake trial staining of different-sized samples of organs of interest from ROSA-26 + controls in order to define how small they must be in order to stain satisfactorily throughout.

The following protocols relate to resolving allozymes of GPI, and staining for *LacZ* activity.

3.3.1. Resolution of Allozymes of GPI

The allozymal combination that is most commonly used is between conceptuses that are *Gpi-1^a/Gpi-1^a* and *Gpi-1^b/Gpi-1^b* in genotype, because other electrophoretically resolvable alleles are relatively unstable. With small tissue samples from embryos, the enzyme can be fully solubilized simply by repeatedly freezing and thawing them in several times their volume of deionized water. Larger samples, particularly from adult mice, may require prior homogenization. Unless the samples are well diluted, clearly separated clean bands may not be found on staining the cellulose acetate plates following electrophoresis. Standard reference samples of blood from animals of the two genotypes should be collected and run on each gel. An advantage of using blood is that the hemoglobin front can be used to monitor the progress of electrophoresis.

Table 3
Stock Solutions for GPI Stain
and Volumes Required for Staining Two Titan Plates

Reagent	Stock	Volume in stain
Tris-citric acid buffer	20.1 g Tris + 8.0 g citric acid in 500 mL H_2O (pH 8.0)	1.0 mL
$MgCl_2$	200 mg/10 mL H_2O	9.0 mL
Fructose-6-phosphate[a]	200 mg/10 mL H_2O	1.0 mL
Nicotinamide[a] adenine dinucleotide phosphate (β-NADP)	27 mg/10 mL H_2O	1.0 mL
Nitroblue tetrazolium[a]	27 mg/10 mL H_2O	1.0 mL
Glucose-6-phosphate dehydrogenase[b]	As supplied	10.0 μL
Phenazine methosulfate[b]	10 mg/1 mL H_2O	30.0 μL

[a]These stocks are conveniently stored at –20°C as 1.0 mL, and the remainder are kept refrigerated at 4°C.
[b]These two components are added immediately before the stain is used.

Details of the necessary buffer stock and other reagents of GPI staining are provided in **Table 3**.

1. Prepare the electrophoresis buffer by dissolving 3.0 g Tris and 14.4 g glycine in 1000 mL of de-ionized water.
2. Presoak cellulose acetate plates in the electrophoresis buffer for 20 min, having introduced them into it carefully to avoid trapping any air bubbles on their surface.
3. Load the samples onto the plates without allowing them to dry out. Loading can be done in two parallel rows using the dispensers provided, which can apply eight samples of 0.5 μL as 5-mm-wide stripes. Alternatively, if the samples are very small, they can be dispensed as dots from pulled Pasteur pipets.
4. Fill the electrophoresis tank with the Tris-glycine buffer and insert the plates using once-folded pieces of Whatman No. 1 filter paper of appropriate size as wicks.
5. Ensure that the polarity is set so that the direction of migration of the allozymes, which is toward the cathode (–), is appropriate for the orientation of the plates. Set the voltage to 210 V and run for 45–60 min. The current should register approx 2 mA for each plate in the apparatus.
6. Remove the plates and keep moist until placed in containers small enough to ensure that approx 6.5 mL of staining solution is sufficient to cover them. Incubate for several minutes at room temperature in the dark.
7. Once the plates have stained adequately, fix by immersion for 30 min in 5% glacial acetic acid in deionized water.
8. Finally, stand plates in a rack until completely dry before handling for scoring visually or for use in a suitable scanner.

3.3.2. Staining for E. coli LacZ Activity (31)

Details of the stock solutions required are given in **Tables 4** and **5**. Fixative: 2% paraformaldehyde/0.2% glutaraldehyde in 0.1 *M* phosphate buffer (pH 7.3).

1. Rinse specimens briefly in PBS before transferring them to the fixative solution and refrigerate for up to 10 min.

Table 4
Solutions for Fixing, Washing, and Staining Tissue for *LacZ* Activity

X-gal wash	Final concentration	Per 100 mL
Phosphate buffer, pH 7.3	0.10 M	100.0 mL
$MgCl_2$	2.00 mM	0.0406 g
Na deoxycholate	0.10%	0.10 g
NP40	0.02%	20.0 mL
BSA	0.05%	0.05 g

Table 5
Solutions for Staining Tissue for *LacZ* Activity

X-gal stain	Stock concentration	Per 10 mL
H_2O		7.3 mL
Na_2HPO_4	1 M	800 μL
NaH_2PO_4	1 M	200 μL
$MgCl_2$	1 M	13 μL
X-Gal	20 mg/mL	500 μL
$K_3Fe(CN)_6$	50 mM	600 μL
$K_4Fe(CN)_6$	50 mM	600 μL

2. Transfer the specimens into wash solution at room temperature for approx 20 min.
3. Repeat **step 2** two times before transferring specimens to stain. Place the staining dishes in a moist plastic sandwich box, which is then sealed with tape and incubated at 37°C in the dark for up to 48 h.
4. Rinse and examine intact (possibly obtaining a photographic record) before refixing preparatory to embedding in wax or resin for final examination. Care should be taken to minimize solubilization of the X-gal reaction product during such further processing, particularly during dehydration.

4. Notes

1. Cutting up the oviduct and then teasing it apart is more likely than flushing to result in damaged morulae and should therefore be used only as a last resort. Some recommend that AT exposure should be at 37°C though this is not essential for rapid dissolution of the ZP *(14)*. The higher temperature increases the risk of the blastocyst sticking even to siliconized pipets or to the floor of the dish, notwithstanding the presence of polyvinylpyrrolidone in the AT. Monitoring the progress of ZP dissolution is facilitated by slightly tilting the substage mirror of the dissecting microscope from the position that is optimal for transillumination. Because mutual adhesion of morulae occurs more readily and firmly while the morulae are warm, having an appropriately heated stage on the dissecting microscope is advantageous. Alternatively or additionally, pre-exposing one member of each pair to phytohemagglutinin *(21)* and then rinsing it before incubation, will aid initial adhesion. Finally, since the viability of conceptuses is reduced to some extent under all in vitro conditions yet devised, they should not be kept in culture longer than necessary.
2. Photographs of stages of blastocyst transfer can be found in **refs. *1*** and *4*. The latter publication also illustrates vasectomy of males required to induce pseudopregnancy. A plenti-

ful supply of heat-sterilized transfer pipets should be available and should be reused only if the tip remains free of debris or can be fully cleared of it. Where different strains are housed together, there may be some risk of pheromonal blockage of pregnancy by foreign males *(37)*. This can be countered by keeping the females with the vasectomized males with whom they mated before blastocyst transfer and thereafter returning them to the same soiled cages, having first removed the males. Where postnatal yield of chimeras is critical, there may be a case for using parous females that have successfully raised a litter as recipients.

3. To avoid risk of damaging the microinstruments when they are being brought in and out of the manipulation chamber, it is important to develop the habit of always positioning them differently in the microscope field before withdrawing them. Thus the recurved needles are moved to the front, the injection pipet to the middle and the blastocyst-holding pipet to the rear of the field.

4. It appears that almost any rabbit-antimouse serum is effective, and that the source of complement is the main factor in determining success or failure to achieve complete lysis of the trophectoderm. In the authors' experience, NRS works most satisfactorily. To obtain clean, undamaged ICMs it is important to allow at least 45 min incubation in pre-equilibrated α + fcs after complement exposure. It is also important to ensure that ICMs are completely devoid of trophectodermal debris, because presence of debris can make them more difficult to transplant.

5. Blastocyst reconstitution is aided by having the needles for excising the host ICM carried on a second Leitz micromanipulator assembly. This avoids the necessity of completing the entire series of donor ICM transplantations before replacing the five instruments with a pair of needles for this final stage in the operation.

6. It appears to be the preparation of recurved needles that mainly discourages people from adopting the technique for injecting dissociated cells into the blastocyst. However, the investment of time needed to attain competence in preparing such needles is no greater than for making good beveled pipets, and the advantages they offer are considerable. First, they enable use of a smooth, straight-tipped injection pipet, which is not only much easier to load with a tight column of cells than a beveled one, but rarely needs to be replaced. In addition, they permit ready injection of very early or poorly cavitated blastocysts as well as those that have hatched from the zona. This is an important consideration in making best use of inbred strains like C57BL/6 in which blastocyst stage and quality are very variable as hosts.

7. The necessity for positioning the tip of the front recurved needle to the left of the back one when they are apposed and for pushing them into the blastocyst well to the rear of its center are both because of the way in which the scissor movement operates. Being pivotal, this movement causes the tip of the front recurved needle to describe an arc and thus move toward the right and hence gradually out of the blastocoele as it is parted from the back one. Because, furthermore, the scissor action only moves the front needle, it tends to swing the blastocyst to one side of the holding pipet, a problem that is largely avoided if the blastocyst is penetrated initially toward its opposite side.

8. Difficulty in pushing tissue into the blastocoele once a triangular hole has been created with the three needles is usually attributable to one of two causes. One is that the vertical distance of the third needle above the injection pipet is either too great or too small. The second is that the aperture of the injection pipet is eccentric so that the tissue fragment is poorly aligned. This can easily occur if the wall thickness of the injection pipet is uneven radially, resulting from insufficient rotation during pulling, or it is not broken perpendicularly to its long axis before heat-polishing.

Acknowledgments

We wish to thank Ann Yates for invaluable help in preparing the manuscript and both the Royal Society and the Wellcome Trust for support.

References

1. Bradley, A. (1987) Production and analysis of chimaeric mice, in *Teratocarcinomas and Embryonic Stem Cells, a Practical Approach* (Robertson, E. J., ed.), IRL, Oxford, UK, pp. 113–151.
2. Wang, Z.-Q., Kiefer, F., Urbanek, P., and Wagner, E. F. (1997) Generation of completely embryonic stem cell-derived mutant mice using tetraploid blastocysts. *Mech. Dev.* **62,** 137–145.
3. Tokunaga, T. and Tsunoda, Y. (1992) Efficacious production of viable germ-line chimeras between embryonic stem (ES) cells and 8-cell stage embryos. *Dev. Growth Differ.* **34,** 561–566.
4. Stewart, C. L. (1993) Production of chimeras between embryonic stem cells and embryos. *Methods Enzymol.* **225,** 823–855.
5. Wood, S. A., Pascoe, W. S., Schmidt, C., Kemler, R., Evans, M. J., and Allen, N. D. (1993) Simple and efficient production of embryonic stem cell-embryo chimeras by co-culture. *Proc. Natl. Acad. Sci. USA* **90,** 4582–4585.
6. Varlet, I., Collignon, J., and Robertson, E. J. (1997) Nodal expression in the primitive endoderm is required for specification of the anterior axis during gastrulation. *Development* **124,** 1033–1044.
7. Wilson, V., Manson, L., Skarnes, W. C., and Beddington, R. S. P. (1995) The T gene is required for normal mesodermal morphogenetic cell movements during gastrulation. *Development* **121,** 877–886.
8. McLaren, A. (1976) *Mammalian Chimaeras.* Cambridge University Press, Cambridge, UK.
9. Eicher, E. M. and Hoppe, P. C. (1973) Use of chimeras to transmit lethal genes in the mouse and to demonstrate allelism of the two X-linked lethal genes *jp* and *msd. J. Exp. Zool.* **183,** 181–184.
10. Mullen, R. J. and Whitten, W. K. (1971) Relationship of genotype and degree of chimerism in coat color to sex ratios and gametogenesis in chimeric mice. *J. Exp. Zool.* **178,** 165–176.
11. Graham, C. F. and Deussen, Z. A. (1978) Features of cell lineage in preimplantation mouse development. *J. Embryol. Exp. Morphol.* **48,** 53–72.
12. Surani, M. A. H. and Barton, S. C. (1984) Spatial distribution of blastomeres is dependent on cell division order and interactions in mouse morula. *Dev. Biol.* **102,** 335–343.
13. James, R. M., Klerkx, A. H. E. M., Keighren, M., Flockhart, J. H., and West, J. D. (1995) Restricted distribution of tetraploid cells in mouse tetraploid-diploid chimeras. *Dev. Biol.* **167,** 213–226.
14. Nicholson, G. L., Yanagimachi, R., and Yanagimachi, H. (1975) Ultrastructural localization of lectin-binding sites on the zonae pellucidae and plasma membranes of mammalian eggs. *J. Cell Biol.* **66,** 263–274.
15. Gardner, D. K. and Sakkas, D. (1993) Mouse embryo cleavage, metabolism and viability: Role of medium composition. *Hum. Reprod.* **8,** 288–295.
16. Lawitts, J. A. and Biggers, J. D. (1993) Culture of preimplantation embryos. *Methods Enzymol.* **225,** 153–164.
17. Biggers, J. D., Whitten, W. K., and Whittingham, D. G. (1971) Culture of mouse embryos in vitro, in *Methods in Mammalian Embryology* (Daniel, J. C., Jr., ed.), Freeman, San Francisco, pp. 86–116.

18. Whittingham, D. G. and Wales, R. G. (1969) Storage of 2-cell mouse embryos in vitro. *Aust. J. Biol. Sci.* **22,** 1065–1068.

19. Stanners, C. P., Eliceiri, G. L., and Green, H. (1971). Two types of ribosome in mouse-hamster hybrid cells. *Nature New Biol.* **230,** 52–54.

20. Gardner, R. L. (1996) Can developmentally significant spatial patterning of the egg be discounted in mammals? *Hum. Reprod. Update* **2,** 3–27.

21. Mintz, B., Gearhart, J. D., and Guymont, A. O. (1973) Phytohemagglutinin-mediated blastomere aggregation and development of allophenic mice. *Dev. Biol.* **31,** 195–199.

22. McLaren, A. and Michie, D. (1956) Studies on the transfer of fertilized mouse eggs to uterine foster-mothers. I. Factors affecting the implantation and survival of native and transferred eggs. *J. Exp. Biol.* **33,** 394–416.

23. Papaioannou, V. E. (1990) In utero manipulations, in *Postimplantation Mammalian Embryos: A Practical Approach* (Copp, A. J. and Cockroft, D. L., eds.), IRL, Oxford, UK, pp. 61–80.

24. Babinet, C. (1980) A simplified method for mouse blastocyst injection. *Exp. Cell Res.* **130,** 15–19.

25. Gardner, R. L. (1978) Production of chimeras by injecting cells or tissues into the blastocyst, in *Methods in Mammalian Reproduction* (Daniel, J. C., Jr., ed.), Freeman, San Francisco, pp. 137–165.

26. Gardner, R. L. (1985) Regeneration of endoderm from primitive ectoderm in the mouse embryo: fact or artifact? *J. Embryol. Exp. Morphol.* **88,** 303–326.

27. Cockroft, D. L. and Gardner, R. L. (1987) Clonal analysis of the developmental potential of 6th and 7th day visceral endoderm cells in the mouse. *Development* **101,** 143–155.

28. Rossant, J., Gardner, R. L., and Alexandre, H. L. (1978) Investigation of the potency of cells from the postimplantation mouse embryo by blastocyst injection: A preliminary report. *J. Embryol. Exp. Morphol.* **48,** 239–247.

29. Gardner, R. L., Lyon, M. F., Evans, E. P., and Burtenshaw, M. D. (1985) Clonal analysis of X-chromosome inactivation and the origin of the germline in the mouse embryo. *J. Embryol. Exp. Morphol.* **88,** 349–363.

30. Solter, D. and Knowles, B. B. (1975) Immunosurgery of mouse blastocyst. *Proc. Natl. Acad. Sci. USA* **72,** 5099–5102.

31. Papaioannou, V. E. (1982) Lineage analysis of the inner cell mass and trophectoderm using microsurgically reconstituted blastocysts. *J. Embryol. Exp. Morphol.* **68,** 199–209.

32. Beddington, R. S. P. and Lawson, K. A. (1990) Clonal analysis of cell lineages, in *Postimplantation Mammalian Embryos: A Practical Approach* (Copp, A. J. and Cockroft, D. L., eds.), Oxford University Press: Oxford, pp. 267–292.

33. Chapman, V. M., Whitten, W. K., and Ruddle, F. H. (1972) Expression of paternal glucose phosphate isomerase-1 (Gpi-1) in preimplantation stages of mouse development. *Dev. Biol.* **26,** 153–158.

34. Gardner, R. L. (1984) An *in situ* marker for clonal analysis of development of the extraembryonic endoderm in the mouse. *J. Embryol. Exp. Morphol.* **80,** 251–288.

35. Thomson, J. A. and Solter, D. (1988) Transgenic markers for mammalian chimeras. *Roux's Arch. Dev. Biol.* **197,** 63–65.

36. Friedrich, G. and Soriano, P. (1991) Promoter traps in embryonic stem cells—A genetic screen to identify and mutate developmental genes in the mouse. *Genes Dev.* **5,** 1513–1523.

37. Bruce, H. M. (1960) A block to pregnancy in the mouse caused by proximity of strange males. *J. Reprod. Fertil.* **1,** 96–103.

39

Cell Grafting and Fate Mapping of the Early-Somite-Stage Mouse Embryo

Simon J. Kinder, Seong-Seng Tan, and Patrick P. L. Tam

1. Introduction

The development of techniques for culturing postimplantation mouse embryos has opened up the possibility of continuous observation of embryonic development in vitro. Current methods have supported extensive growth and morphogenesis of the mouse embryo from the stage of gastrulation (at approx 6.5 d postcoitum [dpc]) to as far as late organogenesis (10.5 dpc). Our capability to grow the postimplantation embryo outside the confine of the uterine environment has also offered an unique opportunity to study development through direct microsurgical manipulations, or, in short, micro-manipulation (1). These manipulations generally fall into two types, the removal of cells or tissues from the embryo and the introduction by transplantation (or grafting) of cells or tissues to the embryo (2). The outcome of cell or tissue ablation provides an assessment of the role played by these cells or tissues in the development of the embryo and the ability of the remaining cells to compensate for the loss of the tissues (3). The addition of cells is achieved by microinjection of single cells or groups of cells using micropipets. Large fragments can also be transplanted by grafting parts of an embryo to defined regions of another embryo. In both cases, the cells or tissues are procured from the donor embryo and transplanted to the host embryo. The transplantation can be made to sites in the host embryo that match the sites in the donor embryo from which the transplanted tissues were obtained. This is termed as orthotopic transplantation. By doing so, the cells or tissues are placed in the "original" environment. Their behavior in the host embryo may therefore reflect their normal fate and differentiation characteristics. On the contrary, the site of transplantation could be totally different from the donor site and this constitutes the heterotopic transplantation. The cells or tissues are now confronted with a new environment, and their behavior reflects the flexibility of differentiation; in other words, plasticity of developmental fate (2,3). In the situation where a large tissue fragment is transplanted, the outcome of the experiment may reveal not only the ability of the tissue to undergo autonomous differentiation as a discrete cellular community but also the ability of the graft to exert any inductive effect on the differentiation of surrounding host tissues.

From: *Methods in Molecular Biology, Vol. 135: Developmental Biology Protocols, Vol. I*
Edited by: R. S. Tuan and C. W. Lo © Humana Press Inc., Totowa, NJ

In this chapter, we describe the protocols employed for making micromanipulation instruments, preparation of culture media, isolating experimental embryos, dissection of cells for transplantation, setting up the micromanipulator, and the grafting of cells to the early-somite-stage mouse embryo. We have focused on the technique of orthotopic grafting as used in experiments for mapping the developmental fate of the cells, but the protocols can be adapted readily for heterotopic grafting.

2. Materials

2.1. Making Instruments for Micromanipulation

1. Wire polishing unit (acid bath with 85% orthophosphoric acid) and orthodontic metal alloy wire (DCA, Hagerstown, MD).
2. Diamond tip pen (ProSciTech, Thuringowa, Australia).
3. "Microcaps" micropipets (cat. no. 100 005 00; Drummond Scientific, Broomall, PA) for making solid glass needles.
4. Thin-walled glass capillaries (od: 1.0 mm, id: 0.75 mm, cat. no. 900 021 61 Drummond) for making grafting pipets.
5. Thick-walled glass capillaries (od: 1.0 mm, id: approx 0.60 mm, cat. no. 520-119; Leica Microsystems, Deerfield, IL) for making holding pipets.
6. Horizontal micropipet puller (Model P-79 Advanced Flaming/Brown Puller; Sutter Instrument Co., Novato, CA).
7. 9-in. Pasteur pipets (cat. no. 93; Chase Instruments, Rockwood, TN).
8. Spirit lamp (CC Thomas).
9. Bunsen burner attached to butane gas bottle.
10. Microforge (Narishige, Tokyo, Japan).

2.2. Preparation of Media for Embryo Handling and Culture

1. Dulbecco's Modified Eagles Medium (DMEM) (cat. no. 12100-103 with enriched glucose content at 4 g/L; Gibco-BRL, Gaithersburg, MD), supplemented with glutamine 200 μM (Trace Biosciences, Castle Hill, Australia) and penicillin/streptomycin mixture, 5000 μg/mL (Trace Biosciences).
2. Chemical ingredients of PB1 medium (*see* **Table 1** for the formulation): KCl (Fisons, Loughborough, Leicestershire, UK), $CaCl_2 \cdot 2H_2O$ (BDH Chemicals, London, UK), KH_2PO_4 (BDH), $MgCl_2 \cdot 6H_2O$ (Ajax Chemicals), Na_2HPO_4 (Ajax), NaCl (BDH), Penicillin (Sigma Chemical Co., St. Louis, MO), phenol red (Sigma), sodium pyruvate (Sigma), bovine serum albumin (BSA; Sigma) and glucose (Sigma).
3. Rat serum: prepared as immediately centrifuged serum from nonheparinized blood collected by exsanguination from dorsal aorta of anesthetized rats, stored at –20°C. Fetal calf serum (FCS; Trace Biosciences). Aliquoted from stock, stored at –20°C. Both sera are heat inactivated for 30 min at 56°C on the day of use.

2.3. Dissection of Early-Somite-Stage Embryo

1. Dissecting instruments: iridectomy scissors (cat. no. 1301011; Fine Scientific Tools, Tuttlingen, Germany), fine scissors (Aesculap, Vancouver, Canada, BC50), standard scissors (BC 374; Aesculap), fine forceps (BD 331; Aesculap), watchmaker's forceps (Fine Scientific Tools).
2. Tissue culture wares: tissue culture dishes (60 mm, cat no. 25010; Corning Glassworks, Corning, NY), 4-well chamber slides (cat. no. 177437; Nalgene Nunc International, Rochester, NY).

Table 1
The Formulation of the PB1 Medium
for Handling Embryos and Tissue Fragments

Components	Concentration (g/L)
NaCl	8.0
KCl	0.2
$Na_2HPO_4 \cdot 12H_2O$	2.88
KH_2PO_4	0.2
$CaCl_2 \cdot 2H_2O$	0.13
$MgCl_2 \cdot 6H_2O$	0.1
Sodium pyruvate	0.036
Phenol Red	0.01
Penicillin	0.06
Bovine serum albumin	4.0
Glucose	1.0
pH	7.3–7.4
Osmolarity	286–292 mOsmol/L

3. Mouth-controlled embryo transfer pipet made from Pasteur pipet connected to the mouth-piece by flexible rubber tubing.

2.4. Preparation of Donor Cells

1. Enzyme solution containing 0.5% trypsin (Trace Biosciences), 2.5% pancreatin (Roche Molecular Biochemicals, Indianapolis, IN), 0.2% glucose (Sigma), and 0.1% polyvinyl pyrollidone dissolved in calcium- and magnesium-free phosphate-buffered saline (PBS; Flow Laboratories, McLean, VA).
2. PBS complete with calcium and magnesium (Flow Labs).

2.5. Grafting Cells to the Host Embryo

1. Dissecting microscope (Wild M3Z, MDG 17) with ergo wedge attachment for eyepieces (cat. no. 10446123; Leica Microsystems, Deerfield, IL).
2. Micromanipulation apparatus (Leica): base plate with fixture points (cat. no. 335 520 139) to hold two manipulators, left and right manipulators (cat. no. 335 520 137, cat. no. 330 520 138), instrument holders (cat. no. 335 520 142, cat. no. 335 520 143), instrument sleeves (cat. no. 335 520 143).
3. Micrometer syringe (e.g., IM5A/5B, IM88, Narishige, Tokyo, Japan).
4. De Fonbrune syringe (cat. no. 4095; Alcatel, Malakoff, France).
5. Light paraffin oil used in the manipulation dish and heavy paraffin oil (BDH) used in the micromanipulation apparatus, both cleansed by mixing with PBS to remove water-soluble impurities. Decant the paraffin oil after it separates from the PBS.

2.6. Embryo Culture

1. Water-jacketed CO_2 incubator (Model 3336; Forma Scientific, Marietta, OH).
2. Embryo culture apparatus (B.T.C. Engineering, Milton, Cambridge, UK): precision incubator, rotating drum for culture bottles, glass culture bottles (15-mL capacity; B.T.C. Engineering).
3. 5% CO_2, 5% O_2, 90% N_2 gas mixture and gas regulator (Model cat. no. HPT500-125-4F-4F; BOC Gases, Sydney, Australia).

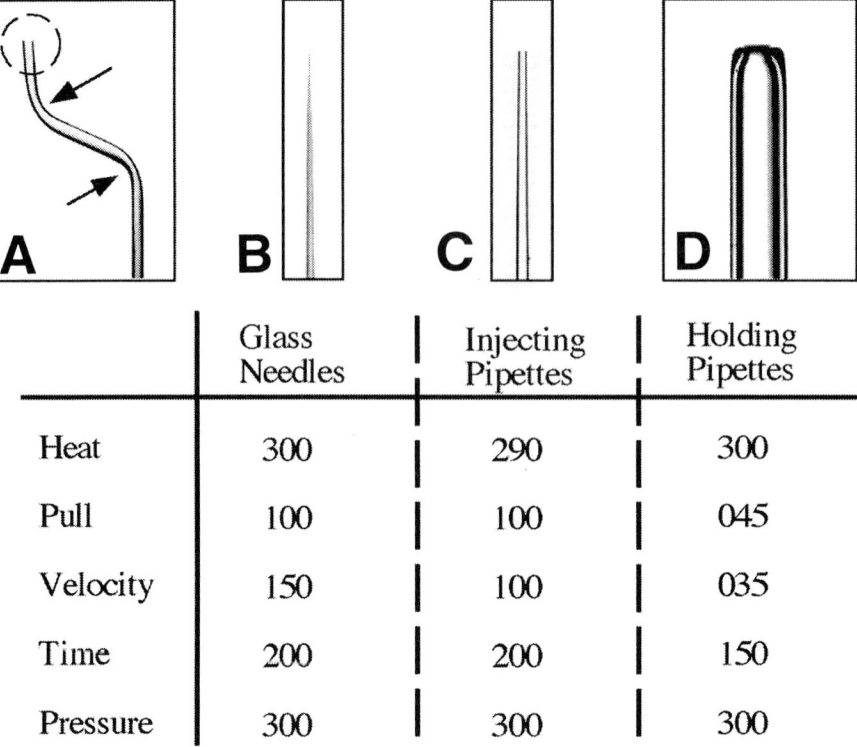

	Glass Needles	Injecting Pipettes	Holding Pipettes
Heat	300	290	300
Pull	100	100	045
Velocity	150	100	035
Time	200	200	150
Pressure	300	300	300

Fig. 1. The settings for making instruments using the Flaming/Brown micropipet puller used in our laboratory. Pressure refers to the value for air pressure. The values may vary with the quality and the configuration of the heating element and also the lapse of working time and should be taken as a general guide only. Pullers of other design will have a different set of parameters. (**A**) shows the two bends (*arrows*) made to the shaft of the micropipet so that the tip (*circled*) may reach the bottom of the manipulation dish. Examples of (**B**) the glass needle, (**C**) grafting pipet, and (**D**) holding pipet prepared using the setting listed are shown at the same magnification (o.d. of the holding pipet is approx 150 μm).

3. Methods

3.1. Making Instruments for Micromanipulation (4,5)

3.1.1. Cleaning the Glass Capillaries

1. Soak the glass capillaries in acetone overnight.
2. Rinse in 5–6 changes of water.
3. Oven dry at 37°C and then place in a clean measuring cylinder.

3.1.2. Making Glass Needles, Grafting Pipets, and Holding Pipets

1. Program the Flaming/Brown horizontal pipet puller with the parameter settings listed in **Fig. 1** (*6*; *see* **Note 1**).
2. Test pull several pipets with the settings to check the consistency of pull.
3. Choose the program settings for pulling glass needles (e.g., **Fig. 1**, col. 1) and place the Microcaps capillaries in the holders. The pull is done automatically after activating the program.

4. Examine the needles under a dissecting microscope and discard broken or blunt ones. The ideal shape of the glass needle is shown in **Fig. 1B**. Make at least 10–15 needles for each experiment.

5. Choose the program settings for pulling grafting pipets (**Fig. 1**, col. B) and place the thin-walled glass capillaries in the holders. The pull is done automatically after activating the program (*see* **Note 2**).

6. Place the pulled pipet on the microforge holder and focus the microscope on the end of the pipet.

7. Move the heating element (with a glass bead melted onto it beforehand) into the field of view. Position the element below the pipet and bring it into the same focus as the pipet.

8. Move the pipet so that the glass bead on the heating element is situated directly beneath where the shaft is to be broken, so that the final pipet will have the desired bore size (*see* **Note 3**).

9. Set the heating element to low heat at which the glass bead on the filament begins to melt.

10. Switch the element on and move the pipet down until just touching the bead. Immediately, as the wall of the pipet and the glass bead fuse, switch off the current. The pipet will break at the point of fusion to give a flushed end (**Fig. 1C**; *see* **Note 4**).

11. The pipet needs to be shaped so that the shaft is clear from the edge of the manipulation chamber as it dips to reach the bottom of the chamber. Set up a small burner by attaching a Pasteur pipet (or a syringe with G20 needle) to the Bunsen burner attached to a butane gas cylinder. Adjust the flame to approx 1 cm in height.

12. Hold the pipet above the flame and heat a point about 1.5–2.0 cm from the pipet tip. The pipet will bend under gravity as the glass melts. Allow the pipet to bend approx 75–80°.

13. Make a second bend in the opposite direction at a position approx 2 cm from the first bend. It is important to make sure both bends are aligned on the same plane (**Fig. 1A**). Make approx 10–15 grafting pipets for each experiment.

14. Program the settings for pulling holding pipets (e.g., Fig. 1, col. D) and place the thick-walled glass capillaries in the holders. The pull is done automatically after activating the program.

15. Break the shaft of the pipet to produce a tip of the desired diameter using the microforge as described above.

16. Position the pipet in the holder and focus on its tip. Polish and reduce the internal diameter of the holding pipet using the microforge (*see* **Note 5**).

17. Bring the heating element into the field of view and align it end-on with the tip of the pipet.

18. Switch on the heating element and increase the current until the filament glows hot. Bring the filament close to the tip of the pipet until the glass begins to melt.

19. Monitor the progression of the melting of tip under the microscope and switch off the filament when a round lip is formed at the end of the pipet and the internal diameter is approx one third of the external diameter (**Fig. 1D**).

20. Bend the holding pipets as described above for grafting pipets. Make approx 4–5 holding pipets for each experiment.

21. Store the needles and pipets in a large (125-mm) culture dish by pressing the instruments into a strip of plasticine on the dish.

3.1.3. Making Metal Needles for Dissecting Embryonic Fragments

1. Cut two lengths of orthodontic wire to approx 4–6 cm long (*see* **Note 6**).

2. Electrolytically sharpen one end of the wire to a fine point by dipping the needles repeatedly in the acid bath of the wire polishing unit.

3. Check the progress of the sharpening regularly under the dissecting microscope (*see* **Note 7**).
4. After sharpening, wash the needles in distilled water to remove any acid, and dip in 70% ethanol to clean and sterilize. Store the needles in a dry container padded with Kimwipe paper (Kimberly-Clark, Neenah, WI).

3.1.4. Making Transfer Pipets for Embryos and Embryonic Fragments

1. To make the embryo transfer pipet, score the shaft of the Pasteur pipet with a diamond pen and break the pipet to give a bore size of approx 3–5 mm.
2. Polish the broken end by heating in the flame of the spirit lamp and check the polished tip under the microscope. Make approx 5 pipets for each experiment.
3. To make the transfer pipet for tissue fragments, heat the shaft of the Pasteur pipet in a flame until the glass softens.
4. Take the pipet off the flame and pull the molten segments into a thin capillary.
5. Score the capillary in the middle and snap to give a pipet of approx 1–1.5 mm in bore size. Make approx 5 pipets for each experiment.
6. Connect the pipet to a mouthpiece by a flexible rubber tubing.

3.2. Preparation of Media for Embryo Handling and Culture

3.2.1. PB1 Medium (see **Note 8**)

1. Dissolve 85 mg sodium pyruvate in 10 mL of 0.9% saline. Dilute to 1/50.
2. Add 13 mg phenol red and 129 mg $NaHCO_3$ to 10 mL distilled H_2O.
3. Add 599 mg penicillin to 1 mL 0.9% NaCl, dilute to 1/100.
4. Combine the ingredients shown in **Table 1** together and equilibrate with 5% CO_2 in air for 10 min.
5. Dissolve 130 mg glucose and 520 mg BSA in 100 mL medium.
6. Filter-sterilize the medium through a 0.2-μm filter. Store at 4°C. Media are good for 2 wk.

3.2.2. DR75 Embryo Culture Medium

Collection of rat serum:

1. Anesthetize the rat with 5% halothane in oxygen.
2. Open the abdominal cavity and puncture the aorta with a hypodermic needle (20 gage) attached to a nonheparinized syringe. Dispense the blood into 15-mL culture tubes.
3. Spin the blood at 1000g for 10 min.
4. Squeeze the serum from the fibrin clot by winding the fibrin around the shaft of a pair of clean forceps. Leave the fibrin clot in the tube and respin.
5. Collect the serum using sterile Pasteur pipets into 10-mL culture tubes. Store at –20°C (*see* **Note 9**).

Preparing DR75 medium:

1. Dissolve one packet of powdered DMEM in 4.75 L distilled H_2O.
2. Add 11 g $NaHCO_3$ and dissolve.
3. Bring up to 5-L volume with distilled H_2O.
4. Adjust pH to 7.2 with either NaOH or HCl.
5. Sterilize by filtration through 0.2-μm membrane.
6. Test sterility by incubating 50-μL aliquot at 37°C for 5 d; examine for the presence of contaminant.
7. Aliquot 20-mL DMEM stock to 50-mL tube.

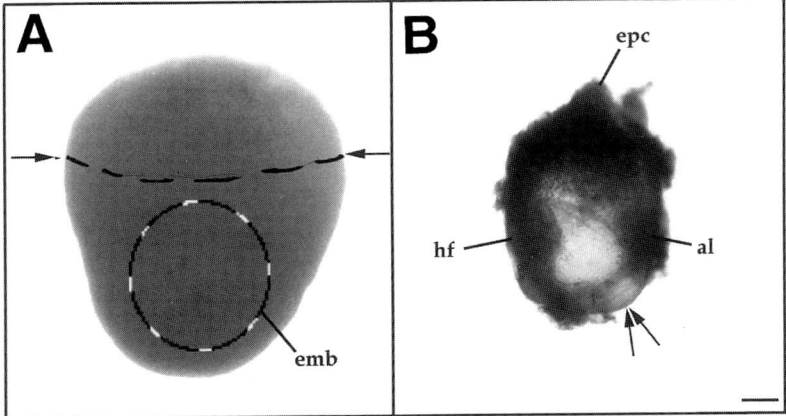

Fig. 2. (**A**) The explanted decidua showing the position of the first cut at the junction of the ectoplacental and embryonic parts (dashed curved line between the arrows). The position of embryo (emb) is outlined by the circle. (**B**) The conceptus dissected out of the decidua showing the Reichert's membrane enveloping the embryo. The *arrows* indicate the site of first tear of the Reichert's membrane. hf; headfold, al; allantois, epc; ectoplacental cone. Bar = 500 μm.

8. Add 20 μ*M* glutamine from stock (stored at –20°C at 200 μ*M* concentration) and 500 μg/mL penicillin/streptomycin (stored at –20°C as 5000 μg/mL stock). Store at 4°C.
9. To make DR75 medium (*see* **Note 10**), heat inactivate an aliquot of rat serum at 56°C for 30 min on the day of use, mix 1 part of DMEM and 3 parts of heat-inactivated serum, equilibrate for at least 2–3 h at 37°C under 5% CO_2 in air (*see* **Note 10**).

3.3. Dissection of Early-Somite-Stage Embryo

1. Sacrifice the pregnant mouse at 8.5 dpc by cervical dislocation.
2. Wet the coat of the abdomen with 70% ethanol.
3. Grasp the skin with blunt forceps and cut through the skin and the abdominal muscle layers with scissors, starting from the pelvic region along the lateral flank to the rib cage on both sides in a V-shaped incision.
4. Reflect the body wall over the thorax, move the viscera to one side to expose the uterus.
5. Sever the cervical and the oviductal connections of the uterine horns and take uterus off the posterior abdominal wall by cutting the mesometrium.
6. Working under a dissecting microscope and using two pairs of watchmaker's forceps, tear open the muscular wall of the uterus to expose the decidua (*see* **Note 11**).
7. Remove the decidua one by one by detaching them from the uterine wall and place them in PB1 medium.
8. Make a shallow cut through the decidual tissue at the junction of the ectoplacental part and the embryo (**Fig. 2A**; *see* **Note 12**) using iridectomy scissors, and extend the cut by nibbling the decidual tissues with watchmaker's forceps along the circumference of the decidua (**Fig. 2A**; *see* **Note 13**).
9. Expose the conceptus by tearing open the decidual tissue along one side of the decidua.
10. Gently lift the conceptus out of the decidua and transfer to fresh PB1 medium.
11. Grasp the Reichert's membrane close to its attachment to the ectoplacental tissue and tear it open down one side of the conceptus (**Fig. 2B**). Trim away the membrane, leaving only a small skirt around the ectoplacental cone (*see* **Note 14**).
12. Culture the embryo in DR75 medium until cell grafting.

3.4. Preparation of Donor Cells

There are two methods by which cells can be prepared for grafting. The tissues may be dissected mechanically to obtain clumps of 5–10 cells *(7,8)*. Alternatively, the tissue layers may be predigested with a mixture of trypsin and pancreatin to aid tissue separation *(7)*. This is then followed by the mechanical isolation of cell clumps. The protocol below (enzyme digestion followed by dissection) can be used for isolating cells from tissues such as neural plate, neural crest, cranial mesoderm, paraxial mesoderm, neural tube and primitive streak *(4,5,7–10)*.

1. Isolate the tissue fragment containing the required cells from the donor embryo by cutting with a pair of metal needles in PB1 + 20% FCS (*see* **Note 15**).
2. Wash the fragment in calcium and magnesium-free phosphate-buffered saline (PBS).
3. Digest the tissue fragment for 15–45 min at 4°C (*see* **Note 16**).
4. Wash the tissue fragment through three changes of PB1 + 20% FCS.
5. Separate the tissue layers into epithelial and mesenchymal portions or cut the fragment into smaller pieces with blunt metal needles (*see* **Note 17**).
6. Cut the tissues further into smaller clumps using fine glass needles and transfer the cell clumps using a mouth-controlled transfer pipet to a drop of PB1 + 20% FCS under paraffin oil (*see* **Note 18**).
7. Repeat the **steps 5** and **6** if cell clumps from other parts of the embryo are to be collected. Keep the cell clumps from different embryonic fragments in separate drops of media.

3.5. Grafting Cells to the Host Embryo

3.5.1. Installing and Aligning Instruments on the Micromanipulator

1. Set up the micromanipulation assembly as shown in **Fig. 3**.
2. Place the de Fonbrune syringe that operates the grafting pipet on the left-hand side of the manipulator assembly and the micrometer syringe that controls the holding pipet on the right-hand side.
3. Bleed the air from the instrument holders by applying positive pressure on the micrometer syringe and de Fonbrune syringe until oil appears at the opening of the holders.
4. Place the grafting pipet in the instrument holder on the right-hand manipulator and the holding pipet on the left-hand manipulator (*see* **Note 19**).
5. Set up the micromanipulation chamber (**Fig. 4A**) by placing 2 drops (50 µL and 100 µL) of PB1 + 20% FCS on the inverted lid of a 60-mm tissue culture dish.
6. Transfer the cell clumps to the smaller off-centre drop. If necessary, further dissection of the cell clumps can be done at this stage using finely drawn glass needles. Cover the drops completely with paraffin oil.
7. Position the dish under a dissecting microscope so that the edge of the larger central drop is at the center of the field of view.
8. Move the holding and grafting pipets into the field of view by tilting the manipulator, adjusting the control knobs, and positioning with the joystick control to bring the pipet tips in focus near the edge of the medium drop (**Fig. 4A**).
9. Adjust the pressure on the syringes so that there is no net movement of oil in or out of the pipets.
10. Zoom in slowly on the microscope and continue to align the pipets so that their tips are always in focus at the center of the field of view even at the highest magnification on the microscope.
11. Take the instruments out of the field of view by retracting the joysticks.

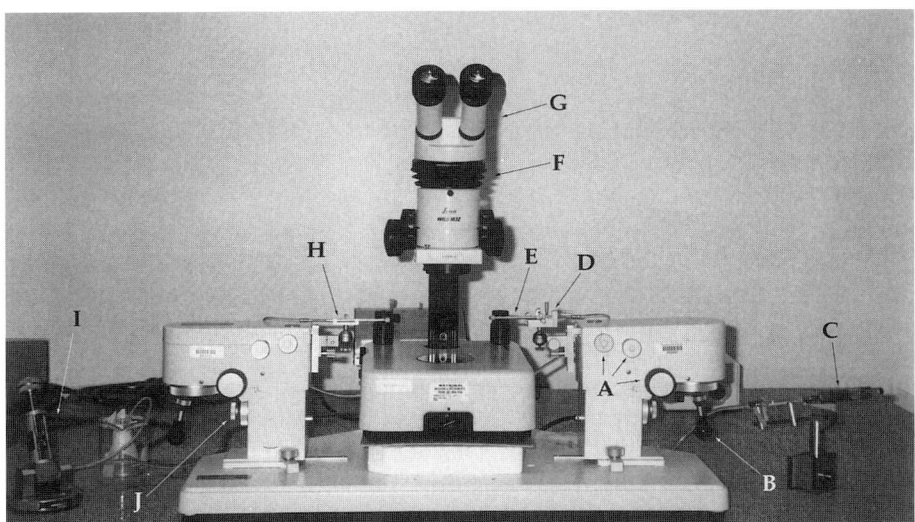

Fig. 3. The micromanipulator assembly. (**A**) Coarse adjustment for positioning the right micromanipulator. (**B**) Joystick for fine positioning of the instrument holder in the horizontal plane. (**C**) Micrometer syringe for the holding pipet. (**D**) Instrument holder for the grafting micropipet. (**E**) Instrument sleeve for the grafting micropipet. (**F**) Ergo-wedge attachment on (**G**) Wild MP3 dissecting microscope. (**H**) Instrument holder for the holding micropipet. (**I**) De Fonbrune oil-pump syringe for operating the grafting micropipet. (**J**) Fine adjustment knob for controlling the elevation of the left micromanipulator.

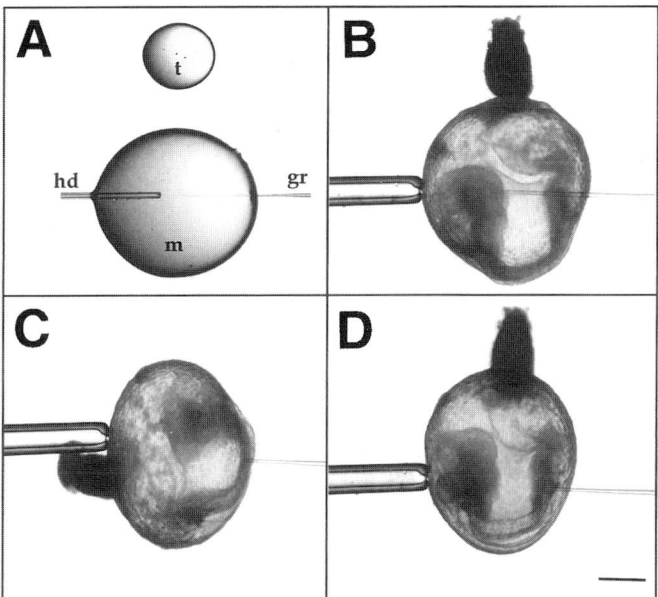

Fig. 4. (**A**) A view under the dissecting microscope of the arrangement of media drops for holding tissue fragments (t) and recipient embryos (m) and the alignment of grafting micro-pipet (gr) and the holding micropipet (hp) in the manipulation dish. The position of the embryo and the alignment of instruments for grafting cells to (**B**) the neural crest region of the head folds, (**C**) the somite, and (**D**) the primitive streak of early-somite-stage (8.5-d) mouse embryos. Bar = 500 µm.

3.5.2. Cell Grafting

1. Transfer 4–5 host embryos to the large drop in the manipulation chamber using the embryo transfer pipet.
2. Bring the holding pipet and grafting pipet back to the field of view and realign the pipets so that the embryos and the instruments are all at the same focal plane (*see* **Note 20**).
3. Take up a short column (approx 20–50 μm) of media into both pipets (*see* **Note 21**).
4. Move the holding pipet out of the field of view but keep the grafting pipet in sight. Move the manipulation chamber to bring the medium drop containing the cell clumps into view.
5. Focus the microscope on the cell clumps, then bring the tip of the grafting pipet in focus close to a chosen clump by tilting the manipulator and adjusting the control knobs. Bring the tip toward the cell clump so that it is just touching the pipet opening.
6. By exerting a negative pressure on the de Fonbrune syringe, draw the cell clump into the pipet following a small quantity of medium so that the clump is held at approx 50 μm from the tip of the pipet (*see* **Note 21**).
7. Move the dish to bring the medium containing the embryos into view. Bring the holding pipet back into the drop.
8. By pushing and rolling the embryo with the holding and grafting pipets, orient the embryo to align the sites of holding and grafting with the instruments (*see* **Note 22**).
9. Hold the embryo by applying suction through the micrometer syringe to the holding pipet so that a small area of the yolk sac membrane or the embryo is drawn into the pipet. Move the embryo gently back and forth to ensure that it has been held tightly.
10. Focus the microscope on the site to be grafted.
11. Bring the grafting pipet toward the embryo and bear it gently against the spot on the surface of the conceptus directly in line of the site of grafting. Using the joystick control, push the grafting pipet through the extraembryonic membranes and the embryonic tissue to reach the site of grafting. It may be necessary to push through the tissue slightly beyond the site of grafting to create the space to accommodate the cell clump.
12. While withdrawing the pipet slowly, gently expel the cell clump from the pipet by exerting a positive pressure on the de Fonbrune syringe. The cell clump should be just out of the pipet as its tip passes the site of grafting. Stop the movement of the fluid in the grafting pipet once the clump has left the pipet (*see* **Note 23**). Check the position of the clump and the content of the pipet after the grafting pipet is completely out of the embryo.
13. Release the embryo from the holding pipet. Transfer the embryos to the DR75 culture medium after grafting is complete (*see* **Note 24**).

4. Notes

1. Different degrees of heating are required when glass of different properties are used for the capillaries. The P-79 Advanced Flaming/Brown Puller has an useful function that conducts a RAMP (heating) test to determine the appropriate filament setting to melt the glass. The RAMP test should be run regularly to correct for variations in glass properties and the working condition of the filament.
2. The puller will break the glass capillaries into two micropipets. Usually both can be used, but it is necessary to routinely check if they are of consistent size and shape. After pulling, the capillary tapers into a fine tip that is too small for transplanting clumps of cells. To fabricate pipets with a desired bore size, the capillary must be broken at its shaft using a microforge. An alternative method is to break the pipet by cutting the shaft with a scalpel blade on a piece of rubber. This has less control of the size and shape of the pipet tip but good pipets can be made.

3. The size of the pipet is defined by the external and internal diameters. The internal diameter is the more critical characteristic, and it should match the size of the cell clumps to be transplanted. For measuring the size of the pipet, an eyepiece ocular micrometer is installed in the eyepiece of the microscope.

4. Some allowance in the positioning of the filament to the micropipet is necessary because the filament will expand on heating. To produce a flushed end on breaking, the filament should be maintained at a temperature that is hot enough to melt the wall of the pipet to the bead but not enough to overheat the glass to cause the bending of the pipet. The breaking force is generated by the contraction of the filament on the pipet as it cools down after the current is turned off.

5. When polishing holding pipets, heating should be stopped when the internal diameter is approx one third the external diameter. A small bore reduces the area that may be held by suction and may lead to an unstable hold, whereas a large diameter may cause tissue damage when a large part of the embryo is drawn into the pipet.

6. The wires can be used as they are, but for ease of handling, they can be glued to wooden handles.

7. Orthodontic wires are available in different thickness and cross-sectional shape and some experimentation is needed to find out the most suitable type for making the needles considering the manipulation to be performed. Thick wire produces blunt needles good for tissue layer separation and trimming of tissues. Thin wire gives thin needles good for cutting up tissue fragments into smaller clumps.

8. PB1 medium is a PBS medium that can be used under normal atmosphere. An alternative medium is M2 solution *(7)*.

9. Rat serum is good for 2–3 mo if stored at –20°C.

10. DR75 should be prepared on the day of use. Do not use if over 24 h old.

11. Care must be taken not to apply pressure to the decidua when splitting the uterine wall and detaching the decidua as this can damage the embryo within.

12. The border of the embryo and the ectoplacental tissue can be discerned by the change in opacity of the decidua from red in color to a translucent area corresponding to the yolk sac.

13. At 8.5 dpc the embryo has a fluid filled yolk sac and amniotic cavity. Exceptional care must be taken to prevent compression or puncture of these fluid-filled cavities. The decidual tissue is extremely sticky and adheres tenaciously to the Reichert's membrane, so when removing the decidual tissue, be careful not to pressure the underlying tissue or puncture the yolk sac.

14. Trimming the Reichert's membrane and ectoplacental cone greatly facilitates the positioning of the embryo during grafting.

15. Dissection from the decidua of donor embryos can be done more expeditiously because there is no need to keep the extraembryonic membranes. As long as the embryo is intact, it is fine for use as a donor. Embryonic fragments can be cut with a pair of metal needles in a shearing action. One needle is used to fix the specimen to the dish and the other is slid along the shaft of the first one to cut up the tissue.

16. The duration of enzyme digestion varies with the size of the fragment and the amount of extracellular matrix. The tissue should be examined regularly during digestion. Tissues are ready for further dissection when tissue layers at the edge of the fragment begin to separate.

17. The tissue layers are separated by first fixing the specimen to the bottom of a plastic culture dish with one needle then pushing in the other needle between the tissue layers and lifting the top tissue layer by swaying this needle back and forth between the layers.

18. Cell clumps can be kept in this medium at ambient temperature for up to 2 h without any significant loss of viability, but they will become sticky with time and are more difficult to handle with pipets. However, it is advisable to complete the grafting within 1 h after tissue dissection and to perform several dissections of donor tissues to stagger with the grafting experiment. Undissected embryonic fragments can be kept in drops of DR75 culture medium under light paraffin oil in a CO_2 incubator at 37°C, 5% CO_2.

19. The movement of the cell clumps is determined by the movement of the fluid in the pipet, which is regulated by the fluid pressure generated by the syringes. The movement of fluid in and out of the pipet is monitored by the movement of the oil-medium interface, which should be kept in sight throughout the grafting procedure. The translation of pressure applied to the syringe to fluid movement is effective only when the whole system is completely bled of air. The presence of air in the system is indicated by a lag in fluid movement when pressure is applied or the continued movement of fluid even when there is no change in syringe pressure. The contralateral arrangement of the syringe and the manipulator (**Fig. 3**) is necessary for simultaneous control of fluid movement and the position of the instrument.

20. Alignment of instruments is done by bringing them into the same plane of focus. This is achieved by first focusing the microscope on the object in the manipulation chamber (cell clumps or embryos) and then adjusting the elevation and the horizontal position of the instruments. Misalignment of instruments results in the rolling of the embryo when the grafting pipet is pushed against the embryo.

21. The column of medium between the cell clumps and the oil-medium interface is to avoid the adhesion of cells to the oil and to stop the oil from getting into embryonic tissues during the expulsion of the cell clumps. The column of medium that keeps the cell clumps away from the tip prevent the loss of the cell clumps when the pipet is taken in and out of the medium drops and during the orientation of the embryo.

22. Orientation of embryos can be achieved in two ways. First, gross orientation can be achieved by using a steel needle and gently rolling the embryo to the desired position. Second, the embryo can be moved by flipping it with the grafting and holding pipets. The ideal orientation of the embryo and the positioning of the instruments vary with the site of grafting (**Fig. 4B–D**). The principle is to hold the embryo either closest to the embryonic structure in order to stabilize it for grafting (e.g., head folds, **Fig. 4B**) or on the side diametrically opposite to the site of grafting (e.g., trunk region or primitive streak, **Fig. 4C,D**). The grafting pipet may have to penetrate through all tissue layers and the amniotic cavity or just through one or two tissues layers to reach the site of grafting.

23. The cell clump may adhere to the end of the pipet, especially after it has been standing in the PB1 medium for too long or when the tip of the pipet is coated with cellular debris and exudate after several graftings. If this happens during grafting, the pipet with the clump partially out at the tip can be left in the embryo for 30–45 s to allow the clump to adhere to the host tissues. The pipet is then jerked away from the embryo to dislodge the clump. Tapping the base plate of the manipulator assembly has the same effect of pulling back the pipet. Change to a new pipet for the next grafting.

24. Up to five early-somite-stage embryos can be cultured in 5 mL of the medium in a 15-mL rotating bottles in an embryo culture incubator (BTC Engineering) under 5% CO_2, 5% O_2, and 90% N_2. The embryos can be cultured for any duration up to approx 2.5 d from the early-somite-stage. A description of the culture system is beyond the scope of this chapter but can be found in **ref. 7**.

Acknowledgment

We thank Anne Camus, Bruce Davidson, Peter Rowe, and Tania Tsang for reading the manuscript. Our work is supported by the National Health and Medical Research Council (NH&MRC) of Australia (S. S. T. and P. T.) and Mr. James Fairfax (for P. T.'s laboratory at CMRI).

References

1. Tam, P. P. L. (1998) Postimplantation mouse development: whole embryo culture and micro-manipulation. *Int. J. Dev. Biol.* **42,** 845–902.
2. Beddington, R. S. P. and Lawson, K. A. (1990) Clonal analysis of cell lineages, in *Postimplantation Mammalian Embryos. A Practical Approach* (Copp, A. J. and Cockroft, D. L., eds.), IRL, Oxford, UK, pp. 267–292.
3. Davidson, B. P., Camus, A., and Tam, P. P. L. (1998) Cell fate and lineage specification in the gastrulating mouse embryo, in *Cell Fate and Lineage Determination* (Moody, S. A., ed.), Academic, London, pp. 491–504.
4. Tam, P. P. L. (1990) Studying development in embryo fragments, in *Postimplantation Mammalian Embryos. A Practical Approach* (Copp, A. J. and Cockroft, D. L., eds.), IRL, Oxford, UK, pp. 317–337.
5. Beddington, R. S. P. (1987) Isolation, culture and manipulation of post-implantation mouse embryos, in *Mammalian Development. A Practical Approach.* (Monk, M., ed.), IRL, Oxford, UK, pp. 43–70.
6. Brown, K. T. and Flaming, D. G. (1986) Advanced micropipette technology for cell physiology, *IBRO Handbook Series: Methods in the Neurosciences*, vol. 9. Wiley, Chichester, UK.
7. Sturm, K. A. and Tam, P. P. L. (1993) Isolation and culture of whole postimplantation embryos and germ layer derivatives. *Methods Enzymol.* **225,** 164–190.
8. Trainor, P. A., Tan, S.-S., and Tam, P. P. L. (1994) Cranial paraxial mesoderm: Regionalisation of cell fate and impact on craniofacial development in mouse embryos. *Development* **120,** 2397–2408.
9. Chan, W. Y. and Tam, P. P. L. (1988) A morphological and experimental study of the mesencephalic neural crest cells in the mouse embryo using wheat germ agglutinin-gold conjugate as the cell marker. *Development* **102,** 427–442.
10. Tam, P. P. L. and Beddington, R. S. P. (1987) The formation of mesodermal tissues in the mouse embryo during gastrulation and early organogenesis. *Development* **99,** 109–126.

40

Interspecific Chimeras

Transplantation of Neural Crest Between Mouse and Chick Embryos

Margaret L. Kirby, Harriett Stadt, Donna Kumiski, and Vlad Herlea

1. Introduction

Several labs have published data using tissue from genetically altered mouse embryos grafted into chick embryos *(1–4)*. The advantages gained from this technique include the ability to analyze the behavior of a population of genetically altered cells among cells with a genetically normal background. In addition, it allows analysis of a particular gene function at stages beyond the normal viability of some strains of homozygous mutant mice. Finally, it provides a marked cell population for study. Two other labs have studied mouse neural crest or neural tube grafts in chick embryos *(3,4)*. Our experience has been with grafting and tracking the cardiac neural crest. This population of cells presents some unique problems because the hosts must be prepared by removal of the endogenous cardiac neural crest. In the absence of the endogenous cardiac neural crest cells, it is imperative that a critical number of grafted mouse neural crest cells migrate into the pharyngeal region and outflow tract or the host embryo develops a severe cardiac outflow-tract malformation that is highly lethal *(5)*. This necessitates extreme care in handling the graft and ensuring that attachment of the graft to the host is rapid.

Mouse–chick chimeras are prepared by homotypic transplantation of the neural folds from rhombomeres *(6–8)*, which are the origin of the cardiac neural crest *(6)* (*see* **Note 1**). In each of the transplantations, the donor and host are at comparable stages of development at the time of the transplantation (3–12 somites). The chick embryo is the host and the mouse the donor. Analysis of the consequences of the graft can be done at various times after constructing the chimera depending on the process under scrutiny. For migration, the chimeras can be used at 24–72 h after surgery; for aortic arch artery anomalies, from 3–5 d after surgery (*see* **Note 2**); for outflow-tract anomalies and cell lineages, from 8 d total incubation to hatching; for glandular derivatives of the caudal pharyngeal arches, from 10 d total incubation to hatching (*see* **Note 3**).

2. Materials

1. Fertilized unincubated chick eggs and breeding age mice.
2. Forced-draft incubator.

From: *Methods in Molecular Biology, Vol. 135: Developmental Biology Protocols, Vol. I*
Edited by: R. S. Tuan and C. W. Lo © Humana Press Inc., Totowa, NJ

3. Neutral red slides are made by heating 1% agar in distilled water to boiling, allowing the solution to cool, and adding 0.01% neutral red. Pipet onto 1" × 3" glass histological slides before the agar solidifies. Allow to dry and store at 4°C, desiccated until use.

4. Tungsten needles: A length of 0.005"-diameter tungsten wire is secured into a glass handle with hot-melt glue. These are sharpened electrolytically by applying a DC potential of 10 V across the wire in saturated potassium chloride.

5. Pancreatin (1%; Gibco-BRL, Gaithersburg, MD, optional) in chick Tyrode's solution.

6. Phosphate-buffered saline (PBS), pH 7.4: Composition per liter, 8 g NaCl, 0.2 g KCl, 1.15 g Na_2HPO_4, and 0.2 g KH_2PO_4.

7. Mitotracker (Molecular Probes Inc., Eugene, OR) as a 0.05-μM solution in Tyrode's. Because Mitotracker is insoluble in water, a stock can be made in absolute ethanol, diluted 1:100 in Tyrode's for use.

8. Multiwell slides (VWR Scientific Products, West Chester, PA). These are used for digestion, dissection, and staining the mouse neural folds.

9. B2 repetitive element probe or any other suitable marker for mouse cells that does not cross-react with avian cells.

10. 4% paraformaldehyde in PBS.

3. Methods

1. Host embryos are made using fertilized Arbor Acre chicken eggs (Seaboard Hatcheries, Athens, GA), incubated for approx 30 h (stages 9–10) at 37°C and 70% humidity in a forced-draft incubator *(7)*. The eggs are windowed and lightly stained with neutral red and the vitelline membrane torn using Dumont No. 5 forceps (Moria Instruments, Paris, France). The chick embryo is prepared by removal of the bulk of the neural folds between the midotic placode and somite 3 (rhombomeres 6–8).

2. The mice on a 12-h light/dark cycle are paired late in one light cycle and the females examined for copulation plugs in the early part of the next light cycle, which is designated as 0.5 d postcoitum (dpc). At 8.5 dpc, pregnant mice are sacrificed by cervical dislocation and the uterine horns rapidly removed and placed into sterile PBS. Each embryo is removed from the implantation sac and placed into sterile phosphate buffered saline. The embryo can be used if it has at least one somite and fewer than 12 somites. For best cell viability and migration, use of enzyme digestion is not recommended *(see* **Note 4***)*. If a digestion is necessary, an explant of all three germ layers can be made from the region corresponding approximately to the midotic placode to somite 5. The otic placode is not visible at these stages so the cranial level of the incision is judged to be 1.5 somite lengths above somite 1. When the embryo has fewer than 5 somites, the caudal incision can also be approximated. The explanted piece is transferred to 50–100 μL of pancreatin (1%; Gibco-BRL) in chick Tyrode's in a multiwell slide for 5–8 min. It is then transferred into 100 μL of Tyrode's containing neutral red or Mitotracker (0.05 μ*M*; Molecular Probes) and the neural tube dissected free of adherent cells. The neural folds are removed from the cleaned neural tube/plate and allowed to stain for 3–5 min or until the tissue appears pink.

3. The mouse neural folds are washed in PBS (**Note 1**), transferred into a host embryo and oriented carefully for dorsoventral, left–right, and craniocaudal axes. After placement of the folds, the chimeras are observed for several hours until the grafts are completely incorporated and do not move in the developing chick embryo when it is flooded with 305 μL of saline. Once the graft is incorporated the eggs are sealed with cellophane tape and returned to the incubator until the embryos are harvested.

4. To examine early migration of the mouse neural crest in its chick host, the fluorescent tag Mitotracker can be used. The chimeras are dissected free of the egg at 12, 24, 48, and 72 h

after surgery and immersion-fixed in 4% phosphate-buffered paraformaldehyde. At 12 and 24 h, the embryos are examined as whole mounts using confocal microscopy. Embryos at all of the stages can be embedded in paraffin, serially sectioned, and mounted. The sections are examined using a rhodamine filter. Because Mitotracker is visible for only a limited time after construction of the chimera (24–72 h), an alternate method must be used to identify mouse cells in the chick environment.

5. Our lab has used a B2 repetitive element from a mouse embryonic cDNA library. The B2 repeat is subcloned into the *Xba*I site of pBluescript II (KS) (Stratagene, La Jolla, CA) to generate plasmid pB2-link5 and contains 280 bp corresponding to the reported sequence of the B2 repeat with 10 nucleotide differences *(8)*. For the preparation of an *in situ* hybridization probe, plasmid pB2-link5 is linearized with *Hind*III. This linear DNA is used to synthesize a riboprobe from the T7 promoter, incorporating digoxygenin-UTP to generate a transcript 335 nucleotides in length (the probe represents the antisense transcript relative to the sequence referenced above, accession number M55334). This probe is then used under standard conditions for *in situ* hybridization in either whole-mount or paraffin sections to detect nuclei of murine origin.

6. Analysis and photography of the heart and great vessels is done at day 8 of incubation. The embryos are decapitated and perfused via the left ventricle with 0.9% saline followed by Carnoy's fixative. The thorax is immersion-fixed overnight in 10% neutral buffered formalin. Macroscopic dissection of the outflow tract and great arteries allows analysis of the patterns of these vessels. The right ventricular free wall is removed for visualization of the ventricular septum. After analysis and photography, the thoracic region is embedded in paraffin and sectioned for more detailed analysis of the outflow tract.

7. Analysis of glandular derivatives of the pharynx can be done at day 12 of incubation (*see* **Note 3**). The chimeras are fixed essentially as previously described, and the heart, great arteries, thymus, thyroid, ultimobranchial, and parathyroid glands are analyzed *in situ* macroscopically. The glands are removed, embedded in paraffin, serially sectioned at 7 μm and stained with hematoxylin and eosin. To assess glandular function of the ultimobranchial glands (parafollicular cells, C cells), rabbit anticalcitonin antibody (Zymed Laboratories, South San Francisco, CA), diluted 1:100 can be used.

4. Notes

1. The cardiac neural crest arises from approximately the same level in the mouse as in the chick *(9)*, and the normal migratory pattern is similar.

2. Even though the initial pattern of the pharyngeal arch arteries and outflow tract are very similar, some differences become apparent as these structures are repatterned to their adult configuration. In the chick, the aortic arch is derived from the 4th aortic arch artery on the right, whereas it is on the left in mammals. The left 4th in the chick forms briefly and closes down while the right 4th in the mouse is incorporated into the base of the brachiocephalic artery. The ductus arteriosus is formed from the 6th aortic arch arteries in both mouse and chick, but in the chick it remains bilaterally patent until hatching while it is present only on the left in the mouse and closes after birth *(10)*.

3. There are differences in the development of the pharyngeal endocrine organs in mouse and chicken. In both, the thyroid forms from the thyroglossal duct evaginating from the endodern in the midline at the level of pharyngeal arch 2 and moving caudally to the base of the neck. In the chick, the thyroid divides to form paired organs situated on the ventral surface of the common carotid arteries near the origin of the subclavian in the thorax. In contrast, in the mouse, the thyroid forms a mustache-shaped bar in the neck that later fuses with the parathyroid and ultimobranchial bodies. The chick has two pairs of

thymus and parathyroid glands derived from the third pouch, whereas the mouse forms a pair of each primordium from the third pouch. In the chick, the descent of the thymus glands follows the course of the jugular veins and the vagus nerves through the thoracic inlet where they form an elongated, lobulated gland, whose enlarged cranial end lies in the vicinity of the third cervical segment with the caudal end extending downward to the thoracic cavity. The thyroid and parathyroid glands are caudal to the thymus *(11)*. In the mouse, the two thymic primordia (one on each side) also form from the third pouch, migrate medially and posteriorly where the two structures fuse to become a single gland with two lobes ventral to the trachea in the anterior mediastinum. Ultimobranchial evaginations form from the caudal wall of the sixth pouch in the chick and from the posterior pharynx in the mouse. In the chick, all these evaginations detach from the pharynx and come to lie in the thorax along the carotid arteries as separate organs that surround the thyroid lobes. In the mouse, the parathyroids and ultimobranchial glands fuse with the lateral thyroid lobes in the neck.

4. We have found that mouse tissue is much more sensitive to enzymatic digestion than avian tissue. Where an absolute definition of the grafted cell population is not necessary, the graft is much more robust if the enzyme digestion is eliminated.

References

1. Auda-Boucher, G., Jarno, V., Fournier-Thibault, C., et al. (1997) Acetylcholine receptor formation in mouse-chick chimera. *Exper. Cell Res.* **236,** 29–42.
2. Fonatine-Pérus, J., Fournier Le Ray, C., and Paulin, D. (1995) Mouse chick chimera: A new model to study the in ovo developmental potentialities of mammalian somites. *Development* **121,** 1705–1718.
3. Fontaine-Pérus, J., Halgand, P., Cheraud, Y., et al. (1997) Mouse-chick chimera: A developmental model of murine neurogenic cells. *Development* **124,** 3025–3036.
4. Serbedzija, G. N. and McMahon, A. P. (1997) Analysis of neural crest cell migration in Splotch mice using a neural crest-specific LaxZ reporter. *Dev. Biol.* **185,** 139–147.
5. Kirby, M. L. and Creazzo, T. L. (1995) Cardiovascular development. Neural crest and new perspectives. *Cardiovasc. Rev.* **3,** 226–235.
6. Kirby, M. L. (1987) Cardiac morphogenesis: Recent research advances. *Ped. Res.* **21,** 219–224.
7. Hamburger, V. and Hamilton, H. L. (1951) A series of normal stages in the development of the chick embryo. *J. Morphol.* **88,** 49–92.
8. Bladon, T. S., Fregeau, C. J., and McBurney, M. W. (1990) Synthesis and precessing of small B2 transcripts in mouse embryonal carcinoma cells. *Mol. Cell. Biol.* **10,** 4058–4067.
9. Fukiishi, Y. and Morriss-Kay, G. M. (1992) Migration of cranial neural crest cells to the pharyngeal arches and heart in rat embryos. *Cell Tissue Res.* **268,** 1–8.
10. Waldo, K. and Kirby, M. L. (1998) Development of the great arteries, in *Living Morphogenesis of the Heart* (De la Cruz, M. V. and Markwald, R. R., eds.), Birkhäuser, Boston, MA.
11. Romanoff, A. L. (1960) *The Avian Embryo: Structural and Functional Development.* Macmillan, New York.

41

Interspecific Mouse-Chick Chimeras

Josiane Fontaine-Pérus

1. Introduction

Embryonic skeletal muscles originate from metameric mesodermal structures called somites, which are formed according to a craniocaudal gradient from the paraxial mesoderm. In addition to providing all of the striated musculature of the body, somites give rise to the dorsal dermis and the axial skeleton. In birds, microsurgical experiments have been performed on newly formed somites to study their derivatives *(1,2)*. It has been demonstrated that the somite contains a dorsal compartment providing the dermomyotome from which striated muscle and dermis originate and a ventral compartment forming the sclerotome from which the axial skeleton develops (for review, *see* **ref. *3***).

An important aspect of muscular development is the relation between nerve and muscle. Most experiments in which innervation was suppressed have shown neural dependence during the different stages of myogenesis *(4–6)*. Information about the nature of interactions between neuroblasts and myogenic cells can be obtained through experimental alterations of the spatial relations normally existing among these cells. This procedure is easy to carry out in the avian embryo *in ovo* but not in the mammalian fetus *in utero*. However, this difficulty can be overcome through the use of a technique developed by us, which consists in implanting xenografts of fetal mouse tissue into the avian embryo. This procedure is of particular interest because it allows the murine fetal cell to be studied by the classic methods of experimental embryology in terms of what happens to the avian embryo cell.

We have developed a technique for transplanting somites and neural tubes of the mouse fetus into the chick embryo. In the case of somite grafts, the implanted cells participate in forming dermis, cartilage, and axial and peripheral muscles in the host *(7,8)*. In the case of neural tube grafts, the ventral area of the tube gives rise to motor axons and the associated crests to peripheral ganglia. Thanks to our methodology, we were able to combine the techniques of experimental embryology performed in birds with those for labeling mouse cells by homologous recombination. The *LacZ* desmin transgenic mouse was used *(9)*, in which insertion of a 280-base-pair sequence of the human desmin gene coupled to the *LacZ* reporter gene allowed myogenic changes in grafted somitic cells to be monitored by simple blue staining *(7)*. The mouse

From: *Methods in Molecular Biology, Vol. 135: Developmental Biology Protocols, Vol. I*
Edited by: R. S. Tuan and C. W. Lo © Humana Press Inc., Totowa, NJ

construction in which the *LacZ* gene was inserted at the level of tenascin gene transcription provided the means of tracing the development of motor axons, owing to X-gal reactivity to Schwann cells *(10)*.

This technical approach associating detailed knowledge acquired about avian cellular embryology with the possibilities presented by the mouse relative to genetic mutations is quite promising. Our mouse–chick chimera makes this system accessible throughout embryogenesis and allows the investigation of many developmental mechanisms that were previously inaccessible in vivo in mammals.

2. Materials

1. Tyrode's solution: 40 g NaCl, 1 g KCl, 1.325 g $CaCl_2 \cdot 2H_2O$, 0.5 g $MgCl_2 \cdot 6H_2O$, 0.28 g $NaH_2PO_4 \cdot 2H_2O$, 5 g glucose, 5 g $NaHCO_3$ in 1 L distilled water. The solution is dispensed in 100-mL aliquots and sterilized by filtration through a 0.45-μm millipore filter and stored at + 4°C.
2. MacEven's solution: 76 g NaCl, 4.2 g KCl, 2.4 g $CaCl_2$, 1.4 g NaH_2PO_4, 0.5 g MCl_2 in 500 mL distilled water. The solution is dispensed in 50-mL aliquots and sterilized by autoclaving. In each aliquot, 2 g glucose and 2 g $NaHCO_3$ are added together with 0.9 L distilled water. The solution is sterilized by filtration through a 0.45-μm millipore filter and stored at +4°C.
3. Pancreatin stock solution: pancreatin 4% NF prepared normal saline (Gibco-BRL, Gaithersburg, MD) stored at –20°C in 20-mL aliquots. Prior to use, pancreatin is mixed at a 0.1% ratio (w/v) in Tyrode's solution.
4. Fetal bovine serum (FBS; Gibco-BRL) stored at –20°C in 20-mL aliquots. Prior to use, FBS is diluted at a 10% ratio (w/v) in Tyrode's solution.
5. Drawing ink (black Pelikan) is diluted (50% in Tyrode's solution) and brought to room temperature immediately before use.

3. Methods

3.1. Preparation of Donor Somites and Neural Tubes from 9-Day Post-Coitum Mouse Fetus

1. Pregnant mice are killed by decapitation. After washing the body with 70% ethanol, abdominal skin is cut off and the uterus rinsed in MacEven's solution. Mouse embryos are then isolated and stored at 4° in MacEven's solution for a maximum of 3 h before grafting (*see* **Note 1**).
2. The truncal part of the embryo is dissected out and immersed in pancreatin diluted in Tyrode's solution for 5 min on ice before dissociation. The ectoderm is then mechanically extirpated using very-thin stainless-steel needles, and the somites and neural tube are isolated from surrounding tissues (i.e., lateral and intermediate plates, mesenchyme, notochord, and endoderm). These structures, once transferred into Tyrode's solution to which fetal calf serum is added, can be kept on ice until 10 min before grafting (*see* **Notes 2** and **3**).

3.2. Preparation of 2-Day-Old Chick Host Embryo

1. Albumen (0.5 mL) is extirpated from the chick embryo by aspiration through a small hole pierced at the top of the egg, which is then obturated with a drop of sealing wax. This allows the upper part of the shell to be cut off without damaging the embryo developing at the bottom of the yolk sac.
2. After the shell is opened, the embryonic area is rinsed with a few drops of Tyrode's solution. A mixture of Tyrode's solution and drawing ink is then microinjected beneath the

embryonic area to allow its visualization. Somite number is estimated and the embryo operated at the 15–18-somite stage.

3. The vitelline membrane is removed. A few drops of Tyrode's solution are placed on the embryo to prevent dehydration.

   a. Somite ablation. A slit is made in the ectoderm between the neural tube and the somite area to be extirpated (generally the last five somites), which is then pushed aside. Using a micropipet, a drop of pancreatin diluted 1:2 in Tyrode's solution is laid on. Host somites are then mechanically extirpated, and the surgical zone is rapidly washed with a few drops of Tyrode's solution to which 20% fetal calf serum is added.

   b. Somite replacement. Mouse donor somites are transferred by micropipet into chick embryos. The orientation of the transplanted somites is determined with respect to that of adventitial tissues.

   c. Neural tube ablation. The neural tube region to be isolated is delimited in the chick host by making two longitudinal slits between the neural tube and the somites and two transverse slits at the anterior and posterior limits of the graft level. The tube is then mechanically extirpated from surrounding tissues.

   d. Neural tube replacement. Mouse neural-tube fragments are transferred by micropipet into chick embryos. The dorsolateral and anteroposterior orientation is respected. The chick host is rinsed with a few drops of Tyrode's solution.

4. Notes

1. The maintenance conditions for mouse fetus are crucial if optimal graft survival is to be ensured. Accordingly, the fetus is isolated from the uterus and immediately stored on ice in MacEven's solution. Mouse structures (somite, paraxial mesoderm, neural tube) are dissociated just prior to *in ovo* grafting.

2. Enzymatic dissociation of mouse tissues should not exceed 5 min.

3. Dissociated mouse tissues should not be stored more than 10 min.

Acknowledgments

I am grateful to Gwenola Auda-Boucher and Yvonnick Chéraud for their assistance. This work was supported by the Association Française contre les Myopathies.

References

1. Chevallier, A., Kieny, M., and Mauger A. (1977) Limb-somite relationship: Origin of the limb musculature. *J. Embryol. Exp. Morphol.* **41,** 245–258.

2. Christ, B., Jacob, H., and Jacob, M. (1978) On the formation of the myotomes in avian embryos. An experimental and scanning electron microscope study. *Experientia* **34,** 514–516.

3. Christ, B. and Ordhal, C. (1995) Early stages of chick somite development. *Anat. Embryol.* **191,** 381–396.

4. Fredette, B. J., and Landmesser, L. T. (1991) A reevaluation of the role of innervation in primary and secondary myogenesis in developing chick muscle. *Dev. Biol.* **143,** 19–33.

5. Wilson, S. J. and Harris, A. J. (1993) Formation of myotubes in aneural rat muscles. *Dev. Biol.* **156,** 509–518.

6. Ashby, P. R., Pinçon Raymond, M., and Harris, A. J. (1993) Regulation of myogenesis in paralyzed muscles in the mouse mutants peroneal-muscular-atrophy and muscular-dys-genesis. *Dev. Biol.* **156,** 2, 529–536.

7. Fontaine-Pérus, J., Jarno, V., Fournier le Ray, C., Li, Z., and Paulin, D. (1995) Mouse chick chimera: A new model to study the in ovo developmental potentialities of mamma-lian somites. *Development* **121,** 1705–1718.

8. Auda-Boucher, G., Jarno, V., Fournier-Thibault, C., Butler-Browne, G., and Fontaine-Pérus, J. (1997) Acetylcholine receptor formation in mouse-chick chimera. *Exp. Cell Res.* **236,** 29–42.

9. Li, Z., Marchand, P., Babinet C., and Paulin, D. (1993) Desmin sequence elements regulating skeletal muscle-specific expression in transgenic mice. *Development* **117,** 947–959.

10. Fontaine-Pérus, J., Halgand, P., Chéraud, Y., Rouaud, T., Velasco, M. E., Cifuentes Diaz, C., and Rieger, F. (1997) Mouse-chick chimera: A developmental model of murine neurogenic cells. *Development* **124,** 3025–3036.

42

Mosaic Analysis in *Caenorhabditis elegans*

John Yochem, Meera Sundaram, and Elizabeth A. Bucher

1. Introduction

The complete description of its nearly invariant cell lineage *(1,2)* and the growing availability of cloned genes and markers for the cell lineage make *Caenorhabditis elegans* particularly favorable for mosaic analysis, and the literature is rich in examples that prove the usefulness of this approach. Because genetic mosaic analysis in *C. elegans* has recently been reviewed by Herman *(3)* who developed many of the techniques *(3–6)*, this review will be more concerned with recent technical advances rather than with an extensive background of the approach. First we shall present a brief summary of the principles of mosaic analysis as it is typically performed in *C. elegans*. This will be followed by a description of markers that indicate mosaicism and by a discussion of a hypothetical analysis of an essential gene.

Genetic mosaics are individuals that have both genotypically mutant and genotypically wild-type cells. The analysis of mosaics can reveal the cells that need to inherit a functional copy of a gene to prevent a mutant phenotype. In this regard, mosaics can be particularly useful for understanding most, if not all, of the cellular requirements of genes, including those that result in early lethality when mutant. For instance, severe loss-of-function mutations in *let-60* (*C. elegans* nomenclature is summarized in **Table 1**), the *C. elegans* gene that encodes the small G protein Ras, invariably result in the death of first-stage larvae *(7,8)*. Mosaic analysis revealed that the lethality results from the failure of one cell to be properly specified during embryogenesis *(9)*. Not only was this insightful in its own right, but it also meant that viable mosaics could be isolated to investigate other functions of *let-60* later in development. These were mosaics in which the vital part (the "lethal focus") of the cell lineage had inherited wild-type *let-60* activity, but other parts of the cell lineage had lost *let-60* activity.

Mosaic analysis can also reveal whether the effect of a mutation is cell autonomous or nonautonomous with respect to cellular phenotype and hence can provide further insight into the function of a gene product. A simple example with relevance to developmental studies is the case of two interacting cells: one that produces an extracellular signal and another that receives the signal (via a receptor) and then responds by adopting a particular fate. The genes encoding the receptor and the signal may have the same mutant phenotype (failure of the receiving cell to adopt the correct fate). However, a

From: *Methods in Molecular Biology, Vol. 135: Developmental Biology Protocols, Vol. I*
Edited by: R. S. Tuan and C. W. Lo © Humana Press Inc., Totowa, NJ

Table 1
Nomenclature[a]

Designation	Explanation
ncl-1(e1865) III	Name of a gene and in parentheses the designation of a mutation in that gene; letter(s) before the number indicate the laboratory in which the allele was isolated; roman numerals indicate the linkage group (one of the six chromosomes) to which the gene maps
Ncl	Phenotype
NCL-1	Product encoded by a gene
Dp(III;f)	Genetic duplication; the roman numeral indicates the linkage group from which the duplication was derived: *f* (free) indicates that the duplication is extrachromosomal
Ex	Extrachromosomal array
SUR-5GFP(NLS)	Fusion between SUR-5 and a version of the green fluorescent protein containing a nuclear localization signal
kuEx77 [SUR-5GFP (NLS) unc-36(+)]	Designation of an array; letter(s) in front of the *Ex* indicate the laboratory of origin; the number following the *Ex* is a serial number in that lab; in brackets is the genotype of the array, in this case one having copies of the plasmid that encodes SUR-5GFP(NLS) and copies of a plasmid containing wild-type *unc-36* DNA

[a]Additional information regarding nomenclature can be found in **ref. 22**.

mutation in the gene encoding the receptor should behave cell autonomously in genetic mosaics; inheritance of a functional copy of the gene should only matter for the cell that expresses the phenotype when an animal is completely mutant. In contrast, a mutation in the gene encoding the signal should behave cell nonautonomously. The cell expressing the phenotype is the cell that fails to receive the signal, and its inheritance of the functional gene is not relevant; what is relevant is inheritance of the functional gene by the cell that produces the signal. The *lin-12* gene is a noteworthy example of the usefulness of this type of analysis; it furnished the first evidence that proteins of the Notch family are receptors for extracellular signals *(10)*.

1.1. Generating Mosaic C. elegans

Although mosaic worms can be generated in a number of ways (*see* **ref. 3**), most approaches exploit the fact that chromosomes in *C. elegans* are holocentric: Rather than binding to a single kinetochore, metaphase spindle fibers attach at different points along the length of each chromosome *(11)*. Consequently, extrachromosomal DNA can be maintained as a minichromosome that for the most part mimics the native chromosomes. Mitotic segregation of the extrachromosomal DNA, however, occurs with less fidelity than for the chromosomes. If a worm has inherited extrachromosomal DNA that contains wild-type copies of a gene whose chromosomal copies have mutations, the random failure of the extrachromosomal DNA to disjoin when a cell divides during development generates a genotypically mutant clone of cells and therefore a mosaic individual *(5,12)*.

In order to recognize mosaic individuals, one must use an appropriate marker to indicate cells that lack the extrachromosomal DNA. Two useful markers, *ncl-1* and SUR-5GFP(NLS), are discussed in detail in **Subheading 4.** Once genotypically mutant cells have been identified, the invariant cell lineage of *C. elegans* makes it possible to deduce which ancestral cell division(s) suffered mitotic nondisjunction of the extrachromosomal DNA. A cell that establishes a mutant clone is often referred to as the point of loss of the extrachromosomal DNA. A cell whose inheritance of the extrachromosomal DNA affects the mutant phenotype of the gene being analyzed is often called a focus for that particular aspect of the phenotype.

1.2. Extrachromosomal DNAs for Mosaic Analysis

There are two types of extrachromosomal DNA that are suitable for mosaic analysis in *C. elegans*: genetic duplications and transgenic arrays. Most extrachromosomal duplications (more commonly referred to as "free" duplications to distinguish them from genetic duplications that have become attached to one of the six chromosomes) have been isolated following treatments with ionizing radiation. To date, approximately one half of the genome has been duplicated, making it likely that a gene of interest will be present on one or more of the many free duplications *(3,5,12)*. Although here we shall be more concerned with transgenic arrays, free duplications can have some advantages for mosaic analysis *(3)*. Furthermore, methods are available to add desired genes to existing free duplications or to fuse two separate duplications *(3)*. Thus, an analysis need not be limited to the genes present on a particular duplication. The significance of this flexibility will become more apparent in the section concerning lineage and cellular markers.

As first demonstrated by Lackner et al. *(13)*, the use of transgenic arrays can greatly facilitate mosaic analysis in *C. elegans*. This type of extrachromosomal DNA arises following microinjection of a mixture of cloned DNA into the gonad *(14,15)*. The mixture usually contains DNA encoding a marker for the establishment of transgenic lines, DNA encoding a marker for the detection of mosaicism, and DNA that complements ("rescues") the recessive alleles of the gene being analyzed. Recombination within the germline between homologous parts of the cloned DNAs—including common vector sequences, such as the gene for ampicillin resistance—can create an extrachromosomal array that contains copies of each of the three input DNAs.

Extrachromosomal arrays can have several advantages over free duplications for mosaic analysis *(16)*. For one, an analysis need not be limited by the preexistence of a free duplication. Also, arrays tend to segregate less efficiently than free duplications during mitosis, resulting in a higher frequency of mosaic segregants. Furthermore, there is greater flexibility in the choice of marker for detecting mosaicism, because it is encoded by one of the DNA clones that is injected. Although revolutionary, the use of arrays has some limitations. The approach requires that the gene of interest be cloned. The level of gene expression from a particular array can be inappropriately high or low, complicating the analysis. Also, a confusing degree of mosaicism can result from arrays that are too poorly transmitted during mitosis. Because the properties of arrays are dependent on the concentrations of the DNAs injected *(14,15)*, transgenic lines that are optimal for mosaic analysis may need to be determined empirically following a series of injections in which the amounts of the input DNAs are varied. Nevertheless,

Table 2
Markers for Individual Cells

Name	What is scored	Scorable cells	Ref.
ncl-1 (III)[a]	The Ncl phenotype, a cell-autonomous enlargement of nucleoli visible with Nomarski optics	Most postembryonic cells; exceptions are the intestine and the germline	**4,13,16**
SUR-5GFP(NLS)[b]	Bright, cell-autonomous localization of green fluorescent protein (GFP) to nuclei	Many nuclei that descend from early parts of the cell lineage; exceptions are the germline and most cells derived from M	**17**

[a]*ncl-1*(+) is contained on cosmid C33C3 (*16*) (available from the *C. elegans* genome sequencing project; *see* the *C. elegans* website at http://eatworms.swmed.edu/).
[b]*SUR-5GFP(NLS)* is encoded by pTG96 (*17*).

the use of arrays is one of the most significant advances for mosaic analysis in *C. elegans*. Because of their growing use, the examples described herein will be based on them.

1.3. Lineage and Cellular Markers for Mosaic Analysis

Mosaic analysis is more informative if the extrachromosomal DNA contains a gene that indicates—or marks—which parts of the cell lineage have inherited this minichromosome. Inheritance of the wild-type copy of the gene being tested can then be inferred, because it is also part of this minichromosome. In general, there are two classes of markers used to detect mosaicism: those that identify individual cells (**Table 2**) and those that implicate parts of the cell lineage based on gross phenotype (**Table 3**). Markers of the former class are essential for a high level of resolution during mosaic analyses, because they identify specific cells that have inherited the extrachromosomal DNA. A traditional cellular marker is the mutation *ncl-1(e1865)*, which confers a cell autonomous nucleolar phenotype that can be detected when a specimen is examined at high magnification with a compound microscope (*4*). The lineage markers, on the other hand, confer phenotypes that can be scored with a dissecting microscope; they allow convenient identification of rare genetic mosaics among thousands of worms, but do not unambiguously identify which specific cells have inherited the extrachromosomal DNA. (Detailed discussion of the lineage markers is beyond the scope of this review, but the salient features of and references for many of them are presented in **Table 3**.) Used in combination, the two classes of markers can provide speed and certainty to mosaic analysis (*3*). Recently, a new marker, SUR-5GFP(NLS), has been described, which can have the advantages of both classes; it can be used to detect certain genetic mosaics with a dissecting microscope and to establish cellular genotypes (*17*).

1.4. Relative Advantages
of the Cell Autonomous Markers ncl-1 and SUR-5GFP(NLS)

Both *ncl-1* and SUR-5GFP(NLS) can be scored in living animals and allow a high degree of confidence when scoring specific cells for inheritance of extrachromosomal

Table 3
Cell Lineage Markers[a]

Gene name (linkage group)	Cells in which loss of function will result in mutant phenotype	Phenotype of recessive mutations[b]	Ref.
ace-1 (X)	P_1	Unc: uncoordinated movement	*23,24*
ced-3 (IV)	Throughout lineage: mostly in AB and MS descendants	Ced: survival of 131 cells that normally die (DIC phenotype)	*25*
ced-4 (III)	*ibid*	*ibid*	*25*
daf-6 (X)	AB, ABp, ABpl, ABpr	Dye-filling defect in 8 pairs of sensilla (partial defect when lost in ABp descendants)	*5,26*
dpy-4[c] (IV)	AB	Dpy	*25*
dpy-17[c] (III)	P_1, P_2	Dpy	*27*
dpy-18[c] (III)	P_1	Dpy	*28*
emo-1 (V)	P_1, P_2, P_3, P_4	Sterile	*29*
glp-1 (III)	P_1, P_2, P_3, P_4	Sterile	*30*
let-23 (II)	AB, ABp, ABpl, ABpr	Vul: vulvaless	*31,32*
lin-31 (II)	AB, ABp, ABpl, ABpr	Vul or partial defective vulva	*16*
mab-5 (III)	Throughout lineage	Mab: male abnormal (defects in tail)	*27*
mab-21 (III)	AB, ABa	Mab	*33*
mpk-1 (III)	AB, ABp	Suppression of Multivulva phenotype of *let-60(gain-of-function)*	*13*
osm-1 (X)	AB, ABp	Cell-autonomous dye-filling defect similar to *daf-6*	*26,34*
pag-3 (X)	AB, ABp	Unc	*35*
sup-10 (X)	P_1	Sup: suppression of Unc: loss of *sup-10(+)* suppresses Unc phenotype of *unc-93* *(gain-of-function)*	*5,26*
unc-3 (X)	AB, ABp, ABpl, ABpr	Unc (semi-Unc when lost in ABp descendants)	*5,34*
unc-7 (X)	AB, ABp, ABpl, ABpr	Unc (semi-Unc when lost in ABp descendants)	*36*
unc-26 (IV)	AB, ABp, ABpl, ABpr	Unc (semi-Unc when lost in ABp descendants)	*25*
unc-29 (I)	P_1, EMS, MS, C	Unc (semi-Unc when lost in EMS, MS, or C)	*16*
unc-30 (IV)	AB, ABp, ABpl, ABpr	Unc (semi-Unc when lost in ABp descendants)	*25*
unc-36 (III)	AB, ABp, ABpl, ABpr	Unc (semi-Unc when lost in ABp descendants)	*27*

[a]Only markers that are considered convenient and easy for mosaic analysis, and for which published descriptions exist, have been listed. Information regarding more recently described markers may be found by searching abstracts of unpublished work on the *C. elegans* website (http://eatworms.swmed.edu/) and then contacting individual investigators. Most of the genes on this list have been cloned (although the exact gene identity is not always known), thereby permitting the creation of extrachromosomal arrays. Information regarding the cloned DNA can be obtained either by searching ACeDB (a *C. elegans* data base; information about ACeDB can be found at the aforementioned website) or by following the Genes and Genomics link at the *C. elegans* website. Cosmids and other clones can be obtained by contacting one of the *C. elegans* genome sequencing centers. Smaller, more convenient clones can often be obtained by contacting the laboratory in which the gene was originally cloned. Strains bearing mutations can usually be obtained from the Caenorhabditis Genetics Center (CGC); a search of gene mutations and free duplications is available through its Gopher Server, which is listed on the *C. elegans* website.

[b]Phenotypic abbreviations are defined the first time they are used.

[c]These gene products are thought to be required in hyp7, a syncytium that forms most of the epidermis. hyp7 is generated by the fusions of cells derived from AB and from P_1 *(2)*. Some of these gene products are required primarily in the AB-derived cells, and others are required primarily in the P_1-derived cells, as indicated. However, losses in either AB or P_1 may cause a detectable semi-Dpy phenotype.

DNA. Also, neither marker has drastic effects on viability, fertility, development, body morphology, or movement.

The *ncl-1(e1865)* and *ncl-1(e1942)* mutations result in enlarged nucleoli (the Ncl phenotype) that can be detected in nearly all cells by Nomarski differential interference contrast microscopy *(4)*. Importantly, the alleles are recessive to one copy of the wild-type gene. Thus, cells with two mutant copies show the Ncl phenotype, whereas cells with two mutant and one or more wild-type copies (present on a free duplication or extrachromosomal array) have nucleoli of normal size (or even of smaller size in the case of arrays containing many wild-type copies) as shown in **Fig. 1**. With the exception of the nuclei of the intestine and germline, the Ncl phenotype can be distinguished from the wild type in nearly all cells of the postembryonic stages, with the phenotype becoming more obvious in later-stage worms *(4)*. A strong maternal contribution, however, makes a scoring of the phenotype difficult in embryos.

The SUR-5GFP(NLS) mosaic marker *(17)* makes use of the green fluorescent protein (GFP) from *Aequorea victoria (18)*. The marker is based on a translational fusion between SUR-5, the product of a gene that, when mutant, suppresses certain dominant-negative mutations in *let-60ras (19)*, and a version of GFP that has a nuclear localization signal (A. Fire, J. Ahnn, G. Seydoux, and S. Xu, personal communication). SUR-5GFP(NLS) is strongly expressed in many cells, and for unknown reasons it has an unusually high affinity for nuclei (**Fig. 2**). Consequently, many fluorescent nuclei can be detected when worms are examined at low magnifications with a dissecting microscope that has been equipped with a Hg lamp.

The SUR-5GFP(NLS) marker has two major advantages over *ncl-1*. First, the marker itself can be used for the detection and establishment of transgenic lines. Second, SUR-5GFP(NLS) can be used as both a lineage marker and a cellular marker. SUR-5GFP(NLS) can be used as a marker for many parts of the cell lineage because the fluorescing nuclei descend from different early blastomeres. Mosaic worms lacking fluorescence in subsets of these nuclei are easy to detect among hundreds of worms growing on a Petri dish. These candidates can then be picked from the plates and the exact point of loss of the array deduced by using a compound microscope to assess fluorescence in a larger set of individual nuclei. This ability to examine normal growth plates for certain mosaics can help minimize the construction of strains by obviating the need for additional lineage markers (although their use may still be desired).

SUR-5GFP(NLS) may be most useful for analyses of early losses because such events are easiest to detect with a dissecting microscope. In particular, worms having potential losses of an array in P_1 (the posterior daughter of the zygote), EMS, or E are easy to detect because of exceptionally bright fluorescence of the nuclei of the cells of the intestine and the nature by which these cells arise in the cell lineage; all of these nuclei and only these nuclei descend from E, a blastomere that is removed from P_1 by only two cell divisions. Thus, mosaic individuals lacking fluorescence of all of the intestinal cells have a one-third chance of having had losses in P_1. Furthermore, distinguishing candidate losses in P_1, EMS, or E with the dissecting microscope is easy because many of the descendants from MS can also be seen at low magnification.

Another feature of the examination of normal growth plates is the potential to detect very rare, nonmutant mosaics that have inherited SUR-5GFP(NLS) in restricted parts

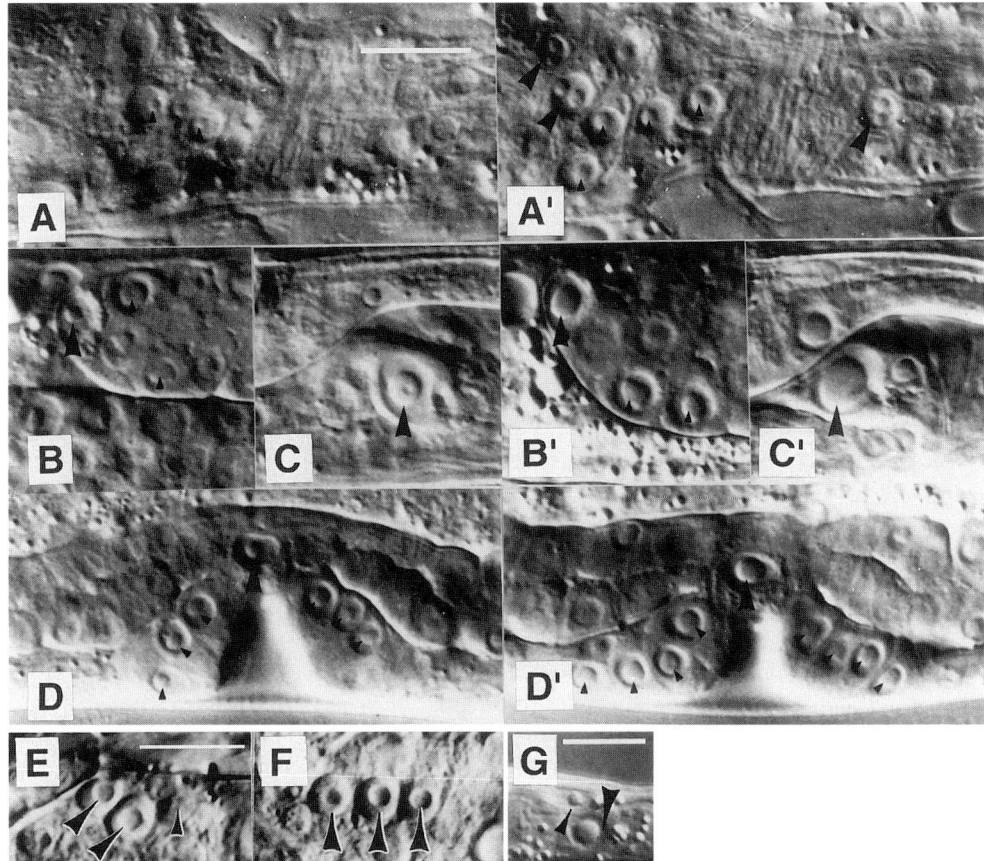

Fig. 1. Phenotype of the *ncl-1* marker in individual cells. The enlarged nucleoli conferred by *ncl-1* mutations can be distinguished from the wild-type by Nomarski microscopy. (A)–(D) Paired panel: left, wild-type; right, *ncl-1(e1865)* homozygotes, of images from right lateral aspects of young L4 hermaphrodites. **(A)** In wild-type, a small fused nucleolus (small arrowheads) is visible in each neuronal nucleus in this view of the nerve ring ganglion. In *ncl-1*, these neurons have either a large, fused nucleolus (*small arrowhead*) or, less frequently, two distinct nucleoli (*large arrowhead*). **(B)** The nucleoli (*large arrowhead*) of distal tip cells of the gonad are compared. Nuclei (*small arrowheads*) of the germline are shown for reference. **(C)** Nucleoli (*large arrowheads*) of the excretory cell are compared. **(D)** Nucleoli of vulval cells (*small arrowheads*) and of the uterine anchor cell (*large arrowheads*) exhibit the phenotype. (E)–(G) The Ncl phenotype in genetic mosaics from a strain of the genotype *dpy-1(e1) ncl-1 (e1865); sDp3(III;f)*. **(E)** The nucleoli of the neurons ASKL (from ABa), ADLL (from ABa), and ASIL (from ABp) (left to right here and in [F]) in an ABa(-) mosaic. Note the enlarged nucleoli in the left two nuclei in contrast to the wild-type nucleolus of ASIL. **(F)** For comparison, the nucleolus of ASIL is enlarged in a *ncl-1* homozygote. **(G)** Two nuclei in the tail indicate mosaicism. The Ncl cell is hyp9 (from ABp) and the non-Ncl cell above it is hyp11 (from C). Bar, 10 μm. Reproduced with permission from **ref. 4**.

of the cell lineage because of consecutive losses of an array. Such cases are easily noticed because of a striking intensity of fluorescence of a few nuclei in otherwise dark animals. These cases can be of much use in further defining the minimal requirement

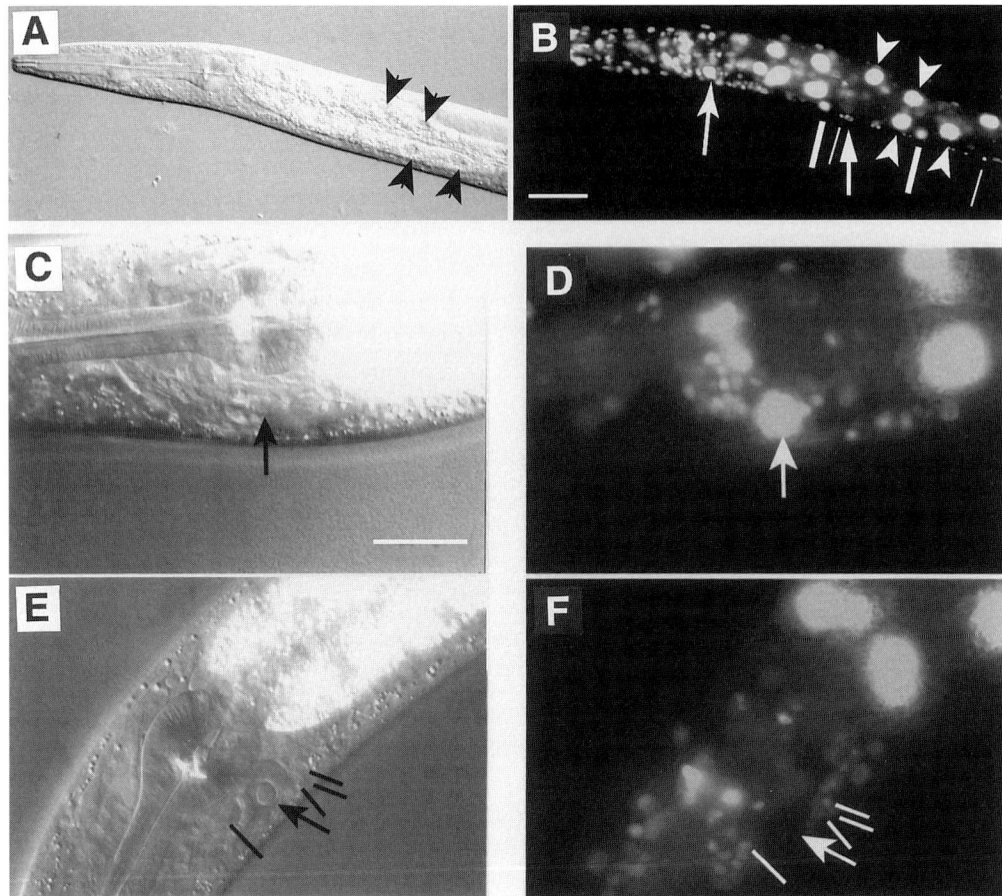

Fig. 2. Fluorescence of SUR-5GFP(NLS) in individual cells. **(A)** Nomarski and **(B)** Fluorescence optics images, respectively, of a longitudinal view (median plane; anterior to the left,
dorsal up) demonstrating bright nuclear fluorescence in nuclei of many cells throughout the
body of an animal transgenic for an array that expresses SUR-5GFP(NLS). The brightest nuclei
are those of the intestine, four of which are indicated with *arrowheads* in both panels. Also very
bright is the nucleus of the excretory cell (anterior-most *arrow*). The nuclei of two (slightly
out-of-focus) body wall muscles (*thick line*), of two neurons in the ventral cord (*thin lines*), and
of a Pn.p cell (posterior *arrow*) are also indicated. (C)–(F) Correlation of the Ncl phenotype
and lack of expression of SUR-5GFP(NLS) in animals of the genotype *ncl-1(e1865) unc-
119(ed3); kuEx78[SUR-5GFP(NLS); ncl-1(+); unc-119(+)]*. **(C)** The Nomarski image of a non-
Ncl excretory cell indicates inheritance of the array. **(D)** The same nucleus (*arrow*) intensely
fluoresces. **(E)** Nomarski image showing the enlarged nucleolus (the Ncl phenotype) present in
the nucleus (*arrow*) of the excretory cell in a mosaic segregant. **(F)** The same nucleus lacks
fluorescence. The nuclei of several neurons (*lines*) serve as reference points. In (A), bar = 10 μm.
Reproduced with permission from **ref. *17***.

for a gene's activity in the cell lineage. This convenient increase in intensity is probably a consequence of a doubling of the copy number of the extrachromosomal DNA at
each nondisjunction (for discussion, *see* **refs. *3*** and ***4***).

SUR-5GFP(NLS) can also facilitate a reciprocal approach, the analysis of very rare segregants that are fully mutant even though they inherited the extrachromosomal DNA. Such cases can refine the analysis because the nuclei that fluoresce should not be relevant to the focus of activity being studied (although one should be aware that insufficient expression of the rescuing DNA on an array can confuse interpretations of this kind of mosaic). This approach is often difficult with a compound microscope because extrachromosomal DNA is not transmitted well during meiosis, resulting in most fully mutant segregants simply not inheriting the DNA.

A disadvantage of SUR-5GFP(NLS) is a failure of all cells to fluoresce *(17)*. For example, the cells that derive from M, a postembryonic blastomere, are nonfluorescent or faint except for the two coelomocytes. If this region of the cell lineage is the object of inquiry, it would be better to use *ncl-1(e1865)*. Because SUR-5GFP(NLS) is limited to the detection of early losses with a dissecting microscope, it may be better to create arrays having both *ncl-1(+)* and SUR-5GFP(NLS) or to create two lines that are individually transgenic for these markers. Another disadvantage of SUR-5GFP(NLS) is a requirement that worms be of a certain size—either the L3 or L4 stage larvae or young adults—for the detection of some early mosaicism with a dissecting microscope. Also, the signal can be faint in older adults, but it is still easier to score in them than the Ncl phenotype. Finally, SUR-5GFP(NLS) may have SUR-5 activity that may make the marker inappropriate for mosaic analyses of the genes encoding members of the Ras pathway *(19)*.

1.5. Hypothetical Genetic Mosaic Analysis of a Lethal Mutation Using SUR-5GFP(NLS) as a Marker

To give the reader an idea of what an actual experiment entails, a hypothetical mosaic analysis of a mutation that causes death of first-stage larvae soon after they hatch is presented in **Figs. 3** and **4**. This example illustrates how the lethal focus of the gene (*let-x*) can be deduced and also how a surprise can be in store that permits greater insight into the gene's function.

The strain constructed for the analysis is simple: It contains mutant copies of *let-x* on the sister chromosomes and an extrachromosomal array that expresses *let-x(+)* and SUR-5GFP(NLS) (**Fig. 3**). Given that the mutation in *let-x* results in the death of homozygotes at an early stage, how was the strain constructed in the first place? In general, construction of transgenic lines for mosaic analyses of lethal mutations can require several steps. First, a strain must exist in which the lethal mutation is balanced by the presence of wild-type DNA. For example, if a recessive mutation that causes uncoordinated movement maps very near *let-x*, a strain can be created with the genotype *let-x/unc-y*. This balanced heterozygote can be propagated by selecting viable progeny that have wild-type movement. The heterozygotes are injected with a mixture of *let-x(+)* and SUR-5GFP(NLS) DNAs, and transgenic lines are established by picking viable green fluorescent progeny that are fully coordinated. Individual segregants of the genotype *let-x; Ex[SUR-5GFP(NLS); let-x(+)]* are then identified. They will always exhibit green fluorescence, will always segregate dead progeny, and will never segregate uncoordinated progeny. Once constructed, such strains maintain themselves and are ready for mosaic analysis.

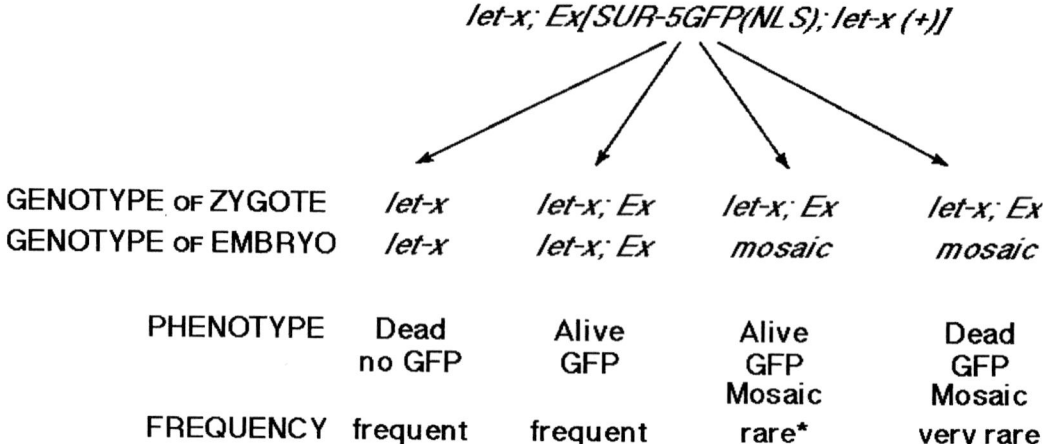

Fig. 3. Possible genotypes and phenotypes of individuals segregating from a strain having a lethal mutation and an extrachromosomal array that rescues the lethal mutation. A hypothetical strain contains a chromosomal recessive lethal mutation, *let-x*, and a complementing extrachromosomal array expressing *let-x*(+) and the marker SUR-5GFP(NLS). Expected classes of nonmosaic and mosaic segregants are indicated. The relative frequencies of these segregant classes will depend on how well the array is transmitted during meiosis and mitosis. Suitable arrays will have mitotic loss rates of 0.1×10^{-3} to 5×10^{-3} losses per cell division *(4,13,16)*, meaning that only 0.8–4% of array-bearing zygotes will experience array loss during one of the eight embryonic cell divisions shown in **Fig. 4**. Thus, mosaic segregants should be rare. If mosaic segregants are not rare, the array may be too unstable for reliable interpretations to be made; this important caution is indicated with an asterisk.

A common class of segregants from the strain are nonfluorescent dead larvae that result from a failure of zygotes to inherit the array (**Fig. 3**). Fluorescent viable progeny that are not mosaic should also be common. Because these have the same genotype as the parent, they can be used to propagate the line. The mosaic individuals are those in which the array was not faithfully transmitted to all daughter cells during embryogenesis. Because we are not using lineage markers that depend on gross phenotype (**Table 3**), our knowledge of mosaicism is derived from the pattern of green fluorescent nuclei revealed by dissecting and compound microscopes. Because the viable mosaic segregants should be rare, we must examine the patterns of many hundreds of worms.

The results of our mosaic analysis are shown in **Fig. 4**. We found several viable mosaic animals in which P_1, ABa, or ABpr had failed to inherit the array (**Fig. 4**, mosaic classes 1, 5, and 6). In contrast, we did not find any viable mosaic animals in which ABpl failed to inherit the array. These data suggest that *let-x*'s activity is vital within ABpl itself or within one or more of its descendants. By examining the cell lineage, we can identify candidate descendants (for example, the excretory cell) that may require *let-x*'s function. In subsequent experiments, we could test these candidate foci by analyzing mosaic animals with late losses within the ABpl lineage. For these subsequent experiments, we might choose to use the *ncl-1* marker.

In our example, the mosaic analysis has also revealed an unexpected aspect of *let-x*'s function. Viable mosaic animals lacking the array in P_1, P_2, or P_3 (mosaic classes 1, 2,

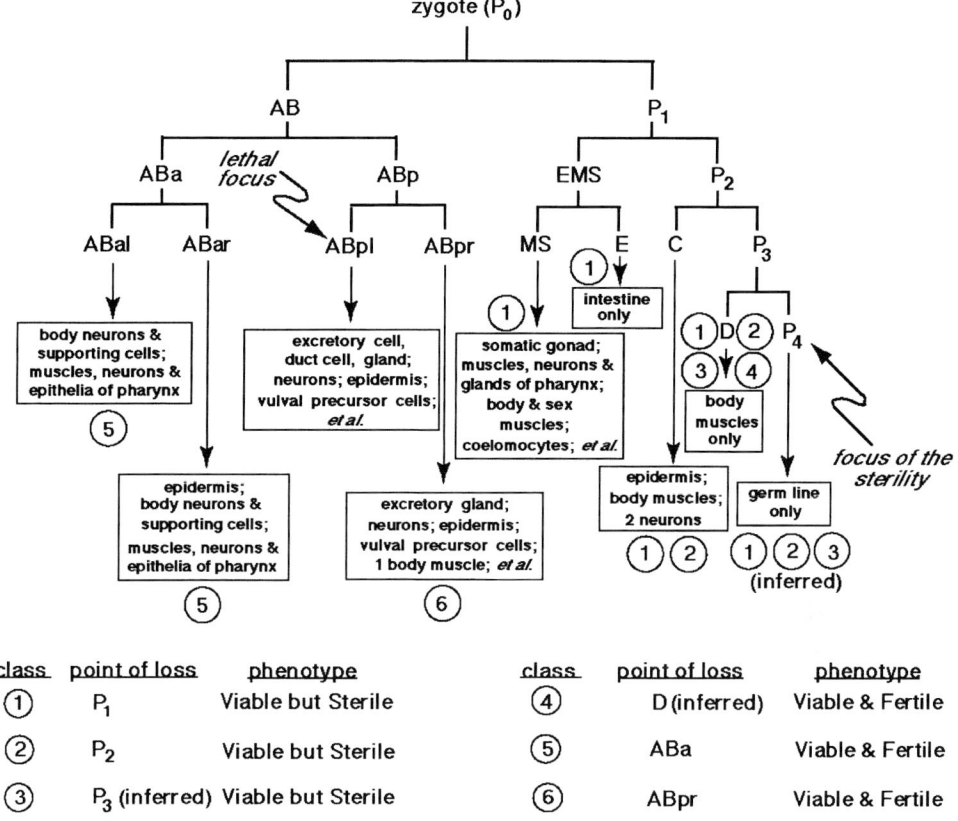

Fig. 4. Hypothetical genetic mosaic analysis of a lethal mutation using SUR-5GFP(NLS) as a marker. Larval and adult cells are indicated in boxes beneath their progenitor in this schematic of the earliest divisions of an embryo (based on **ref. 2**). The indicated cells were examined for expression of the SUR-5GFP(NLS) marker, in viable larvae segregating from the *let-x; Ex[let-x(+); SUR-5GFP(NLS)]* strain introduced in **Fig. 3**. Identification of each fluorescing or nonfluorescing cell was based on the highly reproducible position and morphology of its nucleus (diagrams are available in **refs. 1** and **2**). Nonfluorescence indicated that a cell lacked the extrachromosomal array and was therefore genotypically mutant for *let-x*. Because the lineal relationships of all cells are known in *C. elegans*, we could deduce the common ancestor of all the genotypically mutant cells in a mosaic animal and infer that this ancestor experienced the original loss of the array. We identified many mosaic animals, representing six types of array loss. The phenotype and the point of loss deduced for each type are listed at the bottom with a corresponding number. The genotypically mutant cells observed in each mosaic type are indicated with circled numbers next to the cells. Our examination of viable mosaics means that we are only seeing those losses that are compatible with survival. A failure to see certain types of losses (after an examination of many mosaics) suggests that those losses result in lethality. In this example ABpl losses were never seen, indicating that this is the lethal focus for *let-x*.

and 3) are sterile, suggesting a requirement for *let-x* gene activity for development or viability of the germ cells. Until now, the early lethality of the *let-x* mutation had precluded knowledge of this later requirement. We can infer that *let-x* activity is required within P_4 (the germline progenitor cell), because mosaic animals lacking the array in D (mosaic class 4) are fertile. In subsequent experiments, we could study the sterile *let-x* mosaics in detail in order to learn more about the role of the gene in the germline.

2. Materials

1. Purified cloned DNA(s) for microinjection *(14)* (*see* **Table 3**).
2. Anesthetics: levamisole and sodium azide are available from Sigma Chemical Co. (St. Louis, MO).
3. Stereoscopic, dissecting microscopes: Several companies, including Leica (Leica Microsystems Inc., Deerfield, IL), Olympus (America Inc., Melville, NY), and Carl Zeiss (Jena, Germany), provide models that have been equipped with a Hg vapor lamp, a filter set for viewing GFP, and a base with transmitted light for normal observations. Those that can zoom to ×100 are the most useful, although analyses can be performed with those that only magnify ×40. Because worms will be picked during mosaic analyses and when using GFP for the establishment of transgenic lines, an important consideration is the space between the Petri plate and the objective. Keep in mind that greater working distances can be achieved with an objective of lower power. Initial observations of SUR-5GFP(NLS) were made with a Leica model MZ12 with an KSC 190-807 filter set (excitation 470–490 nm; emission ≥500 nm) and a 1.0 objective (part no. 445819). A more recent Leica instrument, model MFLIII, is reported to have a significant increase in illumination. In combination with optimized GFP filters, this can result in more-sensitive detection of GFP.
4. Media: Growth plates contained NGM *(20)* with either Sigma (A7002) or Difco Bacto agar (Becton-Dickinson, Sparks, MD); autofluorescence of these agars did not interfere with SUR-5GFP(NLS) analyses.
5. Inverted compound microscope and micromanipulator for microinjection of worms *(14,15)*.
6. Upright compound microscope equipped for Nomarski and fluorescence observations. A ×100 plan apo oil objective is required for high-resolution cellular analysis of genetic mosaics.

3. Methods

1. Marker DNA: SUR-5GFP(NLS) is encoded by pTG96 *(17)*. Inject pTG96 (at 100 μg/mL or higher) into the syncytial gonad *(14)* together with cloned DNA that rescues the relevant chromosomal mutation (a lineage marker may also be useful). Alternatively, inject cloned DNAs encoding *ncl-1*(+) *(16)*, other lineage markers if desired (**Table 3**), and the gene that rescues the relevant chromosomal mutation (and perhaps cloned DNA that serves to establish transgenic lines if a lineage marker is not used that also serves this purpose). Generation of strains for mosaic analysis will be very specific (for a discussion, *see* **ref. 14**). It will depend on the phenotype and genetic linkage of the mutation being studied, the relative ease in generating a strain having all of the desired markers, and the material and equipment that are available.
2. With a dissecting microscope equipped with a Hg lamp, examine normal growth plates containing the injected worms for green fluorescence of embryos, indicating successful formation of extrachromosomal arrays. Green eggs are often evident within 24 h after injecting. Alternatively, if this type of microscope is not available, examine growth plates for a behavioral or morphological marker to identify transgenic F_1 animals.
3. Pick transgenic F_1s as described above and establish stable transgenic lines by identifying F_1s that segregate transgenic progeny (e.g., **Fig. 3**). Arrays with germline transmission frequencies of 40–66% have been successfully used for mosaic analysis *(13,16)*. Use of green fluorescence itself as the basis of screening for transgenic F_1 and F_2 progeny can eliminate the need for other markers *(17)*.

4. Identify mosaic animals with early losses by screening growth plates containing hundreds of worms. Use SUR-5GFP(NLS) and a dissecting microscope to identify worms having mosaic patterns of GFP fluorescence *(17)*. For example, if P_1, EMS, or E can be a point of loss, you will find rare semifluorescent worms in which all of the intestinal nuclei are nonfluorescent. Alternatively, use lineage markers (described in **Table 3**) and a standard dissecting microscope to identify mosaic worms based on their phenotype. Once you find mosaic animals, you will determine the specific division at which the extrachromosomal DNA was lost by using a compound microscope to examine individual nuclei for GFP expression, for the Ncl phenotype, or for both. Suitable extrachromosomal arrays will have mitotic loss rates of 0.1×10^{-3} to 5×10^{-3} losses per cell division *(4,13,16)*, meaning that only 0.8–4% of array-bearing zygotes will experience array loss during one of the eight embryonic cell divisions shown in **Fig. 4**. If only using *ncl-1*, proceed to **step 8**; **steps 5–7** relate to specific examinations of SUR-5GFP(NLS) with a dissecting microscope. Also, **steps 8–10** will be presented from the point of view of SUR-5GFP(NLS) as the marker.

5. Examine the mosaics lacking fluorescence in all of their intestinal nuclei for whether they have had losses in P_1, in EMS, or in E. Because the uterus and much of the anterior bulb of the pharynx derive from MS, this is straightforward if E is the point of loss, as fluorescence of the nuclei of these cells can be easily detected with the dissecting microscope. Distinguishing candidate losses in P_1 from those in EMS is based on the body muscles; in P_1 losses, all but one of the body muscles will not fluoresce (the exceptional body muscle descends from AB and is located in the right ventral part of the midbody near the vulva of hermaphrodites).

6. Search for candidate mosaics with losses in AB. A general presence or absence of fluorescence of the neurons of the ventral cord and of the ring and lumbar ganglia is often apparent at higher magnifications, and the nuclei of hyp8, hyp9, and the excretory cell can be easily scored even at lower power.

7. Other early losses may be evident with the dissecting microscope. Candidate losses in MS but not in EMS can be detected by examining the uterus and the terminal bulb of the pharynx, and candidate losses in C or D can be detected by examining those parts of body that contain only body muscles from these parts of the cell lineage.

8. Mount mosaic candidates for examination with the compound microscope. Prepare a slide having a flat pad of 5% Difco Noble Agar (Becton-Dickinson) as described in *(2)*. A candidate can be transferred in one step into 3–5 μL of M9 buffer on the surface of the pad and the drop then covered with a glass cover slip. Examinations may be more convenient if the worms have been immobilized by means of an agent added to the drop of M9 buffer. Two possibilities are 5–10 m*M* sodium azide and 1 m*M* levamisole *(21)*.

9. Candidates are examined at ×1000 with the compound microscope for the true nature of their losses. A record of the positive and negative nuclei of each candidate can be made. Some of the cells in which SUR-5GFP(NLS) is expressed (and their origin in the cell lineage) are as follows: CANL, RID, CANR, ALA, and RMED (from ABalap); m3L and m3VL (from ABalpa); ASKL and ADLL (from ABalpp); m3R and m3VR (from ABarap); ALML, BDUL, ALMR, and BDUR (from ABarpp); ASIL (from ABplaa); PLML, ALNL, HSNL, and the vulva (or its precursor cells) (from ABplap); excretory gland L, excretory cell, and rectal epithelium D (from ABplpa); hyp8/9, rectal epithelium VL, and anal depressor muscle (from ABplpp); ASKR, ADLR, and ASIR (from ABpraa); PLMR, ALNR, HSNR, and the vulva (or its precursor cells) (from ABprap); excretory gland R (from ABprpa); hyp8/9, rectal epithelium VR, DVA, and body muscle (from ABprpp); m7D, m8, and body muscles (from MSaa); left coelomocytes, posterior distal tip cell, dorsal coelomocytes, and those body muscles not from M (from MSap); m6D and body

muscles (from MSpa); right coelomocytes, anterior distal tip cell, and body muscles (from MSpp); all intestinal cells (from E); DVC and body muscles (from Ca); hyp11 and body muscles (from Cp); body muscles from D. With the exception of the intestinal cells, these cells and others can be scored for the Ncl phenotype *(4)*.

4. Notes

1. High concentrations of pTG96 are necessary, and it may be advantageous to pick the most intensely fluorescing worms for the establishment of transgenic lines.

2. If a dissecting microscope with a Hg lamp is not available, plates can be examined with a ×5 or ×10 objective of the compound microscope. Although the image will be reversed, it is also possible to pick worms. The striking rolling behavior of worms transgenic for the common marker pRF4 is often used to establish transgenic lines *(15)*. A twisted body, however, can make the identification of cells difficult during the mosaic analysis. Although there are methods for suppressing the abnormal shape, it is often best to use another marker. One of several possibilities is *unc-36* (**Table 3**); it can be used to establish transgenic lines and to mark part of the cell lineage.

3. Because first-generation transgenic animals are often very mosaic, it may be more expedient for the establishment of stable lines if F_1s are chosen in which cells that derive from P_1 fluoresce; P_4, the precursor of the germline, is the great-granddaughter of P_1 *(2)*.

4. For P_1, EMS, or E losses, it is possible to examine several hundred worms on one plate in less than 15 min. Simultaneously illuminating the plates with a low level of transmitted light can also be useful.

5. The nucleus of the excretory cell is one of the brightest, but this cell (ABplpappaap; *see* **ref.** *2*) is quite removed from AB, and there are therefore many chances for an array to be lost before this cell is born. Nevertheless, this cell can be one of the first examined to identify losses from this side of the cell lineage.

6. Once you gain experience in determining mosaicism with a compound microscope, it can become more efficient to have more than one candidate on a slide. Because the drop can evaporate while the search continues for additional candidates, you can pick candidates as you find them to a normal growth plate; the collection can then be transferred later to a pad. A more-refined approach during the initial observation is to segregate candidates to different plates based on their presumed losses. It is often necessary to determine empirically an optimal concentration of the anesthetics such that paralysis is achieved without detrimental effects. Sodium azide, normally one of the best agents for immobilizing worms, can rapidly destroy the fluorescence of GFP if the initial signal is weak. Although SUR-5GFP(NLS) can be examined in the presence of sodium azide, investigators not yet adept at scoring may prefer to use levamisole. The ideal situation is to avoid anesthetics because of their effects on viability and morphology. This is of further advantage when one hopes to rescue worms from the slides in order to ascertain the effects of mosaicism on subsequent growth, development, or fertility.

7. For those familiar with the anatomy as revealed by Nomarski optics, it can be useful in the beginning to switch back and forth between the Nomarski and fluorescence images until one is comfortable scoring the various nuclei for GFP. It is also possible to simultaneously observe the fluorescent and Nomarski images of certain nuclei. For those unfamiliar with the anatomy, a study of both images in certain types of mosaics can be useful for learning the positions of nuclei. For example, with the exception of the left and right CAN cells, most of the nuclei that descend from ABa are located quite anterior. In a mosaic in which only ABa inherited SUR-5GFP(NLS), the CAN cell nuclei will therefore be the posterior-most nuclei that fluoresce. Learning the Nomarski image of these nuclei is then simply a

matter of switching the image. Indeed, the initial limitation of any mosaic analysis is that it requires a detailed knowledge of *C. elegans* anatomy, which even in its relative simplicity, requires study. Other issues that must be considered when undertaking mosaic analysis as described here include issues of perdurance, possible effects of high-copy-number arrays, the documented variable expression of genes present on arrays, excessive instability of some arrays, and the limitations of the cell lineage and possible multiple foci. All of these issues are discussed thoroughly in *(3,14)*.

Acknowledgments

We thank Simon Tuck for comments on the manuscript, Bob Herman for providing figures for reproduction, and Mike Stiffler and Susien Kim for assistance in preparation of the figures. J. Y. is grateful to Simon Tuck for support and to Min Han and Trent Gu for making mosaic analysis with SUR-5GFP(NLS) possible.

References

1. Sulston, J. E. and Horvitz, H. R. (1977) Post-embryonic cell lineages of the nematode, *Caenorhabditis elegans. Dev. Biol.* **56,** 110–156.
2. Sulston, J. E., Schierenberg, E., White, J. G., and Thomson, J. N. (1983) The embryonic cell lineage of the nematode *Caenorhabditis elegans. Dev. Biol.* **100,** 64–119.
3. Herman, R. (1995) Mosaic analysis in *Caenorhabditis elegans*: Modern biological analysis of an organism, in *Methods in Cell Biology, vol. 48* (Epstein, H. F. and Shakes, D. C., eds.), Academic, San Diego, CA, pp. 123–146.
4. Hedgecock, E. M. and Herman, R. K. (1995) The *ncl-1* gene and genetic mosaics of *Caenorhabditis elegans. Genetics* **141,** 989–1006.
5. Herman, R. K. (1984) Analysis of genetic mosaics of the nematode *Caenorhabditis elegans. Genetics* **108,** 165–180.
6. Herman, R. K. (1989) Mosaic analysis in the nematode *Caenorhabditis elegans. J. Neurogenet.* **5,** 1–24.
7. Han, M. and Sternberg, P. W. (1990) *let-60*, a gene that specifies cell fates during *C. elegans* vulval induction, encodes a *ras* protein. *Cell* **63,** 921–931.
8. Han, M., Aroian, R. V., and Sternberg, P. W. (1990) The *let-60* locus controls the switch between vulval and nonvulval cell fates in *Caenorhabditis elegans. Genetics* **126,** 899–913.
9. Yochem, J., Sundaram, M., and Han, M. (1997) Ras is required for a limited number of cell fates and not for general proliferation in *Caenorhabditis elegans. Mol. Cell. Biol.* **17,** 2716–2722.
10. Seydoux, G. and Greenwald, I. (1989) Cell autonomy of *lin-12* function in a cell fate decision in *C. elegans. Cell* **57,** 1237–1245.
11. Albertson, D. G. and Thomson, J. N. (1982) The kinetochores of *Caenorhabditis elegans. Chromosoma* **86,** 409–428.
12. Herman, R. K., Madl, J. E., and Kari, C. K. (1979) Duplications in *Caenorhabditis elegans. Genetics* **92,** 419–435.
13. Lackner, M. R., Kornfeld, K., Miller, L. M., Horvitz, H. R., and Kim, S. K. (1994) A MAP kinase homolog, *mpk-1*, is involved in *ras*-mediated induction of vulval cell fates in *Caenorhabditis elegans. Genes Dev.* **8,** 160–173.
14. Mello, C. and Fire, A. (1995) DNA transformation, in *Methods in Cell Biology, vol. 48* (Epstein, H. F. and Shakes, D. C., eds.), Academic, San Diego, CA, pp. 451–482.
15. Mello, C. C., Kramer, J. M., Stinchcomb, D., and Ambros, V. (1991) Efficient gene transfer in *C. elegans*: Extrachromosomal maintenance and integration of transforming sequences. *EMBO J.* **10,** 3959–3970.

16. Miller, L. M., Waring, D. A., and Kim, S. K. (1996) Mosaic analysis using a *ncl-1* (+) extrachromosomal array reveals that *lin-31* acts in the Pn.p cells during *Caenorhabditis elegans* vulval development. *Genetics* **143**, 1181–1191.

17. Yochem, J., Gu, T., and Han, M. (1998) A new marker for mosaic analysis in *Caenorhabditis elegans* indicates a fusion between hyp6 and hyp7, two major components of the hypodermis. *Genetics* **149**, 1323–1334.

18. Chalfie, M., Tu, Y., Euskirchen, G., Ward, W. W., and Prasher, D. C. (1994) Green fluorescent protein as a marker for gene expression. *Science* **263**, 802–805.

19. Gu, T., Orita, S., and Han, M. (1998) *Caenorhabditis elegans* SUR-5, a novel but conserved protein, negatively regulates LET-60 Ras activity during vulval induction. *Mol. Cell Biol.* **18**, 4556–4564.

20. Brenner, S. (1974) The genetics of *Caenorhabditis elegans*. *Genetics* **77,** 71–94.

21. Burdine, R. D., Branda, C. S., and Stern, M. J. (1998) EGL-17(FGF) expression coordinates the attraction of the migrating sex myoblasts with vulval induction in *C. elegans*. *Development* **125**, 1083–1093.

22. Hodgkin, J. (1997) Genetics, in *C. elegans II* (Riddle, D. L., Blumenthal, T., Meyer, B. J., and Priess, J. R., eds.), Cold Spring Harbor Laboratory, Plainview, NY, pp. 881–1047.

23. Herman, R. K. and Kari, C. K. (1985) Muscle-specific expression of a gene affecting actylcholinesterase in the nematode *Caenorhabditis elegans*. *Cell* **40,** 509–514.

24. Johnson, C. D., Rand, J. B., Herman, R. K., Stern, B. D., and Russell, R. L. (1988) The acetylcholinesterase genes of *C. elegans*: Identification of a third gene (*ace-3*) and mosaic mapping of a synthetic lethal phenotype. *Neuron* **1,** 165–173.

25. Yuan, J. Y. and Horvitz, H. R. (1990) The *Caenorhabditis elegans* genes *ced-3* and *ced-4* act cell autonomously to cause programmed cell death. *Dev. Biol.* **138,** 33–41.

26. Villeneuve, A. M. and Meyer, B. J. (1990) The role of *sdc-1* in the sex determination and dosage compensation decisions in *Caenorhabditis elegans*. *Genetics* **124,** 91–114.

27. Kenyon, C. (1986) A gene involved in the development of the posterior body region of *C. elegans*. *Cell* **46,** 477–487.

28. Hunter, C. P. and Wood, W. B. (1990) The *tra-1* gene determines sexual phenotype cell-autonomously in *C. elegans*. *Cell* **63,** 1193–1204.

29. Iwasaki, K., McCarter, J., Francis, R., and Schedl, T. (1996) *emo-1*, a *Caenorhabditis elegans* Sec61p gamma homologue, is required for oocyte development and ovulation. *J. Cell Biol.* **134,** 699–714.

30. Austin, J. and Kimble, J. (1987) *glp-1* is required in the germ line for regulation of the decision between mitosis and meiosis in *C. elegans*. *Cell* **51,** 589–599.

31. Koga, M. and Ohshima, Y. (1995) Mosaic analysis of the *let-23* gene function in vulval induction of *Caenorhabditis elegans*. *Development* **121,** 2655–2666.

32. Simske, J. S. and Kim, S. K. (1995). Sequential signalling during *Caenorhabditis elegans* vulval induction. *Nature* **375,** 142–146.

33. Chow, K. L., Hall, D. H., and Emmons, S. W. (1995) The *mab-21* gene of *Caenorhabditis elegans* encodes a novel protein required for choice of alternate cell fates. *Development* **121,** 3615–3626.

34. Herman, R. K. (1987) Mosaic analysis of two genes that affect nervous system structure in *Caenorhabditis elegans*. *Genetics* **116,** 377–388.

35. Jia, Y., Xie, G., McDermott, J. B., and Aamodt, E. (1997) The *C. elegans* gene *pag-3* is homologous to the zinc finger proto-oncogene *gfi-1*. *Development* **124,** 2063–2073.

36. Starich, T. A., Herman, R. K., and Shaw, J. E. (1993) Molecular and genetic analysis of *unc-7*, a *Caenorhabditis elegans* gene required for coordinated locomotion. *Genetics* **133,** 527–541.

VII

Experimental Manipulation of Embryos

43

Local Application of Bone Morphogenic Protein on Developing Chick Embryos Using Fibrous Glass Matrix as a Carrier

Akira Nifuji and Masaki Noda

1. Introduction

Proper cell–cell interactions are essential for controlling embryonic tissue development. The communications from cell to cell are transmitted through direct cell contact via adhesion molecules or short- and/or long-range signaling by diffusible molecule. The molecular basis for these events is the subject of interest in developmental biology. One of the approaches to elucidate the process of cell–cell signaling by diffusible molecules is the local application of carriers containing the molecule by means of microsurgical operation. One clear example is the study on retinoic acid signaling, which has been shown to play an important role during limb development *(1,2)*. When retinoic acid is locally applied to the anterior margin of a limb bud, duplication of limb is observed, showing that signaling by retinoic acid regulates A-P pattern formation of the limb bud. Another example is that local implantation with recombinant protein into the embryonic tissue exhibited alteration of skeletal patterning. We showed that implantation of carrier matrix, fibrous glass matrix (FGM), containing BMP protein, into early somitogenesis, perturbed axial skeletal formation, suggesting involvement of BMP signaling in axial skeletogenesis *(3,4)*.

Here we describe the method of local application of diffusible molecule in order to investigate effects of the molecule at the specific site of embryos. We use the chick embryos as model animal because they are easily manipulated for the local implantation of the BMP-containing carriers.

2. Materials

1. Buffer: Phosphate-buffered saline (PBS), pH 7.4.
2. India ink: Drawing ink A (Pelikan, Hannover, Germany).
3. FGM (Advantec, Tokyo, Japan).
4. Other instruments: Micropipets are prepared by pulling the thin end of a Pasteur pipet over a small Bunsen burner. Micropipets are connected to a mouthpiece by silicone rubber hosing. Microscalpels are manufactured by sharpening sewing steel needles on an Arkansas oil stone. Microscalpels are useful to cut or dissociate the tissues and to precisely place

From: *Methods in Molecular Biology, Vol. 135: Developmental Biology Protocols, Vol. I*
Edited by: R. S. Tuan and C. W. Lo © Humana Press Inc., Totowa, NJ

carriers containing diffusible molecules into tissues. Dental file holders are convenient to hold microscalpels.

3. Methods

3.1. Preparation of Chick Embryos

Chicken eggs are incubated with their long axis in the horizontal plane at 37–39°C for the desired number of days. The egg shells are wiped with 70% ethanol. A small pointed hole is opened at the end of an egg using the tip of the scissors. Then 2 mL albumin is sucked out through the hole to make an air chamber over the embryos and vitelline membrane. The hole is closed with a small piece of tape. A window of egg shell is opened with a diameter of 3 cm on the upper side of the egg by using curved scissors. Then, a small amount of India ink diluted with an equal volume of PBS is injected under the embryos using a micropipet. Embryos are visualized clearly against a black background and embryonic stages can be now determined according to the staging of Hamburger and Hamilton (*5*).

3.2. Preparation of Microcarrier
Containing the Diffusible Molecules (see Notes)

There are several materials that can be used for microcarriers to deliver bioactive diffusible molecules. Those include anion exchange beads (e.g., AG1-X2 beads), Affi-Gel Blue beads (Bio-Rad Laboratories, Hercules, CA), and FGM (*6–8*). For the choice of carriers, the nature of matrix has to be considered (*see* **Notes**). FGM has been shown to be a more effective delivery substance for BMP than the guanidine HCl-insoluble bone matrix (*7*) in in vivo bone induction assay. Therefore, for the experiment of application of BMP into chick embryos, FGM is used as a carrier matrix.

FGM is made of unwoven glass fibrils composed chiefly of silica. Each fibril has a diameter of <10 µm. This matrix itself does not exert any effect on embryonic development.

Recombinant human BMP4 or BMP2 is diluted in PBS with 0.1% bovine serum albumin (BSA) (*9*). Small pieces, approx 500-µm diameter, of FGM are soaked in 0.5 µL of the solution supplemented with BMP in a small plastic dish on ice. Keep the plastic dish containing FGM on ice for at least 30 min to adsorb BMP into FGM. This volume is sufficient for more than 100 implantation experiments. Therefore, a soaking solution of 1 ng/µL BMP is equivalent to 0.5 ng adsorbed in a piece of FGM, 5 pg BMP per one implantation experiment, assuming 100% adsorption occurs. A chip of the matrix in a volume corresponding to 40–80 µm of one somite size is picked up with No. 5 forceps under a stereomicroscope.

3.3. Implantation of Microcarrier (Fig. 1)

The vitelline membrane is cut with a microscalpel. A longitudinal incision is made into the dorsal ectoderm and somite. A small chip of FGM soaked in PBS plus 0.1% BSA or the solution containing appropriate growth factors is placed into the slit of incision using forceps. Then the matrix can be pushed down into the tissue with a microscalpel. The hole of the egg shell is sealed with transparent tape after implantation, and the eggs are returned to the incubator for reincubation at 37–39°C.

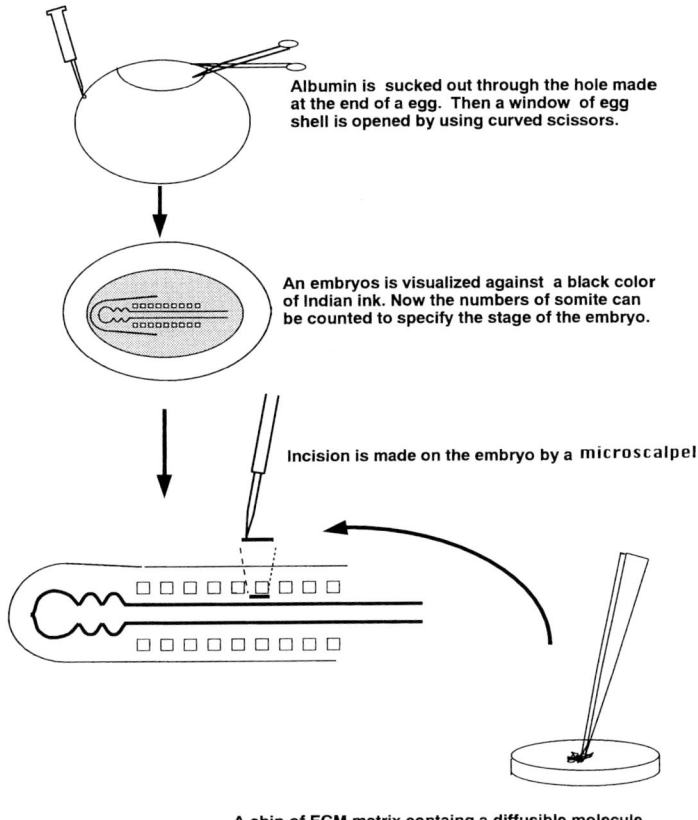

Albumin is sucked out through the hole made at the end of a egg. Then a window of egg shell is opened by using curved scissors.

An embryos is visualized against a black color of Indian ink. Now the numbers of somite can be counted to specify the stage of the embryo.

Incision is made on the embryo by a microscalpel.

A chip of FGM matrix containg a diffusible molecule is picked up. Then transfer it into the slit of incision.

Fig. 1. Outline of FGM implantation procedure.

3.4. Whole-Mount Skeletal Preparations

Skeletons of control and operated embryos are stained *in toto* on E9 or E10. The embryos are removed from the shells and washed in PBS and eviscerated. After washing in PBS, embryos are fixed in 80% ethanol, 20% acetic acid with 0.02% alcian blue overnight, dehydrated in 100% ethanol for 5 d. Embryos are then cleared in 0.5% KOH until the skeletal structure is visible and stained in 0.01% alizarin red in 0.5% KOH for 3–5 h. Then the specimens are cleared through a graded series of KOH and glycerin solutions (10% glycerin in 1% KOH, 20, 50, and 80% glycerin).

Implantation procedures are outlined in **Fig. 1**. Implantation of FGM, which contains BMP, into E2 embryos, results in skeletal anomalies of axial skeletons. One week later, additional processes such as structures extending from thoracic vertebrae and malformed ribs extending distally or dorsally are observed (*arrow*, **Fig. 2**).

3.5. Whole-Mount In Situ *Hybridization Analysis*

To examine the effects of implantation on endogenous gene expression near the site of implantation, whole-mount *in situ* hybridization analysis is performed before embryonic day 5. For later embryos, *in situ* hybridization analysis on tissue sections is

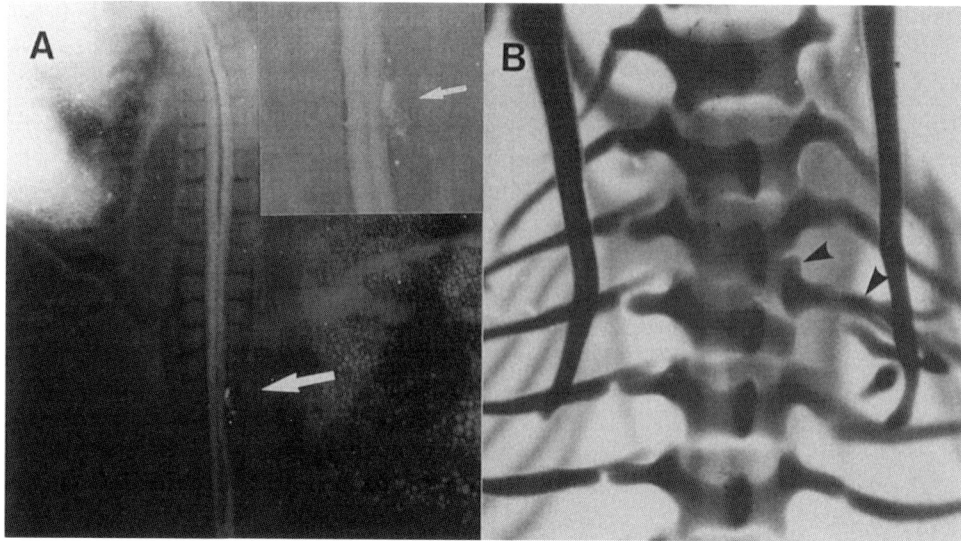

Fig. 2. (*See* color plate 7 appearing after p. 258.) Implantation of FGM containing BMP2 protein into E2 chick embryos perturbed rib and vertebral skeletal formation. (**A**) Dorsal view of the E2 embryos. FGM was implanted into the medial part of newly formed somite (*arrow*). Inset shows a higher magnification view. (**B**) One week later, additional process-like structures and malformed ribs were observed (*arrowhead*).

preferable (*see* protocols for *in situ* hybridization in this volume). In our previous study, endogeneous gene expression of BMP4 and Msx2 was up-regulated at the site of implantation on E4, indicating alteration of cell behaviors prior to overt formation of skeletal tissues (data not shown; *see* **ref. 4**).

4. Notes

Several materials have been reported to deliver biological active compounds. For retinoic acids, positively charged ion-exchange resins were reported to be most suitable (*6*). Affi-Gel-Blue agarose beads were used for FGF, TGF-beta, and BMP implantation (*7,10*). Heparin acrylic beads were also shown to be effective to deliver FGF2 and FGF8 (*11*). In our hands, Affi-Gel Blue beads are also effective to deliver these growth factors. However, FGM matrix is easier to handle than Affi-Gel Blue beads when used for this kind of embryonic manipulation, because this matrix can be separated into fine small pieces.

It was reported that in adult mouse, the efficiency of bone induction observed by BMP depends on the nature of the carrier matrix (*12*). The most common and efficient material to deliver BMP protein is bone residue after extraction by guanidine-HCl. Probably this is because BMPs are trapped at high concentration in the extracellular matrix of bone in the physiological condition (*12*). Kuboki et al. reported that FGM may be a more suitable carrier for BMPs as a stable delivery system than is insoluble bone matrix residue (*8*). The prerequisite nature of carriers as releasing biological materials are (a) stable adsorption of biological substances, (b) gradual and prolonged release of the substances, (c) easy handling, and (d) no toxicity for embryonic development. Hayek et al. *(7)* reported that FGF bound to agarose beads when slowly released over the 3 d after adsorption. Extensive studies on the release of retinoic acids from ion-exchange beads have also been reported. In these studies, it was shown that the release rate of the molecules depended on the nature of the beads. Although the nature of FGM has not yet been extensively analyzed, implantation of FGM containing BMP resulted in the induction of a higher amount of cartilage tissue than insoluble bone matrix-BMP, suggesting that FGM may be one of the best materials to apply BMP in vivo *(8)*. Our data also showed that a very low concentration of BMP was effective to cause perturbation of rib and vertebral formation in chick embryos *(3)*.

Because BMPs have been shown to have diverse effects on cellular differentiation and embryonic development, this local application of FGM containing BMPs into embryos may offer another choice in the study of their roles in development.

References

1. Tickle, C. (1991) Retinoic acid and chick limb bud development. *Development* **1(Suppl)**, 113–121.
2. Tabin, C. (1991) Retinoids, homeobox, and growth factors: Toward molecular models for limb development. *Cell* **66**, 199–217.
3. Nifuji, A., Kellermann, O., Kuboki, Y., Wozney, J. M., and Noda, M. (1997) Perturbation of BMP signaling in somitogenesis resulted in vertebral and rib malformations in the axial skeletal formation. *J. Bone Min. Res.* **12**, 332–342.
4. Takahashi, Y., Tonegawa, A., Matsumoto, K., Ueno, N., Kuroiwa, A., Noda, M., and Nifuji, A. (1996) BMP-4 mediates interacting signals between the neural tubes and skin along the dorsal midline. *Genes cells* **1**, 775–783.
5. Hamburger V. and Hamilton, H. (1951) A series of normal stages in the development of the chick embryo. *J. Morphol.* **88**, 49–92.
6. Eichele, G., Tickle, C., and Alberts, B. M. (1984) Microcontrolled release of biologically active compounds in chick embryos: Beads of 200-mm diameter for the local release of retinoids. *Anal. Biochem.* **142**, 542–555.
7. Hayek, A., Culler, F. L., Beattie, G. M., Lopez, A. D., Cuevas, P., and Baird, A. (1987) An in vivo model for study of the angiogenic effects of basic fibroblast factor. *Biochem. Biophys. Res. Commun.* **147**, 876–880.
8. Kuboki, Y., Saito, T., Murata, M., Takita, H., Mizuno, M., Inoue, T., Nagai, N., and Poole, A. R. (1995) Two distinctive BMP-carriers induce zonal chondrogenesis and membranous ossification, respectively: Geometrical factors of matrices for cell-differentiation. *Connect. Tissue Res.* **32**, 219–226.
9. Wozney, J. M., Rosen, V., Celeste, A. J., Mitsock, L. M., Whitters, M. J., Kitz, R. W., Hewick, R. M., and Wang, E. A. (1988) Novel regulators of bone formation: molecular clones and activities. *Science* **242**, 1528–1534.
10. Hayamizu, T. F., Sessions, S. K., Wanek, N., and Bryant, S. V. (1991) Effects of localized application of transforming growth factor b1 on developing chick limbs. *Dev. Biol.* **145**, 164–173.
11. Vogel, A., Rodriguez, C., and Izpisua-Belmonte, J. C. (1996) Involvement of FGF-8 in initiation, outgrowth and patterning of the vertebrate limb. *Development* **122**, 1737–1750.
12. Reddi, A. H., Wientroub, M. D., and Muthukumaran, N. (1987) Biologic principles of bone induction. *Orthop. Clin. North Am.* **18**, 207–212.

44

Laser Ablation and Fate Mapping

Margaret L. Kirby, Donna Kumiski, Harriett Stadt, and Greg Hunter

1. Introduction

Much of the information about the fate and function of embryonic cell populations was gained in the early years of experimental embryology through ablation studies. With the advent of more sophisticated marking techniques using interspecific chimeras and molecular markers, ablations were generally abandoned because of the problems of misinterpretation of data when unforeseen reconstitution of cell populations occurred following ablation. This is reminiscent of the problems with interpretation of the first generation of gene knockouts where other genes in a family surreptitiously reconstitute the function of the disabled gene. However, even with the problem of data interpretation, useful information can be obtained from ablation experiments when they are properly controlled and conservatively interpreted as has happened for the global null gene mutations. Our group has particularly benefited from ablation technology, because it has provided a clinically relevant model system to explore with a number of interesting approaches. Because of our group's need for a large number of embryos with particular ablations, we have established a largely automated system for production of uniform neural crest ablations in chick embryos. The system, which is relatively simple, consists of a stereomicroscope-mounted laser and motorized micropositioner, which are both controlled by a computer (*see* **Fig. 1**). Very precisely placed ablations can be done by relatively unskilled lab personnel with a minimum of actual hands-on time. In addition, it is possible to graft tissue or cultured cells into the ablation site for tracing or cell lineage studies (*see* **Note 1**).

2. Materials

1. Eggs.
2. Horizontal sander.
3. Neutral red slides made by heating 1% agar in distilled water to boiling. Allow the solution to cool and add 0.01% neutral red. Pipet onto 1" × 3" glass histological slides before the agar solidifies. Allow to dry and store at 4°C, desiccated until use.
4. Fine forceps.
5. Pulsed nitrogen/dye laser (VSL-337/DLM-110; Laser Science, Inc., Franklin, MA).
6. Wide-zoom stereomicroscope.

From: *Methods in Molecular Biology, Vol. 135: Developmental Biology Protocols, Vol. I*
Edited by: R. S. Tuan and C. W. Lo © Humana Press Inc., Totowa, NJ

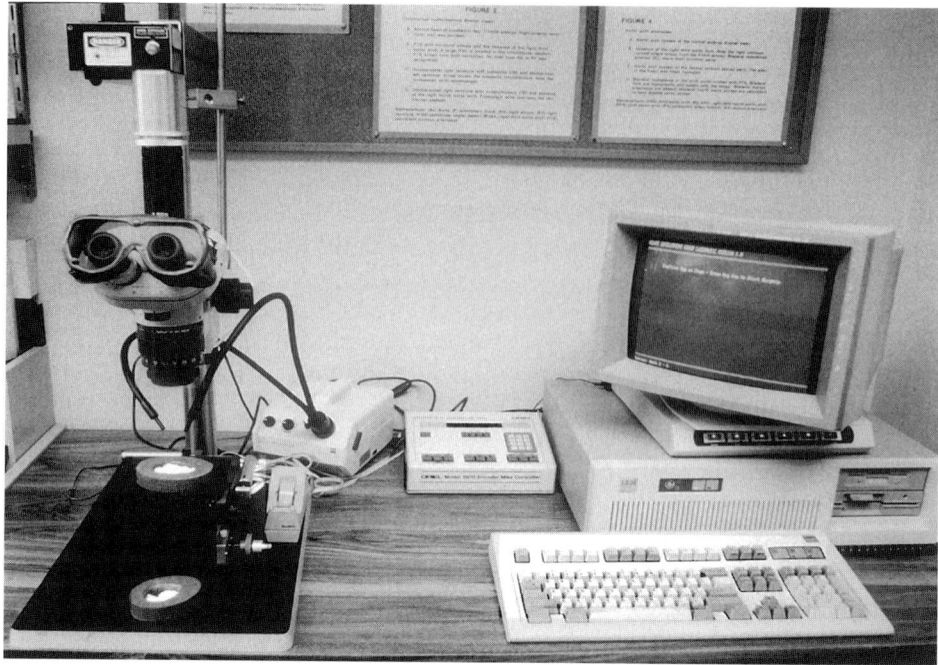

Fig. 1. The laser ablation system consists of an Olympus stereozoom microscope (left) with the laser mounted horizontally through the phototube. Focusing is accomplished by controlling the height (*z*-axis) of the micropositioner on which the egg is placed. The *x*- and *y*-axes of the micropositioner as well as the laser shutter are controlled by an IBM-PC (right).

7. Motorized micropositioner with encoder mike controller (Model 18011; Oriel Instruments, Stratford, CT),
8. IBM PC.
9. Egg holder.

3. Method

1. Chicken eggs are incubated on their sides at 37°C and 90% humidity for approx 30 h. One centimeter oval windows are sanded in the side of the egg. The window is swabbed with 70% ethanol and the shell membrane carefully removed. The embryos are stained with neutral red and appropriate stages chosen for surgery (*1*). The egg containing the embryo to be lesioned is supported by an egg carrier placed on the motorized micropositioner, and the laser is focused at the level of the neural fold. Stage coordinates are preprogrammed and controlled by the IBM PC operating through an encoder mike controller (Oriel). The micropositioner is set at a velocity of 30 µm/s and the tissue is irradiated with a laser beam at a power density of 10^9 W/cm². With the pulse rate of the laser at 20 Hz, this translates to a cumulative power density of 13×10 W/cm². After the program has run, a visual check is used to confirm that all remnants of the neural fold are ablated. The system can be activated manually for final touchup.
2. In our experiments, neural folds are collected by extirpation of the neural folds using electrolytically sharpened tungsten needles. The folds are centrifuged at 2500 rpm (487.5*g*) for 3 min, and then washed and spun 2X in neural crest medium with 1000 U/mL Esgro™ (Gibco-BRL, Gaithersburg, MD). Neural crest medium is made with 15% heat-inactivated horse serum, 10% chick embryo extract, and 0.00005% fungizone in

Dulbecco's Modified Eagle Medium. Esgro is a recombinant form of leukemia inhibitory factor. It is the primary factor in fibroblast feeder layers that inhibits embryonic stem cell differentiation in cell culture. The neural folds are plated in 96-well plates with 50–100 μL of neural crest medium containing Esgro and incubated at 37°C in 5% CO_2 for 3 d. By 24 h after plating, the neural crest cells have migrated away from the neural fold, which can then be removed to enhance the enrichment of the neural crest population.

3. We have used DiI or Mitotracker, fluorescent, vital dyes for tracing cells (Molecular Probes, Eugene, OR). On the day the cells are to be transferred to a host embryo, a working solution of DiI is made by adding 16 μL of a stock solution of DiI (2.5 mg/mL 100% ethanol), to each milliliter of neural crest medium. This solution is filtered through a 0.2-μm syringe filter, and 1 mL is placed over the cultured cells and incubated for 1 h at 37°C (*2*). The DiI-containing medium is removed and 1 mL of trypsin-EDTA added to the cells. After 1 min of incubation, 1.0 mL of neural crest medium is added to each culture dish to dilute the trypsin, and the cells are washed and spun 3X (2500 rpm [487.5*g*] for 3 min) to remove excess unbound dye.

4. The cells are drawn into a capillary micropipet, which is inserted through a small hole made in the vitelline membrane as far laterally and away from the embryo as possible, but in the blastodisc. Taking care not to tear the vitelline membrane any further, the tip of the micropipet is placed over the ablation site, and the cells are gently ejected using positive pressure. Because the vitelline membrane is not damaged during the ablation, the grafted cells or tissue are held in place by slight tension generated under it. The eggs are resealed with cellophane tape and incubated at 37°C in 90% humidity for 24–72 h.

5. The embryos can be fixed or dissected for biochemical or molecular biological techniques at differing ages depending on the system under study.

4. Notes

1. We have found that this is not the method of choice for making solid tissue grafts, because the laser-generated heat partially seals the wound site, making attachment of the graft problematic. Even so, cultured cells can be placed into the lesion site with moderately high success for migration of cells into the host embryo (*3* and unpublished observations). We have used neural crest cells cultured up to 3 d and found all of the neural crest-related cell lineages from the grafted cells in the host embryo. A number of different ablations can be programmed into the computer for studying different developmental events. The programming is done so that the embryo is positioned under a mark in the microscope eyepiece that is centered with respect to the laser beam. The shape of the lesion is determined by movement of the micropositioner from the initial position, and the depth and width is controlled by frequency of the laser shutter opening. We have described the use of this system in ablating neural folds, although other ablations are possible because the plane of the lesion is determined by the plane of focus through the microscope. Even though the initial expense of this equipment is moderately high, savings are realized very quickly, in that relatively unskilled operators can use it accurately and efficiently.

References

1. Hamburger, V. and Hamilton, H. L. (1951) A series of normal stages in the development of the chick embryo. *J. Morphol.* **88,** 49–92.
2. Honig, M. G. and Hume, R. I. (1986) Fluorescent carbocyanine dyes allow living neurons of identified origin to be studied in long-term cultures. *J. Cell Biol.* **103,** 171–187.
3. Kirby, M. L., et al. (1993) Back transplantation of chick cardiac neural crest cells cultured in LIF rescues heart development. *Dev. Dyn.* **198,** 296–311.

45

Photoablation of Cells Expressing β-Galactosidase

Sheila Nirenberg

1. Introduction

Cell ablation can be a powerful technique for studying development. By systematically ablating cells from a developing tissue, one can gain insight into how the cells interact with each other to form the mature structure. Here we describe a new method for performing ablations. This method can be applied to individual cells or whole classes of cells and works well both in vitro and in vivo.

The general procedure is to genetically engineer the cells of interest to express the gene for the enzyme β-galactosidase (β-gal). The β-gal-expressing cells are then labeled with a dye (fluorescein) by treating them with a fluorogenic β-gal substrate, fluorescein-di-β-D-galacto pyranoside (FDG) (Molecular Probes, Inc., Eugene, OR). Once labeled, the cells can be photoablated by illuminating the tissue in the presence of a sensitizing agent, 3-amino-9-ethyl-carbazole (AEC) (Sigma Chemical Co., St. Louis, MO). Light activation of the fluorescein in the presence of the sensitizing agent triggers a chemical reaction that both kills the cells and marks them with a colored product.

This method has been tested on several different β-gal-expressing cell types in the mouse retina and cerebral cortex both in vitro and in vivo and found to ablate >90% of the targeted cells with <2% nonspecific cell death (*1*). Two tests were used to assess cell death. First, cells were scored for the selective uptake of dyes excluded from live cells, ethidium homodimer, and trypan blue (**Fig. 1A–C**) (*2,3*). Second, the tissue containing the β-gal-expressing cells was examined several days after the ablation treatment to assess the loss of the cells (**Fig. 1D**).

This chapter outlines protocols for ablating cells in the mouse retina in vitro and in vivo and in the cerebral cortex in vivo. It also includes methods for examining the effectiveness of the treatment. These protocols should be applicable to most other tissues as well.

2. Materials

1. Saline solutions:
   a. Phosphate-buffered saline (PBS), pH 7.4: 8 g NaCl, 0.2 g KCl, 1.15 g Na_2HPO_4, 0.2 g KH_2PO_4 in 1 L distilled water.
   b. L-15 medium (Sigma Chemical Co., St. Louis, MO)
   c. Ringer's: 110 mM NaCl, 2.5 mM KCl, 1 mM $CaCl_2$, 1.6 mM $MgCl_2$, 10 mM D-glucose, buffered with 22 mM $NaHCO_3$, 5%CO_2/95%O_2, pH 7.4.

From: *Methods in Molecular Biology, Vol. 135: Developmental Biology Protocols, Vol. I*
Edited by: R. S. Tuan and C. W. Lo © Humana Press Inc., Totowa, NJ

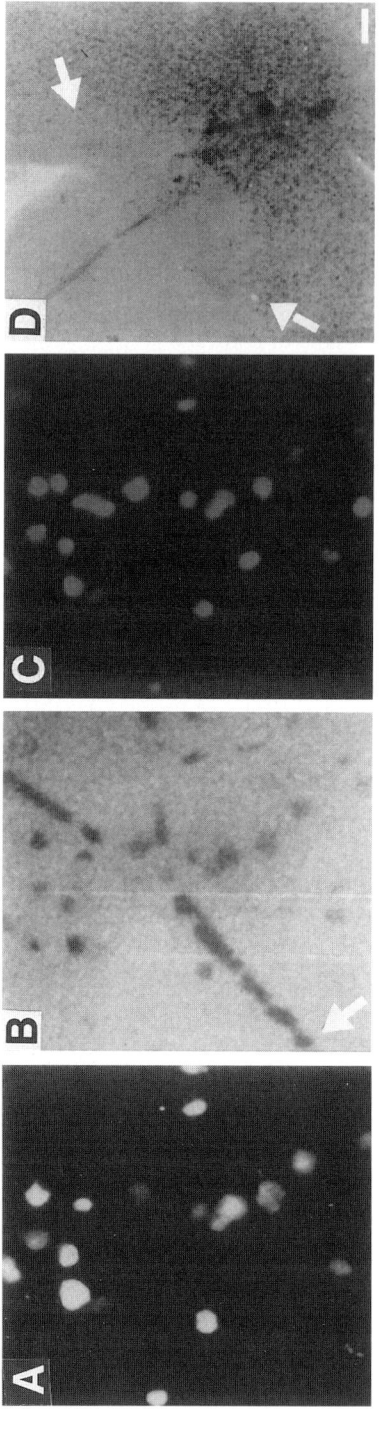

Fig. 1. (*See* color plate 12 appearing after p. 258.) Assays for cell ablation: ethidium homodimer assay and X-gal assay. (A)–(C) View of a retina whole mount from a transgenic mouse line that expressed β-gal in a population of retinal interneurons. (**A**) The retina after treatment with FDG/AEC solution. The FDG was cleaved in the β-gal-expressing cells, rendering them fluorescent. (**B**) The same retina after 5 min of illumination. The colored AEC product appeared in the fluorescein-labeled cells. *Arrow* indicates a blood vessel (to be distinguished from β-gal-expressing cells). (**C**) The same retina after treatment with ethidium homodimer. Approximately 4 h after ablation treatment, the AEC-labeled cells showed uptake of ethidium homodimer. (**D**) View of a retina whole mount stained with X-gal 1 wk after ablation treatment was applied to a region of the retina in vivo. *Arrows* indicate the borders of the region, which show a clear absence of X-gal-stained cells.

476

2. Fluorescein-di-β-D-galactopyranoside (FDG; Molecular Probes) stock solution: 130 mg/mL in dimethyl sulfoxide (DMSO). Store in 5–10-μL aliquots at –20°C.
3. 3-aminoethyl-9-ethylcarbazole (AEC; Sigma) stock solution: 20 mg/mL in DMSO. Store in 50-μL aliquots in foil-covered tubes at –20°C.
4. FDG/AEC solution #1: 1 part FDG stock, 1 part AEC stock, 15 parts DMSO, 3 parts L-15 or Ringer's, which gives a final concentration of 6.5 mg/mL FDG, 1 mg/mL AEC in 85% DMSO/15% L-15. Make fresh before use.
5. FDG/AEC solution #2: 1 part FDG stock and 1 part AEC stock, which gives a final concentration of 65 mg/mL FDG and 10 mg/mL AEC. Make fresh before use.
6. Ethidium homodimer (Molecular Probes): 30 μg/mL in PBS. Can be stored at –20°C for <1 yr.
7. Paraformaldehyde fixative: Add 4 g of paraformaldehyde (e.g., Sigma) to 80 mL distilled water with 50–100 μL of 10 *N* NaOH (used to help dissolve the paraformaldehyde). Heat to 60°C, then cool to room temperature, add 10 mL 10X PBS and bring final volume to 100 mL with distilled water. Bring pH to 7.4 with HCl. Can be stored at 4°C for approx 3 d.
8. Glutaraldehyde fixative: 0.5% in 1X PBS buffer. Dilute from a 25% stock (Sigma). Stock can be thawed and frozen many times. Make dilution immediately before use.
9. X-Gal detection buffer (Sigma): 35 m*M* potassium ferrocyanide, 35 m*M* potassium ferricyanide, 2 m*M* MgCl$_2$, 0.02% Nonidet P-40, which is also called Igepal CA-630 (made from a 10% stock solution), and 0.01% Na deoxycholate (made from a 10% stock solution) in PBS. X-Gal detection buffer can be stored for <1 yr at room temperature in foil-covered container.
10. X-Gal stock (50X): 40 mg/mL 5-bromo-4-chloro-3-indolyl-β-D-galactopyranoside (X-gal) in dimethyl formamide. Store at –20°C in a glass container covered with foil.
11. X-Gal reaction mix: Add 1:49 X-gal stock:X-gal detection buffer. Make fresh before use.

3. Methods
3.1. Photoablating β-Gal-Expressing Cells in Retina Whole Mounts In Vitro

1. Loading β-gal-expressing cells with FDG and AEC: Remove retina from animal and place on a glass cover slip in a small volume of L-15 medium or Ringer's. Use just enough to cover the tissue. Remove all excess liquid to make retina lie flat (*see* **Note 1**). Apply a 2–3-μL drop of FDG/AEC solution #1 on the tissue, then immediately dilute the FDG/AEC solution with a 50-fold vol (100–150 μL) of L-15. Wash 3X with L-15, using a 100 μL vol each time, then leave the retina in 100 μL L-15 for 10 min (*see* **Notes 2** and **3**). After 10 min, you should be able to see the β-gal cells labeled with fluorescein. They can be visualized using a standard fluorescence microscope with a fluorescein filter set (e.g., a Zeiss Axiophot with a 100-W mercury short arc lamp and a filter set containing a 450–490-nm exciter filter, a 510-nm dichroic filter, and a 520–560-nm barrier filter.)
2. Photoablating the cells: Illuminate the region of the tissue that contains the β-gal-expressing cells of interest. Focus on the cells with a 10X, 0.3 N.A. objective and illuminate for 3–15 min. Light activation of the FDG triggers a reaction that causes the colorless AEC to form a brown precipitate in the cells. This precipitate should be visible under brightfield illumination (*see* **Notes 4** and **5**). Once the cells have formed the brown product, return the tissue to a culture dish with fresh media.

3.2. Photoablating β-Gal-Expressing Cells in the Retina In Vivo

1. Loading β-gal-expressing cells with FDG and AEC: Inject the eye of an anesthetized mouse with 0.5–1.0 μL of FDG/AEC solution #2. Assuming the vitreal volume is 5–10 μL,

the solution will be diluted 10–20-fold in the eye. To make the injection, first fill a 10-µL blunt-tipped 33-gage Hamilton syringe with the FDG/AEC solution. Then create a small hole in the cornea to allow the syringe to enter the eye. This can be done by puncturing the cornea with the tip of a 30-gage needle. Then insert the syringe needle into the hole and inject the dye into the vitreous humor. The fluorescein-labeled cells can be visualized by placing the mouse on the microscope stage so that the microscope objective is focused through the optics of the eye onto the retina. You can reduce the corneal refraction by placing a glass cover slip on the animal's cornea (*see* **Note 6**).

2. Photoablating the cells: Focus on the labeled cells and illuminate the retina for 8–10 min with the 10X, 0.3 N.A. objective, as described above in **Subheading 3.1.**

3.3. Photoablating β-Gal-Expressing Cells in the Cortex In Vivo

1. Loading β-gal-expressing cells with FDG and AEC: Cut open the skin above the brain area in the anesthetized mouse. Reflect back or remove the bone. Using a 30-gage needle, make tiny perforations in the dura to permit drug entry into the tissue. Apply FDG/AEC solution #1 to the surface of the brain and allow 10 min for penetration.
2. Photoablating the cells: Focus on the labeled cells and illuminate the retina for 8–10 min with the 10X, 0.3 N.A. objective as described in **Subheading 3.1.** (*see* **Note 7**). Replace the bone and close the skin as described in NIH guidelines for survival surgery.

3.4. Ethidium Assay for Cell Ablation

1. Ethidium homodimer loading: This is for use on live tissue. After ablation treatment, place retina on cover slip and remove all excess liquid to make retina lie flat. Apply 2–3 µL of ethidium homodimer solution on the retinal surface (to drive the dye into the tissue (*see* **Note 1**), then immerse the retina in 2–3 mL of ethidium homodimer solution and place at 37°C for several hours (*see* **Note 8**).
2. Viewing ablated cells: Examine whether the cells that contain the AEC precipitate also show ethidium-labeling. View the AEC-label with brightfield and the ethidium fluorescence with a 630–640-nm excitation filter (*see* **Note 9**).

3.5. X-Gal Assay for Presence (or Loss) of β-Gal-Expressing Cells

1. Fixing retina: Fix retina flat on a cover slip by immersing it in a small volume (just enough to cover tissue) of 4% paraformaldehyde solution for 5 min. Transfer retina to a small dish (e.g., a well in a 24-well dish) and continue to fix for 10 min.
2. Washing retina: Remove fix and wash 5X, each time for greater than 5 min with 2 mL of PBS.
3. Detecting β-gal activity using X-gal: Remove PBS from last wash and add 0.5–1 mL of X-gal reaction mix, incubate at 37°C for several hours or overnight. Remove X-gal reaction mix and wash 3X, each time with 2 mL of PBS. Examine for the loss of X-gal-stained cells by microscopy using standard brightfield illumination (*see* **Note 10**).

4. Notes

1. This flattening of the tissue may be necessary only for the retina and may be skipped when loading dyes into other tissues. Retina whole mounts are notoriously difficult to penetrate with dyes, presumably because of the inner limiting membrane.
2. Note the high concentrations of DMSO used to dissolve the FDG and AEC. These are the concentrations we used in our initial work with this method. Although no DMSO-induced cell death was detected in any of the tissues we examined (*1*), high DMSO concentrations have been reported to be teratogenic in some systems (*see* **Note 4**). Recently, we have had

success with much lower concentrations (1.5 mg/mL FDG, 0.035 mg/mL AEC in 2.5% DMSO/97.5% Ringer's) *(4)*.

3. It is worth mentioning that this procedure works well for loading β-gal-expressing cells with FDG *in tissue*. We have had much less success loading cultured cells using a variety of procedures, including osmotic shock and varying the concentrations of FDG and DMSO.

4. It is difficult to provide a general set of conditions for effective and selective ablation, because cell types vary in their β-gal enzyme activity, position within the tissue, and so forth. For this reason, we are simply reporting conditions under which the ablation treatment was effective and produced little or no nonspecific cell death. These conditions can then serve as a guideline.

   A convenient feature of the method is that AEC forms a visible precipitate during the ablation. Thus, one can vary conditions and use the formation of the AEC product as an assay for cell death. If the β-gal-expressing cells are relatively superficial in the tissue, the formation of the AEC precipitate can be watched as it occurs during the treatment. Alternatively, it can be seen by sectioning the tissue after treatment. We suggest a 4% paraformaldehyde fixation, followed by cryostat-sectioning. The AEC label fades within a few hours of fixation, so it is critical to examine the tissue quickly.

5. Note that fluorescein can cross membranes and accumulate in the tissue. In our experiments, the concentration of fluorescein in the tissue surrounding the β-gal-expressing cells was extremely low compared with the concentration inside the β-gal-expressing cells; thus, we saw no cell death in neighboring cells. However, fluorescein leak is a potential problem. It is a function of several factors, including promoter strength and cell density, so the extent of leak will vary with different β-gal-expressing cells. The effects of leakage can be minimized or avoided by illuminating the cells shortly after the application of FDG—before appreciable leak occurs. The leakage of fluorescein *after* illumination poses no threat, because fluorescein itself is nontoxic.

6. A drop of 1% atropine sulfate (Steris Laboratories, Phoenix, AZ) can be used to dilate the pupil to facilitate viewing the fluorescein-labeled cells.

7. One limitation of this technique is that cells located deep in tissue may be difficult to target using a fluorescence microscope as a light source. The reason for this is that the peak excitation wavelength of fluorescein is absorbed by endogenous tissue chromophores, such as flavins and hemoglobin, which prevent the light from reaching deep into tissue *(5,6)*. Thus far, we have successfully targeted cells as deep as halfway through the cortical plate, approx 300–400 μm below the surface. Cells located deeper in the brain might be targeted with fiber optics inserted into the ventricles.

   Light accessibility is not limiting for embryos of transparent organisms, such as zebrafish. In preliminary experiments with zebrafish embryos, β-gal-expressing cells in deep structures, such as the heart, were readily ablated *(1)*.

8. Ethidium homodimer labeling of AEC-labeled cells can take several hours. It is not clear whether or not this means that the cells take several hours to die. Rather, the slow labeling may reflect slow penetration of ethidium homodimer into the tissue (*see* **Note 1**). Support for the latter possibility was obtained from ablations performed in zebrafish, where uptake of ethidium homodimer into AEC-labeled cells occurred within minutes.

9. The 630–640 nm bandpass filter is used to view the ethidium fluorescence rather than the standard rhodamine barrier filter (590-nm long-pass), because fluorescein fluorescence can sometimes be detected through the rhodamine filter. The 630–640-nm filter completely blocks the fluorescein emission.

10. This assay should be used >36 h after ablation (when ablations are performed in vivo), because β-gal activity can remain for several hours to a day after cells are physiologically dead.

Acknowledgment

The author thanks Peter Latham for comments on the manuscript.

References

1. Nirenberg, S. and Cepko C. L. (1993) Targeted ablation of diverse cell classes in the nervous system in vivid. *J. Neurosci.* **13,** 3238–3251.
2. Lees, G. J. (1989) *In vivo* and *in vitro* staining of acidophilic neurons as indicative of cell death following kainic acid-induced lesions in rat brain. *Acta Neuropathologica* **77,** 519–524.
3. Beletsky, I. P. and Umansky, S. R. (1990) New assay for cell death. *J. Immunol. Methods* **134,** 201–205.
4. Nirenberg, S. and Meister, M. (1997) The light response of retinal ganglion cells is truncated by a displaced amacrine circuit. *Neuron* **18,** 637–650.
5. Anderson, R. R. and Parrish, J. A. (1983) Selective photothermolysis: Precise microsurgery by selective absorption of pulsed radiation. *Science* **220,** 524–527.
6. Batey, D. W. and Eckhert. C. D. (1991) Analysis of flavins in ocular tissues of the rabbit. *Invest. Ophthalmol. Vis. Sci.* **32,** 1981–1985.

46

Exo utero Surgery

Valerie Ngo-Muller and Ken Muneoka

1. Introduction

The study of developmental mechanisms in postimplantation mammalian embryos can be performed for a short term using whole-embryo culture systems (*see* Chapters 7 and 8), or for an extended period using *in utero* *(1,2)* or *exo utero* manipulations *(3)*. The usefulness of both *in utero* and *exo utero* surgical manipulations is restricted because of the fragility and accessibility of the embryo, and the key to selecting the appropriate surgical approach depends largely on the design of the experiment. *In utero* surgery is useful for manipulating late, but not early, embryonic stages, because uterine muscular contractions, in the absence of the hydrostatic buffer that the amniotic cavity provides, are lethal to the embryo until it has developed to a point where it can withstand such contractions. The *exo utero* surgical approach avoids the problem of uterine contractions and is useful for manipulating earlier embryonic stages; however, embryonic manipulation using *exo utero* surgery is largely restricted to stages after resorption of the decidua (from E11.5 on; but *see* **ref. 4**). Prior to E11.5, the decidua inhibits visualizing the embryo, and damage to the decidual layer results in the death of the embryo. After E11.5, *exo utero* surgery allows for direct experimental access to the embryo/fetus and a wide range of experimental perturbations can be successfully performed, including lesioning, grafting, bead implantation, and targeted microinjection. This experimental approach has been used effectively to explore molecular and cellular aspects of mammalian development *(5–17)*.

Exo utero surgery refers to the manipulation of embryos that maintain a placental attachment to the uterus but outside of the confines of the uterine environment. *Exo utero* surgical procedures were developed based on the discovery that embryogenesis to term is not altered after the uterus is surgically opened to release the embryos *(18)*. Under these conditions, embryos develop within the abdominal cavity of the mother as long as the placental attachment to the uterus is maintained. In many ways, *exo utero* development within the abdomen of the mother is analogous to the phenomenon of ectopic pregnancy in humans *(19)*. Surgical procedures involve exposing embryos by opening the uterine wall and operating on embryos within the abdominal cavity of the mother. Access to the embryo itself involves opening the extraembryonic membranes, and these membranes must be sutured after embryonic manipulation. After embryo surgery, the

From: *Methods in Molecular Biology, Vol. 135: Developmental Biology Protocols, Vol. I*
Edited by: R. S. Tuan and C. W. Lo © Humana Press Inc., Totowa, NJ

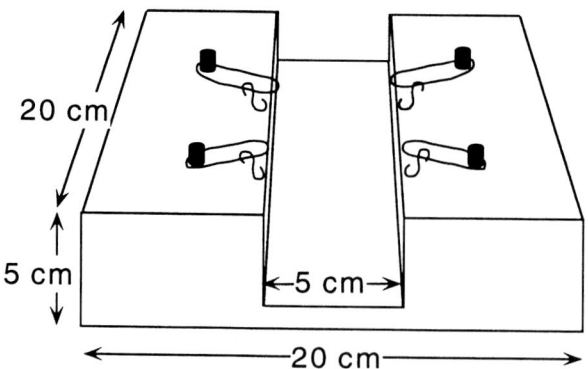

Fig. 1. Surgery operating stage, S-shaped hooks, rubber bands, and pins. The stage is made out of 1-cm-thick foam panels, one large panel for the base measuring 20 × 20 cm, and eight small panels for the upper levels measuring 20 × 7.5 cm. Four small panels are glued on the top of each other, the same thing is repeated with the four other panels, then the two resulting blocks are glued on each side of the large panel. The stage is then covered with tape to allow easy cleaning. S-shaped hooks are made out of paper clips, which are cut and reshaped using pliers.

abdomen of the mother is sutured and embryonic development continues *exo utero*. The entire procedure takes approx 1 h. This chapter outlines the protocol developed in our lab for mouse embryos, but these techniques may also be applicable to embryos of other species.

2. Materials

1. Anesthetics
   a. Nembutal
      i. Stock solution: 50 mg/mL pentobarbital sodium (Abbott Laboratories, North Chicago, IL)
      ii. Working solution 1:9 dilution in sterile water, to be made fresh every week and kept at room temperature (RT).
   b. Fentanyl/Droperidol
      i. Stock solution: 0.4 mg/mL fentanyl (Sigma Chemical Co., St. Louis, MO), 20 mg/mL droperidol (Sigma), 1.8 mg/mL methylparaben (Sigma), 0.2 mg/mL propylparaben (Sigma), adjust pH to 3.1 ± 0.4 with lactic acid, in sterile water. Light sensitive, stored at RT.
      ii. Working solution 1:10 dilution in sterile water. Light sensitive, to be made fresh every week and stored at RT.
2. Hair clipper (rodent; Andis, Racine, WI).
3. Operand® Povidone-Iodine solution USP (Betadine surgical scrub).
4. Surgery operation stage (to be constructed, *see* **Fig. 1**).
5. Surgery microscope (Olympus America, Melville, NY).
6. Fiber optics illuminator (Fisher Scientific Co., Pittsburgh, PA).
7. S-shaped tissue retraction hooks.
8. Microsurgical instruments:
   a. Fine forceps (Dumont #5, SPI Supplies, West Chester, PA).
   b. Blunt forceps.
   c. Microdissection scissors (Roboz Surgical Instrument Co., Rockville, MD).

    d. Iridectomy straight scissors (Roboz).

    e. Iridectomy 45° angled scissors (Roboz).

9. Six-inch cotton-tipped applicators (sterilized by autoclaving).

10. 1-cm diameter cotton balls, made with surgical cotton and sterilized with UV-irradiation (30 min, *see* **Note 1**).

11. Sterile lactated Ringer's solution, 100–200 mL per mouse (Baxter Scientific, Muskegon, MI).

12. Vacuum line and trap (*see* **Note 2**).

13. Sutures

    a. 10.0 monofilament (Sharpoint, Inc., Reading, PA, or Ethicon, Inc., Somerville, NJ).

    b. 6.0 Mersilene (Ethicon).

14. Heating pad.

15. Zephiran (benzalkonium chloride).

3. Methods

3.1. Preparing the Mouse

Preparation of the mouse should be done in an area different from the one where the surgery will be performed. Weigh the pregnant female and anesthetize with an intraperitoneal injection of Nembutal working solution (*see* **Note 3**), using 10 µL/g body wt (i.e., 0.06 mg/g body wt). Holding the tail, let the mouse grab the cage grid with its front limbs, and keep its hind feet suspended. The injection should be made in the lower abdominal region, the angle as parallel to the body axis as possible in order to avoid hitting organs. This technique minimizes the trauma associated with the injection and results in more complete anesthesia of the mouse. It should take <5 min for the mouse to reach deep anesthesia (*see* **Note 4**). Once the mouse is anesthetized, make a subcutaneous injection of 0.05 mL of a fentanyl/droperidol working solution. The injection can be made in the back by lifting the skin.

Place the mouse on its back and shave the abdomen extending from the sternum to approx 5 mm anterior to the vaginal opening. Shave in a posterior-to-anterior direction (i.e., opposite the direction of hair growth). Minimize damage to the skin, and special care should be made not to damage the teats that lie laterally on the abdomen. Shaving can also be done with a razor blade; however, it requires some experience to avoid lacerating the skin. After shaving, take care to remove all hair clippings. Scrub the shaved area with 70% ethanol, then Operand®, then 70% ethanol again, and wipe dry (*see* **Note 5**).

3.2. Exposing the Embryos

Surgery should be done in a clean environment that has little to no traffic and away from heating or cooling vents (i.e., no major air flow; *see* **Note 6**). Place the mouse in the lower level of the operation stage, on a few layers of paper towels so that the surface of the abdomen reaches the upper level of the stage. Ensure that the mouse is properly anesthetized by checking for a response after pinching the tail with forceps. Using microdissection scissors, make a mid-ventral incision of the skin (2- to 3-cm long), and then make a similar incision of the peritoneum (*see* **Note 7**). Use blunt forceps when manipulating the skin and fine forceps when manipulating the peritoneum. Retract both the skin and the peritoneum using S-shaped hooks attached to rubber bands tacked to the upper level of the operating stage.

The two horns of the uterus are located in the lateral regions of the abdominal cavity and join to form the common uterus located at a mid-posterior location. The intestines take up the majority of the abdominal cavity and should be carefully pushed with a cotton-tipped applicator toward the anterior to facilitate access to the uterine horns. The uterus should be relaxed at this point. Embryos are exposed by cutting the uterus on the antiplacental side (i.e., 180° opposite the placenta) where there are no major vessels. This procedure and all subsequent parts of the surgery are done under a surgery microscope. To make this cut, grasp the uterus at the junction of uterine horns with fine forceps and lift the tissue away from the underlying embryo. Make an incision through the myometrium using 45°-angled iridectomy scissors, making sure that the tip of the scissors is directed away (upward) from the underlying embryo. Care should be taken not to rupture the yolk sac of the embryo when making this incision. There are two approaches that we have used to expose embryos. The first is to make a continuous longitudinal incision of the uterus, allowing the uterine tissue to contract to the base of the placenta (*see* **Note 8**). An alternative approach is to make multiple single incisions for each embryo, allowing the embryo to protrude through the incision with the uterine tissues contracting down onto the placenta (*see* **Note 9**).

3.3. Removing Unwanted Embryos (see Note 10)

Embryos can be removed from the uterus without influencing the viability of those remaining. Embryo removal is not necessary but recommended because it makes room in the abdominal cavity for the operation. To remove an embryo, roll a dry cotton-tipped applicator between the placenta and the uterus. The placental attachment is very fragile, so the embryo will detach very easily and will generally adhere to the cotton-tipped applicator. Some bleeding may occur from the uterus, but this bleeding can be controlled by applying pressure for a few seconds with the applicator at the site where the placenta was attached. When all unwanted embryos are removed and any bleeding from the uterus has stopped, flush the abdominal cavity with lactated Ringer's solution until no blood or tissue debris is visible (*see* **Note 11**). Leave the abdominal cavity filled with lactated Ringer's solution to prevent drying of the embryo during the operation.

3.4. Positioning the Embryos to Be Manipulated

Proper positioning of embryos within the abdominal cavity is critical for the success of the embryo manipulation. Position of embryos must be done with care so as not to damage the placental-uterine connection (*see* **Note 12**). The position of the embryo will depend on the region of the embryo to be operated on and the type of manipulation required. Transfer 4–5 sterile cotton balls in a small dish (*see* **Note 13**). Wash them with lactated Ringer's solution to get rid of lint, and dry them using the aspirator setup. Using fine forceps, grasp the uterine muscle wall where there are no blood vessels, and position the embryo in a favorable position. Place the cotton balls in the abdominal cavity around the embryos to hold it in position. A certain amount of movement of the embryo will occur during the operation because of periodic uterine contractions and also because of contractions of intestinal tissue. These movements are inherent to the surgical procedure and must be accommodated by the operator. Once the operation is completed, the embryo may be moved to position the next embryo and this must be done carefully. A cotton-tipped applicator can be used to press the embryo toward the uterus while repositioning the uterus.

3.5. Embryo Manipulation

A variety of embryonic manipulations are possible using this surgical approach. Those involving targeted injection through the extraembryonic membranes or microinjection of substances directly into the embryonic vasculature via the yolk sac vessels do not require opening and suturing of the extra-embryonic membranes and are relatively quick and easy. Because the embryo is visible through the extraembryonic membranes, targeted injections can be carried out with some degree of accuracy (*see* **Note 14**). One consideration is to keep track of operated vs unoperated embryos, so some sort of identifying mark should be placed on each operated embryos. An alternative is to videotape each operation for accurate documentation of each embryonic manipulation.

Extraembryonic membranes have to be opened and subsequently sutured for many manipulations. Grasp the extraembryonic membranes away from major yolk sac vessels with fine forceps and make a small incision in the yolk sac and the amnion using straight iridectomy scissors (*see* **Note 15**). The size of the incision will depend on the type of manipulation, because the incision defines the space within which the operation must be completed (*see* **Note 16**). It is possible to make multiple incisions, but each incision must be sutured separately. In some cases, it is possible to have the part of the embryo to be manipulated protruding through the incision, for example, the developing limb, so the incision can be relatively small. Use a blunt glass probe to position the embryo with respect to the incision. The types of manipulations that have been successful include lesions (*see* **Note 17**), grafting of microcarrier beads (*see* **Note 18**) or tissue (*see* **Note 19**), and injections (*see* **Note 20**). These techniques are routine for other more accessible embryos, such as the chick, and we recommend using the late-stage chick embryo for training and for piloting new projects.

After completing surgery on the embryo, the extraembryonic membranes must be sutured with 10.0 monofilament suture, passing the suture through both the yolk sac and the amnion on both sides of the incision (*see* **Note 21**). Tie a square knot and leave a short length of thread (1 mm). The presence of a suture is a useful mark to identify experimental embryos.

3.6. Closing Up

Once the embryonic operations are completed, carefully remove all cotton balls and reposition the embryos in the lateral regions of the abdomen. Extensively flush the abdominal cavity with lactated Ringer's solution until the abdomen is free of blood or tissue debris. Leave some Ringer's solution in the abdomen, and release the skin and peritoneum from the S-shaped retractor hooks. Using a 6.0 suture, close the peritoneum and the skin with running sutures. We typically suture from the sternum in a posterior direction. While suturing the peritoneum be careful to avoid damaging organs or adipose tissues by lifting the peritoneum while suturing. When the abdominal incision is closed, place the mouse in a cage previously warmed up on a heating pad (*see* **Note 22**). The mouse should wake up within 1 h after surgery, and it is important to monitor the mouse to make sure that there are no postsurgical complications (*see* **Note 23**).

3.7. Cleaning Up

Clean all the surgery area and the dissections tools, including the S-shaped hooks and remaining sutures, with Zephiran (Sanofi Winthrop Laboratories, New York, NY) and then 70% ethanol. Dissection tools should be periodically cleaned in a sonicator. Sterilizing by autoclaving or exposure to ultraviolet light is recommended but not absolutely necessary.

3.8. Collecting Embryos

To collect experimental embryos, euthanize the female mouse and reopen the abdominal incision. Typical survival rates range between 50 and 100% depending on the stage at which embryos were operated (*see* **Note 24**). Embryo survival depends largely on the invasiveness of the operation, although other factors, such as the formation of abdominal adhesions, can cause poor survival. To minimize the occurrence of adhesions, we emphasize the importance of minimizing tissue damage during the operation and flushing the abdominal cavity with lactated Ringer's solution to remove blood and tissue debris (*see* **Note 10**).

If development to postnatal stages is required, fetuses that have been removed from the operated female can be fostered by a female that has recently delivered her litter. For fostering, it is important to wear gloves and to minimize handling of the newborns. Newborns should be transferred to a Petri dish, cleaned of blood and extraembryonic membranes and dried using cotton-tipped applicators. At this point, they should be stimulated to begin breathing by gently rubbing their tail with a cotton-tipped applicator. Use a cotton-tipped applicator to absorb any fluids that come out of their mouths as they begin to breath. Place the newborns on a heating pad or under a lamp and monitor them until they are breathing regularly and moving about. Remove the litter from the foster female and transfer her into a dark room. Remove some of her bedding and put it with the experimental newborns. Transfer the experimental newborns into the foster female's cage and leave them undisturbed (*see* **Note 25**).

4. Notes

1. Using surgical cotton is crucial because it is made of long strands of fibers and will release less lint. Commercial cotton balls for make-up are inappropriate because they enhance postsurgical abdominal adhesions that compromise embryonic survival. Cotton balls are sterilized with ultraviolet light and kept in a covered dish next to the operating area.
2. A vacuum line and trap at the operation station is very important for quickly flushing the abdominal cavity with lactated Ringer's solution during the operation. The trap can be made with a sidearm flask (1000 mL) and flexible tubing that extends to the operating stage. To control the suction we attach to the end of the flexible tubing the barrel of a 1.0-mL syringe (plastic), which has a hole in the side of the barrel that can be used to control suction at the tip.
3. Intraperitoneal injection is preferable to subcutaneous injection for the administration of Nembutal. Subcutaneous injections are evidenced by the formation of a subcutaneous lump at the injection site, and the mouse requires a longer time to reach deep anesthesia. Subcutaneous injections usually require the administration of additional Nembutal for complete anesthesia. Follow-up injections are made intraperitoneal in aliquots of 0.05–0.10 mL. If several animals are to be operated on at one time, it is best to administer the initial anesthetics to each mouse in isolation. If you notice that fluid comes out of the vagina or

the anus following an intraperitoneal injection, then it is likely that the injection was made in the bladder, the uterus, or the intestines. Subsequent injections of Nembutal will be necessary to anesthetize the mouse.

4. Deep anesthesia occurs when complete response to painful stimulus is lost and can be verified by pinching the base of the tail without generating a response. Because pentobarbital is a respiratory depressant, the respiratory rate also becomes very slow. Care must be taken not to overdose the animal when using pentobarbital.

5. This scrubbing is crucial to disinfect the abdomen and to get rid of any residual hair that could later contaminate the abdominal cavity.

6. The surgical area should be clean and away from ventilation. We have had good results working within a plastic hood enclosure that was built with a plastic pipe frame and covered with sheets of clear plastic. Working within a hood enclosure is recommended but not absolutely necessary. It is important to keep the surgery area as clean and as dust free as possible.

7. Cut the peritoneum along the median suture that appears as a white line. This tissue is devoid of blood vessels so little to no bleeding will result. Incisions lateral to the midline can result in considerable bleeding that is difficult to control. Lift the peritoneum with the forceps when making the incision to avoid hitting any organ.

8. Exposing the embryos in the uterine horns is one of the most important and critical steps in this procedure because of the potential for damaging embryos. The uterine muscles should be relaxed from the action of fentanyl/droperidol, making it relatively easy to cut the myometrium without damaging the embryos. One approach is to open each uterine horn with a single continuous longitudinal incision on the antiplacental side of the uterus. After contraction of the myometrium, the tubular uterus is everted with a chain of embryos attached. Each embryo is attached to the uterus by the placenta, and because the mouse placenta is hemoendothelial, this attachment site is very fragile. It is important to keep in mind the importance of maintaining the placenta-uterus association while moving the uterus to position each embryo for operation. When manipulating early-stage embryos (E11.5 and E12.5), the Reichert's membrane may still be surrounding the yolk sac, and maintaining this membrane intact when exposing the embryos will help to stabilize the placental attachment. A tightly contracted uterine musculature is indicative of poor anesthesia and/or excessive stress to the mouse in the administration of the anesthetics. The degree of uterine contractions is directly proportional to the extent of embryo damage resulting from cutting the uterus, thus the handling and administration of anesthetics is vital to the success of this surgical approach. If the uterus is contracted, then cutting the uterus will result in the underlying embryo as well as neighboring embryos being forced through the incision, and this will damage the embryos. Cutting the uterus rapidly can minimize the effect of the contractions, however, making a rapid longitudinal cut of the entire uterine horn difficult.

9. By making discontinuous incisions of the uterus, each embryo will bulge out of the uterus. In this case, the uterus will remain as a closed tube and is contracted onto the placenta. The idea behind this approach is to use the uterine contractions to help maintain the attachment of the placenta to the uterus, at the same time still making the embryo accessible to manipulation. A secondary benefit of this approach is that it minimizes overall tissue damage that is expected to reduce the occurrence of postsurgical adhesions associated with the operation. It is important that the uterus be relaxed for this procedure. Instead of a continuous incision, make a small incision over each embryo on the uterine horn. Each incision should be large enough to allow the embryo to pop out undamaged but small enough so that the uterine muscle wall will retract over the placenta and protect it. This

procedure must be done fairly rapidly because the uterus initiates contraction following each incision, and it is possible that neighboring embryos may be pushed toward a single incision. This procedure can be used for the embryos that will be kept for operation and for the embryos to be discarded as well, for it will help reduce bleeding from the placental attachment sites where embryos are removed and reduce the overall surface of exposed wounds in the abdomen. An alternative approach to removing embryos has been described that involves using a large-gage needle and syringe to remove embryos without opening the uterus wall *(20)*.

10. Deep anesthesia typically lasts approx 1 h, and during that time we typically operate on 2–4 embryos. It is good to keep some nonoperated embryos that can serve as back-ups in case some operated embryos get damaged during the operating procedure. We typically keep 4–6 embryos, i.e., 2–3 on each horn, preferably the ones that are located close to the ovaries.

11. Lactated Ringer's is administered from a sterile bag (or bottle) via a drip tube controlled by the operator with a compression clip. Fluid flow is determined by gravity, so the height of the bag will determine the rate of flow. We generally hang the lactated Ringers about 24 in. above the operating table to one side of the operator. The aspirator is set up on the other side so that the operator can simultaneously control the flow of Ringer's solution with one hand and the aspiration of abdominal fluid with the other hand. Ringer's solution may be warmed up to 37°C if one is working in a very cool room. Under deep anesthesia, the mouse temperature drops, and flushing the abdominal cavity with cold fluid may induce further temperature drop, which may interfere with the animal's recovery. Lactated Ringer's solution can be warmed up by placing a portion of the tubing in a 37°C water bath.

12. When working with E11.5 and E12.5 embryos, if the placenta starts to bleed, it is very likely that the embryo will die shortly after surgery.

13. It is a good idea to try to use the same number of cotton balls each time or to keep a record of how many cotton balls were placed in the abdominal cavity, in order to avoid leaving any in the abdominal cavity once the surgery is completed. Leaving any cotton ball in the abdominal cavity will invariably lead to tissue adhesions and embryo death.

14. The embryo is very visible through the extraembryonic membranes; however, the amniotic fluid can distort the view of the embryo and can influence targeting of the microinjection needle. To avoid this, we recommend puncturing the extraembryonic membranes and draining the amniotic fluid prior to injection. Draining the amniotic fluid also makes it possible to grasp the extraembryonic membranes with fine forceps during the injection, making it easier to pierce the amnion (*see* **Note 15**).

15. The incision of the yolk sac should avoid the major vessels in the yolk sac; however, sometimes this is unavoidable because of the region of the embryo under investigation. It is possible to cut major vessels of the yolk sac by first using blunt forceps to apply pressure to the vessel until there is no bleeding. The yolk sac is opaque and highly vascularized, but the amnion is almost transparent and nonvascularized. Cutting the yolk sac does not always result in cutting the amnion. The amnion is very thin but very resilient, making it difficult to cut or pierce. For example, it is possible to inject a glass needle into the embryo without breaking the amnion, that is, the amniotic layer forms a barrier between the needle and the embryonic tissue. The best way to determine whether it is cut is to use a blunt glass probe to see whether there is direct access to the embryo. If the amnion is not cut, grasp it with fine forceps and make a cut similar to the yolk sac incision. It is important to keep track of the edges of the amnion, as the amnion must be sutured closed at the end of the operation (*see* **Note 21**).

16. The incision must not be larger than approx one third of the diameter of the yolk sac, otherwise the incision will tend to tear and the embryo will come out of the yolk sac. It is interesting to note that late-stage embryos can survive outside of the amnion and yolk sac in the abdominal cavity; however, we have not explored this phenomenon in any detail.

17. Experimental lesioning of the embryo/fetus has been valuable for investigating changes in wound repair and regenerative capabilities with development. Lesioning studies are performed using iridectomy scissors, sharpened needles, or laser ablation *(20,21)*. Bleeding is the predominant concern in these studies. Bleeding can be controlled using direct pressure applied with a blunt glass probe or applying fibrinogen directly to the wound site or both. Fibrinogen does not seem to be as effective for early-stage embryos. Early-stage embryos have a low blood volume and are more likely to die from excessive bleeding than older embryos/fetuses.

18. Grafting of microcarrier beads that act as slow-release vehicles for signaling molecules has become a valuable experimental approach for studying the action of growth factors and hormones. Techniques for grafting microcarrier beads largely depends on the nature of the bead. Some beads, such as heparin conjugated beads or ion exchange beads, tend to stick to tissues and can be grafted into lesion where they readily adhere *(16)*. Other beads, such as agarose beads, are nonadherent and tend to pop out of a graft site. We have success grafting agarose beads by pinning the bead onto a sharpened tungsten needle, and letting it air dry for a few seconds so that it dehydrates and shrinks onto the tip of the needle. A graft site is created by using another sharpened needle to "tunnel" through the tissue to the graft site. The dehydrated bead is then inserted into the tunnel and positioned at the graft site and held there until it hydrates, swelling back to its original size and trapping the bead within the tissue. The bead is gently released from the needle using forceps or a blunt glass probe. The result of such a graft is shown in **Figs. 2A–C**.

19. Tissue grafting is used to investigate the developmental potential of cells placed in the embryonic environment. For the later stage embryo/fetus, grafting is relatively easy, because the integrity of the host tissues at the graft site holds grafted tissue in place during the early stages of wound healing. For example, tunneling beneath the skin to create a subcutaneous graft site results in excellent healing of grafted tissues. Grafting into early-stage embryos is difficult, because the host tissues lack the integrity required to hold tissues in place long enough to allow for healing and integration. To compensate for this, it is necessary to tack the graft to the graft site to hold it in place. Grafted tissues and graft site should have matching shapes with the graft slightly larger than the graft site. In all tissue grafting studies, it is advised that a cell-autonomous marker (e.g., the *lacZ*-expressing Rosa 26 transgenic mouse strain, Jackson Laboratories, Bar Harbor, ME) be used to ensure graft survival and participation in the response (**Fig. 2D,E**).

20. Experiments involving the injection of vital dyes, viral vectors, or cultured cells have been performed to investigate cell fate and the response of cells to the embryonic environment. Injections can be performed using pulled-glass micropipets and targeting the injection site by hand *(21)*.

21. The 10.0 monofilament suture is very fine, and the needle is very small. It tends to "jump" out of the forceps if it is held too tight. It is extremely easy to lose and almost impossible to find. The thread is also very fragile and thus must not be pulled too hard; otherwise it will break. It is crucial to suture the amnion. The amnion heals very fast, and if it is not sutured, it may constrict around parts of the embryo, causing deformities. Suturing with 10.0 monofilament suture is very difficult, especially for the novice. Because closing the extraembryonic membranes is crucial for the success of the operation, we can suggest an alternative approach that entails making a purse-string suture outlining the surgical region

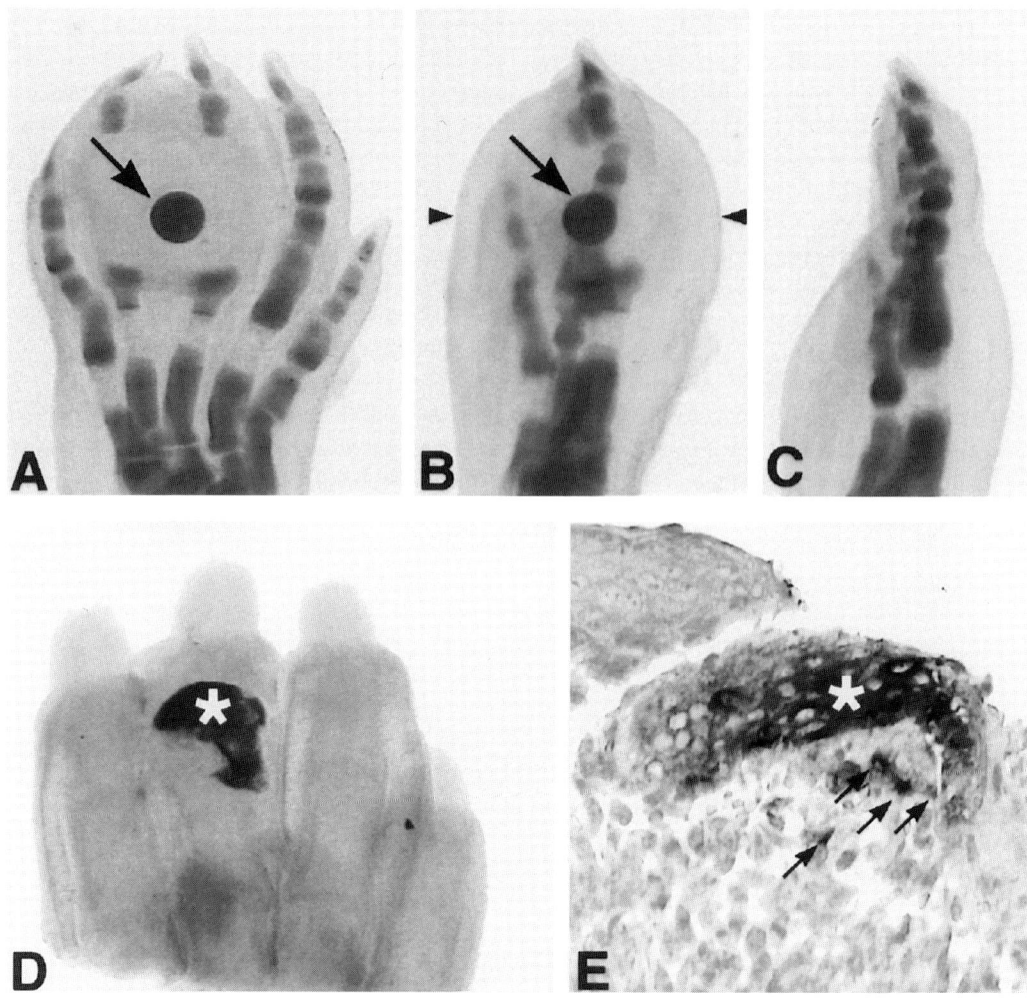

Fig. 2. (*See* color plate 11 appearing after p. 258.) (**A**) Example of microcarrier bead grafting experiment showing the effect of TGFβ-1 on digit formation in CD1 mouse embryo. Agarose Affi Gel® Blue beads (150–300 μL, Biorad Laboratories, Richmond, CA) were soaked in a solution of 50 μg/mL porcine TGFβ-1 (20 beads in a 2-μL vol for 2 h at 4°C) and grafted at the tip of digit III of the left hindlimb at day 12.5 as described in **Note 18**. The operated embryo was collected at term (E18.5) and processed for whole-mount staining with Victoria blue. TGFβ-1 containing microcarrier bead (*arrow*) locally inhibits the formation of skeletal elements without altering skeletal patterning in other regions of the autopodium. (**B**) A lateral view of the limb shown in (A) displaying the hypertrophic effect of the TGFβ-1 bead (*arrow*) on dorsal and ventral tissues of the digits (*arrowheads*). (**C**) Lateral view of a control limb showing normal dorsal-ventral flattening associated with digit formation. (**D**) Example of tissue graft in CD1 mouse embryo. Adult nail bed tissue from the transgenic mouse strain B6,129-TgR(ROSA26)26Sor (Jackson Laboratories), which harbors the *lacZ* transgene, was grafted subcutaneously in the hindlimb of a day 15.5 CD1 embryo described (*see* **Note 19**). The operated embryo was collected 1 d later and processed for *lacZ* expression *(22)*. The stained limb was cleared in methyl salicylate to reveal the grafted tissue (*), then paraffin-embedded and serially sectioned. (**E**) Histological section showing B6,129-TgR(ROSA26)26Sor nail bed tissue graft (*) and individual labeled cells (*arrows*) that have migrated into the host tissues.

prior to opening the extraembryonic membranes. The purse-string suture is made by making two antiparallel passes and cutting the suture without tying it off. The suture will form a circle and the extraembryonic membranes can be cut within the circle. When the operation is completed, the incision can be closed rapidly by tying off the suture. Finally, care must be taken not to involve cotton thread from the cotton used to position the embryo when suturing the membranes, otherwise, removing the cotton at the end of the surgery can cause damage to the embryo.

22. Operated animals should be placed in a separate cage until they wake up or at least placed in a cage with only animals operated on the same day. Placing a heating pad (37°C) under the cage at the beginning of the surgery will ensure that the cage is warm when the mouse is returned to the cage. Warming the mouse up considerably enhances the animal's recovery, for the surgery procedure induces a dramatic decrease in the body temperature. The heating pad should, however, be placed under only half of the cage to allow the animal to choose between a warm or a cool spot.

23. The first signs of waking up are an increase in respiration rate and shivering of the front limbs. If the mouse's lips turn blue, it is a sign that the mouse is in too-deep anesthesia, and special care should be taken to keep the animal warm if it is to survive. The first day after surgery, the mouse usually is hunched over and looks poorly groomed, but the second day it should look normal. The suture should be checked every day for integrity, but usually the skin begins healing within 24 h. We noticed that mice recover much better when they are kept in groups with other operated mice, thus it is better to operate on several animals on the same day rather than one animal each day for several days.

24. Surgery performed on E12.5 embryos or earlier stages usually yields approx 50% survival, whereas embryos operated at stages after E14.5 will yield survival rates approaching 100%.

25. The foster female will be accepting the newborns as long as they look healthy, which can be enhanced by letting them develop slightly longer in the operated female. However, do not exceed one half day longer than the regular term, or the fetuses may detach from the placenta. Another factor for successful fostering is the number of newborns that are left to the foster mother. The number should not be less than four, to ensure sufficient breast stimulation. If less than four operated fetuses are available, keep the weakest newborns of the foster mother's litter. When fostering a combination of experimental and nonexperimental newborns, it is important to mark them for later identification. Injecting carbon particles subcutaneously provides a permanent mark that is easily distinguished.

Acknowledgment

This work is supported by NIH Grant no. HD 35245. V. N. M. is supported by a postdoctoral fellowship from the Association pour la Recherche contre le Cancer.

References

1. Papaioannou, V. E. (1990) In utero manipulations, in *Post Implantation Mammalian Embryos, A Practical Approach* (Copp, A. J. and Cockroft, D. L.), IRL, Oxford University Press, Oxford, UK, pp. 61–80.

2. Brüstle, O., Maskos, U., and McKay, R. D. G. (1995) Host-guided migration allows targeted introduction of neurons into the embryonic brain. *Neuron* **15,** 1275–1285.

3. Muneoka, K., Wanek, N., Trevino, C., and Bryant, S. V. (1990) *Exo utero* surgery, in *Post Implantation Mammalian Embryos, A Practical Approach* (Copp, A. J. and Cockroft, D. L., eds.), IRL, Oxford University Press, Oxford, UK, pp. 41–59.

4. Muneoka K., Wanek, N., and Bryant, S. V. (1989) Mammalian limb bud development: In situ fate maps of early hindlimb buds. *J. Exp. Zool.* **249,** 50–54.

5. Tsutsui, Y., Kashiwai, A., Kawamura, N., Magahama, M., Mizutani, A., and Naruse, I. (1989) Susceptibility of brain cells to murine cytomegalovirus infection in the developing mouse brain. *Acta Neuropathol.* **79,** 262–270.

6. Turner, D. L., Snyder, E. Y., and Cepko, C. L. (1990) Lineage-independent determination of cell type in the embryonic mouse retina. *Neuron* **4,** 833–845.

7. Snyder, D. C., Coltman, B. W., Muneoka, K., and Ide, C. F. (1991) Mapping the early development of projections from the entorhinal cortex in the embryonic mouse using prenatal surgery techniques. *J. Neurobiol.* **22,** 897–906.

8. Serbedzija, G. N., Bronner-Fraser, M., and Fraser, S. E. (1992) Vital dye analysis of cranial neural crest cell migration in the mouse embryo. *Development* **116,** 297–307.

9. Trevino, C., Anderson, R., and Muneoka, K. (1993) 3T3 cell integration and differentiative potential during limb development in the mouse. *Dev. Biol.* **155,** 38–45.

10. Naruse, I. and Keino, H. (1993) Induction of agenesis of the corpus callosum by destruction of anlage of the olfactory bulb using fetal laser surgery *exo utero* in mice. *Brain Res. Dev. Brain Res.* **71,** 69–74.

11. Hopkinson-Woolley, J., Hughes, D., Gordon, S., and Martin, P. (1994) Macrophage recruitment during limb development and wound healing in the embryonic and foetal mouse. *J. Cell Sci.* **107,** 1159–2267.

12. Naruse, I., Keino, H., Taniguchi, M., and Masaki, S. (1994) The role of apoptosis in the manifestation of polydactyly and arhinencephaly in genetic mutant mouse *Pdn/Pdn. Cong. Anom.* **34,** 321–328.

13. Tamagawa, M., Keino, H., Taniguchi, M., Morita, J., and Naruse, I. (1994) Induction of the hydrocephalus by the intraventricular injection of aurintricarboxylic acid in mice *exo utero. Cong. Anom.* **34,** 345–352.

14. Reginelli, A. D., Wang, Y.-Q., Sassoon, D., and Muneoka, K. (1995) Digit tip regeneration correlates with regions of *Msx1* (*Hox 7*) expression in fetal and newborn mice. *Development* **121,** 1065–1076.

15. Naruse, I. and Keino, H. (1995) Apoptosis in the developing CNS. *Prog. Neurobiol.* **47,** 135–155.

16. Iseki, S., Wilkie, A. O. M., Heath, J. K., Ishimaru, T., Eto, K., and Morriss-Kay, G. M. (1997) *Fgfr2* and *osteopontin* domains in the developing skull vault are mutually exclusive and can be altered by locally applied FGF2. *Development* **124,** 3375–3384.

17. Sekimoto, H., Hatta, T., Moriyama, K., Otani, H., Moritake, K., and Tanaka, O. (1997) Developmental analysis of chlorambucil–induced occipital blebs in mice. *Cong. Anom.* **37,** 31–45.

18. Muneoka, K., Wanek, N., and Bryant, S. V. (1986) Mouse embryos develop normally *exo utero. J. Exp. Zool.* **239,** 289.

19. Larsen W. J. (1993) *Human Embryology* (Larsen, W. J., ed.), Churchill Livingstone, New York.

20. Naruse, I., Keino, H., and Taniguchi, M. (1996) Fetal laser surgery *exo utero* in mice. *Cong. Anom.* **36,** 107–113.

21. Copp, A. J. (1990) Studying developmental mechanisms in intact embryos, in *Post Implantation Mammalian Embryos, A Practical Approach* (Copp, A. J. and Cockroft, D. L., eds.), IRL, Oxford University Press, Oxford, UK, pp. 293–316.

22. Murti, R. and Schimenti, J. C. (1991) Microwave-accelerated fixation and *lacZ* activity staining of testicular cells in transgenic mice. *Anal. Biochem.* **198,** 92–96.

VIII

APPLICATION OF VIRAL VECTORS IN THE ANALYSIS OF DEVELOPMENT

47

Methods for Constructing and Producing Retroviral Vectors

Andrea Gambotto, Seon Hee Kim, Sunyoung Kim, and Paul D. Robbins

1. Introduction

Vectors derived from murine retroviruses have been used extensively for gene transfer in both preclinical and clinical studies. Retroviruses are small RNA viruses that replicate through a double-stranded DNA intermediate. The ability of retroviral vectors to integrate efficiently into the host DNA of infected cells, resulting in stable gene expression, makes them well suited for certain gene therapy applications. The majority of the applications using retroviral vectors have involved ex vivo methods in which cells are genetically modified in culture prior to introduction into the animal (*1,2*). However, the recent improvement in methods for production and concentration of retroviruses now allows for direct, in vivo applications (*see* **Notes 1** and **2**). This chapter will give a brief background on Moloney Murine Leukemia Virus- (MLV) based retroviruses and describe how to construct, produce, and titer replication-defective retroviral vectors.

Retroviruses infect target cells through a specific interaction between the viral envelope protein and a cell surface receptor on the target cell. The virus then is internalized where it is uncoated and the RNA reversed transcribed into proviral double-stranded DNA using the virally encoded Pol gene. The double-stranded DNA is then transported to the nucleus where it is integrated into the host DNA using a virally encoded integrase. The integrated provirus is transcribed, producing RNAs encoding the viral Gag, Pol, and Env proteins, which allow for packaging of the full-length unspliced viral RNA containing the psi site; the spliced message does not contain the psi sites and thus is not packaged. At least one of the Gag polypeptides binds directly to the viral RNA to facilitate packaging into the virion that, upon assembly, is released from the cell by budding from the Env-coated cell membrane. An important feature of the virus life cycle is that the infected cell stably produces viruses for the life of the cell without dramatically altering the growth properties of the cell. Thus stable viral producer lines can be generated that allow for continued recombinant virus production.

Most of the retroviral vectors currently in clinical use are based on MLV, a well-studied and characterized retrovirus. The MLV genome encodes for three polyproteins, Gag, Pol, and Env, which are required in *trans* for viral replication and packaging.

From: *Methods in Molecular Biology, Vol. 135: Developmental Biology Protocols, Vol. I*
Edited by: R. S. Tuan and C. W. Lo © Humana Press Inc., Totowa, NJ

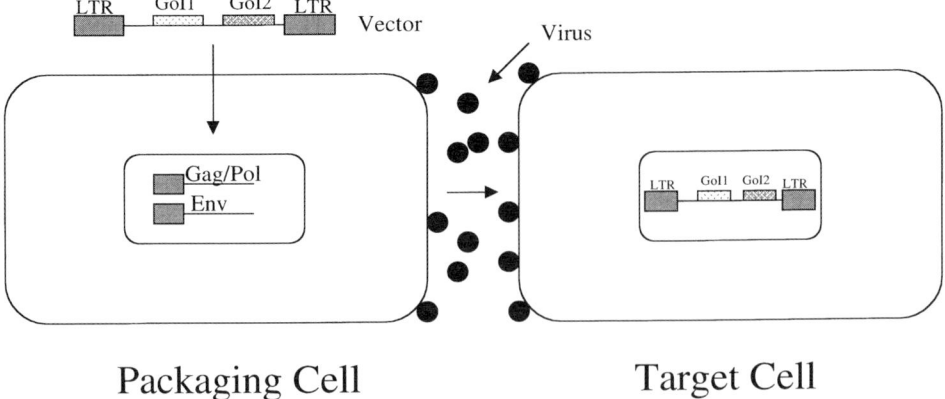

Fig. 1. Generation of infectious, replication-defective retroviral vectors. The retroviral vector carrying the genes of interest (*Go*I1 and *Go*I2) is introduced by transfection into the packaging line that stably expresses the retroviral trans-acting proteins, Gag, Pol, and Env. The vector RNA is packaged into infectious virus that is released from the cell by budding. The infectious virus is able to infect the target cells where it integrates into the host DNA. However, the virus is unable to replicate because the viral proteins are not present.

However, all that is required for viral replication in *cis* are the 5' and 3' long-terminal repeats (LTRs) that contain promoter, poly-adenylation, and integration sequences, a packaging site termed psi, and a tRNA binding site as well as several additional sequences involved in reverse transcription. The genes encoding the three viral proteins can be removed and heterologous genes and transcriptional regulatory sequences inserted (**Fig. 1**). In order to produce infectious, replication-defective retroviral vectors, cell lines that express the three viral proteins, Gag, Pol, and Env, have been generated, termed packaging cells. The vector can then be introduced into the packaging cells either transient or stably by transfection, resulting in the production of infectious virus. However, following infection of a target cell and integration of the vector into the host DNA, the virus is unable to replicate because of the absence of the viral *trans*-acting proteins.

There are three general classes of retroviral vectors *(3–9)*. The first class of vector is the LTR-based vector, where the gene of interest is expressed directly from the LTR (**Fig. 2A–E**). In the LTR-based vectors, two or more genes can be expressed by either differential splicing (**Fig. 2D**) or through the use of an internal ribosome entry site (IRES) (**Fig. 2E**). The second class of vector is the internal promoter vector where the gene of interest is driven from an internal promoter (**Fig. 2F–H**). In this type of vector, the 3' LTR can be mutated so that after infection, LTR-based expression is abolished (**Fig. 2D**) or the LTR can be maintained. The presence of a wild-type LTR allows for expression of a second gene from the LTR (**Fig. 2F**). Finally, there is a so-called reverse orientation vector where the gene of interest is expressed from its own promoter in reverse orientation (**Fig. 2I**). This type of vector is used for insertion of genomic sequences with introns, allowing for more appropriate regulation of gene expression.

Retroviral vectors have the advantage of stably integrating into the host DNA in the infected cell and expressing a therapeutic gene for the life of that cell. Furthermore, the

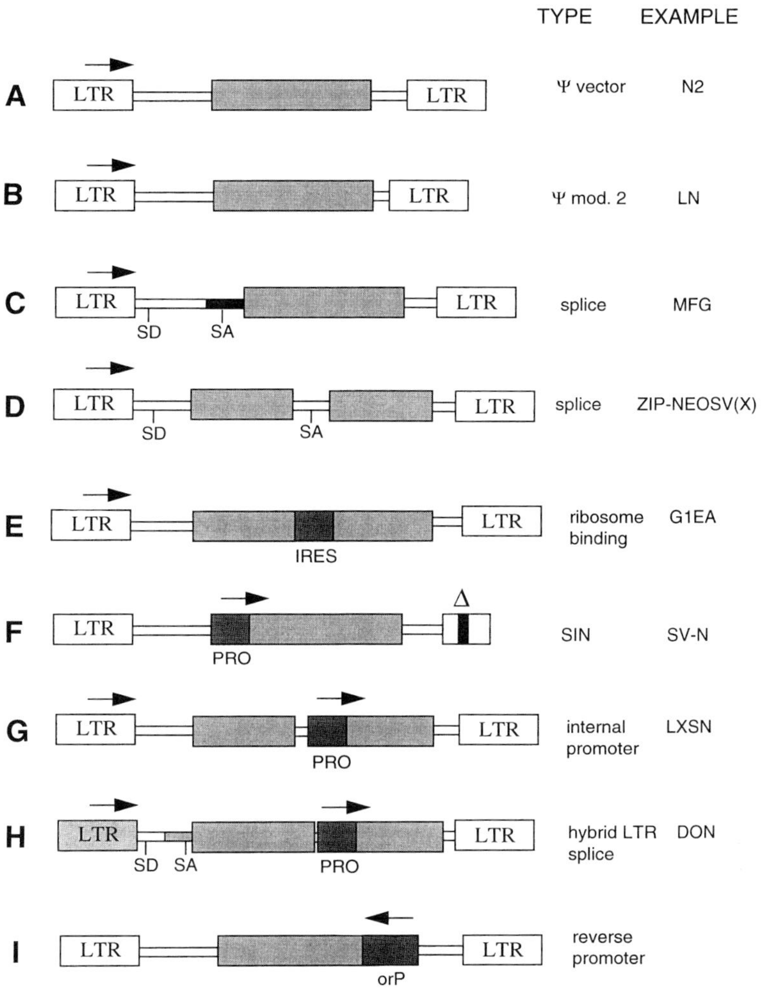

Fig. 2. Retroviral vectors. Shown are diagrams of various types of retroviral vectors designed to express single or multiple genes. (**A**) The ψ modification vectors use the retroviral LTR to drive the expression of a gene inserted 3' to the extended packaging signal (the N2 vector is an example of this type of vector). (**B**) The ψ modification vectors, are similar to the modification 1 vectors but have further deleted 3' sequences derived from the Mo-MLV env region (LN is an example of this type of vector). (**C**) The splicing vectors use the native splice donor and acceptor region from Mo-MLV to express an inserted gene in a similar manner to the natural RNA coding for the env gene (MFG is an example of this class). (**D**) Splicing vectors use the LTR to express both genes, one directly and one via splicing (ZIP-NeoSV(X) is an example of this type of vector). (**E**) The internal ribosome entry site (IRES) vectors use the LTR to drive the expression of a single gene, the second cistron of which is translated via internal ribosome binding (the G1EA vector is an example of this type of vector). (**F**) Self-inactivating vectors, SIN, contain deletions in the 3' LTR that are translated to the 5' LTR during reverse transcription. Expression in SIN vectors is mediated by an internal promoter (SV-N is an example of this vector class). (**G**) Internal promoters use two (or more) promoters to express independent genes (the LXSN-type vectors are examples of this class). (**H**) Hybrid LTR and substitution of splicing acceptor promote strong gene expression. Deletion of all viral structural genes including *gag*, *pol*, and *env* in vector can be used safe in clinical study without helper virus (the DON vector is one of the recently improved vector). (**I**) Reverse orientation vector, where the genomic sequence including promoter, introns, and poly adenylation site, is inserted in the reverse orientation.

provirus is maintained during subsequent mitotic division; thus a retrovirally trans-duced cell can be clonally expanded. Packaging lines that can produce moderate titers (10^6–10^7) of the replication-defective vectors, without giving rise to replication-competent helper virus through recombination, have been generated and used clinically (*see* **Notes 3** and **4**). The disadvantage to the MLV-based retroviral vectors is that they require cell division for integration, in particular, mitosis. Thus the current retroviral vectors are better suited for ex vivo gene therapy where isolated cells can be propagated in culture, genetically modified by retroviral infection, and then transplanted into the recipient. Retroviral vectors also may be used to infect certain rapidly dividing cells in vivo, such as hepatocytes following partial hepatechtomy, rapidly dividing tumor cells, or proliferating synovial cells lining inflamed joints (*see* **Note 5**). Recent advances in producing higher titer viruses and the further development of lentivirus vectors, which can infect nondividing cells, may allow for the general application of retroviral vectors to in vivo gene therapy (*see* **Note 6**).

2. Materials

2.1. Buffers for Cloning

1. PCR reaction: 50 µL of reaction mixture containing template DNA, 4 µL of 2.5 m*M* dNTP mix, 4 µL of 25 m*M* MgCl$_2$, 100 ng of upstream primer, 100 ng of downstream primer, 5 µL of 10X *Taq* buffer, 2.5 U of *Taq* polymerase.
2. TAE agarose gel running buffer: 0.04 *M* Tris-acetate and 0.001 *M* EDTA, 50X stock composition per liter; 242 g Tris base, 57.1 mL glacial acetic acid, 100 mL of 0.5 *M* EDTA, pH 8.0.
3. DNA sample loading buffer: 0.25% bromophenol blue, 0.25% xylene cyanol FF, 30% glycerol in water.
4. Agarose block melting buffer: composition per 100 mL, 90.8 g NaI 1.5 g Na$_2$SO$_3$ in 100 mL distilled H$_2$O.
5. Glass bead washing buffer: 1 *M* NaCl, 50 m*M* MOPS, 50% ethanol.

2.2. Cell Culture Reagents

1. Medium: Dulbecco's Modified Eagle's Medium (DMEM; Gibco-BRL, Gaithersburg, MD) supplemented with 120 µg/mL penicillin G (Sigma, St. Louis, MO, 1690 U/mg) and 200 µg/mL streptomycin sulfate (Sigma, 750 U/mg) containing 10% fetal bovine serum (FBS) or calf serum (CS).
2. Phosphate-buffered saline (PBS) adjusted to pH 7.4 with phosphoric acid or sodium hydroxide): composition per liter, 0.2 g KCl, 0.2 g KH$_2$PO$_4$ (monobasic), 8.0 g NaCl, 1.14 g Na$_2$HPO$_4$.
3. Trypsin-EDTA: composition per liter, 0.5 g Trypsin, 0.38 g EDTA.4Na, and 0.85 g NaCl in 1 L of PBS.

2.3. Transfection and Transduction

1. CaCl$_2$ solution: 2 *M* CaCl$_2$ composition per 100 mL, 29.4 g in 100 mL distilled water.
2. 2X HEPES buffered saline (HBS) solution: 280 m*M* NaCl, 10 m*M* KCl, 1.5 m*M* Na$_2$HPO$_4$, 12 m*M* dextrose, 50 m*M* HEPES. After filtration, store at –20°C.
3. Chloroquine: 25 µg/mL chloroquine (Sigma), 1000X stock composition, 250 mg chloroquine in 10 mL PBS. After filtration, keep in –20°C.

4. Polybrene: 8 µg/mL hexadimethrine bromide (Sigma), 1000X stock composition, 80 mg hexadimethrine bromide in 10 mL of PBS. After filtration, store at –20°C.
5. G418: 1 mg/mL Geneticin (Gibco-BRL), 200X stock composition, 1 g Geneticin (G418-sulfate) in 50 m*M* HEPES or serum-free media. After filtration, store at –20°C.

2.4. Titering of Virus Stocks

1. Sodium azide fixative: 0.1% sodium azide in 1X PBS, 10 g sodium azide in 1 L PBS. After filtration, store at 4°C.
2. Paraformaldehyde: 4% paraformaldehyde in 1X PBS: Add 4 g of paraformaldehyde to 80 mL of distilled water. Add 20 µL of a 10 *M* NaOH solution. Stir and heat to 60°C until totally dissolved. Add 10 mL of 10X PBS. Bring final volume to 100 mL with distilled water. Keep at 4°C.
3. Crystal violet: 0.3% crystal violet in 70% methanol.
4. TE (pH 7.4): 10 m*M* Tris-HCl (pH 7.4), 1 m*M* EDTA (pH 8.0).
5. TES: 0.7% sodium dodecyl sulfate in TE.
6. Depurination solution: 0.25 *M* HCl.
7. Denaturing solution: 0.5 *M* NaOH, 1.5 *M* NaCl.
8. Neutralizing solution: 1.0 *M* Tris-HCl (pH 8.0), 1.5 *M* NaCl.
9. Standard saline citrate: 20X solution; Dissolve 175.3 g NaCl, 88.2 g sodium citrate in 800 mL of H_2O. After adjusting to pH 7.4 with 10 *N* NaOH, add distilled water to final 1 L.

3. Methods

The method for constructing a recombinant retroviral vector involves the use of standard recombinant DNA techniques. The gene of interest is inserted in the correct orientation into one of usually several cloning sites in the vector backbone. First, the gene has to contain the appropriate 5' and 3' restriction sites on each end of the gene to be inserted. The appropriate restriction sites can be introduced by PCR amplifying the DNA fragment using primers containing the correct restriction sites. Alternatively, small double-stranded oligonucleotides, termed linkers, which contain the required restriction site, can be ligated to the ends of the DNA fragment. The construction of a recombinant vector depends on the vector to be used and the sequence of the gene to be inserted. Therefore, several examples of how to insert specific genes into two different vectors are detailed below.

The retroviral vector MFG (**Fig. 2C**) expresses the inserted gene from the LTR using the MLV-envelope spliced message. The start site of translation (ATG) for the inserted gene is fused to the envelope ATG at an *Nco*I site. Because the majority of the genes to be inserted into a vector do not contain an *Nco*I site at the start site of translation, the genes have to be engineered by PCR to contain the appropriate restriction sites. One example of how a gene (*Neo*^{r}) can be engineered by PCR for insertion into MFG is shown in **Fig. 3A**. Because the *Neo* gene contains an internal *Nco*I site, the gene can only be inserted into MFG using a two-step process (detailed in **Subheading 3.1.**) or using a three-part ligation strategy. The strategy for insertion of the gene encoding enhanced green fluorescent protein (*eGFP*) into the vector DONAI is also shown (**Fig. 3B**). DONAI has two restriction sites for insertion of the required gene, *Bam*HI, which gives a "sticky" end, and *Hpa*I, which gives a blunt end. Thus the *eGFP* gene has to be engineered so that it has a 5' *Bam*HI site and a 3' blunt end.

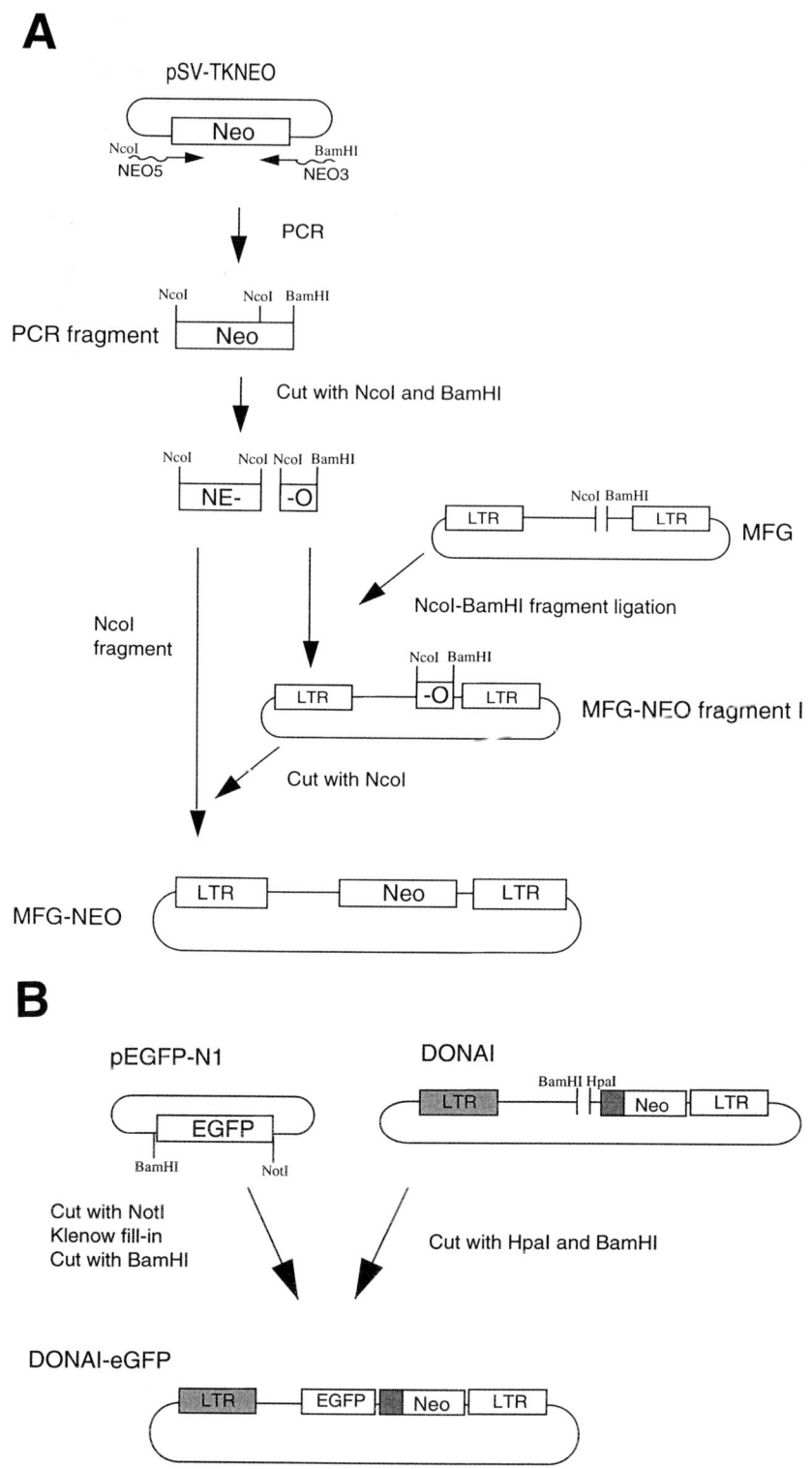

Fig. 3. Construction of retroviral vectors. (**A**) Schematic for the construction of MFG-Neo. (**B**) Schematic for construction of DONAI-*eGFP*.

3.1. Insertion of the Neomycin Resistance Gene into the MFG Retroviral Vector: MFG-Neo

1. *Neo*[r] sequence containing *Nco*I and *Bam*HI sites can be cloned by PCR using the following primers from the plasmid pSV-TKNEO (Stratagene Cloning Systems, La Jolla, CA):

NEO-5 (upstream primer, 5' to 3') <u>CCATGG</u>GATCGGCCATTGAAC
*Nco*I

NEO-3 (downstream primer, 5' to 3') <u>GGATCC</u>TCAGAAGAACTCGTC
*Bam*HI

2. The amplified DNA fragment is then inserted into the *Nco*I/*Bam*HI cloning sites of the MFG vector. Because the Neo[r] sequence has an *Nco*I site, it needs to be cloned into MFG through a two-step process (*see* **Fig. 2**). The *Nco*I/*Bam*HI fragment (232 bp) is first cloned into the *Nco*I/*Bam*HI cloning site of MFG and the *Nco*I fragment (571 bp) subsequently cloned into the *Nco*I site.
3. The amplified *Neo* fragment and the MFG vector are digested with *Nco*I and *Bam*HI restriction enzymes.
4. The desired DNA fragments, *Neo Nco*I-*Nco*I, *Neo Nco*I-*Bam*HI, and MFG *Nco*I-*Bam*HI, can be isolated by a number of different methods, including glass-bead purification. A brief description of how to isolate DNA from following electrophoresis on an agarose gel is given.
    a. The DNA fragments are separated on 0.8% agarose gel in tris-acetate buffer.
    b. The desired DNA fragments are visualized under UV light and the bands cut out of the gel with a clean razor blade.
    c. The gel slices are transferred to Eppendorf tubes and a threefold volume of sodium iodide solution. The tubes incubated at 50°C for 5 min until the agarose blocks are melted.
    d. Ten microliters of the glass-bead slurry is then added to the melted agarose and the tubes incubated on ice for 5 min.
    e. The glass beads containing the bound DNA fragments are then washed 3X with 0.5 mL of glass bead washing buffer.
    f. After drying the residual wash solution, the DNA fragments are eluted in TE buffer at 50°C for 5 min.
5. The purified *Neo Nco*I-*Bam*HI DNA fragment and the MFG *Nco*I and *Bam*HI vector were ligated with T4 DNA ligase in ligation buffer at 16°C water bath for 4 h. Usually a 100:1 ratio of insert to vector is used (100 ng:1 ng) in the ligation.
6. Ligation mixture is then transformed into an *E. coli* competent cell strain such as DH5α that are available commercially and the ampicillin-resistant transformants screened for selection of the colonies containing the MFG-3' NEO plasmid by restriction digest (*see* **Note 7**).

3.2. Insertion of the cDNA for Enhanced Green Fluorescent Protein (eGFP) into DONAI: DONAI-EGFP

1. The EGFP cDNA can be isolated from pEGFP-N1 (Clontech, Palo Alto, CA) by restriction digestion with *Bam*HI and *Not*I enzymes. However, in order to generate a 3' blunt end that will be compatible with the *Hpa*I cloning site in the DONAI vector, pEGFP-N1 is digested first with the *Not*I enzyme followed by filling in the 5' overhang with Klenow DNA polymerase. After the Klenow fill-in reaction, DNA is then cut with *Bam*HI enzyme.
2. DONAI vector was linearized by *Bam*HI and *Hpa*I enzymes and both ends of DONAI dephosphorylated with calf-intestinal phosphatase.

Vector   Plasmid   Transfection   Collecting   Transduction for
Construction   Preparatin   to Packaging cell line Cell-free   Titeration and
                                                               Viral Supernatant   Test of Gene Expression

                                                               OR

                                                               Selection
                                                               for Stable Line

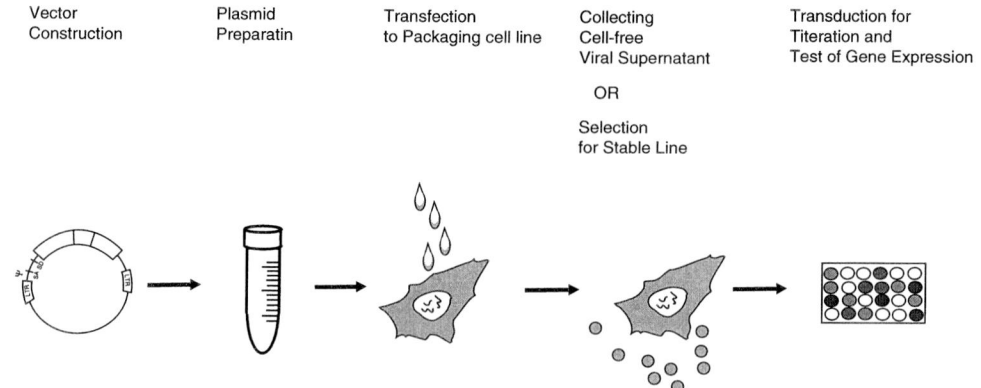

Fig. 4. Flowchart for generation of infectious, replication-defective virus. The cDNA is first inserted into the retroviral vector, the recombinant plasmid propagated, transfected into the appropriate packaging cell line and virus then used for infection of the appropriate target cell.

3. The EGFP fragment and vector were purified by glass-bead extraction and ligated using the same conditions as described above.
4. The *E. coli* transformants were screened for the correct DONAI-eGFP plasmid.

3.3. Preparing Viral Supernatants

Once the recombinant virus is constructed and confirmed by restriction analysis and sequencing to contain that the construct is correct, the next step is to generate infectious virus. A flowchart for the generation of a recombinant retroviral vector and production of infectious virus is shown in **Fig. 4**. Following a large-scale plasmid preparation using standard techniques, the vector DNA can be transfected into packaging cell lines for generation of stable producer clones or production of virus transiently. Producer lines are cell lines that stably express the three viral genes, *gag, pol,* and *env,* encoding the *trans*-acting proteins required for virus production. There are a number of different packaging lines available, derived from murine, canine, and human cells that can be used for virus production. The strengths and weaknesses of the different cells lines and the reasons for choosing one particular cell lines are outlined in **Subheading 4.**

3.3.1. Transient Virus Production

1. 293 derived packaging cell line (e.g., BOSC23, BING, PHOENIX, etc.) need to be passed at 1:3–1:4 to prevent cell clumping that occurs when the cells are passed at low density or when they are allowed to become overconfluent. Pass the cells by rinsing once with PBS and then trypsinizing for about 30 s (*see* **Note 8**).
2. Plate 2×10^6 packaging cells/plate 18–24 h prior to transfection in 4 mL of 10% FCS DMEM. The transfection efficiency will be higher if the cells appear as single cells rather than clumps. All conditions are for 60-mm plates (*see* **Note 9**).
3. Change media to 4 mL of DMEM 10% FCS containing 25 μM chloroquine (*see* **Note 10**).
4. Transfect by diluting 6–10 μg DNA with sterile water to a final volume of 438 μL in a 5-mL Falcon® 2054 polystyrene tube (Falcon, Los Angeles, CA). Add 62 μL of 2 M CaCl$_2$ and 500 μL of 2X HBS (pH 7.05) by bubbling (**Fig. 3**). Immediately (within 1–2 min) add this solution to the cells. Gently rock the plate and incubate at 37°C under 5% CO$_2$ (*see* **Note 11**).

5. At 10 h, suck off the media and replace with 4 mL of DMEM 10% FCS complete media. It is important to not leave the chloroquine longer than 12 h. This will cause a large decrease in titer. The range for chloroquine treatment is 7–11 h (*see* **Note 12**).

6. The cells should be nearly confluent by 24 h after transfection (*see* **Note 13**). Harvest the supernatant at 48–72 h posttransfection. The viral supernatants must be filtered through a 45-µL filter. Store the supernatants at –80°C in 2-mL aliquots.

3.3.2. Stable Producer Cell Line

The capability of a packaging cell line to become a stable producer cell line depends on the possibility to select stable clones with an antibiotic resistance gene present in the retroviral construct or that can be cotransfected with the retroviral vector. For instance, BOSC 23, BING, and PHOENIX packaging cell lines already contain a neomycin resistance gene and thus are not suitable for making a stable producer line for retrovirus carrying this resistance gene. Instead, the zeomycin resistance or related gene needs to be used. In contrast, the CRIP packaging cell line does not contain the neomycin resistance gene, making this packaging cell line suitable when neomycin gene is the retroviral resistance gene. It is also possible to use genes that express fluorescent proteins such as *eGFP* to sort for transient or stably transfected cells.

In order to generate a stable packaging line, the vector can either be introduced by infection or by transfection. For infection, virus can be generated by transfection of a packaging line expressing the amphotropic envelope and then used for infection of a packaging line expressing the ecotropic envelope. It is important to note that a virus carrying a specific envelope protein cannot infect packaging cells expressing that same class of viral envelope. If possible, stable packaging lines should be generated by infection, because this usually allows for higher titers. In addition, a stable producer can be infected multiple times with a virus in order to increase the copy number, resulting in a higher titer virus. If a stable producer needs to be generated by transfection, the method to be used should be similar to that described above for transient transfection, but the cells should be split into selection media 48 h posttransfection. If the virus does not carry the antibiotic resistance gene, then a plasmid expressing the appropriate marker gene can be cotransfected at a ratio of 20:1, vector to marker plasmid.

3.3.3. Generation of Stable Producer by Viral Infection

1. On day 0, plate 5×10^5 packaging cells (e.g., CRIP) in a 10-cm culture dish.
2. On day 1, aspirate medium from the plate (the plate should now contain approx 10^6 cells).
3. Add 2 mL viral supernatant and 2 µL polybrene to the plate (polybrene stock 8 mg/mL) and rock the plate gently (*see* **Note 14**).
4. Incubate the plate for 2–6 h at 37°C under 5% CO_2 and rock it every 15–30 min.
5. After incubation is completed, add 8 mL media. (Viral supernatant may be removed if packaging cells media is different from it.)
6. On day 4, split the cells 1:10 and/or 1:20 and start the selection using the specific antibiotics at specific concentrations (*see* **Note 15**).
7. Once the cells are completely selected (10–30 d), plate 2×10^6 cells in a 10-cm dish in 7 mL of complete media (*see* **Note 16**).
8. Harvest the supernatant after 48–96 h and filter it through a 0.45-µm filter and store at –80°C in 2-mL aliquots (*see* **Note 17**).

3.4. Methods for Titering

There are many methods available to titer viruses. The important point is to be able to measure the amount of infectious virus and to have a virus of known titer to use as a standard. Described below are methods for titering using neo selection, GFP expression, Southern blotting, and PCR.

3.4.1. Neo Resistance (for Viruses Carrying the Gene for neo^r)

1. The target cells for transduction, NIH3T3 cells (5×10^5), are plated in a 6-cm plate the day before infection.
2. The viral supernatants are harvested from the producer lines and passed through a 0.45-μm filter.
3. Three serial 10-fold dilutions of the original virus stock are made in 1-mL total volume and added to the plate with 8 μg/mL polybrene. Be sure to dilute the virus sufficiently (1:10^5) in case the titer of the virus is greater than 10^6 in order to be able to count the resulting colonies.
4. The plates are then placed at 37°C (5% CO_2) for 4 h, swirled occasionally, and then 2 mL of fresh medium added for 20 h.
5. The transduced cells are split 1:10 and seeded onto 100-mm dishes in a medium containing 1 mg/mL of G418. The medium is replaced every 3–4 d.
6. The resulting G418-resistant colonies are counted 10–12 d postinfection by fixing and staining with 0.3% crystal violet in 70% methanol.
7. The virus titers are determined by dividing the total number of colonies by the volume (in milliliters) of the retrovirus stock used and multiplying by 10 to correct for the 1:10 split.

3.4.2. GFP Expression for eGFP Expressing Viruses

1. Supernatant containing the *eGFP* recombinant retrovirus is generated from producer cells for 24 or 48 h in appropriate culture medium.
2. The culture supernatant is subsequently removed and passed through a 0.45-μm filter.
3. For transduction of NIH3T3 fibroblast, supernatant containing eGFP retrovirus is supplemented with polybrene at 8 μg/mL and layered onto subconfluent cell monolayers.
4. The next day, the cells are fed fresh culture medium and grown for 48 h before analysis by flow cytometry.
5. Transduced cells are trypsinized, washed with PBS, and resuspended with 0.3 mL of 2% paraformaldehyde.
6. Fixed cells can be stored at 4°C or analyzed directly by FACS (Becton Dickinson, Sunnyvale, CA) on 200,000 target cells resuspended in PBS with 1% bovine serum albumin and 0.1% sodium azide. Green fluorescence was measured through a 530–nm/30–nm bandpass filter after illumination with the 488-nm line of an argon ion laser.
7. The titer is calculated from the volumes corresponding to the linear slope of the curve according to the following formula:

$$\text{viral titer} = \frac{\text{NIH3T3 cell no. X\% of fluorescent cells}}{\text{volume of supernatant (mL)}}$$

3.4.3. Southern Blotting

3.4.3.1. ISOLATION OF GENOMIC DNA FROM CULTURED CELLS

1. Transduced cells are washed 2X with 10 mL of PBS and 3 mL of TES (TE + 0.7% SDS) solution was added (for 100-mm plate).

2. The homogenized cells are transferred to 15-mL tube, treated with proteinase K (400 µg/mL), and incubated at 50°C for 4 h to overnight.
3. The solution is extracted gently with 2 mL of phenol/chloroform until the white interface significantly diminished.
4. After removing the supernatant to a fresh tube, 2 vol of ethanol was added.
5. The genomic DNA was spooled out by toothpick into a microtube, washed with 70% ethanol, and dissolve in TE buffer.

3.4.3.2. DNA BLOT HYBRIDIZATION

1. The genomic DNAs is digested with *Nhe*I or with other restriction enzymes that cut within the LTRs of the proviral DNA, run in 0.8% agarose gel, and blotted onto nitrocellulose membrane.
2. The gel is depurinated by soaking in 0.25 *M* HCl for 15 min, treated with denaturing solution (0.5 *M* NaOH, 1.5 *M* NaCl) for 15 min, and immersed in neutralizing solution (1.0 *M* Tris-HCl, pH 8.0, 1.5 *M* NaCl).
3. The DNA is transferred to nitrocellulose membrane (Hybond-C; Amersham, Arlington Heights, IL) by capillary action in 20X SSC (3 *M* NaCl, 0.3 *M* sodium citrate).
4. After checking the complete transfer, the membrane is baked at 80°C in a vacuum for 1 h.
5. The blot is hybridized with a ^{32}P-labeled probe specific for the gene carried by the virus and washed with 0.2X SSC containing 0.1% SDS, 1X at room temperature for 20 min and 2X at 50°C for 15 min.
6. The hybridized blot is analyzed after exposing either to phosphoimager plate or X-ray film. The band intensity was compared with control of which the copy number is known. Usually the control consists of genomic DNA spiked with different amounts of proviral plasmid DNA.

3.4.4. PCR Amplification

1. The viral titer can also be estimated by semiquantitative PCR. Vector copy number per cell standards are prepared by diluting GFP-virus producer cells, which contain a single copy of the provirus, with empty-virus producer cells (100%, 50%, 25%, 5%, and 0% of GFP-virus producer cells).
2. DNA is extracted from the cell mixtures and used for PCR reaction. Five micrograms of genomic DNA from NIH3T3 cells transduced both with different volumes of viral supernatant and from the standards is amplified under the below-described condition.
   a. 50 µL of reaction mixture contained genomic DNA, 4 µL of 2.5 m*M* dNTP mix, 4 µL of 25 m*M* MgCl$_2$, 100 ng of upstream primer, 100 ng of downstream primer, 5 µL of 10X *Taq* buffer, and 2.5 U of *Taq* polymerase.
   b. An equal volume of mineral oil is overlayed onto the reaction mixture to prevent evaporation.
   c. Samples are subjected to 24 cycles of amplification. Each cycle included a denaturation step at 94°C for 1 min, a primer annealing step at 50°C for 1 min, and a DNA extension step at 72°C for 1.5 min.
   d. After the last amplification cycle, the samples are incubated at 72°C for 7 min to complete the elongation of the PCR intermediate products.
3. Ten microliters of the PCR products were resolved on 1% agarose gel containing 0.5 µg/mL ethidium bromide in 1X TAE (40 m*M* Tris-acetate, 1 m*M* EDTA) and visualized under UV light.

4. Notes

1. There are numerous types of retroviral vectors that can be used for gene transfer. The choice of which vector to be use depends on the specific application of the vector. For in vivo experiments, the vector should have been demonstrated to be able to infect the target tissue and express for extended periods of time. For in vitro experiments, the viral titers or duration of gene expression are not as important as in vivo. One vector that has been used extensively to express a number of different therapeutic genes is the vector MFG, where the gene of interest is driven from the LTR and expressed from the normal spliced Env message. This vector also has been used clinically for treatment of cancer, arthritis, and Gaucher disease. Multiple genes can be expressed in MFG using internal ribosome entry sites, allowing for coexpression of marker genes.

2. If the vector is to be used for animal experiments, the presence of an immunogenic marker gene such as *Neo* or *eGFP* may result in clearance of the infected cells. In addition, the *Neo* gene has been shown to decrease transcription from the viral vector. Therefore, it is recommended that retroviral vectors to be used for in vivo experiments not contain a selectable marker.

3. There are several different types of packaging cells that can be used to make virus. The murine-based packaging lines, PA317 and CRIP, are able to produce virus at titers between 10^6 and 10^7 without giving rise to replication competent virus. However, it has been difficult to concentrate virus from the murine packaging lines, because there appear to be factors, able to inhibit viral infection, that are also coconcentrated. Packaging lines based on either human 293 cells (BOSC23, BING, PHOENIX), human fibrosarcoma cells (FLY), or canine cells (DA) appear to produce higher titer virus that can be concentrated. The 293 based-packaging lines also can be transfected with high efficiency, allowing for transient production of high titer virus. However, the human-based packaging lines have not been used for clinical trials.

4. The choice of packaging lines also depends on which type of viral envelope is needed *(13–20)*. For infection of murine cells, packaging lines expressing the ecotropic receptor (CRE, BOSC23) can be used. However, for infection of human cells, packaging lines expressing the amphotropic receptor (PHOENIX, BING, CRIP, and PA317) or possibly other receptors (GP13) need to be utilized. The envelope from Gibbon ape leukemia virus (GALV) has been shown to allow for better infection of human hematopoietic cells than the amphotropic envelope. Therefore, the murine-based packaging line GP13, which expresses the GALV envelope, may be useful for certain applications.

5. Retroviral vectors can be pseudotyped using the G protein from vesicular stomatitis virus (VSV) to greatly enhance the host range of the virus. An advantage to using the G-protein of VSV is that the virus can be easily concentrated by centrifugation in contrast to the MLV ecotropic and amphotropic receptors. Indeed, viral titers of 1010 per milliliter can be generated using the VSV G protein as the viral envelope. However, expression of VSV G protein is toxic to cells, so 293-based packaging lines expressing G protein from an inducible promoter have been generated. Alternatively, it is possible to generate virus by transient cotransfection of 293 cells with the vector plasmid and Gag-Pol and VSV G protein expression vectors.

6. Vectors based on lentiviruses such as human immunodeficiency virus or equine infectious anemia virus have recently been generated *(21,22)*. The advantage to lentiviruses compared to MLV-based vectors is that they can infect certain nondividing cells with high efficiency. Thus they are well suited for in vivo gene therapy applications, such as gene transfer to liver, brain, and muscle. The production of replication-defective lentiviral vectors is similar to the production of replication-defective MLV-based vectors described in this chapter.

7. In order to insert the 5' *Nco*I-*Nco*I fragment of *neo*, the correct MFG-3'*Neo* is digested with *Nco*I and treated briefly with calf intestinal phosphatase enzyme to remove the 5' phosphate. This will prevent religation of the vector in the subsequent cloning step. The linearized fragment is then isolated from an agarose gel as previously described and used for the ligation reaction with the *Neo* 5' fragment. The resulting ampicillin-resistant bacteria clones obtained following transformation with the ligation mixture need to be screened for the presence of the *Nco*I fragment in the correct orientation by restriction digest.

8. Plating the cells may be the most important step in obtaining high titers. It is extremely important that the cells not be clumped and are the correct density. Unlike most adherent cell lines, the 293 derived packaging cells do not form nice monolayers. Instead they tend to clump before confluence (at which time the media will become acidic). To overcome the clumping, the cells can be split 1:1 1 or 2 d before plating them for transfection. This may need to be repeated if the cells do not spread well. After the 1:1 splits, it is best to pass the cells 1:2 for 1 or 2 passages and then at 1:3 or 1:4. The cells grow much slower than 293 cells, and the 1:3 split should take about 3–4 d to reach confluence. Transfections are better if you plate the cells for transfection before the plate become confluent.

9. 2×10^6 Cells/plate should result in a plate that is approx 80% confluent prior to transfection. It is also important to count the cells rather than estimating the split. Try to plate at a density so that the cells are 95–100% confluent at 24 h after transfection.

10. Roswell Park Memorial Institute media (RPMI-1640) cannot be used for $CaPO_4$ transfection. The excess positive charge in this media will cause formation of dense precipitate that is toxic to the cell.

11. You can add up to 20 μg of DNA without toxicity; polystyrene tubes yield superior results than tubes manufactured with polypropylene tubes; it is very important that the cells are nearly confluent prior to transfection. If the cells are not very confluent, there will be significant cell death.

12. In order to increase the relative titer/milliliter you may want to change the volume of media to 2.5–3 mL.

13. If the cells are not confluent at this time, you should play with the conditions so that they are confluent by 48 h.

14. If using a 6-cm plate for the infection, use only 1 mL of viral sup. and 1 μL of polybrene.

15. The antibiotic should be the same for which you have the resistance in your viral construct (e.g., G418/Neo; Zeocin/Zeo, etc.). The concentration may vary for the different packaging cell lines.

16. If using a 6-cm plate 10^6 cells, use 4.5 mL of complete media.

17. At this time, freeze the selected producer cell line in 10% DMSO. Thaw and repeat **steps 7** and **8** for preparing more retroviral supernatants.

References

1. Miller, A. D. (1992) Human gene therapy comes of age. *Nature* **357,** 455–460.
2. Mulligan, R. C. (1993) The basic science of gene therapy. *Science* **260,** 926.
3. Kim, S. H., Yu, S. S., Park, J. S., Robbins, P. D., An, C. S., and Kim, S. (1997) Construction of retroviral vectors with improved safety, gene expression, and versatility. *J. Virol.* **72,** 994–1004.
4. Byun, J., Kim, S.-H., Kim, J. M., Robbins, P. D., Yim, J., and Kim, S. (1996) Analysis of the relative level of gene expression from different retroviral vectors used for gene therapy. *Gene Ther.* **3,** 780–788.
5. Guild, B. C., Finer, M. H., Housman, D. E., and Mulligan, R. C. (1988) Development of retrovirus vectors useful for expressing genes in cultured murine embryonal cells and hematopoietic cells in vivo. *J. Virol.* **62,** 3795–3801.

6. Dranoff, G., Jaffee, E., Lazenby, A., Golumbek, P., Levitsky, H., Brose, K., et al. (1993) Vaccination with irradiated tumor cells engineered to secrete murine granulocyte-macrophage colony-stimulating factor stimulates potent, specific, and long-lasting anti-tumor immunity. *Proc. Natl. Acad. Sci. USA* **90,** 3539–3543.

7. Byun, J., Kim, J. M., Kim, S.-H., Yim, J., Robbins, P. D., and Kim, S. (1996) Simple and rapid method for the determination of the recombinant retroviruses titer by G418 selection. *Gene Ther.* **3,** 1018–1020.

8. Miller, A. D. (1992) Retroviral vectors. *Curr. Top. Microb. Immunol.* **158,** 1–24.

9. Miller, A. D., Miller, D. G., Garcia, J. V., and Lynch, C. M. (1993) Use of retroviral vectors for gene transfer and expression. *Methods Enzymol.* **217,** 581–599.

10. Zitvogel, L., Tahara, H., Robbins, P. D., Storkus, W. J., Clarke, M. R., Nalesnik, M. A., et al. (1995) Cancer immunotherapy of established tumors with IL-12. Effective delivery by genetically engineered fibroblasts. *J. Immunol.* **155,** 1393–1403.

11. Riviere, I., Brose, K., and Mulligan, R. C. (1995) Effects of retroviral vector design on expression of human adenosine deaminase in murine bone marrow transplant recipients engrafted with genetically modified cells. *Proc. Natl. Acad. Sci. USA* **92,** 6733–6737.

12. Robbins, P. D., Tahara, H., Mueller, G., Hung, G., Bahnson, A., Zitvogel, L., et al. (1994) Retroviral vectors for use in human gene therapy for cancer, Gaucher disease, and arthritis. *Ann. New York Acad. Sci.* **716,** 72–88.

13. Miller, A. D. and Wolgamot, G. (1997) Murine retroviruses use at least six different receptors for entry into mus dunni cells. *J. Virol.* **71,** 4531–4535.

14. Miller, A. D. (1990) Retrovirus packaging cells. *Hum. Gene Ther.* **1,** 5–14.

15. Pear, W. S., Nolan, G. P., Scott, M. L., and Baltimore, D. (1993) Production of high titer helper free retroviruses by transient transfection. *Proc. Natl. Acad. Sci. USA* **90,** 8392–8396.

16. Kinsella, T. M. and Nolan, G. P. (1996) Episomal vectors rapidly and stably produce high-titer recombinant retrovirus. *Hum. Gene Ther.* **7,** 1405–1413.

17. Miller, A. D., Garcia, J. V., von Suhr, N., Lynch, C. M., Wilson, C., and Eiden, M. V. (1991) Construction and properties of retrovirus packaging lines based on gibbon ape leukemia virus. *J. Virol.* **65,** 2220–2224.

18. Cosset, F. L., Takeuchi, Y., Battini, J. L., Weiss, R. A., and Collins, M. K. (1995) High titer packaging cells producing recombinant retroviruses resistant to human serum. *J. Virol.* **69,** 7430–7436.

19. Danos, O. and Mulligan, R. C. (1988) Safe and efficient generation of recombinant retroviruses with amphotropic and ecotropic host range. *Proc. Natl. Acad. Sci. USA* **85,** 6460–6464.

20. Ory, D. S., Neugeboren, B. A., and Mulligan, R. C. (1996) A stable human-derived packaging cell line for production of high titer retrovirus/vesicular stomatitis vikrus G pseudotypes. *Proc. Natl. Acad. Sci. USA* **93,** 11,400–11,406.

21. Naldini, L., Blomer, U., Gage, F. H., Trono, D., and Verma, I. M. (1996) Efficient transfer, integration, and sustained long-term expression of the transgene in adult rat brains injected with a lentiviral vector. *Proc. Natl. Acad. Sci. USA* **93,** 11,382–11,388.

22. Naldini, L., Blomer, U., Gallay, P., Ory, D., Mulligan, R., Gage, F. H., et al. (1996) In vivo gene delivery and stable transduction of nondividing cells by a lentiviral vector. *Science* **272,** 263–267.

48

Retroviral Gene Transduction
in Limb Bud Micromass Cultures

N. Susan Stott and Cheng-Ming Chuong

1. Introduction

Since Solursh and his coworkers set up the limb bud micromass cultures in 1977 *(1)*, this procedure has become a major model to analyze cellular and molecular event involved in chondrogenesis *(2–7)*. Recently, RCAS retroviral vectors have been developed that can infect chicken embryos *in ovo*. These have advanced our understanding in limb bud patterning remarkably *(8)*. Here we describe a newly developed protocol to merge these two procedures, which will make micromass culture an even more powerful model for the analysis of chondrogenic mechanisms.

Micromass culture has been used to study the role of growth factors, enzyme modulators, drugs, and adhesion molecules in chondrogenesis *(3–7)*. These studies have been facilitated by the development of serum-free media *(9)*. However, because some signaling molecules are intracellular, it has been difficult to dissect the chondrogenic pathways at a molecular level. To manipulate gene expression in micromass cultures, we have used electroporation to introduce exogenous genes into limb bud cells *(7)*. However, this leads to a high mortality rate of cells, so we have been continuing to search for better procedures.

With the advent of gene therapy *(10)*, retroviral mediated gene delivery has become a more mature technology. It has several advantages. The initial infection is not toxic to the cell, the gene expression is more stable than other procedures, and the percentage of infected cells will increase over time if the virus is replication competent. It takes approx 18 h after retrovirus infection for the gene to be expressed. Because precartilage condensation formation in micromass cultures starts within 3 h of plating, we would like to have most, if not all, cells expressing the transgene at this stage. To achieve this goal, we have devised a novel two-stage micromass culture, with a low-density plating window allowing retroviral gene transduction into primary limb bud cells, and a regular high-density plating allowing cells to differentiate (*see* **Note 1**). This strategy was used successfully to demonstrate the dual action of SHH in chondrogenesis (*11*; *see* **Note 2**) and the arrest effect of Wnt 7a at early stages of precartilage condensation formation *(16)*.

From: *Methods in Molecular Biology, Vol. 135: Developmental Biology Protocols, Vol. I*
Edited by: R. S. Tuan and C. W. Lo © Humana Press Inc., Totowa, NJ

2. Materials

1. Fertilized pathogen-free chicken eggs: (SPAFAS Connecticut Hatchery, Preston, CT). These chicken embryos are susceptible to infection by retroviruses carrying the A envelope subgroup *(15)*. Chicken embryos were staged as described *(12)*.
2. Media:
   a. Hank's buffered saline solution (HBSS; Gibco-BRL, Gaithersburg, MD).
   b. Dulbecco's modified Eagle's medium (DMEM) (Gibco-BRL) supplemented with 2% fetal calf serum and gentamicin (1:1000).
   c. Calcium-magnesium free saline (CMF 10X): 80 g of 1.37 M NaCl, 3 g of 0.04 M KCl, 0.5 g of 0.004 M NaH$_2$PO$_4$, 0.25 g of 2 M KH$_2$PO$_4$, 10 g of 0.12 M NaHCO$_3$, 20 g of 0.1 M glucose, in 1000 mL distilled water, pH 7.3.
   d. Defined medium for micromass cultures. Based on Paulsen et al. *(9)*. 60% Ham's F-12 (BioWhittaker, Walkersville, MD), 40% DMEM, 5 µg/mL insulin (Sigma Chemical Co., St. Louis, MO), 5 µg/mL transferrin (Sigma), 50 µg/mL ascorbic acid (Sigma), 100 nM hydrocortisone (Sigma).
3. Enzymes: Trypsin, collagenase (Worthington Biochemicals, Freehold, NJ), type-I collagen (UBI).

3. Methods

The outline of the procedure is shown in **Fig. 1**.

1. SPAFAS pathogen-free eggs are used. The distal one third of limb buds from stage 23-24 chicken embryos *(13)* are dissected and pooled. After soaking the limb buds in ice-cold 2X calcium- and magnesium-free medium (CMF) containing 0.25% EDTA, epithelia are removed and discarded. The pooled limb bud mesenchymal cells are subjected to mild dissociation conditions in HBSS containing 0.006% of trypsin and collagenase for 10 min at 37°C. The digestion is stopped with fetal calf serum. The cells are gently triturated, then centrifuged (150g) at 15°C and resuspended to a concentration of 2×10^7/mL in ice-cold micromass defined medium (DM).
2. For infection with retrovirus, the dissociated limb bud cells are incubated with retrovirus-containing medium (10^6 cells/mL of retroviral medium) at 4°C with gentle shaking for 2 h. After a gentle centrifugation (150g, 3 min), the cell pellets are resuspended in defined medium (*see* **Notes 2**, **4**, and **6**).
3. Tissue culture dishes are precoated with 100 µg/mL of type-I collagen in 0.02 N acetic acid for 1 h at 37°C and then neutralized with HBSS. For low-density plating, cells are plated at 5.5×10^4 cells/cm^2 on these dishes in defined medium and cultured at 37°C, 5% CO$_2$/95% air for 2 d.
4. Cells are washed 2X with HBSS and trypsinized again with 0.006% trypsin and collagenase in HBSS without Ca^{2+} and Mg^{2+}. Digestion is stopped with fetal calf serum and the dissociated cells passed through cell microsieve netting (20 µm) to ensure a single-cell suspension. The cells are plated in 10 µL drops at a density of 2×10^7/mL on collagen type I precoated 35-mm tissue culture dishes. The cells are allowed to attach for 1.5 h and then 1.5 mL of defined medium is added.
5. From this point on, the cultures proceed similar to normal micromass cultures (*see* **Note 3**). Cells form visible precartilage condensations in 1–2 d and cartilage nodules in 3–4 d. These cartilage nodules have been shown to be positive for alcian blue and collagen II immunostaining *(14)*.

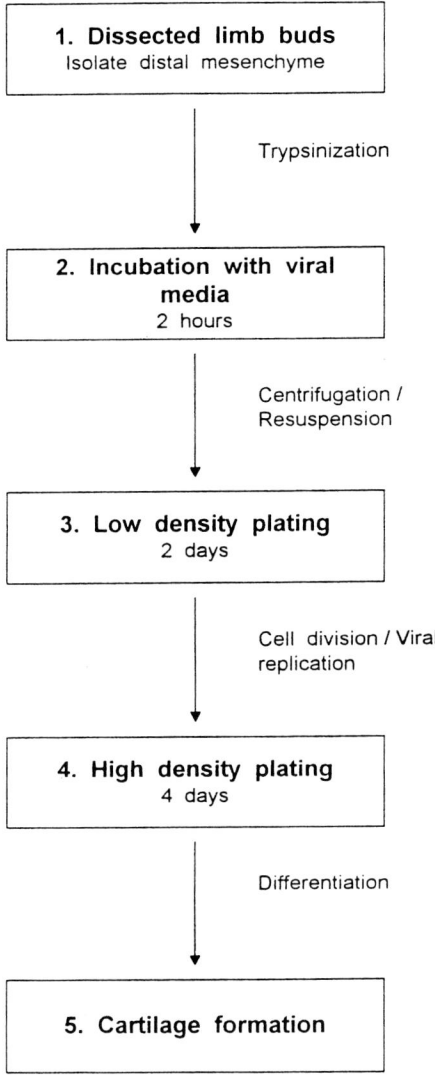

Fig. 1. Flowchart of the protocol.

4. Notes

1. Rationale for the two-stage procedures: The low-density window balances two needs: the need for rapid proliferation and retroviral integration and the need to maintain chondrogenic competence. Low-density culture provides a time window for cell proliferation and viral infection. Fetal calf serum can induce chondrocyte maturation, so we used serum-free medium during this time. To resume chondro-differentiation, we plate cells at high density and with type-I collagen as the substrate. Type-I collagen is used because it is highly expressed in limb bud in vivo prior to the precartilaginous condensation phase and may facilitate cell attachment and enhance chondrogenic competence.

2. Retroviral infectivity: Retroviral medium is produced and titered according to published methods *(8)*. Retroviral media is filtered with a 0.45-μm surfactant-free cellulose acetate

filter (Nalge Nunc International, Rochester, NY) and stored in aliquots at –70°C. The infectivity, judged by staining of antibodies to viral gag, reaches 60–75% of cells after 3 d of high-density plating *(13)*.

3. Effect of this procedure and RCAS on micromass cultures: Infection with RCAS vector did not significantly alter cell numbers as judged by DNA content and ^{3}H thymidine incorporation *(13)*. Both virus-free and RCAS-infected cultures showed a minor and similar level of reduction in overall chondrogenic differentiation compared with regular, traditional one-stage micromass culture *(1)*. This may be a result of the manipulation during the two-stage protocol. RCAS virus infection per se does not decrease the pattern of chondro-differentiation as judged by Alcian blue staining and immunostaining with antibodies to type-II collagen *(13)*.

4. Optimization of virus mediated gene transduction: We have tried to vary some parameters to achieve better results. The following are our experiences. These are the parameters that could be altered when applying similar procedures to other organotypic cultures.

    a. The time for virus infection was varied from 1 to 4 h. During this period, infectivity gradually increases but cell viability drops. Two hours of incubation with virus is a good middle point.

    b. Plating densities between 10^4 and 10^6 cells/cm^2 were tested for cell survival and growth together with different substrates. A density too high allowed the cells to become confluent and triggered premature chondrogenic differentiation, whereas a density too low decreased cell survival and growth. A plating density of 5.5×10^4 cells/cm^2 provided the best middle point for cell growth without differentiation.

    c. We also have tried to vary the length of low-density cultures. This period was varied from 1 to 3 d. With shorter time, infectivity is not high enough and the cell number is low. With longer time, infectivity is high, but cell-survival rate drops and the competence to form cartilage nodules after high-density plating also drops. Thus, 2 d of low-density plating provides the best balance between maintenance of chondrogenic competence and high levels of retroviral gene expression.

5. Because genes in retrovirus take about 18 h to be expressed, direct addition of retroviral media to the micromass cultures would not lead to exogenous gene expression until 18 h later. If the events to be perturbed are late events, one can still get results. This appears to be the case of the suppressive effect of Wnt 7a *(14)*.

6. We typically used RCASBP (A) for limb bud micromass cultures. However, other subgroups of RCAS are available that carry different envelope glycoproteins *(8,15)*. It is also possible to use cells from different specific chicken lines *(15)* to make a chimeric micromass cultures so that viral resistant cells and viral susceptible cells can be mixed. This allows the analysis for autonomous and nonautonomous effects.

7. Another strategy is to use limb buds from transgenic mice. This would be better as all cells then should contain the transgenes. However, the logistics of timing the mice to be at the right developmental stages and checking the genotype makes this approach less suitable. In comparison, it is easier to get large numbers of chicken embryos of the appropriate stages.

8. It is also possible to apply similar strategies to other organotypic cultures. Differences in cell surface receptor availability *(8,10,15)* and cellular properties may require some adjustment of various parameters. It is also possible to test different RCAS strains *(15)*.

Acknowledgment

This work was supported by grants from the National Institutes of Health, National Science Foundation, and the Wright Foundation/University of Southern California (C.-M. C.).

References

1. Ahrens, P. B., Solursh, M., and Reiter, R. S. (1977) Position-related capacity for differentiation of limb mesenchyme in cell culture. *Dev. Biol.* **60,** 69–82.
2. Daniels, K., Reiter, R., and Solursh. M. (1996) Micromass cultures of limb and other mesenchyme, in *Methods in Avian Embryology* (Bronner-Fraser, M., ed.), Academic, New York, pp. 237–247.
3. Hassell, J. R. and Horrigan, E. A. (1982) Chondrogenesis: A model developmental system for measuring teratogenic potential of compounds *Teratog. Carcinog. Mutagen.* **2,** 325–331.
4. Kosher, R. A. and Walker, K. H. (1983) The effect of prostaglandins on in vitro limb cartilage differentiation. *Exp. Cell Res.* **145,** 145–153.
5. Jiang, T. X., Yi, J. R., Ying, S. Y., and Chuong, C. M. (1993) Activin enhances chondrogenesis of limb bud cells: Stimulation of precartilaginous mesenchymal condensations and expression of NCAM. *Dev. Biol.* **155,** 545–557.
6. Lee, Y. S. and Chuong, C. M. (1996) Activation of protein kinase A is a pivotal step involved in both BMP-2 and cyclic AMP-induced chondrogenesis. *J. Cell Physiol.* **170,** 153–165.
7. Widelitz, R. B., Jiang, T. X., Murray, B. A., and Chuong, C. M. (1993) Adhesion molecules in skeletogenesis: II. Neural cell adhesion molecules mediate precartilaginous condensations and enhance chondrogenesis. *J. Cell Physiol.* **156,** 399–411.
8. Morgan, B. A. and Fekete, D. M. (1996) Manipulating gene expression with replication competent viruses. *Methods Cell Biol.* **51,** 185–218.
9. Paulsen, D. F., Chen, W.-D., Pang, L., Johnson, B., and Okello, D. (1994) Stage and region-dependent chondrogenesis and growth of chick wing-bud mesenchyme in serum-containing and defined tissue culture media. *Dev. Dyn.* **200,** 39–52.
10. Gordon, E. M. and Anderson, W. F. (1994) Gene therapy using retroviral vectors. *Curr. Opin. Biotechnol.* **5,** 611–616.
11. Stott, N. S. and Chuong, C. M. (1997) Dual action of sonic hedgehog on chondrocyte hypertrophy: retrovirus mediated ectopic sonic hedgehog expression in limb bud micromass culture induces novel cartilage nodules that are positive for alkaline phosphatase and type X collagen. *J. Cell Sci.* **110,** 2691–2701.
12. Hamburger, V. and Hamilton, H. L. (1951) A series of normal stages in development of the chick embryo. *J. Morphol.* **88,** 49–91.
13. Stott, S., Lee, Y. S., and Chuong, C. M. (1998) Retroviral gene transfer in chondrogenic limb bud micromass cultures. *BioTechniques* **24,** 660–666.
14. Rudnicki, J. A. and Brown, A. M. C. (1997) Inhibition of chondrogenesis by *Wnt* gene expression *in vivo* and *in vitro. Dev. Biol.* **185,** 104–118.
15. Fekete, D. M. and Cepko, C. L. (1993) Retroviral infection coupled with tissue transplantation limits gene transfer in the chicken embryo. *Proc. Natl. Acad. Sci. USA* **90,** 2350–2354.
16. Stott, S., Jiang, T.-X., and Chuong, C.-M. (1999) Successive formative stages of precartilage mesenchymal condensations *in vitro*: modulation of cell adhesion by wnt-7a and BMP-2. *J. Cell Physiol.*, in press.

49

Construction of Adenoviral Vectors

Alan R. Davis, Nelson A. Wivel, Joseph L. Palladino, Luan Tao, and James M. Wilson

1. Introduction

Human adenovirus is a double-stranded DNA virus with a 36-kb linear genome that contains four early transcription units (E1–E4), active at early (3–12 h) times after infection and five late transcription units (L1–L5) active thereafter and producing the structural genes of the virus. For further details on human adenovirus and its life cycle, the reader is referred to **ref. _1_**, which provides a description of the genome structure, and types of these viruses.

Recombinant adenoviruses with insertions in the E1 region are traditionally produced by overlap recombination after cotransfection of 293 cells with a plasmid shuttle vector and a large right-end restriction fragment of viral DNA _(2)_. This shuttle vector typically contains a cassette for a transgene placed in region E1 and flanking sequences from adenovirus for recombination. A high background of parental virus results because of the difficulties in completely separating the right-end restriction fragment of adenoviral DNA from uncut DNA. In addition, mixed plaques can arise containing both parental and recombinant adenoviruses. This necessitates not only extensive screening of initial plaques but also laborious rescreening following subsequent plaque purification.

Others have attempted to overcome these difficulties by using _E. coli (3,4)_ and yeast _(5)_ DNAs instead of viral DNA. These systems lose the transfection efficiency attained with authentic viral DNA and are labor intensive, requiring the isolation of an intact infectious clone of the desired vector in the shuttle system for each desired vector. Another strategy is to develop a selection system, using β-galactosidase as a negative selection system _(6)_ or using a parental vector expressing thymidine kinase coupled with the use of ganciclovir _(7)_. Finally, Cre-lox recombination recently has been described _(8)_ to accomplish the desired recombination in vitro.

In this chapter, we describe a negative selection system based on the traditional cotransfection method using viral DNA from an E1-deleted adenoviral recombinant expressing green fluorescent protein (GFP). _In situ_ fluorescent microscopy distinguishes the recombinant plaques (nonfluorescent or white) from background (fluorescent or green) plaques that are derived from the parental virus. In addition, contamination with parental virus is easily detected at later stages when production lots of the recombinant adenovirus are being made.

From: _Methods in Molecular Biology, Vol. 135: Developmental Biology Protocols, Vol. I_
Edited by: R. S. Tuan and C. W. Lo © Humana Press Inc., Totowa, NJ

The general strategy is to incorporate a simple and efficient screening method into the traditional cotransfection approach. Our previous experience with cotransfection in 293 cells is that the success rate is high and throughput is potentially fast. The problem is that the efficiency of homologous recombination to create a recombinant is often very low compared to the background plaques that emerge from incompletely restricted parent viral DNA or relegated viral DNA. Screening plaques by structural analysis of the viral DNA or expression of most transgenes requires isolation and characterization of individual plaques, which constitutes approx 90% of the effort in isolating new recombinants. Furthermore, the plaques containing recombinant virus are often contaminated with the input virus, requiring multiple rounds of plaque purification.

2. Materials

1. Plasmids: The basic adenoviral transfer vector, pAdCMVlink *(2)* is shown in **Fig. 1** together with landmarks. It is necessary to first clone the transgene DNA into this plasmid. If other promoters or regulatory elements are to be used, a plasmid lacking all regulatory elements (pAdlink) is also available.
2. Viruses: H5.000CMVEGFP, H5.010CMVEGFP, H5.030CMVEGFP, and H5.001 CMVEGFP *(9)*. In addition, the virus H5.000CMVEGFP4 (unpublished) can be used to reduce background (*see* **Notes 1–4**).
3. Cell lines: The 293 cell line was obtained from the American Type Culture Collection (Manassas, VA) *(10)* and used below passage 50.
4. Culture medium:
   a. For cell growth: Dulbecco's modified Eagle's medium (DMEM) with 10% fetal bovine serum (FBS) and 0.01 vol penicillin/streptomycin (Sigma Chemical Co., St. Louis, MO).
   b. For infection: DMEM with 2% FBS and 0.01 vol penicillin/streptomycin (Sigma).
5. Buffers:
   a. 20 mM HEPES buffer, pH 7.8, 150 mM NaCl.
   b. HEPES-buffered saline (HBS)/10% glycerol: 9 vol HBS and 1 vol glycerol.
   c. 2X HBS. For 10 mL: 0.5 mL 5 M NaCl, 1.0 mL 0.5 M HEPES, pH 7.1, 0.15 mL 0.1 M NaPO$_4$ · 7H$_2$O, pH 7.0, 8.5 mL H$_2$O. Combine and adjust to between pH 7.07 and 7.12 with NaOH. Filter sterilize and store at –20°C.
   d. Tris-EDTA: 10 mM Tris-HCl, pH 7.4, 1 mM EDTA.
   e. Light cesium chloride 100 mL: Dissolve 22.39 g cesium chloride in 77.61 mL of 10 mM Tris-HCl, pH 8.0. Autoclave. Density = 1.25 g/mL.
   f. Heavy cesium chloride, 100 mL: Dissolve 42.23 g cesium chloride in 57.77 mL of 10 mM Tris-HCl pH 8.0. Autoclave. Density = 1.45 g/mL.
   g. Proteinase K solution: 1 mg/mL proteinase K in 0.05 M Tris-HCl, pH 7.4, 1 mM EDTA, 0.5% SDS.
   h. Agar overlay mix: 2.0% agarose (Seaplaque agarose; FMC Bioproducts, Rockland, ME), 2X minimal essential medium (MEM; BioWhittaker, Walkersville, MD), FBS.
   i. Penicillin/streptomycin: 10,000 U penicillin and 10 mg streptomycin per milliliter in 0.9% NaCl (Sigma).
   j. Remelt agar solution in microwave. Meanwhile, mix together 100 mL of 2X MEM and 18.75 mL FBS, and 2 mL of the penicillin/streptomycin mixture. Hold solution in 45°C water bath. Cool melted agar to 45°C and add an equal volume of the agar and the MEM mixture. Mix well.

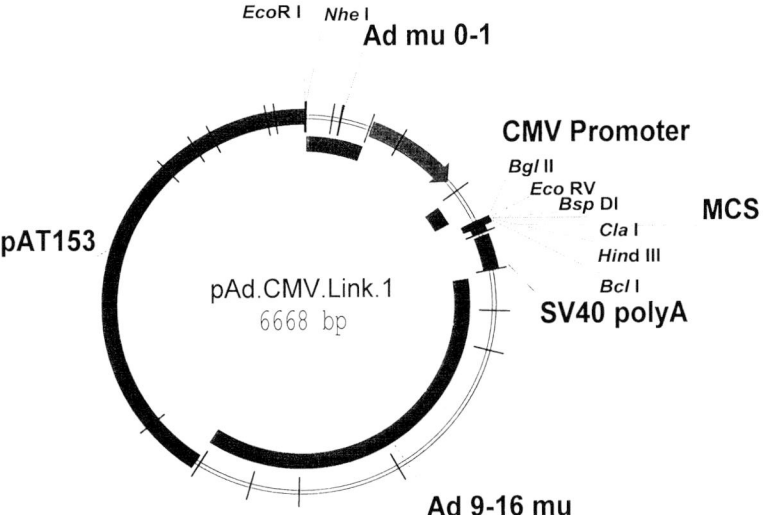

Fig. 1. Map of the shuttle plasmid pAdCMVlink. This plasmid contains a multiple cloning site with the unique restriction sites shown. The MCS is preceded by a CMV promoter and an intron. An SV40 polyadenylation signal is found downstream of the MCS. Additionally, Ad5 map units 0–1 and 9–16 flank the expression cassette. Homologous recombination with the backbone DNA occurs between 9 and 16 map units. The *Eco*RI and *Nhe*I sites in this plasmid are at positions corresponding to the left end of Ad5 DNA.

3. Methods

1. Transfection: Viral DNA (1 μg) is mixed with plasmid DNA (10 μg) and 37.5 μL of 2 *M* $CaCl_2$ and adjusted to a final volume of 300 μL with water. This solution is then added dropwise to 300 μL of 2X HBS solution while vortexing. After standing 30 min, 600 μL of the precipitate is added dropwise to a 60-mm dish of 293 cells growing in DMEM with 10% FBS that is approx 70–80% confluent. The medium is changed on these dishes approx 4 h before transfection. After overnight incubation, the monolayer is washed with phosphate buffered saline (PBS) without calcium or magnesium, followed by an agar overlay (5 mL) consisting of 1% agarose in MEM with 10% FBS. Dishes are overlaid with 3 mL of MEM with 10% FBS/1% agarose every 3–4 d until plaques appear.

2. Picking plaques: Remove plates from the incubator and hold plate up to the light observing monolayer from the bottom of the dish. Plaques will appear as clear translucent areas amid a darker monolayer. Adenovirus plaques formed after this transfection can also be viewed using either bright field or fluorescence microscopy. Cells in liquid culture as well as cells expressing GFP within plaques are viewed using a Nikon Diaphot TMD-EF inverted fluorescence microscope. Blue light is produced for excitation by utilization of a DM510 dichoric mirror and a BA520 barrier filter. If only a few plaques are obtained, each plaque is circled with a marking pen and observed. In this way it is easy to select the white (potential recombinant) plaques. If a large number of plaques are obtained they are scanned using the lowest objective of the microscope and white plaques are marked. The white plaques are picked by aspirating the agar plug from a marked area of the plate with a plugged Pasteur pipet. The plug is placed in 500 μL of DMEM/2% FBS and frozen at –80°C.

3. Plaque purification: After picking white plaques they are purified once by plaquing before any further analysis. A minimum of three white plaques should be used for each recombinant adenovirus.

For plaque purification, one can assume that a plaque contains between 10^3 and 10^4 PFU/mL, and dilution schemes should be based so that there are between 1 and 10 plaques per plate. Typically, dilutions of 5×10^{-2}, 5×10^{-3}, and 5×10^{-4} are made when repurifying a plaque.

a. Day 0
  i. Prepare in advance 6-well plates of 293 cells that are 80–90% confluent.
  ii. Freeze-thaw each plaque 3X. Use a 37°C water bath and an ethanol-dry ice bath for this purpose.
  iii. Prepare dilutions in a 24-well plate using 1 mL DMEM/2% for each dilution.
  iv. Aspirate the media from each well of 293 cells. Aspirate the medium from the cells and replace with 1 mL DMEM/2% FBS. Add 1 mL diluted virus to each.
  v. Incubate plates overnight at 37°C.

b. Day 1
  i. Prepare agar overlay mixture.
  ii. Aspirate media from wells.
  iii. Overlay each well with 2 mL agar mixture by slowly pipeting the solution down the side of the well, taking care not to dislodge the cells. Allow agar to solidify at room temperature (RT) for a minimum of 30 min, then return dishes to the 37°C incubator.

c. Day 3
  i. Prepare agar overlay solution.
  ii. Overlay with 2 mL of agar mixture on top of the existing agar in each well and allow to solidify at room temperature for 30 min. Return plates to incubator. The additional overlays are to feed cells to maintain monolayer integrity.

d. Days 6–9
  For each of the three potential recombinant adenovirus find the dilution with the lowest number of plaques. Circle several plaques. Check these using the inverted immunofluorescent microscope to ensure that they are white. Pick these plaques as described above and store them at –80°C.

4. Expansion of plaques
  a. Prepare 60-mm dishes of 293 cells that are 80–90% confluent.
  b. Freeze-thaw each plaque 3X as described in **step 3**.
  c. Dilute 200 µL of each plaque into 1 mL DMEM/20% FBS.
  d. Aspirate the medium from each dish and replace with 2 mL DMEM/2% FBS.
  e. Add the diluted virus to the dish and incubate overnight at 37°C.
  f. The next morning add 2 mL DMEM/2% FBS.
  g. Continue incubation until cytopathic effect is observed. This is noted by the appearance of grape-like clusters of rounded cells. Also check using the inverted immunofluorescence microscope that there are no green cells present.
  h. Pipet medium and cells into a 15-mL plastic centrifuge tube. Centrifuge 5 min at 400g.
  i. Aspirate the supernatant and resuspend the pellet in a microcentrifuge tube in 1 mL PBS without calcium or magnesium.
  j. Centrifuge 2 min at 850g in an IEC Micromax microcentrifuge. Aspirate the supernatant and resuspend in 1 mL of 10 mM Tris-HCl, pH 8.0.
  k. Remove 300 µL of the resuspended cells into a separate microcentrifuge tube for DNA structure analysis. Store the other 700 µL at –80°C.

5. Analysis of potential recombinant adenoviruses: At this juncture, one of several methods can be used to analyze the potential adenovirus recombinants. These methods mostly detect the expressed transgene and include enzymatic, immunological, or functional tests.

However, a simple first step toward characterization is to prepare Hirt DNA from the infected cell lysate and perform restriction enzyme analysis. PCR can also be performed, but one must keep in mind the sensitivity of this latter test and the potential for false-positive readouts. Preparation of Hirt DNA is performed essentially as described (*11*):

a. Add 33.3 µL of 10% SDS to the 300 µL cell suspension.

b. Incubate 30 min at 37°C.

c. Add 18 µL of 2 mg/mL pronase in 10 m*M* Tris-HCl, pH 8.0, predigested 2 h at 37°C and 7 µL of 1 mg/mL proteinase K.

d. Incubate 2 h at 37°C.

e. Add 250 µL of a solution of 0.67 *M* acetic acid, 3 *M* CsCl, and 1 *M* KOAc.

f. Incubate 15 min at 4°C.

g. Centrifuge 15 min at full speed in a microcentrifuge.

h. Load the supernatants onto a Qiagen Qiaprep spin column (Qiagen, Valencia, CA).

i. Wash 2X by centrifugation with 750 µL 10 m*M* Tris-HCl, pH 7.5, 80 m*M* KOAc, 40 µ*M* EDTA, and 60% (v/v) ethanol. After the washes, centrifuge once for 30 s to dry resin.

j. Elute the DNA using 50 µL Tris-EDTA that has been warmed to 65°C. A small amount (10 µL) of this DNA is sufficient for each restriction enzyme digestion performed.

6. Expansion of positive recombinant adenoviruses:

a. Prepare five 15-cm plates of 293 cells that are 80–90% confluent.

b. After the correct recombinant adenovirus is identified, freeze-thaw the remaining 700 µL of lysate 3X.

c. Centrifuge 5 min in a microcentrifuge.

d. Add the supernatant directly to 100 mL DMEM/2% FBS.

e. Add 20 mL to each of the five plates and incubate overnight at 37°C.

f. The next morning add 0.6 mL FBS.

g. Incubate the culture until complete cytopathic effect appears.

h. Harvest cells and medium in a 250-mL centrifuge bottle. Centrifuge 20 min at 1500*g*.

i. Aspirate the supernatant and resuspend the pellet in 5 mL of 10 m*M* Tris-HCl, pH 8.0. Store frozen at –80°C.

7. Further expansion and preparation of purified recombinant adenovirus:

a. Prepare one Nunc cell factory (Nalgene Nunc, Rochester, NY) and five 15-cm dishes.

b. Freeze-thaw the 5-mL cell suspension 3X.

c. Centrifuge for 10 min at 3500*g*.

d. Add the supernatant to 1.1 L DMEM/2% FBS.

e. Infect the cell factory with 1 L of virus and the five plates with 20 mL each.

f. Monitor the five plates for cytopathic effect.

g. When complete cytopathic effect is achieved (usually 48 h after infection), harvest the cell factory and the five plates into 1-L centrifuge bottles.

h. Again centrifuge at 1500*g* for 30 min.

i. Resuspend the pellet to a final volume of 40 mL with 10 m*M* Tris-HCl, pH 8.0.

j. Freeze–thaw 3X.

k. Centrifuge for 20 min at 3500*g* and collect the supernatant.

8. CsCl gradient purification of the recombinant adenovirus:

a. Pipet 10 mL of light cesium chloride solution into an ultracentrifuge tube for the Beckman SW 28 rotor. Take up 10 mL of heavy cesium chloride solution into a pipet and insert the tip of the pipet to the bottom of the ultracentrifuge tube and carefully dispense the heavy cesium. The light cesium chloride will float atop the heavy layer—the interface between the layers should be very sharp.

b. Carefully layer the viral supernatant on top of the cesium gradient. A maximum of 18 mL of supernatant can be loaded per bucket. The volume of the viral supernatant can be adjusted with 10 mM Tris-HCl, pH 8.1, to fill up the ultracentrifuge tube.

c. Load the tubes into the SW 28 rotor, being certain that the tubes are balanced. Centrifuge for 2 h at 72,000g (RCF$_{MAX}$).

d. Remove tubes from centrifuge and clamp onto a ring stand above a beaker of bleach. Virus will appear as a narrow opaque white band 2/3 down the heavy/light cesium gradient. Puncture tube either under band to pull it in a syringe or at the bottom so cesium can drip out of the tube until band is collected. Do not collect any bands above it.

e. Dilute collected band with equal volume of 10 mM Tris-HCl, pH 8.0.

f. Pipet 4 mL of light cesium chloride solution into an ultracentrifuge tube for the SW 40 rotor and pipet 4 mL of heavy cesium chloride solution beneath it. Then, carefully layer the diluted viral band on top of the gradient (a maximum of 4 mL can be loaded on this small gradient).

g. Load tube into SW 40 rotor and centrifuge overnight at 72,000g (RCF$_{MAX}$) 4°C. A minimum of 4 h is required for this centrifugation.

h. Collect the band by puncturing the side of the tube with a needle and syringe. Transfer virus suspension to a 15-mL conical tube and place on ice.

i. Dilute a sample of the virus 1:50 and take an OD260 reading. OD260 × dilution × 10^{12} = particles/mL.

j. Dilute to 10^{12} particles/mL in 50% glycerol in 10 mM Tris-HCl, pH 8.1 and 100 mM NaCl and aliquot to store at –80°C or proceed with desalting column purification. This diluted virus is best used as a seed stock for future infections. However, for other applications where CsCl may interfere or cause toxicity, such as experiments in animals, it is best to desalt the preparation of purified recombinant adenovirus.

k. Desalting column purification:
   i. Clamp the column (BioRad Econo-Pac® 10DG disposable chromatography columns, 10 mL; BioRad, Hercules, CA) onto a ring stand, remove the tip, and fit end with a stopcock.
   ii. Drain off the buffer on the column and rinse the resin bed with 30 mL HBS.
   iii. When the last of the HBS has reached the top of the resin, add the virus collected from the second cesium chloride gradient. (This size column will desalt a maximum of 3 mL of virus). Allow the virus to enter the resin and close the stopcock.
   iv. Add 10 mL HBS to the column without disturbing the resin. Open the stopcock and allow 2 mL of buffer to drip from the column. Sometimes the virus can be seen as a milky band moving through the column; if this is visible, begin collecting 1-mL fractions as this band exits the column. If virus is not visible, begin collecting 0.5-mL fractions immediately after the 2 mL of buffer has flowed through the column. Keep collected fractions on ice and add glycerol to a final concentration of 10% to each aliquot.
   v. Dilute a sample of each fraction 1:50 and take an OD$_{260}$ reading. Particle concentration of each fraction is calculated as follows:
   vi. OD$_{260}$ × dilution factor × 10^{12} particles/mL = concentration (particles/mL).
   vii. Storage concentration should not exceed 5 × 10^{12} particles/mL. Dilute with less-concentrated fractions or with additional HBS/10% glycerol. Aliquot into amounts suitable for a single experiment. Experiments using animals generally require between 1 × 10^{10} and 1 × 10^{12} particles per dose whereas in vitro experiments require less vector. Repeated freezing and thawing of the virus can cause a drop in viability.
   viii. The viral titer should also be determined by plaque assay as described next.

9. Plaque assay of recombinant adenoviruses:
   a. Prepare two 6-well plates of 293 cells that are 80–90% confluent.
   b. Twenty-four hours prior to beginning the assay, seed each well with 5×10^5 cells. The growth medium is DMEM/2% FBS.
   c. The initial concentration (in particles/mL) can be determined from the OD_{260} spectrophotometer reading previously taken. With this information, one can determine an appropriate range of five 10-fold dilutions based on the concentration. For example, if the stock concentration is 10^{13} particles/mL, an appropriate dilution range would be from 10^{-8} to 10^{-12}. Dilute the virus in DMEM/2% FBS so that there is a minimum final volume of 2 mL for each of the final five dilutions. Aspirate the medium from each well of 293 cells. Add 1 mL of diluted virus to each well so that each dilution is run in duplicate. Incubate the plates for 1 h, add 100 μL FBS to each well, and then incubate for 16–20 h. Do not exceed 20 h.
   d. On day 1, aspirate medium from the wells. Overlay each well with 4 mL of agar mixture by slowly pipeting the solution down the side of the well, taking care not to dislodge the cells. The agar mixture is made by adding 23 mL of 2X MEM, 2 mL of FBS, 0.62 mL of 1 M $MgCl_2$, 0.5 mL of Pen/Strep, and 26 mL of 1.6% Nobel agar. Allow agar to solidify at RT for approx 30 min, and then return the plates to the incubator until day 5.
   e. On day 5, overlay each well with 1.5 mL of the agar mixture. Allow agar to solidify at RT for approx 30 min, and then return plates to the incubator until day 9.
   f. On day 9, add 500 μL of 1.4% neutral red per 50 mL of agar that has been freshly prepared. Overlay each well with 2 mL of the neutral red/agar mixture. Allow agar to solidify as described previously and return plates to the incubator.
   g. On day 10, remove the plates from the incubator and hold up to the light, observing the monolayer from the bottom of the plate. Count the plaques in each well; they will appear as clear pale orange areas among the darker reddish-orange monolayer.

4. Notes

1. Preparation of viral DNA: Viral DNA is needed for the preparation of restriction enzyme cleaved viral DNA backbones. The procedure for preparation of viral DNA is described below: Adenovirus DNA is isolated from purified virions by digestion of the capsid proteins with proteinase K (1 mg/mL) in the presence of SDS (1 mg/mL proteinase K 50 mM Tris-HCl, pH 7.4, 1 mM, 1% SDS). This is followed by deproteinization with phenol followed by ethanol precipitation.
   a. From the adenovirus preparation, collect band from the second cesium chloride gradient in a 15-mL conical tube. Add to this a 4X vol of the above proteinase K solution.
   b. Mix by inversion and incubate tube at 37°C for l h.
   c. Extract 2X with an equal volume of phenol:chloroform (50:50). Extract 1X with chloroform.
   d. Add 1/10 vol of 3 M Na acetate, pH 5.2, mix well, and precipitate with 2 vol of 95% ethanol (–20°C). Centrifuge for 30 min at 3500g and decant ethanol. Add 5 mL 70% (–20°C) ethanol and centrifuge 15 min at 3500g. Dry pellet briefly.
   e. Resuspend DNA pellet in 200–500 μL of Tris-EDTA.
   f. Repeat **steps c–e**.
   g. Determine DNA concentration by reading the absorbance at 260 nm.
2. Preparation of adenoviral backbone DNA by restriction enzyme digestion:
   a. H5.000CMVEGFP, H5.010CMVEGFP, and H5.001CMVEGFP DNA are digested within a 200 μL reaction with 10 U *Cla*I for every 1 μg of viral DNA for 18 h at 37°C.

Another 1 U of *Cla*I is added per microgram viral DNA and digestion continued for 2 h at 37°C. H5.000CMVEGFP4 DNA is placed in a 200-μL reaction containing 10 U of *Cla*I and *Bst*BI per microgram of viral DNA. Digestion is carried out for 2 h at 37°C and 2 h at 65°C.

   b. After digestion, the reaction mixture is extracted 1X with phenol:chloroform:isoamyl alcohol (24:24:1) and 1X with chloroform.

   c. Add 1/10 vol of 3 *M* Na acetate, pH 5.2, mix well, and precipitate with 2 vol of 95% ethanol (−20°C). Centrifuge for 30 min at 3500*g* and decant ethanol. Add 1 mL of 70% ethanol (−20°C) for every 200 μL of original reaction and centrifuge 15 min at 3500*g*. Dry pellet briefly.

   d. Resuspend DNA pellet in 100 μL of Tris-EDTA.

3. Size limitations: One important factor in choice of these backbones is the size of the transgene cassette to be incorporated. Adenoviruses generally cannot package more than 5–6% of their genome or approx 1800 bp. The E1 deletion formed by cloning in AdCMVlink is approx 2900 bp. Therefore, the total cassette should not exceed approx 4700 bp. Because cassette regulatory elements (i.e., CMV promoter and SV40 poly-adenylation signal) occupy approx 1300 bp, this leaves room for a transgene of approx 3400 bp. If the transgene is of a size larger than this, backbones deleted in the E3 region can be used. The E3 region of adenovirus codes for a family of genes involved in escape of the virus from host immune response. This region can be deleted without effect on the growth of the virus in cell culture. Viral backbones with E3 region deletions include sub360 (substitution of sequence resulting in a net deletion of 300 bp), dl327 (1800 bp deletion), and dl7001 (3200 bp) deletion.

4. Adenoviral vectors that express green fluorescent protein: The adenoviral vectors H5.000CMVEGFP, H5.010CMVEGFP, H5.001CMVEGFP, and H5.030CMVEGFP have been described. The first two have little (010) or no (000) deletion within region E3 and thus can accommodate inserts (including regulatory elements) of approx 3000 bp. The third has a deletion in region E4 and the fourth has an 1800 bp deletion in region E3 and can accommodate an insert of 5000 bp and in practice up to 6000 bp. Each of the viral DNA of these vectors has a unique *Cla*I site just after the green fluorescent protein-coding region. Therefore cleavage with *Cla*I should render the DNA noninfectious, unless it recombines with the shuttle plasmid. Uncut DNAs form green plaques. In order to reduce background, a new vector, H5.000CMVEGFP4, has been developed in that its DNA has not only a *Cla*I but also a *Bst*BI site just after the coding region of green fluorescent protein. Cleavage with both enzymes substantially decreases the background of green plaques.

References

1. Ginsberg, H. S. (1984) *The Adenoviruses*. Plenum, New York.
2. Davis, A. and Wilson, J. (1996) Adenovirus vectors. *Curr. Protocols Hum. Genet.* 12.4.1–12.4.18.
3. Chartier, C., Degryse, E., Gantzer, M., and Dieterle, A. (1996) Efficient generation of recombinant adenovirus vectors by homologous recombination in Escherichia coli. *J. Virol.* **70,** 4805–4810.
4. Bett, A., Haddara, W., Prevec, I., and Graham, F. L. (1994) An efficient and flexible system for construction of adenovirus vectors with insertions or deletions in early regions 1 and 3. *Proc. Natl. Acad. Sci. USA* **91,** 8802–8806.
5. Ketner, G., Spencer, F., Tugendreich, S., Connelly, C., and Hieter, P. (1994) Efficient manipulation of the human adenovirus genome as an infectious yeast artificial chromosome clone. *Proc. Natl. Acad. Sci. USA* **91,** 6186–6190.

6. Schaack, J., Langer, S., and Guo, X. (1995) Efficient selection of recombinant adeno-viruses by vectors that express β-galactosidase. *J. Virol.* **69,** 3920–3923.

7. Imler, J. L., Chartier, C., Dieterle, A., et al. (1995) An efficient procedure to select and recover recombinant adenovirus vectors. *Gene Ther.* **2,** 263–268.

8. Hardy, S., Kitamura, M., and Harris-Stancil, T. (1997) Construction of adenovirus vectors through Cre-lox recombination. *J. Virol.* **71,** 1842–1849.

9. Davis, A. R., Meyers, K., and Wilson, J. M. (1998) High throughput method for creating and screening recombinant adenoviruses. *Gene Ther.* **5,** 1148–1152.

10. Graham, F. L., Smiley, J., Russel, W. C., and Nairn, R. (1977) Characteristics of a human cell line transformed by DNA from human adenovirus type 5. *J. Gen. Virol.* **36,** 59–74.

11. Arad, U. (1998) Modified Hirt procedure for rapid purification of extrachromosomal DNA from mammalian cells. *BioTechniques* **24,** 760–762.

50

Application of Adenoviral Vectors

Analysis of Eye Development

Jean Bennett, Yong Zeng, Abha R. Gupta, and Albert M. Maguire

1. Introduction

Recent technological advances in the generation of recombinant adenovirus vectors have been applied to a wide range of studies in developmental biology. These vectors have allowed analyses of the regulation of expression and function of specific genes as well as evaluation of patterns of cell migration during development. Recombinant adenovirus-mediated transgene delivery has several advantages over standard transgenic approaches (such as transgenesis achieved through pronuclear injection). Adenoviruses can be used to deliver transgenes to specific tissues (of a variety of different animal species) at specified developmental time points through surgical application. Such methods are likely to be less costly, use fewer animals, and be more time efficient than more traditional transgenic delivery systems.

Recombinant adenoviruses do have limitations for use in developmental biology studies, however. They do not target certain cells/tissues efficiently and, in adult mammals, both the adenovirus and cells transduced with adenovirus can be subject to a rapid immune-mediated clearance. The eye, however, has several features that minimize these difficulties. Many different ocular cell types are efficiently transduced with adenovirus. In addition, the unique structure of the eye permits localized application of a high concentration of the vector. The fact that the eye enjoys a relatively immune-privileged status minimizes the possibility of toxicity and rapid immune-mediated loss of the transgenes. Finally, it is possible in the eye to follow transgene expression over time in vivo in a noninvasive fashion through direct visualization. This chapter outlines the protocols developed for purification and application of recombinant adenoviral vectors to the postnatal mammalian (murine) eye. Methods are described for in vivo and in vitro assessment of two reporter genes that are commonly used: *lacZ*, the bacterial gene encoding β-galactosidase and *GFP*, the jellyfish-derived gene encoding the bioluminescent green fluorescent protein. The protocols should be adaptable for use with fetal eyes and with eyes of other (nonmammalian) species.

From: *Methods in Molecular Biology, Vol. 135: Developmental Biology Protocols, Vol. I*
Edited by: R. S. Tuan and C. W. Lo © Humana Press Inc., Totowa, NJ

2. Materials

2.1. Virus Purification

1. Media/cells and recombinant adenovirus:
   a. DMEM/10%: DMEM with high glucose (cat. no. 11965-084; Gibco-BRL, Gaithersburg, MD) + 10% fetal bovine serum (FBS) + 1% penicillin/streptomycin.
   b. DMEM/2%: DMEM + 2% FBS + 1% penicillin/streptomycin.
   c. 293 (human embryonic kidney) cells are available from ATCC (cat. no. CRL-1573; American Tissue Culture Co., Rockville, MD). Cells should be 80–90% confluent for infection. To prepare 50 plates: split one 150-mm plate into seven plates. Cells will take 3–4 d to reach confluence. When confluent, expand the cells further into fifty 150-mm plates. Cells should be grown in DMEM/10%.
   d. Starter virus: For infection of fifty 150-mm plates, 5×10^{11} particles of E1-deleted virus are needed (or 10^{10} particles per plate). Use virus obtained from "seed stock" (media from infected 293 cells) if possible. Alternatively, frozen virus can be used (*see* **Subheading 3.1., step 3**).
   e. Disinfectant: 10% bleach (0.525 g sodium hypochlorite/100 mL).
2. Buffers: Tris(8.1): 10 mM Tris-Cl, pH 8.1.
3. Solutions for virus purification:
   a. Centrifugation: Light cesium: dissolve 22.39 g cesium chloride in 77.61 mL of 10 mM Tris-HCl, pH 8.0, Heavy cesium: dissolve 42.23 g of cesium chloride in 57.77 mL of 10 mM Tris-HCl, pH 8.0, autoclave and store cesium chloride solutions at room temperature.
   b. Tris-NaCl: 10 mM Tris-HCl/100 mM NaCl, pH 8.1.
   c. HEPES-buffered saline (HBS): 20 mM HEPES, 150 mM NaCl, pH 7.8, filter with 0.22-μM filter.
   d. HBS-3%S: HBS with 3% sucrose.
   e. Agar overlay mixture: Mix together 25 mL of 2X Basal Medium Eagles (BME) medium, 0.62 mL of 1 M MgCl$_2$, and 1 mL FBS and hold solution in 45°C water bath. Melt agar solution (1.6% Difco Agar Noble, cell-culture tested) in a microwave and cool it to 45°C. Add 25 mL of the melted agar to the BME mixture and stir.
   f. Neutral red agar overlay mixture: Add 1.4 g neutral red to 100 mL distilled water. Mix well and autoclave. Store in amber bottle at 4°C. Add 500 μL of 1.4% neutral red to 50 mL of the agar overlay mixture.

2.2. Surgical Procedures

1. Sterile phosphate-buffered saline (PBS). PBS is prepared by dilution of a 10X stock. Composition per liter of 10X stock: 80 g NaCl, 2 g KCl, 21.4 g Na$_2$HPO$_4 \cdot$ 7H$_2$O, 3.5 g KH$_2$PO$_4$, pH 7.4.
2. PBS-20%S: Sterile 1X PBS containing 20% w/v sucrose.
3. Avertin: Stock solution: dissolve 10 g 2,2,2-tribromoethanol (Avertin; Aldrich Chemical, Milwaukee, WI) in 10 mL 2-methyl,2-butanol. Working solution: To 1.25 mL of avertin stock solution, add PBS to a final volume of 50 mL. Store both stock and working solution wrapped in foil at 4°C.

2.3. Transgene Expression

1. Paraformaldehyde fixative: 4% paraformaldehyde in 1X PBS. Add 4 g of paraformaldehyde (e.g., Fisher purified grade) to 60–70 mL of distilled water. Add 20 μL of 10 M NaOH solution (helps paraformaldehyde to dissolve). Stir and heat to 60°C until dissolved.

Add 10 mL of 10X PBS. Adjust pH to 7.4. Bring final volume to 100 mL with distilled water. Use within 24 h with storage at 4°C. Paraformaldehyde is toxic and a carcinogen. This solution should be prepared in a chemical fumehood, and gloves should be worn.

2. β-galactosidase (β-gal) substrate: X-Gal buffer: 10 mM K$_3$Fe(CN)$_6$, 10 mM K$_4$Fe(CN)$_6$ · 3H$_2$O, 2 mM MgCl$_2$, 0.01% sodium deoxycholate, 0.02% NP40 in PBS, pH 7.5. Store at room temperature for 3 mo; X-Gal concentrate: 40 mg/mL of 5-bromo-4-chloro-3-indolyl galactopyranoside (Sigma Chemical Co., St. Louis, MO) in dimethylformamide. Store at –20°C. When ready to perform X-Gal staining, add X-Gal concentrate to X-Gal mixer to a final concentration of 1 mg/mL.

3. Methods
3.1. Purification of Recombinant Adenovirus (see Note 4.1.1.)

1. Infection of 293 cells: For infection of fifty 150-mm plates of 293 cells, suspend 5×10^{11} particles of virus in 1 L of DMEM/2%. Aspirate media from the 50 dishes of 293 cells. Slowly add 20 mL of media/virus solution to the side of each plate. Incubate plates in incubator at 37°C for approx 40 h. At this time, cytopathic effect (CPE) will be detectable. (The cells will appear round and refractile and will begin to lift off the plate as the virus is amplified). Using a 10-mL pipet, pipet the media in each dish over the cells to dislodge the cells from the plate.

2. Cell harvest (*see* **Note 4.1.2.**): Collect the cells and media into 500-mL or 1-L autoclaved centrifuge bottles. Centrifuge at 4000g for 25 min at 4°C. Remove as much supernatant as possible with a 5-mL pipet and discard it into 10% bleach. Be careful not to dislodge the cell pellet. Resuspend the pellet in 0.5 mL per 150-mL plate (or 25 mL for 50 plates) of Tris(8.1). Place the suspension in a 50-mL Sorvall tube. At this point the suspension can be stored at –70°C.

3. Purification of virus from cell lysate (*see* **Note 4.1.3.**): Freeze-thaw the cell suspension 3X by sequential immersions in a 95% ethanol/dry ice bath and a 37°C water bath in order to release the virus from the cells. Spin lysates 20 min at 3000g at 4°C in a Sorvall centrifuge. Avoiding the pellet, harvest the supernatant, which will be cloudy. Pipet 10 mL light cesium into an ultracentrifuge tube (ultra-clean 1 × 3.5 in., reorder no. 344058; Beckman, Fullerton, CA). Pipet 10 mL heavy cesium with the pipet tip touching the bottom of the light cesium-containing ultracentrifuge tube. This will result in a sharp interface between the two layers. Load the viral supernatant (a maximum of 18 mL) on top of the cesium gradient. Place the tubes into a SW 28 (Beckman) rotor, being certain that the tubes are balanced. Spin in an ultracentrifuge for 2 h at 60,000g at 4°C. Remove the tubes from the centrifuge and clamp onto a ring stand above a beaker of 10% bleach. Virus will appear as a narrow opaque white band two thirds down the heavy/light cesium gradient (**Fig. 1**). Puncture the tube at the bottom so that cesium can drip out of the tube until the lowest band can be collected (in a sterile 15-mL tube). Do not collect any of the higher bands. Dilute the collected band with an equal volume of Tris(8.1). Pipet 4 mL of light cesium into an ultracentrifuge tube (ultra-clean 9/16 × 3.5 in., reorder no. 344059; Beckman). Pipet 4 mL of heavy cesium beneath this layer. Load the diluted viral band (a maximum of 4 mL) on top of the gradient. Place the tubes into a SW 41Ti (Beckman) rotor, being certain that the tubes are balanced. Spin overnight (or a minimum of 4 h) at 50,000g at 4°C in an ultracentrifuge. Remove the tubes and collect the band as described above. Transfer the virus suspension to a sterile 15-mL conical tube (approx 0.5 mL/tube) and place on ice. Dilute an aliquot of the virus 1:50 and take an OD$_{260}$ reading. Particles/mL can be calculated by multiplying: OD$_{260}$ × dilution × 10^{12}. Dilute to 10^{12} particles/mL in 50% glycerol in Tris-NaCl and aliquot to store as stock for future cell infections (*see* **Subheading 2.1.**, **step 1d**) or proceed with desalting column purification.

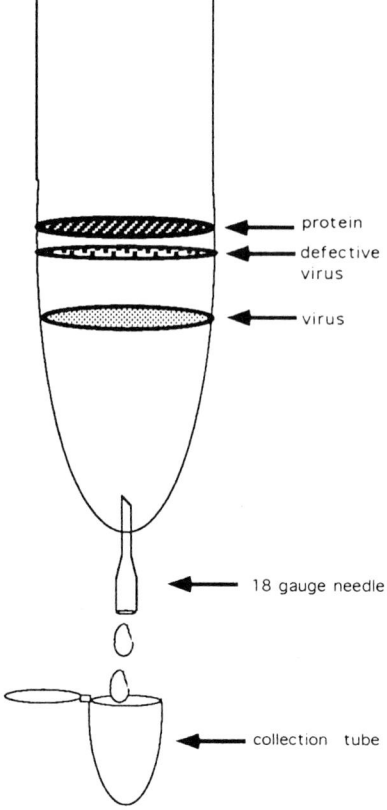

Fig. 1. Appearance of bands in cesium chloride gradient after ultracentrifugation to purify recombinant adenovirus.

4. Desalting of the virus and storage (*see* **Note 4.1.4.**): Clamp a DG-10 Econo-pack column (cat. #732-21010; Bio-Rad, Hercules, CA) onto a ring stand, remove the tip, and fit the end with a stopcock. Drain off the buffer in the column and rinse the resin bed with 30 mL HBS3%S. When the last of the HBS3%S has reached the top of the resin, add the virus (maximum of 3 mL) collected from the second cesium chloride gradient. Allow the virus to enter the resin and close the stopcock. Add 10 mL HBS-3%S to the column without disturbing the resin. Open the stopcock and allow 2 mL of buffer to drip from the column. The virus may be visible as a milky band moving through the column. If it is visible, begin collecting 1-mL fractions as this band exits the column. If the virus is not visible, begin collecting 0.5-mL fractions immediately after the 2 mL of buffer have flowed through the column. Keep collected fractions on ice. Dilute a sample of each fraction 1:50 and measure OD_{260}. Particle concentration of each fraction is calculated as follows:

$$OD_{260} \times \text{dilution factor} \times 10^{12} \text{ particles/mL} = x \text{ particles/mL}$$

Storage concentration should not exceed 5×10^{12} particles/mL. Dilute with less-concentrated fractions or with additional HBS-3%S. Aliquot into amounts appropriate for a single experiment. Freeze on dry ice and store at –80°C.

5. Adenovirus plaque assay: The number of infectious particles per total number of viral particles should be determined for each new virus preparation. This is generally expressed

as plaque forming units (pfu) on 293 cells (i.e., 1 pfu indicates approx one live virus). Number of pfus is determined by exposing replicate 293 cell plates to a range of doses of the adenovirus and counting the number of plaques per unit area. To two 6-well plates of 80–90% confluent 293 cells, add 1 mL of diluted virus in DMEM without serum (for a total of five dilutions) to each well. A duplicate should be run for each dilution. An appropriate dilution range would be from 10^{-8} to 10^{-12} in logarithmic increments. Control wells (exposed to media without virus) should also be included. Incubate the plates at 37°C for 1 h. Add 100 µL FBS to each well and incubate plates at 37°C for 16–20 h. On the next day (day 1), aspirate media from the wells and overlay each well with 4 mL of agar overlay mixture. The mixture should be pipetted slowly down the side of the well, taking care not to dislodge the cells. The agar should be allowed to solidify at room temperature (RT) for a minimum of 30 min and then the dishes should be returned to the 37°C incubator. On day 5, feed the cells by overlaying the existing agar mixture with 2–4 mL of new agar mixture and allow to solidify at RT for 30 min. Return plates to the incubator. On day 9, overlay each well with 2–4 mL of neutral red agar mixture and allow to solidify at RT for 30 min. Return plates to incubator. On day 10, remove plates from incubator and hold plate up to light, observing the monolayer from the bottom of the dish. Plaques will appear as clear pale-orange areas among a darker reddish-orange monolayer. Count the plaques in each well. Determine plaque counts for each dilution by averaging the duplicate wells. This average will determine the number of plaque forming units (pfu) per milliliter of adenovirus (*see* **Note 4.1.5.**).

3.2. Intraocular Injection of Recombinant Adenovirus

1. Preparation of virus for injection: Immediately before surgery, an aliquot of purified virus (**Subheading 3.1.4.**) is thawed on wet ice. The virus is diluted as desired with sterile PBS or PBS-20%S, and the diluted virus is stored until use at 4°C. Generally, application of 1×10^7 pfu in a volume of 1 µL results in high levels of transduction in the mouse eye. Therefore, if the virus titer is 1×10^{11} pfu, it should be diluted 1:10. A range of doses should be tried in a pilot experiment to determine the optimal dosage of the particular virus (*see* **Note 4.2.1.**).

2. Surgical procedures (*see* **Note 4.2.2.**): Surgical procedures are provided for mice; these can be adapted for use with other animal species. Mice are anesthetized with avertin through an intraperitoneal (ip) injection of 0.5–0.6 g/kg of avertin working solution. Adult mice generally require approx 0.5 mL of working strength avertin. Pupils are dilated with topical application of 1 drop of Mydriacyl (1.0% tropicamide; Alcon, Humacao, Puerto Rico). The anesthetized mouse is positioned on the base of a dissecting microscope (e.g., Nikon SMZU microscope; Optical Apparatus, Ardmore, PA) with the eye to be injected under view (approx ×15 magnification). Illumination is provided with a fiberoptics light source (e.g., Footec model 3115 PS-12W-B30, Optical Apparatus). A drop of betadine is placed in the fornix and ocular adnexa (i.e., on the cornea adjacent to the lower eyelid) in order to minimize contamination from resident bacteria.

   For subretinal injections, Vannas iridotomy scissors (e.g., cat. no. RS-5610; Roboz Surgical, Rockville, MD) and jeweler's forceps (Dumont #5, Roboz Surgical) are used to incise the conjunctiva just posterior to the cornea. The conjunctiva and underlying episcleral tissue are opened circumferentially (i.e., conjunctival peritomy) to expose the underlying sclera (**Fig. 2A**). The conjunctiva adjacent to the cornea is grasped with forceps, providing traction to rotate the globe and allowing optimal surgical exposure. The tip of a sterile 30-gauge needle is passed obliquely through the sclera, choroid, and retina into the vitreous avoiding the lens (**Fig. 2B**). A bead of clear vitreous will prolapse through

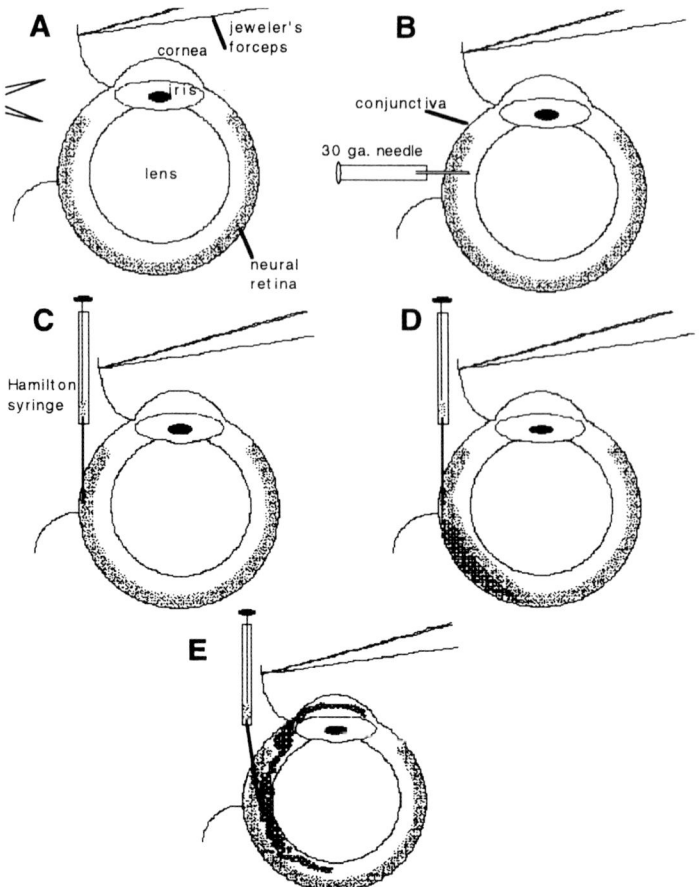

Fig. 2. Surgical approach for injection of recombinant adenovirus in the mouse eye.
(**A**) The conjunctiva is incised at the temporal aspect of the eye to expose the underlying sclera.
(**B**) A 30-gauge needle is passed through the sclera, choroid, retinal pigment epithelium, and
neural retina. (**C**) The tip of the needle mounted to the Hamilton syringe is introduced into the
incision tangential to the neural retina. (**D**) 1 µL of virus is microinjected into the subretinal
space, resulting in a localized retinal detachment. (**E**) Intravitreal injection is achieved by
inserting the needle through the retina and into the vitreous, being careful not to touch the lens.

this scleral incision, and occasionally the translucent edge of the retina can be seen internally. At this point, the globe is gently retracted to expose the sclerotomy. The tip of a
33-gauge blunt needle (Hamilton #79633; Baxter Scientific Products, Edison, NJ) mounted
on a 10-µL Hamilton syringe (Hamilton #801RN; Baxter Scientific) is introduced into the
incision oriented tangentially to the neural retina (**Fig. 2C**). Guidance is aided by use of
jeweler's forceps in the opposite hand. The 33-gauge needle passes through the sclera and
the choroid (an external transscleral transchoroidal approach) and then terminates in the
subretinal space. Under visual guidance, approx 1.0 µL of virus is injected (**Fig. 2D**). Two
individuals are necessary for this procedure: one to hold the syringe and the other to inject.
Direct visualization is used to determine the accuracy of the subretinal injection. The
retina in the region of the injection will appear pale and raised, reflecting the localized site
of detachment (**Fig. 3A**). After visualization, ointment (preferably PredG; Allergan

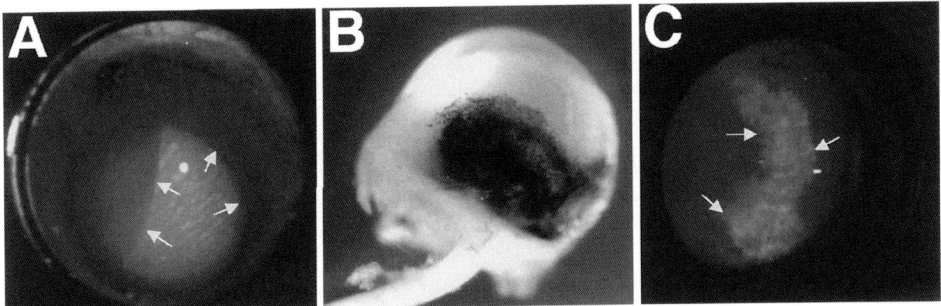

Fig. 3. Assessment of the retina/reporter gene expression after intraocular adenovirus administration. (A) Appearance of the retina after accurate subretinal injection. Shown is a retinal detachment (borders demarcated by *arrows*) viewed by ophthalmoscopy. The choroidal/retinal vessels cannot be resolved easily in the detached region of the retina, whereas they are identifiable in the untreated region. Reporter gene expression can be assessed (B) in vitro through analysis of eyes reacted in whole-mount fashion for transgenes such as *lacZ* or (C) in vivo using ophthalmoscopy for transgenes such as *GFP* (*arrows*).

Pharmaceuticals, Irvine, CA) should be applied to the corneas to minimize drying of this tissue while the animal is anesthetized (and does not blink spontaneously).

Some modifications of the procedure may be necessary in order to perform intraocular injections on neonatal mice. It is often difficult to achieve adequate anesthesia with Avertin because of combinations of toxicity and difficulty of delivering adequate volumes of anesthetic to the neonatal mouse peritoneum. Anesthesia by hypothermia can often be achieved reliably and without toxicity in neonatal mice—it may be useful to consult with the local animal care and use committee regarding this procedure. In addition, in neonatal mice, it will be necessary to open the eyelids, which do not open spontaneously until approx 10–11 d after birth. The eyelids are opened by carefully incising the skin between the upper and lower lid with the beveled edge of a 30-gauge needle. The line of skin can be identified by its lack of hair follicles and its smooth surface. Once a small opening is made, the tips of a jeweler's forcep are placed in the space between the lids and are carefully spread open to expose the eye. While exerting gentle fingertip pressure, the lids should be slipped behind the eye, allowing the entire globe to prolapse forward. Forceps are used to strip any conjunctiva or episcleral tissue away from the injection site, just posterior to the cornea. The scleral incision is made with only the tip of the bevel of a 30-gauge needle. Using the entire bevel creates too large an incision. The 33-gauge blunt tip needle mounted on the Hamilton syringe is then placed just inside the eye, tangential to the sclera, and the virus is injected toward the curve of the globe. It is useful to mark the needle approx 0.5 mm from the tip in order to identify the length of needle placed inside the eye. It is also advisable for the surgeon to steady the hand holding the Hamilton syringe on a roll of towels or other fixed surface. Once the needle is positioned just inside the eye, an assistant should inject 1 µL of virus, while the surgeon maintains the position of the needle. After injection, the needle is carefully removed and the conjunctiva is gently repositioned. The incision is not closed, as the suture needle would result in holes as large as the 30-gauge incision. Neonatal mice must be restored to their mothers (or to foster mothers) after treatment.

Intravitreal injections proceed similarly as subretinal injections except that the 33-gauge needle is guided through the hole in the neural retina created by the lateral canthotomy in a direction parallel with the lens capsule (i.e., avoiding damage to the lens) (**Fig. 2E**).

3. Assessment of the health of the retina can be performed using indirect ophthalmoscopy, if desired. To accomplish this, the eye is dilated with 1.0% tropicamide. An assistant is needed to hold the mouse. In a dark room, the light source from the ophthalmoscope (e.g., Keeler, Broomall, PA) is fixed on the dilated eye. A 90-diopter lens is slowly moved toward the eye in the line of the light beam until the retinal vessels appear in view (**Fig. 3A**). With practice, this can be performed reliably in approx 10 s.

3.3. Tissue Examination

1. Some reporter gene products can be visualized in the eye in vivo. GFP, for example, can be identified by ophthalmoscopy by illuminating the eye with a blue light (**Fig. 3C**) *(1)*. The procedures described for ophthalmoscopy in **Subheading 3.2.3.** are followed except that a gelatin Wratten filter (Wratten #47B gelatin excitation filter, cat. #149-5795; Eastman Kodak, Rochester, NY) is taped over the light source. Use of this filter results in transmission of blue light at 450–490 nm, which corresponds to excitatory wavelengths for the GFP (and the enhanced GFP, EGFP) chromophore. Illumination of the GFP-expressing eye with this blue light will result in emission of green fluorescence. Photographs can be made using a Kowa camera (Keeler Instruments, Broomall, PA) equipped with the same Wratten excitation filter as described above (*see* **Note 4.3.1.**).
2. After killing the mouse, enucleation of the mouse eye is accomplished by cutting the optic nerve, blood vessels, and muscles with curved iridectomy scissors (*see* **Note 4.3.2.**). The method of fixation depends on the tests that are to be performed on the tissue. Many forms of immunohistochemistry and transgene expression studies can be performed on paraformaldehyde-fixed eyes. GFP expression can be easily identified by fluorescence microscopy of sections of eyes fixed for 2 h with 4% paraformaldehyde *(1)*.

    For identification of *lacZ*-expressing cells, the eyes should be fixed for 1 h in fresh 4% paraformaldehyde, washed in PBS, and then reacted overnight in X-Gal buffer plus X-Gal concentrate at 37°C. After injection in albino mice, it should then be possible to identify *lacZ*-expressing cells in whole-mount preparations through identification of the blue X-Gal reaction product (**Fig. 3B**) *(2–5)*. The tissue should then be postfixed in 4% paraformaldehyde for 12 h. If desired, the lens can be removed at this time by slitting the cornea with iridectomy scissors and gently teasing the lens out of the eye with Dumont forceps. Samples can then be further processed as desired. To obtain good histology in frozen sections it may be desirable to cryoprotect the tissue through sequential immersion in PBS containing 10% sucrose and PBS containing 30% sucrose (for 12 h each at 4°C). The tissue can then be immersed in Optimal Tissue Compound (O.T.C.; Baxter Scientific) and frozen in a dry ice bath containing 2-methylbutane.

4. Notes

4.1. Purification of Recombinant Adenovirus

1. Consult with local Environmental Health and Safety Office before initiating experiments using recombinant adenovirus, and follow their guidelines. Appropriate (biohazard level 2) precautions should be used in working with recombinant adenovirus. These include no mouth pipetting, use of gloves and protective apparel, and appropriate containment of animals. All materials coming in contact with adenovirus should be decontaminated with 10% bleach, and waste materials should be autoclaved and disposed of appropriately. Infection of 293 cells: Early passage cells should be used for optimal results. Generally, the earlier passage cells grow more slowly than those of later passage. These cells must be 80–90% confluent to infect. All solutions should be prepared using Milli-Q-derived water

(cat. #CPMQ004D2, Millipore, Bedford, MA) and should be sterilized by autoclaving or passing through a 0.22-μm filter. Fetal bovine serum should not be autoclaved.

2. Cell harvest: If the cell pellet is loose after centrifugation of bottles, resuspend it in a small volume of supernatant, place it in a 50-mL Sorvall tube and spin at 3000*g* for 10 min at 4°C.

3. Purification of virus from cell lysate: Approximately 10^{13}–10^{14} particles will be collected at this point.

4. Desalting of the virus, storage, and evaluation: Use a sterile 18-gauge 1.5-in. needle to collect the fractions. Recombinant adenovirus transduction efficiency decreases markedly with each successive freeze–thaw cycle, so these should be kept to a minimum.

    Polymerase chain reaction (PCR) can be used to screen for contaminating wild-type adenovirus: PCR primers (which amplify the adenovirus E1a region that should be missing in E1-deleted adenovirus recombinants) are as follows *(1)*: forward: 5'-TCGAAG AGGTACTGGCTGAT-3'; reverse: 5'-TGACAAGACCTGCAACCGTG-3'. PCR should be done with a total of 35 cycles with an initial denaturation of 7 min at 95°C, denaturation temperature of 94°C for 1 min, annealing temperature of 55°C for 1 min, extension temperature of 72°C for 1 min, and a final extension of 10 min at 72°C. All purified recombinant stocks used should lack the E1a region. An alternative (less-sensitive) assessment of contamination with wild-type replication-competent adenovirus would be to infect HeLa cells at a multiplicity of infection (MOI) of 10, passage them for 30 d, and examine them for cytopathic effects. All stocks used should lack such effects.

5. Typically, the ratio of particles to pfu is 20–50:1.

4.2. Intraocular Injection of Recombinant Adenovirus

1. Dilution with PBS-20%S may be preferable to dilution with PBS, as use of the denser solution facilitates monitoring of the accuracy of injection.

2. Consult with the local Institutional Animal Care and Use Committee before initiating experiments using animals, and follow their guidelines. It is essential to follow procedures that minimize stress and pain and involve use of the minimal number of animals to obtain statistically reliable data. Avertin provides fast-onset, short-term (approx 1 h) anesthesia for mice. However, use of old solutions can result in toxicity. Avertin should be stored at 4°C. It should be warmed to 37°C immediately before use. Younger mice are more sensitive to avertin, so the lowest dose possible should be used. To avoid hypothermia, maintain animals under a heat lamp until they have recovered their ability to roll over or blink. This takes approx 1 h for avertin-induced anesthesia (or longer, if higher doses are used). Complete recovery takes 2–4 h. Anesthetics other than avertin may be better suited for other animal species. Pentobarbital (25–40 mg/kg ip) is effective with rats, for instance. Pupils can be dilated while waiting for anesthetic to take effect. Pupils should dilate within 5 min and stay dilated for up to 2 h.

    Consult with Environmental Health and Safety Office regarding appropriate biohazard containment of animals. Precautions to be used in working with animals exposed to recombinant adenovirus include use of gloves and protective apparel and appropriate containment of animals. Animals should be "clean" (i.e., free of known pathogens). Animals should be monitored for untoward reactions (such as inflammation, infection, or discomfort) after the procedure. Preparations should be in place to treat such reactions or to terminate the experiment.

    The surgical field and all reagents and surgical tools that come in contact with the eye should be sterile. Surgical instruments should be autoclaved; fresh (sterile) needles should be used for injections. Hamilton syringes can crack if autoclaved. These should be sterilized by washing 6–7X with sterile water, incubating in 70% ethanol for 2 d, washing again in sterile water 6–7X, and air-drying on a sterile surface. It is critical to have surgical tools that are in good condition. The Dumont forceps are particularly fragile. Care should be taken in storage of these instruments. They are very expensive and fragile. If they become damaged/misaligned, they can be repaired by following the instructions in the tweezer repair kit (Ernest F. Fullam, Inc., Latham, NY).

A transscleral transchoroidal approach is necessary in the mouse because of the very large relative size of the mouse lens (and the likelihood of damaging the lens through use of an anterior approach). An anterior approach can be used in larger animals with relatively small lenses (e.g., rabbits). It may be easier to learn how to perform and assess accuracy of subretinal injections in albino mice (such as CD-1 mice) rather than pigmented mice (such as C57Bl/6 mice). This is because the injection needle can be visualized through the lightly pigmented sclera in the albino mice. Note that material injected into the eye can reflux. In such a situation, the refluxed material should be blotted gently with sterile gauze and discarded. Larger volumes of virus solution can be injected into larger eyes than the murine eye. For example, a rat eye will accommodate 2–4 µL administered to the subretinal space, and a rabbit eye will accommodate 100 µL.

Accuracy of subretinal injections should be determined soon after injection, as the detachments resolve spontaneously within a few hours. It will take a great deal of experience to be able to identify detachments reliably. All injections in which blebs are not successfully raised should be noted and analyzed separately.

The topical ointment PredG reduces postsurgical inflammation. However, because it reduces inflammation, it may alter immune response. This should be kept in mind in the experimental design. Retinas can be examined days or weeks after injection to verify that the detachment has resolved with ophthalmoscopy (*see* **Subheading 3.2.3.**).

A challenge in working with neonatal mice is to prevent cannibalism after treatment. Treated neonatal mice should not be returned to their mothers until they have completely recovered from anesthesia. Choice of strains may minimize the occurrence of cannibalism. In our experience, CD-1 mice have low rates of cannibalism, whereas C57Bl/6 mice have high rates.

Similar to subretinal injections, approx 1.0 µL can be administered with intravitreal injections. Intravitreal injections result in exposure not only to the inner retina (ganglion cell layer) and lens, but also to the trabecular meshwork, iris, and corneal endothelium *(3,4)*. Anterior chamber (intracameral) injections through the thin cornea are also possible but are not recommended as they often lead to iris prolapse and lens capsule rupture.

4.3. Assessment of Transgene Expression

1. It is extremely difficult to obtain good photographs of the mouse retina using the Kowa camera. This procedure is recommended only for experts in indirect ophthalmoscopy. In some instances (e.g., if the cornea or lens have been damaged during the injection), it will be difficult to clearly visualize the retina. For experiments incorporating use of GFP-containing adenoviruses, it may be possible to observe retinal fluorescence with ophthalmoscopy without focusing with the 90-diopter lens (i.e., fluorescence may be visible through illumination with blue light alone).

2. Enucleation and fixation techniques: The retina degrades rapidly after death. Therefore, plans must be made to harvest eyes immediately after the animal is killed. A mark should be made on the anterior portion of the eye before enucleation. This will allow identification of the region of the eye that has been injected and proper orientation of the tissue during processing. The mark can be made with non–water-soluble ink. It should be remembered, however, that some processing steps (such as the steps required for paraffin embedding) will eradicate this mark. To facilitate orientation of tissue in paraffin blocks,

a sclerotomy can be performed in the region opposite the injection site. This can be accomplished by removing a small piece of the sclera of fixed eyes with iridectomy scissors and Dumont forceps. In some cases, a 27-gauge needle puncture mark can be made for orientation purposes. In this case, care should be taken to avoid penetrating the lens capsule. Higher-gauge needles should not be used to puncture the mouse eye, as they lead to great histological damage. Solutions for fixing and reacting the tissues need not be sterile. O.C.T.-immersed tissue can be frozen in liquid nitrogen prior to cryosectioning; however, methylbutane is preferred as this minimizes cracking of the tissue.

Acknowledgments

The authors thank Jim Wilson and members of the University of Pennsylvania Institute for Human Gene Therapy Vector Core facility for advice regarding purification of recombinant adenovirus vectors. This work was supported by a Research to Prevent Blindness Career Developmental Award, NEI RO1 EY10820; a Center Grant from Foundation Fighting Blindness, Inc.; the Pennsylvania Lions Sight Conservation and Eye Research Foundation; the Paul and Evanina Mackall Foundation Trust, and the F. M. Kirby Foundation.

References

1. Bennett, J., Duan, D., Engelhardt, J. F., and Maguire, A. M. (1997) Real-time non-invasive *in vivo* assessment of adeno-associated virus-mediated retinal transduction. *Invest. Ophthalmol. Vis. Sci.* **38,** 2857–2863.
2. Bennett, J., Wilson, J., Sun, D., Forbes, B., and Maguire, A. M. (1994) Adenovirus vector-mediated in vivo gene transfer into adult murine retina. *Invest. Ophthalmol. Vis. Sci.* **35,** 2535–2542.
3. Budenz, D., Bennett, J., Alonso L., and Maguire, A. (1995) In vivo gene transfer into murine trabecular meshwork and corneal endothelial cells. *Invest. Ophthalmol. Vis. Sci.* **36,** 2211–2215.
4. Hoffman, L. M., Maguire, A. M., and Bennett, J. (1997) Cell-mediated immune response and stability of intraocular transgene expression after adenovirus-mediated delivery. *Invest. Ophthalmol. Vis. Sci.* **38,** 2224–2233.
5. Bennett, J., Pakola, S., Zeng, Y., and Maguire, A. M. (1996) Humoral antibody response after administration of E1-deleted adenoviruses: Immune privilege of the subretinal space. *Hum. Gene Ther.* **7,** 1763–1769.

51

The Application of Adenoviral Vectors in the Study of Mammalian Cardiovascular Development

Craig S. Mickanin and H. Scott Baldwin

1. Introduction

The development of the cardiovascular system represents one of the earliest critical processes in mammalian development (*1*). Historically, cardiovascular development has been studied by utilizing in vitro cell- and organ-based systems or by use of avian-based culture models. Even though a great deal of information has been gained in these studies, the possibility that experimental results may be altered by artifacts of the in vitro system or that there may be discrepancies between avian and mammalian development cannot be overlooked. The advent of transgenic technology has provided a wealth of information in the study of cardiovascular development, and tremendous expansion in our knowledge of the molecular basis of heart development has come from the use of transgenics. There are drawbacks using this approach as well. Typical transgenic overexpression or "gain of function" studies are limited in the availability of promoter elements that are specific for the cell types being studied. Likewise, the use of targeted null mutations has highlighted the importance of the cardiovascular system, but null mutations often result in early embryonic lethality, prohibiting the study of latter events in cardiac morphogenesis. To overcome some of these limitations in the study of cardiovascular development, we have developed a technique whereby high-titer recombinant adenoviral stocks are microinjected directly into the developing cardiovascular system of cultured postimplantation mammalian embryos. This allows for precise spatial and temporal gene expression in the heart and blood vessels, and we have elucidated that gene expression can be made cell-type-specific for either endothelium or myocardium, the two cell types present during early cardiac morphogenesis, depending on the choice of otherwise cell-type-nonspecific promoter elements used to express the exogenous gene of interest (*see* **ref. 2**; **Fig. 1**). This technique could also be adapted for use in "loss-of-function" experiments in which dominant-negative or antisense gene constructs are utilized.

The technique outlined in this chapter comprises three distinct steps. First, whole embryo culture of early postimplantation mouse embryos has been described in detail in numerous places (*see* **ref. 3** and **4**), and thus are not detailed for the purpose of this chapter. Second, the generation of high-titer recombinant adenovirus stocks has also

From: *Methods in Molecular Biology, Vol. 135: Developmental Biology Protocols, Vol. I*
Edited by: R. S. Tuan and C. W. Lo © Humana Press Inc., Totowa, NJ

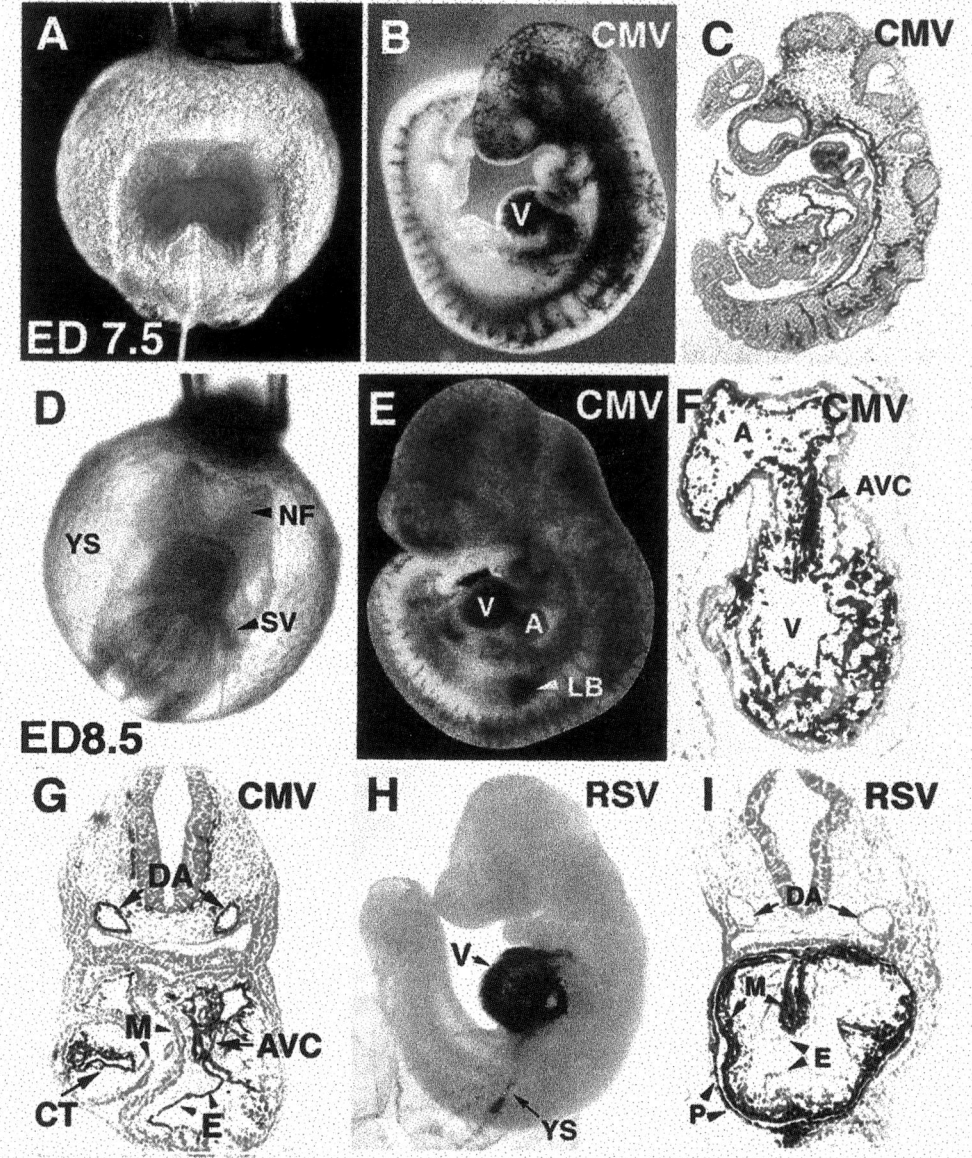

Fig. 1. (*See* color plate 13 appearing after p. 258.) (**A**) While being stabilized by retraction of the ectoplacental cone (EPC) into a holding pipet, viral constructs containing the *lacZ* reporter gene were injected into the pericardial coelom of an E 7.5 (2–3-somite) embryo via a micropipet inserted through the anterior intestinal portal (AIP) or developing foregut. Diffusion of phenol red solution is seen within the neural folds and was subsequently detected within the yolk sac (YS) but not in amniotic fluid. Following culture for 36 h, the extraembryonic membranes were removed and the embryos processed to detect β-galactosidase activity. (**B**) Embryos injected with 5×10^4–5×10^5 particles of virus appeared developmentally normal, and β-galactosidase activity is detected throughout the developing vasculature including the atria (A) and ventricle (V) of the heart, the first (1) and second (2) pharyngeal arches, and the dorsal aorta (DA). (**C**) A sagittal section through the embryo shown in (B) confirms that β-galactosidase activity was restricted to the endothelium of the developing embryos as seen in the endocardium of the atrium, ventricle,

been described in detail (*see* **refs. 5** and **6**). The third is the microinjection of purified adenovirus into the developing heart and vascular system of mouse embryos, which will be the focus of this chapter. Ultimate success in the application of this technique to the study of cardiovascular development requires technical expertise in all of the steps outlined above but is a powerful tool in elucidating the molecular basis of cardiovascular development.

2. Materials

The materials used for the techniques of embryo culture and adenovirus purification will not be detailed herein, but may be found elsewhere in this volume.

1. Stainless steel watchmaker's forceps.
2. Sterile Petri dishes.
3. Embryo culture medium (75% heat-inactivated rat serum/25% Tyrode's solution) (*see* **refs. 3–6**).
4. Sterile PB1 Medium: filter sterilized, prewarmed to 37°C, and supplemented with penicillin and streptomycin: 8 g/L NaCl, 0.2 g/L KCL, 0.1 g/L MgCl$_2$ · 6H$_2$O, 2.88 g/L Na$_2$HPO$_4$ · 12H$_2$O, 0.2 g/L KH$_2$PO$_4$, 0.1 g/L CaCl$_2$, 1.0 g/L Glucose, 0.036 g/L Na Pyruvate.
5. Bench Warmer (Model #TWW100W; Thermolyne Inc., Dubuque, IA).

(*Figure 1 caption continued*) and the lining of the dorsal aorta. No β-galactosidase activity was detected within the neural epithelium lining the developing forebrain (FB), in the somites (S), or in other nonendothelial cells. (**D**) E 8.75 (8–10-somite) embryos were injected with 1×10^7– 1×10^8 particles of recombinant virus containing the *lacZ* reporter gene via a micropipet inserted into the sinus venosus (SV). Injected solution was easily detected circulating through the heart (HT) and embryonic circulation and was subsequently detected in the yolk sac vasculature. (**E**) Following culture for 36 h and removal of extraembryonic membranes, β-galactosidase activity could be detected throughout the vasculature of the developing embryo, including the atria and ventricle of the developing heart, as well as the endothelium of the dorsal aorta, limb bud (LB), and liver primordia (LV). (**F**) A sagittal section through the heart of the embryo shown in (H) demonstrated that β-galactosidase activity was restricted to the endocardium of the atrium, ventricle, and atrioventricular canal (AVC) of the developing heart and was not detected within the myocardium. Of particular interest, the endothelial cells that had undergone epithelial/mesenchymal transformation and migrated into the extracellular matrix of the atrioventricular canal continued to express β-galactosidase (*). To determine if the pattern of β-galactosidase expression was affected by promoter specificity, E 7.75 (3–4-somite) embryos were injected into the pericardial coelom, similar to those described in (A), with 1×10^4–1×10^5 particles of virus containing the *lacZ* reporter gene driven by either the CMV or RSV promoter. Embryos were then cultured for 36 h and processed for β-galactosidase activity. (**G**) As with all other experiments utilizing the CMV promoter, a cross section through the heart demonstrated β-galactosidase activity in the endocardium of the heart and conotruncus (CT) and endothelium of the dorsal aorta. No activity was detected within myocardium of either the atrium or ventricle. (**H**) However, whole-mount preparations of embryos injected with virus containing the RSV-*lacZ* construct demonstrated intense β-galactosidase activity throughout the ventricle as well as some staining within the vasculature of the developing yolk sac. (**I**) A cross section through the heart confirmed β-galactosidase activity throughout the pericardium (PC) and myocardium with minimal expression detected within the endocardium and no expression within the endothelium of the dorsal aorta.

6. Dissecting microscope (Nikon SMZ-U Zoom 1:10; Tokyo, Japan).
7. Reflected light base for microscope (Nikon SMZ-U Diascopic Stand).
8. Lightweight Micromanipulators (Model M-152; Narishige, Tokyo, Japan).
9. Microcapillary pipet puller (Model P-87 Flaming-Brown Type; Sutter Instrument Co., Novato, CA).
10. Microcapillary pipets (Standard wall aluminosilicate, AF-100-53-10; Sutter Instrument Co.).
11. Microloader pipet (Varipette 4710/4810; Eppendorf, Inc., Hamburg, Germany).
12. Microloader pipet tips (cat. no. 5242 956.003; Eppendorf).
13. Microinjector (Transjector 5246; Eppendorf).

3. Method

1. Mouse embryos are dissected from timed pregnant females and cultured for 4–18 h. This time period allows for damaged embryos to be eliminated from the experiment. Because embryos within a litter can range significantly in age, embryos may be of very different stages of development and can thus be grouped according to developmental stage or discarded. A typical experiment in our laboratory will consist of 4–5 experimental groups of 5–10 embryos per group, bringing the total of embryos needed to 20–50. We routinely discard up to one half of the embryos put in culture after the initial culture period because of slight differences in experimental stage, and thus, a typical experiment will begin with 3–5 pregnant females. A strain of mouse that has a large average litter size, such as CD-1 or NIH, should be chosen. We have injected embryos as early as E7.0 and as late as E.11. The exact stage of embryo will vary depending on the specific stage or morphological event that the researcher intends to investigate.

2. Pipets are pulled prior to beginning the injection of embryos. Care must be taken to ensure that embryos remain in culture at 37°C until immediately before injection. Aluminosilicate pipets are pulled on a pipet puller, and as many as 50 individual pipets can be pulled in advance. Aluminosilicate pipets provide a more rigid tip than standard borosilicate glass pipets, and, unlike quartz pipets, can be pulled on a standard Flaming-Brown type pipet puller. We have tested many brands of prepulled microinjection pipets, and none have proven useful for this technique. Pipets are pulled such that a long (0.5–1 cm) taper is achieved, and the diameter of the bore should be as small as possible. A 3–4-mm-wide trough filament yields the most reliable tip for our purpose, although other filaments will suffice. For the holding pipet, we have found that the microloading pipet tips used to backload the injection pipet work well if cut with a razor blade to give a flattened end. As no suction is applied to the embryo with the holding pipet, any instrument that provides a smooth flattened surface on which to essentially "push" the embryo will suffice.

3. The adenovirus stock is thawed on ice for 30 min prior to injection. We routinely dilute frozen stocks of adenovirus 1:1 with PB1 if the stocks are cryopreserved in 3% sucrose (*see* Notes). These dilutions will minimize the teratological effects of the preservatives but will significantly decrease the concentration of the virus. Thus, virus stocks must be of the highest titer and activity and used within 2–3 mo of freezing. The microinjection capillary pipet is backloaded using an Eppendorf microloader pipet tip with 5–10 µL of the diluted adenovirus stock. This volume has proven sufficient for injected as many as 50 embryos in a single experiment because the volume of virus injected into a single embryo is quite small (10–100 pL for E7.5 embryo). Great care must be taken to avoid bubbles in the injection pipet, as these will not be eliminated from the tip of the pipet during the injection. The filled pipet is secured to a pressurized microinjector and mounted onto a micromanipulator several centimeters above the surface of the base of a dissecting microscope. Our laboratory has designed a modified microscope base with removable

micromanipulator supports on either side of the viewing area, which provides a convenient and flexible system for positioning the injection pipet and the holding pipet in a position in which the injections can be done (*see* Notes).

4. Embryos are removed from culture immediately prior to injection and placed into a Petri dish filled with 37°C PB1 medium. The dish is centered on the base of the microscope, and the injection pipet and holding pipet are crudely positioned on either side of the embryo. Using fine watchmaker's forceps, the tip of the injection pipet is gently broken as close to the end as possible. We normally do not bevel the injection pipets, but this is at the discretion of the researcher. The tip of the injection pipet is then lowered beneath the surface of the PB1 medium and inspected visually for standing flow rate and injection flow rate. The holding pipet is brought close to the embryo directly opposite to the injection pipet, and every effort must be made to align the injection pipet, the embryo, and the holding pipet in a single plane. This will minimize the movement of the embryo during the injection and will allow the embryo to be kept somewhat fixed during the initial "piercing" of the vessel wall or myocardium.

5. While the microinjector is exerting a constant low holding pressure, the injection pipet is brought into contact with the embryo at the site of the injection (**Fig. 1**). The embryo will move slightly and provide resistance to the pipet, and care must be taken not to apply excessive force to the embryo with the injection pipet, which will cause the pipet to pierce the embryo abruptly, and the pipet will be pushed deep into the embryo proper. The blood vessels at E8–E11 are very thin, and because the blood of the embryo will not clot, any tearing of the vessel wall will result in almost complete exsanguination of the embryo. The virus can be injected using gradually increasing injection pressure until an optimal pressure is discovered. This optimal pressure varies with the size of the injection pipet bore, the viscosity of the solution being injected, and the capacity of the space (e.g., pericardial cavity, ventricular lumen) to be injected. We have found that an injection pressure of 100 hPa is near optimal for most applications but have used pressures as low as 10 hPa for the injection of the head-fold mesenchyme of E7.5 embryos. (*See* Notes for details pertaining to injection of specific stages of development.)

6. When injections are complete, the embryos are placed back in culture for an additional 6–36 h and then assayed for both morphological and histological activity of the virus. We include as controls a group of embryos injected with a virus containing a reporter gene, such as β-galactosidase (*lacZ*), to control for the injection, and a group of uninjected embryos to control for the culture conditions.

4. Notes

Although the method described herein is not complicated, the application must be carefully optimized to achieve experimentally significant results. We have injected embryos at a variety of developmental stages and anatomical locations to determine the ability of the embryo to develop normally under a variety of injection protocols. We have primarily attempted to confine our injections to two locations. At E7.5–8.0, the pericardial cavity can be filled with high-titer virus, and the endocardium and myocardium are in essence bathed with virus. One advantage of this stage is the relative absence of red blood cells, and exsanguination as a result of tearing of the vascular plexus is not a major concern. We have made the unexpected observation that a transgene expressed under control of the cytomegalovirus (CMV) promoter will be selectively expressed in the endothelium of the heart, whereas gene expression driven by the Rous Sarcoma Virus LTR is specific to the heart muscle or myocardium (*2*).

Even though we have not established a mechanism for the unexpected cell-type restricted pattern, it offers a convenient system for directing gene expression spatially in the heart. Newer methods of adenoviral construction include the presence of a CMV-driven GFP cassette in the viral backbone that should help in identifying cells both infected by and expressing injected transgene *(6)*. We have also focused efforts in optimizing intravascular injections of various stages of development, but primarily E9–E9.5. At this stage, embryos will develop normally with the injection of up to several microliters of virus, and gene expression can be seen at sites throughout the embryonic vasculature in the embryo proper. Of note, we have discovered that it is not necessary to insert the tip of the injection pipet into the vessel lumen to achieve intravenous injection. If the injection pipet is brought into contact with the vessel to be injected, the pressure of the injection will literally "force" the virus solution through the vessel wall, without creating a tear that will allow red blood cells within the vessel to leak out. Injection at this stage is particularly useful for the study of epithelial-mesenchymal transformation, a process whereby endothelial cells of the atrioventricular canal or outflow tract delaminate from the endothelium and change phenotype to constitute the cellular aspect of the heart valves.

Even though we have not focused on the preparation of the adenovirus in this chapter, it is a very critical parameter of ultimate success. Unlike retroviruses, adenovirus vectors do not integrate, and, therefore, concentration of adenovirus must be high to provide gene transduction through the several cycles of cell division that will occur during the culture period. Therefore, the virus solution should be diluted as little as possible prior to injection. Although most investigators cryopreserve adenovirus stock solutions in 10% glycerol solution, we have found that a glycerol concentration of even 1% is mildly teratogenic when injected directly into a developing mouse embryo. When the adenovirus stock is diluted to the point at which the glycerol concentration is no longer teratogenic, the virus will be below the concentration at which gene expression is detectable. Our studies have shown, however, that a sucrose solution of up to 1.5% has no teratogenic effect on the developing embryo. Therefore, we routinely use 3% sucrose as a cryopreservative and dilute the solution 1:1 with PB1 immediately prior to injection. This also provides for a somewhat viscous final solution, which will aid the researcher in the visualization of the injection. It does, however, necessitate that viral stocks be utilized within 3 mo of preparation, as viral activity appears to diminish significantly after this period.

Perhaps the most difficult part of this protocol is the visualization of the injection. Whereas the experienced embryologist will develop a "feel" for the injection, the novice will certainly require several attempts at injecting to gain the necessary experience. We have found that the addition of dyes (such as Janus Green or Neutral Red) will aid in the visualization of the injection but may interfere with the activity of the virus. We have found this to be especially true of Neutral Red. When injecting into the pericardial cavity, the investigator will clearly see the increase in volume of the cavity. Intravascular injections are more difficult to visualize because of the flow of hemoglobinized red blood cells through the vessel. We have found that under a relatively high injection pressure, the flow of virus solution from the injection pipet will momentarily impede the flow of blood in the vessel. Red blood cells will appear to "stand still" and even flow in the direction opposite to blood flow in the vessel under this condition.

Even though standard micromanipulators work well for this protocol, we have designed a system that provides an additional level of flexibility for use in other applications. The vertical supports for the Narishige Model M-152 Micromanipulators are fastened directly into a small square metal block, which is attached to the microscope base via thumbscrews that secure into machined, reinforced holes machined into the top surface of the microscope base. The entire micromanipulator apparatus can thus be easily removed from the microscope base by the loosening of a single thumbscrew. This allows for the microscope base to be clear of obstructions during the initial embryo dissection, which is of tremendous advantage to the investigator.

The duration of the postinjection culture period will be determined by the goal of the experiment undertaken, but we normally allow embryos to develop for 24–48 h. We have detected exogenous protein expression in embryos in as little as 3 h after injection, and, in the case of endothelium, transduction of cells proceeds for as long as 24–36 h. Therefore, this system can be adapted for many studies during various periods of cardiovascular development.

References

1. Copp, A. J. (1995) Death before birth: clues from gene knockouts and mutations. *Trends Genet.* **11,** 87–93.
2. Baldwin, H. S., Mickanin, C., and Buck, C. A. (1997) *Adenovirus*-mediated gene transfer during initial organogenesis in the mammalian embryo is promoter-dependent and tissue-specific. *Gene Ther.* **4,** 1142–1149.
3. Cockroft, D. L. (1990) Dissection and culture of postimplantation embryos. *Postimplantation Mammalian Embryos: A Practical Approach* (Copp, A. J. and Cockroft, D. L., eds.), IRL Press at Oxford University Press, Oxford, UK, pp. 15–40.
4. Sturne, K. and Tam, P. P. L. (1993) Isolation and culture of whole postimplantation embryos and germ layer derivatives, in *Guide to Techniques in Moose Development. Methods in Enzymology, vol. 225* (Wassarman, P. M. and DePamphilia, M. L., eds.), Academic, San Diego, CA, pp. 164–190.
5. Graham, F. L. and Prevec, L. (1995) Methods for construction of adenovirus vectors. *Mol. Biotechnol.* **3,** 207–219.
6. HE, T. C., Zhou, S., Da Costa, L. T., Yu, J., Kinzler, D. W., and Vogelstein, B. (1998) A simplified system for generating recombinant adenoviruses. *Proc. Natl. Acad. Sci. USA* **95,** 2509–2514.

Index